Linear Algebra and Optimization for Machine Learning

Charu C. Aggarwal

Linear Algebra and Optimization for Machine Learning

A Textbook

 Springer

Charu C. Aggarwal
Distinguished Research Staff Member
IBM T.J. Watson Research Center
Yorktown Heights, NY, USA

ISBN 978-3-030-40346-1   ISBN 978-3-030-40344-7   (eBook)
https://doi.org/10.1007/978-3-030-40344-7

This Springer imprint is published by the registered company Springer Nature Switzerland AG.
The registered company address is: Gewerbestrasse 11, 6330 Cham, Switzerland

To my wife Lata, my daughter Sayani,
and all my mathematics teachers

Contents

1 Linear Algebra and Optimization: An Introduction   **1**
   1.1   Introduction   1
   1.2   Scalars, Vectors, and Matrices   2
      1.2.1   Basic Operations with Scalars and Vectors   3
      1.2.2   Basic Operations with Vectors and Matrices   8
      1.2.3   Special Classes of Matrices   12
      1.2.4   Matrix Powers, Polynomials, and the Inverse   14
      1.2.5   The Matrix Inversion Lemma: Inverting the Sum of Matrices . . .   17
      1.2.6   Frobenius Norm, Trace, and Energy   19
   1.3   Matrix Multiplication as a Decomposable Operator   21
      1.3.1   Matrix Multiplication as Decomposable Row and Column
              Operators   21
      1.3.2   Matrix Multiplication as Decomposable Geometric Operators . . .   25
   1.4   Basic Problems in Machine Learning   27
      1.4.1   Matrix Factorization   27
      1.4.2   Clustering   28
      1.4.3   Classification and Regression Modeling   29
      1.4.4   Outlier Detection   30
   1.5   Optimization for Machine Learning   31
      1.5.1   The Taylor Expansion for Function Simplification   31
      1.5.2   Example of Optimization in Machine Learning   33
      1.5.3   Optimization in Computational Graphs   34
   1.6   Summary   35
   1.7   Further Reading   35
   1.8   Exercises   36

2 Linear Transformations and Linear Systems   **41**
   2.1   Introduction   41
      2.1.1   What Is a Linear Transform?   42
   2.2   The Geometry of Matrix Multiplication   43

	2.3	Vector Spaces and Their Geometry	51
	2.3.1	Coordinates in a Basis System	55
	2.3.2	Coordinate Transformations Between Basis Sets	57
	2.3.3	Span of a Set of Vectors	59
	2.3.4	Machine Learning Example: Discrete Wavelet Transform	60
	2.3.5	Relationships Among Subspaces of a Vector Space	61
	2.4	The Linear Algebra of Matrix Rows and Columns	63
	2.5	The Row Echelon Form of a Matrix	64
	2.5.1	LU Decomposition	66
	2.5.2	Application: Finding a Basis Set	67
	2.5.3	Application: Matrix Inversion	67
	2.5.4	Application: Solving a System of Linear Equations	68
	2.6	The Notion of Matrix Rank	70
	2.6.1	Effect of Matrix Operations on Rank	71
	2.7	Generating Orthogonal Basis Sets	73
	2.7.1	Gram-Schmidt Orthogonalization and QR Decomposition	73
	2.7.2	QR Decomposition	74
	2.7.3	The Discrete Cosine Transform	77
	2.8	An Optimization-Centric View of Linear Systems	79
	2.8.1	Moore-Penrose Pseudoinverse	81
	2.8.2	The Projection Matrix	82
	2.9	Ill-Conditioned Matrices and Systems	85
	2.10	Inner Products: A Geometric View	86
	2.11	Complex Vector Spaces	87
	2.11.1	The Discrete Fourier Transform	89
	2.12	Summary	90
	2.13	Further Reading	91
	2.14	Exercises	91

3	**Eigenvectors and Diagonalizable Matrices**		**97**
	3.1	Introduction	97
	3.2	Determinants	98
	3.3	Diagonalizable Transformations and Eigenvectors	103
	3.3.1	Complex Eigenvalues	107
	3.3.2	Left Eigenvectors and Right Eigenvectors	108
	3.3.3	Existence and Uniqueness of Diagonalization	109
	3.3.4	Existence and Uniqueness of Triangulization	111
	3.3.5	Similar Matrix Families Sharing Eigenvalues	113
	3.3.6	Diagonalizable Matrix Families Sharing Eigenvectors	115
	3.3.7	Symmetric Matrices	115
	3.3.8	Positive Semidefinite Matrices	117
	3.3.9	Cholesky Factorization: Symmetric LU Decomposition	119
	3.4	Machine Learning and Optimization Applications	120
	3.4.1	Fast Matrix Operations in Machine Learning	121
	3.4.2	Examples of Diagonalizable Matrices in Machine Learning	121
	3.4.3	Symmetric Matrices in Quadratic Optimization	124
	3.4.4	Diagonalization Application: Variable Separation for Optimization	128
	3.4.5	Eigenvectors in Norm-Constrained Quadratic Programming	130

3.5   Numerical Algorithms for Finding Eigenvectors   131
        3.5.1   The QR Method via Schur Decomposition   132
        3.5.2   The Power Method for Finding Dominant Eigenvectors   133
3.6   Summary   135
3.7   Further Reading   135
3.8   Exercises   135

4   Optimization Basics: A Machine Learning View   141
4.1   Introduction   141
4.2   The Basics of Optimization   142
        4.2.1   Univariate Optimization   142
                4.2.1.1   Why We Need Gradient Descent   146
                4.2.1.2   Convergence of Gradient Descent   147
                4.2.1.3   The Divergence Problem   148
        4.2.2   Bivariate Optimization   149
        4.2.3   Multivariate Optimization   151
4.3   Convex Objective Functions   154
4.4   The Minutiae of Gradient Descent   159
        4.4.1   Checking Gradient Correctness with Finite Differences   159
        4.4.2   Learning Rate Decay and Bold Driver   159
        4.4.3   Line Search   160
                4.4.3.1   Binary Search   161
                4.4.3.2   Golden-Section Search   161
                4.4.3.3   Armijo Rule   162
        4.4.4   Initialization   163
4.5   Properties of Optimization in Machine Learning   163
        4.5.1   Typical Objective Functions and Additive Separability   163
        4.5.2   Stochastic Gradient Descent   164
        4.5.3   How Optimization in Machine Learning Is Different   165
        4.5.4   Tuning Hyperparameters   168
        4.5.5   The Importance of Feature Preprocessing   168
4.6   Computing Derivatives with Respect to Vectors   169
        4.6.1   Matrix Calculus Notation   170
        4.6.2   Useful Matrix Calculus Identities   171
                4.6.2.1   Application: Unconstrained Quadratic Programming . . .   173
                4.6.2.2   Application: Derivative of Squared Norm   174
        4.6.3   The Chain Rule of Calculus for Vectored Derivatives   174
                4.6.3.1   Useful Examples of Vectored Derivatives   175
4.7   Linear Regression: Optimization with Numerical Targets   176
        4.7.1   Tikhonov Regularization   178
                4.7.1.1   Pseudoinverse and Connections to Regularization   179
        4.7.2   Stochastic Gradient Descent   179
        4.7.3   The Use of Bias   179
                4.7.3.1   Heuristic Initialization   180
4.8   Optimization Models for Binary Targets   180
        4.8.1   Least-Squares Classification: Regression on Binary Targets   181
                4.8.1.1   Why Least-Squares Classification Loss Needs Repair . . .   183

      4.8.2   The Support Vector Machine 184
            4.8.2.1   Computing Gradients 185
            4.8.2.2   Stochastic Gradient Descent 186
      4.8.3   Logistic Regression 186
            4.8.3.1   Computing Gradients 188
            4.8.3.2   Stochastic Gradient Descent 188
      4.8.4   How Linear Regression Is a Parent Problem in Machine
            Learning 189
  4.9   Optimization Models for the MultiClass Setting 190
      4.9.1   Weston-Watkins Support Vector Machine 190
            4.9.1.1   Computing Gradients 191
      4.9.2   Multinomial Logistic Regression 192
            4.9.2.1   Computing Gradients 193
            4.9.2.2   Stochastic Gradient Descent 194
  4.10   Coordinate Descent 194
      4.10.1   Linear Regression with Coordinate Descent 196
      4.10.2   Block Coordinate Descent 197
      4.10.3   K-Means as Block Coordinate Descent 197
  4.11   Summary 198
  4.12   Further Reading 199
  4.13   Exercises 199

5 Advanced Optimization Solutions   205
  5.1   Introduction 205
  5.2   Challenges in Gradient-Based Optimization 206
      5.2.1   Local Optima and Flat Regions 207
      5.2.2   Differential Curvature 208
            5.2.2.1   Revisiting Feature Normalization 209
      5.2.3   Examples of Difficult Topologies: Cliffs and Valleys 210
  5.3   Adjusting First-Order Derivatives for Descent 212
      5.3.1   Momentum-Based Learning 212
      5.3.2   AdaGrad 214
      5.3.3   RMSProp 215
      5.3.4   Adam 215
  5.4   The Newton Method 216
      5.4.1   The Basic Form of the Newton Method 217
      5.4.2   Importance of Line Search for Non-quadratic Functions 219
      5.4.3   Example: Newton Method in the Quadratic Bowl 220
      5.4.4   Example: Newton Method in a Non-quadratic Function 220
  5.5   Newton Methods in Machine Learning 221
      5.5.1   Newton Method for Linear Regression 221
      5.5.2   Newton Method for Support-Vector Machines 223
      5.5.3   Newton Method for Logistic Regression 225
      5.5.4   Connections Among Different Models and Unified Framework . . . 228
  5.6   Newton Method: Challenges and Solutions 229
      5.6.1   Singular and Indefinite Hessian 229
      5.6.2   The Saddle-Point Problem 229

		5.6.3	Convergence Problems and Solutions with Non-quadratic Functions	231
			5.6.3.1 Trust Region Method	232
5.7		Computationally Efficient Variations of Newton Method		233
	5.7.1	Conjugate Gradient Method		233
	5.7.2	Quasi-Newton Methods and BFGS		237
5.8		Non-differentiable Optimization Functions		239
	5.8.1	The Subgradient Method		240
		5.8.1.1 Application: L_1-Regularization		242
		5.8.1.2 Combining Subgradients with Coordinate Descent		243
	5.8.2	Proximal Gradient Method		244
		5.8.2.1 Application: Alternative for L_1-Regularized Regression		245
	5.8.3	Designing Surrogate Loss Functions for Combinatorial Optimization		246
		5.8.3.1 Application: Ranking Support Vector Machine		247
	5.8.4	Dynamic Programming for Optimizing Sequential Decisions		248
		5.8.4.1 Application: Fast Matrix Multiplication		249
5.9	Summary			250
5.10	Further Reading			250
5.11	Exercises			251

6 Constrained Optimization and Duality — **255**

6.1	Introduction			255
6.2	Primal Gradient Descent Methods			256
	6.2.1	Linear Equality Constraints		257
		6.2.1.1 Convex Quadratic Program with Equality Constraints		259
		6.2.1.2 Application: Linear Regression with Equality Constraints		261
		6.2.1.3 Application: Newton Method with Equality Constraints		262
	6.2.2	Linear Inequality Constraints		262
		6.2.2.1 The Special Case of Box Constraints		263
		6.2.2.2 General Conditions for Projected Gradient Descent to Work		264
		6.2.2.3 Sequential Linear Programming		266
	6.2.3	Sequential Quadratic Programming		267
6.3	Primal Coordinate Descent			267
	6.3.1	Coordinate Descent for Convex Optimization Over Convex Set		268
	6.3.2	Machine Learning Application: Box Regression		269
6.4	Lagrangian Relaxation and Duality			270
	6.4.1	Kuhn-Tucker Optimality Conditions		274
	6.4.2	General Procedure for Using Duality		276
		6.4.2.1 Inferring the Optimal Primal Solution from Optimal Dual Solution		276
	6.4.3	Application: Formulating the SVM Dual		276
		6.4.3.1 Inferring the Optimal Primal Solution from Optimal Dual Solution		278

6.4.4   Optimization Algorithms for the SVM Dual   279
        6.4.4.1   Gradient Descent   279
        6.4.4.2   Coordinate Descent   280
6.4.5   Getting the Lagrangian Relaxation of Unconstrained Problems . .   281
        6.4.5.1   Machine Learning Application: Dual of Linear Regression   283
6.5   Penalty-Based and Primal-Dual Methods   286
  6.5.1   Penalty Method with Single Constraint   286
  6.5.2   Penalty Method: General Formulation   287
  6.5.3   Barrier and Interior Point Methods   288
6.6   Norm-Constrained Optimization   290
6.7   Primal Versus Dual Methods   292
6.8   Summary   293
6.9   Further Reading   294
6.10 Exercises   294

7   Singular Value Decomposition   299
7.1   Introduction   299
7.2   SVD: A Linear Algebra Perspective   300
  7.2.1   Singular Value Decomposition of a Square Matrix   300
  7.2.2   Square SVD to Rectangular SVD via Padding   304
  7.2.3   Several Definitions of Rectangular Singular Value Decomposition .   305
  7.2.4   Truncated Singular Value Decomposition   307
        7.2.4.1   Relating Truncation Loss to Singular Values   309
        7.2.4.2   Geometry of Rank-k Truncation   311
        7.2.4.3   Example of Truncated SVD   311
  7.2.5   Two Interpretations of SVD   313
  7.2.6   Is Singular Value Decomposition Unique?   315
  7.2.7   Two-Way Versus Three-Way Decompositions   316
7.3   SVD: An Optimization Perspective   317
  7.3.1   A Maximization Formulation with Basis Orthogonality   318
  7.3.2   A Minimization Formulation with Residuals   319
  7.3.3   Generalization to Matrix Factorization Methods   320
  7.3.4   Principal Component Analysis   320
7.4   Applications of Singular Value Decomposition   323
  7.4.1   Dimensionality Reduction   323
  7.4.2   Noise Removal   324
  7.4.3   Finding the Four Fundamental Subspaces in Linear Algebra   325
  7.4.4   Moore-Penrose Pseudoinverse   325
        7.4.4.1   Ill-Conditioned Square Matrices   326
  7.4.5   Solving Linear Equations and Linear Regression   327
  7.4.6   Feature Preprocessing and Whitening in Machine Learning   327
  7.4.7   Outlier Detection   328
  7.4.8   Feature Engineering   329
7.5   Numerical Algorithms for SVD   330
7.6   Summary   332
7.7   Further Reading   332
7.8   Exercises   333

8   Matrix Factorization   **339**
  8.1   Introduction   339
  8.2   Optimization-Based Matrix Factorization   341
        8.2.1   Example: K-Means as Constrained Matrix Factorization   342
  8.3   Unconstrained Matrix Factorization   342
        8.3.1   Gradient Descent with Fully Specified Matrices   343
        8.3.2   Application to Recommender Systems   346
                8.3.2.1   Stochastic Gradient Descent   348
                8.3.2.2   Coordinate Descent   348
                8.3.2.3   Block Coordinate Descent: Alternating Least Squares . .   349
  8.4   Nonnegative Matrix Factorization   350
        8.4.1   Optimization Problem with Frobenius Norm   350
                8.4.1.1   Projected Gradient Descent with Box Constraints   351
        8.4.2   Solution Using Duality   351
        8.4.3   Interpretability of Nonnegative Matrix Factorization   353
        8.4.4   Example of Nonnegative Matrix Factorization   353
        8.4.5   The I-Divergence Objective Function   356
  8.5   Weighted Matrix Factorization   356
        8.5.1   Practical Use Cases of Nonnegative and Sparse Matrices   357
        8.5.2   Stochastic Gradient Descent   359
                8.5.2.1   Why Negative Sampling Is Important   360
        8.5.3   Application: Recommendations with Implicit Feedback Data   360
        8.5.4   Application: Link Prediction in Adjacency Matrices   360
        8.5.5   Application: Word-Word Context Embedding with GloVe   361
  8.6   Nonlinear Matrix Factorizations   362
        8.6.1   Logistic Matrix Factorization   362
                8.6.1.1   Gradient Descent Steps for Logistic Matrix
                          Factorization   363
        8.6.2   Maximum Margin Matrix Factorization   364
  8.7   Generalized Low-Rank Models   365
        8.7.1   Handling Categorical Entries   367
        8.7.2   Handling Ordinal Entries   367
  8.8   Shared Matrix Factorization   369
        8.8.1   Gradient Descent Steps for Shared Factorization   370
        8.8.2   How to Set Up Shared Models in Arbitrary Scenarios   370
  8.9   Factorization Machines   371
  8.10 Summary   375
  8.11 Further Reading   375
  8.12 Exercises   375

9   The Linear Algebra of Similarity   **379**
  9.1   Introduction   379
  9.2   Equivalence of Data and Similarity Matrices   379
        9.2.1   From Data Matrix to Similarity Matrix and Back   380
        9.2.2   When Is Data Recovery from a Similarity Matrix Useful?   381
        9.2.3   What Types of Similarity Matrices Are "Valid"?   382
        9.2.4   Symmetric Matrix Factorization as an Optimization Model   383
        9.2.5   Kernel Methods: The Machine Learning Terminology   383

9.3   Efficient Data Recovery from Similarity Matrices 385
     9.3.1   Nyström Sampling 385
     9.3.2   Matrix Factorization with Stochastic Gradient Descent 386
     9.3.3   Asymmetric Similarity Decompositions 388
9.4   Linear Algebra Operations on Similarity Matrices 389
     9.4.1   Energy of Similarity Matrix and Unit Ball Normalization 390
     9.4.2   Norm of the Mean and Variance 390
     9.4.3   Centering a Similarity Matrix 391
         9.4.3.1   Application: Kernel PCA 391
     9.4.4   From Similarity Matrix to Distance Matrix and Back 392
         9.4.4.1   Application: ISOMAP 393
9.5   Machine Learning with Similarity Matrices 394
     9.5.1   Feature Engineering from Similarity Matrix 395
         9.5.1.1   Kernel Clustering 395
         9.5.1.2   Kernel Outlier Detection 396
         9.5.1.3   Kernel Classification 396
     9.5.2   Direct Use of Similarity Matrix 397
         9.5.2.1   Kernel K-Means 397
         9.5.2.2   Kernel SVM 398
9.6   The Linear Algebra of the Representer Theorem 399
9.7   Similarity Matrices and Linear Separability 403
     9.7.1   Transformations That Preserve Positive Semi-definiteness 405
9.8   Summary 407
9.9   Further Reading 407
9.10   Exercises 407

10 The Linear Algebra of Graphs   **411**
10.1   Introduction 411
10.2   Graph Basics and Adjacency Matrices 411
10.3   Powers of Adjacency Matrices 416
10.4   The Perron-Frobenius Theorem 419
10.5   The Right Eigenvectors of Graph Matrices 423
     10.5.1   The Kernel View of Spectral Clustering 423
         10.5.1.1   Relating Shi-Malik and Ng-Jordan-Weiss Embeddings . . 425
     10.5.2   The Laplacian View of Spectral Clustering 426
         10.5.2.1   Graph Laplacian 426
         10.5.2.2   Optimization Model with Laplacian 428
     10.5.3   The Matrix Factorization View of Spectral Clustering 430
         10.5.3.1   Machine Learning Application: Directed Link
                 Prediction 430
     10.5.4   Which View of Spectral Clustering Is Most Informative? 431
10.6   The Left Eigenvectors of Graph Matrices 431
     10.6.1   PageRank as Left Eigenvector of Transition Matrix 433
     10.6.2   Related Measures of Prestige and Centrality 434
     10.6.3   Application of Left Eigenvectors to Link Prediction 435
10.7   Eigenvectors of Reducible Matrices 436
     10.7.1   Undirected Graphs 436
     10.7.2   Directed Graphs 436

10.8   Machine Learning Applications 439
        10.8.1 Application to Vertex Classification 440
        10.8.2 Applications to Multidimensional Data 442
10.9   Summary 443
10.10   Further Reading 443
10.11   Exercises 444

11 Optimization in Computational Graphs   **447**
11.1   Introduction 447
11.2   The Basics of Computational Graphs 448
        11.2.1 Neural Networks as Directed Computational Graphs 451
11.3   Optimization in Directed Acyclic Graphs 453
        11.3.1 The Challenge of Computational Graphs 453
        11.3.2 The Broad Framework for Gradient Computation 455
        11.3.3 Computing Node-to-Node Derivatives Using Brute Force 456
        11.3.4 Dynamic Programming for Computing Node-to-Node Derivatives . 459
                11.3.4.1 Example of Computing Node-to-Node Derivatives 461
        11.3.5 Converting Node-to-Node Derivatives into Loss-to-Weight
                Derivatives 464
                11.3.5.1 Example of Computing Loss-to-Weight Derivatives 465
        11.3.6 Computational Graphs with Vector Variables 466
11.4   Application: Backpropagation in Neural Networks 468
        11.4.1 Derivatives of Common Activation Functions 470
        11.4.2 Vector-Centric Backpropagation 471
        11.4.3 Example of Vector-Centric Backpropagation 473
11.5   A General View of Computational Graphs 475
11.6   Summary 478
11.7   Further Reading 478
11.8   Exercises 478

Bibliography   **483**

Index   **491**

Preface

"Mathematics is the language with which God wrote the universe." – Galileo

A frequent challenge faced by beginners in machine learning is the extensive background required in linear algebra and optimization. One problem is that the existing linear algebra and optimization courses are not specific to machine learning; therefore, one would typically have to complete more course material than is necessary to pick up machine learning. Furthermore, certain types of ideas and tricks from optimization and linear algebra recur more frequently in machine learning than other application-centric settings. Therefore, there is significant value in developing a view of linear algebra and optimization that is better suited to the specific perspective of machine learning.

It is common for machine learning practitioners to pick up missing bits and pieces of linear algebra and optimization via "osmosis" while studying the solutions to machine learning applications. However, this type of unsystematic approach is unsatisfying, because the primary focus on machine learning gets in the way of learning linear algebra and optimization in a generalizable way across new situations and applications. Therefore, we have inverted the focus in this book, with linear algebra and optimization as the primary topics of interest and solutions to machine learning problems as the *applications* of this machinery. *In other words, the book goes out of its way to teach linear algebra and optimization with machine learning examples.* By using this approach, the book focuses on those aspects of linear algebra and optimization that are more relevant to machine learning and also teaches the reader how to apply them in the machine learning context. As a side benefit, the reader will pick up knowledge of several fundamental problems in machine learning. At the end of the process, the reader will become familiar with many of the basic linear-algebra- and optimization-centric algorithms in machine learning. Although the book is not intended to provide exhaustive coverage of machine learning, it serves as a "technical starter" for the key models and optimization methods in machine learning. Even for seasoned practitioners of machine learning, a systematic introduction to fundamental linear algebra and optimization methodologies can be useful in terms of providing a fresh perspective.

The chapters of the book are organized as follows:

1. *Linear algebra and its applications:* The chapters focus on the basics of linear algebra together with their common applications to singular value decomposition, matrix factorization, similarity matrices (kernel methods), and graph analysis. Numerous machine learning applications have been used as examples, such as spectral clustering,

kernel-based classification, and outlier detection. The tight integration of linear algebra methods with examples from machine learning differentiates this book from generic volumes on linear algebra. The focus is clearly on the most relevant aspects of linear algebra for machine learning and to teach readers how to apply these concepts.

2. *Optimization and its applications:* Much of machine learning is posed as an optimization problem in which we try to maximize the accuracy of regression and classification models. The "parent problem" of optimization-centric machine learning is least-squares regression. Interestingly, this problem arises in both linear algebra and optimization and is one of the key connecting problems of the two fields. Least-squares regression is also the starting point for support vector machines, logistic regression, and recommender systems. Furthermore, the methods for dimensionality reduction and matrix factorization also require the development of optimization methods. A general view of optimization in computational graphs is discussed together with its applications to backpropagation in neural networks.

This book contains exercises both within the text of the chapter and at the end of the chapter. The exercises within the text of the chapter should be solved as one reads the chapter in order to solidify the concepts. This will lead to slower progress, but a better understanding. For in-chapter exercises, hints for the solution are given in order to help the reader along. The exercises at the end of the chapter are intended to be solved as refreshers after completing the chapter.

Throughout this book, a vector or a multidimensional data point is annotated with a bar, such as $\overline{X}$ or $\overline{y}$. A vector or multidimensional point may be denoted by either small letters or capital letters, as long as it has a bar. Vector dot products are denoted by centered dots, such as $\overline{X} \cdot \overline{Y}$. A matrix is denoted in capital letters without a bar, such as R. Throughout the book, the $n \times d$ matrix corresponding to the entire training data set is denoted by D, with n data points and d dimensions. The individual data points in D are therefore d-dimensional row vectors and are often denoted by $\overline{X}_1 \ldots \overline{X}_n$. Conversely, vectors with one component for each data point are usually n-dimensional column vectors. An example is the n-dimensional column vector $\overline{y}$ of class variables of n data points. An observed value y_i is distinguished from a predicted value $\hat{y}_i$ by a circumflex at the top of the variable.

Yorktown Heights, NY, USA   Charu C. Aggarwal

Acknowledgments

I would like to thank my family for their love and support during the busy time spent in writing this book. Knowledge of the very basics of optimization (e.g., calculus) and linear algebra (e.g., vectors and matrices) starts in high school and increases over the course of many years of undergraduate/graduate education as well as during the postgraduate years of research. As such, I feel indebted to a large number of teachers and collaborators over the years. This section is, therefore, a rather incomplete attempt to express my gratitude.

My initial exposure to vectors, matrices, and optimization (calculus) occurred during my high school years, where I was ably taught these subjects by S. Adhikari and P. C. Pathrose. Indeed, my love of mathematics started during those years, and I feel indebted to both these individuals for instilling the love of these subjects in me. During my undergraduate study in computer science at IIT Kanpur, I was taught several aspects of linear algebra and optimization by Dr. R. Ahuja, Dr. B. Bhatia, and Dr. S. Gupta. Even though linear algebra and mathematical optimization are distinct (but interrelated) subjects, Dr. Gupta's teaching style often provided an integrated view of these topics. I was able to fully appreciate the value of such an integrated view when working in machine learning. For example, one can approach many problems such as solving systems of equations or singular value decomposition either from a linear algebra viewpoint or from an optimization viewpoint, and both perspectives provide complementary views in different machine learning applications. Dr. Gupta's courses on linear algebra and mathematical optimization had a profound influence on me in choosing mathematical optimization as my field of study during my PhD years; this choice was relatively unusual for undergraduate computer science majors at that time. Finally, I had the good fortune to learn about linear and nonlinear optimization methods from several luminaries on these subjects during my graduate years at MIT. In particular, I feel indebted to my PhD thesis advisor James B. Orlin for his guidance during my early years. In addition, Nagui Halim has provided a lot of support for all my book-writing projects over the course of a decade and deserves a lot of credit for my work in this respect. My manager, Horst Samulowitz, has supported my work over the past year, and I would like to thank him for his help.

I also learned a lot from my collaborators in machine learning over the years. One often appreciates the true usefulness of linear algebra and optimization only in an applied setting, and I had the good fortune of working with many researchers from different areas on a wide range of machine learning problems. A lot of the emphasis in this book to specific aspects of linear algebra and optimization is derived from these invaluable experiences and collaborations. In particular, I would like to thank Tarek F. Abdelzaher, Jinghui Chen, Jing Gao, Quanquan Gu, Manish Gupta, Jiawei Han, Alexander Hinneburg, Thomas Huang, Nan Li, Huan Liu, Ruoming Jin, Daniel Keim, Arijit Khan, Latifur Khan, Mohammad M. Masud, Jian Pei, Magda Procopiuc, Guojun Qi, Chandan Reddy, Saket Sathe, Jaideep Srivastava, Karthik Subbian, Yizhou Sun, Jiliang Tang, Min-Hsuan Tsai, Haixun Wang, Jianyong Wang, Min Wang, Suhang Wang, Wei Wang, Joel Wolf, Xifeng Yan, Wenchao Yu, Mohammed Zaki, ChengXiang Zhai, and Peixiang Zhao.

Several individuals have also reviewed the book. Quanquan Gu provided suggestions on Chapter 6. Jiliang Tang and Xiaorui Liu examined several portions of Chapter 6 and pointed out corrections and improvements. Shuiwang Ji contributed Problem 7.2.3. Jie Wang reviewed several chapters of the book and pointed out corrections. Hao Liu also provided several suggestions.

Last but not least, I would like to thank my daughter Sayani for encouraging me to write this book at a time when I had decided to hang up my boots on the issue of book writing. She encouraged me to write this one. I would also like to thank my wife for fixing some of the figures in this book.

Author Biography

Charu C. Aggarwal is a Distinguished Research Staff Member (DRSM) at the IBM T. J. Watson Research Center in Yorktown Heights, New York. He completed his undergraduate degree in Computer Science from the Indian Institute of Technology at Kanpur in 1993 and his Ph.D. from the Massachusetts Institute of Technology in 1996.

He has worked extensively in the field of data mining. He has published more than 400 papers in refereed conferences and journals and authored more than 80 patents. He is the author or editor of 19 books, including textbooks on data mining, recommender systems, and outlier analysis. Because of the commercial value of his patents, he has thrice been designated a Master Inventor at IBM. He is a recipient of an IBM Corporate Award (2003) for his work on bioterrorist threat detection in data streams, a recipient of the IBM Outstanding Innovation Award (2008) for his scientific contributions to privacy technology, and a recipient of two IBM Outstanding Technical Achievement Awards (2009, 2015) for his work on data streams/high-dimensional data. He received the EDBT 2014 Test of Time Award for his work on condensation-based privacy-preserving data mining. He is also a recipient of the IEEE ICDM Research Contributions Award (2015) and the ACM SIGKDD Innovation Award (2019), which are the two highest awards for influential research contributions in data mining.

He has served as the general cochair of the IEEE Big Data Conference (2014) and as the program cochair of the ACM CIKM Conference (2015), the IEEE ICDM Conference (2015), and the ACM KDD Conference (2016). He served as an associate editor of the IEEE Transactions on Knowledge and Data Engineering from 2004 to 2008. He is an associate editor of the IEEE Transactions on Big Data, an action editor of the Data Mining and Knowledge Discovery Journal, and an associate editor of the Knowledge and Information Systems Journal. He serves as the editor-in-chief of the ACM Transactions on Knowledge Discovery from Data as well as the ACM SIGKDD Explorations. He serves on the advisory board of the Lecture Notes on Social Networks, a publication by Springer. He has served as the vice president of the SIAM Activity Group on Data Mining and is a member of the SIAM Industry Committee. He is a fellow of the SIAM, ACM, and IEEE, for "contributions to knowledge discovery and data mining algorithms."

Chapter 1

Linear Algebra and Optimization: An Introduction

"No matter what engineering field you're in, you learn the same basic science and mathematics. And then maybe you learn a little bit about how to apply it."–Noam Chomsky

1.1 Introduction

Machine learning builds mathematical models from data containing multiple *attributes* (i.e., variables) in order to predict some variables from others. For example, in a cancer prediction application, each data point might contain the variables obtained from running clinical tests, whereas the predicted variable might be a binary diagnosis of cancer. Such models are sometimes expressed as linear and nonlinear relationships between variables. These relationships are discovered in a data-driven manner by *optimizing* (maximizing) the "agreement" between the models and the observed data. This is an optimization problem.

Linear algebra is the study of linear operations in *vector spaces*. An example of a vector space is the infinite set of all possible *Cartesian coordinates* in two dimensions in relation to a fixed point referred to as the *origin*, and each *vector* (i.e., a 2-dimensional coordinate) can be viewed as a member of this set. This abstraction fits in nicely with the way data is represented in machine learning as points with multiple dimensions, albeit with dimensionality that is usually greater than 2. These dimensions are also referred to as attributes in machine learning parlance. For example, each patient in a medical application might be represented by a vector containing many attributes, such as age, blood sugar level, inflammatory markers, and so on. It is common to apply linear functions to these high-dimensional vectors in many application domains in order to extract their analytical properties. The study of such linear transformations lies at the heart of linear algebra.

While it is easy to visualize the spatial geometry of points/operations in 2 or 3 dimensions, it becomes harder to do so in higher dimensions. For example, it is simple to visualize

© Springer Nature Switzerland AG 2020
C. C. Aggarwal, *Linear Algebra and Optimization for Machine Learning*,
https://doi.org/10.1007/978-3-030-40344-7_1

a 2-dimensional rotation of an object, but it is hard to visualize a 20-dimensional object and its corresponding rotation. This is one of the primary challenges associated with linear algebra. However, with some practice, one can transfer spatial intuitions to higher dimensions. Linear algebra can be viewed as a generalized form of the geometry of Cartesian coordinates in d dimensions. Just as one can use analytical geometry in two dimensions in order to find the intersection of two lines in the plane, one can generalize this concept to any number of dimensions. The resulting method is referred to as *Gaussian elimination* for solving systems of equations, and it is one of the fundamental cornerstones of linear algebra. Indeed, the problem of *linear regression*, which is fundamental to linear algebra, optimization, and machine learning, is closely related to solving systems of equations. This book will introduce linear algebra and optimization with a specific focus on machine learning applications.

This chapter is organized as follows. The next section introduces the definitions of vectors and matrices and important operations. Section 1.3 closely examines the nature of matrix multiplication with vectors and its interpretation as the composition of simpler transformations on vectors. In Section 1.4, we will introduce the basic problems in machine learning that are used as application examples throughout this book. Section 1.5 will introduce the basics of optimization, and its relationship with the different types of machine learning problems. A summary is given in Section 1.6.

1.2   Scalars, Vectors, and Matrices

We start by introducing the notions of scalars, vectors, and matrices, which are the fundamental structures associated with linear algebra.

1. *Scalars:* Scalars are individual numerical values that are typically drawn from the real domain in most machine learning applications. For example, the value of an attribute such as *Age* in a machine learning application is a scalar.

2. *Vectors:* Vectors are arrays of numerical values (i.e., arrays of scalars). Each such numerical value is also referred to as a *coordinate*. The individual numerical values of the arrays are referred to as *entries*, *components*, or *dimensions* of the vector, and the number of components is referred to as the vector *dimensionality*. In machine learning, a vector might contain components (associated with a data point) corresponding to numerical values like *Age*, *Salary*, and so on. A 3-dimensional vector representation of a 25-year-old person making 30 dollars an hour, and having 5 years of experience might be written as the array of numbers $[25, 30, 5]$.

3. *Matrices:* Matrices can be viewed as rectangular arrays of numerical values containing both *rows* and *columns*. In order to an access an element in the matrix, one must specify its row index and its column index. For example, consider a data set in a machine learning application containing d properties of n individuals. Each individual is allocated a row, and each property is allocated in column. In such a case, we can define a data matrix, in which each row is a d-dimensional vector containing the properties of one of the n individuals. The *size* of such a matrix is denoted by the notation $n \times d$. An element of the matrix is accessed with the pair of indices (i, j), where the first element i is the row index, and the second element j is the column index. The row index increases from top to bottom, whereas the column index increases from left to right. The value of the (i, j)th entry of the matrix is therefore equal to the jth property of the ith individual. When we define a matrix $A = [a_{ij}]$, it refers to the fact

that the (i, j)th element of A is denoted by a_{ij}. Furthermore, defining $A = [a_{ij}]_{n \times d}$ refers to the fact that the size of A is $n \times d$. When a matrix has the same number of rows as columns, it is referred to as a *square matrix*. Otherwise, it is referred to as a *rectangular matrix*. A rectangular matrix with more rows than columns is referred to as *tall*, whereas a matrix with more columns than rows is referred to as *wide* or *fat*.

It is possible for scalars, vectors, and matrices to contain complex numbers. This book will occasionally discuss complex-valued vectors when they are relevant to machine learning.

Vectors are special cases of matrices, and scalars are special cases of both vectors and matrices. For example, a scalar is sometimes viewed as a 1×1 "matrix." Similarly, a d-dimensional vector can be viewed as a $1 \times d$ matrix when it is treated as a *row vector*. It can also be treated as a $d \times 1$ matrix when it is a *column vector*. The addition of the word "row" or "column" to the vector definition is indicative of whether that vector is naturally a row of a larger matrix or whether it is a column of a larger matrix. By default, vectors are assumed to be column vectors in linear algebra, unless otherwise specified. We always use an overbar on a variable to indicate that it is a vector, although we do not do so for matrices or scalars. For example, the row vector $[y_1, \ldots, y_d]$ of d values can be denoted by $\overline{y}$ or $\overline{Y}$. In this book, scalars are always represented by lower-case variables like a or δ, whereas matrices are always represented by upper-case variables like A or Δ.

In the sciences, a vector is often geometrically visualized as a quantity, such as the velocity, that has a magnitude as well as a direction. Such vectors are referred to as *geometric vectors*. For example, imagine a situation where the positive direction of the X-axis corresponds to the eastern direction, and the positive direction of the Y-axis corresponds to the northern direction. Then, a person that is simultaneously moving at 4 meters/second in the eastern direction and at 3 meters/second in the northern direction is really moving in the north-eastern direction in a straight line at $\sqrt{4^2 + 3^2} = 5$ meters/second (based on the *Pythagorean theorem*). This is also the length of the vector. The vector of the velocity of this person can be written as a directed line from the origin to $[4, 3]$. This vector is shown in Figure 1.1(a). In this case, the *tail* of the vector is at the origin, and the *head* of the vector is at $[4, 3]$. Geometric vectors in the sciences are allowed to have arbitrary tails. For example, we have shown another example of the same vector $[4, 3]$ in Figure 1.1(a) in which the tail is placed at $[1, 4]$ and the head is placed at $[5, 7]$. In contrast to geometric vectors, only vectors that have tails at the origin are considered in linear algebra (although the mathematical results, principles, and intuition remain the same). This does not lead to any loss of expressivity. *All vectors, operations, and spaces in linear algebra use the origin as an important reference point.*

1.2.1 Basic Operations with Scalars and Vectors

Vectors of the same dimensionality can be added or subtracted. For example, consider two d-dimensional vectors $\overline{x} = [x_1 \ldots x_d]$ and $\overline{y} = [y_1 \ldots y_d]$ in a retail application, where the ith component defines the volume of sales for the ith product. In such a case, the vector of aggregate sales is $\overline{x} + \overline{y}$, and its ith component is $x_i + y_i$:

$$\overline{x} + \overline{y} = [x_1 \ldots x_d] + [y_1 \ldots y_d] = [x_1 + y_1 \ldots x_d + y_d]$$

Vector subtraction is defined in the same way:

$$\overline{x} - \overline{y} = [x_1 \ldots x_d] - [y_1 \ldots y_d] = [x_1 - y_1 \ldots x_d - y_d]$$

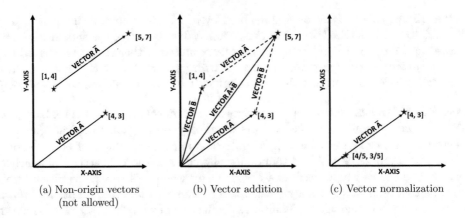

(a) Non-origin vectors   (b) Vector addition   (c) Vector normalization
(not allowed)

Figure 1.1: Examples of vector definition and basic operations

Vector addition is commutative (like scalar addition) because $\overline{x} + \overline{y} = \overline{y} + \overline{x}$. When two vectors, $\overline{x}$ and $\overline{y}$, are added, the origin, $\overline{x}$, $\overline{y}$, and $\overline{x} + \overline{y}$ represent the vertices of a parallelogram. For example, consider the vectors $\overline{A} = [4, 3]$ and $\overline{B} = [1, 4]$. The sum of these two vectors is $\overline{A} + \overline{B} = [5, 7]$. The addition of these two vectors is shown in Figure 1.1(b). It is easy to show that the four points $[0, 0]$, $[4, 3]$, $[1, 4]$, and $[5, 7]$ form a parallelogram in 2-dimensional space, and the addition of the vectors is one of the diagonals of the parallelogram. The other diagonal can be shown to be parallel to either $\overline{A} - \overline{B}$ or $\overline{B} - \overline{A}$, depending on the direction of the vector. Note that vector addition and subtraction follow the same rules in linear algebra as for geometric vectors, except that the tails of the vectors are always origin rooted. For example, the vector $(\overline{A} - \overline{B})$ should no longer be drawn as a diagonal of the parallelogram, but as an origin-rooted vector with the same direction as the diagonal. Nevertheless, the diagonal abstraction still helps in the computation of $(\overline{A} - \overline{B})$. One way of visualizing vector addition (in terms of the velocity abstraction) is that if a platform moves on the ground with velocity $[1, 4]$, and if the person walks on the platform (relative to it) with velocity $[4, 3]$, then the overall velocity of the person relative to the ground is $[5, 7]$.

It is possible to multiply a vector with a scalar by multiplying each component of the vector with the scalar. Consider a vector $\overline{x} = [x_1, \ldots x_d]$, which is scaled by a factor of a:

$$\overline{x}' = a\overline{x} = [a\, x_1 \ldots a\, x_d]$$

For example, if the vector $\overline{x}$ contains the number of units sold of each product, then one can use $a = 10^{-6}$ to convert units sold into number of *millions of* units sold. The scalar multiplication operation simply scales the *length* of the vector, but does not change its *direction* (i.e., relative values of different components). The notion of "length" is defined more formally in terms of the norm of the vector, which is discussed below.

Vectors can be multiplied with the notion of the *dot product*. The dot product between two vectors, $\overline{x} = [x_1, \ldots, x_d]$ and $\overline{y} = [y_i, \ldots y_d]$, is the sum of the element-wise multiplication of their individual components. The dot product of $\overline{x}$ and $\overline{y}$ is denoted by $\overline{x} \cdot \overline{y}$ (with a dot in the middle) and is formally defined as follows:

$$\overline{x} \cdot \overline{y} = \sum_{i=1}^{d} x_i y_i \tag{1.1}$$

Consider a case where we have $\overline{x} = [1, 2, 3]$ and $\overline{y} = [6, 5, 4]$. In such a case, the dot product of these two vectors can be computed as follows:

$$\overline{x} \cdot \overline{y} = (1)(6) + (2)(5) + (3)(4) = 28 \tag{1.2}$$

The dot product is a special case of a more general operation, referred to as the *inner product*, and it preserves many fundamental rules of Euclidean geometry. The space of vectors that includes a dot product operation is referred to as a *Euclidean space*. The dot product is a commutative operation:

$$\overline{x} \cdot \overline{y} = \sum_{i=1}^{d} x_i y_i = \sum_{i=1}^{d} y_i x_i = \overline{y} \cdot \overline{x}$$

The dot product also inherits the distributive property of scalar multiplication:

$$\overline{x} \cdot (\overline{y} + \overline{z}) = \overline{x} \cdot \overline{y} + \overline{x} \cdot \overline{z}$$

The dot product of a vector, $\overline{x} = [x_1, \ldots x_d]$, with itself is referred to as its squared *norm* or Euclidean norm. The norm defines the vector length and is denoted by $\| \cdot \|$:

$$\|\overline{x}\|^2 = \overline{x} \cdot \overline{x} = \sum_{i=1}^{d} x_i^2$$

The norm of the vector is the Euclidean distance of its coordinates from the origin. In the case of Figure 1.1(a), the norm of the vector $[4, 3]$ is $\sqrt{4^2 + 3^2} = 5$. Often, vectors are *normalized* to unit length by dividing them with their norm:

$$\overline{x}' = \frac{\overline{x}}{\|\overline{x}\|} = \frac{\overline{x}}{\sqrt{\overline{x} \cdot \overline{x}}}$$

Scaling a vector by its norm does not change the relative values of its components, which define the direction of the vector. For example, the Euclidean distance of $[4, 3]$ from the origin is 5. Dividing each component of the vector by 5 results in the vector $[4/5, 3/5]$, which changes the length of the vector to 1, but not its direction. This shortened vector is shown in Figure 1.1(c), and it overlaps with the vector $[4, 3]$. The resulting vector is referred to as a *unit vector*.

A generalization of the Euclidean norm is the L_p-norm, which is denoted by $\| \cdot \|_p$:

$$\|\overline{x}\|_p = \left(\sum_{i=1}^{d} |x_i|^p \right)^{(1/p)} \tag{1.3}$$

Here, $| \cdot |$ indicates the absolute value of a scalar, and p is a positive integer. For example, when p is set to 1, the resulting norm is referred to as the Manhattan norm or the L_1-norm.

The (squared) Euclidean distance between $\overline{x} = [x_1, \ldots x_d]$ and $\overline{y} = [y_1, \ldots, y_d]$ can be shown to be the dot product of $\overline{x} - \overline{y}$ with itself:

$$\|\overline{x} - \overline{y}\|^2 = (\overline{x} - \overline{y}) \cdot (\overline{x} - \overline{y}) = \sum_{i=1}^{d} (x_i - y_i)^2 = \text{Euclidean}(\overline{x}, \overline{y})^2$$

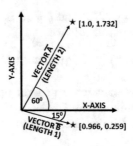

Figure 1.2: The angular geometry of vectors $\overline{A}$ and $\overline{B}$

Dot products satisfy the *Cauchy-Schwarz inequality*, according to which the dot product between a pair of vectors is bounded above by the product of their lengths:

$$|\sum_{i=1}^{d} x_i y_i| = |\overline{x} \cdot \overline{y}| \leq \|\overline{x}\| \, \|\overline{y}\| \tag{1.4}$$

The Cauchy-Schwarz inequality can be proven by first showing that $|\overline{x} \cdot \overline{y}| \leq 1$ when $\overline{x}$ and $\overline{y}$ are unit vectors (i.e., the result holds when the arguments are unit vectors). This is because both $\|\overline{x} - \overline{y}\|^2 = 2 - 2\overline{x} \cdot \overline{y}$ and $\|\overline{x} + \overline{y}\|^2 = 2 + 2\overline{x} \cdot \overline{y}$ are nonnegative. This is possible only when $|\overline{x} \cdot \overline{y}| \leq 1$. One can then generalize this result to arbitrary length vectors by observing that the dot product scales up linearly with the norms of the underlying arguments. Therefore, one can scale up both sides of the inequality with the norms of the vectors.

Problem 1.2.1 (Triangle Inequality) *Consider the triangle formed by the origin, $\overline{x}$, and $\overline{y}$. Use the Cauchy-Schwarz inequality to show that the side length $\|\overline{x} - \overline{y}\|$ is no greater than the sum $\|\overline{x}\| + \|\overline{y}\|$ of the other two sides.*

A hint for solving the above problem is that both sides of the triangle inequality are non-negative. Therefore, the inequality is true if and only if it holds after squaring both sides.

The Cauchy-Schwarz inequality shows that the dot product between a pair of vectors is no greater than the product of vector lengths. In fact, the ratio between these two quantities is the cosine of the angle between the two vectors (which is always less than 1). For example, one often represents the coordinates of a 2-dimensional vector in *polar form* as $[a, \theta]$, where a is the length of the vector, and θ is the counter-clockwise angle the vector makes with the X-axis. The Cartesian coordinates are $[a\cos(\theta), a\sin(\theta)]$, and the dot product of this Cartesian coordinate vector with $[1, 0]$ (the X-axis) is $a\cos(\theta)$. As another example, consider two vectors with lengths 2 and 1, respectively, which make (counter-clockwise) angles of $60°$ and $-15°$ with respect to the X-axis in a 2-dimensional setting. These vectors are shown in Figure 1.2. The coordinates of these vectors are $[2\cos(60), 2\sin(60)] = [1, \sqrt{3}]$ and $[\cos(-15), \sin(-15)] = [0.966, -0.259]$.

The cosine function between two vectors $\overline{x} = [x_1 \ldots x_d]$ and $\overline{y} = [y_i, \ldots y_d]$ is algebraically defined by the dot product between the two vectors after scaling them to unit norm:

$$\cos(\overline{x}, \overline{y}) = \frac{\overline{x} \cdot \overline{y}}{\sqrt{\overline{x} \cdot \overline{x}} \sqrt{\overline{y} \cdot \overline{y}}} = \frac{\overline{x} \cdot \overline{y}}{\|\overline{x}\| \, \|\overline{y}\|} \tag{1.5}$$

The algebraically computed cosine function over $\overline{x}$ and $\overline{y}$ has the normal trigonometric interpretation of being equal to $\cos(\theta)$, where θ is the angle between the vectors $\overline{x}$ and $\overline{y}$.

For example, the two vectors $\overline{A}$ and $\overline{B}$ in Figure 1.2 are at an angle of 75° to each other, and have norms of 1 and 2, respectively. Then, the algebraically computed cosine function over the pair $[\overline{A}, \overline{B}]$ is equal to the expected trigonometric value of $\cos(75)$:

$$\cos(\overline{A}, \overline{B}) = \frac{0.966 \times 1 - 0.259 \times \sqrt{3}}{1 \times 2} \approx 0.259 \approx \cos(75)$$

In order to understand why the algebraic dot product between two vectors yields the trigonometric cosine value, one can use the *cosine law* from Euclidean geometry. Consider the triangle created by the origin, $\overline{x} = [x_1, \ldots, x_d]$ and $\overline{y} = [y_1, \ldots, y_d]$. We want to find the angle θ between $\overline{x}$ and $\overline{y}$. The Euclidean side lengths of this triangle are $a = \|\overline{x}\|$, $b = \|\overline{y}\|$, and $c = \|\overline{x} - \overline{y}\|$. The cosine law provides a formula for the angle θ in terms of side lengths as follows:

$$\cos(\theta) = \frac{a^2 + b^2 - c^2}{2ab} = \frac{\|\overline{x}\|^2 + \|\overline{y}\|^2 - \|\overline{x} - \overline{y}\|^2}{2\left(\|\overline{x}\|\right)\left(\|\overline{y}\|\right)} = \frac{\overline{x} \cdot \overline{y}}{\sqrt{\overline{x} \cdot \overline{x}}\sqrt{\overline{y} \cdot \overline{y}}}$$

The second relationship is obtained by expanding $\|\overline{x} - \overline{y}\|^2$ as $(\overline{x} - \overline{y}) \cdot (\overline{x} - \overline{y})$ and then using the distributive property of dot products. Almost all the wonderful geometric properties of Euclidean spaces can be algebraically traced back to this simple relationship between the dot product and the trigonometric cosine. *The simple algebra of the dot product operation hides a lot of complex Euclidean geometry.* The exercises at the end of this chapter show that many basic geometric and trigonometric identities can be proven very easily with algebraic manipulation of dot products.

A pair of vectors is *orthogonal* if their dot product is 0, and the angle between them is 90° (for non-zero vectors). The vector $\overline{0}$ is considered orthogonal to every vector. A *set* of vectors is *orthonormal* if each pair in the set is mutually orthogonal and the norm of each vector is 1. Orthonormal directions are useful because they are employed for transformations of points across different orthogonal coordinate systems with the use of 1-dimensional *projections*. In other words, a new set of coordinates of a data point can be computed with respect to the changed set of directions. This approach is referred to as *coordinate transformation* in analytical geometry, and is also used frequently in linear algebra. The 1-dimensional projection operation of a vector $\overline{x}$ on a unit vector is defined the dot product between the two vectors. It has a natural geometric interpretation as the (positive or negative) distance of $\overline{x}$ from the origin *in the direction of the unit vector*, and therefore it is considered a coordinate in that direction. Consider the point $[10, 15]$ in a 2-dimensional coordinate system. Now imagine that you were given the orthonormal directions $[3/5, 4/5]$ and $[-4/5, 3/5]$. One can represent the point $[10, 15]$ in a new coordinate system defined by the directions $[3/5, 4/5]$ and $[-4/5, 3/5]$ by computing the dot product of $[10, 15]$ with each of these vectors. Therefore, the new coordinates $[x', y']$ are defined as follows:

$$x' = 10 * (3/5) + 15 * (4/5) = 18, \quad y' = 10 * (-4/5) + 15 * (3/5) = 1$$

One can express the original vector using the new axes and coordinates as follows:

$$[10, 15] = x'[3/5, 4/5] + y'[-4/5, 3/5]$$

These types of transformations of vectors to new representations lie at the heart of linear algebra. In many cases, transformed representations of data sets (e.g., replacing each $[x, y]$ in a 2-dimensional data set with $[x', y']$) have useful properties, which are exploited by machine learning applications.

1.2.2   Basic Operations with Vectors and Matrices

The transpose of a matrix is obtained by flipping its rows and columns. In other words, the (i, j)th entry of the transpose is the same as the (j, i)th entry of the original matrix. Therefore, the transpose of an $n \times d$ matrix is a $d \times n$ matrix. The transpose of a matrix A is denoted by A^T. An example of a transposition operation is shown below:

$$\begin{bmatrix} a_{11} & a_{12} \\ a_{21} & a_{22} \\ a_{31} & a_{32} \end{bmatrix}^T = \begin{bmatrix} a_{11} & a_{21} & a_{31} \\ a_{12} & a_{22} & a_{32} \end{bmatrix}$$

It is easy to see that the transpose of the transpose $(A^T)^T$ of a matrix A is the original matrix A. Like matrices, row vectors can be transposed to column vectors, and vice versa.

Like vectors, matrices can be added only if they have exactly the same sizes. For example, one can add the matrices A and B only if A and B have exactly the same number of rows and columns. The (i, j)th entry of $A+B$ is the sum of the (i, j)th entries of A and B, respectively. The matrix addition operator is commutative, because it inherits the commutative property of scalar addition of its individual entries. Therefore, we have:

$$A + B = B + A$$

A *zero matrix* or *null matrix* is the matrix analog of the scalar value of 0, and it contains only 0s. It is often simply written as "0" even though it is a matrix. It can be added to a matrix of the same size without affecting its values:

$$A + 0 = A$$

Note that matrices, vectors, and scalars all have their own definition of a zero element, which is required to obey the above *additive identity*. For vectors, the zero element is the vector of 0s, and it is written as "$\overline{0}$" with an overbar on top.

It is easy to show that the transpose of the sum of two matrices $A = [a_{ij}]$ and $B = [b_{ij}]$ is given by the sum of their transposes. In other words, we have the following relationship:

$$(A + B)^T = A^T + B^T \tag{1.6}$$

The result can be proven by demonstrating that the (i, j)th element of both sides of the above equation is $(a_{ji} + b_{ji})$.

An $n \times d$ matrix A can either be multiplied with a d-dimensional column vector $\overline{x}$ as $A\overline{x}$, or it can be multiplied with an n-dimensional row vector $\overline{y}$ as $\overline{y}A$. When an $n \times d$ matrix A is multiplied with d-dimensional column vector $\overline{x}$ to create $A\overline{x}$, an element-wise multiplication is performed between the d elements of each row of the matrix A and the d elements of the column vector $\overline{x}$, and then these element-wise products are added to create a scalar. Note that this operation is the same as the dot product, except that one needs to transpose the rows of A to column vectors to rigorously express it as a dot product. This is because dot products are defined between two vectors of the same type (i.e., row vectors or column vectors). At the end of the process, n scalars are computed and arranged into an n-dimensional column vector in which the ith element is the product between the ith row of A and $\overline{x}$. An example of a multiplication of a 3×2 matrix $A = [a_{ij}]$ with a 2-dimensional column vector $\overline{x} = [x_1, x_2]^T$ is shown below:

$$\begin{bmatrix} a_{11} & a_{12} \\ a_{21} & a_{22} \\ a_{31} & a_{32} \end{bmatrix} \begin{bmatrix} x_1 \\ x_2 \end{bmatrix} = \begin{bmatrix} a_{11}x_1 + a_{12}x_2 \\ a_{21}x_1 + a_{22}x_2 \\ a_{31}x_1 + a_{32}x_2 \end{bmatrix} \tag{1.7}$$

One can also post-multiply an n-dimensional row vector with an $n \times d$ matrix $A = [a_{ij}]$ to create a d-dimensional row vector. An example of the multiplication of a 3-dimensional row vector $\overline{v} = [v_1, v_2, v_3]$ with the 3×2 matrix A is shown below:

$$[v_1, v_2, v_3] \begin{bmatrix} a_{11} & a_{12} \\ a_{21} & a_{22} \\ a_{31} & a_{32} \end{bmatrix} = [v_1 a_{11} + v_2 a_{21} + v_3 a_{31}, v_1 a_{12} + v_2 a_{22} + v_3 a_{32}] \quad (1.8)$$

It is clear that the multiplication operation between matrices and vectors is not commutative.

The multiplication of an $n \times d$ matrix A with a d-dimensional column vector $\overline{x}$ to create an n-dimensional column vector $A\overline{x}$ is often interpreted as a *linear transformation* from d-dimensional space to n-dimensional space. The precise mathematical definition of a linear transformation is given in Chapter 2. For now, we ask the reader to observe that the result of the multiplication is a weighted sum of the columns of the matrix A, where the weights are provided by the scalar components of vector $\overline{x}$. For example, one can rewrite the matrix-vector multiplication of Equation 1.7 as follows:

$$\begin{bmatrix} a_{11} & a_{12} \\ a_{21} & a_{22} \\ a_{31} & a_{32} \end{bmatrix} \begin{bmatrix} x_1 \\ x_2 \end{bmatrix} = x_1 \begin{bmatrix} a_{11} \\ a_{21} \\ a_{31} \end{bmatrix} + x_2 \begin{bmatrix} a_{12} \\ a_{22} \\ a_{32} \end{bmatrix} \quad (1.9)$$

Here, a 2-dimensional vector is mapped into a 3-dimensional vector as a weighted combination of the columns of the matrix. Therefore, the $n \times d$ matrix A is occasionally represented in terms of its ordered set of n-dimensional columns $\overline{a}_1 \ldots \overline{a}_d$ as $A = [\overline{a}_1 \ldots \overline{a}_d]$. This results in the following form of matrix-vector multiplication using the columns of A and a column vector $\overline{x} = [x_1 \ldots x_d]^T$ of coefficients:

$$A\overline{x} = \sum_{i=1}^{d} x_i \overline{a}_i = \overline{b}$$

Each x_i corresponds to the "weight" of the ith direction $\overline{a}_i$, which is also referred to as the ith *coordinate* of $\overline{b}$ using the (possibly non-orthogonal) directions contained in the columns of A. This notion is a generalization of the (orthogonal) Cartesian coordinates defined by d-dimensional vectors $\overline{e}_1 \ldots \overline{e}_d$, where each $\overline{e}_i$ is an axis direction with a single 1 in the ith position and remaining 0s. For the case of the Cartesian system defined by $\overline{e}_1 \ldots \overline{e}_d$, the coordinates of $\overline{b} = [b_1 \ldots b_d]^T$ are simply $b_1 \ldots b_d$, since we have $\overline{b} = \sum_{i=1}^{d} b_i \overline{e}_i$.

The dot product between two vectors can be viewed as a special case of matrix-vector multiplication. In such a case, a $1 \times d$ matrix (row vector) is multiplied with a $d \times 1$ matrix (column vector), and the result is the same as one would obtain by performing a dot product between the two vectors. However, a subtle difference is that the dot product is defined between two vectors of the same type (typically column vectors) rather than between the matrix representation of a row vector and the matrix representation of a column vector. In order to implement a dot product as a matrix-matrix multiplication, we would first need to convert one of the column vectors into the matrix representation of a row vector, and then perform the matrix multiplication by ordering the "wide" matrix (row vector) before the "tall" matrix (column vector). The resulting 1×1 matrix contains the dot product. For example, consider the dot product in matrix form, which is obtained by matrix-centric multiplication of a row vector with a column vector:

$$\overline{v} \cdot \overline{x} = [v_1, v_2, v_3] \begin{bmatrix} x_1 \\ x_2 \\ x_3 \end{bmatrix} = [v_1 x_1 + v_2 x_2 + v_3 x_3]$$

The result of the matrix multiplication is a 1×1 matrix containing the dot product, which is a scalar. It is clear that we always obtain the same 1×1 matrix, irrespective of the order of the arguments in the dot product, as long as we transpose the *first* vector in order to place the "wide" matrix before the "tall" matrix:

$$\overline{x} \cdot \overline{v} = \overline{v} \cdot \overline{x}, \quad \overline{x}^T \overline{v} = \overline{v}^T \overline{x}$$

Therefore, dot products are commutative.

However, if we order the "tall" matrix before the "wide" matrix, what we obtain is the *outer product* between the two vectors. The outer product between two 3-dimensional vectors is a 3×3 matrix! In vector form, the outer product is defined between two column vectors $\overline{x}$ and $\overline{v}$ and is denoted by $\overline{x} \otimes \overline{v}$. However, it is easiest to understand the outer product by using the matrix representation of the vectors for multiplication, wherein the first of the vectors is converted into a column vector representation (if needed), and the second of the two vectors is converted into a row vector representation (if needed). In other words, the "tall" matrix is always ordered before the "wide" matrix:

$$\overline{x} \otimes \overline{v} = \overline{x}\,\overline{v}^T = \begin{bmatrix} x_1 \\ x_2 \\ x_3 \end{bmatrix} [v_1, v_2, v_3] = \begin{bmatrix} v_1 x_1 & v_2 x_1 & v_3 x_1 \\ v_1 x_2 & v_2 x_2 & v_3 x_2 \\ v_1 x_3 & v_2 x_3 & v_3 x_3 \end{bmatrix}$$

Unlike dot products, outer products can be performed between two vectors of different lengths. Conventionally, outer products are defined between two column vectors, and the second vector is *transposed* into a matrix containing a single row before matrix multiplication. In other words, the jth component of the second vector (in d dimensions) becomes the $(1, j)$th element of the second matrix (of size $1 \times d$) in the multiplication. The first matrix is simply a $d \times 1$ matrix derived from the column vector. Unlike dot products, the outer product is not commutative; the *order* of the operands matters not only to the values in the final matrix, but also to the size of the final matrix:

$$\overline{x} \otimes \overline{v} \neq \overline{v} \otimes \overline{x}, \quad \overline{x}\,\overline{v}^T \neq \overline{v}\,\overline{x}^T$$

The multiplication between vectors, or the multiplication of a matrix with a vector, are both special cases of multiplying two matrices. However, in order to multiply two matrices, certain constraints on their sizes need to be respected. For example, an $n \times k$ matrix U can be multiplied with a $k \times d$ matrix V only because the number of columns k in U is the same as the number of rows k in V. The resulting matrix is of size $n \times d$, in which the (i, j)th entry is the dot product between the vectors corresponding to the ith row of U and the jth column of V. Note that the dot product operations within the multiplication require the underlying vectors to be of the same sizes. The outer product between two vectors is a special case of matrix multiplication that uses $k = 1$ with arbitrary values of n and d; similarly, the inner product is a special case of matrix multiplication that uses $n = d = 1$, but some arbitrary value of k. Consider the case in which the (i, j)th entries of U and V are u_{ij} and v_{ij}, respectively. Then, the (i, j)th entry of UV is given by the following:

$$(UV)_{ij} = \sum_{r=1}^{k} u_{ir} v_{rj} \tag{1.10}$$

An example of a matrix multiplication is shown below:

$$\begin{bmatrix} u_{11} & u_{12} \\ u_{21} & u_{22} \\ u_{31} & u_{32} \end{bmatrix} \begin{bmatrix} v_{11} & v_{12} & v_{13} \\ v_{21} & v_{22} & v_{23} \end{bmatrix} = \begin{bmatrix} u_{11}v_{11} + u_{12}v_{21} & u_{11}v_{12} + u_{12}v_{22} & u_{11}v_{13} + u_{12}v_{23} \\ u_{21}v_{11} + u_{22}v_{21} & u_{21}v_{12} + u_{22}v_{22} & u_{21}v_{13} + u_{22}v_{23} \\ u_{31}v_{11} + u_{32}v_{21} & u_{31}v_{12} + u_{32}v_{22} & u_{31}v_{13} + u_{32}v_{23} \end{bmatrix} \tag{1.11}$$

Note that both the two earlier matrix-to-vector and vector-to-matrix multiplications can be viewed as special cases of this more general operation. This is because a d-dimensional row vector can be treated as an $1 \times d$ matrix and a n-dimensional column vector can be treated as a $n \times 1$ matrix. For example, if we multiply this type of special $n \times 1$ matrix with a $1 \times d$ matrix, we will obtain an $n \times d$ matrix with some special properties.

Problem 1.2.2 (Outer Product Properties) *Show that if an $n \times 1$ matrix is multiplied with a $1 \times d$ matrix (which is also an outer product between two vectors), we obtain an $n \times d$ matrix with the following properties: (i) Every row is a multiple of every other row, and (ii) every column is a multiple of every other column.*

It is also possible to show that matrix products can be broken up into the sum of simpler matrices, each of which is an outer product of two vectors. We have already seen that each entry in a matrix product is itself an *inner* product of two vectors extracted from the matrix. What about outer products? It can be shown that the entire matrix is the sum of as many outer products as the common dimension k of the two multiplied matrices:

Lemma 1.2.1 (Matrix Multiplication as Sum of Outer Products) *The product of an $n \times k$ matrix U with a $k \times d$ matrix V results in an $n \times d$ matrix, which can be expressed as the sum of k outer-product matrices; each of these k matrices is the product of an $n \times 1$ matrix with a $1 \times d$ matrix. Each $n \times 1$ matrix corresponds to the ith column U_i of U and each $1 \times d$ matrix corresponds to the ith row V_i of V. Therefore, we have the following:*

$$UV = \sum_{r=1}^{k} \underbrace{U_r V_r}_{n \times d}$$

Proof: Let u_{ij} and v_{ij} be the (i,j)th entries of U and V, respectively. It can be shown that the rth term in the summation on the right-hand side of the equation in the statement of the lemma contributes $u_{ir}v_{rj}$ to the (i,j)th entry in the summation matrix. Therefore, the overall sum of the terms on the right-hand side is $\sum_{r=1}^{k} u_{ir}v_{rj}$. This sum is exactly the same as the definition of the (i,j)th term of the matrix multiplication UV (cf. Equation 1.10). ∎
In general, matrix multiplication is *not* commutative (except for special cases). In other words, we have $AB \neq BA$ in the general case. This is different from scalar multiplication, which is commutative. A concrete example of non-commutativity is as follows:

$$\begin{bmatrix} 1 & 1 \\ 0 & 0 \end{bmatrix} \begin{bmatrix} 1 & 0 \\ 1 & 0 \end{bmatrix} = \begin{bmatrix} 2 & 0 \\ 0 & 0 \end{bmatrix} \neq \begin{bmatrix} 1 & 0 \\ 1 & 0 \end{bmatrix} \begin{bmatrix} 1 & 1 \\ 0 & 0 \end{bmatrix} = \begin{bmatrix} 1 & 1 \\ 1 & 1 \end{bmatrix}$$

In fact, if the matrices A and B are not square, it might be possible that one of the products, AB, is possible to compute based on the sizes of A and B, whereas BA might not be computable. For example, it is possible to compute AB for the 4×2 matrix A and the 2×5 matrix B. However, it is not possible to compute BA because of mismatching dimensions.

Although matrix multiplication is not commutative, it is associative and distributive:

$$A(BC) = (AB)C, \qquad\qquad\qquad\qquad \text{[Associativity]}$$
$$A(B+C) = AB + AC, \quad (B+C)A = BA + CA, \qquad \text{[Distributivity]}$$

The basic idea for proving each of the above results is to define variables for the dimensions and entries of each of $A = [a_{ij}]$, $B = [b_{ij}]$, and $C = [c_{ij}]$. Then, an algebraic expression can be computed for the (i, j)th entry on both sides of the equation, and the two are shown to be equal. For example, in the case of associativity, this type of expansion yields the following:

$$[A(BC)]_{ij} = [(AB)C]_{ij} = \sum_k \sum_m a_{ik} b_{km} c_{mj}$$

These properties also hold for matrix-vector multiplication, because all vectors are special cases of matrices. The associativity property is very useful in ensuring efficient matrix multiplication by carefully selecting from the different choices allowed by associativity.

Problem 1.2.3 *Express the matrix ABC as the weighted sum of outer products of vectors extracted from A and C. The weights are extracted from matrix B.*

Problem 1.2.4 *Let A be an 1000000×2 matrix. Suppose you have to compute the 2×1000000 matrix $A^T A A^T$ on a computer with limited memory. Would you prefer to compute $(A^T A)A^T$ or would you prefer to compute $A^T (A A^T)$?*

Problem 1.2.5 *Let D be an $n \times d$ matrix for which each column sums to 0. Let A be an arbitrary $d \times d$ matrix. Show that the sum of each column of DA is also zero.*

The key point in showing the above result is to use the fact that the sum of the rows of D can be expressed as $\overline{e}^T D$, where $\overline{e}$ is a column vector of 1s.

The transpose of the product of two matrices is given by the product of their transposes, but the order of multiplication is reversed:

$$(AB)^T = B^T A^T \tag{1.12}$$

This result can be easily shown by working out the algebraic expression for the (i, j)th entry in terms of the entries of $A = [a_{ij}]$ and $B = [b_{ij}]$. The result for transposes can be easily extended to any number of matrices, as shown below:

Problem 1.2.6 *Show the following result for matrices $A_1 \ldots A_n$:*

$$(A_1 A_2 A_3 \ldots A_n)^T = A_n^T A_{n-1}^T \ldots A_2^T A_1^T$$

The multiplication between a matrix and a vector also satisfies the same type of transposition rule as shown above.

1.2.3   Special Classes of Matrices

A *symmetric matrix* is a square matrix that is its own transpose. In other words, if A is a symmetric matrix, then we have $A = A^T$. An example of a 3×3 symmetric matrix is shown below:

$$\begin{bmatrix} 2 & 1 & 3 \\ 1 & 4 & 5 \\ 3 & 5 & 6 \end{bmatrix}$$

Note that the (i, j)th entry is always equal to the (j, i)th entry for each $i, j \in \{1, 2, 3\}$.

Problem 1.2.7 *If A and B are symmetric matrices, then show that AB is symmetric if and only if AB = BA.*

The *diagonal* of a matrix is defined as the set of entries for which the row and column indices are the same. Although the notion of diagonal is generally used for square matrices, the definition is sometimes also used for rectangular matrices; in such a case, the diagonal starts at the upper-left corner so that the row and column indices are the same. A square matrix that has values of 1 in all entries along the diagonal and 0s for all non-diagonal entries is referred to as an *identity matrix*, and is denoted by I. In the event that the non-diagonal entries are 0, but the diagonal entries are different from 1, the resulting matrix is referred to as a *diagonal matrix*. Therefore, the identity matrix is a special case of a diagonal matrix. Multiplying an $n \times d$ matrix A with the identity matrix of the appropriate size in any order results in the same matrix A. One can view the identity matrix as the analog of the value of 1 in scalar multiplication:

$$AI = IA = A \tag{1.13}$$

Since A is an $n \times d$ matrix, the size of the identity matrix I in the product AI is $d \times d$, whereas the size of the identity matrix in the product IA is $n \times n$. This is somewhat confusing, because the same notation I in Equation 1.13 refers to identity matrices of two different sizes. In such cases, ambiguity is avoided by subscripting the identity matrix to indicate its size. For example, an identity matrix of size $d \times d$ is denoted by I_d. Therefore, a more unambiguous form of Equation 1.13 is as follows:

$$AI_d = I_n A = A \tag{1.14}$$

Although diagonal matrices are assumed to be square by default, it is also possible to create a relaxed definition[1] of a diagonal matrix, which is not square. In this case, the diagonal is aligned with the *upper-left corner* of the matrix. Such matrices are referred to as *rectangular diagonal matrices*.

Definition 1.2.1 (Rectangular Diagonal Matrix) *A rectangular diagonal matrix is an $n \times d$ matrix in which each entry (i, j) has a non-zero value if and only if $i = j$. Therefore, the diagonal of non-zero entries starts at the upper-left corner of the matrix, although it might not meet the lower-right corner.*

A *block diagonal matrix* contains square blocks $B_1 \ldots B_r$ of (possibly) non-zero entries along the diagonal. All other entries are zero. Although each block is square, they need not be of the same size. Examples of different types of diagonal and block diagonal matrices are shown in the top row of Figure 1.3.

A generalization of the notion of a diagonal matrix is that of a *triangular matrix*.

Definition 1.2.2 (Upper and Lower Triangular Matrix) *A square matrix is an **upper triangular matrix** if all entries (i, j) below its main diagonal (i.e., satisfying $i > j$) are zeros. A matrix is **lower triangular** if all entries (i, j) above its main diagonal (i.e., satisfying $i < j$) are zeros.*

Definition 1.2.3 (Strictly Triangular Matrix) *A matrix is said to be **strictly** triangular if it is triangular **and** all its diagonal elements are zeros.*

---

[1] Instead of referring to such matrices as rectangular diagonal matrices, some authors use a quotation around the word *diagonal*, while referring to such matrices. This is because the word "diagonal" was originally reserved for square matrices.

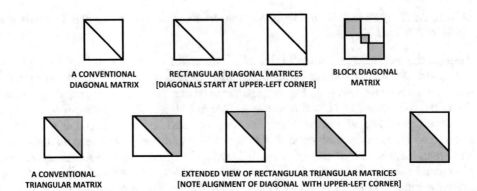

Figure 1.3: Examples of conventional/rectangular diagonal and triangular matrices

We make an important observation about operations on pairs of upper-triangular matrices.

Lemma 1.2.2 (Sum or Product of Upper-Triangular Matrices) *The sum of upper-triangular matrices is upper triangular. The product of upper-triangular matrices is upper triangular.*

Proof Sketch: This result is easy to show by proving that the scalar expressions for the (i, j)th entry in the sum and the product are both 0, when $i > j$.   ∎
The above lemma naturally applies to lower-triangular matrices as well.

Although the notion of a triangular matrix is generally meant for square matrices, it is sometimes used for rectangular matrices. Examples of different types of triangular matrices are shown in the bottom row of Figure 1.3. The portion of the matrix occupied by non-zero entries is shaded. Note that the number of non-zero entries in rectangular triangular matrices heavily depends on the shape of the matrix. Finally, a matrix A is said to be *sparse*, when most of the entries in it have 0 values. It is often computationally efficient to work with such matrices.

1.2.4   Matrix Powers, Polynomials, and the Inverse

Square matrices can be multiplied with themselves without violating the size constraints of matrix multiplication. Multiplying a square matrix with itself many times is analogous to raising a scalar to a particular power. The nth power of a matrix is defined as follows:

$$A^n = \underbrace{AA \ldots A}_{n \text{ times}} \qquad (1.15)$$

The zeroth power of a matrix is defined to be the identity matrix of the same size. When a matrix satisfies $A^k = 0$ for some integer k, it is referred to as *nilpotent*. For example, all strictly triangular matrices of size $d \times d$ satisfy $A^d = 0$. Like scalars, one can raise a square matrix to a fractional power, although it is not guaranteed to exist. For example, if $A = V^2$, then we have $V = A^{1/2}$. Unlike scalars, it is not guaranteed that $A^{1/2}$ exists for an arbitrary matrix A, even after allowing for complex-valued entries in the result (see Exercise 14). In general, one can compute a polynomial function $f(A)$ of a square matrix in much the same way as one computes polynomials of scalars. Instead of the constant term used in a scalar polynomial, multiples of the identity matrix are used; the identity matrix is the matrix analog of the scalar value of 1. For example, the matrix analog of the scalar polynomial $f(x) = 3x^2 + 5x + 2$, when applied to the $d \times d$ matrix A, is as follows:

$$f(A) = 3A^2 + 5A + 2I$$

All polynomials of the same matrix A always commute with respect to the multiplication operator.

Observation 1.2.1 (Commutativity of Matrix Polynomials) *Two polynomials $f(A)$ and $g(A)$ of the same matrix A will always commute:*

$$f(A)g(A) = g(A)f(A)$$

The above result can be shown by expanding the polynomial on both sides, and showing that the same polynomial is reached with the distributive property of matrix multiplication.

Can we raise a matrix to a negative power? The *inverse* of a square matrix A is another square matrix denoted by A^{-1} so that the multiplication of the two matrices (in any order) will result in the identity matrix:

$$AA^{-1} = A^{-1}A = I \tag{1.16}$$

A simple formula exists for inverting 2×2 matrices:

$$\begin{bmatrix} a & b \\ c & d \end{bmatrix}^{-1} = \frac{1}{ad - bc} \begin{bmatrix} d & -b \\ -c & a \end{bmatrix} \tag{1.17}$$

An example of two matrices that are inverses of each other is shown below:

$$\begin{bmatrix} 8 & 3 \\ 5 & 2 \end{bmatrix} \begin{bmatrix} 2 & -3 \\ -5 & 8 \end{bmatrix} = \begin{bmatrix} 2 & -3 \\ -5 & 8 \end{bmatrix} \begin{bmatrix} 8 & 3 \\ 5 & 2 \end{bmatrix} = \begin{bmatrix} 1 & 0 \\ 0 & 1 \end{bmatrix}$$

The inverse of a 1×1 matrix containing the element a is simply the 1×1 matrix containing the element $1/a$. Therefore, a matrix inverse naturally generalizes a scalar inverse. Not all matrices have inverses, just as an inverse does not exist for the scalar $a = 0$. A matrix for which an inverse exists is referred to as *invertible* or *nonsingular*. Otherwise, it is said to be *singular*. For example, if the rows in Equation 1.17 are proportional, we would have $ad - bc = 0$, and therefore, the matrix would not be invertible. An example of a matrix that is not invertible is as follows:

$$A = \begin{bmatrix} 1 & 1 \\ 2 & 2 \end{bmatrix}$$

Note that multiplying A with any 2×2 matrix B will always result in a 2×2 matrix AB in which the second row is twice the first. This is not the case for the identity matrix, and, therefore, an inverse of A does not exist. The fact that the rows in the non-invertible matrix A are related by a proportionality factor is not a coincidence. As you will learn in Chapter 2, matrices that are invertible always have the property that a non-zero linear combination of the rows does not sum to zero. In other words, each vector direction in the rows of an invertible matrix must contribute new, non-redundant "information" that cannot be conveyed using sums, multiples, or linear combinations of other directions. The second row of A is twice its first row, and therefore the matrix A is not invertible.

When the inverse of a matrix A does exist, it is unique. Furthermore, the product of a matrix with its inverse is always commutative and leads to the identity matrix. A natural consequence of these facts is that the inverse of the inverse $(A^{-1})^{-1}$ is the original matrix A. We summarize these properties of inverses in the following two lemmas.

Lemma 1.2.3 (Commutativity of Multiplication with Inverse) *If the product AB of $d \times d$ matrices A and B is the identity matrix I, then BA must also be equal to I.*

Proof: We present a restricted proof by making the assumption that a matrix C always exists so that $CA = I$. Then, we have:

$$C = CI = C(AB) = (CA)B = IB = B$$

∎

The commutativity of the product of a matrix and its inverse can be viewed as an extension of the statement in Observation 1.2.1 that the product of a matrix A with any polynomial of A is always commutative. A fractional or negative power of a matrix A (like A^{-1}) also commutes with A.

Lemma 1.2.4 *When the inverse of a matrix exists, it is always unique. In other words, if B_1 and B_2 satisfy $AB_1 = AB_2 = I$, we must have $B_1 = B_2$.*

Proof: Since $AB_1 = AB_2$, it follows that $AB_1 - AB_2 = 0$. Therefore, we have $A(B_1 - B_2) = 0$. One can pre-multiply the relationship with B_1 to obtain the following:

$$\underbrace{B_1 A}_{I}(B_1 - B_2) = 0$$

This proves that $B_1 = B_2$.   ∎

The negative power A^{-r} for $r > 0$ represents $(A^{-1})^r$. Any polynomial or negative power of a diagonal matrix is another diagonal matrix in which the polynomial function or negative power is applied to each diagonal entry. All diagonal entries of a diagonal matrix need to be non-zero for it to be invertible or have negative powers. The polynomials and inverses of triangular matrices are also triangular matrices of the same type (i.e., lower or upper triangular). A similar result holds for block diagonal matrices.

Problem 1.2.8 (Inverse of Triangular Matrix Is Triangular) *Consider the system of d equations contained in the rows of $R\overline{x} = \overline{e}_k$ for the $d \times d$ upper-triangular matrix R, where $\overline{e}_k$ is a d-dimensional column vector with a single value of 1 in the kth entry and 0 in all other entries. Discuss why solving for $\overline{x} = [x_1 \ldots x_d]^T$ is simple in this case by solving for the variables in the order $x_d, x_{d-1}, \ldots x_1$. Furthermore, discuss why the solution for $R\overline{x} = \overline{e}_k$ must satisfy $x_i = 0$ for $i > k$. Why is the solution $\overline{x}$ equal to the kth column of the inverse of R? Discuss why the inverse of R is also upper-triangular.*

Problem 1.2.9 (Block Diagonal Polynomial and Inverse) *Suppose that you have a block diagonal matrix B, which has blocks $B_1 \ldots B_r$ along the diagonal. Show how you can express the polynomial function $f(B)$ and the inverse of B in terms of functions on block matrices.*

The inverse of the product of two square (and invertible) matrices can be computed as a product of their inverses, but with the order of multiplication reversed:

$$(AB)^{-1} = B^{-1}A^{-1} \tag{1.18}$$

Both matrices must be invertible for the product to be invertible. We can use the associativity property of matrix multiplication to show the above result:

$$(AB)(B^{-1}A^{-1}) = A((BB^{-1})A^{-1}) = A((I)A^{-1}) = AA^{-1} = I$$

One can extend the above results to show that $(A_1 A_2 \ldots A_k)^{-1} = A_k^{-1} A_{k-1}^{-1} \ldots A_1^{-1}$. Note that the individual matrices A_i must be invertible for their product to be invertible. Even if one of the matrices A_i is not invertible, the product will not be invertible (see Exercise 52).

Problem 1.2.10 *Suppose that the matrix B is the inverse of matrix A. Show that for any positive integer n, the matrix B^n is the inverse of matrix A^n.*

The inversion and the transposition operations can be applied in any order without affecting the result:
$$(A^T)^{-1} = (A^{-1})^T \tag{1.19}$$
This result holds because $A^T (A^{-1})^T = (A^{-1} A)^T = I^T = I$. One can similarly show that $(A^{-1})^T A^T = I$. In other words, $(A^{-1})^T$ is the inverse of A^T.

An *orthogonal matrix* is a square matrix whose inverse is its transpose:
$$AA^T = A^T A = I \tag{1.20}$$
Although such matrices are formally defined in terms of having orthonormal *columns*, the commutativity in the above relationship implies the remarkable property that they contain *both* orthonormal columns and orthonormal rows.

A useful property of invertible matrices is that they define uniquely solvable systems of equations. For example, the solution to $A\overline{x} = \overline{b}$ exists and is uniquely defined as $\overline{x} = A^{-1}\overline{b}$ when A is invertible (cf. Chapter 2). One can also view the solution $\overline{x}$ as a new set of coordinates of $\overline{b}$ in a different (and possibly non-orthogonal) coordinate system defined by the vectors contained in the columns of A. Note that when A is orthogonal, the solution simplifies to $\overline{x} = A^T \overline{b}$, which is equivalent to evaluating the dot product between $\overline{b}$ and each column of A to compute the corresponding coordinate. In other words, we are *projecting $\overline{b}$* on each orthonormal column of A to compute the corresponding coordinate.

1.2.5 The Matrix Inversion Lemma: Inverting the Sum of Matrices

Is it possible to compute the inverse of the sum of two matrices as a function of polynomials or inverses of the individual matrices? In order to answer this question, note that it is not possible to easily do this even for scalars a and b (which are special cases of matrices). For example, it is not possible to easily express $1/(a+b)$ in terms of $1/a$ and $1/b$. Furthermore, the sum of two matrices A and B need not be invertible even when A and B are invertible. In the scalar case, we might have $a + b = 0$, in which case it is not possible to compute $1/(a+b)$. Therefore, it is not easy to compute the inverse of the sum of two matrices.

Some special cases are easier to invert, such as the sum of A with the identity matrix. In such a case, one can generalize the scalar formula for $1/(1+a)$ to matrices. The scalar formula for $1/(1+a)$ for $|a| < 1$ is that of an infinite geometric series:
$$\frac{1}{1+a} = 1 - a + a^2 - a^3 + a^4 + \ldots + \text{Infinite Terms} \tag{1.21}$$

The absolute value of a has to be less than 1 for the infinite summation not to blow up. The corresponding analog is the matrix A, which is such that raising it to the nth power causes all the entries of the matrix to go to 0 as $n \Rightarrow \infty$. In other words, the *limit* of A^n as $n \Rightarrow \infty$ is the zero matrix. For such matrices, the following result holds:
$$(I + A)^{-1} = I - A + A^2 - A^3 + A^4 + \ldots + \text{Infinite Terms}$$
$$(I - A)^{-1} = I + A + A^2 + A^3 + A^4 + \ldots + \text{Infinite Terms}$$

The result can be used for inverting triangular matrices (although more straightforward alternatives exist):

Problem 1.2.11 (Inverting Triangular Matrices) *A $d \times d$ triangular matrix L with non-zero diagonal entries can be expressed in the form $(\Delta + A)$, where Δ is an invertible diagonal matrix and A is a* **strictly** *triangular matrix. Show how to compute the inverse of L using only diagonal matrix inversions and matrix multiplicatons/additions. Note that strictly triangular matrices of size $d \times d$ are always nilpotent and satisfy $A^d = 0$.*

It is also possible to derive an expression for inverting the sum of two matrices in terms of the original matrices under the condition that one of the two matrices is "compact." By compactness, we mean that one of the two matrices has so much structure to it that it can be expressed as the product of two much smaller matrices. The *matrix-inversion lemma* is a useful property for computing the inverse of a matrix after incrementally updating it with a matrix created from the outer-product of two vectors. These types of inverses arise often in iterative optimization algorithms such as the *quasi-Newton method* and for incremental linear regression. In these cases, the inverse of the original matrix is already available, and one can cheaply update the inverse with the matrix inversion lemma.

Lemma 1.2.5 (Matrix Inversion Lemma) *Let A be an invertible $d \times d$ matrix, and $\overline{u}$ and $\overline{v}$ be non-zero d-dimensional column vectors. Then, $A + \overline{u}\,\overline{v}^T$ is invertible if and only if $\overline{v}^T A^{-1} \overline{u} \neq -1$. In such a case, the inverse is computed as follows:*

$$(A + \overline{u}\,\overline{v}^T)^{-1} = A^{-1} - \frac{A^{-1}\overline{u}\,\overline{v}^T A^{-1}}{1 + \overline{v}^T A^{-1}\overline{u}}$$

Proof: If the matrix $(A + \overline{u}\,\overline{v}^T)$ is invertible, then the product of $(A + \overline{u}\,\overline{v}^T)$ and A^{-1} is invertible as well (as the product of two invertible matrices). Post-multiplying $(A + \overline{u}\,\overline{v}^T)A^{-1}$ with $\overline{u}$ yields a non-zero vector, because of the invertibility of the former matrix. Otherwise, we can further pre-multiply the resulting equation $(A + \overline{u}\,\overline{v}^T)A^{-1}\overline{u} = 0$ with the inverse of $(A + \overline{u}\,\overline{v}^T)A^{-1}$ in order to yield $\overline{u} = \overline{0}$, which is against the assumptions of the lemma. Therefore, we have:

$$(A + \overline{u}\,\overline{v}^T)A^{-1}\overline{u} \neq 0$$
$$\overline{u} + \overline{u}\,\overline{v}^T A^{-1}\overline{u} \neq 0$$
$$\overline{u}(1 + \overline{v}^T A^{-1}\overline{u}) \neq 0$$
$$1 + \overline{v}^T A^{-1}\overline{u} \neq 0$$

Therefore, the precondition of invertibility is shown.

Conversely, if the precondition $1 + \overline{v}^T A^{-1}\overline{u} \neq 0$ holds, we can show that the matrix $P = A^{-1} - \frac{A^{-1}\overline{u}\,\overline{v}^T A^{-1}}{1 + \overline{v}^T A^{-1}\overline{u}}$ is a valid inverse of $Q = (A + \overline{u}\,\overline{v}^T)$. Note that the matrix P is well defined only when the precondition holds. In such a case, expanding both PQ and QP algebraically yields the identity matrix. For example, expanding PQ yields the following:

$$PQ = I + A^{-1}\overline{u}\,\overline{v}^T - \frac{A^{-1}\overline{u}\,\overline{v}^T + A^{-1}\overline{u}\,[\overline{v}^T A^{-1}\overline{u}]\,\overline{v}^T}{1 + \overline{v}^T A^{-1}\overline{u}}$$
$$= I + A^{-1}\overline{u}\,\overline{v}^T - \frac{A^{-1}\overline{u}\,\overline{v}^T(1 + [\overline{v}^T A^{-1}\overline{u}])}{1 + \overline{v}^T A^{-1}\overline{u}}$$
$$= I + A^{-1}\overline{u}\,\overline{v}^T - A^{-1}\overline{u}\,\overline{v}^T = I$$

Although matrix multiplication is not commutative in general, the above proof uses the fact that the scalar $\bar{v}^T A^{-1} \bar{u}$ can be moved around in the order of matrix multiplication because it is a scalar. ∎

Variants of the matrix inversion lemma are used in various types of iterative updates in machine learning. A specific example is *incremental* linear regression, where one often wants to invert matrices of the form $C = D^T D$, where D is an $n \times d$ data matrix. When a new d-dimensional data point $\bar{v}$ is received, the size of the data matrix becomes $(n+1) \times d$ with the addition of row vector $\bar{v}^T$ to D. The matrix C is now updated to $D^T D + \bar{v}\bar{v}^T$, and the matrix inversion lemma comes in handy for updating the inverted matrix in $O(d^2)$ time. One can even generalize the above result to cases where the vectors $\bar{u}$ and $\bar{v}$ are replaced with "thin" matrices U and V containing a small number k of columns.

Theorem 1.2.1 (Sherman–Morrison–Woodbury Identity) *Let A be an invertible $d \times d$ matrix and let U, V be $d \times k$ non-zero matrices for some small value of k. Then, the matrix $A + UV^T$ is invertible if and only if the $k \times k$ matrix $(I + V^T A^{-1} U)$ is invertible. Furthermore, the inverse is given by the following:*

$$(A + UV^T)^{-1} = A^{-1} - A^{-1} U (I + V^T A^{-1} U)^{-1} V^T A^{-1}$$

This type of update is referred to as a *low-rank update*; the notion of rank will be explained in Chapter 2. We provide some exercises relevant to the matrix inversion lemma.

Problem 1.2.12 *Suppose that I and P are two $k \times k$ matrices. Show the following result:*

$$(I + P)^{-1} = I - (I + P)^{-1} P$$

A hint for solving this problem is to check what you get when you left multiply both sides of the above identity with $(I + P)$. A closely related result is the *push-through identity:*

Problem 1.2.13 (Push-Through Identity) *If U and V are two $n \times d$ matrices, show the following result:*

$$U^T (I_n + VU^T)^{-1} = (I_d + U^T V)^{-1} U^T$$

Use the above result to show the following for any $n \times d$ matrix D and scalar $\lambda > 0$:

$$D^T (\lambda I_n + DD^T)^{-1} = (\lambda I_d + D^T D)^{-1} D^T$$

A hint for solving the above problem is to see what happens when one left-multiplies *and* right-multiplies the above identities with the appropriate matrices. The push-through identity derives its name from the fact that we push in a matrix on the left and it comes out on the right. This identity is very important and is used repeatedly in this book.

1.2.6 Frobenius Norm, Trace, and Energy

Like vectors, one can define norms of matrices. For the rectangular $n \times d$ matrix A with (i, j)th entry denoted by a_{ij}, its *Frobenius norm* is defined as follows:

$$\|A\|_F = \|A^T\|_F = \sqrt{\sum_{i=1}^{n} \sum_{j=1}^{d} a_{ij}^2} \tag{1.22}$$

Note the use of $\|\cdot\|_F$ to denote the Frobenius norm. The squared Frobenius norm is the sum of squares of the norms of the row-vectors (or, alternatively, column vectors) in the matrix. It is invariant to matrix transposition. The *energy* of a matrix A is an alternative term used in machine learning community for the squared Frobenius norm.

The *trace* of a square matrix A, denoted by $\text{tr}(A)$, is defined by the sum of its diagonal entries. The energy of a rectangular matrix A is equal to the trace of either AA^T or $A^T A$:

$$\|A\|_F^2 = \text{Energy}(A) = \text{tr}(AA^T) = \text{tr}(A^T A) \tag{1.23}$$

More generally, the trace of the product of two matrices $C = [c_{ij}]$ and $D = [d_{ij}]$ of sizes of $n \times d$ is the sum of their entrywise product:

$$\text{tr}(CD^T) = \text{tr}(DC^T) = \sum_{i=1}^{n} \sum_{j=1}^{d} c_{ij} d_{ij} \tag{1.24}$$

The trace of the product of two matrices $A = [a_{ij}]_{n \times d}$ and $B = [b_{ij}]_{d \times n}$ is invariant to the order of matrix multiplication:

$$\text{tr}(AB) = \text{tr}(BA) = \sum_{i=1}^{n} \sum_{j=1}^{d} a_{ij} b_{ji} \tag{1.25}$$

Problem 1.2.14 *Show that the Frobenius norm of the outer product of two vectors is equal to the product of their Euclidean norms.*

The Frobenius norm shares many properties with vector norms, such as *sub-additivity* and *sub-multiplicativity*. These properties are analogous to the triangle inequality and the Cauchy-Schwarz inequality, respectively, in the case of vector norms.

Lemma 1.2.6 (Sub-additive Frobenius Norm) *For any pair of matrices A and B of the same size, the triangle inequality $\|A + B\|_F \leq \|A\|_F + \|B\|_F$ is satisfied.*

The above result is easy to show by simply treating a matrix as a vector and creating two long vectors from A and B, each with dimensionality equal to the number of matrix entries.

Lemma 1.2.7 (Sub-multiplicative Frobenius Norm) *For any pair of matrices A and B of sizes $n \times k$ and $k \times d$, respectively, the sub-multiplicative property $\|AB\|_F \leq \|A\|_F \|B\|_F$ is satisfied.*

Proof Sketch: Let $\bar{a}_1 \ldots \bar{a}_n$ correspond to the rows of A, and $\bar{b}_1 \ldots \bar{b}_d$ contain the transposed columns of B. Then, the (i, j)th entry of AB is $\bar{a}_i \cdot \bar{b}_j$, and the squared Frobenius norm of the matrix AB is $\sum_{i=1}^{n} \sum_{j=1}^{d} (\bar{a}_i \cdot \bar{b}_j)^2$. Each $(\bar{a}_i \cdot \bar{b}_j)^2$ is less than $\|\bar{a}_i\|^2 \|\bar{b}_j\|^2$ according to the Cauchy-Schwarz inequality. Therefore, we have the following:

$$\|AB\|_F^2 = \sum_{i=1}^{n} \sum_{j=1}^{d} (\bar{a}_i \cdot \bar{b}_j)^2 \leq \sum_{i=1}^{n} \sum_{j=1}^{d} \|\bar{a}_i\|^2 \|\bar{b}_j\|^2 = (\sum_{i=1}^{n} \|\bar{a}_i\|^2)(\sum_{j=1}^{d} \|\bar{b}_j\|^2) = \|A\|_F^2 \|B\|_F^2$$

Computing the square-root of both sides yields the desired result.   ■

Problem 1.2.15 (Small Matrices Have Large Inverses) *Show that the Frobenius norm of the inverse of an $n \times n$ matrix with Frobenius norm of ϵ is at least $\sqrt{n}/\epsilon$.*

1.3 Matrix Multiplication as a Decomposable Operator

Matrix multiplication can be viewed as a vector-to-vector function that maps one vector to another. For example, the multiplication of a d-dimensional column vector $\overline{x}$ with the $d \times d$ matrix A maps it to another d-dimensional vector, which is the output of the function $f(\overline{x})$:

$$f(\overline{x}) = A\overline{x}$$

One can view this function as a vector-centric generalization of the univariate linear function $g(x) = a\,x$ for scalar a. This is one of the reasons that matrices are viewed as *linear operators* on vectors. Much of linear algebra is devoted to understanding this transformation and leveraging it for efficient numerical computations.

One issue is that if we have a large $d \times d$ matrix, it is often hard to interpret what the matrix is really doing to the vector in terms of its individual components. This is the reason that it is often useful to interpret a matrix as a product of simpler matrices. Because of the beautiful property of the associativity of matrix multiplication, one can interpret a product of simple matrices (and a vector) as the *composition of simple operations on the vector*. In order to understand this point, consider the case when the above matrix A can be decomposed into the product of simpler $d \times d$ matrices $B_1, B_2, \ldots B_k$, as follows:

$$A = B_1 B_2 \ldots B_{k-1} B_k$$

Assume that each B_i is simple enough that one can intuitively interpret the effect of multiplying a vector $\overline{x}$ with B_i easily (such as rotating the vector or scaling it). Then, the aforementioned function $f(\overline{x})$ can be written as follows:

$$f(\overline{x}) = A\overline{x} = [B_1 B_2 \ldots B_{k-1} B_k]\overline{x}$$
$$= B_1(B_2 \ldots [B_{k-1}(B_k \overline{x})]) \quad \text{[Associative Property of Matrix Multiplication]}$$

The nested brackets on the right provide an order to the operations. In other words, we first apply the operator B_k to $\overline{x}$, then apply B_{k-1}, and so on all the way down to B_1. Therefore, *as long as we can decompose a matrix into the product of simpler matrices, we can interpret matrix multiplication with a vector as a sequence of simple, easy-to-understand operations on the vector.* In this section, we will provide two important examples of decomposition, which will be studied in greater detail throughout the book.

1.3.1 Matrix Multiplication as Decomposable Row and Column Operators

An important property of matrix multiplication is that the rows and columns of the product can be manipulated by applying the corresponding operations on one of the two matrices. In a product AX of two matrices A and X, interchanging the ith and jth *rows* of the *first* matrix A will also interchange the corresponding rows in the product (which has the same number of rows as the first matrix). Similarly, if we interchange the *columns* of the *second* matrix, this interchange will also occur in the product (which has the same number of columns as the second matrix). There are three main elementary operations, corresponding to interchange, addition, and multiplication. The elementary row operations on matrices are defined as follows:

- *Interchange operation:* The ith and jth rows of the matrix are interchanged. The operation is fully defined by two indices i and j in any order.

- *Addition operation:* A scalar multiple of the jth row is added to the ith row. The operation is defined by two indices i, j in a specific order, and a scalar multiple c.

- *Scaling operation:* The ith row is multiplied with scalar c. The operation is fully defined by the row index i and the scalar c.

The above operations are referred to as *elementary row operations*. One can define exactly analogous operations on the columns with *elementary column operations*.

An *elementary matrix* is a matrix that differs from the identity matrix by applying a single row or column operation. Pre-multiplying a matrix X with an elementary matrix corresponding to an interchange results in an interchange of the rows of X. In other words, if E is the elementary matrix corresponding to an interchange, then a pair of rows of $X' = EX$ will be interchanged with respect to X. A similar result holds true for other operations like row addition and row scaling. Some examples of 3×3 elementary matrices with the corresponding operations are illustrated in the table below:

Interchange	Addition	Scaling
$\begin{array}{ccc} 0 & 1 & 0 \\ 1 & 0 & 0 \\ 0 & 0 & 1 \end{array}$	$\begin{array}{ccc} 1 & c & 0 \\ 0 & 1 & 0 \\ 0 & 0 & 1 \end{array}$	$\begin{array}{ccc} 1 & 0 & 0 \\ 0 & c & 0 \\ 0 & 0 & 1 \end{array}$
(a) Interchange rows 1, 2	(b) Add $c \times$ (row 2) to row 1	(c) Multiply row 2 by c

These matrices are also referred to as *elementary matrix operators* because they are used to apply specific row operations on arbitrary matrices. The scalar c is always non-zero in the above matrices, because all elementary matrices are invertible and are different from the identity matrix (albeit in a minor way). Pre-multiplication of X with the appropriate elementary matrix can result in a row exchange, addition, or row-wise scaling being applied to X. For example, the first and second rows of the matrix X can be exchanged to create X' as follows:

$$\underbrace{\begin{bmatrix} 0 & 1 & 0 \\ 1 & 0 & 0 \\ 0 & 0 & 1 \end{bmatrix}}_{\text{Operator}} \underbrace{\begin{bmatrix} 1 & 2 & 3 \\ 4 & 5 & 6 \\ 7 & 8 & 9 \end{bmatrix}}_{X} = \underbrace{\begin{bmatrix} 4 & 5 & 6 \\ 1 & 2 & 3 \\ 7 & 8 & 9 \end{bmatrix}}_{X'}$$

The first row of the matrix can be scaled up by 2 with the use of the appropriate scaling operator:

$$\underbrace{\begin{bmatrix} 2 & 0 & 0 \\ 0 & 1 & 0 \\ 0 & 0 & 1 \end{bmatrix}}_{\text{Operator}} \underbrace{\begin{bmatrix} 1 & 2 & 3 \\ 4 & 5 & 6 \\ 7 & 8 & 9 \end{bmatrix}}_{X} = \underbrace{\begin{bmatrix} 2 & 4 & 6 \\ 4 & 5 & 6 \\ 7 & 8 & 9 \end{bmatrix}}_{X'}$$

Post-multiplication of matrix X with the following elementary matrices will result in exactly analogous operations on the columns of X to create X':

Interchange	Addition	Scaling
$\begin{array}{ccc} 0 & 1 & 0 \\ 1 & 0 & 0 \\ 0 & 0 & 1 \end{array}$	$\begin{array}{ccc} 1 & 0 & 0 \\ c & 1 & 0 \\ 0 & 0 & 1 \end{array}$	$\begin{array}{ccc} 1 & 0 & 0 \\ 0 & c & 0 \\ 0 & 0 & 1 \end{array}$
(a) Interchange col. 1, 2	(b) Add $c \times$ (col. 2) to col. 1	(c) Multiply col. 2 by c

Only the elementary matrix for the addition operation is slightly different between row and column operations (although the other two matrices are the same). In the following, we show an example of how post-multiplication with the appropriate elementary matrix can result in a column exchange operation:

$$
\underbrace{\begin{bmatrix} 1 & 2 & 3 \\ 4 & 5 & 6 \\ 7 & 8 & 9 \end{bmatrix}}_{X}
\underbrace{\begin{bmatrix} 0 & 1 & 0 \\ 1 & 0 & 0 \\ 0 & 0 & 1 \end{bmatrix}}_{\text{Operator}}
=
\underbrace{\begin{bmatrix} 2 & 1 & 3 \\ 5 & 4 & 6 \\ 8 & 7 & 9 \end{bmatrix}}_{X'}
$$

Note that this example is very similar to the one provided for row interchange, except that the corresponding elementary matrix is *post-multiplied* in this case.

Problem 1.3.1 *Define a 4×4 operator matrix so that pre-multiplying any matrix X with this matrix will result in addition of c_i times the ith row of X to the 2nd row of X for each $i \in \{1, 2, 3, 4\}$ in one shot. Show that this matrix can be expressed as the product of three elementary addition matrices and a single elementary multiplication matrix.*

These types of elementary matrices are always invertible. The inverse of the interchange matrix is itself. The inverse of the scaling matrix is obtained by replacing the entry c with $1/c$. The inverse of the row or column addition matrix is obtained by replacing c with $-c$. We make the following observation:

Observation 1.3.1 *The inverse of an elementary matrix is another elementary matrix.*

Keeping the inverses of elementary matrices in mind can sometimes be useful. Therefore, the reader is encouraged to work out the details of these matrices using the exercise below:

Problem 1.3.2 *Write down one example of each of the three types [i.e., interchange, multiplication, and addition] of elementary matrices for performing row operations on a matrix of size 4×4. Work out the inverse of these matrices. Repeat this result for each of the three types of matrices for performing column operations.*

The following exercises are examples of the utility of the inverses of elementary matrices:

Problem 1.3.3 *Let A and B be two matrices. Let A_{ij} be the matrix obtained by exchanging the ith and jth columns of A, and B_{ij} be the matrix obtained by exchanging the ith and jth rows of B. Write each of A_{ij} and B_{ij} in terms of A or B, and an elementary matrix. Now explain why $A_{ij}B_{ij} = AB$.*

Problem 1.3.4 *Let A and B be two matrices. Let matrix A' be created by adding c times the jth column of A to its ith column, and matrix B' be created by subtracting c times the ith row of B from its jth row. Explain using the concept of elementary matrices why the matrices AB and $A'B'$ are the same.*

It is also possible to apply elementary operations to matrices that are not square. For an $n \times d$ matrix, the pre-multiplication operator matrix will be of size $n \times n$, whereas the post-multiplication operator matrix will be of size $d \times d$.

Permutation Matrices

An elementary row (or column) interchange operator matrix is a special case of a *permutation matrix*. A permutation matrix contains a single 1 in each row, and a single 1 in each column. An example of a permutation matrix P is shown below:

$$P = \begin{bmatrix} 0 & 0 & 1 & 0 \\ 1 & 0 & 0 & 0 \\ 0 & 0 & 0 & 1 \\ 0 & 1 & 0 & 0 \end{bmatrix}$$

Pre-multiplying any matrix with a permutation matrix shuffles the rows, and post-multiplying any matrix with a permutation matrix shuffles the columns. For example, pre-multiplying any four-row matrix with the above matrix P reorders the rows as follows:

$$\text{Row } 3 \Rightarrow \text{Row } 1 \Rightarrow \text{Row } 4 \Rightarrow \text{Row } 2$$

Post-multiplying any four-column matrix with P reorders the columns, albeit in the reverse order:

$$\text{Column } 2 \Rightarrow \text{Column } 4 \Rightarrow \text{Column } 1 \Rightarrow \text{Column } 3$$

It is noteworthy that a permutation matrix and its transpose are inverses of one another because they have orthonormal columns. Such matrices are useful in reordering the items of a data matrix, and applications will be shown for graph matrices in Chapter 10. Since one can shuffle the rows of a matrix by using a sequence of row interchange operations, it follows that *any permutation matrix is a product of row interchange operator matrices*.

Applications of Elementary Operator Matrices

The row manipulation property is used to compute the inverses of matrices. This is because a matrix A and its inverse X are related as follows:

$$AX = I$$

Row operations are applied on A to convert the matrix to the identity matrix. A systematic approach to perform such row operations to convert A to the identity matrix is the *Gaussian elimination method* discussed in Chapter 2. These operations are mirrored on the right-hand side so that the identity matrix is converted to the inverse. As the final result of the row operations, we obtain the following:

$$IX = A^{-1}$$

Elementary matrices are fundamental because *one can decompose any square and invertible matrix into a product of elementary matrices*. In fact, if one is willing to augment the set of elementary multiplication operators to allow the scalar c on the diagonal to be zero (which is traditionally not the case), then one can express any square matrix as a product of augmented elementary matrices.

Finally, we discuss the important application of finding a solution to the system of equations $A\overline{x} = \overline{b}$. Here, A is an $n \times d$ matrix, $\overline{x}$ is d-dimensional column vector, and $\overline{b}$ is an n-dimensional row vector. Note that a feasible solution might not exist to this system of equations, especially when some groups of equations are mutually inconsistent. For example, the equations $\sum_{i=1}^{100} x_i = +1$ and $\sum_{i=1}^{100} x_i = -1$ are mutually inconsistent.

The matrix-centric methodology for solving such a system of linear equations derives its inspiration from the well-known methodology of eliminating variables from systems of equations in multiple variables. For example, if we have a pair of linear equations in x_1 and x_2, we can create an equation without one of the variables by subtracting an appropriate multiple of one equation from the other. This operation is identical to the elementary row addition operation discussed in this chapter. This general principle can be applied to systems containing any number of variables, so that the rth equation is defined only in terms of $x_r, x_{r+1}, \ldots x_d$. This is equivalent to converting the original system $A\overline{x} = \overline{b}$ into a new system $A'\overline{x} = \overline{b}'$ where A' is triangular. Therefore, if we apply a sequence $E_1 \ldots E_k$ of elementary row operations to the system of equations, we obtain the following relationship:

$$\underbrace{E_k E_{k-1} \ldots E_1 A}_{A'} \overline{x} = \underbrace{E_k E_{k-1} \ldots E_1 \overline{b}}_{\overline{b}'}$$

A triangular system of equations is solved by first processing equations with fewer variables and iteratively backsubstituting these values to reduce the system to fewer variables. These methods will be discussed in detail in Chapter 2. It is noteworthy that the problem of solving linear equations is a special case of the fundamental machine learning problem of *linear regression*, in which the best-fit solution is found to an inconsistent system of equations. Linear regression serves as the "parent problem" to many machine learning problems like *least-squares classification*, *support-vector machines*, and *logistic regression*.

1.3.2 Matrix Multiplication as Decomposable Geometric Operators

Aside from decompositions involving elementary matrices, other forms of decompositions are based on matrices with *geometric* interpretations, such as rotation, reflection, and scaling. For example, a 90° counter-clockwise rotation of the vector $[2, 1]$ transforms it to $[-1, 2]$. A reflection of the point $[2, 1]$ across the X-axis yields $[2, -1]$; a scaling along the X-axis and Y-axis by respective factors of 2 and 3 yields $[4, 3]$. All these simple transformations on a vector in two dimensions can be defined by pre-multiplication of the corresponding column vector with a 2×2 matrix (or post-multiplication of a row vector with the transpose of this 2×2 matrix). For example, consider the column vector representation of a point with polar coordinates $[a, \alpha]$ and Cartesian coordinates $[a \cos(\alpha), a \sin(\alpha)]$. The point has magnitude a and makes a counter-clockwise angle of α with the X-axis. Then, one can multiply it with the *rotation matrix* shown below to yield a counter-clockwise rotation of the vector with angle θ:

$$\begin{bmatrix} \cos(\theta) & -\sin(\theta) \\ \sin(\theta) & \cos(\theta) \end{bmatrix} \begin{bmatrix} a\cos(\alpha) \\ a\sin(\alpha) \end{bmatrix} = \begin{bmatrix} a[\cos(\alpha)\cos(\theta) - \sin(\alpha)\sin(\theta)] \\ a[\cos(\alpha)\sin(\theta) + \sin(\alpha)\cos(\theta)] \end{bmatrix} = \begin{bmatrix} a\cos(\alpha + \theta) \\ a\sin(\alpha + \theta) \end{bmatrix}$$

The final result is obtained by using a standard trigonometric identity for the cosines and sines of the sums of angles, and the Cartesian coordinates shown on the right-hand side are equivalent to the polar coordinates $[a, \alpha + \theta]$. In other words, the original coordinates $[a, \alpha]$ have been rotated counter-clockwise by angle θ. The basic geometric operations like rotation, reflection, and scaling can be performed by post-multiplication with appropriately chosen matrices. We list these matrices below, which are defined for pre-multiplying column vectors:

Rotation	Reflection	Scaling
$\begin{array}{cc} \cos(\theta) & -\sin(\theta) \\ \sin(\theta) & \cos(\theta) \end{array}$	$\begin{array}{cc} 1 & 0 \\ 0 & -1 \end{array}$	$\begin{array}{cc} c_1 & 0 \\ 0 & c_2 \end{array}$
(a) Rotate counter-clockwise by θ	(b) Reflect across X-axis	(c) Scale x and y by factors of c_1 and c_2

The above matrices are also referred to as *elementary* matrices for geometric operations (like the elementary matrices for row and column operations). It is possible for the diagonal entries of the scaling matrix to be negative or 0. Strictly speaking, the elementary reflection matrix can be considered a special case of the scaling matrix by setting the different values of c_i to values drawn from $\{-1, 1\}$.

Problem 1.3.5 *The above list of matrices for rotation, reflection, and scaling is designed to transform a column vector $\overline{x}$ using the matrix-to-vector product $A\overline{x}$. Write down the corresponding matrices for the case when you want to transform a row vector $\overline{u}$ as $\overline{u}B$.*

The matrix for a sequence of transformations can be computed by multiplying the corresponding matrices. This is easy to show by observing that if we have $A = A_1 \ldots A_k$, then successively pre-multiplying a column-vector $\overline{x}$ with $A_k \ldots A_1$ is the same as the expression $A_1(A_2(\ldots(A_k\overline{x})))$. Because of the associativity of matrix multiplication, one can express this matrix as $(A_1 \ldots A_k)\overline{x} = A\overline{x}$. Conversely, if a matrix can be expressed as a product of simpler matrices (like the geometric ones shown above), then multiplication of a vector with that matrix is equivalent to a sequence of the above geometric transformations.

A fundamental result of linear algebra is that any square matrix can be shown to be a product of rotation/reflection/scaling matrices by using a technique called *singular value decomposition*. In other words, *all linear transformations of vectors defined by matrix multiplication corresponding to the application of a sequence of rotations, reflections, and scaling on the vector*. Chapter 2 generalizes the 2×2 matrices in the above table to any number of dimensions by using $d \times d$ matrices. These concepts are sometimes more complex in higher dimensions — for example, it is possible to use an arbitrarily oriented axis of rotation in higher dimensions unlike in the case of two dimensions. The decomposition of a matrix into geometrically interpretable matrices can also be used for computing inverses.

Problem 1.3.6 *Suppose that you are told that any invertible square matrix A can be expressed as a product of elementary rotation/reflection/scaling matrices as $A = R_1 R_2 \ldots R_k$. Express the inverse of A in terms of the easily computable inverses of $R_1, R_2, \ldots, R_k$.*

It is also helpful to understand the row addition operator, discussed in the previous section. Consider the 2×2 row-addition operator:

$$A = \begin{bmatrix} 1 & c \\ 0 & 1 \end{bmatrix}$$

This operator *shears* the space along the direction of the first coordinate For example, if vector $\overline{z}$ is $[x, y]^T$, then $A\overline{z}$ yields the new vector $[x+cy, y]^T$. Here, the y-coordinate remains unchanged, whereas the x-coordinate gets sheared in proportion to its height. The shearing of a rectangle into a parallelogram is shown in Figure 1.4. An elementary row operator matrix is a very special case of a triangular matrix; correspondingly, a triangular matrix *with unit diagonal entries* corresponds to a sequence of shears. This is because one can convert an identity matrix into any such triangular matrix with a sequence of elementary row addition operations.

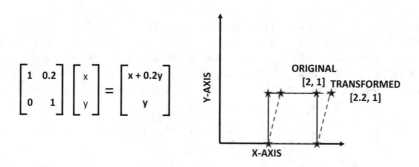

Figure 1.4: An elementary row addition operator can be interpreted as a shear transform

As discussed earlier in this section, a linear transformation can be viewed as a succession of simpler transforms. This simpler sequence of transforms is obtained by *decomposing* a matrix A into the product of simpler matrices $B_1 \ldots B_k$ as follows:

$$f(\overline{x}) = A\overline{x} = B_1(B_2 \ldots [B_{k-1}(B_k\overline{x})])$$

Each B_i is typically a group of similar transforms, such as orthogonal matrices (sequence of rotations), diagonal matrices (sequence of scalings), or triangular matrices with unit diagonal entries (sequence of shears). There is considerable flexibility in terms of how these decompositions can be performed. For example, this book discusses the *LU decomposition*, the *QR decomposition*, and the *singular value decomposition*.

1.4 Basic Problems in Machine Learning

Machine learning is about constructing models on observed examples in the rows of data matrices, and using these models to make predictions about missing entries of previously unseen examples. This process is also referred to as *learning*, which is where "machine learning" derives its name. Throughout this book, we assume that we have an $n \times d$ data matrix D, which contains n examples of d-dimensional data points in its rows. A *dimension* or *attribute* is one of the d properties of a data point, and a column of D contains this property for all data instances. For example, in a medical application, each row of the data matrix D might correspond to a patient, and the d dimensions might represent the different attributes garnered from the patient, such as their height, weight, test results, and so on. Machine learning uses these examples for various applications, such as that of predicting the value of a particular dimension in the data, finding anomalous patients, or grouping similar patients. These correspond to classical problems in machine learning, such as classification, anomaly detection, and clustering. This section will introduce these classical problems.

1.4.1 Matrix Factorization

Matrix factorization is an alternative term for matrix decomposition, and it usually refers to an optimization-centric view of decomposition. Matrix factorization decomposes an $n \times d$ matrix D into two *factor* matrices U and V of respective sizes $n \times k$ and $d \times k$, so that $UV^T \approx D$. Here, $k \ll \min\{n, d\}$ is a parameter referred to as the *rank* of the factorization. The notion of rank is introduced formally in Chapter 2. The rank controls the "conciseness" of the factorization because the total number of entries in U and V is $k(n+d)$, which is much smaller than the original number of entries in D. Matrix factorization is a generalization of (real-valued) scalar factorization to matrices. There are an infinite number of factors of the same matrix D, just as a scalar can be factored in an infinite number of pairs of real values. For example, the scalar 6 can be written as 2×3, as 1.5×4, or as $\sqrt{2} \times \sqrt{18}$. An example of a matrix factorization of a 3×3 matrix into two smaller matrices is shown below:

$$\begin{bmatrix} 1 & -1 & 1 \\ -1 & 1 & -1 \\ 2 & -2 & 2 \end{bmatrix} = \begin{bmatrix} 1 \\ -1 \\ 2 \end{bmatrix} [1, -1, 1]$$

In the above case, the factorization is exact, although it is often allowed to be *approximately* true in order to minimize the sizes of the factor matrices U and V. If one is willing to allow for a reasonable level of approximation, the value of k can be quite small.

A common approach for matrix factorization is to set up the following optimization problem:

$$\text{Minimize } J = \|D - UV^T\|_F^2 \tag{1.26}$$

Here, $\| \cdot \|_F^2$ refers to the squared Frobenius norm, which is the sum of the squares of the entries in the *residual* matrix $(D - UV^T)$. The objective function J is minimized with the use of gradient descent on the parameter matrices U and V, whose entries are variables of this optimization problem. By minimizing this objective function, one will ensure that the matrix $(D - UV^T)$ will have entries that are small in magnitude, and therefore $D \approx UV^T$. These types of objective functions are also referred to as *loss functions*, because they measure how much information UV^T "loses" with respect to the original matrix D.

One can even factorize an incompletely specified matrix D by formulating the optimization objective function only with the observed entries. This basic principle serves as the foundation of *recommender systems*. For example, consider a setting in which we have n users and d ratings; the (i, j)th entry of D provides the rating of the user i for item j. Most of the entries of D are unobserved, because users typically rate only a small subset of items. In such a case, the objective function $\|D - UV^T\|_F^2$ will need to be modified, so that we sum up the squared errors only over the *observed* entries in D. This is because the values of the remaining entries in $(D - UV^T)$ are unknown. Setting up an optimization problem only in terms of a subset of entries allows us to learn fully specified matrices U and V. Therefore, UV^T provides a prediction of the *fully reconstructed* matrix D. This application will be discussed in greater detail in Chapter 8.

1.4.2   Clustering

The problem of clustering is that of partitioning the rows of the $n \times d$ data matrix D into groups of similar rows. For example, imagine a setting where one has data records in which the rows of D correspond to different individuals, and the different dimensions (columns) of D correspond to the number of units of each product bought in a supermarket. Then, a clustering application might try to segment the data set into groups of similar individuals with particular types of buying behavior. The number of clusters might either be specified by the analyst up front, or the algorithm might use a heuristic to set the number of "natural" clusters in the data. One can often use the segmentation created by clustering as a preprocessing step for other analytical goals. For example, on closer examination of the clusters, one might learn that particular individuals are interested in household articles in a grocery store, whereas others are interested in fruits. This information can be used by the supermarket to make recommendations. Various clustering algorithms like *k-means* and *spectral clustering* are introduced in Chapters 8, 9, and 10.

1.4.3 Classification and Regression Modeling

The problem of classification is closely related to clustering, except that more guidance is available for grouping the data with the use of the notion of *supervision*. In the case of clustering, the data is partitioned into groups without any regard for the types of clusters we wish to find. In the case of classification, the *training* data are already partitioned into specific types of groups. Therefore, in addition to the $n \times d$ data matrix D, we have an $n \times 1$ array of labels denoted by $\overline{y}$. The ith entry in $\overline{y}$ corresponds to the ith row in the data matrix D, and the former is a categorical label defining a semantic name for the cluster (or *class*) to which the ith row of D belongs. In the case of the grocery example above, we might decide up front that we are interested in the classes $\mathcal{L} = \{$ *fruits, poultry, all else* $\}$. Note that these classes might often be clustered in the data in terms of the similarity of the rows in the data matrix D, although this is not always necessary. For example, clusters that are clearly distinct might be located in a single class. Furthermore, it might be possible that other distinct clusters might exist that are corresponding to specific sub-categories within the *all else* label. This might be the case because the end-user (e.g., merchant) might not have any interest in identifying items in the *all else* category, whereas the other labels might help the merchant identify candidate customers for a promotion. Therefore, in the classification problem, the training data defines the clusters of *interest* with the use of examples. The actual segmentation of the rows is done on a separate $n_t \times d$ *test* data matrix D_t, in which the labels are not specified. Therefore, for each row of D_t, one needs to map it one of the labels from the set $\mathcal{L}$. This mapping is done with the use of a classification *model* that was constructed on the training data. The test data is *unseen* during the process of model construction, as the rows of D and D_t are not the same.

A common setting in classification is that the label set is *binary* and only contains two possible values. In such a case, it is common to use the label set $\mathcal{L}$ from $\{0, 1\}$ or from $\{-1, +1\}$. The goal is to *learn* the ith entry y_i in $\overline{y}$ as a function of the ith row $\overline{X}_i$ of D:

$$y_i \approx f(\overline{X}_i)$$

The function $f(\overline{X}_i)$ is often parameterized with a weight vector $\overline{W}$. Consider the following example of binary classification into the labels $\{-1, +1\}$:

$$y_i \approx f_{\overline{W}}(\overline{X}_i) = \text{sign}\{\overline{W} \cdot \overline{X}_i\}$$

Note that we have added a subscript to the function to indicate its parametrization. How does one compute $\overline{W}$? The key idea is to penalize any kind of mismatching between the *observed value* y_i and the predicted value $f(\overline{X}_i)$ with the use of carefully constructed *loss function*. Therefore, many machine learning models reduce to the following optimization problem:

$$\text{Minimize}_{\overline{W}} \sum_i \text{Mismatching between } y_i \text{ and } f_{\overline{W}}(\overline{X}_i)$$

Once the weight vector $\overline{W}$ has been computed by solving the optimization model, it is used to predict the value of the class variable y_i for instances in which the class variable is not known. Classification is also referred to as *supervised learning*, because it uses the training data to build a model that performs the classification of the test data. In a sense, the training data serves as the "teacher" providing supervision. The ability to use the knowledge in the training data in order to classify the examples in *unseen* test data is referred to as *generalization*. There is no utility in classifying the examples of the training data again, because their labels have already been observed.

Regression

The label in classification is also referred to as *dependent* variable, which is categorical in nature. In the regression modeling problem, the $n \times d$ training data matrix D is associated with an $n \times 1$ vector $\overline{y}$ of dependent variables, which are *numerical*. Therefore, the only difference from classification is that the array $\overline{y}$ contains numerical values (rather than categorical ones), and can therefore be treated as a vector. The dependent variable is also referred to as a *response variable, target variable*, or *regressand* in the case of regression. The independent variables are also referred to as *regressors*. Binary response variables are closely related to regression, and some models solve binary classification *directly* with the use of a regression model (by pretending that the binary labels are numerical). This is because binary values have the flexibility of being treated as either categorical or as numerical values. However, more than two classes like $\{Red, Green, Blue\}$ cannot be ordered, and are therefore different from regression.

The regression modeling problem is closely related to linear algebra, especially when a *linear optimization model* is used. In the linear optimization model, we use a d-dimensional column vector $\overline{W} = [w_1 \ldots w_d]^T$ to represent the weights of the different dimensions. The ith entry y_i of $\overline{y}$ is obtained as the dot product of the ith row $\overline{X}_i$ of D and $\overline{W}$. In other words, the function $f(\cdot)$ to be learned by the optimization problem is as follows:

$$y_i = f(\overline{X}_i) = \overline{X}_i \overline{W}$$

One can also state this condition across all training instances using the full $n \times d$ data matrix D:

$$\overline{y} \approx D\overline{W} \tag{1.27}$$

Note that this is a matrix representation of n linear equations. In most cases, the value of n is much greater than d, and therefore, this is an *over-determined* system of linear equations. In over-determined cases, there is usually no solution for $\overline{W}$ that *exactly* satisfies this system. However, we can minimize the sum of squares of the errors to get as close to this goal as possible:

$$J = \frac{1}{2}\|D\overline{W} - \overline{y}\|^2 \tag{1.28}$$

On solving the aforementioned optimization problem, it will be shown in Chapter 4 that the solution $\overline{W}$ can be obtained as follows:

$$\overline{W} = (D^T D)^{-1} D^T \overline{y} \tag{1.29}$$

Then, for each row $\overline{Z}$ of the test data matrix D_t, the dot product of $\overline{W}^T$ and $\overline{Z}$ is the corresponding prediction of the real-valued dependent variable.

1.4.4   Outlier Detection

In the outlier detection problem, we have an $n \times d$ data matrix D, and we would like find rows of D that are very different from most of the other rows. This problem has a natural relationship of complementarity with the clustering problem, in which the aim is to find groups of similar rows. In other words, outliers are rows of D that do not naturally fit in with the other rows. Therefore, clustering methods are often used to find outliers. Matrix factorization methods are also used often for outlier detection. This book will introduce various outlier detection methods as applications of linear algebra and optimization.

1.5 Optimization for Machine Learning

Much of machine learning uses optimization in order to define *parameterized* models for learning problems. These models treat dependent variables as functions of independent variables, such as Equation 1.27. It is assumed that some examples are available containing observed values of both dependent and independent variables for training. These problems define *objective functions* or *loss functions*, which penalize differences between predicted and observed values of dependent variables (such as Equation 1.28). Therefore, the training phase of machine learning methods requires the use of optimization techniques.

In most cases, the optimization models are posed in minimization form. The most basic condition for optimality of the function $f(x_1, \ldots, x_d)$ at $[x_1 \ldots x_d]$ is that each *partial derivative* is 0:

$$\frac{\partial f(x_1, \ldots, x_d)}{\partial x_r} = \lim_{\delta \to 0} \frac{f(x_1, \ldots, x_r + \delta, \ldots, x_d) - f(x_1, \ldots, x_r, \ldots, x_d)}{\delta} = 0, \quad \forall r$$

The basic idea is that the rate of change of the function in any direction is 0, or else one can move in a direction with negative rate of change to further improve the objective function. This condition is necessary, but not sufficient, for optimization. More details of relevant optimality conditions are provided in Chapter 4.

The d-dimensional vector of partial derivatives is referred to as the *gradient*:

$$\nabla f(x_1, \ldots x_d) = \left[\frac{\partial f(\cdot)}{\partial x_1} \cdots \frac{\partial f(\cdot)}{\partial x_d} \right]^T$$

The gradient is denoted by the symbol ∇, and putting it in front of a function refers to the vector of partial derivatives with respect to the argument.

1.5.1 The Taylor Expansion for Function Simplification

Many objective functions in machine learning are very complicated in comparison with the relatively simple structure of polynomial functions (which are much easier to optimize). Therefore, if one can approximate complex objective functions with simpler polynomials (even within restricted regions of the space), it can go a long way toward solving optimization problems in an iterative way.

The Taylor expansion expresses any smooth function as a polynomial (with an infinite number of terms). Furthermore, if we only want an approximation of the function in a small locality of the argument, a small number of polynomial terms (typically no more than 2 or 3) will often suffice. First, consider the univariate function $f(w)$. This function can be expanded about any point a in the domain of the function by using the following expansion:

$$f(w) = f(a) + (w - a)f'(a) + \frac{(w - a)^2 f''(a)}{2!} + \ldots + \frac{(w - a)^r}{r!} \left[\frac{d^r f(w)}{d w^r} \right]_{w=a} + \ldots$$

Here, $f'(a)$ is the first derivative of $f(w)$ at a, $f''(w)$ is the second derivative, and so on. Note that $f(w)$ could be an arbitrary function, such as $\sin(w)$ or $\exp(w)$, and the expansion expresses it as a polynomial with an infinite number of terms. The case of $\exp(w)$ is particularly simple, because the nth order derivative of $\exp(w)$ is itself. For example, $\exp(w)$ can be expanded about $w = 0$ as follows:

$$\exp(w) = \exp(0) + \exp(0)w + \exp(0)\frac{w^2}{2!} + \exp(0)\frac{w^3}{3!} + \ldots + \exp(0)\frac{w^n}{n!} \ldots \quad (1.30)$$

$$= 1 + w + \frac{w^2}{2!} + \frac{w^3}{3!} + \ldots + \frac{w^n}{n!} \ldots \quad (1.31)$$

In other words, the exponentiation function can be expressed as an infinite polynomial, in which the trailing terms rapidly shrink in size because $\lim_{n\to\infty} w^n/n! = 0$. For some functions like $\sin(w)$ and $\exp(w)$, the Taylor expansion converges to the true function by including an increasing number of terms (irrespective of the choice of w and a). For other functions like $1/w$ or $\log(w)$, a converging expansion exists in restricted ranges of w at any particular value of a. More importantly, the Taylor expansion almost always provides a very good approximation of any smooth function *near* $w = a$, and the approximation is exact at $w = a$. Furthermore, higher-order terms tend to vanish when $|w - a|$ is small, because $(w - a)^r/r!$ rapidly converges to 0 for increasing r. Therefore, one can often obtain good quadratic approximations of a function near $w = a$ by simply including the first three terms.

In practical settings like optimization, one is often looking to change the value w from the current point $w = a$ to a "nearby" point in order to improve the objective function value. In such cases, using only up to the quadratic term of the Taylor expansion about $w = a$ provides an excellent simplification in the neighborhood of $w = a$. In gradient-descent algorithms, one is often looking to move from the current point by a relatively small amount, and therefore lower-order Taylor approximations can be used to guide the steps in order to improve the polynomial approximation rather than the original function. It is often much easier to optimize polynomials than arbitrarily complex functions.

One can also generalize the Taylor expansion to multivariable functions $F(\overline{w})$ with d-dimensional arguments of the form $\overline{w} = [w_1 \ldots w_d]^T$. The Taylor expansion of the function $F(\overline{w})$ about $\overline{w} = \overline{a} = [a_1 \ldots a_d]^T$ can be written as follows:

$$F(\overline{w}) = F(\overline{a}) + \sum_{i=1}^{d}(w_i - a_i)\left[\frac{\partial F(\overline{w})}{\partial w_i}\right]_{\overline{w}=\overline{a}} + \sum_{i=1}^{d}\sum_{j=1}^{d}\frac{(w_i - a_i)(w_j - a_j)}{2!}\left[\frac{\partial^2 F(\overline{w})}{\partial w_i \partial w_j}\right]_{\overline{w}=\overline{a}} +$$

$$+ \sum_{i=1}^{d}\sum_{j=1}^{d}\sum_{k=1}^{d}\frac{(w_i - a_i)(w_j - a_j)(w_k - a_k)}{3!}\left[\frac{\partial^3 F(\overline{w})}{\partial w_i \partial w_j \partial w_k}\right]_{\overline{w}=\overline{a}} + \ldots$$

In the multivariable case, we have $O(d^2)$ second-order interaction terms, $O(d^3)$ third-order interaction terms, and so on. One can see that the number of terms becomes unwieldy very quickly. Luckily, we rarely need to go beyond second-order approximations in practice. Furthermore, the above expression can be rewritten using the gradients and matrices compactly. For example, the second-order approximation can be written in vector form as follows:

$$F(\overline{w}) \approx F(\overline{a}) + [\overline{w} - \overline{a}]^T \nabla F(\overline{w}) + [\overline{w} - \overline{a}]^T H(\overline{a})[\overline{w} - \overline{a}]$$

Here, $\nabla F(\overline{W})$ is the gradient, and $H(\overline{a}) = [h_{ij}]$ is the $d \times d$ matrix of all second-order derivatives of the following form:

$$h_{ij} = \left[\frac{\partial^2 F(\overline{w})}{\partial w_i \partial w_j}\right]_{\overline{w}=\overline{a}}$$

A third-order expansion would require the use of a *tensor*, which is a generalization of the notion of a matrix. The first- and second-order expansions will be used frequently in this book for developing various types of optimization algorithms, such as the Newton method.

Problem 1.5.1 (Euler Identity) *The Taylor series is valid for complex functions as well. Use the Taylor series to show the Euler identity $e^{i\theta} = cos(\theta) + i\,sin(\theta)$.*

1.5.2 Example of Optimization in Machine Learning

An example of a parameterized model discussed in an earlier section is that of linear regression, in which we want to determine a d-dimensional vector $\overline{W} = [w_1 \ldots w_d]^T$ so that we can predict the n-dimensional dependent variable vector $\overline{y}$ as a function $\overline{y} = D\overline{W}$ of the $n \times d$ matrix D of the observed values. In order to minimize the difference between predicted and observed values, the following objective function is minimized:

$$J = \frac{1}{2}\|D\overline{W} - \overline{y}\|^2 \tag{1.32}$$

Here, D is an $n \times d$ data matrix, whereas $\overline{y}$ is an n-dimensional column vector of dependent variables. Therefore, this is a simple optimization problem in d parameters. Finding the optimal solution requires techniques from differential calculus. The simplest approach is to set the partial derivative with respect to each parameter w_i to 0, which provides a necessary (but not sufficient) condition for optimality:

$$\frac{\partial J}{\partial w_i} = 0, \quad \forall i \in \{1 \ldots d\} \tag{1.33}$$

The partial derivatives can be shown to be the following (cf. Section 4.7 of Chapter 4):

$$\left[\frac{\partial J}{\partial w_1} \cdots \frac{\partial J}{\partial w_d}\right]^T = D^T D\overline{W} - D^T\overline{y} \tag{1.34}$$

For certain types of *convex objective functions* like linear regression, setting the vector of partial derivatives to the zero vector is both necessary and sufficient for minimization (cf. Chapters 3 and 4). Therefore, we have $D^T D\overline{W} = D^T\overline{y}$, which yields the following:

$$\overline{W} = (D^T D)^{-1} D^T\overline{y} \tag{1.35}$$

Linear regression is a particularly simple problem because the optimal solution exists in *closed form*. However, in most cases, one cannot solve the resulting optimality conditions in such a form. Rather, the approach of *gradient-descent* is used. In gradient descent, we use a computational algorithm of initializing the parameter set $\overline{W}$ randomly (or a heuristically chosen point), and then change the parameter set in the direction of the negative derivative of the objective function. In other words, we use the following updates repeatedly with step-size α, which is also referred to as the *learning rate*:

$$[w_1 \ldots w_d]^T \Leftarrow [w_1 \ldots w_d]^T - \alpha \left[\frac{\partial J}{\partial w_1} \cdots \frac{\partial J}{\partial w_d}\right]^T = \overline{W} - \alpha[D^T D\overline{W} - D^T\overline{y}] \tag{1.36}$$

The d-dimensional vector of partial derivatives is referred to as the *gradient vector*, and it defines an instantaneous direction of best rate of improvement of the objective function at the current value of the parameter vector $\overline{W}$. The gradient vector is denoted by $\nabla J(\overline{W})$:

$$\nabla J(\overline{W}) = \left[\frac{\partial J}{\partial w_1} \cdots \frac{\partial J}{\partial w_d}\right]^T$$

Therefore, one can succinctly write gradient descent in the following form:

$$\overline{W} \Leftarrow \overline{W} - \alpha \nabla J(\overline{W})$$

The size of the step is defined by the learning rate α. Note that the best rate of improvement is only over a step of infinitesimal size, and does not hold true for larger steps of finite size. Since the gradients change on making a step, one must be careful not to make steps that are too large or else the effects might be unpredictable. These updates are repeatedly executed to *convergence*, when further improvements become too small to be useful. Such a situation will occur when the gradient vector contains near-zero entries. Therefore, this computational approach will also (eventually) reach a solution approximately satisfying the optimality conditions of Equation 1.33. As we will show in Chapter 4, the gradient descent method (and many other optimization algorithms) can be explained with the use of the Taylor expansion.

Using gradient descent for optimization is a tricky exercise, because one does not always converge to an optimal solution for a variety of reasons. For example, even the wrong step-size, α, might result in unexpected numerical overflows. In other cases, one might terminate at suboptimal solutions, when the objective function contains multiple minima relative to specific local regions. Therefore, there is a significant body of work on designing optimization algorithms (cf. Chapters 4, 5, and 6).

1.5.3   Optimization in Computational Graphs

Many machine learning problems can be represented as the process of learning a function of the inputs that matches the observed variables in the data. For example, the least-squares optimization problem can be represented as the following sequence of operations:

Input (d variables) $\Rightarrow$ Dot product with parameter vector $\overline{W}$ $\Rightarrow$ Prediction $\Rightarrow$ Squared loss

A graphical representation of these types of operations on the inputs is presented in Figure 1.5(a). This model has d input nodes containing the features $x_1 \ldots x_d$ of the data, and a single (computational) output node creating the dot product $\sum_{i=1}^{d} w_i x_i$. The weights $[w_1 \ldots w_d]$ are associated with the edges. Therefore, each node computes a function of its inputs, and the edges are associated with the parameters to be learned. By choosing a more complex topology of the computational graph with more nodes, one can create more powerful models, which often do not have direct analogs in traditional machine learning

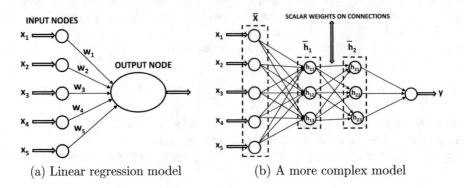

(a) Linear regression model   (b) A more complex model

Figure 1.5: The computational graph view of machine learning

(cf. Figure 1.5(b)). Each node of this graph can compute a function of its incoming nodes and the edge parameters. The overall function is potentially extremely complex, and often cannot be expressed compactly in closed form (like the simple relationship $y = \sum_{i=1}^{d} w_i x_i$ in a linear regression model). A model with many layers of nodes is referred to as a *deep learning model*. Such models can learn complex, nonlinear relationships in the data.

How does one compute gradients with respect to edge parameters in computational graphs? This is achieved with the use of a technique referred to as *backpropagation*, which will be introduced in Chapter 11. The backpropagation algorithm yields exactly the same gradient as is computed in traditional machine learning. For example, since Figure 1.5(a) models linear regression, the backpropagation algorithm will yield exactly the same gradient as computed in the previous section. The main difference is that the backpropagation algorithm will also be able to compute gradients in more complex cases like Figure 1.5(b). Almost all the well-known machine learning models (based on gradient descent) can be represented as relatively simple computational graphs. Therefore, computational graphs are extremely powerful abstractions, as they include traditional machine learning as special cases. We will discuss the power of such models and the associated algorithms in Chapter 11.

1.6 Summary

Linear algebra and optimization are intimately related because many of the basic problems in linear algebra, such as finding the "best" solution to an over-determined system of linear equations, are solved using optimization techniques. Many optimization models in machine learning can also be expressed as objective functions and constraints using matrices/vectors. A useful technique that is used in many of these optimization problems is to decompose these matrices into simpler matrices with specific algebraic/geometric properties. In particular, the following two types of decomposition are commonly used in machine learning:

- Any square and invertible matrix A can be decomposed into a product of elementary matrix operators. If the matrix A is not invertible, it can still be decomposed with a *relaxed* definition of matrix operators, which are allowed to be non-invertible.

- Any square matrix A can be decomposed into a product of two rotation matrices and one scaling (diagonal) matrix in the particular order of rotation, scaling, and rotation. This idea is referred to as *singular value decomposition* (cf. Chapter 7).

An alternative view of machine learning expresses predictions as *computational graphs*; this idea also forms the basis for the field of *deep learning*.

1.7 Further Reading

Several basic books on linear algebra are available, such as those by Strang [122, 123], David Lay [77], and Hoffman-Kunze [62]. These books are, however, generic books on linear algebra, and the focus is not specifically on machine learning topics. Some recent books have focused on a machine learning perspective [23, 119, 122, 125]. The classic matrix computation book by Golub and Van Loan [52] provides an overview of fundamental numerical algorithms. A closely related field to linear algebra is that of optimization. Several books are available on optimization from a generic perspective [10, 15, 16, 22, 99], whereas others focus on machine learning [1–4, 18, 19, 39, 46, 53, 56, 85, 94, 95].

1.8   Exercises

1. For any two vectors $\overline{x}$ and $\overline{y}$, which are each of length a, show that (i) $\overline{x} - \overline{y}$ is orthogonal to $\overline{x} + \overline{y}$, and (ii) the dot product of $\overline{x} - 3\overline{y}$ and $\overline{x} + 3\overline{y}$ is negative.

2. Consider a situation in which you have three matrices A, B, and C, of sizes 10×2, 2×10, and 10×10, respectively.

   (a) Suppose you had to compute the matrix product ABC. From an efficiency perspective, would it computationally make more sense to compute $(AB)C$ or would it make more sense to compute $A(BC)$?

   (b) If you had to compute the matrix product CAB, would it make more sense to compute $(CA)B$ or $C(AB)$?

3. Show that if a matrix A satisfies $A = -A^T$, then all the diagonal elements of the matrix are 0.

4. Show that if we have a matrix satisfying $A = -A^T$, then for any column vector $\overline{x}$, we have $\overline{x}^T A \overline{x} = 0$.

5. Suppose we have an $n \times n$ matrix A that can be written as $A = D^T$ for some $n \times d$ matrix D. Show that $\overline{x}^T A \overline{x} \geq 0$ for any n-dimensional column vector $\overline{x}$.

6. Show that the matrix product AB remains unchanged if we scale the ith column of A and the ith row of B by respective factors that are inverses of each other.

7. Show that any matrix product AB can be expressed in the form $A' \Delta B'$, where A' is a matrix in which the sum of the squares of the entries in each column is 1, B' is a matrix in which the sum of the squares of the entries in each row is 1, and Δ is an appropriately chosen diagonal matrix with nonnegative entries on the diagonal.

8. Discuss how a permutation matrix can be converted to the identity matrix using at most d elementary row operations of a single type. Use this fact to express A as the product of at most d elementary matrix operators.

9. Suppose that you reorder all the columns of an invertible matrix A using some random permutation, and you know A^{-1} for the original matrix. Show how you can (simply) compute the inverse of the reordered matrix from A^{-1} without having to invert the new matrix from scratch. Provide an argument in terms of elementary matrices.

10. Suppose that you have approximately factorized an $n \times d$ matrix D as $D \approx UV^T$, where U is an $n \times k$ matrix and V is a $d \times k$ matrix. Show how you can derive an infinite number of alternative factorizations $U'V'^T$ of D, which satisfy $UV^T = U'V'^T$.

11. Either prove each of the following statements or provide a counterexample:

    (a) The order in which you apply two elementary row operations to a matrix does not affect the final result.

    (b) The order in which you apply an elementary row operation and an elementary column operation does not affect the final result.

    It is best to think of these problems in terms of elementary matrix operations.

12. Discuss why some power of a permutation matrix is always the identity matrix. [Hint: Think in terms of the finiteness of the number of permutations.]

13. Consider the matrix polynomial $\sum_{i=0}^{t} a_i A^i$. A straightforward evaluation of this polynomial will require $O(t^2)$ matrix multiplications. Discuss how you can reduce the number of multiplications to $O(t)$ by rearranging the polynomial.

14. Let $A = [a_{ij}]$ be a 2×2 matrix with $a_{12} = 1$, and 0s in all other entries. Show that $A^{1/2}$ does not exist even after allowing complex-valued entries.

15. **Parallelogram law:** The parallelogram law states that the sum of the squares of the sides of a parallelogram is equal to the sum of the squares of its diagonals. Write this law as a vector identity in terms of vectors $\overline{A}$ and $\overline{B}$ of Figure 1.1. Now use vector algebra to show why this vector identity must hold.

16. Write the first four terms of the Taylor expansion of the following univariate functions about $x = a$: (i) $\log_e(x)$; (ii) $\sin(x)$; (iii) $1/x$; (iv) $\exp(x)$.

17. Use the multivariate Taylor expansion to provide a quadratic approximation of $\sin(x + y)$ in the vicinity of $[x, y] = [0, 0]$. Confirm that this approximation loses its accuracy with increasing distance from the origin.

18. Consider a case where a $d \times k$ matrix P is initialized by setting all values randomly to either -1 or $+1$ with equal probability, and then dividing all entries by $\sqrt{d}$. Discuss why the columns of P will be (roughly) mutually orthogonal for large values of d of the order of 10^6. This trick is used frequently in machine learning for rapidly generating the random projection of an $n \times d$ data matrix D as $D' = DP$.

19. Consider the perturbed $d \times d$ matrix $A_\epsilon = A + \epsilon B$, where the value of ϵ is small. Show the following useful approximation for approximating A_ϵ^{-1} from A^{-1}:

$$A_\epsilon^{-1} \approx A^{-1} - \epsilon A^{-1} B A^{-1}$$

20. Suppose that you have a 5×5 matrix A, in which the rows/columns correspond to people in a social network in the order John, Mary, Jack, Tim, and Robin. The entry (i, j) corresponds to the number of times person i sent a message to person j. Define a matrix P, so that PAP^T contains the same information, but with the rows/columns in the order Mary, Tim, John, Robin, and Jack.

21. Suppose that the vectors $\overline{x}$, $\overline{y}$, and $\overline{x} - \overline{y}$ have lengths 2, 3, and 4, respectively. Find the length of $\overline{x} + \overline{y}$ using only vector algebra (and no Euclidean geometry).

22. Show that the inverse of a symmetric matrix is symmetric.

23. Let $A_1, A_2, \ldots A_d$ be $d \times d$ matrices that are strictly upper triangular. Then, the product of $A_1, A_2, \ldots A_d$ is the zero matrix.

24. **Apollonius's identity:** Let ABC be a triangle, and AD be the median from A to BC. Show the following using only vector algebra and no Euclidean geometry:

$$AB^2 + AC^2 = 2(AD^2 + BD^2)$$

[Hint: Orient your triangle properly with respect to the origin.]

25. **Sine law:** Express the sine of the *interior* angle between $\bar{a}$ and $\bar{b}$ (i.e., the angle not greater than 180 degrees) purely in terms of $\bar{a} \cdot \bar{a}$, $\bar{b} \cdot \bar{b}$, and $\bar{a} \cdot \bar{b}$. You are allowed to use $\sin^2(x) + \cos^2(x) = 1$. Consider a triangle, two sides of which are the vectors $\bar{a}$ and $\bar{b}$. The *opposite* angles to these vectors are A and B, respectively. Show the following using only vector algebra and no Euclidean geometry:

$$\frac{\|\bar{a}\|}{\sin(A)} = \frac{\|\bar{b}\|}{\sin(B)}$$

26. **Trigonometry with vector algebra:** Consider a unit vector $\bar{x} = [1, 0]^T$. The vector $\bar{v}_1$ is obtained by rotating $\bar{x}$ counter-clockwise by angle θ_1, and $\bar{v}_2$ is obtained by rotating $\bar{x}$ clockwise by θ_2. Use the rotation matrix to obtain the coordinates of unit vectors $\bar{v}_1$ and $\bar{v}_2$, and then show the following well-known trigonometric identity:

$$\cos(\theta_1 + \theta_2) = \cos(\theta_1)\cos(\theta_2) - \sin(\theta_1)\sin(\theta_2)$$

27. **Coordinate geometry with matrix algebra:** Consider the two lines $y = 3x + 4$ and $y = 5x + 2$ in the 2-dimensional plane. Write the equations in matrix form for appropriately chosen A and $\bar{b}$:

$$A \begin{bmatrix} x \\ y \end{bmatrix} = \bar{b}$$

Find the intersection coordinates (x, y) of the two lines by inverting matrix A.

28. Use the matrix inversion lemma to invert a 10×10 matrix with 1s in each entry other than the diagonal entries, which contain the value 2.

29. **Solid geometry with vector algebra:** Consider the origin-centered hyperplane in 3-dimensional space that is defined by the equation $z = 2x + 3y$. This equation has infinitely many solutions, all of which lie on the plane. Find two solutions that are not multiples of one another and denote them by the 3-dimensional column vectors $\bar{v}_1$ and $\bar{v}_2$, respectively. Let $V = [\bar{v}_1, \bar{v}_2]$ be a 3×2 matrix with columns $\bar{v}_1$ and $\bar{v}_2$. Geometrically describe the set of all vectors that are linear combinations of $\bar{v}_1$ and $\bar{v}_2$ with real coefficients c_1 and c_2:

$$\mathcal{V} = \left\{ V \begin{bmatrix} c_1 \\ c_2 \end{bmatrix} : c_1, c_2 \in \mathcal{R} \right\}$$

Now consider the point $[x, y, z]^T = [2, 3, 1]^T$, which does not lie on the above hyperplane. We want to find a point $\bar{b}$ on the hyperplane for which $\bar{b}$ is as close to $[2, 3, 1]^T$ as possible. How is the vector $\bar{b} - [2, 3, 1]^T$ geometrically related to the hyperplane? Use this fact to show the following condition on $\bar{b}$:

$$V^T \left(\bar{b} - \begin{bmatrix} 2 \\ 3 \\ 1 \end{bmatrix} \right) = \begin{bmatrix} 0 \\ 0 \end{bmatrix}$$

Find a way to eliminate the 3-variable vector $\bar{b}$ from the above equation and replace with the 2-variable vector $\bar{c} = [c_1, c_2]^T$ instead. Substitute numerical values for entries in V and find $\bar{c}$ and $\bar{b}$ with a 2×2 matrix inversion.

30. Let A and B be two $n \times d$ matrices. One can partition them columnwise as $A = [A_1, A_2]$ and $B = [B_1, B_2]$, where A_1 and B_1 are $n \times k$ matrices containing the first k columns of A and B, respectively, in the same order. Let A_2 and B_2 contain the remaining columns. Show that the matrix product AB^T can be expressed as follows:

$$AB^T = A_1 B_1^T + A_2 B_2^T$$

31. Matrix centering: In machine learning, a common centering operation of an $n \times n$ similarity matrix S is the update $S \Leftarrow (I - U/n)S(I - U/n)$, where U is an $n \times n$ matrix of 1s. Use the associative property of matrix multiplication to implement this update efficiently. [Hint: Express U as a product of smaller matrices.]

32. Energy preservation in orthogonal transformations: Show that if A is an $n \times d$ matrix and P is a $d \times d$ orthogonal matrix, then we have $\|AP\|_F = \|A\|_F$.

33. Tight sub-multiplicative case: Suppose that $\overline{u}$ and $\overline{v}$ are column vectors (of not necessarily the same dimensionality). Show that the matrix $\overline{u}\overline{v}^T$ created from the outer product of $\overline{u}$ and $\overline{v}$ has Frobenius norm of $\|\overline{u}\| \, \|\overline{v}\|$.

34. Frobenius orthogonality and Pythagorean theorem: Two $n \times d$ matrices A and B are said to be Frobenius orthogonal if the sum of entry-wise products of their corresponding elements is zero [i.e., $\text{tr}(AB^T) = 0$]. Show the following:

$$\|A + B\|_F^2 = \|A\|_F^2 + \|B\|_F^2$$

35. Let $\overline{x}$ and $\overline{y}$ be two orthogonal column vectors of dimensionality n. Let $\overline{a}$ and $\overline{b}$ be two arbitrary d-dimensional column vectors. Show that the outer products $\overline{x}\,\overline{a}^T$ and $\overline{y}\,\overline{b}^T$ are Frobenius orthogonal (see Exercise 34 for definition of Frobenius orthogonality).

36. Suppose that a sequence of row and column operations is performed on a matrix. Show that as long as the ordering among row operations and the ordering among column operations is maintained, the way in which the row sequence and column sequence are merged does not change the final result matrix. [Hint: Use operator matrices.]

37. Show that any orthogonal upper-triangular matrix is a diagonal matrix.

38. Consider a set of vectors $\overline{x}_1 \ldots \overline{x}_n$, which are known to be *unit normalized*. You do not have access to the vectors, but you are given all pairwise squared Euclidean distances in the $n \times n$ matrix Δ. Discuss why you can derive the $n \times n$ pairwise dot product matrix by adding 1 to each entry of the matrix $-\frac{1}{2}\Delta$.

39. We know that every matrix commutes with its inverse. We want to show a generalization of this result. Consider the polynomial functions $f(A)$ and $g(A)$ of the square matrix A, so that $f(A)$ is invertible. Show the following commutative property:

$$[f(A)]^{-1}g(A) = g(A)[f(A)]^{-1}$$

40. Give an example of a 2×2 matrix A and a polynomial function $f(\cdot)$, so that A is invertible, but $f(A)$ is not invertible. Give an example of a matrix A, so that A is not invertible, but $f(A)$ is invertible. Note that the constant term in the polynomial corresponds to a multiple of the identity matrix.

41. Let A be a rectangular matrix and $f(\cdot)$ be a polynomial function. Show that $A^T f(AA^T) = f(A^T A)A^T$. Assuming invertibility of $f(AA^T)$ and $f(A^T A)$, show:

$$[f(A^T A)]^{-1} A^T = A^T [f(AA^T)]^{-1}$$

Interpret the push-through identity as a special case of this result.

42. Discuss why one cannot generalize the formula for the scalar binomial expansion $(a+b)^n$ to the matrix expansion $(A+B)^n$. Also discuss why generalization is possible in cases where $B = f(A)$ for some polynomial function $f(\cdot)$.

43. Suppose that A is a $d \times d$ matrix satisfying $A^4 = 0$. Derive an algebraic expression for $(I + A)^{-1}$ as a matrix polynomial in A.

44. Compute the inverse of the following triangular matrix by expressing it as the sum of two carefully chosen matrices (cf. Section 1.2.5):

$$A = \begin{bmatrix} 1 & 0 & 0 \\ 2 & 1 & 0 \\ 1 & 3 & 1 \end{bmatrix}$$

45. Express a $d \times d$ matrix M of 1s as the outer product of two d-dimensional vectors. Use the matrix inversion lemma to compute an algebraic expression for $(I + M)^{-1}$.

46. Show that if A and B commute, the matrix polynomials $f(A)$ and $g(B)$ commute.

47. Show that if invertible matrices A and B commute, A^k and B^s commute for all integers $k, s \in [-\infty, \infty]$. Show the result of Exercise 46 for an extended definition of "polynomials" with both positive and negative integer exponents included.

48. Let $U = [u_{ij}]$ be an upper-triangular $d \times d$ matrix. What are the diagonal entries of the matrix polynomial $f(U)$ as scalar functions of the matrix entries u_{ij}?

49. **Inverses behave like matrix polynomials:** The *Cayley-Hamilton theorem* states that a finite-degree polynomial $f(\cdot)$ always exists for any matrix A satisfying $f(A) = 0$. Use this fact to prove that the inverse of A is also a finite-degree polynomial.

50. Derive the inverse of a 3×3 row addition operator by inverting the sum of matrices.

51. For any non-invertible matrix A, show that the infinite summation $\sum_{k=0}^{\infty} (I - A)^k$ cannot possibly converge to a finite matrix. Give two examples to show that if A is invertible, the summation might or might not converge.

52. The chapter shows that the product, $A_1 A_2 \ldots A_k$, of invertible matrices is invertible. Show the converse that if the product $A_1 A_2 \ldots A_k$ of square matrices is invertible, each matrix A_i is invertible. [Hint: You need only the most basic results discussed in this chapter for the proof.]

53. Show that if a $d \times d$ diagonal matrix Δ with distinct diagonal entries $\lambda_1 \ldots \lambda_d$ commutes with A, then A is diagonal.

54. What fraction of 2×2 binary matrices with 0-1 entries are invertible?

Chapter 2

Linear Transformations and Linear Systems

"You can't criticize geometry. It is never wrong."– Paul Rand

2.1 Introduction

Machine learning algorithms work with data matrices, which can be viewed as collections of row vectors or as collections of column vectors. For example, one can view the rows of an $n \times d$ data matrix D as a set of n points in a space of dimensionality d, and one can view the columns as features. These collections of row vectors and column vectors define *vector spaces*. In this chapter, we will introduce the basic properties of vector spaces and their connections to solving linear systems of equations. This problem is also a special case of the problem of linear regression, which is one of the fundamental building blocks of machine learning.

We will also study matrix multiplication as a linear operator with geometric interpretation. As discussed in Section 1.3.2 of Chapter 1, multiplying a matrix with a vector can be used to implement rotation, scaling, and reflection operations on the vector. In fact, *a multiplication of a vector with a matrix can be shown to be some combination of rotation, scaling, and reflection being applied to the vector.* Much of linear algebra draws inspirations from Cartesian geometry. However, Cartesian geometry is often studied in only 2 or 3 dimensions. On the other hand, linear algebra is naturally defined in spaces of any dimensionality.

This chapter is organized as follows. The remainder of this section introduces the concept of linear transformations. The next section provides a provides a basic understanding of the geometric properties of linear transformations. The basics of linear algebra are introduced in Section 2.3. The linear algebra of row spaces and column spaces is introduced in Section 2.4. The problem of solving systems of linear equations is discussed in Section 2.5. The notion of matrix rank is introduced in Section 2.6. Different methods for generating orthogonal basis sets are introduced in Section 2.7. In Section 2.8, we show that solving

© Springer Nature Switzerland AG 2020
C. C. Aggarwal, *Linear Algebra and Optimization for Machine Learning*,
https://doi.org/10.1007/978-3-030-40344-7_2

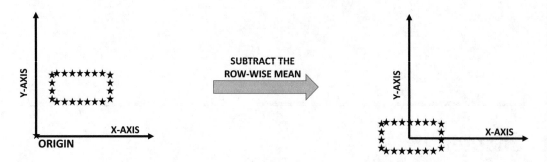

Figure 2.1: Mean-centering: a translation operation

systems of linear equations is a special case of least-squares regression, which is one of the fundamental building blocks of machine learning. The issue of ill-conditioned matrices and ill-conditioned systems of equations is discussed in Section 2.9. Inner products are introduced in Section 2.10. Complex vector spaces are introduced in Section 2.11. A summary is given in Section 2.12.

2.1.1   What Is a Linear Transform?

Linear transformations are at the heart of operations performed on vectors in linear algebra, and they are typically accomplished by multiplying matrices and vectors. A linear transform is defined as follows:

Definition 2.1.1 (Linear Transform) *A vector-to-vector function $f(\overline{x})$ defines a linear transform of $\overline{x}$, if the following conditions are satisfied for any scalar c:*

$$f(c\overline{x}) = c \cdot f(\overline{x}), \quad \forall \overline{x} \ in \ domain \ of \ f(\cdot)$$
$$f(\overline{x} + \overline{y}) = f(\overline{x}) + f(\overline{y}), \quad \forall \overline{x}, \overline{y} \ in \ domain \ of \ f(\cdot)$$

A vector-to-vector function is a generalization of the notion of scalar functions, and it maps a d-dimensional vector to an n-dimensional vector for some d and n. Consider the function $f(\overline{x}) = A\overline{x}$, which pre-multiplies the d-dimensional column vector $\overline{x}$ with the $n \times d$ matrix A to create an n-dimensional column vector. This function satisfies the conditions of Definition 2.1.1, and is therefore a linear transform.

On the other hand, the *translation* operator is *not* a linear transform. Consider the translation of the d-dimensional vector $\overline{x}$ with the vector $\overline{b} = [b_1 \dots b_d]^T$ as follows:

$$f(\overline{x}) = \overline{x} + \overline{b}$$

This transform does not obey the additive and multiplicative properties. The translation operation is often used in machine learning for *mean-centering* the data, where a constant mean vector is subtracted from each row of the data set. As a result, the mean value of each column of the transformed data set becomes 0. An example of the effect of mean-centering on the scatter plot of a 2-dimensional data set is illustrated in Figure 2.1.

Translation is a special case of the class of *affine transforms*, which includes any transform of the form $f(\overline{x}) = A\overline{x} + \overline{c}$, where A is an $n \times d$ matrix, $\overline{x}$ is d-dimensional vector, and $\overline{c}$ is an n-dimensional column vector. Stated simply, *an affine transform is a combination of a linear transform with a translation*. One can define an affine transform as follows:

Definition 2.1.2 (Affine Transform) *A vector-to-vector function $f(\overline{x})$ defines an affine transform of $\overline{x}$, if the following condition is satisfied for any scalar λ:*

$$f(\lambda\overline{x} + [1 - \lambda]\overline{y}) = \lambda f(\overline{x}) + [1 - \lambda]f(\overline{y}), \quad \forall \overline{x}, \overline{y} \text{ in domain of } f(\cdot)$$

All linear transforms are special cases of affine transforms, but not vice versa. There is considerable confusion and ambiguity in the use of the terms "linear" and "affine" in mathematics. Many subfields of mathematics use the terms "linear" and "affine" interchangeably. For example, the simplest univariate function $f(x) = m \cdot x + b$, which is widely referred to as "linear," allows a non-zero translation b; this would make it an affine transform. However, the notion of linear transform from the *linear algebra* perspective is much more restrictive, and it does not even include the univariate function $f(x) = m \cdot x + b$, unless the *bias term* b is zero. The class of linear transforms (from the linear algebra perspective) can always be geometrically expressed as a sequence of one or more rotations, reflections, and dilations/contractions about the origin. The origin always maps to itself after these operations, and therefore translation is not included. Unfortunately, the use of the word "linear" in machine learning almost always allows translation (with copious use of bias terms), which makes the terminology somewhat confusing. In this book, the words "linear transform" or "linear operator" will be used in the context of linear algebra (where translation is not allowed). Terms such as "linear function" will be used in the context of machine learning (where translation is allowed).

2.2 The Geometry of Matrix Multiplication

The discussion in the previous section already shows that the multiplication of a d-dimensional vector with an $n \times d$ matrix is an example of a linear transformation. It turns out that *the converse is also true*:

Lemma 2.2.1 (Linear Transformation Is Matrix Multiplication) *Any linear mapping $f(\overline{x})$ from d-dimensional vectors to n-dimensional vectors can be represented as the matrix-to-vector product $A\overline{x}$ by constructing A as follows. The columns of the $n \times d$ matrix A are $f(\overline{e}_1) \ldots f(\overline{e}_d)$, where $\overline{e}_i$ is the ith column of the $d \times d$ identity matrix.*

Proof: The result $f(\overline{e}_i) = A\overline{e}_i$ holds, because $A\overline{e}_i$ returns the ith column of A, which is $f(\overline{e}_i)$. Furthermore, one can express $f(\overline{x})$ for any vector $\overline{x} = [x_1 \ldots x_d]^T$ as follows:

$$f(\overline{x}) = f(\sum_{i=1}^{d} x_i \overline{e}_i) = \sum_{i=1}^{d} x_i f(\overline{e}_i) = \sum_{i=1}^{d} x_i [A\overline{e}_i] = A[\sum_{i=1}^{d} x_i \overline{e}_i] = A\overline{x}$$

Therefore, the linear transformation $f(\overline{x})$ can always be expressed as $A\overline{x}$. ∎

Setting A to the scalar m yields a special case of the scalar-to-scalar linear function $f(x) = m \cdot x + b$ (with $b = 0$). For vector-to-vector transformations, one can either transform a row vector $\overline{y}$ as $\overline{y}V$ or (equivalently) transform the column vector $\overline{x} = \overline{y}^T$ as $V^T\overline{x}$:

$$f(\overline{y}) = \overline{y}V \quad \text{[Linear transform on row vector } \overline{y}]$$
$$g(\overline{x}) = V^T\overline{x} \quad \text{[Same transform on column vector } \overline{x} = \overline{y}^T]$$

One can also treat a matrix-to-matrix multiplication between $n \times d$ matrix D and $d \times d$ matrix V as a linear transformation of the rows of the first matrix. In other words, the ith row of the $n \times d$ matrix $D' = DV$ is the transformed representation of the ith row of the original matrix D. Data matrices in machine learning often contain multidimensional points in their rows.

Matrix transformations can be broken up into geometrically interpretable sequences of transformations by expressing matrices as products of simpler matrices (cf. Section 1.3 of Chapter 1):

Observation 2.2.1 (Matrix Product as Sequence of Geometric Transformations)
The geometric transformation caused by multiplying a vector with $V = V_1 V_2 \ldots V_r$ can be viewed a sequence of simpler geometric transformations by regrouping the product as follows:

$$\underbrace{\overline{y}\, V = ([[(\overline{y}V_1)V_2] \ldots V_r)}_{\textit{For row vector } \overline{y}}, \qquad \underbrace{V^T \overline{x} = (V_r^T [V_{r-1}^T \ldots (V_1^T \overline{x})])}_{\textit{For column vector } \overline{x} = \overline{y}^T}$$

Note the groupings of the expressions using parentheses so that simple geometric operations corresponding to matrices $V_1 \ldots V_r$ are sequentially applied to the corresponding vectors. In the following, we discuss some important geometric operators. We start with orthogonal operators.

Orthogonal Transformations

The orthogonal 2×2 matrices V_r and V_c that respectively rotate 2-dimensional row and column vectors by θ degrees in the counter-clockwise direction are as follows:

$$V_r = \begin{bmatrix} \cos(\theta) & \sin(\theta) \\ -\sin(\theta) & \cos(\theta) \end{bmatrix}, \quad V_c = \begin{bmatrix} \cos(\theta) & -\sin(\theta) \\ \sin(\theta) & \cos(\theta) \end{bmatrix} \qquad (2.1)$$

If we have an $n \times 2$ data matrix D, then the product DV_r will rotate each *row* of D using V_r, whereas the product $V_c D^T$ will equivalently rotate each *column* of D^T. One can also view a data rotation DV_r in terms of projection of the original data on a rotated axis system. Counter-clockwise rotation of the data with a fixed axis system is the same as clockwise rotation of the axis system with fixed data. In essence, the two columns of the transformation matrix V_r represent the mutually orthogonal unit vectors of a new axis system that is rotated *clockwise* by θ. These two new columns are shown on the left of Figure 2.2 for a counter-clockwise rotation of $30°$. The transformation returns the coordinates DV_r of the data points on these column vectors, because we are computing the dot product of each row of D with the (unit length) columns of V_r. In this case, the columns of V_r (orthonormal directions in new axis system) make counter-clockwise angles of $-30°$ and $60°$ with the vector $[1, 0]$. Therefore, the corresponding matrix V_r is obtained by populating the columns with vectors of the form $[\cos(\theta), \sin(\theta)]^T$, where θ is the angle each new orthonormal axis direction makes with the vector $[1, 0]$. This results in the following matrix V_r:

$$V_r = \begin{bmatrix} \cos(-30) & \cos(60) \\ \sin(-30) & \sin(60) \end{bmatrix} = \begin{bmatrix} \cos(30) & \sin(30) \\ -\sin(30) & \cos(30) \end{bmatrix} \qquad (2.2)$$

After performing the projection of each data point on the new axes, we can reorient the figure so that the new axes are aligned with the original X- and Y-axes (as shown in the left-to-right transition of Figure 2.2). It is easy to see that the final result is a counter-clockwise rotation of the data points by $30°$ about the origin.

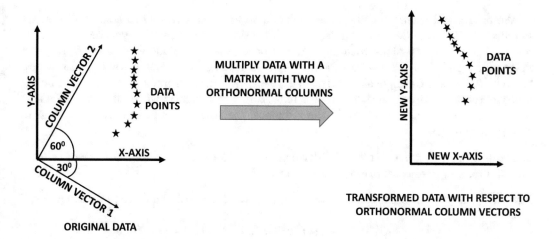

Figure 2.2: An example of counter-clockwise rotation with 30° with matrix multiplication. The two columns of the transformation matrix are shown in the figure on the left

Orthogonal matrices might include reflections. Consider the following matrix:

$$V = \begin{bmatrix} 0 & 1 \\ 1 & 0 \end{bmatrix} \tag{2.3}$$

For any 2-dimensional data set contained in the $n \times 2$ matrix D, the transformation DV of the rows of D simply flips the two coordinates in each row of D. The resulting transformation cannot be expressed purely as a rotation. This is because this transformation changes the *handedness* of the data — for example, if the scatter plot of the n rows of the $n \times 2$ matrix D depicts a right hand, the scatter plot of the $n \times 2$ matrix DV will depict a left hand. Intuitively, when you look at your reflection in the mirror, your left hand appears to be your right hand. This implies that a reflection needs to performed somewhere. The key point is that V can be expressed as the product of a counter-clockwise rotation of 90°, followed by a reflection across the vector $[0, 1]$:

$$V = \begin{bmatrix} \cos(90) & \sin(90) \\ -\sin(90) & \cos(90) \end{bmatrix} \begin{bmatrix} -1 & 0 \\ 0 & 1 \end{bmatrix}$$

When a row of D is post-multiplied with V, it is first rotated counter-clockwise with 90° because of multiplication with the first matrix, and then its first coordinate is multiplied with -1 (i.e., reflection across the Y-axis $[0, 1]$) because of multiplication with the second matrix. An example of the above transformation can be elucidated by post-multiplying the 2-dimensional row vector $[3, 4]$ with V:

$$[3, 4]\,V = \underbrace{[3, 4] \begin{bmatrix} \cos(90) & \sin(90) \\ -\sin(90) & \cos(90) \end{bmatrix}}_{\text{Rotate } 90° \text{ counter-clockwise}} \begin{bmatrix} -1 & 0 \\ 0 & 1 \end{bmatrix} = \underbrace{[-4, 3] \begin{bmatrix} -1 & 0 \\ 0 & 1 \end{bmatrix}}_{\text{Reflect}} = [4, 3]$$

Note that the intermediate result $[-4, 3]$ is indeed a 90° rotation of $[3, 4]$. The decomposition of an orthogonal matrix into rotations and reflections is not unique. For example, if we reflected across $[1, 0]$ instead of $[0, 1]$ in the above example, then a 270° counter-clockwise rotation will do the same job.

An orthogonal matrix might correspond to a *sequence* of rotations in a space of dimensionality greater than 3. For example, if a 4-dimensional object in the $xyzw$-axis system is rotated once in the xy-plane with angle α and once in the zw-plane with angle β, the two independent rotations cannot be expressed by a single angle or plane of rotation. However, the resulting 4×4 orthogonal matrix is still called a "rotation matrix" (in spite of being a sequence of rotations). In some cases, reflections are included with rotations. When a compulsory reflection is included in the sequence, the resulting matrix is referred to as a *rotreflection* matrix.

Lemma 2.2.2 (Closure Under Multiplication) *The product of any number of orthogonal matrices is always an orthogonal matrix.*

Proof: For any set of orthogonal matrices $A_1, A_2, \ldots A_n$, we can show the following:

$$(A_1 A_2 \ldots A_n)(A_1 A_2 \ldots A_n)^T = A_1 A_2 \ldots A_n A_n^T A_{n-1}^T \ldots A_1^T = I$$

One obtains the final result by repeatedly grouping pairs of adjacent orthogonal matrices like $A_n A_n^T$, and replacing it with the identity matrix. Since the transpose of the product matrix $A_1 A_2 \ldots A_n$ is also its inverse, it follows that the product matrix is orthogonal. ∎
What about the commutativity of the product of orthogonal matrices? At first glance, one might mistakenly assume that the product of rotation matrices is commutative. After all, it should not matter whether you first rotate an object $50°$ and then $30°$ or vice versa. However, this type of 2-dimensional visualization of commutativity breaks down in higher dimensions (or when reflection is combined with rotation even in two dimensions). In other words, the product of orthogonal matrices is not necessarily commutative. The main issue is that rotations in higher dimensions are associated with a vector referred to as the *axis of rotation*. Orthogonal matrices that do not correspond to the same axis of rotation may not be commutative; for example, if we successively rotate a sphere by $90°$ about two mutually perpendicular axes, the point on the sphere closest to us will land at different places depending on which rotation occurs first. In order to understand this point, consider the following two 3×3 matrices $R_{[1,0,0]}$ and $R_{[0,1,0]}$, which can perform counter-clockwise rotations of angles α, β about $[1,0,0]$ and $[0,1,0]$, respectively:

$$R_{[1,0,0]} = \begin{bmatrix} 1 & 0 & 0 \\ 0 & \cos(\alpha) & \sin(\alpha) \\ 0 & -\sin(\alpha) & \cos(\alpha) \end{bmatrix}, \quad R_{[0,1,0]} = \begin{bmatrix} \cos(\beta) & 0 & \sin(\beta) \\ 0 & 1 & 0 \\ -\sin(\beta) & 0 & \cos(\beta) \end{bmatrix} \tag{2.4}$$

In order to understand the nature of orthogonal matrices in more than two dimensions, we ask the reader to convince themselves of the following facts:

1. Post-multiplication of row vector $[x, y, z]$ with matrix $R_{[1,0,0]}$ only rotates the vector about $[1,0,0]$ (without changing the first coordinate), whereas the matrix $R_{[0,1,0]}$ rotates this vector about $[0,1,0]$ (without changing the second coordinate).

2. The matrix $R_{[1,0,0]} R_{[0,1,0]}$ is a matrix with orthonormal rows and columns (which can be verified algebraically).

3. The product of $R_{[1,0,0]}$ and $R_{[0,1,0]}$ is sensitive to the order of multiplication. Therefore, the order of rotations matters.

All 3-dimensional rotation matrices can be geometrically expressed as a single rotation, albeit with an arbitrary axis of rotation.

Givens Rotations and Householder Reflections

It is not possible to express a rotation matrix using a single angle in dimensionalities greater than 3 — in such cases, independent rotations of different angles might be occurring in unrelated planes (e.g., xy-plane and zw-plane). Therefore, one must express a rotation transformation as a *sequence* of *elementary* rotations, each of which occurs in a 2-dimensional plane. One natural choice for defining an elementary rotation is the *Givens rotation*, which is a generalization of Equation 2.4 to higher dimensions. A $d \times d$ *Givens rotation* always selects two coordinate axes and performs the rotation in that plane, so that post-multiplying a d-dimensional row vector with that rotation matrix changes only two coordinates. The $d \times d$ Givens rotation matrix is different from the $d \times d$ identity matrix in only 2×2 relevant entries; these entries are the same as those of a 2×2 rotation matrix. For example, the 4×4 Givens rotation matrix $G_r(2,4,\alpha)$ below rotates only the second and fourth coordinates counter-clockwise by α when post-multiplied to a row vector, and its transpose $G_c(2,4,\alpha)$ can be pre-multiplied to a column vector to achieve the same result:

$$G_r(2,4,\alpha) = \begin{bmatrix} 1 & 0 & 0 & 0 \\ 0 & \cos(\alpha) & 0 & \sin(\alpha) \\ 0 & 0 & 1 & 0 \\ 0 & -\sin(\alpha) & 0 & \cos(\alpha) \end{bmatrix}, \quad G_c(2,4,\alpha) = \begin{bmatrix} 1 & 0 & 0 & 0 \\ 0 & \cos(\alpha) & 0 & -\sin(\alpha) \\ 0 & 0 & 1 & 0 \\ 0 & \sin(\alpha) & 0 & \cos(\alpha) \end{bmatrix}$$

$$\underbrace{}_{\text{For row vectors}} \qquad \underbrace{}_{\text{For column vectors}}$$

The notations $G.(\cdot,\cdot,\cdot)$ for row-wise and column-wise transformation matrices are respectively subscripted by either "r" or "c." All orthogonal matrices can be decomposed into Givens rotations, although a reflection might also be needed. We state the following result [52], although a formal proof is omitted:

Lemma 2.2.3 (Givens Geometric Decomposition) *All $d \times d$ orthogonal matrices can be shown to be products of at most $O(d^2)$ Givens rotations and at most a single elementary reflection matrix (obtained by negating one diagonal element of the identity matrix).*

The Givens rotation has many useful applications in numerical linear algebra [52].

Problem 2.2.1 *Show that you can express a $d \times d$ elementary row interchange matrix as the product of a $90°$ Givens rotation and an elementary reflection.*

So far we have introduced only diagonal reflection matrices that flip the sign of a vector component. The *Householder reflection matrix* is an orthogonal matrix that reflects a vector $\overline{x}$ in any "mirror" hyperplane of arbitrary orientation; such a hyperplane passes through the origin and its orientation is defined by an *arbitrary* normal vector $\overline{v}$ (of unit length). Assume that both $\overline{x}$ and $\overline{v}$ are column vectors. First, note that the distance of $\overline{x}$ from the "mirror" hyperplane is $c = \overline{x} \cdot \overline{v}$. An object and its mirror image are separated by twice this distance along $\overline{v}$. Therefore, to perform the reflection of $\overline{x}$ and create its mirror image $\overline{x}'$, one must subtract *twice* of $c\overline{v}$ from $\overline{x}$:

$$\overline{x}' \Leftarrow \overline{x} - 2\,(\overline{x} \cdot \overline{v})\overline{v} = \overline{x} - 2\,(\overline{v}^T\overline{x})\overline{v} = \overline{x} - 2\,\overline{v}(\overline{v}^T\overline{x}) = \overline{x} - 2\,(\overline{v}\overline{v}^T)\overline{x} = \underbrace{\left(I - 2\overline{v}\,\overline{v}^T\right)}_{\text{Householder}} \overline{x}$$

For any unit (column) vector $\overline{v}$, the matrix $(I - 2\overline{v}\,\overline{v}^T)$ is an elementary reflection matrix in the hyperplane perpendicular to $\overline{v}$ and passing through the origin. This matrix is referred to as the *Householder reflection matrix*. Any orthogonal matrix can be represented with fewer Householder reflections than Givens rotations; therefore, the former is a more expressive transform.

Lemma 2.2.4 (Householder Geometric Decomposition) *Any orthogonal matrix of size $d \times d$ can be expressed as the product of at most d Householder reflection matrices.*

Problem 2.2.2 (Reflection of a Reflection) *Verify algebraically that the square of the Householder reflection matrix is the identity matrix.*

Problem 2.2.3 *Show that the elementary reflection matrix, which varies from the identity matrix only in terms of flipping the sign of the ith diagonal element, is a special case of the Householder reflection matrix.*

Problem 2.2.4 (Generalized Householder) *Show that a sequence of k mutually orthogonal Householder transformations can be expressed as $I - 2QQ^T$ for a $d \times k$ matrix Q containing orthonormal columns. Which $(d - k)$-dimensional plane is this a reflection in?*

Rigidity of Orthogonal Transformations

Dot products and Euclidean distances between vectors are unaffected by multiplicative transformations with orthogonal matrices. This is because an orthogonal transformation is a sequence of rotations and reflections, which does not change lengths and angles. This fact can also be shown algebraically. Consider two d-dimensional row vectors $\overline{x}$ and $\overline{y}$ that are respectively transformed to $\overline{x}V$ and $\overline{y}V$ using the $d \times d$ orthogonal matrix V. Then, the dot product between these transformed vectors is as follows:

$$[\overline{x}V] \cdot [\overline{y}V] = [\overline{x}V][\overline{y}V]^T = [\overline{x}V][V^T \overline{y}^T] = \overline{x}(VV^T)\overline{y}^T = \overline{x}(I)\overline{y}^T = \overline{x} \cdot \overline{y}$$

This equivalence for dot products naturally carries over to Euclidean distances and angles, which are functions of dot products. This also means that orthogonal transformations preserve the sum of squares of Euclidean distances of the data points (i.e., rows of a data matrix D) about the origin, which is also the (squared) Frobenius norm or energy of the $n \times d$ matrix D. When the $n \times d$ matrix D is multiplied with the $d \times d$ orthogonal matrix V, the Frobenius norm of DV can be expressed in terms of the trace operator as follows:

$$\|DV\|_F^2 = \mathrm{tr}[(DV)(DV)^T] = \mathrm{tr}[D(VV^T)D^T] = \mathrm{tr}(DD^T) = \|D\|_F^2$$

Transformations that preserve distances between pairs of points are said to be *rigid*. Rotations and reflections not only preserve distances between points but also absolute distances of points from the origin. Translations (which are not linear transforms) are also rigid because they preserve distances between pairs of transformed points. However, translations usually do not preserve distances from the origin.

Scaling: A Non-rigid Transformation

In general, multiplication of a vector $\overline{x}$ with an arbitrary matrix V might change its length. If such a matrix can be decomposed into simpler geometric operator matrices as $V = V_1 V_2 \ldots V_r$, it means that there must be some fundamental geometric transformation V_i among these operator matrices that does not preserve distances. This fundamental transformation is that of *dilation/contraction* (or, more generally, *scaling*). The basic form of this transformation scales the ith dimension of the vector $\overline{x}$ by a scaling factor λ_i. Such a transformation can be achieved by post-multiplying row vector $\overline{x}$ with a $d \times d$ diagonal matrix Δ in which the ith diagonal entry is λ_i. Note that it is possible for the entries to be negative, in which case the *reflection* operation (along the corresponding axis direction) is combined with dilation/contraction. When the scaling factors across different dimensions are different, the scaling is said to be *anisotropic*. An example of a 2×2 matrix Δ corresponding to anisotropic scaling is as follows:

$$\Delta = \begin{bmatrix} 2 & 0 \\ 0 & 0.5 \end{bmatrix}$$

Multiplying a 2-dimensional vector with this matrix scales the first coordinate by 2 and the second coordinate by 0.5. This transformation is not rigid because of non-unit scaling factors in various directions. Furthermore, if we flip the sign of the first diagonal entry by changing it from 2 to -2, then this transformation will combine positive dilation/contraction with reflection via the following decomposition:

$$\begin{bmatrix} -2 & 0 \\ 0 & 0.5 \end{bmatrix} = \underbrace{\begin{bmatrix} 2 & 0 \\ 0 & 0.5 \end{bmatrix}}_{\text{Stretching}} \underbrace{\begin{bmatrix} -1 & 0 \\ 0 & 1 \end{bmatrix}}_{\text{Reflection}}$$

Thus, a reflection matrix is a special case of a scaling (diagonal) matrix.

General Case: Combining Orthogonal and Scaling Transformations

Multiplying an $n \times d$ data matrix D with a diagonal matrix Δ to create $D\Delta$ results in scaling of the ith dimension (column) of the data matrix D with the ith diagonal entry of Δ. This is an example of axis-parallel scaling, where the directions of scaling are aligned with the axes of representation. Just as axis-parallel scalings are performed with diagonal matrices, scalings along arbitrary directions are performed with *diagonalizable* matrices (cf. Chapter 3).

Consider the case in which we want to scale each 2-dimensional row of an $n \times 2$ data matrix in the direction $[\cos(-30), \sin(-30)]$ by a factor of 2, and in the direction $[\cos(60), \sin(60)]$ by a factor of 0.5. This can be achieved by (i) first rotating the data set D by an angle $30°$ by multiplying D with orthogonal matrix V to create DV, (ii) then multiplying the resulting matrix DV with diagonal matrix Δ with diagonal entries 2 and 0.5 to create $(DV)\Delta$, and (iii) finally rotating the data set in the reverse direction (i.e., by angle $-30°$) by multiplying $DV\Delta$ with V^T to create $(DV\Delta)V^T$. The resulting transformation can be regrouped using the associativity property of matrix multiplication as follows:

$$D' = D(V\Delta V^T)$$

Such transformations of the form $V\Delta V^T$ will be discussed in Chapter 3.

The matrix for performing the aforementioned anisotropic scaling along the two orthogonal vector directions $[\cos(-30), \sin(-30)]$ and $[\cos(60), \sin(60)]$ at scale factors of 2 and 0.5 can be obtained by defining V and Δ as follows:

$$V = \begin{bmatrix} \cos(-30) & \cos(60) \\ \sin(-30) & \sin(60) \end{bmatrix} = \begin{bmatrix} \cos(30) & \sin(30) \\ -\sin(30) & \cos(30) \end{bmatrix}, \qquad \Delta = \begin{bmatrix} 2 & 0 \\ 0 & 0.5 \end{bmatrix}$$

Therefore, we obtain the following transformation matrix $A = V\Delta V^T$:

$$A = \begin{bmatrix} \cos(30) & \sin(30) \\ -\sin(30) & \cos(30) \end{bmatrix} \begin{bmatrix} 2 & 0 \\ 0 & 0.5 \end{bmatrix} \begin{bmatrix} \cos(30) & -\sin(30) \\ \sin(30) & \cos(30) \end{bmatrix} = \begin{bmatrix} 1.625 & -0.650 \\ -0.650 & 0.875 \end{bmatrix}$$

Consider a square with coordinates at $[0, 0]$, $[0, 1]$, $[1, 0]$, and $[1, 1]$. What happens to these coordinates after post-multiplication with the above matrix A? The origin is always transformed to the origin by a linear transformation, and therefore we only need to worry about

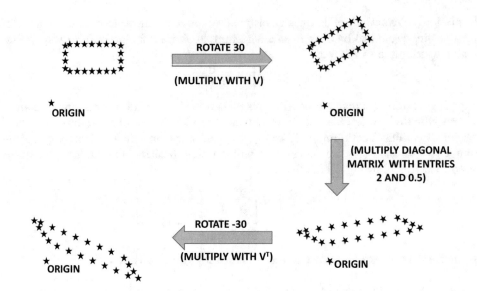

Figure 2.3: An example of anisotropic scaling along two mutually orthogonal directions

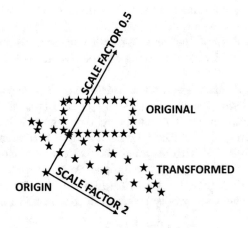

Figure 2.4: The transformation of Figure 2.3 as shown in terms of scaling along two directions

the other three points by stacking them up into a 3×2 matrix denoted by matrix D. The resulting transformed matrix $D' = DA$ is as follows:

$$D' = DA = \begin{bmatrix} 1 & 0 \\ 0 & 1 \\ 1 & 1 \end{bmatrix} \begin{bmatrix} 1.625 & -0.650 \\ -0.650 & 0.875 \end{bmatrix} = \begin{bmatrix} 1.625 & -0.650 \\ -0.650 & 0.875 \\ 0.975 & 0.225 \end{bmatrix}$$

It is also helpful to understand the nature of the distortion pictorially. An example of the sequence of transformations in terms V, Δ, V^T (for a rectangular scatterplot) are shown in Figure 2.3. The corresponding data set $D' = D(V\Delta V^T)$ and the scaling are shown in a concise way in Figure 2.4. One can also generalize this intuition to higher dimensions.

Not all transformations can be expressed in the form $V\Delta V^T$, as shown above. However, all is not lost. A beautiful result, referred to as *singular value decomposition* (cf. Chapter 7), states that any square matrix A can be expressed in the form $A = U\Delta V^T$, where U and V are both orthogonal matrices (which might be different) and Δ is a *nonnegative* scaling matrix. Therefore, *all linear transformations defined by matrix multiplication can be expressed as a sequence of rotations/reflections, together with a single anisotropic scaling.* This result can even be extended to rectangular matrices.

2.3 Vector Spaces and Their Geometry

A *vector space* is an infinite *set* of vectors satisfying certain types of *set closure* properties under addition and scaling operations. One of the most important vector spaces in linear algebra is the set of all n-dimensional vectors:

Definition 2.3.1 (Space of n-Dimensional Vectors) *The space $\mathcal{R}^n$ consists of the set of all column vectors with n real components.*

By convention, the vectors in $\mathcal{R}^n$ are assumed to be column vectors in linear algebra. *Geometrically, all vectors in $\mathcal{R}^n$ are assumed to have tails at the origin.* This is unlike vectors in many scientific fields like physics, where the vector $\overline{x}$ might have a tail at $\overline{a}$ and head at $\overline{x} + \overline{a}$. The space $\mathcal{R}^n$ contains an infinite set of vectors, because any real-valued component of an n-dimensional vector can have an infinite number of possible values. Furthermore, we can scale any vector from $\mathcal{R}^n$ or add two vectors from $\mathcal{R}^n$ to still stay in $\mathcal{R}^n$. This is the defining property of a vector space, which might contain a *subset* $\mathcal{V}$ of vectors from $\mathcal{R}^n$:

Definition 2.3.2 (Vector Space in $\mathcal{R}^n$) *A subset of vectors $\mathcal{V}$ from $\mathcal{R}^n$ is a vector space, if it satisfies the following properties:*

1. *If $\overline{x} \in \mathcal{V}$ then $c\overline{x} \in \mathcal{V}$ for any scalar $c \in \mathcal{R}$.*

2. *If $\overline{x}, \overline{y} \in \mathcal{V}$, then $\overline{x} + \overline{y} \in \mathcal{V}$.*

The zero vector, denoted by $\overline{0}$, is included in all vector spaces, and always satisfies the additive identity $\overline{x} + \overline{0} = \overline{x}$. A singleton set containing the zero vector can also be considered a vector space (albeit a rather simple one), because it satisfies both the above properties.

Consider the subset of vectors from $\mathcal{R}^3$, such that the head of each vector lies on a 2-dimensional hyperplane passing through the origin (and the tail is the origin). This set of vectors is a vector space because adding or scaling vectors on an origin-centered hyperplane leads to other vectors on the same hyperplane. Furthermore, all multiples of an arbitrary vector like $[2, 1, 3]^T$ (i.e., all points on an infinite line in $\mathcal{R}^3$) also form a vector space, which is also a special case of a hyperplane. In general, vector spaces that are subsets of $\mathcal{R}^n$ correspond to vectors sitting on an origin-centered hyperplane of dimensionality at most n. Therefore, vector spaces in $\mathcal{R}^n$ can be nicely mapped to our geometric understanding of lower-dimensional hyperplanes. The *origin-centered nature of these hyperplanes is important*; the set of vectors with tails at the origin and heads on a hyperplane that is not origin-centered does not define a vector space, because this set of vectors is not closed under scaling and addition. Another example of a set of vectors that is not a vector space is the set of all vectors with only non-negative components in $\mathcal{R}^3$, because it is not closed under multiplication with negative scalars. Other than the zero vector space, all vector spaces contain an infinite set of vectors.

Finally, we observe that a fixed linear transformation of each element of a vector space results in another vector space, because of the way in which linear transformations preserve the properties of addition and scalar multiplication (cf. Definition 2.1.1). For example, multiplying all vectors on an origin-centered hyperplane with the same matrix results in a set of vectors sitting on another origin-centered hyperplane after undergoing a set of geometrically interpretable linear transformations (like origin-centered rotation and scaling).

Definition 2.3.2 seems somewhat restrictive at first glance, because we have required all vector spaces to be subsets of $\mathcal{R}^n$. The modern notion of a vector space is more general than vectors from $\mathcal{R}^n$, because it allows all kinds of abstract objects to be considered "vectors" and infinite sets of such objects to be considered vector spaces (along with appropriately defined vector addition and scalar multiplication operations on these objects). For example, the space of all upper-triangular matrices of a specific size is a vector space, although the addition operation now corresponds to element-wise addition of the matrices. Similarly, the space of all polynomial functions of a specific maximum degree is a vector space, and the addition operation corresponds to addition of constituent monomial coefficients. In each case, the nature of the addition and multiplication operations, and the definition of the zero vector (such as the zero matrix or zero polynomial) depends on the type of object being considered. It is also possible for the components of vectors and the scalar c in Definition 2.3.2 to be drawn from the complex domain (or other sets of values[1] satisfying a set of properties known as the *field axioms*). Most of this book works with real-valued vector spaces, although we will occasionally consider vectors drawn from $\mathcal{C}^n$, where $\mathcal{C}$ corresponds to the field of complex numbers (cf. Section 2.11).

The assumption that vector spaces are subsets of $\mathcal{R}^n$ is not as restrictive as one might think, because we can indirectly represent most vector spaces over a real field by mapping them to $\mathcal{R}^n$. For example, the vector space of $m \times m$ upper-triangular matrices can be represented indirectly by populating a vector from $\mathcal{R}^{[m(m+1)/2]}$ with matrix entries. Similarly, polynomials with a pre-defined maximum degree can be represented as finite-length vectors containing the coefficients of various monomials that constitute the polynomial. It can be formally shown that large classes of vector spaces over the real field can be *indirectly* represented using $\mathcal{R}^n$, via the process of *coordinate representation* (cf. Section 2.3.1). Furthermore, staying in $\mathcal{R}^n$ has the distinct advantage of being able to work with easily understandable operations over matrices and vectors.

Problem 2.3.1 *Let $\overline{x} \in \mathcal{R}^d$ be a vector and A be an $n \times d$ matrix. Is each of the following a vector space? (a) All $\overline{x}$ satisfying $A\overline{x} = \overline{0}$; (b) All $\overline{x}$ satisfying $A\overline{x} \geq \overline{0}$; (c) All $\overline{x}$ satisfying $A\overline{x} = \overline{b}$ for some non-zero $\overline{b} \in \mathcal{R}^n$; (d) All $n \times n$ matrices in which the row sums and column sums are the same for a particular matrix (but not necessarily across matrices).*

A subset of the vector space, which is itself a vector space, is referred to as a *subspace*:

Definition 2.3.3 (Subspace) *A vector space $\mathcal{S}$ is a subspace of another vector space $\mathcal{V}$, if any vector $\overline{x} \in \mathcal{S}$ is also present in $\mathcal{V}$. In addition, when $\mathcal{V}$ contains vectors not present in $\mathcal{S}$, the subspace $\mathcal{S}$ is a **proper** subspace of $\mathcal{V}$.*

The set notation "$\subseteq$" is used to denote a subspace as in $\mathcal{S} \subseteq \mathcal{V}$. The notation "$\subset$" denotes a proper subspace of the parent space. The requirement that subspaces are vector spaces ensures that subspaces of $\mathcal{R}^n$ contain vectors residing on hyperplanes in n-dimensional space

---

[1]The field axioms are the properties of associativity, commutativity, distributivity, identity, and inverses. For example, real numbers, complex numbers, and rational numbers form a field. However, integers do not form a field. Refer to http://mathworld.wolfram.com/Field.html. Therefore, one can define vectors over the set of real numbers, complex numbers, or rational numbers. Although one can define vectors more restrictively over the set of integers, such vectors will not satisfy some fundamental rules of linear algebra required for them to be considered a vector space.

passing through the origin. When the hyperplane defining the subspace has dimensionality strictly less than n, the corresponding subspace is a proper subspace of $\mathcal{R}^n$ because non-hyperplane vectors in $\mathcal{R}^n$ are not members of the subspace. For example, the set of all scalar multiples of the vector $[2, 1, 5]^T$ defines a proper subspace of $\mathcal{R}^3$, and it contains all vectors lying on a 1-dimensional hyperplane passing through the origin. However, vectors that do not lie on this 1-dimensional hyperplane are not members of the subspace. Similarly, the vectors $[1, 0, 0]^T$ and $[1, 2, 1]^T$ can be used to define a 2-dimensional hyperplane $\mathcal{V}_1$, each point on which is a linear combination of this pair of vectors. The set of vectors sitting on this hyperplane also define a proper subspace of $\mathcal{R}^3$. Both the vectors $[5, 4, 2]^T$ and $[0, 2, 1]^T$ lie in this subspace because of the following:

$$\begin{bmatrix} 5 \\ 4 \\ 2 \end{bmatrix} = 3 \begin{bmatrix} 1 \\ 0 \\ 0 \end{bmatrix} + 2 \begin{bmatrix} 1 \\ 2 \\ 1 \end{bmatrix}, \qquad \begin{bmatrix} 0 \\ 2 \\ 1 \end{bmatrix} = \begin{bmatrix} 1 \\ 2 \\ 1 \end{bmatrix} - \begin{bmatrix} 1 \\ 0 \\ 0 \end{bmatrix}$$

All scalar multiples of $[5, 4, 2]^T$ also define a vector space $\mathcal{V}_2$ that is a proper subspace of $\mathcal{V}_1$, because the line defining $\mathcal{V}_2$ sits on the hyperplane corresponding to $\mathcal{V}_1$. In other words, we have $\mathcal{V}_2 \subset \mathcal{V}_1 \subset \mathcal{R}^3$. For the vector space $\mathcal{R}^3$, examples of proper subspaces could be the set of vectors sitting on (i) any 2-dimensional plane passing through the origin, (ii) any 1-dimensional line passing through the origin, and (iii) the zero vector. Furthermore, subspace relationships might exist among the lower-dimensional hyperplanes when one of them contains the other (e.g., a 1-dimensional line sitting on a plane in $\mathcal{R}^3$).

A set of vectors $\{\bar{a}_1 \ldots \bar{a}_d\}$ is *linearly dependent* if a non-zero linear combination of these vectors sums to zero:

Definition 2.3.4 (Linear Dependence) *A set of non-zero vectors $\bar{a}_1 \ldots \bar{a}_d$ is linearly dependent, if a set of d scalars $x_1 \ldots x_d$ can be found so that at least some of the scalars are non-zero, and the following condition is satisfied:*

$$\sum_{i=1}^{d} x_i \bar{a}_i = \bar{0}$$

We emphasize the fact that all scalars $x_1 \ldots x_d$ cannot be zero. Such a coefficient set is said to be *non-trivial*. When no such set of non-zero scalars can be found, the resulting set of vectors is said to be *linearly independent*. It is relatively easy to show that a set of vectors $\bar{a}_1 \ldots \bar{a}_d$ that are mutually orthogonal must be linearly independent. If these vectors are linearly dependent, we must have non-trivial coefficients $x_1 \ldots x_d$, such that $\sum_{i=1}^{d} x_i \bar{a}_i = \bar{0}$. However, taking the dot product of the linear dependence condition with each $\bar{a}_i$ and setting each $\bar{a}_i \cdot \bar{a}_j = 0$ for $i \neq j$ yields each $x_i = 0$, which is a *trivial* coefficient set.

Consider the earlier example of three linearly dependent vectors $[0, 2, 1]^T$, $[1, 2, 1]^T$, and $[1, 0, 0]^T$, which lie on a 2-dimensional hyperplane passing through the origin. These vectors satisfy the following linear dependence condition:

$$\begin{bmatrix} 0 \\ 2 \\ 1 \end{bmatrix} - \begin{bmatrix} 1 \\ 2 \\ 1 \end{bmatrix} + \begin{bmatrix} 1 \\ 0 \\ 0 \end{bmatrix} = \bar{0}$$

Therefore, the coefficients x_1, x_2, and x_3 of the linear dependence condition are $+1$, -1, and $+1$ in this case. The key point is that one only needs two of these three vectors to define the hyperplane on which all the vectors lie. This minimal set of vectors is also referred to as a *basis*, and is defined as follows:

Definition 2.3.5 (Basis) *A basis (or basis set) of a vector space* $V \subseteq \mathcal{R}^n$ *is a* **minimal** *set of vectors* $\mathcal{B} = \{\bar{a}_1 \ldots \bar{a}_d\} \subseteq V$, *so that all vectors in* V *can be expressed as linear combinations of* $\bar{a}_1 \ldots \bar{a}_d$. *In other words, for any vector* $\bar{v} \in V$, *we can find scalars* $x_1 \ldots x_d$ *so that* $\bar{v} = \sum_{i=1}^{d} x_i \bar{a}_i$, *and one cannot do this for any proper subset of* $\mathcal{B}$.

It is helpful to think of a basis geometrically as a coordinate system of directions or *axes*, and the scalars $x_1 \ldots x_d$ as coordinates in order to express vectors. For example, the two commonly used axis directions in the classical 2-dimensional plane of Cartesian geometry are $[1, 0]^T$ and $[0, 1]^T$, although we could always rotate this axis system by θ to get a new set of axes $\{[\cos(\theta), \sin(\theta)]^T, [-\sin(\theta), \cos(\theta)]^T\}$ and corresponding coordinates. Furthermore, the representative directions need not even be mutually orthogonal. For example, every point in $\mathcal{R}^2$ can be expressed as a linear combination of $[1, 1]^T$ and $[1, 2]^T$. Clearly, the basis set is not unique, just as coordinate systems are not unique in classical Cartesian geometry.

Note that the vectors in a basis must be linearly independent. This is because if the vectors in the basis $\mathcal{B}$ are linearly dependent, we can drop any vector occurring in the linear dependence condition from $\mathcal{B}$ without losing the ability to express all vectors in V in terms of the remaining vectors. Furthermore, if the linear combination of a set of vectors $\mathcal{B}$ cannot express a particular vector in $\bar{v} \in V$, one can add $\bar{v}$ to the set $\mathcal{B}$ without disturbing its linear independence. This process can be continued until all vectors in V are expressed by a linear combination of the set $\mathcal{B}$. Therefore, an alternative definition of the basis as follows:

Definition 2.3.6 (Basis: Alternative Definition) *A basis (or basis set) of a vector space* V *is a* **maximal** *set of* **linearly independent** *vectors in it.*

Both definitions of the basis are equivalent and can be derived from one another. An interesting artifact is that the vector space containing only the zero vector has an empty basis. A vector space containing non-zero vectors always has an infinite number of possible basis sets. For example, if we select any three linearly independent vectors in $\mathcal{R}^3$ (or even scale the vectors in a basis set), the resulting set of vectors is a valid basis of $\mathcal{R}^3$. An important result, referred to as the *dimension theorem of vector spaces*, states that the size of every basis set of a vector space must be the same:

Theorem 2.3.1 (Dimension Theorem for Vector Spaces) *The number of members in every possible basis set of a vector space* V *is always the same. This value is referred to as the* **dimensionality** *of the vector space.*

Proof: Suppose that we have two basis sets $\bar{a}_1 \ldots \bar{a}_d$ and $\bar{b}_1 \ldots \bar{b}_m$ so that $d < m$. In such a case, we will prove that a subset of the vectors in $\bar{b}_1 \ldots \bar{b}_m$ must be linearly dependent, which is a contradiction with the pre-condition of the lemma.

Each vector $\bar{b}_i$ is a linear combination of the basis vectors $\bar{a}_1 \ldots \bar{a}_d$:

$$\bar{b}_i = \sum_{j=1}^{d} \beta_{ij} \bar{a}_j \quad \forall i \in \{1 \ldots m\} \tag{2.5}$$

A key point is that we have $m > d$ linear dependence conditions (see Equation 2.5), and we can eliminate each of the d vectors $\bar{a}_1 \ldots \bar{a}_d$ at the cost of reducing one equation. For example, we can select a linear dependence condition in which $\bar{a}_1$ occurs with a non-zero coefficient, and express $\bar{a}_1$ as a linear combination of $\bar{a}_2 \ldots \bar{a}_d$ and at least one of $\bar{b}_1 \ldots \bar{b}_m$. This linear expression for $\bar{a}_1$ is substituted in all the other linear dependence conditions. The linear dependence condition that was originally selected in order to create the expression for $\bar{a}_1$ is dropped. This process reduces the number of linear dependence conditions and the number of vectors from the basis set $\{\bar{a}_1 \ldots \bar{a}_d\}$ by 1. One can repeat this process with each of $\bar{a}_2 \ldots \bar{a}_d$, and in each case, the corresponding vector is eliminated while reducing the number of linear dependence conditions by 1. Therefore, after all the vectors $\bar{a}_1 \ldots \bar{a}_d$ have been eliminated, we will be left with $(m - d) > 0$ linear conditions between $\bar{b}_1 \ldots \bar{b}_m$. This implies that $\bar{b}_1 \ldots \bar{b}_m$ are linearly dependent. ∎

The notion of subspace dimensionality is identical to that of geometric dimensionality of hyperplanes in $\mathcal{R}^n$. For example, any set of n linearly independent directions in $\mathcal{R}^n$ can be used to create a basis (or coordinate system) in $\mathcal{R}^n$. For subspaces corresponding to lower-dimensional hyperplanes, we only need as many linearly independent vectors sitting on the hyperplane as are needed to uniquely define it. This value is the same as the geometric dimensionality of the hyperplane. This leads to the following result:

Lemma 2.3.1 (Matrix Invertibility and Linear Independence) *An $n \times n$ square matrix A has linearly independent columns/rows if and only if it is invertible.*

Proof: An $n \times n$ square matrix with linearly independent columns defines a basis for all vectors in $\mathcal{R}^n$ in its columns. Therefore, we can find n coefficient vectors $\bar{x}_1, \ldots, \bar{x}_n \in \mathcal{R}^n$ so that $A\bar{x}_i = \bar{e}_i$ for each i, where $\bar{e}_i$ is the ith column of the identity matrix. These conditions can be written in matrix form as $A[\bar{x}_1 \ldots \bar{x}_n] = [\bar{e}_1 \ldots \bar{e}_n] = I_n$. Since A and $[\bar{x}_1 \ldots \bar{x}_d]$ multiply to yield the identity matrix, we have $A^{-1} = [\bar{x}_1 \ldots \bar{x}_n]$. Conversely, if the matrix A is invertible, multiplication of $A\bar{x} = \bar{0}$ with A^{-1} shows that $\bar{x} = \bar{0}$ is the only solution (which implies linear independence). One can show similar results with the rows. ∎

When vector spaces contain abstract objects like degree-p polynomials of the form $\sum_{i=0}^{p} c_i t^i$, the basis contains simple instantiations of these objects like $\{t^0, t^1, \ldots t^p\}$. Choosing a basis like this allows as to use the coefficients $[c_0 \ldots c_p]^T$ of each polynomial as the new vectors space $\mathcal{R}^{p+1}$. Carefully chosen basis sets *allow us to automatically map all d-dimensional vector spaces over real fields to $\mathcal{R}^d$ for finite values of d.* For example, $\mathcal{V}$ might be a d-dimensional subspace of $\mathcal{R}^n$ (for $d < n$). However, once we select d basis vectors, the set of d-dimensional combination coefficients for these vectors themselves create the "nicer" vector space $\mathcal{R}^d$. Therefore, we have a one-to-one *isomorphic* mapping between any d-dimensional vector space $\mathcal{V}$ and $\mathcal{R}^d$.

2.3.1 Coordinates in a Basis System

Let $\bar{v} \in \mathcal{V} \subset \mathcal{R}^n$ be a vector drawn from a d-dimensional vector space $\mathcal{V}$ for $d < n$. In other words, the vector space contains all vectors sitting on a d-dimensional hyperplane in $\mathcal{R}^n$. The coefficients $x_1 \ldots x_d$, in terms of which the vector $\bar{v} = \sum_{i=1}^{d} x_i \bar{a}_i$ is represented in a particular basis are referred to as its coordinates. A particular basis set of the vector space $\mathcal{R}^n$, referred to as the *standard basis*, contains the n-dimensional column vectors $\{\bar{e}_1, \ldots \bar{e}_n\}$, where each $\bar{e}_i$ contains a 1 in the ith entry and a value of 0 in all other entries. The standard basis set is often chosen by default, where the scalar components of vectors are the same as their coordinates. However, scalar components of vectors are not the same as their coordinates for arbitrary basis sets. The standard basis is restrictive because it cannot be used as the basis of a *proper* subspace of $\mathcal{R}^n$.

An important result is that the coordinates of a vector in any basis must be unique:

Lemma 2.3.2 (Uniqueness of Coordinates) *The coordinates $\bar{x} = [x_1, \ldots, x_d]^T$ of any vector $\bar{v} \in \mathcal{V}$ in terms of a basis set $\mathcal{B} = \{\bar{a}_1 \ldots \bar{a}_d\}$ are always unique.*

Proof: Suppose that the coordinates are not unique, and we have two distinct sets of coordinates $x_1 \ldots x_d$ and $y_1 \ldots y_d$. Then, we have $\bar{v} = \sum_{i=1}^{d} x_i \bar{a}_i = \sum_{i=1}^{d} y_i \bar{a}_i$. Therefore, we have $\sum_{i=1}^{d}(x_i - y_i)\bar{a}_i = \bar{v} - \bar{v} = \bar{0}$. This implies that the vectors $\bar{a}_1 \ldots \bar{a}_d$ are linearly dependent. This results in the contradiction from the statement of the lemma that $\mathcal{B}$ is a basis (unless the coordinate sets $x_1 \ldots x_d$ and $y_1 \ldots y_d$ are identical). ∎

How can one find these unique coordinates? When $\bar{a}_1 \ldots \bar{a}_d$ correspond to an orthonormal basis of $\mathcal{V}$, the coordinates are simply the dot products of $\bar{v}$ with these vectors. By taking the dot product of both sides of $\bar{v} = \sum_{i=1}^{d} x_i \bar{a}_i$ with each $\bar{a}_j$ and using orthonormality, it is easy to show that $x_j = \bar{v} \cdot \bar{a}_j$. For example, if $\bar{a}_1 = [1, 1, 1]^T/\sqrt{3}$ and $\bar{a}_2 = [1, -1, 0]^T/\sqrt{2}$ constitute the orthonormal basis set of vector space $\mathcal{V}$ containing all points in the plane of these vectors, the vector $[2, 0, 1]^T \in \mathcal{V}$ can be shown to have coordinates $[\sqrt{3}, \sqrt{2}]^T$ (using the dot product method). Even though the basis vectors are drawn from $\mathcal{R}^3$, the vector space $\mathcal{V}$ is a 2-dimensional plane, and it will have only two coordinates.

It is much trickier to find the coordinates of a vector $\bar{v}$ in a non-orthogonal basis system. The general problem is that of solving the system of equations $A\bar{x} = \bar{v}$ for $\bar{x} = [x_1 \ldots x_d]^T$, where the n-dimensional columns of the $n \times d$ matrix A contain the (linearly independent) basis vectors. The problem boils down to finding a solution to the system of equations $A\bar{x} = \bar{v}$, where $A = [\bar{a}_1 \ldots \bar{a}_d]$ contains the basis vectors of the d-dimensional vector space $\mathcal{V} \subseteq \mathcal{R}^n$. Note that the basis vectors are themselves represented using n components like the vectors of $\mathcal{R}^n$, even though the vector space $\mathcal{V}$ is a d-dimensional subspace of $\mathcal{R}^n$ and the coordinate vector $\bar{x}$ lies in $\mathcal{R}^d$. If $d = n$, and the matrix A is square, the solution is simply $\bar{x} = A^{-1}\bar{v}$. However, when A is not square, one may not be able to find valid coordinates, if $\bar{v}$ does not lie in $\mathcal{V} \subset \mathcal{R}^n$. This occurs when $\bar{v}$ does not geometrically lie on the hyperplane H_A defined by all possible linear combinations of the columns of A. However, one can find the *best fit* coordinates $\bar{x}$ by observing that *the line joining the closest linear combination $A\bar{x}$ of the columns of A to $\bar{v}$ must be orthogonal to the hyperplane H_A*, and it is therefore also orthogonal to every column of A. The condition that $(A\bar{x} - \bar{v})$ is orthogonal to every column of A can be expressed as the *normal equation* $A^T(A\bar{x} - \bar{v}) = \bar{0}$. This results in the following:

$$\bar{x} = (A^T A)^{-1} A^T \bar{v} \tag{2.6}$$

The best-fit solution includes the exact solution when it is possible. The matrix $(A^T A)^{-1} A^T$ is referred to as the *left-inverse* of the matrix A with linearly independent columns and we will encounter it repeatedly in this book via different derivations (see Section 2.8).

In order to illustrate the nature of coordinate transformations, we will show the coordinates of the same vector $[10, 15]^T$ in three different basis sets including the standard basis set. The three basis sets correspond to the standard basis set, a basis set $\left\{ \left[\frac{3}{5}, \frac{4}{5}\right]^T, \left[-\frac{4}{5}, \frac{3}{5}\right]^T \right\}$ obtained by rotating each vector in the standard basis counter-clockwise by $\sin^{-1}(4/5)$, and a non-orthogonal basis $\{[1, 1]^T, [1, 2]^T\}$ in which the vectors are not even unit normalized. Each of these basis sets defines a coordinate system for representing $\mathcal{R}^2$, and the non-orthogonal coordinate system seems very different from the conventional system of Cartesian coordinates. The corresponding basis directions are shown in Figure 2.5(a), (b), and (c), respectively. For the case of the standard basis in Figure 2.5(a), the coordinates of the vector $[10, 15]^T$ are the same as its vector components (i.e., 10 and 15). However, this is not the case in any other basis. The coordinates of the vector $[10, 15]^T$ in the orthonormal (rotated) basis of Figure 2.5(b) are $[18, 1]^T$, and the coordinates in the non-orthogonal basis of Figure 2.5(c) are $[5, 5]^T$. The explanation for these values of the coordinates arises from the decomposition of $[10, 15]^T$ in terms of various basis sets:

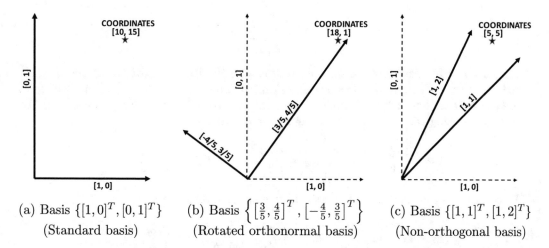

(a) Basis $\{[1,0]^T, [0,1]^T\}$
(Standard basis)

(b) Basis $\left\{\left[\frac{3}{5}, \frac{4}{5}\right]^T, \left[-\frac{4}{5}, \frac{3}{5}\right]^T\right\}$
(Rotated orthonormal basis)

(c) Basis $\{[1,1]^T, [1,2]^T\}$
(Non-orthogonal basis)

Figure 2.5: Examples of different bases in $\mathcal{R}^2$ with corresponding coordinates of the same vector $[10, 15]^T$. A basis set may be non-orthogonal and unnormalized, as in (c)

$$\begin{bmatrix} 10 \\ 15 \end{bmatrix} = \underbrace{10 \begin{bmatrix} 1 \\ 0 \end{bmatrix} + 15 \begin{bmatrix} 0 \\ 1 \end{bmatrix}}_{\text{Standard basis}} = \underbrace{18 \begin{bmatrix} 3/5 \\ 4/5 \end{bmatrix} + 1 \begin{bmatrix} -4/5 \\ 3/5 \end{bmatrix}}_{\text{Basis of Figure 2.5(b)}} = \underbrace{5 \begin{bmatrix} 1 \\ 1 \end{bmatrix} + 5 \begin{bmatrix} 1 \\ 2 \end{bmatrix}}_{\text{Basis of Figure 2.5(c)}}$$

Although the notion of a non-orthogonal coordinate system does exist in analytical geometry, it is rarely used in practice because of loss of visual interpretability of the coordinates. However, such non-orthogonal basis systems are very natural to linear algebra, where some loss of geometric intuition is often compensated by algebraic simplicity.

2.3.2 Coordinate Transformations Between Basis Sets

The previous section discusses how different basis sets correspond to different coordinate systems for the vectors in $\mathcal{R}^n$. A natural question arises as to how one can transform the coordinates $\overline{x}_a$ defined with respect to the n-dimensional basis set $\{\overline{a}_1, \ldots, \overline{a}_n\}$ of $\mathcal{R}^n$ into the coordinates $\overline{x}_b$ defined with respect to the n-dimensional basis set $\{\overline{b}_1, \ldots, \overline{b}_n\}$. The goal is to find an $n \times n$ matrix $P_{a \to b}$ that transforms $\overline{x}_a$ to $\overline{x}_b$:

$$\overline{x}_b = P_{a \to b}\overline{x}_a$$

For example, how might one transform the coordinates in the orthogonal basis set of Figure 2.5(b) into the non-orthogonal system of Figure 2.5(c)? Here, the key point is to observe that the coordinates $\overline{x}_a$ and $\overline{x}_b$ are representations of the same vector, and they would therefore have the same coordinates in the standard basis. First, we use the basis sets to construct two $n \times n$ matrices $A = [\overline{a}_1 \ldots \overline{a}_n]$ and $B = [\overline{b}_1 \ldots \overline{b}_n]$. Since the coordinates $\overline{x}$ of $\overline{x}_a$ and $\overline{x}_b$ must be identical in the standard basis, we have the following:

$$A\overline{x}_a = B\overline{x}_b = \overline{x}$$

We have already established (cf. Lemma 2.3.1) that square matrices defined by linearly independent vectors are invertible. Therefore, multiplying both sides with B^{-1}, we obtain the following:

$$\overline{x}_b = \underbrace{\left[B^{-1}A\right]}_{P_{a \to b}} \overline{x}_a$$

In order to verify that this matrix does indeed perform the intended transformation, let us compute the coordinate transformation matrix from the system in Figure 2.5(b) to the system in Figure 2.5(c). Therefore, our matrices A and B in these two cases can be constructed using the basis vectors in Figure 2.5 as follows:

$$A = \begin{bmatrix} 3/5 & -4/5 \\ 4/5 & 3/5 \end{bmatrix}, \quad B = \begin{bmatrix} 1 & 1 \\ 1 & 2 \end{bmatrix}, \quad B^{-1} = \begin{bmatrix} 2 & -1 \\ -1 & 1 \end{bmatrix}$$

The coordinate transformation matrix can be computed as follows:

$$P_{a \to b} = B^{-1}A = \begin{bmatrix} 2 & -1 \\ -1 & 1 \end{bmatrix} \begin{bmatrix} 3/5 & -4/5 \\ 4/5 & 3/5 \end{bmatrix} = \begin{bmatrix} 2/5 & -11/5 \\ 1/5 & 7/5 \end{bmatrix}$$

In order to check whether this coordinate transformation works correctly, we want to check whether the coordinate $[18, 1]^T$ in Figure 2.5(b) gets transformed to $[5, 5]^T$ in Figure 2.5(c):

$$P_{a \to b} \begin{bmatrix} 18 \\ 1 \end{bmatrix} = \begin{bmatrix} 2/5 & -11/5 \\ 1/5 & 7/5 \end{bmatrix} \begin{bmatrix} 18 \\ 1 \end{bmatrix} = \begin{bmatrix} 5 \\ 5 \end{bmatrix}$$

Therefore, the transformation matrix correctly converts coordinates from one system to another. The main computational work involved in the transformation is in inverting the matrix B. One observation is that when B is an orthogonal matrix, the transformation matrix simplifies to $B^T A$. Furthermore, when the matrix A (i.e., source representation) corresponds to the standard basis, the transformation matrix is B^T. Therefore, working with orthonormal bases simplifies computations, which is why the identification of orthonormal basis sets is an important problem in its own right (cf. Section 2.7.1).

It is also possible to perform coordinate transformations between basis sets that define a particular d-dimensional *subspace* $\mathcal{V}$ of $\mathcal{R}^n$, rather than all of $\mathcal{R}^n$. Let $\overline{a}_1 \ldots \overline{a}_d$ amd $\overline{b}_1 \ldots \overline{b}_d$ be two basis sets for this d-dimensional subspace $\mathcal{V}$, such that each of these basis vectors is expressed in terms of the standard basis of $\mathcal{R}^n$. Furthermore, let $\overline{x}_a$ and $\overline{x}_b$ be two d-dimensional coordinates of the same vector $\overline{v} \in \mathcal{V}$ in terms of the two basis sets. We want to transform the known coordinates $\overline{x}_a$ to the unknown coordinates $\overline{x}_b$ in the second basis set (and find a best fit if the two basis sets represent different vector spaces). As in the previous case, let $A = [\overline{a}_1 \ldots \overline{a}_d]$ and $B = [\overline{b}_1 \ldots \overline{b}_d]$ be two $n \times d$ matrices whose columns contain each of these two sets of basis vectors. Since $\overline{x}_a$ and $\overline{x}_b$ are coordinates of the same vector, and have the same coordinates in the standard basis of $\mathcal{R}^n$, we have $A\overline{x}_a = B\overline{x}_b$. However, since the matrix B is not square, it cannot be inverted in order to solve for $\overline{x}_b$ in terms of $\overline{x}_a$, and we sometimes might have to be content with a best fit. We observe that this best-fit problem is similar to what was derived in Equation 2.6 with the use of the normal equation, and $A\overline{x}_a - B\overline{x}_b$ needs to be orthogonal to every column of B in order to be a best-fit solution. This implies that $B^T(A\overline{x}_a - B\overline{x}_b) = \overline{0}$, and we have the following:

$$\overline{x}_b = \underbrace{(B^T B)^{-1} B^T A}_{P_{a \to b}} \overline{x}_a$$

When B is square and invertible, it is easy to show that this solution simplifies to $B^{-1}A\overline{x}_a$.

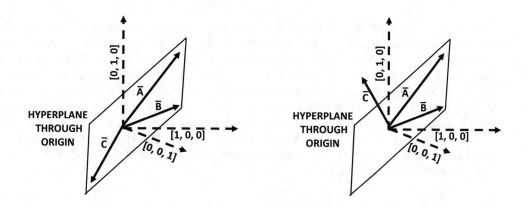

(a) $\text{Span}(\{\overline{A}, \overline{B}\}) = \text{Span}(\{\overline{A}, \overline{B}, \overline{C}\})$ (b) $\text{Span}(\{\overline{A}, \overline{B}\}) \neq \text{Span}(\{\overline{A}, \overline{B}, \overline{C}\})$
$\text{Span}(\{\overline{A}, \overline{B}, \overline{C}\}) = $ All vectors on hyperplane $\text{Span}(\{\overline{A}, \overline{B}, \overline{C}\}) = $ All vectors in $\mathcal{R}^3$

Figure 2.6: The span of a set of linearly dependent vectors has lower dimension than the number of vectors in the set

2.3.3 Span of a Set of Vectors

Even though a vector space is naturally defined by a basis set (which is linearly independent), one can also define a vector space by using a set of linearly dependent vectors. This is achieved with the notion of *span*:

Definition 2.3.7 (Span) *The span of a finite set of vectors $\mathcal{A} = \{\overline{a}_1, \ldots, \overline{a}_d\}$ is the vector space defined by all possible linear combinations of the vectors in $\mathcal{A}$:*

$$Span(\mathcal{A}) = \{\overline{v} : \overline{v} = \sum_{i=1}^{d} x_i \overline{a}_i, \forall x_1 \ldots x_d \in \mathcal{R}\}$$

For example, consider the vector spaces drawn on $\mathcal{R}^3$. In this case, the span of the two vectors $[0, 2, 1]^T$, $[1, 2, 1]^T$ is the set of all vectors lying on the 2-dimensional hyperplane defined by the vectors $[0, 2, 1]^T$ and $[1, 2, 1]^T$. Points that do not lie on this hyperplane do not lie in the span of two vectors. The span of an augmented set of three vectors, which additionally includes the vector $[1, 0, 0]^T$, is no different from the span of the first two vectors; this is because the vector $[1, 0, 0]^T$ is linearly dependent on $[0, 2, 1]^T$ and $]1, 2, 1]^T$. Therefore, adding a vector to a set $\mathcal{A}$ increases its span only when the added vector does not lie in the subspace defined by the span of $\mathcal{A}$. When the set $\mathcal{A}$ contains linearly independent vectors, it is also a basis set of its span.

A pictorial example of what a span captures in $\mathcal{R}^3$ is illustrated in Figure 2.6. In Figure 2.6(a), the three vectors $\overline{A}$, $\overline{B}$, and $\overline{C}$ lie on a hyperplane passing through the origin, although they are *pairwise* linearly independent. Therefore, any pair of them can span the 2-dimensional subspace containing all vectors lying on this hyperplane; however, the span of all three vectors is still this same subspace because of the linear dependence of the three vectors. Adding any number of vectors lying on the hyperplane to the set will not change the span of the set. On the other hand, the three vectors in Figure 2.6(b) are linearly independent, and therefore their span is $\mathcal{R}^3$.

Since the three vectors in Figure 2.6(b) are linearly independent and span $\mathcal{R}^3$, they can be used to create a valid coordinate system to represent any vector in $\mathcal{R}^3$ (albeit a non-orthogonal one). A natural question arises as to what would happen if one tried to use the three linearly dependent vectors $\overline{A}$, $\overline{B}$, and $\overline{C}$ in Figure 2.6(a) to create a "coordinate system" of $\mathcal{R}^3$. First, note that any 3-dimensional vector that does not lie on the hyperplane of Figure 2.6(a) cannot be represented as a linear combination of the three vectors $\overline{A}$, $\overline{B}$, and $\overline{C}$. Therefore, no valid coordinates would exist to represent such a vector. Furthermore, even in cases where $\overline{b}$ does lie on the hyperplane of Figure 2.6(a), the solution to $A\overline{x} = \overline{b}$ may not be unique because of linear dependence of the columns of A, and therefore unique "coordinates" may not exist.

2.3.4   Machine Learning Example: Discrete Wavelet Transform

Basis transformations are used frequently in machine learning of time series. A time-series of length n can be treated as a point in $\mathcal{R}^n$, where each real value represents the series value at a clock tick. For example, a time-series of temperatures collected each second over an hour would result in a vector from $\mathcal{R}^{3,600}$. One common characteristic of time-series is that consecutive values are very similar in most real applications; for example, consecutive temperature readings would typically be the same almost all the time. Therefore, most of the information would be hidden in a few variations across time. The Haar wavelet transformation performs precisely a basis transformation that extracts the important variations. Typically, only a few such differences will be large, which results in a sparse vector. Aside from the space-efficiency advantages of doing so, some predictive algorithms seem to work better with coordinates that reflect trend differences.

For example, consider the series $\overline{s} = [8, 6, 2, 3, 4, 6, 6, 5]^T$ in $\mathcal{R}^8$. The representation corresponds to the values in the standard basis. However, we want a basis in which the differences between *contiguous* regions of the series are emphasized. Therefore, we define the following set of 8 vectors to create a new basis in $\mathcal{R}^8$ together with an interpretation of what their coefficients represent to within a proportionality factor:

Interpretation of basis coefficient	Unnormalized basis vectors	Basis vector norm
Series sum	$[1, 1, 1, 1, 1, 1, 1, 1]^T$	$\sqrt{8}$
Difference between halves	$[1, 1, 1, 1, -1, -1, -1, -1]^T$	$\sqrt{8}$
Difference between quarters	$[1, 1, -1, -1, 0, 0, 0, 0]^T$	$\sqrt{4}$
	$[0, 0, 0, 0, 1, 1, -1, -1]^T$	$\sqrt{4}$
Difference between eighths	$[1, -1, 0, 0, 0, 0, 0, 0]^T$	$\sqrt{2}$
	$[0, 0, 1, -1, 0, 0, 0, 0]^T$	$\sqrt{2}$
	$[0, 0, 0, 0, 1, -1, 0, 0]^T$	$\sqrt{2}$
	$[0, 0, 0, 0, 0, 0, 1, -1]^T$	$\sqrt{2}$

Note that all basis vectors are orthogonal, although they are not normalized to unit norm. We would like to transform the time-series from the standard basis into this new set of orthogonal vectors (after normalization). The problem is simplified by the fact that we have to transform from a standard basis. As discussed at the end of the previous section, we can create an orthogonal matrix B using these vectors, and then simply multiply the time series $\overline{s} = [8, 6, 2, 3, 4, 6, 6, 5]^T$ with B^T to create the transformed representation. Note that the transposed matrix B^T will contain the basis vectors in its *rows* rather than columns. For numerical and computational efficiency, we will not normalize the columns of B to unit norm up front, and simply normalize the coordinates of $\overline{s}$ after multiplying with the *unnormalized* matrix B^T. Therefore, the unnormalized coordinates $\bar{s}_u$ and normalized coordinates $\bar{s}_n$ can be computed as follows:

$$\bar{s}_u = \underbrace{\begin{bmatrix} 1 & 1 & 1 & 1 & 1 & 1 & 1 & 1 \\ 1 & 1 & 1 & 1 & -1 & -1 & -1 & -1 \\ 1 & 1 & -1 & -1 & 0 & 0 & 0 & 0 \\ 0 & 0 & 0 & 0 & 1 & 1 & -1 & -1 \\ 1 & -1 & 0 & 0 & 0 & 0 & 0 & 0 \\ 0 & 0 & 1 & -1 & 0 & 0 & 0 & 0 \\ 0 & 0 & 0 & 0 & 1 & -1 & 0 & 0 \\ 0 & 0 & 0 & 0 & 0 & 0 & 1 & -1 \end{bmatrix}}_{B^T} \underbrace{\begin{bmatrix} 8 \\ 6 \\ 2 \\ 3 \\ 4 \\ 6 \\ 6 \\ 5 \end{bmatrix}}_{\bar{s}} = \begin{bmatrix} 40 \\ -2 \\ 9 \\ -1 \\ 2 \\ -1 \\ -2 \\ 1 \end{bmatrix}, \quad \bar{s}_n = \begin{bmatrix} 40/\sqrt{8} \\ -2/\sqrt{8} \\ 9/\sqrt{4} \\ -1/\sqrt{4} \\ 2/\sqrt{2} \\ -1/\sqrt{2} \\ -2/\sqrt{2} \\ 1/\sqrt{2} \end{bmatrix}$$

The rightmost vector $\bar{s}_n$ contains the normalized wavelet coefficients. In many cases, the dimensionality of the time-series is reduced by dropping those coefficients that are very small in absolute magnitude. Therefore, a compressed representation of the time series can be created. Note that the matrix B is very sparse, and it contains $O(n \log(n))$ non-zero entries for a transformation in $\mathcal{R}^n$. Furthermore, since the matrix only contains values from $\{-1, 0, +1\}$, the matrix multiplication reduces to only addition or subtraction of vector components. In other words, such a matrix multiplication is very efficient.

2.3.5 Relationships Among Subspaces of a Vector Space

In this section, we study the different types of relationships among the subspaces of a vector space. Although this section makes the assumption that all vector spaces are subspaces of $\mathcal{R}^n$ (because of the relevance to machine learning), the underlying results hold even under more general assumptions. First, we discuss the concept of *disjoint vector spaces*:

Definition 2.3.8 (Disjoint Vector Spaces) *Two vector spaces $\mathcal{U} \subseteq \mathcal{R}^n$ and $\mathcal{W} \subseteq \mathcal{R}^n$ are disjoint if and only if the two spaces do not contain any vector in common other than the zero vector.*

If $\mathcal{U}$ and $\mathcal{W}$ are disjoint with basis sets $\mathcal{B}_u$ and $\mathcal{B}_w$, the union $\mathcal{B} = \mathcal{B}_u \cup \mathcal{B}_w$ of these basis sets is a linearly independent set. Otherwise, we can apply the linear dependence condition to $\mathcal{B}$ and place elements from each of the vector spaces on the two sides of the dependence condition to create a vector that lies in both $\mathcal{U}$ and $\mathcal{W}$. This is a contradiction to the pre-condition of disjointedness.

An origin-centered plane in $\mathcal{R}^3$ and an origin-centered line in $\mathcal{R}^3$ represent disjoint vector spaces as long as the line is not subsumed by the plane. However, vector spaces created by any pair of origin-centered planes in $\mathcal{R}^3$ are not disjoint because they intersect along a 1-dimensional line. The hyperplanes corresponding to two disjoint vector spaces *must intersect only at the origin*, which is a 0-dimensional vector space. A special case of *disjointedness* of vector spaces is that of *orthogonality* of the two spaces:

Definition 2.3.9 (Orthogonal Vector Spaces) *Two vector spaces $\mathcal{U} \subseteq \mathcal{R}^n$ and $\mathcal{W} \subseteq \mathcal{R}^n$ are orthogonal if and only if for any pair of vectors $\bar{u} \in \mathcal{U}$ and $\bar{w} \in \mathcal{W}$, the dot product of the two vectors is 0:*

$$\bar{u} \cdot \bar{w} = 0 \tag{2.7}$$

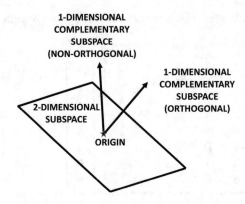

Figure 2.7: Illustration of complementary subspace

Disjoint pairs of vector spaces need not be orthogonal, but orthogonal pairs of vector spaces are always disjoint. One can show this result by contradiction. If the orthogonal vector spaces $\mathcal{U}$ and $\mathcal{W}$ are not disjoint, one can select $\overline{u} \in \mathcal{U}$ and $\overline{w} \in \mathcal{W}$ to be the same non-zero vector (i.e., $\overline{u} = \overline{w} \neq \overline{0}$) from the non-disjoint portion of the space, which cannot satisfy the condition of Equation 2.7 (and this results in a contradiction).

Two orthogonal subspaces, such that the union of their basis sets span all of $\mathcal{R}^n$ are referred to as *orthogonal complementary subspaces*.

Definition 2.3.10 (Orthogonal Complementary Subspace) *Let $\mathcal{U}$ be a subspace of $\mathcal{R}^n$. Then, $\mathcal{W}$ is an orthogonal complementary subspace of $\mathcal{U}$ if and only if it satisfies the following properties:*

- *The spaces $\mathcal{U}$ and $\mathcal{W}$ are orthogonal (and therefore disjoint).*

- *The union of the basis sets of $\mathcal{U}$ and $\mathcal{W}$ forms a basis for $\mathcal{R}^n$.*

The notion of orthogonal complementary subspace is a special case of that of *complementary subspaces*. Two subspaces are complementary when they are disjoint and the union of their basis sets spans all of $\mathcal{R}^n$. However, they need not be orthogonal. For a given subspace, there are an infinite number of complementary subspaces, whereas there is only one orthogonal complementary subspace. Consider the case in which the subspace $\mathcal{U}$ of $\mathcal{R}^3$ is the set of all vectors lying on a 2-dimensional plane passing through the origin. This plane is shown in Figure 2.7. Then any of the infinite number of vectors that emanate from the origin and do not lie on this plane can be used as the singleton basis set to define a complementary 1-dimensional subspace of $\mathcal{U}$. However, there is a unique subspace defined by the vector perpendicular to this plane, which is the *orthogonal* complementary subspace with respect to $\mathcal{U}$.

Problem 2.3.2 *Consider two disjoint vector spaces in $\mathcal{R}^3$ with basis sets $\{[1,1,1]^T\}$ and $\{[1,0,0]^T, [0,1,0]^T\}$, respectively. Express the vector $[0,1,1]^T$ as the sum of two vectors, such that each of them belongs to one of the two spaces. Note that you will have to solve a system of three linear equations to solve this problem.*

Problem 2.3.3 *Let $\mathcal{U} \subset \mathcal{R}^3$ be defined by the basis set $\{[1,0,0]^T, [0,1,0]^T\}$. State the basis sets of two possible complementary subspaces of $\mathcal{U}$. In each case, provide a decomposition of the vector $[1,1,1]^T$ as a sum of vectors from these complementary subspaces.*

Problem 2.3.4 *Let $\mathcal{U} \subset \mathcal{R}^3$ be defined by the basis set $\mathcal{B} = \{[1, 1, 1]^T, [1, -1, 1]^T\}$. Formulate a system of equations to find the orthogonal complementary subspace $\mathcal{W}$ of $\mathcal{U}$. Use the orthogonality of $\mathcal{U}$ and $\mathcal{W}$ to propose a fast method to express the vector $[2, 2, 1]^T$ as a sum of vectors from these complementary subspaces.*

2.4 The Linear Algebra of Matrix Rows and Columns

The rows and columns of an $n \times d$ matrix A span vector spaces, referred to as *row spaces* and *column spaces*, respectively.

Definition 2.4.1 (Row Spaces and Column Spaces) *For an $n \times d$ matrix A, its column space is defined as the vector space spanned by its columns, and it is a subspace of $\mathcal{R}^n$. The row space of A is defined as the vector space spanned by the columns of A^T (which are simply the transposed rows of A). The row space of A is a subspace of $\mathcal{R}^d$.*

A remarkable result in linear algebra is that the dimensionality of the row space (also referred to as *row rank*) and that of the column space (also referred to as *column rank*) of any $n \times d$ matrix A is the same. We will show this result slightly later. We have already shown this equivalence in some special cases where the rows of a square matrix must be linearly independent when the columns are linearly independent, and vice versa (cf. Lemma 2.3.1). Such matrices are said to be of *full rank*. Rectangular matrices are said to be of full rank when *either* the rows *or* the columns are linearly independent. The former is referred to as *full row rank*, whereas the latter is referred to as *full column rank*.

Since the columns of an $n \times d$ matrix A might span only a subspace of $\mathcal{R}^n$ and the (transposed) rows of A might span only a subspace of $\mathcal{R}^d$, how does one characterize the orthogonal complements of these subspaces? This is achieved with the notion of *null spaces*.

Definition 2.4.2 (Null Space) *The null space of a matrix A is the subspace of $\mathcal{R}^d$ containing all column vectors $\overline{x} \in \mathcal{R}^d$, such that $A\overline{x} = \overline{0}$.*

The null space of a matrix A is essentially the orthogonal complementary subspace of the row space of A. The reason is that the condition $A\overline{x} = \overline{0}$ ensures that the dot product of $\overline{x}$ with each transposed row of A (or a linear combination of them) is 0. Note that if $d > n$, the d-dimensional rows of A (after transposition to column vectors) will always span a *proper* subspace of $\mathcal{R}^d$, whose orthogonal complement is non-empty; in other words, the null space of A will be non-empty in this case. For square and non-singular matrices, the null space only contains the zero vector.

The notion of a null space refers to a *right* null space by default. This is because the vector $\overline{x}$ occurs on the right side of matrix A in the product $A\overline{x}$, which must evaluate to the zero vector. Similar to the definition of a right null space, one can define the *left null space* of a matrix, which is the orthogonal complement of the vector space spanned by the columns of the matrix.

Definition 2.4.3 (Left Null Space) *The left null space of an $n \times d$ matrix A is the subspace of $\mathcal{R}^n$ containing all column vectors $\overline{x} \in \mathcal{R}^n$, such that $A^T\overline{x} = \overline{0}$. The left null space of A is the orthogonal complementary subspace of the column space of A.*

Alternatively, the left null space of a matrix A contains all vectors $\overline{x}$ satisfying $\overline{x}^T A = \overline{0}^T$. The row space, column space, the right null space, and the left null space are referred to as the *four fundamental subspaces of linear algebra*.

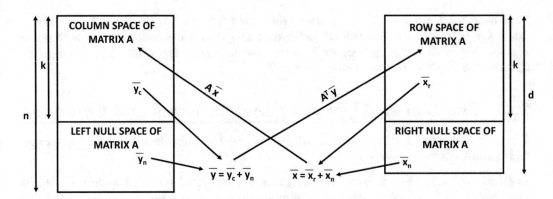

Figure 2.8: The four fundamental subspaces of linear algebra for an $n \times d$ matrix A

In Figure 2.8, we have shown the relationships among the four fundamental subspaces of linear algebra for an $n \times d$ matrix A. In this particular case, the value of n is chosen to be greater than d. Multiplying A with any d-dimensional vector $\overline{x} \in \mathcal{R}^d$ maps to the column space of A (including the zero vector) because the vector $A\overline{x}$ is a linear combination of the columns of A. Similarly, multiplying any n-dimensional vector $\overline{y} \in \mathcal{R}^n$ with A^T to create the vector $A^T\overline{y}$ yields a member of the row space of A, which is a linear combination of the (transposed) rows of A. Another noteworthy point in Figure 2.8 is that *the ranks of the row space and the column space are the same.* The equality is a fundamental result in linear algebra, which will be shown in a later section. The fixed value of the row rank and column rank is also referred to as the rank of the matrix. For example, consider the following 3×4 matrix:

$$A = \begin{bmatrix} 1 & 0 & 1 & 0 \\ 0 & 1 & 0 & 1 \\ 1 & 1 & 1 & 1 \end{bmatrix} \tag{2.8}$$

Note that neither the rows nor the columns of this matrix are linearly independent. The row space has the basis vectors $[1, 0, 1, 0]^T$, and $[0, 1, 0, 1]^T$, whereas the column space has the basis vectors $[1, 0, 1]^T$, and $[0, 1, 1]^T$. Therefore, the row rank is the same as the column rank, which is the same as the matrix rank of 2.

Problem 2.4.1 *Find a basis for each of the right and left null spaces of matrix A in Equation 2.8.*

Problem 2.4.2 *For any $n \times d$ matrix A, show why the matrices $P = A^T A + \lambda I_d$ and $Q = AA^T + \lambda I_n$ always have an empty null space for any $\lambda > 0$.*

A hint for solving the above problem is to show that $\overline{x}^T P \overline{x}$ can never be zero.

2.5   The Row Echelon Form of a Matrix

The row echelon form is useful for transforming matrices to simpler forms with elementary row operations (cf. Section 1.3.1 of Chapter 1) that are *row equivalent* to the original matrix. The material in this section is therefore based on the row operation concepts developed in Section 1.3.1.

Definition 2.5.1 (Row and Column Equivalence) *Two matrices are said to be row equivalent, if one matrix is obtained from the other by a sequence of elementary row operations such as row interchange, row addition, or multiplication of a row with a non-zero scalar. Similarly, two matrices are said to be column equivalent, if one matrix is obtained from the other with a sequence of elementary column operations.*

Note that applying elementary row operations does not change the vector space spanned by the rows of a matrix. This is because row interchange and non-zero scaling operations do not fundamentally change the (normalized) row set of the matrix. Furthermore, the span of any pair of row vectors $\{\overline{r}_i, \overline{r}_j\}$ is the same as that of $\{\overline{r}_i, \overline{r}_i + c\overline{r}_j\}$ for non-zero scalar c because $\overline{r}_j$ can be expressed in terms of the new set of rows as $[(\overline{r}_i + c\overline{r}_j) - \overline{r}_i]/c$. Therefore, any vector in the span of the original set of rows is also in the span of the new set of rows. The converse can also be seen to be true because the new row vectors are directly expressed in terms of the original rows. Similarly, column operations do not change the column space. However, row operations do change the column space, and column operations do change the row space. These results are summarized as follows:

Lemma 2.5.1 *Elementary row operations do not change the vector space spanned by the rows, whereas elementary column operations do not change the vector space spanned by the columns.*

A particularly convenient row-equivalent conversion of the matrix A is the *row echelon form*, which is useful for solving linear systems of the type $A\overline{x} = \overline{b}$. By applying the same row operations to both the matrix A and the vector $\overline{b}$ in the system of equations $A\overline{x} = \overline{b}$, one can simplify the matrix A to a form that makes the system easily solvable. This is exactly the row echelon form, and the procedure is equivalent to the *Gaussian elimination* method for solving systems of equations.

Definition 2.5.2 (Row Echelon Form) *An $n \times d$ matrix A is said to be in row echelon form, if and only if (i) the leftmost non-zero entry in each row is 1, (ii) the column index of the leftmost non-zero entry in each row increases with row index, and (iii) all zero rows (if any) occur at the bottom of the matrix.*

All row echelon matrices are (rectangular) upper-triangular matrices, but the converse is not true. For example, consider the following pair of upper-triangular matrices:

$$A' = \begin{bmatrix} 1 & 7 & 4 & 3 & 5 \\ 0 & 0 & 1 & 7 & 6 \\ 0 & 0 & 0 & 1 & 3 \\ 0 & 0 & 0 & 0 & 1 \end{bmatrix} \qquad B' = \begin{bmatrix} 1 & 7 & 4 & 3 & 5 \\ 0 & 0 & 1 & 7 & 6 \\ 0 & 0 & 1 & 5 & 3 \\ 0 & 0 & 0 & 0 & 1 \end{bmatrix}$$

Here, the matrix A' is in row echelon form, whereas the matrix B' is not. This is because the leftmost non-zero entry of the second and third rows of matrix B' have the same column index. The increasing column index of the leading non-zero entry ensures that *non-zero rows in echelon form are always linearly independent*; adding rows in the order from bottom to top of the matrix to a set S always increases the span of S by 1.

The bulk of the work in Gaussian elimination is to create a matrix in which the column index of the leftmost non-zero entry is *different* for each row; further row interchange operations can create a matrix in which the leftmost non-zero entry has an *increasing* column index, and row scaling operations can change the leftmost entry to 1. The entire process uses three phases:

- **Row addition operations:** We repeatedly identify pairs of rows, so that the column index of the leftmost non-zero entry is the same. For example, the second and third rows of matrix B' in the above example have a tied column index of the leftmost non-zero entry. The elementary row addition operation is applied to the pair so that one of these leftmost entries is set to 0. For example, consider two rows $\overline{r}_1$ and $\overline{r}_2$ with the same leftmost column index. If the leftmost non-zero entries of rows $\overline{r}_1$ and $\overline{r}_2$ have values 3 and 7, respectively, then we can change row $\overline{r}_1$ to $\overline{r}_1 - (3/7)\overline{r}_2$, so that the leftmost entry of $\overline{r}_1$ becomes 0. We could also change $\overline{r}_2$ to $\overline{r}_2 - (7/3)\overline{r}_1$ to achieve a similar effect. *We always choose to perform the operation on the lower of the two rows in order to ensure that the corresponding operator matrix is a lower triangular matrix and the number of leading zeros in the lower row increases by 1.* Since the matrix contains $n \times d$ entries, and each operation increases the number of leading zeros in the matrix, the procedure is guaranteed to succeed in removing column-index ties after $O(nd)$ row addition operations [each of which requires $O(d)$ time]. However, depending on the configuration of the original matrix, one may not be able to reach a matrix in which the column index of the leftmost non-zero entry always increases. For example, a 2×2 matrix with a value of 0 in the top-left corner and a value of 1 in every other entry can never be converted to upper-triangular form with row addition operations.

- **Row interchange operations:** In this phase, we permute the rows of the matrix, so that the column index of the leftmost non-zero entry increases with increasing column index. The permutation of the rows is achieved by interchanging "violating" pairs of rows repeatedly, which do not satisfy the aforementioned condition. Random selection of violating pairs will require $O(d^2)$ interchanges, although more judicious selection can ensure that this is done in $O(d)$ interchanges.

- **Row scaling operations:** Each row is divided by its leading non-zero entry to convert the matrix to row echelon form.

All of the above operations can be implemented with the elementary row operations discussed in Section 1.3.1 of Chapter 1.

2.5.1   LU Decomposition

The goal of LU decomposition is to express a matrix as the product of a (square) lower triangular matrix L and a (rectangular) upper triangular matrix U. However, it is not always possible to create an LU decomposition of a matrix without permuting its rows first. We provide an example in which row permutation is essential:

Observation 2.5.1 *A non-singular matrix $A = [a_{ij}]$ with $a_{11} = 0$ can never be expressed in the form $A = LU$, where $L = [l_{ij}]$ is lower-triangular and $U = [u_{ij}]$ is upper-triangular.*

The above observation can be shown by contradiction by assuming that $A = LU$ is possible. Since $A = LU$, it can be shown that $a_{11} = l_{11}u_{11}$. In order for a_{11} to be zero, either l_{11} or u_{11} must be 0. In other words, either the first row of L is zero or the first column of U is zero. This means that either the first row or the first column of $A = LU$ is zero. In other words, A cannot be non-singular, which is a contradiction.

Let us examine the effect of the first two steps (row addition and interchange steps) of the Gaussian elimination algorithm, which already creates a rectangular upper triangular matrix U. Note that the row addition operations are always lower triangular matrices, because lower rows are always subtracted from upper rows. Furthermore, the sequence of row interchange operations is a permutation of rows, and can therefore be expressed as the permutation matrix P. Therefore, we can express the first two steps of the Gaussian elimination process in terms of a permutation matrix P and the m row-addition operations defined by lower-triangular matrices $L_1 \ldots L_m$:

$$PL_mL_{m-1}\ldots L_1A = U$$

Multiplying both sides with P^T and the inverses of the lower-triangular matrices L_i in the proper sequence, we obtain the following:

$$A = \underbrace{L_1^{-1}L_2^{-1}\ldots L_m^{-1}}_{L} P^T U$$

The inverses and products of lower-triangular matrices are lower triangular (cf. Chapter 1). Therefore, we can consolidate these matrices to obtain a single lower-triangular matrix L of size $n \times n$. In other words, we have the following:

$$A = LP^T U$$

This is, however, not the standard form of the LU decomposition. With some bookkeeping, it is possible to obtain a decomposition in which the permutation matrix P^T occurs before the lower-triangular matrix L (although these matrices would be different when re-ordered):

$$A = P^T LU$$

One can also write this decomposition as $PA = LU$. This is the standard form of LU decomposition.

2.5.2   Application: Finding a Basis Set

The Gaussian elimination method can be used to find a basis set of a bunch of (possibly linearly dependent) vectors. Let $\bar{a}_1 \ldots \bar{a}_n$ be a set of n row vectors, each of which have d dimensions. Then, we can create an $n \times d$ matrix A whose rows are $\bar{a}_1 \ldots \bar{a}_n$. The process discussed in the previous section can be applied to create the row echelon form. The non-zero rows in the reduced matrix are always linearly independent because of the fact that their leading entries have a different column index. In cases where the original rows of A are linearly dependent, and the rank k of the corresponding vector space is strictly less than n, the final $(n - k)$ rows of the row echelon matrix will be zero vectors. The reduced row vectors (which are non-zero) correspond to the linearly independent basis set.

2.5.3   Application: Matrix Inversion

In order to invert a non-singular matrix A, we first perform row operations to convert it to the upper-triangular $d \times d$ matrix $U = [u_{ij}]$ in row echelon form. For invertible/nonsingular matrices like U, it is possible to further convert the matrix U to an identity matrix with the use of only row operations. First, the non-diagonal entries on the $(d - 1)$th row are converted to 0 by subtracting an appropriate multiple [which is $u_{d-1,d}$] of the dth row from it. Then, the non-diagonal entries of the $(d - 2)$th row are converted to 0 by subtracting appropriate multiples [which are $u_{d-2,d-1}$ and $u_{d-2,d}$] of the $(d - 1)$th and dth rows from it. In other words, the rows are processed in order of reducing row index, and at most

$d(d-1)/2$ row operations will be required. This approach works only when the matrix is nonsingular, or else some of the diagonal entries will be 0s. One can obtain the inverse of A by performing the same row operations starting with the identity matrix, as one performs these row operations on A to reach the identity matrix. *A sequence of row operations that transforms A to the identity matrix will transform the identity matrix to $B = A^{-1}$.* The idea is that we perform the same row operations on both sides of the equation $AA^{-1} = I$. The row operations on the left-hand side AA^{-1} can be performed on A until it is transformed to the identity matrix.

2.5.4   Application: Solving a System of Linear Equations

Consider the problem where we want to find all solutions $\overline{x} = [x_1, x_2, \ldots x_d]^T$ that satisfy $A\overline{x} = \overline{b}$, where A is an $n \times d$ matrix and $\overline{b}$ is an n-dimensional column vector. If the columns of the matrix A are $\overline{a}_1 \ldots \overline{a}_d$, $\overline{b}$ needs to be expressed as a linear combination of these columns. This is because the matrix condition $A\overline{x} = \overline{b}$ can be rewritten in terms of the columns of A as follows:

$$\sum_{i=1}^{d} x_i \overline{a}_i = \overline{b} \tag{2.9}$$

Depending on A and $\overline{b}$, three cases arise:

1. If the vector $\overline{b}$ does not occur in the column space of A, then no solution exists to this system of linear equations although *best fits* are possible. This case is studied in detail in Section 2.8.

2. If the vector $\overline{b}$ occurs in the column space of A, and A has linearly independent columns (which implies that the columns form the basis of a d-dimensional subspace of $\mathcal{R}^n$), the solution is unique. This result is based on the uniqueness of coordinates (cf. Lemma 2.3.2). In the special case that A is square, the solution is simply $\overline{x} = A^{-1}\overline{b}$.

3. If the vector $\overline{b}$ occurs in the column space of A and the columns of A are linearly dependent, then an infinite number of solutions exists to $A\overline{x} = \overline{b}$. Note that if $\overline{x}_1$ and $\overline{x}_2$ are solutions, then $\lambda \overline{x}_1 + (1 - \lambda)\overline{x}_2$ is also a solution for any real λ.

The first situation arises very commonly in over-determined systems of linear equations where the number of rows of the matrix is much greater than the number of columns. It is possible for inconsistent systems of equations to occur even in matrices where the number of rows is less than the number of columns. In order to understand this point, consider the case where $\overline{b} = [1, 1]^T$, and a 2×100 matrix A contains two non-zero row vectors, so that the second row vector is twice the first. However, it is impossible to find any non-zero solution to the $A\overline{x} = \overline{b}$ unless the second component of $\overline{b}$ is twice the first. Similarly, the third case occurs more commonly in cases where the number of columns d is greater than the number of rows n, but it is possible to find linearly dependent column vectors even when $d < n$. We present some exercises in order to gain some intuition about these difficult cases:

Problem 2.5.1 *Suppose that no solution exists to the system of equations $A\overline{x} = \overline{b}$, where A is an $n \times d$ matrix and $\overline{b}$ is an n-dimensional column vector. Show that an n-dimensional column vector $\overline{z}$ must exist that satisfies $\overline{z}^T A = \overline{0}$ and $\overline{z}^T \overline{b} \neq \overline{0}$.*

The above practice exercise simply states that if a system of equations is inconsistent, then a weighted combination of the equations can always be found so that the left-hand side adds up to zero, whereas the right-hand side adds up to a non-zero quantity. As a hint to solve the exercise, note that $\bar{b}$ does not fully lie in the column space of A, but can be expressed as a sum of vectors from the column space and left null space of A. The vector $\bar{z}$ can be derived from this decomposition.

Problem 2.5.2 *Express the system of equations $\sum_{i=1}^{5} x_i = 1$, $\sum_{i=1}^{2} x_i = -1$, and $\sum_{i=3}^{5} x_i = -1$ as $A\bar{x} = \bar{b}$ for appropriately chosen A and $\bar{b}$. Informally discuss by inspection why this system of equations is inconsistent. Now define a vector $\bar{z}$ satisfying the conditions of the previous exercise to show that the system is inconsistent.*

The process of row echelon conversion is useful to identify whether a system of equations is inconsistent, and also to characterize the set of solutions to a system of consistent equations. One can use a sequence of row operations to convert the linear system $A\bar{x} = \bar{b}$ to a new system $A'\bar{x} = \bar{b}'$ in which the matrix A' is in row echelon form. Whenever a row operation is performed on A, exactly the same operation is performed on $\bar{b}$. The resulting system $A'\bar{x} = \bar{b}'$ contains a wealth of information about the solutions to the original system. Inconsistent systems will contain zero rows at the bottom of A' after row echelon conversion, but a corresponding non-zero entry in the same row of $\bar{b}'$ (try to explain this using Problem 2.5.1 while recognizing that A' contains linearly independent rows). Such a system can never have a solution because a zero value on the left is being equated with a non-zero value on the right. *All zero rows in A' need to be matched with zero entries in $\bar{b}'$ for the system to have a solution.*

Assuming that the system is not inconsistent, how does one detect systems with unique solutions? *In such cases, each column will contain a leftmost non-zero entry of some row.* It is possible for some of the rows to be zeros. We present two examples of matrices, the first of which satisfies the aforementioned property, and the second does not satisfy the property:

$$M' = \begin{bmatrix} 1 & 7 & 4 \\ 0 & 1 & 2 \\ 0 & 0 & 1 \\ 0 & 0 & 0 \end{bmatrix} \qquad N' = \begin{bmatrix} 1 & 7 & 4 & 3 & 5 \\ 0 & 1 & 9 & 7 & 6 \\ 0 & 0 & 0 & 1 & 3 \\ 0 & 0 & 0 & 0 & 1 \end{bmatrix}$$

Note that the matrix N' does not satisfy the uniqueness condition because the third column (whose entries are in bold) does not contain the leftmost non-zero entry of any row. Such a column is referred to as a *free column* because one can view the variable corresponding to it as a free parameter. If there is no free column, one will obtain a square, triangular, invertible matrix on dropping the zero rows of A' and corresponding zero entries of $\bar{b}'$. For example, one obtains a square, triangular, and invertible matrix on dropping the zero rows of M'. This matrix will be an upper-triangular matrix, which has values of 1 along the diagonal. It is easy to find a unique solution by using *backsubstitution*. One can first set the last component of $\bar{x}$ to the last component of $\bar{b}'$, and substitute it into the system of equations to obtain a smaller upper-triangular system. This process is applied iteratively to find all components of $\bar{x}$.

The final case is one in which some free columns exist, which are not the leading non-zero entries of some row. The variables corresponding to the free columns can be set to any value, and a unique solution for the other variables can always be found. In this case, the solution space contains infinitely many solutions. Consider the following system in row echelon form:

$$\underbrace{\begin{bmatrix} 1 & 2 & 1 & -3 \\ 0 & 0 & 1 & 2 \\ 0 & 0 & 0 & 0 \end{bmatrix}}_{A'} \begin{bmatrix} x_1 \\ x_2 \\ x_3 \\ x_4 \end{bmatrix} = \begin{bmatrix} 3 \\ 2 \\ 0 \end{bmatrix}$$

In this system of equations, the second and fourth columns do not contain any entry that are the leading non-zero entries of any row. Therefore, we can set x_2 and x_4 to arbitrary numerical values (say, α and β) and also drop all the zero rows. Furthermore, setting x_2 and x_4 to numerical values will result in a system of equations with only two variables x_1 and x_3 (because α and β are now constants rather than variables). The vector $\vec{b}'$ on the right-hand size is adjusted to reflect the effect of these numerical constants. After making these adjustments, the aforementioned system becomes the following:

$$\begin{bmatrix} 1 & 1 \\ 0 & 1 \end{bmatrix} \begin{bmatrix} x_1 \\ x_3 \end{bmatrix} = \begin{bmatrix} 3 - 2\alpha + 3\beta \\ 2 - 2\beta \end{bmatrix}$$

This system is a square 2×2 system of equations with a unique solution in terms of α and β. The value of x_3 is set to $2-2\beta$, and then back-substitution is used to derive $x_1 = 1-2\alpha+5\beta$. Therefore, the set of solutions $[x_1, x_2, x_3, x_4]$ is defined as follows:

$$[x_1, x_2, x_3, x_4] = [1 - 2\alpha + 5\beta, \alpha, 2 - 2\beta, \beta]$$

Here, α and β can be set to arbitrary numerical values; therefore, the system has infinitely many solutions.

Problem 2.5.3 (Coordinate Transformations with Row Echelon) *Consider the vector space $V \subset \mathcal{R}^n$ with basis $\mathcal{B} = \{\bar{a}_1 \ldots \bar{a}_d\}$, so that $d < n$. Show how to use the row echelon method to find the d coordinates of $\bar{v} \in V$ in the basis $\mathcal{B}$.*

2.6   The Notion of Matrix Rank

Any matrix can be reduced to a (rectangular) diagonal matrix with only row and column operations. The reason for this is that we can first use row operations to convert a matrix to row echelon form. This matrix is a (rectangular) upper-triangular matrix. Subsequently, we can reduce it to a diagonal matrix using column operations. First, column operations are used to move all free columns to the rightmost end of the matrix. The non-free columns are reduced to a diagonal matrix. This is done in order of increasing column index j by subtracting appropriate multiples of all non-free columns up to index $(j - 1)$ from that column. Then, all free columns are reduced to zero columns by subtracting appropriate multiples of the non-free columns (each of which has only one non-zero entry). This will result in a rectangular diagonal matrix in which all free columns are converted to zero columns. In other words, any $n \times d$ matrix A can be expressed in the following form:

$$RAC = \Delta$$

Here, R is an $n \times n$ matrix that is the product of the elementary row operator matrices, C is a $d \times d$ matrix that is the product of the elementary column operator matrices, and Δ is an $n \times d$ rectangular diagonal matrix.

This result has the remarkable implication that the ranks of the row space and the column space of a matrix are the same.

Lemma 2.6.1 *The rank of the row space of a matrix is the same as that of its column space.*

Proof Sketch: The condition $RA = \Delta C^{-1}$ implies that the row rank of A is the same as the number of non-zero diagonal entries in Δ (since row operations do not change rank of A according to Lemma 2.5.1, and ΔC^{-1} contains as many non-zero, linearly independent rows as the number of non-zero diagonal entries in Δ). Similarly, the condition $AC = R^{-1}\Delta$ implies that the column rank of A is the same as the number of non-zero diagonal entries in Δ. Therefore, the row rank of A is the same as its column rank. ∎

The common value of the rank of the row space and the column space is referred to as the rank of a matrix.

Definition 2.6.1 (Matrix Rank) *The rank of a matrix is equal to the rank of its row space, which is the same as the rank of its column space.*

Two natural corollaries of the above result are the following:

Corollary 2.6.1 *The rank of an $n \times d$ matrix is at most $min\{n, d\}$.*

The matrix A contains d columns and therefore the rank of the column space is at most d. Similarly, the rank of the row space is at most n. Since both ranks are the same, it follows that this value must be at most $min\{n, d\}$.

Corollary 2.6.2 *Consider an $n \times d$ matrix A with rank $k \leq min\{n, d\}$. Then the rank of the null space of A is $d - k$ and the rank of the left null space of A is $n - k$.*

This follows from the fact that rows of A are d-dimensional vectors, and the null space of A is the orthogonal complement of the vector space defined by the (transposed) rows of A. Therefore, the rank of the null space of A must be $d - k$. A similar argument can be made for the left null space of A.

2.6.1 Effect of Matrix Operations on Rank

It is common to use matrix addition and multiplication operations in machine learning. In such cases, it is helpful to understand the effect of matrix addition and multiplication on the rank. In this context, we establish lower and upper bounds on the results obtained using matrix operations.

Lemma 2.6.2 (Matrix Addition Upper Bound) *Let A and B be two matrices with ranks a and b, respectively. Then, the rank of $A + B$ is at most $a + b$.*

Proof: Each row of $A + B$ can be expressed as a linear combination of the rows of A and the rows of B. Therefore, the rank of the row space of $(A + B)$ is at most $a + b$. ∎

One can show a similar result for the lower bound on matrix addition:

Lemma 2.6.3 (Matrix Addition Lower Bound) *Let A and B be two matrices with ranks a and b, respectively. Then, the rank of $A + B$ is at least $|a - b|$.*

Proof: The result follows directly from Lemma 2.6.2, because one can express the relationship $A + B = C$ as $A + (-C) = (-B)$ or as $B + (-C) = (-A)$. Therefore, if A and B have ranks a and b, then the rank of $-C$ must be at least $|a - b|$ from the previous lemma. ∎

One can also derive upper and lower bounds for multiplication operations.

Lemma 2.6.4 (Matrix Multiplication Upper Bound) *Let A and B be two matrices with ranks a and b, respectively. Then, the rank of AB is at most $min\{a, b\}$.*

Proof: Each column of AB is a linear combination of the columns of A, where the linear combination coefficients defining the ith column of AB are provided in the ith column of B. Therefore, the rank of the column space of AB is no greater than that of the column space of A. However, the column space of a matrix is the same as its rank. Therefore, the matrix rank of AB is no greater than the matrix rank of A.

Similarly, each row of AB is a linear combination of the rows of B, where the linear combination coefficients defining the ith row of AB are included in the ith row of A. Therefore, the rank of the row space of AB is no greater than that of the row space of B. However, the row space of a matrix is the same as its rank. Therefore, the matrix rank of AB is no greater than the matrix rank of B. Combining the above two results, we obtain the fact that rank of AB is no greater than $min\{a, b\}$.   ∎

Establishing a lower bound on the rank of the product of two matrices is much harder than establishing an upper bound; a useful bound exists only in some special cases.

Lemma 2.6.5 (Matrix Multiplication Lower Bound) *Let A and B be $n \times d$ and $d \times k$ matrices of ranks a and b, respectively. Then, the rank of AB is at least $a + b - d$.*

We omit a formal proof of this result, which is also referred to as *Sylvester's inequality*. It is noteworthy that d is the shared dimension of the two matrices (thereby allowing multiplication), and the result is not particularly useful when $a + b \leq d$. In such a case, the lower bound on the rank becomes negative, which is trivially satisfied by every matrix and therefore not informative. A useful lower bound can be established when the two matrices have rank close to the shared dimension d (i.e., the maximum possible value). What about the case when one or both matrices are square and are *exactly* of full rank? Some natural corollaries of the above result are the following:

Corollary 2.6.3 *Multiplying a matrix A with a square matrix B of full rank does not change the rank of matrix A.*

Corollary 2.6.4 *Let A and B be two square matrices. Then AB is non-singular if and only if A and B are both non-singular.*

In other words, the product is of full rank if and only if both matrices are of full rank. This result is important from the perspective of the invertibility of the *Gram matrix* $A^T A$ of the column space of A. Note that the Gram matrix often needs to be inverted in machine learning applications like linear regression. In such cases, the inversion of the Gram matrix is part of the closed-form solution (see, for example, Equation 1.29 of Chapter 1). It is helpful to know that the invertibility of the Gram matrix is determined by the linear independence of the columns of the underlying data matrix of feature variables:

Lemma 2.6.6 (Linear Independence and Gram Matrix) *The matrix $A^T A$ is said to be the Gram matrix of the column space of an $n \times d$ matrix A. The columns of the matrix A are linearly independent if and only if $A^T A$ is invertible.*

Proof: Consider the case where $A^T A$ is invertible. This means that the rank of $A^T A$ is d, and therefore the rank of each of the factors of $A^T A$ must also be at least d. This means that A must have rank at least d, which is possible only when the d columns of A are linearly independent.

Now suppose that A has linearly independent columns. Then, for any non-zero vector $\overline{x}$, we have $\overline{x}^T A^T A \overline{x} = \|A\overline{x}\|^2 \geq 0$. This value can be zero only when $A\overline{x} = \overline{0}$. However, we know that $A\overline{x} \neq \overline{0}$ for a non-zero vector $\overline{x}$, because of the linear independence of the columns of A. In other words, $\overline{x}^T A^T A \overline{x}$ is *strictly* positive, which is possible only when $A^T A \overline{x}$ is a non-zero vector. In other words, for any non-zero vector $\overline{x}$ we have $A^T A \overline{x} \neq \overline{0}$, which implies that the square matrix $A^T A$ has linearly independent columns. This is possible only when $A^T A$ is invertible (cf. Lemma 2.3.1). ∎

One can use a very similar approach to show the stronger result that the ranks of the matrices A, $A^T A$, and AA^T are the same (see Exercise 2). The matrix AA^T is the Gram matrix of the *row space* of A, and is also referred to as the *left Gram matrix*.

2.7 Generating Orthogonal Basis Sets

Orthogonal basis sets have many useful properties like ease of coordinate transformations, projections, and distance computation. In this section, we will discuss how to convert a non-orthogonal basis set to an orthogonal basis set with the use of Gram-Schmidt orthogonalization. We also provide an example of a useful orthogonal basis of $\mathcal{R}^n$, which is obtained with the use of the *discrete cosine transform*.

2.7.1 Gram-Schmidt Orthogonalization and QR Decomposition

It is desired to find an orthonormal basis set of the span of the non-orthogonal vectors $\mathcal{A} = \{\overline{a}_1 \ldots \overline{a}_d\}$. We first discuss the simpler case in which the vectors of $\mathcal{A}$ are linearly independent and the basis vectors are unnormalized. We assume that each $\overline{a}_i$ is drawn from $\mathcal{R}^n$ and $n \geq d$ (to ensure linear independence of $\{\overline{a}_1 \ldots \overline{a}_d\}$). Therefore, one is looking for an orthogonal basis of a *subspace* of $\mathcal{R}^n$.

An orthogonal basis $\{\overline{q}_1 \ldots \overline{q}_d\}$ can be found with the use of *Gram-Schmidt* orthogonalization. The basic idea of Gram-Schmidt orthogonalization is to successively remove the projections of previously generated vectors from a vector belonging to $\mathcal{A}$ to iteratively create orthogonal vectors. We start by setting the first basis vector $\overline{q}_1$ to $\overline{a}_1$, and then adjust $\overline{a}_2$ by removing its projection on $\overline{q}_1$ from it in order to create $\overline{q}_2$, which is orthogonal to $\overline{q}_1$. Subsequently, the projections of $\overline{a}_3$ on both $\overline{q}_1$ and $\overline{q}_2$ are removed from $\overline{a}_3$ to create the next basis vector $\overline{q}_3$. This process is iteratively continued till all d basis vectors are generated. If the set $\mathcal{A}$ is a linearly independent set of basis vectors, the generated basis set $\{\overline{q}_1 \ldots \overline{q}_d\}$ will only contain non-zero vectors (or else we will obtain a linear dependence relation between $\overline{a}_1 \ldots \overline{a}_j$ when $\overline{q}_j$ evaluates to $\overline{0}$). These vectors can also be normalized by dividing each with its norm. Gram-Schmidt orthogonalization produces a basis set that depends on the order in which the vectors of $\mathcal{A}$ are processed.

Next, we describe the process formally. In the initial step, $\overline{q}_1$ is generated as follows:

$$\overline{q}_1 = \overline{a}_1 \tag{2.10}$$

Subsequently, an iterative process is used for generating $\overline{q}_i$ after $\overline{q}_1 \ldots \overline{q}_{i-1}$ have been generated. The vector $\overline{q}_i$ is generated by subtracting the projection of $\overline{a}_i$ on the subspace defined by the already generated vectors $\overline{q}_1 \ldots \overline{q}_{i-1}$. Note that the projection of $\overline{a}_i$ onto a previously generated $\overline{q}_r$ $(r < i)$ is simply $\frac{\overline{a}_i \cdot \overline{q}_r}{\|\overline{q}_r\|}$. Therefore, the process of generating $\overline{q}_i$ is as follows:

1. Compute $\overline{q}_i = \overline{a}_i - \sum_{r=1}^{i-1} \frac{(\overline{a}_i \cdot \overline{q}_r)}{\|\overline{q}_r\|} \frac{\overline{q}_r}{\|\overline{q}_r\|} = \overline{a}_i - \sum_{r=1}^{i-1} \frac{(\overline{a}_i \cdot \overline{q}_r)}{\overline{q}_r \cdot \overline{q}_r} \overline{q}_r$.

2. Increment i by 1.

This process is repeated for each $i = 2 \ldots d$. This algorithm is referred to as the *unnormalized Gram-Schmidt method*. In practice, the vectors are scaled to unit norm after the process.

We can show that the resulting vectors are mutually orthogonal by induction. For example, consider the case when we make the inductive assumption that $\bar{q}_1 \ldots \bar{q}_{i-1}$ are orthogonal. Then, we can show that $\bar{q}_i$ is also orthogonal to each $\bar{q}_j$ for $j \in \{1 \ldots i - 1\}$:

$$\bar{q}_j \cdot \bar{q}_i = \bar{q}_j \cdot \underbrace{\left[\bar{a}_i - \sum_{r=1}^{i-1} \frac{(\bar{a}_i \cdot \bar{q}_r)}{\|\bar{q}_r\|} \frac{\bar{q}_r}{\|\bar{q}_r\|} \right]}_{\text{[Drop terms using induction]}} = \bar{q}_j \cdot \bar{a}_i - \frac{(\bar{q}_j \cdot \bar{q}_j)}{\|\bar{q}_j\|^2} (\bar{q}_j \cdot \bar{a}_i) = 0$$

Therefore, the inductive assumption of mutual orthogonality can also be extended to $\bar{q}_i$ from $\bar{q}_1 \ldots \bar{q}_{i-1}$.

Aside from the orthogonality of the generated basis, we need to show that the span of $\bar{q}_1 \ldots \bar{q}_i$ remains the same as that of $\bar{a}_1 \ldots \bar{a}_i$ for all $i \leq d$. This result can be shown by induction. The result is trivially true at $i = 1$. Now, make the inductive assumption that the span of $\bar{q}_1 \ldots \bar{q}_{i-1}$ is the same as that of $\bar{a}_1 \ldots \bar{a}_{i-1}$. In each iterative step, adding $\bar{q}_i$ to the current basis has the same effect as adding $\bar{a}_i$ to the current basis, because $\bar{q}_i$ is adjusted from $\bar{a}_i$ additively using a linear combination of vectors $\{\bar{q}_1 \ldots \bar{q}_{i-1}\}$ already in the basis. Therefore, the span of $\bar{q}_1 \ldots \bar{q}_i$ is the same as that of $\bar{a}_1 \ldots \bar{a}_i$.

What happens when the vectors in $\mathcal{A}$ are not linearly independent? In such cases, some of the generated vectors $\bar{q}_i$ turn out to be zero vectors, and they are discarded as soon as they are computed. In such a case, the Gram-Schmidt method returns fewer than d basis vectors. As a specific example, in the case when $\bar{a}_2 = 3\bar{a}_1$, it is easy to show that $\bar{q}_2 = \bar{a}_2 - 3\bar{q}_1 = \bar{a}_2 - 3\bar{a}_1$ will be the zero vector. In general, when $\bar{a}_i$ is linearly dependent on $\bar{a}_1 \ldots \bar{a}_{i-1}$, the projection of the vector $\bar{a}_i$ on the subspace defined by $\bar{q}_1 \ldots \bar{q}_{i-1}$ is itself; therefore, subtracting this projection of $\bar{a}_i$ from $\bar{a}_i$ will result in the zero vector.

Problem 2.7.1 (A-Orthogonality) *Two n-dimensional vectors $\bar{x}$ and $\bar{y}$ are said to be A-orthogonal, if we have $\bar{x}^T A \bar{y} = 0$ for an $n \times n$ invertible matrix A. Given a set of $d \leq n$ linearly independent vectors from $\mathcal{R}^n$, show how to generate an A-orthogonal basis for them.*

Problem 2.7.2 (Randomized A-Orthogonality) *Propose a method to find a randomized orthogonal basis of $\mathcal{R}^n$ using the Gram-Schmidt method. Now generalize the method to find a randomized A-orthogonal basis of $\mathcal{R}^n$.*

2.7.2   QR Decomposition

We first discuss the QR decomposition of an $n \times d$ matrix with linearly independent columns. Since the columns are linearly independent, we must have $n \geq d$. Gram-Schmidt orthogonalization can be used to decompose an $n \times d$ matrix A with the linearly independent columns into the product of an $n \times d$ matrix Q with orthonormal columns and an upper-triangular $d \times d$ matrix R. In other words, we want to compute the following *QR decomposition*:

$$A = QR \tag{2.11}$$

Consider an $n \times d$ matrix A with linearly independent columns $\bar{a}_1 \ldots \bar{a}_d$. Then, we perform the Gram-Schmidt orthogonalization as discussed above (with the normalization step included), and construct the matrix Q with orthonormal columns $\bar{q}_1 \ldots \bar{q}_d \in \mathcal{R}^n$ obtained from Gram-Schmidt orthogonalization. The columns appear in the same order as obtained by processing $\bar{a}_1 \ldots \bar{a}_d$ by the Gram-Schmidt algorithm. Since the vectors $\bar{a}_1 \ldots \bar{a}_d$ are linearly independent, one would derive a full set of d orthonormal basis vectors. Note that the projection of $\bar{a}_r$ on each $\bar{q}_j$ is $\bar{q}_j \cdot \bar{a}_r$, which provides its jth coordinate in the new orthonormal basis. Therefore, we define a $d \times d$ matrix R, in which the (j, r)th entry is $\bar{q}_j \cdot \bar{a}_r$. For $j > r$, $\bar{q}_j$ is orthogonal to the space spanned by $\bar{a}_1 \ldots \bar{a}_r$, and therefore the value of $\bar{q}_j \cdot \overline{a_r}$ is 0. Therefore, the matrix R is upper triangular. It is easy to see that the rth column of the product QR is the appropriate linear combination of the orthonormal basis defined by Gram-Schmidt orthogonalization (to yield $\bar{a}_r$), and therefore $A = QR$.

What happens when the columns of the $n \times d$ matrix A are not linearly independent? In such a case, the Gram-Schmidt process will yield the vectors $\bar{q}_1 \ldots \bar{q}_d$, which are either unit-normalized vectors or zero vectors. Assume that k of the vectors $\bar{q}_1 \ldots \bar{q}_d$ are non-zero. We can assume that the zero vectors also have zero coordinates in the Gram-Schmidt representation, since the coordinates of zero vectors are irrelevant from a representational point of view. As in the previous case, we create the decomposition QR (including the zero columns in Q and matching zero rows in R), where Q is a $n \times d$ matrix and R is a $d \times d$ upper-triangular (rectangular) matrix. Subsequently, we drop all the zero columns from Q, and also drop the zero rows with matching indices from R. As a result, the matrix Q is now of size $n \times k$ and the matrix R is of size $k \times d$. This provides the most concise, *generalized* QR decomposition of the original $n \times d$ matrix A.

Problem 2.7.3 (Solving Linear Equations) *Show how you can use QR decomposition to solve the system of equations $A\bar{x} = \bar{b}$ with back-substitution. Assume that A is a $d \times d$ matrix with linearly independent columns and $\bar{b}$ is a d-dimensional column vector.*

Leveraging Givens Rotations and Householder Reflections

The following section provides a brief overview of advanced methods for QR decomposition, and the reader may omit this section without loss of continuity. It is possible to perform QR decomposition of any $n \times d$ matrix A by applying $O(nd)$ Givens rotations (defined on page 47) to the columns of A. Pre-multiplying a square matrix A with the Givens rotation matrix can be used to change a single entry below the diagonal to zero (without disturbing the entries already zeroed out), provided that the angle of rotation is properly chosen and the entries are zeroed in the proper order. The basic geometric principle behind zeroing an entry is that it is always possible to rotate a 2-dimensional vector for an appropriate angle until one of its coordinates is zeroed out. Pre-multiplying A with an $n \times n$ Givens rotation matrix performs an operation on each column vector of A. Although the column vectors of A are not 2-dimensional, Givens rotations always perform rotations in 2-dimensional projections without affecting other coordinates, and therefore such an angle always exists.

Given an $n \times d$ matrix A, the approach successively pre-multiplies A with an $n \times n$ Givens rotation matrix, so as to turn one entry below the diagonal to zero (without disturbing the entries that have already been turned to zero). The running matrix after pre-multiplication with orthogonal matrices is denoted by variable R, and this matrix is upper-triangular at the end of the process. Let $Q_1 \ldots Q_s$ be Givens matrices successively chosen in this way, so that we have the following repeated process:

$$A = \underbrace{Q_1^T Q_1}_{I} A = Q_1^T \underbrace{Q_2^T Q_2}_{I} \underbrace{Q_1 A}_{R} = \ldots = \underbrace{(Q_1^T \ldots Q_s^T)}_{\text{Orthogonal } Q} \underbrace{(Q_s Q_{s-1} \ldots Q_1 A)}_{\text{Triangular } R}$$

Therefore, the approach requires at most $O(nd)$ Givens rotations, although far fewer rotations will be required for sparse matrices. Entries (below the diagonal) with the smallest column index j are zeroed first, and those with the same column index are selected in order of decreasing row index i. Based on the notations on page 47, the Givens matrix used for pre-multiplication of the current transformation R of A is $G_c(i-1, i, \alpha)$, where α is chosen to zero out the (i, j)th entry of the current matrix corresponding to running variable R. Multiplication of $G_c(i-1, i, \alpha)$ with R affects only the $(i-1)$th and ith entries of each column of R. If the lower-triangular portions of columns before index j have already been set to 0, then multiplication with the Givens matrix will not affect them (since a rotation of a zero vector is a zero vector). Therefore, work already done on setting earlier column entries to 0 will remain undisturbed. Consider the current column index j, whose entries are being set to 0. If the current matrix R contains entries r_{ij}, then one can pull out the portion of the product of the Givens matrix $G_c(i-1, i, \alpha)$ with R corresponding to the rotation of the 2-dimensional vector $[r_{i-1,j}, r_{ij}]^T$:

$$\begin{bmatrix} \cos(\alpha) & -\sin(\alpha) \\ \sin(\alpha) & \cos(\alpha) \end{bmatrix} \begin{bmatrix} r_{i-1,j} \\ r_{ij} \end{bmatrix} = \begin{bmatrix} \sqrt{r_{i-1,j}^2 + r_{ij}^2} \\ 0 \end{bmatrix}$$

One can verify that the solution to the above system yields the following value of α:

$$\sin(\alpha) = \frac{-r_{ij}}{\sqrt{r_{ij}^2 + r_{i-1,j}^2}}, \quad \cos(\alpha) = \frac{r_{i-1,j}}{\sqrt{r_{ij}^2 + r_{i-1,j}^2}} \tag{2.12}$$

Note that α takes on (absolute) value of $90°$, when $r_{i-1,j}$ is 0 but r_{ij} is not 0. Furthermore, α is 0 or 180 when r_{ij} is already zero, and no rotation needs to be done (since a $180°$ rotation only flips the sign of $r_{i-1,j}$). The ordering of the processing of the $O(nd)$ entries is necessary to ensure that already zeroed entries are not disturbed by further rotations. The pseudocode for the process is as follows:

```
Q ⇐ I; R ⇐ A;
for j = 1 to d − 1 do
  for i = n down to (j + 1) do
    Choose α based on Equation 2.12;
    Q ⇐ Q G_c(i, i − 1, α)^T; R ⇐ G_c(i, i − 1, α) R;
  endfor
endfor
return Q, R;
```

For $n \geq d$ and a matrix A with linearly independent columns, the above approach will create an $n \times n$ matrix Q and an $n \times d$ matrix R. These matrices are larger than the ones obtained with the Gram-Schmidt method. However, the bottom $(n-d)$ rows of R will be zeros, and therefore one can drop the last $(n-d)$ columns of Q and the bottom $(n-d)$ rows of R without affecting the result. This yields a smaller QR decomposition with $n \times d$ matrix Q and $d \times d$ matrix R.

It is also possible to use this approach of iteratively modifying Q and R with Householder reflection matrices instead of Givens rotation matrices. In this case, at most $(d-1)$ reflections will be needed to triangulize the matrix, because each iteration is able to zero out all the entries below the diagonal for a particular column (and the final one can be ignored). The columns are processed in order of increasing column index. The basic geometric principle is that for any n-dimensional coordinate vector (first column of A), it is possible to orient a $(n-1)$-dimensional "mirror" passing through the origin, so that the image of the vector is mapped to a point in which only the first coordinate is non-zero. Such a transformation is defined by multiplication with a Householder reflection matrix. We encourage the reader to visualize a 1-dimensional reflection plane in 2-dimensional space, so that a specific point $[x, y]^T$ is mapped to $[\sqrt{x^2 + y^2}, 0]^T$. This principle also applies more generally to vectors in n-dimensional space, such as the first column $\overline{c}_1$ of A. One can choose $\overline{v}_1$ (normal vector to the "mirror" hyperplane) in the first iteration to be the unit vector joining $\overline{c}_1$ to a column vector $\|\overline{c}_1\|[1, 0, \ldots, 0]^T$ of equal length in which only the first component is non-zero. Therefore, we have $\overline{v}_1 \propto (\overline{c}_1 - \|\overline{c}_1\|[1, 0, \ldots, 0]^T)$, and it is scaled to unit norm. One can then compute the Householder matrix $Q_1 = (I - 2\overline{v}_1\overline{v}_1^T)$. Pre-multiplying A with Q_1 will zero the bottom $(n - 1)$ entries of the first column $\overline{c}_1$ of A. In subsequent iterations, the entries of the first row of the resulting matrix $R = Q_1 A$ remain frozen to their current values, and all modifications are performed only on the bottom $(n - 1)$ rows. Therefore, the $n \times n$ Householder reflection matrix $Q_2 = (I - 2\overline{v}_2\overline{v}_2^T)$ will be chosen in the second iteration so that any changes occur only in the bottom $(n - 1)$ dimensions. The second iteration zeros out the bottom $(n - 2)$ entries of the second column $\overline{c}_2$ of the running matrix R. This is achieved by first copying $\overline{c}_2$ to $\overline{c}_{2,n-1}$, resetting the first entry of $\overline{c}_{2,n-1}$ to zero, evaluating unit vector $\overline{v}_2 \propto \overline{c}_{2,n-1} - \|\overline{c}_{2,n-1}\|[0, 1, 0, \ldots 0]^T$, and then updating $R \Leftarrow R(I - 2\overline{v}_2\overline{v}_2^T)$. In the next iteration, the Householder matrix is computed by defining $\overline{c}_{3,n-2}$ as a partial copy of the vector $\overline{c}_3$ with the first *two* entries set to zero. One can set the unit vector $\overline{v}_3 \propto \overline{c}_{3,n-2} - \|\overline{c}_{3,n-1}\|[0, 0, 1, 0, \ldots 0]^T$, and then update $R \Leftarrow R(I - 2\overline{v}_3\overline{v}_3^T)$. This process is iteratively applied to zero the appropriate number of entries of each column of R. The final orthogonal matrix of the QR decomposition is obtained as $Q_1^T \ldots Q_{d-1}^T$. Careful implementation choices are required to reduce numerical errors. For example, in the first iteration, one can reflect $\overline{c}_1$ to either $\|\overline{c}_1\|[1, 0, \ldots 0]^T$ or to $-\|\overline{c}_1\|[1, 0, \ldots 0]^T$. Selecting the further of the two choices reduces numerical errors.

2.7.3   The Discrete Cosine Transform

The Gram-Schmidt basis does not expose any specific properties of a vector with the help of its coordinates. On the other hand, the wavelet basis discussed in Section 2.3.4 is an orthogonal basis that exposes local variations in a time series. The *discrete cosine transform* uses a basis with trigonometric properties in order to expose periodicity in a time series.

Consider a time-series drawn from $\mathcal{R}^n$, which has n values (e.g., temperatures) drawn at n equally spaced clock ticks. Choosing a basis in which each basis vector contains equally spaced samples of a cosine time-series of a particular periodicity allows a transformation in which the coordinates of the basis vectors can be interpreted as the amplitudes of the different periodic components of the series. For example, a time-series of temperatures over 10 years will have day-night variations as well as summer-winter variations, which will be captured by the coordinates of different basis vectors (periodic components). These coordinates are helpful in many machine learning applications.

Consider a high-dimensional time series of length n, which is represented as a column vector in $\mathcal{R}^n$. The n-dimensional basis vector of this time series with the largest possible periodicity uses n equally spaced samples of the cosine function ranging between 0 and π radians. The samples of the cosine function are spaced at a distance of π/n radians from one another, and a natural question arises as to where one might select the first sample. Although different variations of the discrete cosine transform select the first sample at different points of the cosine function, the most common choice is to ensure that the samples are symmetric about $\pi/2$, and therefore the first sample is chosen at $\pi/2n$. This yields the following basis vector $\overline{b}$:

$$\overline{b} = [\cos(\pi/2n), \cos(3\pi/2n), \ldots, \cos([2n-1]\pi/2n)]^T$$

For a time-series of length n, this is the largest possible level of periodicity, where the entire basis vector is an n-dimensional sample of only half a cosine wave (covering π radians). To address smaller periodicities in the data, we would need more basis vectors in which the n-dimensional sample is drawn from a larger number of cosine waves (i.e., a larger angle than π). In other words, the n samples of the cosine function are obtained by sampling the cosine function at n points between 0 and $(j-1)\pi$ for each value of $j \in \{1, \ldots, n\}$:

$$\overline{b}_j = [\cos([j-1]\pi/2n), \cos(3[j-1]\pi/2n), \ldots, \cos([2n-1][j-1]\pi/2n)]^T$$

Setting $j = 1$ yields $\overline{b}_1$ as a column vector of 1s, which is not periodic, but is a useful basis vector for capturing constant offsets. The case of $j = 2$ corresponds to half a cosine wave as discussed above.

One can create an unnormalized basis matrix $B = [\overline{b}_1 \ldots \overline{b}_n]$ whose columns contain the basis vectors discussed above. Let us assume that the ith component of the jth basis vector $\overline{b}_j$ is denoted by b_{ij}. In other words, the (i, j)th entry of B is b_{ij}, where b_{ij} is defined as follows:

$$b_{ij} = \cos\left(\frac{\pi(2i-1)(j-1)}{2n}\right), \quad \forall i, j \in \{1 \ldots n\}$$

The above basis matrix includes the non-periodic (special) basis vector, and it is unnormalized because the norm of each column is not 1. A key point is the columns of the basis matrix B are orthogonal:

Lemma 2.7.1 (Orthogonality of Basis Vectors) *The dot product of any pair of basis vectors $\overline{b}_p$ and $\overline{b}_q$ of the discrete cosine transform for $p \neq q$ is 0.*

Proof Sketch: We use the identity that $\cos(x)\cos(y) = [\cos(x+y) + \cos(x-y)]/2$. Using this identity, it can be shown that the dot product between $\overline{b}_p$ and $\overline{b}_q$ is as follows:

$$\overline{b}_p \cdot \overline{b}_q = \frac{1}{2}\sum_{i=1}^{n}\cos\left(\frac{[p+q][2i-1]\pi}{2n}\right) + \frac{1}{2}\sum_{i=1}^{n}\cos\left(\frac{[p-q][2i-1]\pi}{2n}\right)$$

The right-hand side can be broken up into the sum of two cosine series with their arguments in arithmetic progression. This is a standard trigonometric identity [73]. Using the formula for the sum of cosine series with arguments in arithmetic progression, these sums can be shown to be proportional to $\sin(n\delta/2)\cos(n\delta/2)/\sin(\delta/2) \propto \sin(n\delta)/\sin(\delta/2)$, where $\delta = (p+q)\pi/n$ in the first cosine series, and $\delta = (p-q)\pi/n$ in the second cosine series. The value of $\sin(n\delta)$ is 0 for both values of δ, and therefore both series sum to 0. ∎

Lemma 2.7.2 (Norms of Basis Vectors) *The norm of the special basis vector $\overline{b}_1$ of the discrete cosine transform is $\sqrt{n}$, whereas the norm of each $\overline{b}_p$ for $p \in \{2, \ldots, n\}$ is $\sqrt{n/2}$.*

Proof Sketch: The proof for $\overline{b}_1$ is trivial. For $p > 1$ the squared norms of $\overline{b}_p$ are the sums of squares of cosines with arguments in arithmetic progression. Here, we can use the trigonometric identity $\cos^2(x) = (1 + \cos(2x))/2$. Therefore, we obtain the following:

$$\|\overline{b}_p\|^2 = \frac{n}{2} + \underbrace{\frac{1}{2}\sum_{i=1}^{n}\cos\left(\frac{p[2i-1]\pi}{n}\right)}_{0}$$

As in the proof of the previous lemma, the cosine series with angles in arithmetic progression sums to 0. The result follows.   ∎

The basis matrix B is orthogonal after matrix normalization. One can normalize the matrix B by dividing all matrix entries with $\sqrt{n}$, and then multiplying columns 2 through n with $\sqrt{2}$. For example, an 8×8 normalized basis matrix for the cosine transform is as follows:

$$B = \frac{1}{2} \begin{bmatrix} \frac{1}{\sqrt{2}} & \cos(\frac{\pi}{16}) & \cos(\frac{2\pi}{16}) & \cos(\frac{3\pi}{16}) & \cos(\frac{4\pi}{16}) & \cos(\frac{5\pi}{16}) & \cos(\frac{6\pi}{16}) & \cos(\frac{7\pi}{16}) \\ \frac{1}{\sqrt{2}} & \cos(\frac{3\pi}{16}) & \cos(\frac{6\pi}{16}) & \cos(\frac{9\pi}{16}) & \cos(\frac{12\pi}{16}) & \cos(\frac{15\pi}{16}) & \cos(\frac{18\pi}{16}) & \cos(\frac{21\pi}{16}) \\ \frac{1}{\sqrt{2}} & \cos(\frac{5\pi}{16}) & \cos(\frac{10\pi}{16}) & \cos(\frac{15\pi}{16}) & \cos(\frac{20\pi}{16}) & \cos(\frac{25\pi}{16}) & \cos(\frac{30\pi}{16}) & \cos(\frac{35\pi}{16}) \\ \frac{1}{\sqrt{2}} & \cos(\frac{7\pi}{16}) & \cos(\frac{14\pi}{16}) & \cos(\frac{21\pi}{16}) & \cos(\frac{28\pi}{16}) & \cos(\frac{35\pi}{16}) & \cos(\frac{42\pi}{16}) & \cos(\frac{49\pi}{16}) \\ \frac{1}{\sqrt{2}} & \cos(\frac{9\pi}{16}) & \cos(\frac{18\pi}{16}) & \cos(\frac{27\pi}{16}) & \cos(\frac{36\pi}{16}) & \cos(\frac{45\pi}{16}) & \cos(\frac{54\pi}{16}) & \cos(\frac{63\pi}{16}) \\ \frac{1}{\sqrt{2}} & \cos(\frac{11\pi}{16}) & \cos(\frac{22\pi}{16}) & \cos(\frac{33\pi}{16}) & \cos(\frac{44\pi}{16}) & \cos(\frac{55\pi}{16}) & \cos(\frac{66\pi}{16}) & \cos(\frac{77\pi}{16}) \\ \frac{1}{\sqrt{2}} & \cos(\frac{13\pi}{16}) & \cos(\frac{26\pi}{16}) & \cos(\frac{39\pi}{16}) & \cos(\frac{52\pi}{16}) & \cos(\frac{65\pi}{16}) & \cos(\frac{78\pi}{16}) & \cos(\frac{91\pi}{16}) \\ \frac{1}{\sqrt{2}} & \cos(\frac{15\pi}{16}) & \cos(\frac{30\pi}{16}) & \cos(\frac{45\pi}{16}) & \cos(\frac{60\pi}{16}) & \cos(\frac{75\pi}{16}) & \cos(\frac{90\pi}{16}) & \cos(\frac{105\pi}{16}) \end{bmatrix}$$

Consider the time-series $\overline{s} = [8, 6, 2, 3, 4, 6, 6, 5]^T$, which is the same example used in Section 2.3.4 on wavelet transformations. This time-series can be transformed to the basis of the discrete cosine transform by solving the system of equations $B\overline{x} = \overline{s}$ in order to compute the coordinates $\overline{x}$. Since B is an orthogonal matrix, the solution $\overline{x}$ is given by $\overline{x} = B^T \overline{s}$. The smaller coefficients can be set to 0 in order to enable space-efficient sparse representations.

The focus on capturing periodicity makes the discrete cosine transform quite different from the wavelet transform. It is closely related to the *discrete Fourier transform* (cf. Section 2.11.1), and the former is the preferred choice in some applications like *jpeg compression*. The discrete cosine transform has many variants depending on how one samples the cosine function to generate the basis vectors. The version presented in this section is referred to as *DCT-II*, and it is the most popular version of the transform [121].

2.8 An Optimization-Centric View of Linear Systems

Linear algebra is closely related to many problems in linear optimization, which recur frequently in machine learning. Indeed, solving a system of linear equations is a special case of one of the most fundamental problems in machine learning, which is referred to as *linear regression*. One way of solving the system of equations $A\overline{x} = \overline{b}$ is to view it as an optimization problem in which we want to minimize the objective function $\|A\overline{x} - \overline{b}\|^2$. This is classical least-squares regression, which is the genesis of a vast array of models in machine learning. Least-squares regression tries to find the *best possible fit* to a system of equations (rather than an exact one). The minimum possible value of the objective function is 0, which occurs when a feasible solution exists for $A\overline{x} = \overline{b}$. However, if the system of equations is inconsistent, the optimization problem will return the best possible fit with a non-zero (positive) optimal value. Therefore, the goal is to minimize the following objective function:

$$J = \underbrace{\|A\overline{x} - \overline{b}\|^2}_{\text{Best Fit}}$$

Although one can use calculus to solve this problem (see Section 4.7 of Chapter 4), we use a geometric argument. The closest approach from a point to a hyperplane is always orthogonal to the hyperplane. The vector $(\overline{b} - A\overline{x}) \in \mathcal{R}^n$, which joins $\overline{b}$ to its closest approximation $\overline{b}' = A\overline{x}$ on the hyperplane defined by the column space of A, must be orthogonal to the hyperplane and therefore to every column of A (see Figure 2.9). Hence, we obtain the *normal equation* $A^T(\bar{b} - A\bar{x}) = \bar{0}$, which yields the following:

$$\bar{x} = (A^T A)^{-1} A^T \bar{b} \tag{2.13}$$

The assumption here is that $A^T A$ is invertible, which can occur only when the columns of A are linearly independent (according to Lemma 2.6.6). This can happen only when A is a "tall" matrix (i.e., $n \geq d$). The matrix $L = (A^T A)^{-1} A^T$ is referred to as the *left-inverse* of the matrix A, which is a generalization of the concept of a conventional inverse to rectangular matrices. In such a case, it is evident that we have $LA = (A^T A)^{-1}(A^T A) = I_d$. Note that the identity matrix I_d is of size $d \times d$. However, AL will be a (possibly larger) $n \times n$ matrix, and it can never be the identity matrix when $n > d$. Therefore, the left-inverse is a one-sided inverse.

An important point is that there are many matrices L' for which $L'A = I_d$, when the matrix A satisfies $d < n$ and has linearly independent columns, although the choice $(A^T A)^{-1} A^T$ is the preferred one. In order to understand this point, let $\bar{z}_1 \ldots \bar{z}_d$ be any set of n-dimensional row vectors such that $\bar{z}_i A = \bar{0}$. As long as the tall matrix A is of rank strictly less than n (i.e., non-empty left null space), such a set of non-zero vectors can be found. Note that even if the rank of the left null space of A is 1, we can find d such vectors that are scalar multiples of one another. We can stack up these d vectors into a $d \times n$ matrix Z, such that the ith row contains the vector $\bar{z}_i$. Then, it can be shown that any $d \times n$ matrix L_z (in which Z is chosen according to the aforementioned procedure) is a left-inverse of L:

$$L_z = (A^T A)^{-1} A^T + Z$$

This is easy to show because we have:

$$L_z A = ((A^T A)^{-1} A^T + Z)A = \underbrace{(A^T A)^{-1}(A^T A)}_{I} + \underbrace{ZA}_{0} = I$$

Using L_z to solve the system of equations as $\bar{x} = L_z \bar{b}$ will provide the same solution as $\bar{x} = (A^T A)^{-1} A^T \bar{b}$, when a consistent solution to the system of equations exists. However, it will not provide an equally good *best-fit* to an *inconsistent* system of equations because it was not derived from the optimization-centric view of linear systems. This is the reason that even though alternative left-inverses exist, only one of them is the preferred one.

What happens when $n < d$ or when $(A^T A)$ is not invertible? In such a case, we have an infinite number of possible best-fit solutions, all of which have the same optimal value (which is typically but not necessarily[2] 0). Although there are an infinite number of best-fit solutions, one can discriminate further using a *conciseness criterion*, according to which we want $\|\bar{x}\|^2$ as small as possible (as a secondary criterion) among alternative minima for $\|A\bar{x} - \bar{b}\|^2$ (which is the primary criterion). The conciseness criterion is a well-known principle in machine learning, wherein simple solutions are preferable over complex ones (see Chapter 4). When the *rows* of A are linearly independent, the most concise solution $\bar{x}$ is the following (see Exercise 31):

$$\bar{x} = A^T (AA^T)^{-1} \bar{b} \tag{2.14}$$

---

[2]When $n < d$, we could have an inconsistent system $A\bar{x} = \bar{b}$ with *linearly dependent* rows *and* columns in A; an example is the equation pair $\sum_{i=1}^{10} x_i = 1$ and $\sum_{i=1}^{10} x_i = -1$. However, linearly independent rows and $n < d$ guarantees an infinite number of consistent solutions.

The matrix $R = A^T(AA^T)^{-1}$ is said to be the *right-inverse* of A, because we have $AR = (AA^T)(AA^T)^{-1} = I_n$. The linear independence of the rows also ensures that the column space of A spans all of $\mathcal{R}^n$ and therefore the system is consistent for *any* vector $\bar{b}$. It is also easy to verify that $A\bar{x} = (AA^T)(AA^T)^{-1}\bar{b} = \bar{b}$.

Problem 2.8.1 *What is the left-inverse of a matrix containing a single column-vector* $[a, b, c]^T$ *?*

The special case in which the matrix A is square and invertible is a "nice" case in which the left- and right-inverses turn out to be the same.

Problem 2.8.2 *If a matrix A is square and invertible, show that its left- and right-inverses both simplify to* A^{-1}.

Problem 2.8.3 *Consider an $n \times d$ matrix A with linearly independent rows and $n < d$. How many matrices R are there that satisfy $AR = I_n$?*

2.8.1 Moore-Penrose Pseudoinverse

How does one solve inconsistent linear systems of the form $A\bar{x} = \bar{b}$, when neither the rows nor the columns of A are linearly independent (and, therefore, neither of $A^T A$ or AA^T is invertible)? Although the following description will require some optimization results developed in later chapters, the goal of this presentation is to give the reader the full picture of different cases associated with linear systems (and the connections with optimization and machine learning). Therefore, at some points in this section, we use some results developed in later chapters (and it is not necessary for the reader to know the details of the underlying derivations at this stage to understand the broader intuition).

A natural approach to addressing inconsistent linear systems in which neither the rows nor the columns of A are linearly independent is to *combine* the idea of finding a *best-fit* solution with a *concise* one. This is achieved by minimizing the following objective function:

$$J = \underbrace{\|A\bar{x} - \bar{b}\|^2}_{\text{Best Fit}} + \underbrace{\lambda(\sum_{i=1}^{d} x_i^2)}_{\text{Concise}}$$

The additional term in the objective function is a *regularization term*, which tends to favor small absolute components of the vector $\bar{x}$. This is precisely the conciseness criterion discussed in the previous section. The value $\lambda > 0$ is the regularization parameter, which regulates the relative importance of the best-fit term and the conciseness term.

We have not yet introduced the methods required to compute the solution to the above optimization problem (which are discussed in Section 4.7 of Chapter 4). For now, we ask the reader to make the leap of faith that this optimization problem has the following alternative forms of the solution:

$$\bar{x} = (A^T A + \lambda I_d)^{-1} A^T \bar{b} \quad \text{[Regularized left-inverse form]}$$
$$\bar{x} = A^T (AA^T + \lambda I_n)^{-1} \bar{b} \quad \text{[Regularized right-inverse form]}$$

It is striking how similar both the above forms are to left- and right-inverses introduced in the previous section, and they are referred to as the regularized left inverses and right inverses, respectively. *Both solutions turn out to be the same* because of the *push-through identity* (cf. Problem 1.2.13 of Chapter 1). An important difference of the regularized form of the solution from the previous section is that both the matrices $(A^T A + \lambda I_d)$ and $(AA^T + \lambda I_n)$ are always invertible for $\lambda > 0$ (see Problem 2.4.2), irrespective of the linear independence of the rows and columns of A. How should be parameter $\lambda > 0$ be selected? If our *primary* goal is to find the best-fit solution, and the (limited) purpose of the regularization term is to only play a tie-breaking role among equally good fits (with the *secondary* conciseness criterion), it makes sense to allow λ to be infinitesimally small.

In the limit that $\lambda \to 0^+$, these (equivalent) matrices are the same as the Moore-Penrose pseudoinverse. This provides the following *limit-based definition*:

$$\lim_{\lambda \to 0^+}(A^T A + \lambda I_d)^{-1} A^T = \lim_{\lambda \to 0^+} A^T (AA^T + \lambda I_n)^{-1} \text{ [Moore-Penrose Pseudoinverse]}$$

Note that λ approaches 0 from the right, and the function can be discontinuous at $\lambda = 0$ in the most general case. The conventional inverse, the left-inverse, and the right-inverse are special cases of the Moore-Penrose pseudoinverse. When the matrix A is invertible, all four inverses are the same. When only the columns of A are linearly independent, the Moore-Penrose pseudoinverse is the left-inverse. When only the rows of A are linearly independent, the Moore-Penrose pseudoinverse is the right-inverse. When neither the rows nor columns of A are linearly independent, the Moore-Penrose pseudoinverse provides a generalized inverse that none of these special cases can provide. Therefore, *the Moore-Penrose pseudoinverse respects both the best-fit and the conciseness criteria like the left- and right inverses.*

The Moore-Penrose pseudoinverse is computed as follows. An $n \times d$ matrix A of rank r has a *generalized* QR decomposition of the form $A = QR$, where Q is an $n \times r$ matrix with orthonormal columns, and R is a *rectangular* $r \times d$ upper-triangular matrix of full row rank. The matrix RR^T is therefore invertible. Then, the pseudoinverse of A is as follows:

$$A^+ = \lim_{\lambda \to 0^+}(R^T R + \lambda I_d)^{-1} R^T Q^T = \lim_{\lambda \to 0^+} R^T (RR^T + \lambda I_n)^{-1} Q^T = R^T (RR^T)^{-1} Q^T$$

We used $Q^T Q = I$ in the first step and the push-though identity in the second step. Another approach using singular value decomposition is discussed in Section 7.4.4.

2.8.2   The Projection Matrix

The optimization-centric solution for solving over-determined systems of equations with $d < n$ is a more general approach (as compared to the row echelon method), because it also provides an approximate solution to the inconsistent system of equations $A\overline{x} = \overline{b}$. The optimization-centric approach recognizes that the linear system of equations is inconsistent when $\overline{b}$ does not lie in the span of the columns of A. Therefore, it is also able to "solve" this inconsistent system by projecting $\overline{b}$ on the hyperplane defined by the columns of A and then using this projection $\overline{b}'$ to solve the modified (and consistent) system $A\overline{x} = \overline{b}'$. After all, $\overline{b}'$ is the closest approximation of $\overline{b}$ within the span of the columns of A. Mapping from $\overline{b}$ to $\overline{b}'$ can also be understood in the context of a linear transformation by a *projection matrix*. In this section, we will examine the nature of the projection matrix, because it turns out to be a useful linear operator in many settings of linear algebra and optimization.

First, we will consider the simple case when the columns of A are orthonormal, and emphasize its orthogonality by using the notation $Q = A$ (which is commonly used for orthogonal matrices). Therefore, the system of equations is $Q\overline{x} = \overline{b}$. The projection of an n-dimensional vector $\overline{b}$ on a d-dimensional orthonormal basis system (for $d < n$) is easy to compute. For example, if the $n \times d$ matrix Q contains d orthonormal columns, then the coordinates of $\overline{b}$ on these vectors are given by the dot products with the columns.

In other words, the coordinates are represented in the d-dimensional vector $\overline{x} = Q^T \overline{b}$. Furthermore, the actual linear combination[3] of the columns of Q with these coordinates is $\overline{b}' = Q\overline{x} = QQ^T\overline{b}$. The vector $\overline{b}'$ is the projection of $\overline{b}$ on the d-dimensional plane created by the columns of Q. Note that if the original matrix Q is square, then its orthonormal columns would imply that $QQ^T = Q^TQ = I$, and therefore $\overline{b}' = QQ^T\overline{b} = \overline{b}$. This is not particularly surprising because the projection of an n-dimensional vector on the full n-dimensional space is itself. For cases in which the columns of Q are orthonormal but the matrix Q satisfies $d < n$, the matrix $P = QQ^T$ is the projection matrix. Projecting a column vector by pre-multiplying with P might result in a different vector; however, projecting again by pre-multiplying with P will not change the projection further. For example, projecting a vector in $\mathcal{R}^3$ on a 2-dimensional plane will result in a "shadow" of the vector on the plane; projecting that smaller vector again on the same plane will not change it. Therefore, projection matrices always satisfy $P^2 = P$:

$$P^2 = (QQ^T)(QQ^T) = Q\underbrace{(Q^TQ)}_{I}Q^T = QQ^T = P \qquad (2.15)$$

This is referred to as the *idempotent property* of projection matrices.

Next, we discuss the projection matrix of a more general $n \times d$ matrix A of full rank. Therefore, if $\overline{x}$ contains the coordinates of $\overline{b}'$ in the basis of the column space of A, we have $\overline{b}' = A\overline{x}$. We want to minimize the squared distance $\|\overline{b}' - \overline{b}\|^2 = \|A\overline{x} - \overline{b}\|^2$, because the projection is always the smallest distance to the plane. This is exactly the same problem as discussed in the optimization-centric view discussed in the previous section. Since we assume linearly independent columns with $d < n$, one can use the left-inverse to obtain the following:

$$\overline{x} = (A^TA)^{-1}A^T\overline{b} \qquad (2.16)$$

Note that $\overline{x}$ corresponds to the coordinate vector in terms of the columns of A, which provides the best approximation $A\overline{x} = \overline{b}'$. The projection of $\overline{b}$ on the plane defined by the d linearly independent columns of A can also be represented in terms of the projection matrix:

$$\overline{b}' = A\overline{x} = \underbrace{A(A^TA)^{-1}A^T}_{P}\overline{b} \qquad (2.17)$$

Therefore, the $n \times n$ projection matrix is $P = A(A^TA)^{-1}A^T$. The projection matrix is always symmetric and satisfies $P^T = P$. When the columns of A are orthonormal and $d < n$, we have $A^TA = I$, and it is easy to show that the projection matrix simplifies to AA^T. Furthermore, the symmetric projection matrix always satisfies $P^2 = P$:

$$P^2 = A\underbrace{(A^TA)^{-1}(A^TA)}_{I}(A^TA)^{-1}A^T = A(A^TA)^{-1}A^T = P \qquad (2.18)$$

In fact, any symmetric matrix satisfying $P^2 = P$ can be shown to be a projection matrix. The projection matrix is useful for finding the closest approximation of an n-dimensional vector $\overline{b}$ on a plane defined by fewer than n vectors, when the point does not lie on the plane. In fact, the classical problem of least-squares regression can be viewed as that of trying to project an n-dimensional column vector of *response variables* to its *concisely*

---

[3]The *columns* of A are orthonormal. For $d < n$, we have $Q^TQ = I_d$ but $QQ^T \neq I_n$. It is only in the case of square matrices that we have $Q^TQ = QQ^T = I$.

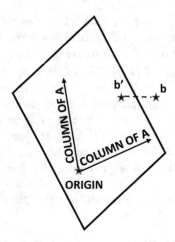

Figure 2.9: The projection of the 3-dimensional vector $\overline{b}$ on to its closest 3-dimensional point $\overline{b}'$ lying on a 2-dimensional plane defined by the columns of the 3×2 matrix A is shown for the inconsistent system $A\overline{x} = \overline{b}$. Multiplying $\overline{b}$ with the 3×3 projection matrix yields $\overline{b}'$

modeled approximation on a d-dimensional plane using a coefficient vector containing the $d \ll n$ coordinates of the projection of the n-dimensional response variables. This situation is shown in Figure 2.9, where we assume that we have a 3×2 matrix A, which is such that the 3-dimensional vector $\overline{b}$ does not lie inside the span of the two columns of A. These two column vectors are shown in Figure 2.9. Multiplying $\overline{b}$ with the 3×3 projection matrix finds the closest approximation $\overline{b}'$ of $\overline{b}$ which does lie in the span of the two columns. Subsequently, one can find the 2-dimensional vector of coordinates $\overline{x}$ of $\overline{b}'$ in terms of the two columns, which is the same as computing the solution to $A\overline{x} = \overline{b}'$. The resulting vector $\overline{x}$ is exactly the solution to least-squares regression (cf. Section 4.7 of Chapter 4).

Multiplying an $n \times d$ matrix A with any non-singular $d \times d$ matrix B creates a matrix AB with the same projection matrix as A, because the projection matrix $(AB)([AB]^T AB)^{-1}(AB)^T$ can be algebraically simplified to the projection matrix of A after canceling B and B^T with their inverses. This is because *the projection matrix of A only depends on the vector space spanned by the columns of A* and post-multiplying A with a non-singular matrix does not change the span of its columns. Therefore, an efficient way of computing the projection matrix and the projection $\overline{b}'$ of $\overline{b}$ is to use QR-decomposition $A = QR$ to compute the projection matrix as $P = QQ^T$. Note that Q is an $n \times d$ matrix like A, and R is a $d \times d$ upper-triangular matrix. The projection $\overline{b}'$ can be computed as $QQ^T \overline{b}$. The best-fit solution $\overline{x}$ to $A\overline{x} = \overline{b}$ is the solution to $QR\overline{x} = \overline{b}'$ as follows:

$$R\overline{x} = Q^T\overline{b}' = Q^T QQ^T \overline{b} = Q^T \overline{b} \tag{2.19}$$

Backsubstitution can be used to solve $R\overline{x} = Q^T \overline{b}$. We provide an example of the use of QR-decomposition to compute the projection matrix:

$$A = \begin{bmatrix} 1 & 2 \\ 0 & 2 \\ 1 & 2 \end{bmatrix} = QR = \begin{bmatrix} 1/\sqrt{2} & 0 \\ 0 & 1 \\ 1/\sqrt{2} & 0 \end{bmatrix} \begin{bmatrix} \sqrt{2} & 2\sqrt{2} \\ 0 & 2 \end{bmatrix}$$

The projection matrix P can be computed as follows:

$$P = A(A^T A)^{-1} A^T = QQ^T = \begin{bmatrix} 1/2 & 0 & 1/2 \\ 0 & 1 & 0 \\ 1/2 & 0 & 1/2 \end{bmatrix}$$

Problem 2.8.4 (Orthogonal Complementary Projections) *Suppose that* $P = QQ^T$ *is a projection matrix, where* Q *is an* $n \times d$ *matrix with orthogonal columns. Show that* $(I - P)$ *is also a projection matrix in the orthogonal complementary vector space to the projection space of* P. *A hint is to show that* $(I - P)$ *can be expressed as* $Q_1 Q_1^T$.

2.9 Ill-Conditioned Matrices and Systems

Ill-conditioned matrices are "almost" singular, and in some cases their non-singularity is only a result of numerical approximations that some algorithm might already have made during the computation of the matrix. Trying to invert such a matrix will result in very large entries, numerical overflows, and all types of round-off errors. In other words, the earlier errors will be greatly magnified. Consider the matrix A and its perturbation A_ϵ:

$$A = \begin{bmatrix} 1 & 1 \\ 1 & 1 \end{bmatrix}, \quad A_\epsilon = \begin{bmatrix} 1 + 10^{-8} & 1 \\ 1 & 1 + 10^{-8} \end{bmatrix}$$

Note that the matrix A is singular, whereas the matrix A_ϵ is invertible. The matrix A_ϵ could easily have been created by computer finite-precision errors during computation of what was intended to be A. The inverse of the matrix can be approximated as follows:

$$A_\epsilon^{-1} \approx \frac{10^8}{2} \begin{bmatrix} 1 + 10^{-8}/2 & -1 + 10^{-8}/2 \\ -1 + 10^{-8}/2 & 1 + 10^{-8}/2 \end{bmatrix} = \frac{10^8}{2} \begin{bmatrix} 1.000000005 & -0.999999995 \\ -0.999999995 & 1.000000005 \end{bmatrix}$$

It is evident that the inverse contains very large entries, and many entries need to be represented to a very high degree of precision in order to perform accurate multiplication with the original matrix. The combination of the two is a deadly cocktail because of the disproportionate effect of round-off errors and the possibility of numerical overflows in some cases. In order to understand the problematic aspects of this type of inversion, consider the case where one tries to solve the system of equations $A\bar{x} = \bar{b}$. One of the properties of A_ϵ is that $A_\epsilon \bar{x}$ is always non-zero (because the matrix A_ϵ is nonsingular), but the value of the norm $\|A_\epsilon \bar{x}\|$ will vary a lot. For example, choosing $\bar{x} = [1, 1]$ will result in $\|A\bar{x}\| \approx \sqrt{2}$. On the other hand, choosing $\bar{x} = [1, -1]^T$ will result in $\|A\bar{x}\| = 10^{-8}\sqrt{2}$.

This type of variation can cause numerical problems in near-singular systems. Since the entries of A_ϵ^{-1} are very large, small changes in $\bar{b}$ can lead to large and unstable changes in the solution $\bar{x}$. The resulting solutions might sometimes not be semantically meaningful, if the non-singularity of A_ϵ was caused by computational errors. For example, one would always be able to find a solution to $A_\epsilon \bar{x} = \bar{b}$, but in some cases the solution might be so large so as to cause a numerical overflow (caused by magnification of a tiny computational error). In the above case, using $\bar{b} = [1, -1]^T$ might lead to numerical problems, where all entries are of the order of 10^8. The problem of ill-conditioning is ubiquitous in matrix operations and linear algebra. One can quantify the ill-conditioning of a square and invertible matrix A with the notion of *condition numbers*:

Definition 2.9.1 (Condition Number) *Let A be a $d \times d$ invertible matrix. Let $\|A\overline{x}\|/\|\overline{x}\|$ be the scaling ratio of vector $\overline{x}$. Then, the condition number of A is defined as the ratio of the largest scaling ratio of A (over all d-dimensional vectors) to the smallest scaling ratio over all d-dimensional vectors.*

The smallest possible condition number of 1 occurs for the identity matrix (or any orthogonal matrix). After all, orthogonal matrices only rotate or reflect a vector without scaling it. Singular matrices have undefined condition numbers, and near-singular matrices have extremely large condition numbers. One can compute the condition number of a matrix using a method called *singular value decomposition* (cf. Section 7.4.4.1 of Chapter 7). The intuitive idea is that singular value decomposition tells us about the various scale factors in a linear transformation (also referred to as *singular values*). Therefore, the ratio of the largest to smallest scale factor gives us the condition number. See Section 7.4.4.1 of Chapter 7 on methods for solving ill-conditioned systems.

2.10   Inner Products: A Geometric View

The dot product is a natural approach for measuring similarity in vector spaces. The *inner product* is a generalization of this concept. In some engineering applications, the similarity between two real-valued vectors is obtained as the dot product after stretching the vectors in some "important" directions with the linear transformation A. Therefore, we first provide a practical and easy-to-visualize definition of inner products that works only for $\mathcal{R}^n$:

Definition 2.10.1 (Inner Products: Restricted Definition) *A mapping from $\overline{x}, \overline{y} \in \mathcal{R}^n$ to $\langle \overline{x}, \overline{y} \rangle \in \mathcal{R}$ is an inner product if and only if $\langle \overline{x}, \overline{y} \rangle$ is always equal to the dot product between $A\overline{x}$ and $A\overline{y}$ for some $n \times n$ non-singular matrix A. The inner product $\langle \overline{x}, \overline{y} \rangle$ can also be expressed using the Gram matrix $S = A^T A$:*

$$\langle \overline{x}, \overline{y} \rangle = (A\overline{x})^T (A\overline{x}) = \overline{x}^T [A^T A] \overline{y} = \overline{x}^T S \overline{y}$$

When the linear transformation A is a rotreflection matrix, the matrix S is the identity matrix, and the inner product specializes to the normal dot product. The inner product also induces cosines and distances with respect to transformation A:

$$\text{cosine}_A(\overline{x}, \overline{y}) = \frac{\langle \overline{x}, \overline{y} \rangle}{\sqrt{\langle \overline{x}, \overline{x} \rangle} \sqrt{\langle \overline{y}, \overline{y} \rangle}} = \frac{\overline{x}^T S \overline{y}}{\sqrt{\overline{x}^T S \overline{x}} \sqrt{\overline{y}^T S \overline{y}}} = \frac{(A\overline{x})^T (A\overline{y})}{\|A\overline{x}\|_2 \|A\overline{y}\|_2}$$

$$\text{distance}_A(\overline{x}, \overline{y})^2 = \langle \overline{x} - \overline{y}, \overline{x} - \overline{y} \rangle = (\overline{x} - \overline{y})^T S (\overline{x} - \overline{y}) = \|A\overline{x} - A\overline{y}\|_2^2$$

It is easy to see that the induced distances and angles correspond to our normal geometric understanding of lengths and angles *after using the matrix A to perform a linear transformation on the vectors*. The value $\sqrt{\langle \overline{x} - \overline{y}, \overline{x} - \overline{y} \rangle}$ is referred to as a *metric*, which satisfies all laws of Euclidean geometry, such as the triangle inequality. This is not particularly surprising, given that it *is* a Euclidean distance in transformed space.

   A more general definition of inner products that works beyond $\mathcal{R}^n$ (e.g., for abstract vector spaces) is based on particular axiomatic rules that need to be followed:

Definition 2.10.2 (Inner-Product: General Definition) *The real value $\langle \overline{u}, \overline{v} \rangle$ is an inner product between $\overline{u}$ and $\overline{v}$, if it satisfies the following axioms for all $\overline{u}$ and $\overline{v}$:*

Additivity: $\langle \overline{u}, \overline{v} + \overline{w} \rangle = \langle \overline{u}, \overline{v} \rangle + \langle \overline{u}, \overline{w} \rangle$, $\langle \overline{v} + \overline{w}, \overline{u} \rangle = \langle \overline{v}, \overline{u} \rangle + \langle \overline{w}, \overline{u} \rangle$

Multiplicativity: $\langle c\overline{u}, \overline{v} \rangle = c\langle \overline{u}, \overline{v} \rangle$, $\langle \overline{u}, c\overline{v} \rangle = c\langle \overline{u}, \overline{v} \rangle$ $\forall c \in \mathcal{R}$

Commutativity: $\langle \overline{u}, \overline{v} \rangle = \langle \overline{v}, \overline{u} \rangle$

Positive definiteness: $\langle \overline{u}, \overline{u} \rangle \geq 0$, *with equality only for the zero vector*

Every finite-dimensional inner product $\langle \overline{x}, \overline{y} \rangle$ in $\mathcal{R}^n$ satisfying the above axioms can be shown to be equivalent to $\overline{x}^T S \overline{y}$ for some carefully chosen Gram matrix $S = A^T A$. Therefore, at least for finite-dimensional vector spaces in $\mathcal{R}^n$, *the linear transformation definition and the axiomatic definition of* $\langle \overline{x}, \overline{y} \rangle$ *are equivalent.* The following exercise shows how such a matrix S can be constructed from the axiomatic definition of an inner product:

Problem 2.10.1 (Axiomatic Inner-Product Is Transformed Dot Product) *Suppose that the inner product* $\langle \overline{x}, \overline{y} \rangle$ *satisfies the axiomatic definition for all pairs* $\overline{x}, \overline{y} \in \mathcal{R}^n$. *Show that the inner product* $\langle \overline{x}, \overline{y} \rangle$ *can also be expressed as* $\overline{x}^T S \overline{y}$, *where the* (i, j)th *entry of* S *is* $\langle \overline{e}_i, \overline{e}_j \rangle$. *Here,* $\overline{e}_i$ *is the ith column of the* $n \times n$ *identity matrix. The next chapter shows that matrices like* S *can always be expressed as* $A^T A$ *for* $n \times n$ *matrix* A *because of the positive definite axiom. Why is* $\langle \overline{x}, \overline{y} \rangle$ *equal to the vanilla dot product between* $A\overline{x}$ *and* $A\overline{y}$?

Problem 2.10.2 *Suppose that you are given all* $n \times n$ *real-valued inner products between pairs drawn from* n *linearly independent vectors in* $\mathcal{R}^n$. *Show how you can compute* $\langle \overline{x}, \overline{y} \rangle$ *for any* $\overline{x}, \overline{y} \in \mathcal{R}^n$ *using the basic axioms of inner products.*

2.11   Complex Vector Spaces

As discussed earlier in this chapter, vector spaces can be defined over any *field* that satisfies the *field axioms*. One such example of a field is the domain of complex numbers. A complex number is a value of the form $a + i\,b$ where $i = \sqrt{-1}$. Complex numbers are often written in the *polar form* $r\left[\cos(\theta) + i\sin(\theta)\right]$, where $r = \sqrt{a^2 + b^2}$ and $\theta = \cos^{-1}(a/r)$. One can also show the following *Euler identity* by comparing the Taylor expansions of the exponential and trigonometric series (see Problem 1.5.1):

$$\exp(i\theta) = \cos(\theta) + i\sin(\theta)$$

The angle θ must be expressed in radians for this formula to hold. Therefore, a complex number may be represented as $r \cdot \exp(i\theta)$. The polar representation is very convenient in the context of many linear algebra operations. This is because the multiplication of two complex numbers is a simple matter of adding angular exponents and multiplying their magnitudes. This property is used in various types of matrix products.

One can define a vector space over the complex domain using the same additive and multiplicative properties over $\mathcal{C}^n$ as in $\mathcal{R}^n$:

Definition 2.11.1 (Vector Space in $\mathcal{C}^n$) *A set of vectors* $\mathcal{V}$ *that correspond to a subset of* $\mathcal{C}^n$ *is a vector space, if it satisfies the following properties:*

1. *If* $\overline{x} \in \mathcal{V}$ *then* $c\overline{x} \in \mathcal{V}$ *for any scalar* $c \in \mathcal{C}$.

2. *If* $\overline{x}, \overline{y} \in \mathcal{V}$, *then* $\overline{x} + \overline{y} \in \mathcal{V}$.

Here, it is important to note that the *multiplicative scalar is drawn from the complex domain*. For example, the value of c could be a number such as $1+i$. This is an important difference from Definition 2.3.2 on real-valued vector spaces. The consequence of this fact is that one can still use the standard basis $\overline{e}_1 \ldots \overline{e}_n$ to represent any vector in $\mathcal{C}^n$. Here, each $\overline{e}_i$ is an n-dimensional vector with a 1 in the ith entry, and a 0 in all other entries. Although $\overline{e}_i$ has real components, all real vectors are special cases of complex-valued vectors. Any vector $\overline{x} = [x_1 \ldots x_d]^T \in \mathcal{C}^n$ can be expressed in terms of standard basis, where the ith coordinate is the complex number x_i. The key point is that the coordinates can also be complex values, since the vector space is defined over the complex field. We need to be able to perform operations such as projections in order to create coordinate representations. This is achieved with the notion of complex inner products.

As in the case of real inner products, one wants to retain geometric properties of Euclidean spaces (like notions of lengths and angles). Generalizing inner products from the real domain to the complex domain can be tricky. In real-valued Euclidean spaces, the dot product of the vector with itself provides the squared norm. *This definition does not work for complex vectors.* For example, a blind computation of the real-valued definition of squared norm of $\overline{v} = [1, 2i]^T$ results in the following:

$$\overline{v}^T \overline{v} = [1, 2i] \begin{bmatrix} 1 \\ 2i \end{bmatrix} = 1^2 + 4i^2 = 1 - 4 = -3 \tag{2.20}$$

We obtain a *negative* value for squared norm, which is intended to be a proxy for the squared length. Therefore, we need modified axioms for the complex-valued inner product $\langle \overline{u}, \overline{v} \rangle$:

Additivity: $\langle \overline{u}, \overline{v} + \overline{w} \rangle = \langle \overline{u}, \overline{v} \rangle + \langle \overline{u}, \overline{w} \rangle$, $\langle \overline{v} + \overline{w}, \overline{u} \rangle = \langle \overline{v}, \overline{u} \rangle + \langle \overline{w}, \overline{u} \rangle$

Multiplicativity: $\langle c\overline{u}, \overline{v} \rangle = c^* \langle \overline{u}, \overline{v} \rangle$, $\langle \overline{u}, c\overline{v} \rangle = c\langle \overline{u}, \overline{v} \rangle$ $\forall c \in \mathcal{C}$

Conjugate symmetry: $\langle \overline{u}, \overline{v} \rangle = \langle \overline{v}, \overline{u} \rangle^*$

Positive definiteness: $\langle \overline{u}, \overline{u} \rangle \geq 0$, with equality only for the zero vector

The superscript '*' indicates the *conjugate* of a complex number, which is obtained by negating the imaginary part of the number. The inner product computation of Equation 2.20 is invalid is because it violates the positive definite property.

For a scalar complex number, its squared norm is defined by its product with its conjugate. For example, the squared norm of $a + ib$ is $(a - ib)(a + ib) = a^2 + b^2$. In the case of vectors, we can combine transposition with conjugation in order to define inner products. The conjugate transpose of a complex vector or matrix is defined as follows:

Definition 2.11.2 (Conjugate Transpose of Vector and Matrix) *The conjugate transpose $\overline{v}^*$ of a complex vector $\overline{v}$ is obtained by transposing the vector and replacing each entry with its complex conjugate. The conjugate transpose V^* of a complex matrix V is obtained by transposing the matrix and replacing each entry with its complex conjugate.*

Therefore, the conjugate transpose of $[1, 2i]^T$ is $[1, -2i]$, and the conjugate transpose of $[1 + i, 2 + 3i]^T$ is $[1 - i, 2 - 3i]$.

A popular way of defining[4] the inner product between vectors $\overline{u}, \overline{v} \in \mathcal{C}^n$, which is the direct analog of the dot product, is the following:

$$\langle \overline{u}, \overline{v} \rangle = \overline{u}^* \overline{v} \tag{2.21}$$

---

[4] Some authors define $\langle \overline{u}, \overline{v} \rangle = \overline{v}^* \overline{u}$ (which is a conjugate of the definition here). The choice does not really matter as long as it is used consistently.

The inner product can be a complex number. Unlike vectors in $\mathcal{R}^n$, the inner product is *not* commutative over the complex domain, because $\langle \overline{u}, \overline{v} \rangle$ is the complex conjugate of $\langle \overline{v}, \overline{u} \rangle$ (i.e., conjugate symmetry property). The squared norm of a vector $\overline{v} \in \mathcal{C}^n$ is defined as $\overline{v}^* \overline{v}$ rather than $\overline{v}^T \overline{v}$; this is the inner product of the vector with itself. Based on this definition, the squared norm of $[1, 2i]^T$ is $[1, -2i][1, 2i]^T$, which is $1^2 + 2^2 = 5$. Similarly, the squared norm of $[1 + i, 2 + 3i]^T$ is $(1 + i)(1 - i) + (2 + 3i)(2 - 3i) = 1 + 1 + 4 + 9 = 15$. Note that both are positive, which is consistent with the positive definite property.

As in the real domain, two complex vectors are orthogonal when their inner product is 0. In such a case, both the complex conjugates $\langle \overline{u}, \overline{v} \rangle$ and $\langle \overline{v}, \overline{u} \rangle$ are zero.

Definition 2.11.3 (Orthogonality in $\mathcal{C}^n$) *Two vectors $\overline{u}$ and $\overline{v}$ from $\mathcal{C}^n$ are orthogonal if and only if $\overline{u}^* \overline{v} = \overline{v}^* \overline{u} = 0$.*

An orthonormal set of vectors in $\mathcal{C}^n$ corresponds to any set of vectors $\overline{v}_1 \ldots \overline{v}_n$, such that $\overline{v}_i^* \overline{v}_j$ is 1 when $i = j$, and 0, otherwise. Note that the standard basis is also orthogonal in $\mathcal{C}^n$. As in the real domain, an $n \times n$ matrix containing orthogonal columns from $\mathcal{C}^n$ is referred to as *orthogonal* or *unitary*.

Definition 2.11.4 (Orthogonal Matrix with Complex Entries) *A matrix V with complex-valued entries is orthogonal or* **unitary** *if and only if $VV^* = V^*V = I$.*

It is relatively easy to compute the inverse of orthogonal matrices by simply computing their conjugate transposes. This idea has applications to the discrete Fourier transform.

2.11.1 The Discrete Fourier Transform

The discrete Fourier transform is closely related to the discrete cosine transform, and it is capable of finding an orthonormal basis for time-series in the complex domain. As a practical matter, it is used as an alternative to the discrete cosine transform (cf. Section 2.7.3) for real-valued series with a high level of periodicity.

Consider a complex-valued time series $\overline{s} \in \mathcal{C}^n$, which we would like to transform into a complex and orthogonal basis. The Fourier basis uses n mutually orthogonal basis vectors $\overline{b}_1 \ldots \overline{b}_n$ from $\mathcal{C}^n$, so that the basis vector $\overline{b}_j$ is defined as follows:

$$\overline{b}_j = [1, \exp(\omega[j-1]i), \ldots, \underbrace{\exp(\omega[k-1][j-1]i)}_{k\text{th component}}, \ldots, \exp(\omega[n-1][j-1]i)]^T / \sqrt{n}$$

Note that the value of i in the above does nor refer to a variable but to the imaginary number $\sqrt{-1}$. The value of ω is $2\pi/n$ in radians, and therefore each complex number is written in polar form. We make the following assertion:

Lemma 2.11.1 (Orthonormality of Fourier Basis) *The basis vectors $\overline{b}_1 \ldots \overline{b}_n$ of the Fourier transform are orthonormal.*

Proof: It is easy to see that $\overline{b}_p^* \overline{b}_p = [\sum_{k=0}^{n-1}(1/n)\exp(0)] = 1$. On computing $\overline{b}_p^* \overline{b}_q$ for $p \neq q$, one can sum a geometric series of exponentials:

$$\overline{b}_p^* \overline{b}_q = \sum_{k=0}^{n-1} \exp(k[q-p]\omega i) = \frac{\exp([n\omega][q-p]i) - 1}{\exp([q-p]\omega i) - 1} = \frac{\overbrace{\exp(2\pi[q-p]i)}^{1} - 1}{\exp([q-p]\omega) - 1} = 0$$

One of the simplifications above uses the fact that $\exp(i\theta)$ is 1 when θ is a multiple of 2π. ∎

One can, therefore, create a basis matrix B whose columns contain the basis vectors $\overline{b}_1 \ldots \overline{b}_n$. For example, the 8×8 basis matrix for transformation of vectors in $\mathcal{C}^8$ is as follows:

$$\frac{1}{\sqrt{8}} \begin{bmatrix} 1 & 1 & 1 & 1 & 1 & 1 & 1 & 1 \\ 1 & \exp(\frac{2\pi i}{8}) & \exp(\frac{4\pi i}{8}) & \exp(\frac{6\pi i}{8}) & \exp(\frac{8\pi i}{8}) & \exp(\frac{10\pi i}{8}) & \exp(\frac{12\pi i}{8}) & \exp(\frac{14\pi i}{16}) \\ 1 & \exp(\frac{4\pi i}{8}) & \exp(\frac{8\pi i}{8}) & \exp(\frac{12\pi i}{8}) & \exp(\frac{16\pi i}{8}) & \exp(\frac{20\pi i}{8}) & \exp(\frac{24\pi i}{8}) & \exp(\frac{28\pi i}{8}) \\ 1 & \exp(\frac{6\pi i}{8}) & \exp(\frac{12\pi i}{8}) & \exp(\frac{18\pi i}{8}) & \exp(\frac{24\pi i}{8}) & \exp(\frac{30\pi i}{8}) & \exp(\frac{36\pi i}{8}) & \exp(\frac{42\pi i}{8}) \\ 1 & \exp(\frac{8\pi i}{8}) & \exp(\frac{16\pi i}{8}) & \exp(\frac{24\pi i}{8}) & \exp(\frac{32\pi i}{8}) & \exp(\frac{40\pi i}{8}) & \exp(\frac{48\pi i}{8}) & \exp(\frac{56\pi i}{8}) \\ 1 & \exp(\frac{10\pi i}{8}) & \exp(\frac{20\pi i}{8}) & \exp(\frac{30\pi i}{8}) & \exp(\frac{40\pi i}{8}) & \exp(\frac{50\pi i}{8}) & \exp(\frac{60\pi i}{8}) & \exp(\frac{70\pi i}{8}) \\ 1 & \exp(\frac{12\pi i}{8}) & \exp(\frac{24\pi i}{8}) & \exp(\frac{36\pi i}{8}) & \exp(\frac{48\pi i}{8}) & \exp(\frac{60\pi i}{8}) & \exp(\frac{72\pi i}{8}) & \exp(\frac{84\pi i}{8}) \\ 1 & \exp(\frac{14\pi i}{8}) & \exp(\frac{28\pi i}{8}) & \exp(\frac{42\pi i}{8}) & \exp(\frac{56\pi i}{8}) & \exp(\frac{70\pi i}{8}) & \exp(\frac{84\pi i}{8}) & \exp(\frac{98\pi i}{8}) \end{bmatrix}$$
$$\underbrace{}_{B}$$

The matrix B is orthogonal, and therefore the basis transformation is *length preserving*:

$$\|B\overline{s}\|^2 = (Bs)^*(Bs) = s^* \underbrace{(B^*B)}_{I} s = \|s\|^2$$

Given a *complex-valued* time-series $\overline{s}$ from $\mathcal{C}^8$, one can transform it to the Fourier basis by solving the system of equations $B\overline{x} = \overline{s}$. The solution to this system is simply $\overline{x} = B^*\overline{s}$, which provides the complex coefficients of the series. As a practical matter, the approach is used for real-valued time series. For example, consider our running example of the time-series $\overline{s} = [8, 6, 2, 3, 4, 6, 6, 5]^T$, which is used in Section 2.3.4 on the wavelet transform. One can simply pretend that this series is a special case of a complex-valued series, and compute the Fourier coefficients as $\overline{x} = B^*\overline{s}$. The main problem with this approach is that it transforms a series from $\mathcal{R}^8$ to $\mathcal{C}^8$, since the coordinates in $\overline{x}$ will have imaginary components. A naïve solution to this problem is to create a representation in R^{16} that contains both real and imaginary parts of each component of $\overline{x}$. Therefore, the Fourier transformation contains twice the number of real-valued coefficients as the original series. This increase is a consequence of treating a real-valued time-series as a special case of a complex-valued series. Because of the real-valued nature of the original series, wasteful redundancy exists in the coordinate vector $\overline{x}$, whose kth component is always the complex conjugate of the $(8 - k)$th component for all k. Therefore, one can keep only the first four components of the vector $\overline{x} \in \mathcal{C}^8$ and unroll the real and imaginary components of these four complex numbers into $\mathcal{R}^8$. Furthermore, one sets the small Fourier coefficients to zero in practice, which leads to space-efficient sparse vector representations.

Problem 2.11.1 *Use the 8×8 Fourier matrix proposed in this section in order to create the Fourier representation of $\overline{s} = [8, 6, 2, 3, 4, 6, 6, 5]^T$.*

2.12   Summary

Machine learning applications often use additive and multiplicative transformations with matrices, which correspond to the fundamental building blocks of linear algebra. These building blocks are utilized for different types of decompositions such as the QR decomposition and the LU decomposition. The decompositions are the workhorses to solution methodologies for many matrix-centric problems in machine learning. Specific examples include solving systems of linear equations and linear regression.

2.13   Further Reading

Fundamental books on linear algebra include those by Strang [122, 123], David Lay [77], and Hoffman-Kunze [62]. The matrix computation book by Golub and Van Loan [52] teaches important numerical methods. A discussion of numerical methods that combine linear algebra and optimization is provided in [99].

2.14   Exercises

1. If we have a square matrix A that satisfies $A^2 = I$, it is always the case that $A = \pm I$. Either prove the statement or provide a counterexample.

2. Show that the matrices A, AA^T, and $A^T A$ must always have the same rank for any $n \times d$ matrix A. Start by showing that $A\overline{x} = \overline{0}$ if and only if $A^T A\overline{x} = \overline{0}$.

3. Provide a geometric interpretation of A^9, where A is a 2×2 rotation matrix at a counter-clockwise angle of $60°$.

4. Consider 6×10 matrices A and B of rank 6. What is the minimum and maximum possible rank of the 6×6 matrix AB^T. Provide examples of A and B in each case.

5. Use each of row reduction and Gram-Schmidt to find basis sets for the span of $\{[1, 2, 1]^T, [2, 1, 1]^T, [3, 3, 2]^T\}$. What are the best-fit coordinates of $[1, 1, 1]^T$ in each of these basis sets? Verify that the best-fit vector is the same in the two cases.

6. Propose a test using Gram-Schmidt orthogonalization to identify whether two sets of (possibly linearly dependent) vectors span the same vector space.

7. A $d \times d$ skew symmetric matrix satisfies $A^T = -A$. Show that all diagonal elements of such a matrix are 0. Show that each $\overline{x} \in \mathcal{R}^d$ is orthogonal to $A\overline{x}$ if and only if A is skew symmetric. What is the difference from a pure rotation by $90°$?

8. Consider the 4×4 Givens matrix $G_c(2, 4, 90)$ based on the notations on page 47. This matrix performs a $90°$ counter-clockwise rotation of a 4-dimensional column vector in the plane of the second and fourth dimensions. Show how to obtain this matrix as the product of two Householder reflection matrices. Think geometrically based on Section 2.2 in order to solve this problem. Is the answer to this question unique?

9. Repeat Exercise 8 for a Givens matrix that rotates a column vector counter-clockwise for $10°$ instead of $90°$.

10. Consider the 5×5 matrices A, B, and C, with ranks 5, 2, and 4, respectively. What is the minimum and maximum possible rank of $(A + B)C$.

11. Solve the following system of equations using the Gaussian elimination procedure:

$$\begin{bmatrix} 0 & 1 & 1 \\ 1 & 1 & 1 \\ 1 & 2 & 1 \end{bmatrix} \begin{bmatrix} x_1 \\ x_2 \\ x_3 \end{bmatrix} = \begin{bmatrix} 2 \\ 3 \\ 4 \end{bmatrix}$$

Now use these row operations to create an LU decomposition. Is it possible to perform an LU decomposition of this matrix without the use of a permutation matrix?

12. Solve the system of equations in the previous exercise using QR decomposition. Use the Gram-Schmidt method for orthogonalization. Use the QR decomposition to compute the inverse of the matrix if it exists.

13. Why must the column space of matrix AB must be a subspace of the column space of A? Show that all four fundamental subspaces of A^{k+1} must be the same as that of A^k for some integer k.

14. Consider a vector space $\mathcal{V} \subset \mathcal{R}^3$ and two of its possible basis sets $\mathcal{B}_1 = \{[1, 0, 1]^T, [1, 1, 0]^T\}$ and $\mathcal{B}_2 = \{[0, 1, -1]^T, [2, 1, 1]^T\}$. Show that $\mathcal{B}_1$ and $\mathcal{B}_2$ are basis sets for the same vector space. What is the dimensionality of this vector space? Now consider a vector $\overline{v} \in \mathcal{V}$ with coordinates $[1, 2]^T$ in basis $\mathcal{B}_1$, where the order of coordinates matches the order of listed basis vectors. What is the standard basis representation of $\overline{v}$? What are the coordinates of $\overline{v}$ in $\mathcal{B}_2$?

15. Find the projection matrix of the following matrix using the QR method:

$$A = \begin{bmatrix} 3 & 6 \\ 0 & 1 \\ 4 & 8 \end{bmatrix}$$

How can you use the projection matrix to determine whether the vector $\overline{b} = [1, 1, 0]^T$ belongs to the column space of A? Find a solution (or best-fit solution) to $A\overline{x} = \overline{b}$.

16. For the problem in Exercise 15, does a solution exist to $A^T \overline{x} = \overline{c}$, where $\overline{c} = [2, 2]^T$? If no solution exists, find the best-fit. If one or more solutions exist, find the one for which $\|\overline{x}\|$ is as small as possible.

17. **Gram-Schmidt with Projection Matrix:** Given a set of $m < n$ linearly independent vectors $\overline{a}_1 \ldots \overline{a}_m$ in $\mathcal{R}^n$, let A_r be the $n \times r$ matrix defined as $A_r = [\overline{a}_1, \overline{a}_2, \ldots, \overline{a}_r]$ for each $r \in \{1 \ldots m\}$. Show that after initializing $\overline{q}_1 = \overline{a}_1$, the unnormalized Gram-Schmidt vectors $\overline{q}_2 \ldots \overline{q}_m$ of $\overline{a}_2 \ldots \overline{a}_m$ can be computed non-recursively using the projection matrix P_s as follows:

$$\overline{q}_{s+1} = [I - A_s(A_s^T A_s)^{-1} A_s^T]\overline{a}_{s+1} = \overline{a}_{s+1} - [P_s \overline{a}_{s+1}] \quad \forall s \in \{1, \ldots, m-1\}$$

18. Consider a $d \times d$ matrix A such that its right null space is identical to its column space. Show that d is even, and provide an example of such a matrix.

19. Show that the columns of the $n \times d$ matrix A are linearly independent if and only if $f(\overline{x}) = A\overline{x}$ is a one-to-one function.

20. Consider an $n \times n$ matrix A. Show that if the length of the vector $A\overline{x}$ is strictly less than that of the vector $\overline{x}$ for all non-zero $\overline{x} \in \mathcal{R}^n$, then $(A - I)$ is invertible.

21. It is intuitively obvious that an $n \times n$ projection matrix P will always satisfy $\|P\overline{b}\| \leq \|\overline{b}\|$ for any $\overline{b} \in \mathcal{R}^n$, since it projects $\overline{b}$ on a lower-dimensional hyperplane. Show **algebraically** that $\|P\overline{b}\| \leq \|\overline{b}\|$ for any $\overline{b} \in \mathcal{R}^n$. [Hint: Express the rank-d projection matrix $P = QQ^T$ for $n \times d$ matrix Q and start by showing $\|QQ^T\overline{b}\| = \|Q^T\overline{b}\|$. What is the geometric interpretation of $Q^T\overline{b}$ and $QQ^T\overline{b}$?]

22. Let A be a 10×10 matrix. If A^2 has rank 6, find the minimum and maximum possible ranks of A. Give examples of both matrices.

23. Suppose that we have a system of equations $A\overline{x} = \overline{b}$ for some $n \times d$ matrix A. We multiply both sides of the above equation with a non-zero, $m \times n$ matrix B to obtain the new system $BA\overline{x} = B\overline{b}$. Provide an example to show that the solution sets to the two systems need not be identical. How are the solution sets related in general? Provide one example of a sufficient condition on a *rectangular* matrix B under which they are identical. [For scalar equations, multiplying both sides by a scalar value does not change the equation unless that value is 0. This exercise shows that multiplying both sides of a vector equation with a matrix can have more intricate effects.]

24. Show that every $n \times n$ Householder reflection matrix can be expressed as $Q_1 Q_1^T - Q_2 Q_2^T$, where concatenating the columns of Q_1 and Q_2 creates an $n \times n$ orthogonal matrix, and Q_2 contains a single column. What is the nature of the linear transformation, when Q_2 contains more than one column?

25. Show that if B^k has the same rank as that of B^{k+1} for a particular value of $k \geq 1$, then B^k has the same rank as B^{k+r} for all $r \geq 1$.

26. Show that if an $n \times n$ matrix B has rank $(n-1)$, and the matrix B^k has rank $(n-k)$, then each matrix B^r for r from 1 to k has rank $(n-r)$. Show how to construct a *chain* of vectors $\overline{v}_1 \ldots \overline{v}_k$ so that $B\overline{v}_i = \overline{v}_{i-1}$ for $i > 1$, and $B\overline{v}_1 = \overline{0}$. [Note: You will encounter a similar but more complex *Jordan chain* in Chapter 3.]

27. Suppose that $B^k \overline{v} = \overline{0}$ for a particular vector $\overline{v}$ for some $k \geq 2$, and $B^r \overline{v} \neq \overline{0}$ for all $r < k$. Show that the vectors $\overline{v}, B\overline{v}, B^2\overline{v}, \ldots, B^{k-1}\overline{v}$ must be linearly independent.

28. **Inverses with QR decomposition:** Suppose you perform QR decomposition of an invertible $d \times d$ matrix as $A = QR$. Show how you can use this decomposition relationship for finding the inverse of A by solving d different triangular systems of linear equations, each of which can be solved by backsubstitution. Show how to compute the left or right inverse of a matrix with QR decomposition and back substitution.

29. **Least-squares error by QR decomposition:** Let $A\overline{x} = \overline{b}$ be a system of equations in which the $n \times d$ matrix A has linearly independent columns. Suppose that you decompose $A = QR$, where Q is an $n \times d$ matrix with orthogonal columns and R is a $d \times d$ upper-triangular matrix. Show that the best-fit error (using the least-squares model) is given by $\|\overline{b}\|^2 - \|Q^T\overline{b}\|^2$. How would you find the least-squares error via QR decomposition in the case that A does not have linearly independent columns or rows? [Hint: Think geometrically in terms of the projection matrix.]

30. Consider a modified least-squares problem of minimizing $\|A\overline{x} - \overline{b}\|^2 + \overline{c}^T\overline{x}$, where A is an $n \times d$ matrix, $\overline{x}, \overline{c}$ are d-dimensional vectors, and $\overline{b}$ is an n-dimensional vector. Show that the problem can be reduced to the standard least-squares problem as long as $\overline{c}$ lies in the row space of A. What happens when $\overline{c}$ does not lie in the row space of A? [Hint: First examine the univariate version of this problem.]

31. **Right-inverse yields concise solution:** Let $\overline{x} = \overline{v}$ be any solution to the consistent system $A\overline{x} = \overline{b}$ with $n \times d$ matrix A containing linearly independent rows. Let $\overline{v}_r = A^T(AA^T)^{-1}\overline{b}$ be the solution given by the right inverse. Then, show the following:

$$\|\overline{v}\|^2 = \|\overline{v} - \overline{v}_r\|^2 + \|\overline{v}_r\|^2 + 2\overline{v}_r^T(\overline{v} - \overline{v}_r) \geq \|\overline{v}_r\|^2 + 2\overline{v}_r^T(\overline{v} - \overline{v}_r)$$

Now show that $\overline{v}_r^T(\overline{v} - \overline{v}_r) = 0$ and therefore $\|\overline{v}\|^2 \geq \|\overline{v}_r\|^2$.

32. Show that any 2×2 Givens rotation matrix is a product of at most two Householder reflection matrices. Think geometrically before wading into the algebra. Now generalize the proof to $d \times d$ matrices.

33. Show algebraically that if two tall matrices of full rank have the same column space, then they have the same projection matrix.

34. Construct 4×3 matrices A and B of rank 2 that are not multiples of one another, but with the same four fundamental subspaces of linear algebra. [Hint: $A = U V$.]

35. Show that any Householder reflection matrix $(I - 2\overline{v}\,\overline{v}^T)$ can be expressed as follows:

$$(I - 2\overline{v}\,\overline{v}^T) = \begin{bmatrix} \cos(\theta) & \sin(\theta) \\ \sin(\theta) & -\cos(\theta) \end{bmatrix}$$

Relate $\overline{v}$ to θ geometrically.

36. Show how any vector $\overline{v} \in \mathcal{R}^n$ can be transformed to $\overline{w} \in \mathcal{R}^n$ as $\overline{w} = c\,H\overline{v}$, where c is a scalar and H is an $n \times n$ Householder reflection matrix. Think geometrically to solve this exercise.

37. A block upper-triangular matrix is a generalization of a block diagonal matrix (cf. Section 1.2.3) that allows non-zero entries above the square, diagonal blocks. Consider a block upper-triangular matrix with invertible diagonal blocks. Make an argument why such a matrix is row equivalent to an invertible block diagonal matrix. Generalize the backsubstitution method to solving linear equations of the form $A\overline{x} = \overline{b}$ when A is block upper-triangular. You may assume that the diagonal blocks are easily invertible.

38. If P is a projection matrix, show that $(P + \lambda I)$ is invertible for any $\lambda > 0$. [Hint: Show that $\overline{x}^T(P + \lambda I)\overline{x} > 0$ for all $\overline{x}$, and therefore $(P + \lambda I)\overline{x} \neq 0$.]

39. If R is a Householder reflection matrix, show that $(R + I)$ is always singular, and that $(R + \lambda I)$ is invertible for any $\lambda \notin \{1, -1\}$.

40. Length-preserving transforms are orthogonal: We already know that if A is an $n \times n$ orthogonal matrix, then $\|A\overline{x}\| = \|\overline{x}\|$ for all $\overline{x} \in \mathcal{R}^n$. Prove the converse of this result that if $\|A\overline{x}\| = \|\overline{x}\|$ for all $\overline{x} \in \mathcal{R}^n$, then A is orthogonal.

41. Let A be a square $n \times n$ matrix so that $(A + I)$ has rank $(n - 2)$. Let $f(x)$ be the polynomial $f(x) = x^3 + x^2 + x + 1$. Show that $f(A)$ has rank at most $(n - 2)$. Furthermore, show that $f(A)$ has rank exactly $(n - 2)$ if A is symmetric.

42. Suppose that a $d \times d$ matrix A exists along with d vectors $\overline{x}_1 \ldots \overline{x}_d$ so that $\overline{x}_i^T A \overline{x}_j$ is zero if and only if $i \neq j$. Show that the vectors $\overline{x}_1 \ldots \overline{x}_d$ are linearly independent. Note that A need not be symmetric.

43. Suppose that a $d \times d$ *symmetric* matrix S exists along with d vectors $\overline{x}_1 \ldots \overline{x}_d$ so that $\overline{x}_i^T S \overline{x}_j$ is zero when $i \neq j$ and positive when $i = j$. Show that $\langle \overline{x}, \overline{y} \rangle = \overline{x}^T S \overline{y}$ is a valid inner product over all $\overline{x}, \overline{y} \in \mathcal{R}^d$. [Hint: The positive definite axiom is the hard part.]

44. Cauchy-Schwarz and triangle inequality for general inner products: Let $\overline{u}$ and $\overline{v}$ be two vectors for which $\langle \overline{u}, \overline{u} \rangle = \langle \overline{v}, \overline{v} \rangle = 1$. Show using only the inner-product axioms that $|\langle \overline{u}, \overline{v} \rangle| \leq 1$. Now show the more general Cauchy-Schwarz inequality by defining $\overline{u}$ and $\overline{v}$ appropriately in terms of $\overline{x}$ and $\overline{y}$:

$$|\langle \overline{x}, \overline{y} \rangle| \leq \sqrt{\langle \overline{x}, \overline{x} \rangle \langle \overline{y}, \overline{y} \rangle}$$

Now use this result (and the inner-product axioms) to prove the triangle inequality for the triangle formed by $\overline{x}$, $\overline{y}$, and the origin:

$$\sqrt{\langle \overline{x}, \overline{x} \rangle} + \sqrt{\langle \overline{y}, \overline{y} \rangle} \geq \sqrt{\langle \overline{x} - \overline{y}, \overline{x} - \overline{y} \rangle}$$

45. If the matrix computed by the polynomial function $f(A) = \sum_{i=0}^{d} c_i A^i$ has rank strictly greater than that of A, is there anything you can say about the coefficients $c_0 \ldots c_d$?

46. Let S be a symmetric matrix and $g(S) = S^3 - S^2 + S$. Without using the results of the next chapter, show that $g(S)$ has the same rank as that of S.

47. Let A be an $n \times m$ matrix and B be a $k \times d$ matrix. Show that the column space of AXB is always a subspace of the column space of A, and the row space of AXB is a subspace of the row space of B for any $m \times k$ matrix X.

48. Suppose that A is an $n \times m$ matrix and B is a $k \times d$ matrix, both of full rectangular rank. You want to find the $m \times k$ matrix X so that $C = AXB$, where C is a known $n \times d$ matrix. What should the shapes of each of A and B be (i.e., tall or wide) for the system of equations to be *guaranteed* to be consistent? Derive a closed-form expression for one solution, X, in terms of A, B, and C in this case. When is this solution unique?

49. Suppose that A is an $n \times m$ matrix and B is a $k \times d$ matrix, both of full rectangular rank. A is tall and B is wide. The system of equations $C = AXB$ is inconsistent. You want to find the *best-fit* $m \times k$ matrix X so that $\|C - AXB\|_F^2$ is as small as possible, where C is a known $n \times d$ matrix. So you model $Y \approx XB$, and first fix Y to the best-fit solution to $\|C - AY\|_F^2$. Then, you find the best-fit solution to $\|Y - XB\|_F^2$ for fixed Y. Use the normal equations to derive closed-form expressions for X and Y. Show that the closed-form solution for X and the best-fit C' to C are as follows:

$$X = \underbrace{(A^T A)^{-1} A^T}_{\text{Left Inverse}} C \underbrace{B^T (BB^T)^{-1}}_{\text{Right inverse}}, \quad C' = \underbrace{A(A^T A)^{-1} A^T}_{\text{Project columns}} C \underbrace{B^T (BB^T)^{-1} B}_{\text{Project rows}}$$

[Note: Sequential optimization of variables (like Y and X) is suboptimal in general, but it works in this case.]

50. **Challenge Problem:** Let A be an $n \times m$ matrix and B be a $k \times d$ matrix. You want to find the $m \times k$ matrix X so that $C = AXB$, where C is a known $n \times d$ matrix. Nothing is known about the linear independence of rows or columns of A, B, and C. Propose a variation of the Gaussian elimination method to solve the system of equations $C = AXB$. How can you recognize inconsistent systems of equations or systems with an infinite number of solutions? [Note: Closed-form solution in Exercise 23 of Chapter 4.]

51. Use the limit-based definition of the Moore-Penrose pseudoinverse to show that $A^T AA^+ = A^T$ and $B^+ BB^T = B^T$. [Note: Proofs based on QR/SVD are simple.]

52. We know that the best-fit solution to $A\overline{x} = \overline{b}$ is given by $\overline{x}^* = A^+ \overline{b}$. For inconsistent systems, we have $A\overline{x}^* = AA^+ \overline{b} \neq \overline{b}$. Use the limit-based definition of A^+ to show that the matrix AA^+ is both symmetric and idempotent (which is an alternative definition of a projection matrix). What type of projection does AA^+ perform here?

Chapter 3

Eigenvectors and Diagonalizable Matrices

"Mathematics is the art of giving the same name to different things." – Henri Poincare

3.1 Introduction

Any square matrix A of size $d \times d$ can be considered a linear operator, which maps the d-dimensional column vector $\overline{x}$ to the d-dimensional vector $A\overline{x}$. A linear transformation $A\overline{x}$ is a combination of operations such as rotations, reflections, and scalings of a vector $\overline{x}$.

A diagonalizable matrix is a special type of linear operator that only corresponds to a simultaneous scaling along d different directions. These d different directions are referred to as *eigenvectors* and the d scale factors are referred to as *eigenvalues*. All such matrices can be decomposed using an invertible $d \times d$ matrix V and a diagonal $d \times d$ matrix Δ:

$$A = V\Delta V^{-1}$$

The columns of V contain d eigenvectors and the diagonal entries of Δ contain the eigenvalues. For any $\overline{x} \in \mathcal{R}^d$, one can geometrically interpret $A\overline{x}$ using the decomposition in terms of a sequence of three transformations: (i) Multiplication of $\overline{x}$ with V^{-1} computes the coordinates of $\overline{x}$ in a (possibly non-orthogonal) basis system corresponding to the columns (eigenvectors) of V, (ii) multiplication of $V^{-1}\overline{x}$ with Δ to create $\Delta V^{-1}\overline{x}$ dilates these coordinates with scale factors in Δ in the eigenvector directions, and (iii) final multiplication with V to create $V\Delta V^{-1}\overline{x}$ transforms the coordinates back to the original basis system (i.e., the standard basis). The overall result is an anisotropic scaling in d eigenvector directions. Linear transformations that can be represented in this way correspond to *diagonalizable* matrices. *A $d \times d$ diagonalizable matrix represents a linear transformation corresponding to anisotropic scaling in d linearly independent directions.*

When the columns of matrix V are orthonormal vectors, we have $V^{-1} = V^T$. In such a case, the scaling is done along mutually orthogonal directions, and the matrix A is always

© Springer Nature Switzerland AG 2020
C. C. Aggarwal, *Linear Algebra and Optimization for Machine Learning*,
https://doi.org/10.1007/978-3-030-40344-7_3

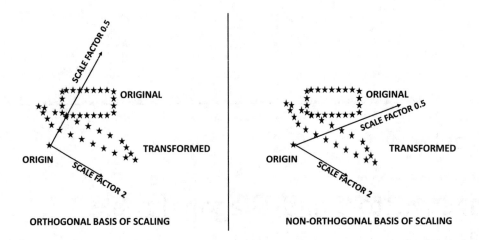

Figure 3.1: Examples of transformations caused by diagonal matrices. The transformation on the left occurs when the matrix A is symmetric

symmetric. This is because we have $A^T = V\Delta^T V^T = V\Delta V^T = A$. The two cases of anisotropic scaling with orthogonal basis systems and non-orthogonal basis systems are shown in Figure 3.1. Here, the scale factors in the two directions are 0.5 and 1, which correspond to contraction and dilation, respectively.

This chapter studies the properties of eigenvectors, diagonalizable matrices, and their applications. The concept of determinant is introduced in Section 3.2. The concepts of diagonalization, eigenvectors, and eigenvalues are discussed in Section 3.3. The special case of symmetric matrices is also discussed in this section. Machine learning applications and examples of symmetric matrices are given in Section 3.4. Numerical algorithms for finding eigenvectors and eigenvalues of diagonalizable matrices are discussed in Section 3.5. A summary is given in Section 3.6.

3.2 Determinants

Imagine a scatter plot of n coordinate vectors $\overline{x}_1 \ldots \overline{x}_n \in \mathcal{R}^d$, which corresponds to the outline of a d-dimensional object. Multiplying these vectors with a $d \times d$ matrix A to create the vectors $A\overline{x}_1 \ldots A\overline{x}_n$ will result in a distortion of the object. When the matrix A is diagonalizable, this distortion is fully described by anisotropic scaling, which affects the "volume" of the object. How can one determine the scale factors of the transformation implied by multiplication with a matrix? To do so, one must first obtain some notion of the effect of a linear transformation on the volume of an object. This is achieved by the notion of the *determinant* of a square matrix, which can be viewed as a quantification of its "volume." A rather loose but intuitive definition of the determinant is as follows:

Definition 3.2.1 (Determinant: Geometric View) *The determinant of a $d \times d$ matrix is the (signed) volume of the d-dimensional parallelepiped defined by its row (or column) vectors.*

The determinant of a matrix A is denoted by $\det(A)$. The above definition is self-consistent because the volume defined by the row vectors and the volume defined by the column vectors of a square matrix can be mathematically shown to be the same. This definition is, however,

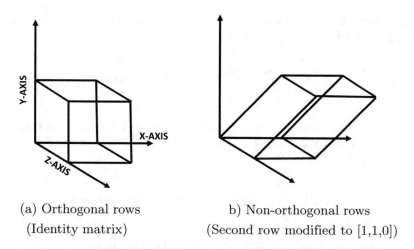

(a) Orthogonal rows
(Identity matrix)

b) Non-orthogonal rows
(Second row modified to [1,1,0])

Figure 3.2: Parallelepipeds before and after a row operation on the 3×3 identity matrix

incomplete because it does not define the sign of $\det(A)$. The sign of the determinant tells us about the effect of multiplication by A on the *orientation* of the basis system. For example, a Householder reflection matrix always has a determinant of -1 because it changes the orientation of the vectors it transforms. It is noteworthy that multiplying an $n \times 2$ data matrix containing the 2-dimensional scatter plot of a right hand (in its rows) with a 2×2 reflection matrix will change the scatter plot to that of a left hand. The sign of the determinant keeps track of this orientation effect of the linear transformation. The geometric view of useful because it provides us an intuitive idea of what the determinant actually computes in terms of absolute values. Consider the following two matrices:

$$A = \begin{bmatrix} 1 & 0 & 0 \\ 0 & 1 & 0 \\ 0 & 0 & 1 \end{bmatrix}, \quad B = \begin{bmatrix} 1 & 0 & 0 \\ 1 & 1 & 0 \\ 0 & 0 & 1 \end{bmatrix} \tag{3.1}$$

The parallelepipeds formed by the rows of each matrix are shown in Figure 3.2(a) and (b), respectively. The determinant of both matrices can be shown to be 1, and both parallelepipeds have a base area of 1 and a height of 1. The first of these matrices is simply the identity matrix, which is an orthogonal matrix. An orthogonal matrix always forms a unit hypercube, and so *the absolute value of its determinant is always 1.*

A matrix needs to be non-singular (i.e., invertible) in order for the determinant to be non-zero. For example, if we have a 3×3 matrix that has a rank of 2, then all three row vectors must lie on a 2-dimensional plane. Therefore, the parallelepiped formed by these three row vectors cannot have a non-zero 3-dimensional volume. The determinant of the $d \times d$ matrix A can also be defined in terms of $(d-1) \times (d-1)$ submatrices of A:

Definition 3.2.2 (Determinant: Recursive View) *Let $A = [a_{ij}]$ be a $d \times d$ matrix and A_{ij} be the $(d-1) \times (d-1)$ matrix formed by dropping the ith row and jth column, while maintaining the relative ordering of retained rows and columns. The determinant $det(A)$ is recursively defined as follows:*

1. If A is a 1×1 matrix, its determinant is equal to the single scalar inside it.

2. *If A is larger than a 1×1 matrix, its determinant is given by the following expression for any fixed value of $j \in \{1 \ldots d\}$:*

$$det(A) = \sum_{i=1}^{d}(-1)^{(i+j)} a_{ij} det(A_{ij}) \quad \text{[Fixed column } j] \tag{3.2}$$

The above computation fixes a column j, and then expands using all the elements of that column. Any choice of j will yield the same determinant. It is also possible to fix a row i and expand along that row:

$$det(A) = \sum_{j=1}^{d}(-1)^{(i+j)} a_{ij} det(A_{ij}) \quad \text{[Fixed row } i] \tag{3.3}$$

The recursive definition implies that some matrices have easily computable determinants:

- **Diagonal matrix:** The determinant of a diagonal matrix is the product of its diagonal entries.

- **Triangular matrix:** The determinant of a triangular matrix is the product of its diagonal entries.

- A matrix containing a row (or column) of 0s will have a determinant of 0.

Consider the following matrix:

$$A = \begin{bmatrix} a & b \\ c & d \end{bmatrix} \tag{3.4}$$

Using the above rule, the determinant of A can be shown to be $ad - bc$ by expanding along the first column. Now, let us consider the slightly larger matrix:

$$A = \begin{bmatrix} a & b & c \\ d & e & f \\ g & h & i \end{bmatrix} \tag{3.5}$$

In this case, we can expand along the first column to obtain the following:

$$det(A) = a \cdot det \begin{bmatrix} e & f \\ h & i \end{bmatrix} - d \cdot det \begin{bmatrix} b & c \\ h & i \end{bmatrix} + g \cdot det \begin{bmatrix} b & c \\ e & f \end{bmatrix}$$

$$= a(ei - hf) - d(bi - hc) + g(bf - ec)$$

$$= aei - ahf - dbi + dhc + gbf - gec$$

An immediate observation is that the determinant contains $3! = 6$ terms, which is the number of possible ways in which three elements can be permuted. In fact, this perspective provides a permutation-centric definition of the determinant, which is also referred to as the *Leibniz formula*:

Definition 3.2.3 (Determinant: Explicit Formula) *Consider a $d \times d$ matrix $A = [a_{ij}]$, and let Σ be the set of all $d!$ permutations of $\{1 \ldots d\}$. In other words, for each $\sigma = \sigma_1 \sigma_2 \ldots \sigma_d \in \Sigma$, the value of σ_i is a permuted integer from $\{1 \ldots d\}$. The sign value (denoted by $sgn(\sigma)$) of a permutation $\sigma \in \Sigma$ is $+1$, if the permutation can be reached from $\{1 \ldots d\}$*

with an even number of element interchanges and it is −1 otherwise. Then, the determinant of A is defined as follows:

$$det(A) = \sum_{\sigma \in \Sigma} \left(sgn(\sigma) \prod_{i=1}^{d} a_{i\sigma_i} \right) \tag{3.6}$$

The permutation-centric definition of a determinant is the most direct one, although it is difficult to use computationally, and it is not particularly intuitive.

Problem 3.2.1 *Suppose that you have a d × d matrix A, which is not invertible. Provide an informal argument with the geometric view of determinants, as to why addition of i.i.d. Gaussian noise with variance λ to each entry of the matrix A will almost certainly make it invertible.*

Useful Properties of Determinants

The recursive and geometric definitions of the determinant imply the following properties:

1. Switching two rows (or columns) of a matrix A flips the sign of the determinant.

2. The determinant of a matrix is the same as that of its transpose.

$$det(A) = det(A^T) \tag{3.7}$$

3. A matrix with two identical rows has a determinant of 0. This also means that adding or subtracting the multiple of row j of the matrix from row i and using the result to replace row i does not change the determinant. Note that we are "shearing" the parallelepiped in the 2-dimensional plane defined by rows i and j (as in Figure 3.2) without changing its volume.

4. Multiplying a single row of the matrix A with c to create the new matrix A' results in multiplication of the determinant of A by a factor of c (because we are scaling the volume of the matrix parallelepiped by c).

$$det(A') = c \cdot det(A) \tag{3.8}$$

A natural corollary of the above result is that multiplying the entire $d \times d$ matrix by c scales its determinant by c^d.

5. The determinant of a matrix A is non-zero only if the matrix is non-singular (i.e., invertible). Geometrically, a parallelepiped of linearly dependent vectors lies in a lower dimensional plane with zero volume.

These results can also be used to derive an important *product-wise property* of determinants.

Lemma 3.2.1 *The determinant of the product of two matrices A and B is the product of their determinants:*

$$det(AB) = det(A) \cdot det(B) \tag{3.9}$$

Proof: Consider two matrices A and B. One can apply the same elementary row addition and interchange operations on A and AB to create matrices A' and $[AB]'$ while maintaining $A'B = [AB]'$. Furthermore, one can apply the same elementary column operations on B and

AB to create matrices B' and $[AB]'$ while maintaining $AB' = [AB]'$. Performing a row addition operation on A or a column addition operation on B has no effect on $\det(A) \cdot \det(B)$, and there is also no effect on $\det(AB)$ when the same row/column operation is performed on AB. Performing a row interchange on A or a column interchange on B has the same negation effect on $\det(A) \cdot \det(B)$ as on $\det(AB)$ when the same operation is performed on AB. By using row addition/interchange operations on A and column addition/interchange operations on B, one can obtain upper-triangular matrices A' and B' (see Chapter 2). Note that $A'B'$ is also upper-triangular since the product of two upper-triangular matrices is upper triangular. Furthermore, each diagonal entry of $A'B'$ is the product of the corresponding diagonal entries of A' and B'. Since the determinant of an upper-triangular matrix is equal to the product of its diagonal entries, it is easy to show that the product of the determinants of A' and B' is equal to the determinant of $A'B'$. The same result, therefore, holds for A, B, and AB, since the sequence of row and column operations to obtain $A'B'$ from AB is the same as the concatenation of the sequence of row operations on A and column operations on B to obtain A' and B', respectively. As we have already discussed, each of these operations has the same effect on $\det(A) \cdot \det(B)$ as on $\det(AB)$. The result follows.   ∎

A corollary of this result is that the determinant of the inverse of a matrix is the inverse of its determinant:

$$\det(A^{-1}) = \frac{\det(I)}{\det(A)} = \frac{1}{\det(A)} \tag{3.10}$$

The product-wise property of determinants can be geometrically interpreted in terms of parallelepiped volumes:

1. Multiplying matrix A with matrix B (in any order) always scales up the (parallelepiped) volume of B with the volume of A. Therefore, even though $AB \neq BA$ (in general), their volumes are always the same.

2. Multiplying matrix A with a diagonal matrix with values $\lambda_1 \ldots \lambda_d$ along the diagonal scales up the volume of A with $\lambda_1 \lambda_2 \ldots \lambda_d$. This is not particularly surprising because we are stretching the axes with these factors, which explains the nature of the scaling of the volume of the underlying parallelepiped.

3. Multiplying A with a rotation matrix simply rotates the parallelepiped, and it does not change the determinant of the matrix.

4. Reflecting a parallelepiped to its mirror image changes its sign without changing its volume. The sign of the determinant tells us a key fact about the orientation of the data created using multiplicative transformation with A. For example, consider an $n \times 2$ data set D containing the 2-dimensional scatter plot of a right hand in its rows. A negative determinant of a 2×2 matrix A means that multiplicative transformation of the $n \times 2$ data set D with A will result in a scatter plot of a right hand in D changing into that of a (possibly stretched and rotated) left hand in DA.

5. Since all linear transformations are combinations of rotations, reflections, and scaling (see Chapter 7), one can compute the absolute effect of a linear transformation on the determinant by focusing only on the scaling portions of the transformation.

The product-wise property of determinants is particularly useful for matrices with special structure. For example, an orthogonal matrix satisfies $A^T A = I$, and therefore we have $\det(A)\det(A^T) = \det(I) = 1$. Since the determinants of A and A^T are equal, it follows that the *square* of the determinant of A is 1.

Lemma 3.2.2 *The determinant of an orthogonal matrix is either $+1$ or -1.*

One can use this result to simplify the determinant computation of a matrix with various types of decompositions containing orthogonal matrices.

Problem 3.2.2 *Consider a $d \times d$ matrix A that is decomposed into the form $A = Q\Sigma P^T$, where Q and P are $d \times d$ orthonormal matrices, and Σ is a $d \times d$ diagonal matrix containing the* **nonnegative** *values $\sigma_1 \ldots \sigma_d$. What is the* **absolute** *value of the determinant of A. Can the sign of the determinant be negative? Why or why not? Does the answer to any of the questions change when $Q = P$?*

Problem 3.2.3 (Restricted Affine Property of Determinants) *Consider two matrices A and B, which differ in* **exactly one row** *(say, the ith row). Show that for any scalar λ, we have $det(\lambda A + [1 - \lambda]B) = \lambda det(A) + [1 - \lambda]det(B)$.*

A hint for solving the above problem is to use the recursive definition of determinants.

Problem 3.2.4 *Work out the determinants of all the elementary row operator matrices introduced in Chapter 1.*

Problem 3.2.5 *How can one compute the determinant from the QR decomposition or the LU decomposition of a square matrix.*

Problem 3.2.6 *Consider a $d \times d$ square matrix A such that $A = -A^T$. Use the properties of determinants to show that if d is odd, then the matrix is singular.*

Problem 3.2.7 *Suppose that you have a $d \times d$ matrix in which the absolute value of every entry is no greater than 1. Show that the absolute value of the determinant is no greater than $(d)^{d/2}$. Provide an example of a 2×2 matrix in which the determinant is equal to this upper bound. [Hint: Think about the geometric view of determinants.]*

3.3 Diagonalizable Transformations and Eigenvectors

We will first define the notion of eigenvectors formally:

Definition 3.3.1 (Eigenvectors and Eigenvalues) *A d-dimensional column vector $\overline{x}$ is said to be an eigenvector of $d \times d$ matrix A, if the following relationship is satisfied for some scalar λ:*

$$A\overline{x} = \lambda\overline{x} \tag{3.11}$$

The scalar λ is referred to as its eigenvalue.

An eigenvector can be viewed as "stretching direction" of the matrix, where multiplying the vector with the matrix simply stretches the former. For example, the vectors $[1, 1]^T$ and $[1, -1]^T$ are eigenvectors of the following matrix with eigenvalues 3 and -1, respectively:

$$\begin{bmatrix} 1 & 2 \\ 2 & 1 \end{bmatrix}\begin{bmatrix} 1 \\ 1 \end{bmatrix} = 3\begin{bmatrix} 1 \\ 1 \end{bmatrix}, \quad \begin{bmatrix} 1 & 2 \\ 2 & 1 \end{bmatrix}\begin{bmatrix} 1 \\ -1 \end{bmatrix} = -1\begin{bmatrix} 1 \\ -1 \end{bmatrix}$$

Each member of the standard basis is an eigenvector of the diagonal matrix, with eigenvalue equal to the ith diagonal entry. All vectors are eigenvectors of the identity matrix.

The number of eigenvectors of a $d \times d$ matrix A may vary, but only diagonalizable matrices represent anisotropic scaling in d linearly independent directions; therefore, *we need to be able to find d linearly independent eigenvectors.* Let $\bar{v}_1 \ldots \bar{v}_d$ be d linearly independent eigenvectors and $\lambda_1 \ldots \lambda_d$ be the corresponding eigenvalues. Therefore, the eigenvector condition holds in each case:

$$A\bar{v}_i = \lambda_i \bar{v}_i, \quad \forall i \in \{1 \ldots d\} \tag{3.12}$$

One can rewrite this condition in matrix form:

$$A[\bar{v}_1 \ldots \bar{v}_d] = [\lambda_1 \bar{v}_1 \ldots \lambda_d \bar{v}_d] \tag{3.13}$$

By defining V to be a $d \times d$ matrix containing $\bar{v}_1 \ldots \bar{v}_d$ in its columns, and Δ to be a diagonal matrix containing $\lambda_1 \ldots \lambda_d$ along the diagonal, one can rewrite Equation 3.13 as follows:

$$AV = V\Delta \tag{3.14}$$

Post-nultiplying with V^{-1}, we obtain the *diagonalization* of the matrix A:

$$A = V\Delta V^{-1} \tag{3.15}$$

Note that V is an *invertible* $d \times d$ matrix containing *linearly independent* eigenvectors, and Δ is a $d \times d$ diagonal matrix, whose diagonal elements contain the eigenvalues of A. The matrix V is also referred to as a *basis change matrix*, because it tells us that the linear transformation A is a diagonal matrix Δ *after changing the basis to the columns of V*.

The determinant of a diagonalizable matrix is defined by the product of its eigenvalues. Since diagonalizable matrices represent linear transforms corresponding to anisotropic scaling in arbitrary directions, a diagonalizable transform should scale up the volume of an object by the product of these scaling factors. It is helpful to think of the matrix A in terms of the transform it performs on the unit parallelepiped corresponding to the orthonormal columns of the identity matrix:

$$A = AI$$

The transformation scales this unit parallelepiped with scaling factors $\lambda_1 \ldots \lambda_d$ in d directions. The ith scaling multiplies the volume of the parallelepiped by λ_i. As a result, the final volume of the parallelepiped defined by the identity matrix (after all the scalings) is the product of $\lambda_1 \ldots \lambda_d$. This intuition provides the following result:

Lemma 3.3.1 *The determinant of a diagonalizable matrix is equal to the product of its eigenvalues.*

Proof: Let A be a $d \times d$ matrix with the following diagonalization:

$$A = V\Delta V^{-1} \tag{3.16}$$

By taking the determinant of both sides, we obtain the following:

$$\det(A) = \det(V\Delta V^{-1}) = \det(V)\det(\Delta)\det(V^{-1}) \quad \text{[Productwise Property]}$$
$$= \det(\Delta) \quad \text{[Since } \det(V^{-1}) = 1/\det(V)]$$

Since the determinant of a diagonal matrix is equal to the product of its diagonal entries, the result follows.   ∎

The presence of a zero eigenvalue implies that the matrix A is singular because its determinant is zero. One can also infer this fact from the observation that the corresponding eigenvector $\bar{v}$ satisfies $A\bar{v} = \bar{0}$. In other words, the matrix A is not of full rank because its null space is nonempty. A nonsingular, diagonalizable matrix can be inverted easily according to the following relationship:

$$(V\Delta V^{-1})^{-1} = V\Delta^{-1}V^{-1} \tag{3.17}$$

Note that Δ^{-1} can be obtained by replacing each eigenvalue in the diagonal of Δ with its reciprocal. Matrices with zero eigenvalues cannot be inverted; the reciprocal of zero is not defined.

Problem 3.3.1 *Let A be a square, diagonalizable matrix. Consider a situation in which we add α to each diagonal entry of A to create A'. Show that A' has the same eigenvectors as A, and its eigenvalues are related to A by a difference of α.*

It is noteworthy that the ith eigenvector $\overline{v}_i$ belongs to the null space of $A - \lambda_i I$ because $(A - \lambda_i I)\overline{v}_i = 0$. In other words, the determinant of $A - \lambda_i I$ must be zero. This polynomial expression that yields the eigenvalue roots is referred to as the *characteristic polynomial* of A.

Definition 3.3.2 (Characteristic Polynomial) *The characteristic polynomial of a $d \times d$ matrix A is the degree-d polynomial in λ obtained by expanding $det(A - \lambda I)$.*

Note that this is a degree-d polynomial, which always has d roots (including repeated or complex roots) according to the *fundamental theorem of algebra*. The d roots of the characteristic polynomial of *any* $d \times d$ matrix are its eigenvalues.

Observation 3.3.1 *The characteristic polynomial $f(\lambda)$ of $d \times d$ matrix A is a polynomial in λ of the following form, where $\lambda_1 \ldots \lambda_d$ are eigenvalues of A:*

$$det(A - \lambda I) = (\lambda_1 - \lambda)(\lambda_2 - \lambda)\ldots(\lambda_d - \lambda) \tag{3.18}$$

Therefore, the eigenvalues and eigenvectors of a matrix A can be computed as follows:

1. The eigenvalues of A can be computed by expanding $det(A - \lambda I)$ as a polynomial expression in λ, setting it to zero, and solving for λ.

2. For each root λ_i of this polynomial, we solve the system of equations $(A - \lambda_i I)\overline{v} = 0$ in order to obtain one or more eigenvectors. The linearly independent eigenvectors with eigenvalue λ_i, therefore, define a basis of the right null space of $(A - \lambda_i I)$.

The characteristic polynomial of the $d \times d$ identity matrix is $(1 - \lambda)^d$. This is consistent with the fact that an identity matrix has d repeated eigenvalues of 1, and every d-dimensional vector is an eigenvector belonging to the null space of $A - \lambda I$. As another example, consider the following matrix:

$$B = \begin{bmatrix} 1 & 2 \\ 2 & 1 \end{bmatrix} \tag{3.19}$$

Then, the matrix $B - \lambda I$ can be written as follows:

$$B - \lambda I = \begin{bmatrix} 1 - \lambda & 2 \\ 2 & 1 - \lambda \end{bmatrix} \tag{3.20}$$

The determinant of the above expression $(1 - \lambda)^2 - 4 = \lambda^2 - 2\lambda - 3$, which is equivalent to $(3 - \lambda)(-1 - \lambda)$. By setting this expression to zero, we obtain eigenvalues of 3 and -1, respectively. The corresponding eigenvectors are $[1, 1]^T$ and $[1, -1]^T$, respectively, which can be obtained from the null-spaces of each $(A - \lambda_i I)$.

We need to diagonalize B as $V \Delta V^{-1}$. The matrix V can be constructed by stacking the eigenvectors in columns. The normalization of columns is not unique, although choosing V to have unit columns (which results in V^{-1} having unit rows) is a common practice. One can then construct the diagonalization $B = V \Delta V^{-1}$ as follows:

$$B = \begin{bmatrix} 1/\sqrt{2} & 1/\sqrt{2} \\ 1/\sqrt{2} & -1/\sqrt{2} \end{bmatrix} \begin{bmatrix} 3 & 0 \\ 0 & -1 \end{bmatrix} \begin{bmatrix} 1/\sqrt{2} & 1/\sqrt{2} \\ 1/\sqrt{2} & -1/\sqrt{2} \end{bmatrix}$$

Problem 3.3.2 *Find the eigenvectors, eigenvalues, and a diagonalization of each of the following matrices:*

$$A = \begin{bmatrix} 1 & 0 \\ -1 & 2 \end{bmatrix}, \quad B = \begin{bmatrix} 1 & 1 \\ -2 & 4 \end{bmatrix}$$

Problem 3.3.3 *Consider a $d \times d$ matrix A such that $A = -A^T$. Show that all non-zero eigenvalues would need to occur in pairs, such that one member of the pair is the negative of the other.*

One can compute a polynomial of a square matrix A in the same way as one computes the polynomial of a scalar — the main differences are that non-zero powers of the scalar are replaced with powers of A and that the scalar term c in the polynomial is replaced by cI. When one computes the characteristic polynomial in terms of its matrix, one always obtains the zero matrix! For example, if the matrix B is substituted in the aforementioned characteristic polynomial $\lambda^2 - 2\lambda - 3$, we obtain the matrix $B^2 - 2B - 3I$:

$$B^2 - 2B - 3I = \begin{bmatrix} 5 & 4 \\ 4 & 5 \end{bmatrix} - 2 \begin{bmatrix} 1 & 2 \\ 2 & 1 \end{bmatrix} - 3 \begin{bmatrix} 1 & 0 \\ 0 & 1 \end{bmatrix} = 0$$

This result is referred to as the *Cayley-Hamilton theorem*, and it is true for all matrices whether they are diagonalizable or not.

Lemma 3.3.2 (Cayley-Hamilton Theorem) *Let A be any matrix with characteristic polynomial $f(\lambda) = det(A - \lambda I)$. Then, $f(A)$ evaluates to the zero matrix.*

The Cayley-Hamilton theorem is true in general for any square matrix A, but it can be proved more easily in some special cases. For example, when A is diagonalizable, it is easy to show the following for any polynomial function $f()$:

$$f(A) = V f(\Delta) V^{-1}$$

Applying a polynomial function to a diagonal matrix is equivalent to applying a polynomial function to each diagonal entry (eigenvalue). Applying the characteristic polynomial to an eigenvalue will yield 0. Therefore, $f(\Delta)$ is a zero matrix, which implies that $f(A)$ is a zero matrix. One interesting consequence of the Cayley-Hamilton theorem is that the inverse of a non-singular matrix can always be expressed as a polynomial of degree $(d - 1)$!

Lemma 3.3.3 (Polynomial Representation of Matrix Inverse) *The inverse of an invertible $d \times d$ matrix A can be expressed as a polynomial of A of degree at most $(d - 1)$.*

Proof: The constant term in the characteristic polynomial is the product of the eigenvalues, *which is non-zero in the case of nonsingular matrices.* Therefore, only in the case of non-singular matrices, we can write the Cayley-Hamilton matrix polynomial $f(A)$ in the form $f(A) = A[g(A)] + cI$ for some scalar constant $c \neq 0$ and matrix polynomial $g(A)$ of degree $(d-1)$. Since the Cayley-Hamilton polynomial $f(A)$ evaluates to zero, we can rearrange the expression above to obtain $A\underbrace{[-g(A)/c]}_{A^{-1}} = I$. ∎

Problem 3.3.4 *Show that any matrix polynomial of a $d \times d$ matrix can always be reduced to a matrix polynomial of degree at most $(d-1)$.*

The above lemma explains why the inverse shows many special properties (e.g., commutativity of multiplication with inverse) shown by matrix polynomials. Similarly, both polynomials and inverses of triangular matrices are triangular. Triangular matrices contain eigenvalues on the main diagonal.

Lemma 3.3.4 *Let A be a $d \times d$ triangular matrix. Then, the entries $\lambda_1 \ldots \lambda_d$ on its main diagonal are its eigenvalues.*

Proof: Since $A - \lambda_i I$ is singular for any eigenvalue λ_i, it follows that at least one of the diagonal values of the triangular matrix $A - \lambda_i I$ must be zero. This can only occur if λ_i is a diagonal entry of A. The converse can be shown similarly. ∎

3.3.1 Complex Eigenvalues

It is possible for the characteristic polynomial of a matrix to have complex roots. In such a case, a real-valued matrix might be diagonalizable with complex eigenvectors/eigenvalues. Consider the case of the rotation transform, which is not diagonalizable with real eigenvalues. After all, it is hard to imagine a real-valued eigenvector that when transformed with a 90° rotation would point in the same direction as the original vector. However, this is indeed possible when working in complex fields! The key point is that multiplication with the imaginary number i rotates a complex vector to an orthogonal orientation. One can verify that the complex vector $\overline{u} = \overline{a} + i\,\overline{b}$ is always orthogonal to the vector $\overline{v} = i[\overline{a} + i\,\overline{b}]$ using the definition of complex inner products (cf. Section 2.11 of Chapter 2).

Consider the following 90° rotation matrix of column vectors:

$$A = \begin{bmatrix} \cos(90) & -\sin(90) \\ \sin(90) & \cos(90) \end{bmatrix} = \begin{bmatrix} 0 & -1 \\ 1 & 0 \end{bmatrix}$$

The characteristic polynomial of A is $(\lambda^2 + 1)$, which does not have any real-valued roots. The two complex roots of the polynomial are $-i$ and i. The corresponding eigenvectors are $[-i, 1]^T$ and $[i, 1]^T$, respectively, and these eigenvectors can be found by solving the linear systems $(A - iI)\overline{x} = 0$ and $(A + iI)\overline{x} = 0$. Solving a system of linear equations on a complex field of coefficients is fundamentally not different from how it is done in the real domain. We verify that the corresponding eigenvectors satisfy the eigenvalue scaling condition:

$$\begin{bmatrix} 0 & -1 \\ 1 & 0 \end{bmatrix} \begin{bmatrix} -i \\ 1 \end{bmatrix} = -i \begin{bmatrix} -i \\ 1 \end{bmatrix}, \quad \begin{bmatrix} 0 & -1 \\ 1 & 0 \end{bmatrix} \begin{bmatrix} i \\ 1 \end{bmatrix} = i \begin{bmatrix} i \\ 1 \end{bmatrix}$$

Each eigenvector is rotated by $90°$ because of multiplication with i or $-i$. One can then put these eigenvectors (after normalization) in the columns of V, and compute the matrix V^{-1}, which is also a complex matrix. The resulting diagonalization of A is as follows:

$$A = V\Delta V^{-1} = \begin{bmatrix} -i/\sqrt{2} & i/\sqrt{2} \\ 1/\sqrt{2} & 1/\sqrt{2} \end{bmatrix} \begin{bmatrix} -i & 0 \\ 0 & i \end{bmatrix} \begin{bmatrix} i/\sqrt{2} & 1/\sqrt{2} \\ -i/\sqrt{2} & 1/\sqrt{2} \end{bmatrix}$$

It is evident that the use of complex numbers greatly extends the family of matrices that can be diagonalized. In fact, one can write the family of 2×2 rotation matrices at an angle θ (in radians) as follows:

$$\begin{bmatrix} \cos(\theta) & -\sin(\theta) \\ \sin(\theta) & \cos(\theta) \end{bmatrix} = \begin{bmatrix} -i/\sqrt{2} & i/\sqrt{2} \\ 1/\sqrt{2} & 1/\sqrt{2} \end{bmatrix} \begin{bmatrix} e^{-i\theta} & 0 \\ 0 & e^{i\theta} \end{bmatrix} \begin{bmatrix} i/\sqrt{2} & 1/\sqrt{2} \\ -i/\sqrt{2} & 1/\sqrt{2} \end{bmatrix} \quad (3.21)$$

From Euler's formula, it is known that $e^{i\theta} = \cos(\theta) + i\sin(\theta)$. It seems geometrically intuitive that multiplying a vector with the mth power of a θ-rotation matrix should rotate the vector m times to create an overall rotation of $m\theta$. The above diagonalization also makes it *algebraically* obvious that the mth power of the θ-rotation matrix yields a rotation of $m\theta$, because the diagonal entries in the mth power become $e^{\pm i\,m\theta}$.

Problem 3.3.5 *Show that all complex eigenvalues of a real matrix must occur in conjugate pairs of the form $a + bi$ and $a - bi$. Also show that the corresponding eigenvectors also occur in similar pairs $\overline{p} + i\overline{q}$ and $\overline{p} - i\overline{q}$.*

3.3.2   Left Eigenvectors and Right Eigenvectors

Throughout this book, we have defined an eigenvector as a *column* vector satisfying $A\overline{x} = \lambda\overline{x}$ for some scalar λ. Such an eigenvector is a right eigenvector because $\overline{x}$ occurs on the right side of the product $A\overline{x}$. When a vector is referred to as an "eigenvector" without any mention of "right" or "left," it refers to a right eigenvector by default.

A left eigenvector is a *row* vector $\overline{y}$, such that $\overline{y}A = \lambda\overline{y}$ for some scalar λ. It is necessary for $\overline{y}$ to be a row vector for $\overline{y}$ to occur on the left-hand side of the product $\overline{y}A$. It is noteworthy that (the transposed representation of) a right eigenvector of a matrix need not be a left eigenvector and vice versa, unless the matrix A is symmetric. If the matrix A is symmetric, then the left and right eigenvectors are transpositions of one another.

Lemma 3.3.5 *If a matrix A is symmetric then each of its left eigenvectors is a right eigenvector after transposing the row vector into a column vector. Similarly, transposing each right eigenvector results in a row vector that is a left eigenvector.*

Proof: Let $\overline{y}$ be a left eigenvector. Then, we have $(\overline{y}A)^T = \lambda\overline{y}^T$. The left-hand side can be simplified to $A^T\overline{y}^T = A\overline{y}^T$. Re-writing with the simplified left-hand side, we have the following:

$$A\overline{y}^T = \lambda\overline{y}^T \quad (3.22)$$

Therefore, $\overline{y}^T$ is a right eigenvector of A. A similar approach can be used to show that each right eigenvector is a left eigenvector after transposition. ∎
This relationship between left and right eigenvectors holds only for symmetric matrices. How about the eigenvalues? It turns out that the left eigenvalues and right eigenvalues are the same irrespective of whether or not the matrix is symmetric. This is because the characteristic polynomial in both cases is $\det(A - \lambda I) = \det(A^T - \lambda I)$.

Consider a diagonalizable $d \times d$ matrix A, which can be converted to its diagonalized matrix Δ as follows:

$$A = V\Delta V^{-1} \tag{3.23}$$

In this case, the right eigenvectors are the d columns of the $d \times d$ matrix V. However, the left eigenvectors are the *rows* of the matrix V^{-1}. This is because the left eigenvectors of A are the right eigenvectors of A^T after transposition. Transposing A yields the following;

$$A^T = (V\Delta V^{-1})^T = (V^{-1})^T \Delta V^T$$

In other words, the right eigenvectors of A^T are the columns of $(V^{-1})^T$, which are the transposed rows of V^{-1}.

Problem 3.3.6 *The right eigenvectors of a diagonalizable matrix $A = V\Delta V^{-1}$ are columns of V, whereas the left eigenvectors are rows of V^{-1}. Use this fact to infer the relationships between left and right eigenvectors of a diagonalizable matrix.*

3.3.3 Existence and Uniqueness of Diagonalization

The characteristic polynomial provides insights into the existence and uniqueness of a diagonalization. In this section, we assume that complex-valued diagonalization is allowed, although the original matrix is assumed to be real-valued. In order to perform the diagonalization, we need d linearly independent eigenvectors. We can then put the d linearly independent eigenvectors in the columns of matrix V and the eigenvalues along the diagonal of Δ to perform the diagonalization $V\Delta V^{-1}$. First, we note that the characteristic polynomial has at least one distinct root (which is possibly complex), and the minimum number of roots occurs when the same root is repeated d times. Given a root λ, the matrix $A - \lambda I$ is singular, since its determinant is 0. Therefore, we can find the vector $\overline{x}$ in the null space of $(A - \lambda I)$. Since this vector satisfies $(A - \lambda I)\overline{x} = 0$, it follows that it is an eigenvector. We summarize this result:

Observation 3.3.2 *A well-defined procedure exists for finding an eigenvector from each* **distinct** *root of the characteristic polynomial. Since the characteristic polynomial has at least one (possibly complex) root, every real matrix has at least one (possibly complex) eigenvector.*

Note that we *might* be able to find more than one eigenvector for an eigenvalue when the root is repeated, which is a key deciding factor in whether or not the matrix is diagonalizable. First, we show the important result that the eigenvectors belonging to distinct eigenvalues are linearly independent.

Lemma 3.3.6 *The eigenvectors belonging to distinct eigenvalues are linearly independent.*

Proof Sketch: Consider a situation where the characteristic polynomial of a $d \times d$ matrix A has $k \leq d$ distinct roots $\lambda_1 \ldots \lambda_k$. Let $\overline{v}_1 \ldots \overline{v}_k$ represent eigenvectors belonging to these eigenvalues.

Suppose that the eigenvectors are linearly dependent, and therefore we have $\sum_{i=1}^{k} \alpha_i \overline{v}_i = 0$ for scalars $\alpha_1 \ldots \alpha_k$ (at least some of which must be non-zero). One can then pre-multiply the vector $\sum_{i=1}^{k} \alpha_i \overline{v}_i$ with the matrix $(A - \lambda_2 I)(A - \lambda_3 I) \ldots (A - \lambda_k I)$ in order to obtain the following:

$$\alpha_1 [\prod_{i=2}^{k}(\lambda_1 - \lambda_i)]\overline{v}_1 = 0$$

Since the eigenvalues are distinct, it follows that $\alpha_1 = 0$. One can similarly show that each of $\alpha_2 \ldots \alpha_k$ is zero. Therefore, we obtain a contradiction to our linear dependence assumption. ∎

In the special case that the matrix A has d distinct eigenvalues, one can construct an invertible matrix V from the eigenvectors. This makes the matrix A diagonalizable.

Lemma 3.3.7 *When the roots of the characteristic polynomial are distinct, one can find d linearly independent eigenvectors. Therefore, a (possibly complex-valued) diagonalization $A = V \Delta V^{-1}$ of a real-valued matrix A with d distinct roots always exists.*

In the case that the characteristic polynomial has distinct roots, one can not only show existence of a diagonalization, but we can also show that the diagonalization can be performed in an almost unique way (with possibly complex eigenvectors and eigenvalues). We use the word "almost" because one can multiply any eigenvector with any scalar, and it still remains an eigenvector with the same eigenvalue. If we scale the ith column of V by c, we can scale the ith row of V^{-1} by $1/c$ without affecting the result. Finally, one can shuffle the order of left/right eigenvectors in V^{-1}, V and eigenvalues in Δ in the same way without affecting the product. By imposing a non-increasing eigenvector order, and a normalization and sign convention on the diagonalization (such as allowing only unit normalized eigenvectors in which the first non-zero component is positive), one can obtain a unique diagonalization.

On the other hand, if the characteristic polynomial is of the form $\prod_i (\lambda_i - \lambda)^{r_i}$, where at least one r_i is strictly greater than 1, the roots are not distinct. In such a case, the solution to $(A - \lambda_i I)\overline{x} = 0$ might be a vector space with dimensionality less than r_i. As a result, we may or may not be able to find the full set of d eigenvectors required to create the matrix V for diagonalization.

The *algebraic multiplicity* of an eigenvalue λ_i is the number of times $(A - \lambda_i I)$ occurs as a factor in the characteristic polynomial. For example, if A is a $d \times d$ matrix, its characteristic polynomial always contains d factors (including repetitions and complex-valued factors). We have already shown that an algebraic multiplicity of 1 for each eigenvalue is the simple case where a diagonalization exists. In the case where the algebraic multiplicities of some eigenvalues are strictly greater than 1, one of the following will occur:

- Exactly r_i linearly independent eigenvectors exist for each eigenvalue with algebraic multiplicity r_i. Any linear combination of these eigenvectors is also an eigenvector. In other words, a vector space of eigenvectors exists with rank r_i, and any basis of this vector space is a valid set of eigenvectors. Such a vector space corresponding to a specific eigenvalue is referred to as an *eigenspace*. In this case, one can perform the diagonalization $A = V \Delta V^{-1}$ by choosing the columns of V in an infinite number of possible ways as the basis vectors of all the underlying eigenspaces.

- If less that r_i eigenvectors exist for an eigenvalue with algebraic multiplicity r_i, a diagonalization does not exist. The closest we can get to a diagonalization is the *Jordan normal form* (see Section 3.3.4). Such a matrix is said to be *defective*.

In the first case above, it is no longer possible to have a unique diagonalization even after imposing a normalization and sign convention on the eigenvectors.

For an eigenvalue λ_i with algebraic multiplicity r_i, the system of equations $(A - \lambda_i I)\overline{x} = 0$ might have as many as r_i solutions. When we have two or more distinct eigenvectors (e.g., $\overline{v}_1$ and $\overline{v}_2$) for the same eigenvalue, any linear combination $\alpha \overline{v}_1 + \beta \overline{v}_2$ will also be an eigenvector for all scalars α and β. Therefore, for creating a diagonalization $A = V \Delta V^{-1}$, one can construct the columns of V in an infinite number of possible ways. The best example of this situation is the identity matrix in which any unit vector is an eigenvector with eigenvalue 1. One can "diagonalize" the (already diagonal) identity matrix I in an infinite number of possible ways $I = V\Delta V^{-1}$, where Δ is identical to I and V is any invertible matrix.

Repeated eigenvalues also create the possibility that a diagonalization might not exist. This occurs when the number of linearly independent eigenvectors for an eigenvalue is less than its algebraic multiplicity. Even though the characteristic polynomial has d roots (including repetitions), one might have fewer than d eigenvectors. In such a case, the matrix is not diagonalizable. Consider the following matrix A:

$$A = \begin{bmatrix} 1 & 1 \\ 0 & 1 \end{bmatrix} \tag{3.24}$$

The characteristic polynomial is $(1 - \lambda)^2$. Therefore, we obtain a single eigenvalue of $\lambda = 1$ with algebraic multiplicity of 2. However, the matrix $(A - \lambda I)$ has rank 1, and we obtain only a single eigenvector $[1, 0]^T$. Therefore, this matrix is not diagonalizable. *Matrices containing repeated eigenvalues and missing eigenvectors of the repeated eigenvalues are not diagonalizable.* The number of eigenvectors of an eigenvalue is referred to as its *geometric multiplicity*, which is at least 1 and at most the algebraic multiplicity.

3.3.4 Existence and Uniqueness of Triangulization

Where do the "missing eigenvectors" of defective matrices go? Consider an eigenvalue with λ with multiplicity k. The characteristic polynomial only tells us that the null space of $(A - \lambda I)^k$ has dimensionality k, but it does not guarantee this for $(A - \lambda I)$. The key point is that the system of equations $(A - \lambda I)^k \overline{x} = 0$ is guaranteed to have k linearly independent solutions, although the system of equations $(A - \lambda I)\overline{x} = 0$ might have anywhere between 1 and k solutions. Can we somehow use this fact to get something close to a diagonalization?

Let the system of equations $(A - \lambda I)\overline{x} = 0$ have $r < k$ solutions. All the k solutions of $(A - \lambda I)^k \overline{x} = 0$ are *generalized* eigenvectors and $r < k$ of them are *ordinary eigenvectors*. It is possible to decompose the set of k generalized eigenvectors into r *Jordan chains*. The ith Jordan chain contains an ordered sequence of $m(i)$ (generalized) eigenvectors out of the k eigenvectors, so that we have $\sum_{i=1}^{r} m(i) = k$. The sequence of generalized eigenvectors for the ith Jordan chain is denoted by $\overline{v}_1 \ldots \overline{v}_{m(i)}$, so that the first eigenvector $\overline{v}_1$ is an ordinary eigenvector satisfying $A\overline{v}_1 = \lambda\overline{v}_1$, and the remaining satisfy the chain relation $A\overline{v}_j = \lambda\overline{v}_j + \overline{v}_{j-1}$ for $j > 1$. Note that these chain vectors are essentially obtained as $\overline{v}_{m(i)-r} = (A - \lambda I)^r \overline{v}_{m(i)}$ for each r from 1 to $m(i) - 1$. A full proof of the existence of Jordan chains is quite complex, and is omitted.

The matrix V contains the generalized eigenvectors in its columns, with eigenvectors belonging to the same Jordan chain occurring consecutively in the same order as their chain relations, and with the ordinary eigenvector being the leftmost of this group of columns. This matrix V can be used to create the *Jordan normal form*, which "almost" diagonalizes the matrix A with an upper-triangular matrix U:

$$A = VUV^{-1} \tag{3.25}$$

The upper-triangular matrix U is "almost" diagonal, and it contains diagonal entries containing eigenvalues in the same order as the corresponding generalized eigenvectors in V. In addition, at most $(d - 1)$ entries, which are just above the diagonal, can be 0 or 1. An entry just above the diagonal is 0 if and only if the corresponding eigenvector is an ordinary eigenvector, and it is 1, if it is not an ordinary eigenvector. It is not difficult to verify that

$AV = VU$ is the matrix representation of all the eigenvector relations (including chain relations), which implies that $A = VUV^{-1}$. Each entry immediately above the diagonal is referred to as a *super-diagonal entry*. A large matrix may sometimes contain only a small number of repeated eigenvalues, and the number of non-zero entries above the diagonal is always bounded above by these repetitions. Therefore, the Jordan normal form contains a small number of super-diagonal 1s in additional to the non-zero entries on the diagonal. In the special case of diagonalizable matrices, the Jordan normal form is the diagonalization of the matrix.

The existence of the Jordan normal form implies that all square matrices are triangulizable, although it is possible for the eigenvectors and eigenvalues to be complex even for real matrices. The triangulizability of a matrix is not unique. One can create different types of triangulizations by imposing different types of constraints on the basis vectors and the triangular matrix. For example, the Jordan normal form has a special structure of the upper-triangular matrix U, but no special structure on the basis vectors in V. Another form of triangulization is the Schur decomposition in which the basis change matrix P is orthogonal, and the upper-triangular matrix U contains the eigenvalues on the diagonal with no other special properties:

$$A = PUP^T \tag{3.26}$$

A Schur decomposition can be found using iterative QR decomposition, and it is one of the methods used for computing the eigenvalues of a matrix (cf. Section 3.5.1). The Schur decomposition of a symmetric matrix is the same as its diagonalization. This is because if we have $A = A^T$, then we must have $PUP^T = PU^TP^T$, which is the same as saying that $P(U - U^T)P^T = 0$. Since P is non-singular, we must have $U = U^T$. This is possible only when U is diagonal. A (possibly complex-valued) Schur decomposition of a real matrix always exists, although it might not be unique (just as the diagonalization is not unique).

Diagonalizability vs Triangulizibility: A Geometric View

How can one geometrically interpret the Jordan normal form? Note that each entry of 1 on the super-diagonal can be zeroed out by using an elementary row addition operator with the row below it, provided that we perform the elementary row addition operations from bottom to top order on consecutive pairs of rows. As we have already discussed, elementary row addition operations correspond to *shear matrices*. Multiplications with shear matrices cause transformations of the type that change the cube in Figure 3.2(a) to the parallelepiped in Figure 3.2(b). In fact, the transformation that would convert Figure 3.2(a) into Figure 3.2(b) is not a diagonalizable one; it cannot be represented purely as a stretching operation along specific directions, because changing a cube to a non-rectangular parallelepiped requires stretching in arbitrary directions, which would also change the directions of the parallelepiped edges from its axis-parallel orientation. See Figure 3.1 for an example of the effect of arbitrarily oriented scaling on axis-parallel edges. Therefore, additional rotations would be needed for re-alignment. *Non-diagonalizable matrices always contain this type of "residual" rotation.*

Diagonalizable transforms are those in which a (possibly non-orthogonal) basis system exists along which one can scale the space. For non-diagonalizable matrices, scaling alone is not adequate. If we are additionally willing to allow some rotation after the scaling, non-diagonalizable transformations can be represented as well. As discussed in Lemma 7.2.2 of Chapter 7, every square matrix can be decomposed into the product of a diagonalizable matrix and the "residual" rotation matrix. This decomposition is referred to as the *polar decomposition* of a matrix. Note that rotation matrices are also diagonalizable, albeit with complex eigenvalues. Therefore, every real matrix can be expressed as the product of at most two diagonalizable matrices (although one might have complex eigenvalues).

3.3.5 Similar Matrix Families Sharing Eigenvalues

Similar matrices are defined as follows:

Definition 3.3.3 *Two matrices A and B are said to be similar when $B = VAV^{-1}$.*

Similarity is a commutative and transitive property. In other words, if A and B are similar, then B and A are similar as well. Furthermore, if A and B are similar, and if B and C are similar, then A and C are also similar. Therefore, similar matrices form a *family* of related matrices.

What do similar matrices mean? When we have two similar matrices A and B, then multiplying a vector with either A or B results in the same transformation of that vector as long as the basis is appropriately chosen in each case. For example, two similar matrices of size 3×3 might each correspond to a $60°$ rotation of a 3-dimensional vector, but the axis of rotation might be different. Similarly, two similar transforms might scale a vector by the same factors in different directions. One can interpret this point in terms of their Jordan normal forms.

Lemma 3.3.8 (Jordan Normal Forms of Similar Matrices) *Let A and B be two similar matrices satisfying $B = VAV^{-1}$. Then, their Jordan normal forms (with possibly complex eigenvalues) will be related:*

$$A = V_1 U V_1^{-1}, \quad B = V_2 U V_2^{-1}$$

The matrix V_2 is related to V_1 as $V_2 = VV_1$.

The above lemma is easy to show by direct substitution of the Jordan form of A in the relationship $B = VAV^{-1}$. An important consequence of the above result is that *similar matrices are have the same eigenvalues (and their corresponding multiplicities).* Furthermore, if one member of a similar family is diagonalizable, then all members are diagonalizable as well, and a diagonal matrix is included in the family.

As introduced in Chapter 2, the sum of the diagonal entries of a matrix is referred to as its trace. The trace of a matrix A *is equal to the sum of its eigenvalues, whether it is diagonalizable or not.*

Lemma 3.3.9 *The traces of similar matrices are equal, and are equal of the sum of the eigenvalues of that family (whether it is diagonalizable or not).*

Proof: Here, we will use the property of the trace that $\text{tr}(GH) = \text{tr}(HG)$ for square matrices G and H. Let A and B be similar matrices such that $A = VBV^{-1}$. Then, we have the following:

$$\text{tr}(A) = \text{tr}(V[BV^{-1}]) = \text{tr}([BV^{-1}]V) = \text{tr}(B[V^{-1}V]) = \text{tr}(B)$$

Therefore, the traces of similar matrices are equal. This also implies that the trace of a matrix is equal to the trace of the upper-triangular matrix in its Jordan normal form (which is equal to the sum of the eigenvalues of the family). ∎

Similar matrices perform similar operations, but in different basis systems. For example, a similar family of **diagonalizable** matrices performs anisotropic scaling with the same factors, albeit in completely different eigenvector directions.

Problem 3.3.7 (Householder Family) *Show that all Householder reflection matrices are similar, and the family includes the elementary reflection matrix that differs from the identity matrix in one element.*

A hint for solving the above problem is that this matrix is diagonalizable.

Problem 3.3.8 (Projection Family) *Section 2.8.2 introduces the $n \times n$ projection matrix $P = A(A^T A)^{-1} A^T$ for $n \times d$ matrix A with full column rank d and $n > d$. Show that all projection matrices P obtained by varying A (but for particular values of n and d) are similar. What is the trace of P? Provide a geometric interpretation of $(I - P)$ and $(I - 2P)$.*

A hint for solving this problem is to first express the projection matrix in the form QQ^T by using QR decomposition of A, where Q is an orthogonal matrix. Now extract the eigenvectors and eigenvalues of the projection matrix by using the properties of Q, and verify that the eigenvalues are always the same for fixed values of n and d.

Problem 3.3.9 (Givens Family) *Show that all Givens matrices with the same rotation angle α are similar, because for any such pair of Givens matrices G_1 and G_2, one can find a permutation matrix P such that $G_2 = P G_1 P^T$. Now consider an orthogonal matrix Q that is not a permutation matrix. Provide a geometric interpretation of $Q G_1 Q^T$.*

For the reader who is familiar with graph adjacency matrices, we recommend the following exercise (or to return to it after reading Chapter 10):

Problem 3.3.10 (Similarity in Graph Theory) *Consider a graph G_A whose adjacency matrix is A. Show that the adjacency matrix B of the **isomorphic** graph G_B obtained by reordering the vertices of G_A is similar to matrix A. What type of matrix is used for the basis transformation between A and B?*

Geometric Interpretability of Trace

Since the trace of a matrix is invariant to similarity transformations, a natural question arises as to whether it can be interpreted in a geometric way. The interpretation of the trace of a square matrix is not a simple one, especially when the underlying matrix is not symmetric. Fortunately, many of the square matrices encountered in machine learning appear in the form of Gram matrices $A^T A$, where A is either an $n \times d$ data set or its transpose. Examples of such matrices include the *regularized graph adjacency matrix*, the *covariance matrix* and the *dot product similarity matrix*. We make the following observation:

Observation 3.3.3 *The trace of the Gram matrix $A^T A$ is equal to the energy in its base matrix A.*

The above observation follows directly from the definition of energy in Equation 1.23 of Chapter 1. One consequence of the observation is that if we apply an orthonormal similarity transformation AP on a data set contained in the $n \times d$ matrix A, its energy, which is equal to the trace of $P^T (A^T A) P$ does not change. This fact can be used to infer the result that the sum of the variances of all dimensions in a mean-centered data set is always the same, irrespective the choice of basis:

Problem 3.3.11 (Covariance Family) *Let D be a **mean-centered** $n \times d$ data set with n rows and d dimensions, and let P be any $d \times d$ orthogonal matrix. Let DP be the transformed $n \times d$ data set in the new orthogonal basis system. A covariance matrix is a $d \times d$ matrix, in which (i, j)th entry is the covariance between dimensions i and j, with diagonal entries representing variances. Show that all covariance matrices of DP over different choices of orthogonal P are similar and they therefore have the same trace.*

3.3.6 Diagonalizable Matrix Families Sharing Eigenvectors

A diagonalizable matrix family that shares eigenvectors (but not eigenvalues) is referred to as *simultaneously diagonalizable*. This idea is complementary to the notion of diagonalizable, similar matrices that share eigenvalues, but not eigenvectors.

Definition 3.3.4 (Simultaneous Diagonalizability) *Two diagonalizable matrices A and B are said to be simultaneously diagonalizable, if a $d \times d$ invertible matrix V exists, such that the columns of V are the eigenvectors of both A and B. Therefore, we have the following:*

$$A = V \Delta_1 V^T$$
$$B = V \Delta_2 V^T$$

Here, Δ_1 and Δ_2 are diagonal matrices.

The geometric interpretation of simultaneously diagonalizable matrices is that *they perform anisotropic scaling in the same set of directions.* However, the scaling factors might be different, since the diagonal matrices are different. Simultaneous diagonalizability is a property that is closely related to matrix commutativity.

Lemma 3.3.10 *Diagonalizable matrices are also simultaneously diagonalizable if and only if they are commutative.*

Problem 3.3.12 *Let A and B be two diagonalizable matrices that share the same set of eigenvectors. Provide a geometric interpretation of why $AB = BA$.*

Problem 3.3.13 (Givens Commutative Family) *The multiplication of rotation matrices in dimensionalities greater than 2 is not commutative in general. However, the $d \times d$ family of Givens rotation matrices $G_c(i, j, \theta)$ is known to commutative over **fixed** dimension pair i, j and varying θ. Provide a geometric interpretation of this commutativity. Now provide an algebraic interpretation in terms of simultaneous diagonalizability by generalizing Equation 3.21 to $d \times d$ matrices.*

3.3.7 Symmetric Matrices

Symmetric matrices arise repeatedly in machine learning. This is because covariance matrices, dot-product matrices, (undirected) graph adjacency matrices, and similarity (kernel) matrices are used frequently in machine learning. Furthermore, many of the applications associated with such matrices require some type of diagonalization. One of the fundamental properties of symmetric matrices is that they are always diagonalizable, and have orthonormal eigenvectors. This result is referred to as the *spectral theorem*:

Theorem 3.3.1 (Spectral Theorem) *Let A be a $d \times d$ symmetric matrix with real entries. Then, A is always diagonalizable with real eigenvalues and has orthonormal, real-valued eigenvectors. In other words, A can be diagonalized in the form $A = V \Delta V^T$ with orthogonal matrix V.*

Proof: First, we need to show that the eigenvalues of A are real. Let $(\overline{v}, \lambda)$ represents a eigenvector-eigenvalue pair of a *real* matrix. We start with the most general assumption that this pair could be complex. Pre-multiplying the equation $A\overline{v} = \lambda\overline{v}$ with the conjugate transpose $\overline{v}^*$ of $\overline{v}$, we obtain $\overline{v}^* A \overline{v} = \lambda \overline{v}^* \overline{v} = \lambda \|\overline{v}\|^2 = \lambda$. In other words we have $\overline{v}^* A \overline{v} = \lambda$. Taking the conjugate transpose of both sides of this 1×1 "matrix," we obtain:

$$\lambda^* = [\overline{v}^* A \overline{v}]^* = \overline{v}^* A^* [\overline{v}^*]^* = \overline{v}^* A^* \overline{v} = \overline{v}^* A \overline{v} = \lambda$$

We used the real and symmetric nature of A in the above derivation. Therefore, the eigenvalue λ is equal to its conjugate, and it is real. The eigenvector $\overline{v}$ is also real because it belongs to the null space of the real matrix $(A - \lambda I)$.

We claim that eigenvalues with multiplicity greater than 1 do not have missing eigenvectors. If there are missing eigenvectors, two non-zero vectors $\overline{v}_1$ and $\overline{v}_2$ must exist in a Jordan chain such that $A \overline{v}_1 = \lambda \overline{v}_1$ and $A \overline{v}_2 = \lambda \overline{v}_2 + \overline{v}_1$ (see Section 3.3.3). Then, we can show that $(A - \lambda I)^2 \overline{v}_2 = 0$, by successively applying the eigenvector condition. Therefore, $\overline{v}_2^T (A - \lambda I)^2 \overline{v}_2$ is zero as well. At the same time, one can show the contradictory result that this quantity is non-zero by using the symmetric nature of the matrix A:

$$\overline{v}_2^T (A - \lambda I)^2 \overline{v}_2 = [\overline{v}_2^T (A^T - \lambda I^T)][(A - \lambda I)\overline{v}_2] = \|(A - \lambda I)\overline{v}_2\|^2 = \|\overline{v}_1\|^2 \neq 0$$

Therefore, we obtain a contradiction, and A is diagonalizable (with no missing eigenvectors).

Next, we need to show that all eigenvectors are mutually orthogonal. Within the eigenspace of a repeated eigenvalue, we can always choose an orthonormal basis of eigenvectors. Furthermore, two eigenvectors $\overline{v}_1$ and $\overline{v}_2$ belonging to distinct eigenvalues λ_1 and λ_2 are also orthogonal. This is because transposing the scalar $\overline{v}_1^T A \overline{v}_2$ results in the same scalar $\overline{v}_2^T A^T \overline{v}_1 = \overline{v}_2^T A \overline{v}_1$. Using this, we can show the following:

$$\overline{v}_1^T \underbrace{[A \overline{v}_2]}_{\lambda_2 \overline{v}_2} = \overline{v}_2^T \underbrace{[A \overline{v}_1]}_{\lambda_1 \overline{v}_1}$$

$$\lambda_1 (\overline{v}_1 \cdot \overline{v}_2) = \lambda_2 (\overline{v}_1 \cdot \overline{v}_2)$$

$$(\lambda_1 - \lambda_2)(\overline{v}_1 \cdot \overline{v}_2) = 0$$

This is possible only when the dot product of the two eigenvectors is zero. ∎

Since the inverse of an orthogonal matrix is its transpose, it is common to write the diagonalization of symmetric matrices in the form $A = V \Delta V^T$ instead of $A = V \Delta V^{-1}$. Multiplying a data matrix D with a symmetric matrix represents anisotropic scaling of its rows along orthogonal axis directions. An example of such a scaling is illustrated on the left-hand side of Figure 3.1.

The eigenvectors of a symmetric matrix A are not only orthogonal but also *A-orthogonal*.

Definition 3.3.5 (A-Orthogonality) *A set of column vectors $\overline{v}_1 \ldots \overline{v}_d$ is A-orthogonal, if and only if $\overline{v}_i^T A \overline{v}_j = 0$ for all pairs $[i, j]$ with $i \neq j$.*

The notion of A-orthogonality is a generalization of orthogonality, and setting $A = I$ reverts the definition to the usual notion of orthogonality. Note that $\overline{v}_i^T A \overline{v}_j$ is simply a different choice of inner product from the vanilla dot product (cf. Definition 2.10.1).

Lemma 3.3.11 *The eigenvectors of a symmetric $d \times d$ matrix A are A-orthogonal.*

Proof: For any pair of eigenvectors $\overline{v}_i$ and $\overline{v}_j$ with eigenvalues λ_i and λ_j, we have the following:

$$\overline{v}_i^T A \overline{v}_j = \overline{v}_i^T [\lambda_j \overline{v}_j] = \lambda_j \overline{v}_i^T \overline{v}_j = 0$$

The result follows. ∎

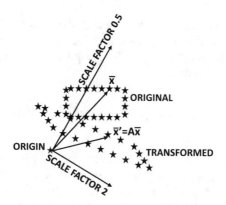

Figure 3.3: Positive semidefinite transforms do not change angular orientations of points by more than 90°

One can use a natural generalization of Gram-Schmidt orthogonalization (cf. Problem 2.7.1) to find A-orthogonal basis sets (which is a more efficient choice than eigenvector computation). In many applications like *conjugate gradient descent*, one is often looking for A-orthogonal directions, where A is the Hessian of the optimization function.

Problem 3.3.14 (Frobenius Norm vs Eigenvalues) *Consider a matrix with real eigenvalues. Show that its squared Frobenius norm is at least equal to the sum of the squares of its eigenvalues, and that strict equality is observed for symmetric matrices. You will find the Schur decomposition helpful.*

3.3.8   Positive Semidefinite Matrices

A symmetric matrix is *positive semidefinite* if and only if all its eigenvalues are non-negative. From a geometric perspective, pre-multiplication of a set of d-dimensional vectors $\overline{x}_1 \ldots \overline{x}_n$ with a $d \times d$ positive semidefinite matrix A to create $A\overline{x}_1 \ldots A\overline{x}_n$ will distort the scatter-plot of the vectors (see Figure 3.3 for $d = 2$), so that the scatter-plot is stretched along all the eigenvector directions with *non-negative* scale factors. For example, the scale factors in Figure 3.3 are 2 and 0.5. The nonnegativity of scale factors ensures that transformed vectors do not have large angles with respect to the original vectors (i.e., angles greater than 90°). The angle between a data vector $\overline{x}$ and its transformed representation $\overline{x}' = A\overline{x}$ is shown in Figure 3.3; this angle is no greater than 90° because of the fact that the scale factors are nonnegative. Since the cosine of any such angle is nonnegative, it follows that the dot product $\overline{x}^T(A\overline{x})$ between any column vector $\overline{x} \in \mathcal{R}^d$ and its transformed representation, $A\overline{x}$, is nonnegative. This observation provides the definition of positive semidefinite matrices:

Definition 3.3.6 (Positive Semidefinite Matrix) *A $d \times d$ symmetric matrix A is positive semidefinite if and only if for any non-zero vector $\overline{x} \in \mathcal{R}^d$, the following is true:*

$$\overline{x}^T A \overline{x} \geq 0 \tag{3.27}$$

Figure 3.3 provides the pictorial intuition as to why Definition 3.3.6 is equivalent to stating that the eigenvalues are nonnegative. In the following, we show this result formally:

Lemma 3.3.12 *Definition 3.3.6 on positive semidefiniteness of a $d \times d$ symmetric matrix A is equivalent to stating that A has nonnegative eigenvalues.*

Proof: According to the spectral theorem, we can always diagonalize a symmetric matrix A as $V\Delta V^T$. Suppose that the eigenvalues $\lambda_1 \ldots \lambda_d$ in Δ are all nonnegative. Then, for any column vector $\overline{x}$, let us denote $\overline{y} = V^T \overline{x}$. Furthermore, let the ith component of $\overline{y}$ be denoted by y_i. Therefore, we have:

$$\overline{x}^T A \overline{x} = \overline{x}^T V \Delta V^T \overline{x} = (V^T \overline{x})^T \Delta (V^T \overline{x}) = \overline{y}^T \Delta \overline{y} = \sum_{i=1}^{d} \lambda_i y_i^2$$

It is clear that the final expression on the right is nonnegative because each λ_i is nonnegative. Therefore, the matrix A is positive semidefinite according to Definition 3.3.6.

To prove the converse, let us assume that A is positive semidefinite according to Definition 3.3.6. Therefore, it is the case that $\overline{x}^T A \overline{x} \geq 0$ for any $\overline{x}$. Then, let us select $\overline{x}$ to be the ith column of V (which is also the ith eigenvector). Then, because of the orthonormality of the columns of V, we have $V^T \overline{x} = \overline{e}_i$, where $\overline{e}_i$ contains a single 1 in the ith position, and 0s in all other positions. As a result, we have the following:

$$\overline{x}^T A \overline{x} = \overline{x}^T V \Delta V^T \overline{x} = (V^T \overline{x})^T \Delta (V^T \overline{x}) = \overline{e}_i^T \Delta \overline{e}_i = \lambda_i$$

Therefore, λ_i needs to be nonnegative because we know that $\overline{x}^T A \overline{x} \geq 0$. The result follows. ∎

A minor variation on the notion of positive semidefinite matrix is that of a *positive definite* matrix, where the matrix A cannot be singular.

Definition 3.3.7 (Positive Definite Matrix) *A $d \times d$ symmetric matrix A is positive definite if and only if for any non-zero vector $\overline{x} \in \mathcal{R}^d$, the following is true:*

$$\overline{x}^T A \overline{x} > 0 \tag{3.28}$$

The eigenvalues of such a matrix need to be strictly positive.

Lemma 3.3.13 *A symmetric matrix $A = V\Delta V^T$ is positive definite, if and only if it has positive eigenvalues.*

Unlike positive semidefinite matrices, positive definite matrices are guaranteed to be invertible. The inverse matrix is simply $V\Delta^{-1}V^T$; here, Δ^{-1} can always be computed because none of the eigenvalues are zero.

One can also define *negative semidefinite matrices* as those matrices in which every eigenvalue is non-positive, and $\overline{x}^T A \overline{x} \leq 0$ for each column vector $\overline{x}$. A negative semidefinite matrix can be converted into a positive semidefinite matrix by reversing the sign of each entry in the matrix. A *negative definite* matrix is one in which every eigenvalue is strictly negative. Symmetric matrices with both positive and negative eigenvalues are said to be *indefinite*.

Any matrix of the form BB^T or $B^T B$ (i.e., Gram matrix form) is always positive semidefinite. The Gram matrix is fundamental to machine learning, and it appears repeatedly in different forms. Note that B need not be a square matrix. This provides yet another definition of positive semidefiniteness.

Lemma 3.3.14 *A $d \times d$ matrix A is positive semi-definite if and only if it can be expressed in the form $B^T B$ for some matrix B.*

Proof: For any non-zero column vector $\overline{x} \in \mathcal{R}^d$, we have:

$$\overline{x}^T B^T B \overline{x} = (B\overline{x})^T (B\overline{x}) = \|B\overline{x}\|^2 \geq 0$$

The result follows.

Conversely, *any* positive semidefinite matrix A can be expressed in the eigendecomposition form $A = Q\Sigma^2 Q^T = (Q\Sigma)(Q\Sigma)^T$. Then, by setting $B = (Q\Sigma)^T$, we obtain the form $A = B^T B$. ∎

Note that we could also have stated this lemma using BB^T instead of $B^T B$, and the proof is similar. We will use the above result extensively for kernel feature engineering in Chapter 9.

Problem 3.3.15 *If C is a positive semidefinite matrix, show that there exists a square-root matrix $\sqrt{C}$ that satisfies the following:*

$$\sqrt{C}\sqrt{C} = C$$

Problem 3.3.16 *If a matrix C is positive definite, then so is C^{-1}.*

A hint for solving the above problems is to examine the eigendecomposition trick used in the proof of Lemma 3.3.14.

3.3.9 Cholesky Factorization: Symmetric LU Decomposition

The fact that positive definite matrices can be symmetrically factorized into Gram matrix form is a useful result for kernel methods in machine learning. The use of eigendecomposition to achieve this goal is a natural choice, but not the only one. Given a factorization the $d \times d$ matrix as $A = BB^T$, one can use any orthogonal $d \times d$ matrix P to create the alternative factorization $A = B(PP^T)B^T = (BP)(BP)^T$. One of these infinite choices of symmetric factorizations of A is one in which B is lower-triangular. In other words, one can express the positive definite matrix A in the form LL^T, where $L = [l_{ij}]$ is some $d \times d$ lower-triangular matrix. This is referred to as the *Cholesky factorization*.

The Cholesky decomposition is a special case of LU decomposition, and it can be used only for positive definite matrices. Although a matrix might have an infinite number of LU decompositions, a positive definite matrix has a unique Cholesky factorization. It is computationally more efficient to compute the Cholesky decomposition for positive definite matrices than the generic LU decomposition.

Let the columns of the matrix $L = [l_{ij}]_{d \times d}$ be denoted by $\overline{l}_1 \ldots \overline{l}_d$. Furthermore, since the matrix $A = [a_{ij}]_{d \times d}$ is symmetric, we will focus only on the lower-triangular entries a_{ij} (with $i \geq j$) to set up a system of equations that can be easily solved using back-substitution. First, note that for any $i \geq j$, we have the following condition:

$$a_{ij} = \underbrace{\sum_{k=1}^{d} l_{ik} l_{jk}}_{A_{ij} = (LL^T)_{ij}} = \underbrace{\sum_{k=1}^{j} l_{ik} l_{jk}}_{\text{Lower-triangular } L}$$

Note that the subscript for k only runs up to j instead of d for lower-triangular matrices and $i \geq j$. This condition easily sets up a simple system of equations for computing the entries in each column of L one-by-one while back substituting the entries already computed, *as long as we do the computations in the correct order.* For example, we can compute the first column of L by setting $j = 1$, and iterating over all $i \geq j$:

$$l_{11} = \sqrt{a_{11}}$$
$$l_{i1} = a_{i1}/l_{11} \quad \forall i > 1$$

We can repeat the same process to compute the second column of L as follows:

$$l_{22} = \sqrt{a_{22} - l_{21}^2}$$
$$l_{i2} = (a_{i2} - l_{i1}l_{21})/l_{22} \quad \forall i > 2$$

A generalized iteration for the jth column yields the pseudocode for Cholesky factorization:

Initialize $L = [0]_{d \times d}$;
for $j = 1$ to d **do**
  $l_{jj} = \sqrt{a_{jj} - \sum_{k=1}^{j-1} l_{jk}^2}$;
  for $i = j + 1$ to d **do**
    $l_{ij} = (a_{ij} - \sum_{k=1}^{j-1} l_{ik}l_{jk})/l_{jj}$;
  endfor
endfor
return $L = [l_{ij}]$;

Each computation of l_{ij} requires $O(d)$ time, and therefore the Cholesky method requires $O(d^3)$ time. The above algorithm works for positive-definite matrices. If the matrix is singular and positive *semi*-definite, then at least one l_{jj} will be 0. This will cause a division by 0 during the computation of l_{ij}, which results in an undefined value. The decomposition is no longer unique, and a Cholesky factorization does not exist in such a case. One possibility is to add a small positive value to each diagonal entry of A to make it positive definite and then restart the factorization. If the matrix A is indefinite or negative semidefinite, it will show up during the computation of at least one l_{jj}, where one will be forced to compute the square-root of a negative quantity. The Cholesky factorization is the preferred approach for testing the positive definiteness of a matrix.

Problem 3.3.17 (Solving a System of Equations) *Show how you can solve the system of equations* $(LL^T)\overline{x} = \overline{b}$ *by successively solving two triangular systems of equations, the first of which is* $L\overline{y} = \overline{b}$. *Use this fact to discuss the utility of Cholesky factorization in certain types of systems of equations. Where does the approach not apply?*

Problem 3.3.18 (Cholesky Factorization from Any Symmetric Factorization) *Suppose that you are already given a symmetric factorization* $B^T B$ *of* $d \times d$ *positive definite matrix* A, *where* B *is a tall matrix with linearly independent columns. Show that the Cholesky factorization of* A *can be extracted by performing the QR-decomposition of* B.

3.4   Machine Learning and Optimization Applications

The linear algebra ideas in this chapter are used frequently in machine learning and optimization. This section will provide an overview of the most important examples, which will be used throughout this book.

3.4.1 Fast Matrix Operations in Machine Learning

Consider a situation, where one wants to compute A^k for some positive integer k. Repeated matrix multiplication can be expensive. Furthermore, there is no way to compute A^k, when k tends to ∞ in the limit. It turns out that diagonalization is very useful, even if it is complex valued. This is because one can express A^k as follows:

$$A^k = V \Delta^k V^{-1} \tag{3.29}$$

Note that it is often easy to compute Δ^k, because we only need to exponentiate the individual entries along the diagonal. By using this approach, one can compute A^k in relatively few operations. As $k \to \infty$, it is often the case that A^k will either vanish to 0 or explode to very large entries depending on whether the largest eigenvalue is less than 1 or whether it is greater than 1. One can easily compute a polynomial function in A by computing a polynomial function in Δ. These types of applications often arise when working with the adjacency matrices of graphs (cf. Chapter 10).

3.4.2 Examples of Diagonalizable Matrices in Machine Learning

There are several positive semidefinite matrices that arise repeatedly in machine learning applications. This section will provide an overview of these matrices.

Dot Product Similarity Matrix

A dot product similarity matrix of an $n \times d$ data matrix D is an $n \times n$ matrix containing the pairwise dot products between the rows of D.

Definition 3.4.1 *Let D be an $n \times d$ data matrix containing d-dimensional points in its rows. Let S be an $n \times n$ similarity matrix between the points, where the (i,j)th entry is the dot product between the ith and jth rows of D. Therefore, the similarity matrix S is related to D as follows:*

$$S = DD^T \tag{3.30}$$

Since the dot product is in the form of a Gram matrix, it is positive semidefinite (cf. Lemma 3.3.14):

Observation 3.4.1 *The dot product similarity matrix of a data set is positive semidefinite.*

A dot product similarity matrix is an alternative way of specifying the data set, because one can recover the data set D from the similarity matrix to within rotations and reflections of the original data set. This is because each computational procedure for performing symmetric factorization $S = D'D'^T$ of the similarity matrix might yield a a different D', which can be viewed as a rotated and reflected version of D. Examples of such computational procedures include eigendecomposition or Cholesky factorization. All the alternatives yield the same dot product. After all, dot products are invariant to axis rotation of the coordinate system. Since machine learning applications are only concerned with the relative positions of points, this type of ambiguous recovery is adequate in most cases. One of the most common methods to "recover" a data matrix from a similarity matrix is to use eigendecomposition:

$$S = Q \Delta Q^T \tag{3.31}$$

The matrix Δ contains only nonnegative eigenvalues of the positive semidefinite similarity matrix, and therefore we can create a new diagonal matrix Σ containing the square-roots of the eigenvalues. Therefore, the similarity matrix S can be written as follows:

$$S = Q\Sigma^2 Q^T = \underbrace{(Q\Sigma)}_{D'}\underbrace{(Q\Sigma)^T}_{D'^T} \tag{3.32}$$

Here, $D' = Q\Sigma$ is an $n \times n$ data set containing n-dimensional representations of the n points. It seems somewhat odd that the new matrix $D' = Q\Sigma$ is an $n \times n$ matrix. After all, if the similarity matrix represents dot products between d-dimensional data points for $d \ll n$, we should expect the recovered matrix D' to be a rotated representation of D in d dimensions. What are the extra $(n - d)$ dimensions? Here, the key point is that if the similarity matrix S was indeed created using dot products on d-dimensional points, then DD^T will also have rank at most d. Therefore, at least $(n - d)$ eigenvalues in Δ will be zeros, which correspond to dummy coordinates.

But what if we did not use dot product similarity to calculate S from D? What if we used some other similarity function? It turns out that this idea is the essence of *kernel methods* in machine learning (cf. Chapter 9). Instead of using the dot product $\overline{x} \cdot \overline{y}$ between two points, one often uses similarity functions such as the following:

$$\text{Similarity}(\overline{x}, \overline{y}) = \exp(-\|\overline{x} - \overline{y}\|^2 / \sigma^2) \tag{3.33}$$

Here, σ is a parameter that controls the sensitivity of the similarity function to distances between points. Such a similarity function is referred to as a *Gaussian kernel*. If we use a similarity function like this instead of the dot product, we might recover a data set that is different from the original data set from which the similarity was constructed. In fact this recovered data set may not have dummy coordinates, and all $n > d$ dimensions might be relevant. Furthermore, the recovered representations $Q\Sigma$ from such similarity functions might yield better results for machine learning applications than the original data set. This type of fundamental transformation of the data to a new representation is referred to as *nonlinear feature engineering*, and it goes beyond the natural (linear) transformations like rotation that are common in linear algebra. In fact, it is even possible to extract multidimensional representations from data sets of *arbitrary* objects between which only similarity is specified. For example, if we have a set of n graph or time-series objects, and we only have the $n \times n$ similarity matrix of these objects (and no multidimensional representation), we can use the aforementioned approach to create a multidimensional representation of each object for off-the-shelf learning algorithms.

Problem 3.4.1 *Suppose you were given a similarity matrix S that was constructed using some arbitrary heuristic (rather than dot products) on a set of n arbitrary objects (e.g., graphs). As a result, the matrix is symmetric but not positive semidefinite. Discuss how you can repair the matrix S by modifying only its self-similarity (i.e., diagonal) entries, so that the matrix becomes positive semidefinite.*

A hint for solving this problem is to examine the effect of adding a constant value to the diagonal on the eigenvalues. This trick is used frequently for applying kernel methods in machine learning, when a similarity matrix is constructed using an arbitrary heuristic.

Covariance Matrix

Another common matrix in machine learning is the *covariance matrix*. Just as the similarity matrix computes dot products between rows of matrix D, the covariance matrix computes (scaled) dot products between *columns* of D *after mean-centering the matrix*. Consider a set of scalar values $x_1 \ldots x_n$. The mean μ and the variance σ^2 of these values are defined as follows:

$$\mu = \frac{\sum_{i=1}^n x_i}{n}$$

$$\sigma^2 = \frac{\sum_{i=1}^n (x_i - \mu)^2}{n} = \frac{\sum_{i=1}^n x_i^2}{n} - \mu^2$$

Consider a data matrix in which two columns have values $x_1 \ldots x_n$ and $y_1 \ldots y_n$, respectively. Also assume that the means of the two columns are μ_x and μ_y. In this case, the covariance σ_{xy} is defined as follows:

$$\sigma_{xy} = \frac{\sum_{i=1}^n (x_i - \mu_x)(y_i - \mu_y)}{n} = \frac{\sum_{i=1}^n x_i y_i}{n} - \mu_x \mu_y$$

The notion of covariance is an extension of variance, because $\sigma_x^2 = \sigma_{xx}$ is simply the variance of $x_1 \ldots x_n$. If the data is mean-centered with $\mu_x = \mu_y = 0$, the covariance simplifies to the following:

$$\sigma_{xy} = \frac{\sum_{i=1}^n x_i y_i}{n} \qquad \text{[Mean-centered data only]}$$

It is noteworthy that the expression on the right-hand side is simply a scaled version of the dot product between the *columns*, if we represent the x values and y values as an $n \times 2$ matrix. Note the close relationship to the similarity matrix, which contains dot products between all pairs of *rows*. Therefore, if we have an $n \times d$ data matrix D, which is mean-centered, we can compute the covariance between the column i and column j using this approach. Such a matrix is referred to as the *covariance matrix*.

Definition 3.4.2 (Covariance Matrix of Mean-Centered Data) *Let D be an $n \times d$ mean-centered data matrix. Then, the covariance matrix C of D is defined as follows:*

$$C = \frac{D^T D}{n}$$

The *unscaled version* of the matrix, in which the factor of n is not used in the denominator, is referred to as the *scatter matrix*. In other words, the scatter matrix is simply $D^T D$. The scatter matrix is the Gram matrix of the column space of D, whereas the similarity matrix is the Gram matrix of the row space of D. Like the similarity matrix, the scatter matrix and covariance matrix are both positive semidefinite, based on Lemma 3.3.14.

The covariance matrix is often used for *principal component analysis* (cf. Section 7.3.4). Since the $d \times d$ covariance matrix C is positive semidefinite, one can diagonalize it as follows:

$$C = P \Delta P^T \tag{3.34}$$

The data set D is transformed to $D' = DP$, which is equivalent to representing each row of the original matrix D in the axis system of directions contained in the columns of P. This new data set has some interesting properties in terms of its covariance structure. One can also write the diagonal matrix as $\Delta = P^T C P$. The diagonal matrix Δ is the new covariance matrix of the transformed data $D' = DP$. In order to see why this is true, note that the transformed data is also mean centered because the sum of its columns can be shown to be 0. The covariance matrix of the transformed data is therefore $D'^T D'/n = (DP)^T(DP)/n = P^T(D^TD)P/n$. This expression simplifies to $P^T C P = \Delta$. In other words, the transformation represents a *decorrelated version of the data.*

The entries on the diagonal of Δ are the variances of the individual dimensions in the transformed data, and they represent the nonnegative eigenvalues of the positive semidefinite matrix C. Typically, only a few diagonal entries are large (in relative terms), which contain most of the variance in the data. The remaining low-variance directions can be dropped from the transformed representation. One can select a small subset of columns from P corresponding to the largest eigenvalues in order to create a $d \times k$ transformation matrix P_k, where $k \ll d$. The $d \times k$ transformed data matrix is defined as $D'_k = DP_k$. Each row is a new k-dimensional representation of the data set. It turns out that this representation has a highly reduced dimensionality, but it still retains most of the data variability (like Euclidean distances between points). For mean-centered data, the discarded $(d - k)$ columns of DP are not very informative because they are all very close to 0. In fact, one can show using optimization methods that this representation provides an optimal reduction of the data in k dimensions (or *principal components*), so that the least amount of variance in the data is lost. We will revisit this problem in Chapters 7 and 8.

3.4.3   Symmetric Matrices in Quadratic Optimization

Many machine learning applications are posed as optimization problems over a squared objective function. Such objective functions are *quadratic*, because the highest term of the polynomial is 2. The simplest versions of these quadratic functions can be expressed as $\overline{x}^T A \overline{x}$, where A is a $d \times d$ matrix and $\overline{x}$ is a d-dimensional column vector of optimization variables. The process of solving such optimization problems is referred to as *quadratic programming.* Quadratic programming is an extremely important class of problems in optimization, because arbitrary functions can be locally approximated as quadratic functions by using the method of *Taylor expansion* (cf. Section 1.5.1 of Chapter 1). This principle forms the basis of many optimization techniques, such as the *Newton method* (cf. Chapter 5).

The shape of the function $\overline{x}^T A \overline{x}$ critically depends on the nature of the matrix A. Functions in which A is positive semidefinite correspond to *convex functions*, which take the shape of a bowl with a minimum but no maximum. Functions in which A is negative semidefinite are *concave*, and they take on the shape of an inverted bowl. Examples of convex and concave functions are illustrated in Figure 3.4. Formally, convex and concave functions satisfy the following properties for any pair of vectors $\overline{x}_1$ and $\overline{x}_2$ and any scalar $\lambda \in (0, 1)$:

$$f(\lambda \overline{x}_1 + (1 - \lambda)\overline{x}_2) \leq \lambda f(\overline{x}_1) + (1 - \lambda)f(\overline{x}_2) \quad \text{[Convex function]}$$
$$h(\lambda \overline{x}_1 + (1 - \lambda)\overline{x}_2) \geq \lambda h(\overline{x}_1) + (1 - \lambda)h(\overline{x}_2) \quad \text{[Concave function]}$$

Functions in which A is neither positive nor negative semidefinite (i.e., A is indefinite) have neither global maxima nor do they have global minima. Such quadratic functions have *saddle points*, which are inflection points looking like *both* maxima or minima, depending on which direction one approaches that point from. An example of an indefinite function is illustrated in Figure 3.6.

Consider the quadratic function $f(x_1, x_2) = x_1^2 + x_2^2$, which is convex and has a single global minimum at $(0, 0)$. If we plot this function in three dimensions with $f(x_1, x_2)$ on the

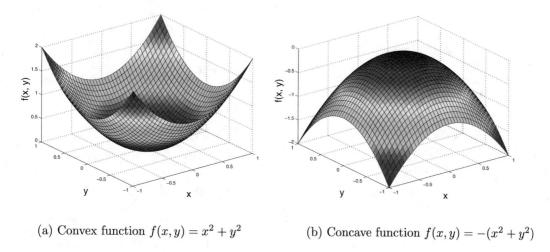

(a) Convex function $f(x, y) = x^2 + y^2$ (b) Concave function $f(x, y) = -(x^2 + y^2)$

Figure 3.4: Illustration of convex and concave functions

vertical axis in addition to the two horizontal axes representing x_1 and x_2, we obtain an upright bowl, as shown in Figure 3.4(a). One can express $f(x, y)$ in matrix form as follows:

$$f(x_1, x_2) = [x_1, x_2] \begin{bmatrix} 1 & 0 \\ 0 & 1 \end{bmatrix} \begin{bmatrix} x_1 \\ x_2 \end{bmatrix}$$

In this case, the function represents a perfectly circular bowl, and the corresponding matrix A for representing the ellipse $\overline{x}^T A \overline{x} = r^2$ is the 2×2 identity matrix, which is a trivial form of a positive semidefinite matrix. We can also use various vertical cross sections of the circular bowl shown in Figure 3.4(a) to create a *contour plot*, so that the value of $f(x_1, x_2)$ at each point on a contour line is constant. The contour plot of the circular bowl in shown in Figure 3.5(a). Note that using the negative of the identity matrix (which is a negative semidefinite matrix) results in an inverted bowl, as shown in Figure 3.4(b). The negative of a convex function is always a concave function, and vice versa. Therefore, maximizing concave functions is almost exactly similar to minimizing convex functions.

The function $f(\overline{x}) = \overline{x}^T A \overline{x}$ corresponds to a *perfectly circular* bowl, when A is set to the identity matrix (cf. Figures 3.4(a) and 3.5(a)). Changing A from the identity matrix leads to several interesting generalizations. First, if the diagonal entries of A are set to different (nonnegative) values, the circular bowl would become elliptical. For example, if the bowl is stretched twice in one direction as compared to the other, the diagonal entries would be in the ratio of $2^2 : 1 = 4 : 1$. An example of such a function is following:

$$f(x_1, x_2) = 4x_1^2 + x_2^2$$

One can represent this ellipse in matrix form as follows:

$$f(x_1, x_2) = [x_1, x_2] \begin{bmatrix} 4 & 0 \\ 0 & 1 \end{bmatrix} \begin{bmatrix} x_1 \\ x_2 \end{bmatrix}$$

The contour plot for this case is shown in Figure 3.5(b). Note that the vertical direction x_2 is stretched even though the x_1 direction has diagonal entry of 4. The diagonal entries are *inverse squares* of stretching factors.

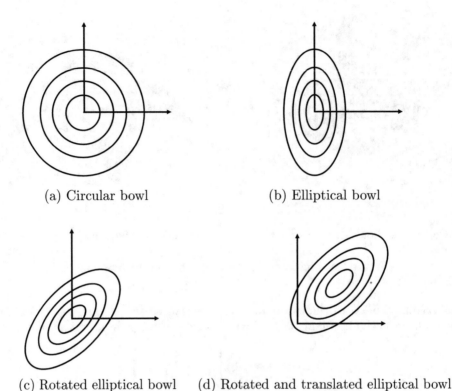

(a) Circular bowl

(b) Elliptical bowl

(c) Rotated elliptical bowl   (d) Rotated and translated elliptical bowl

Figure 3.5: Contour plots of quadratic functions created with 2×2 positive semidefinite matrices

So far, we have only considered quadratic functions in which the stretching occurs along axis-parallel directions. Now, consider the case where we start with the diagonal matrix Δ and rotate using basis matrix P, where P contains the two vectors that are oriented at $45°$ to the axes. Therefore, consider the following rotation matrix:

$$P = \begin{bmatrix} \cos(45) & \sin(45) \\ -\sin(45) & \cos(45) \end{bmatrix} \tag{3.35}$$

In this case, we use $A = P\Delta P^T$ in order to define $\overline{x}^T A \overline{x}$. The approach computes the coordinates of $\overline{x}$ as $\overline{y} = P^T \overline{x}$, and then computes $f(\overline{x}) = \overline{x}^T A \overline{x} = \overline{y}^T \Delta \overline{y}$. Note that we are stretching the coordinates of the *new* basis. The result is a stretched ellipse in the direction of the basis defined by the columns of P (which is a $45°$ clockwise rotation matrix for column vectors). One can compute the matrix A in this case as follows:

$$A = \begin{bmatrix} \cos(45) & \sin(45) \\ -\sin(45) & \cos(45) \end{bmatrix} \begin{bmatrix} 4 & 0 \\ 0 & 1 \end{bmatrix} \begin{bmatrix} \cos(45) & \sin(45) \\ -\sin(45) & \cos(45) \end{bmatrix}^T = \begin{bmatrix} 5/2 & -3/2 \\ -3/2 & 5/2 \end{bmatrix}$$

One can represent the corresponding function as follows:

$$f(x_1, x_2) = [x_1, x_2] \begin{bmatrix} 5/2 & -3/2 \\ -3/2 & 5/2 \end{bmatrix} \begin{bmatrix} x_1 \\ x_2 \end{bmatrix} = \frac{5}{2}(x_1^2 + x_2^2) - 3x_1 x_2$$

The term involving $x_1 x_2$ captures the interactions between the attributes x_1 and x_2. This is the direct result of a change of basis that is no longer aligned with the axis system. The contour plot of an ellipse that is aligned at 45° with the axes is shown in Figure 3.5(c).

All these cases represent situations where the optimal solution to $f(x_1, x_2)$ is at $(0, 0)$, and the resulting function value is 0. How can we generalize to a function with optimum occurring at $\overline{b}$ and an optimum value of c (which is a scalar)? The corresponding function is of the following form:

$$f(\overline{x}) = (\overline{x} - \overline{b})^T A (\overline{x} - \overline{b}) + c \tag{3.36}$$

The matrix A is equivalent to half the *Hessian matrix* of the quadratic function. The $d \times d$ Hessian matrix $H = [h_{ij}]$ of a function of d variables is a symmetric matrix containing the second-order derivatives with respect to each pair of variables.

$$h_{ij} = \frac{\partial^2 f(\overline{x})}{\partial x_i \partial x_j} \tag{3.37}$$

Note that $\overline{x}^T H \overline{x}$ represents the *directional* second derivative of the function $f(\overline{x})$ along $\overline{x}$ (cf. Chapter 4), and it represents the second derivative of the rate of change of $f(\overline{x})$, when moving along direction $\overline{x}$. This value is always nonnegative for convex functions irrespective of $\overline{x}$, which ensures that the value of $f(\overline{x})$ is minimum when the first derivative of the rate of change of $f(\overline{x})$ along each direction $\overline{x}$ is 0. In other words, the Hessian needs to be positive semidefinite. This is a generalization of the condition $g''(x) \geq 0$ in 1-dimensional convex functions. We make the following assertion, which is shown formally in Chapter 4:

Observation 3.4.2 *Consider a quadratic function, whose quadratic term is of the form* $\overline{x}^T A \overline{x}$. *Then, the quadratic function is convex, if and only if the matrix A is positive semidefinite.*

Many quadratic functions in machine learning are of this form. A specific example is the dual objective function of a support vector machine (cf. Chapter 6).

One can construct an example of the general form of the quadratic function by translating the 45°-oriented, origin-centered ellipse of Figure 3.5(c). For example, if we center the elliptical objective function at $[1, 1]$ and add 2 to the optimal values, we obtain the function $(\overline{x}^T - [1, 1]) A (\overline{x} - [1, 1]^T) + 2$. The resulting objective function, which takes an optimal value of 2 at $[1, 1]$ is shown below:

$$f(x_1, x_2) = \frac{5}{2}(x_1^2 + x_2^2) - 2(x_1 + x_2) - 3x_1 x_2 + 4 \tag{3.38}$$

This type of quadratic objective function is common in many machine learning algorithms. An example of the contour plot of a translated ellipse is shown in Figure 3.5(d), although it doe snot show the vertical translation by 2.

It is noteworthy that the most general form of a quadratic function in multiple variables is as follows:

$$f(\overline{x}) = \overline{x}^T A' \overline{x} + \overline{b}'^T \overline{x} + c' \tag{3.39}$$

Here, A' is a $d \times d$ symmetric matrix, $\overline{b}'$ is a d-dimensional column vector, and c' is a scalar. In the 1-dimensional case, A' and $\overline{b}'$ are replaced by scalars, and one obtains the familiar form $ax^2 + bx + c$ of univariate quadratic functions. Furthermore, as long as $\overline{b}'$ belongs to the column space of A', one can convert the general form of Equation 3.39 to the *vertex* form of Equation 3.36. It is important for $\overline{b}'$ to belong to the column space of A' for an optimum to exist. For example, the 2-dimensional function is $G(x_1, x_2) = x_1^2 + x_2$ does not have a minimum because the function is partially linear in x_2. The vertex form of Equation 3.39 considers only strictly quadratic functions in which all cross-sections of the function are quadratic. Only strictly quadratic functions are interesting for optimization, because linear functions usually do not have a maximum or minimum. One can relate the coefficients of Equations 3.36 and 3.39 as follows:

$$A' = A, \quad \bar{b}' = -2A\bar{b}, \quad c' = \bar{b}^T \bar{b} + c$$

Given A', $\bar{b}'$ and c', the main condition for being able to arrive at the vertex form of Equation 3.36 is the second condition $\bar{b}' = -2A\bar{b} = -2A'\bar{b}$ for which a solution will exist only when $\bar{b}'$ occurs in the column space of A'.

Finally, we discuss the case where the matrix A used to create the function $\bar{x}^T A \bar{x}$ is *indefinite*, and has both positive and negative eigenvalues. An example of such a function is the following:

$$g(x_1, x_2) = [x_1, x_2] \begin{bmatrix} 1 & 0 \\ 0 & -1 \end{bmatrix} \begin{bmatrix} x_1 \\ x_2 \end{bmatrix} = x_1^2 - x_2^2$$

The gradient at $(0, 0)$ is 0, which seems to be an optimum point. However, this point behaves like both a maximum and a minimum, when examining second derivatives. If we approach the point from the x_1 direction, it seems like a minimum. If we approach it from the x_2 direction, it seems like a maximum. This is because the directional second derivatives in the x_1 and x_2 directions are simply twice the diagonal entries (which are of opposite sign). The shape of the objective function resembles that of a riding saddle, and the point $(0, 0)$ is referred to as a *saddle point*. An example of this type of objective function is shown in Figure 3.6. Objective functions containing such points are often notoriously hard for optimization.

3.4.4   Diagonalization Application: Variable Separation for Optimization

Consider the quadratic function $f(\bar{x}) = \bar{x}^T A \bar{x} + \bar{b}^T \bar{x} + c$. Unless the symmetric matrix A is diagonal, the resulting function contains terms of the form $x_i x_j$. Such terms are referred to as *interacting terms*. Most real-world quadratic functions contain such terms. It is noteworthy that any multivariate quadratic function can be transformed to an *additively separable* function (without interacting terms) by basis transformation of the input variables of the function. This type of change in basis brings us back to using linear algebra tricks. Additively separable functions are much easier to optimize, because one can decompose the optimization problem into smaller optimization problems on individual variables. For example, a multivariate quadratic function would appear as a simple sum of univariate quadratic functions (each of which is extremely simple to optimize). One can show this simple result by using the linear algebra tricks that we have learned in this chapter. We first define the notion of separable functions:

Definition 3.4.3 (Additively Separable Functions) *A function $F(x_1, x_2, \ldots, x_d)$ in d variables is said to be additively separable, if it can be expressed in the following form for appropriately chosen univariate functions $f_1(\cdot), f_2(\cdot), \ldots f_d(\cdot)$:*

$$F(x_1, x_2, \ldots, x_d) = \sum_{i=1}^{d} f_i(x_i)$$

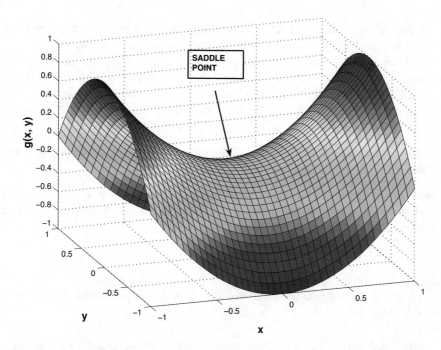

Figure 3.6: Illustration of quadratic function $g(x, y) = x^2 - y^2$ created by indefinite matrix

Consider the following quadratic function defined on a d-dimensional vector $\overline{x} = [x_1, \ldots x_d]^T$.

$$f(\overline{x}) = \overline{x}^T A \overline{x} + \overline{b}^T \overline{x} + c$$

Since A is a $d \times d$ symmetric matrix, one can diagonalize it as $A = V\Delta V^T$, and use the variable transformation $\overline{x} = V\overline{x}'$ (which is the same as $\overline{x}' = V^T\overline{x}$). On performing this transformation one obtains the new function $g(\overline{x}') = f(V\overline{x}')$, which is identical to the original function in a different basis. It is easy to show that the quadratic function may be expressed as follows:

$$f(V\overline{x}') = \overline{x}'^T \Delta \overline{x}' + \overline{b}^T V\overline{x}' + c$$

After this variable transformation, one obtains an additively separable function, because the matrix Δ is diagonal. One can solve for $\overline{x}'$ using d univariate optimizations, and then transform back $\overline{x}'$ to $\overline{x}$ using $\overline{x} = V\overline{x}'$.

Although this approach simplifies optimization, the problem is that eigenvector computation of A can be expensive. However, one can generalize this idea and try to find *any* matrix V (with possibly non-orthogonal columns), which satisfies $A = V\Delta V^T$ for some diagonal matrix Δ. Note that $A = V\Delta V^T$ would not[1] be a true diagonalization of A if the columns of V are not orthonormal. However, it is good enough to create a separable transformation for optimization, which is what we really care about. The columns of such non-orthogonal matrices are computationally much easier to evaluate than true eigenvectors, and the transformed variables are referred to as *conjugate directions*. The columns of V are referred to as *A-orthogonal directions*, because for any pair of (distinct) columns $\overline{v}_i$

---

[1] A true diagonalization must satisfy $V^T = V^{-1}$.

and $\overline{v}_j$, we have $\overline{v}_i^T A \overline{v}_j = \Delta_{ij} = 0$. There are an infinite number of possible ways of creating conjugate directions, and the eigenvectors represent a special case. In fact, a generalization of the Gram-Schmidt method can be used to find such directions (cf. Problem 2.7.1). This basic idea forms the principle of the *conjugate gradient descent* method discussed in Section 5.7.1 of Chapter 5, which can be used even for non-quadratic functions. Here, we provide a conceptual overview of the iterative conjugate gradient method for arbitrary (possibly non-quadratic) function $h(\overline{x})$ from current point $\overline{x} = \overline{x}_t$:

1. Create a quadratic approximation $f(\overline{x})$ of non-quadratic function $h(\overline{x})$ using the second-order Taylor expansion of $h(\overline{x})$ at $\overline{x} = \overline{x}_t$.

2. Compute the optimal solution x^* of the quadratic function $f(\overline{x})$ using the separable variable optimization approach discussed above as a set of d univariate optimization problems.

3. Set $\overline{x}_{t+1} = \overline{x}^*$ and $t \Leftarrow t + 1$. Go back to step 1.

The approach is iterated to convergence. The aforementioned algorithm provides the conceptual basis for the conjugate gradient method. The detailed method is provided in Section 5.7.1 of Chapter 5.

3.4.5   Eigenvectors in Norm-Constrained Quadratic Programming

A problem that arises frequently in different types of machine learning settings is one in which we wish to optimize $\overline{x}^T A \overline{x}$, where $\overline{x}$ is constrained to unit norm. Here, A is a $d \times d$ *symmetric* data matrix. This type of problem arises in many feature engineering and dimensionality reduction applications like *principal component analysis*, *singular value decomposition*, and *spectral clustering*. Such an optimization problem is posed as follows:

$$\text{Optimize } \overline{x}^T A \overline{x}$$
$$\text{subject to:}$$
$$\|\overline{x}\|^2 = 1$$

The optimization problem can be in either minimization or maximization form. Constraining the vector $\overline{x}$ to be the unit vector fundamentally changes the nature of the optimization problem. Unlike the previous section, it is no longer important whether the matrix A is positive semidefinite or not. One would have a well-defined optimal solution, even if the matrix A is indefinite. Constraining the norm of the vector helps in avoiding vectors with unbounded magnitudes or trivial solutions (like the zero vector), even when the matrix A is indefinite.

Let $\overline{v}_1 \ldots \overline{v}_d$ be the d orthonormal eigenvectors of the symmetric matrix A. Note that the set of eigenvectors creates a basis for $\mathcal{R}^d$, and therefore any d-dimensional vector $\overline{x}$ can be expressed as a linear combination of $\overline{v}_1 \ldots \overline{v}_d$ as follows:

$$\overline{x} = \sum_{i=1}^{d} \alpha_i \overline{v}_i \qquad (3.40)$$

We will re-parameterize this optimization problem in terms of the parameters $\alpha_1 \ldots \alpha_d$ by substituting for $\overline{x}$ in the optimization problem. By making this substitution, and setting each $A\overline{v}_i = \lambda_i \overline{v}_i$, we obtain the following re-parameterized optimization problem:

$$\text{Optimize } \sum_{i=1}^{d} \lambda_i \alpha_i^2$$

subject to:

$$\sum_{i=1}^{d} \alpha_i^2 = 1$$

The expression $\|\overline{x}\|^2$ in the constraint is simplified to $(\sum_{i=1}^{d} \alpha_i \overline{v}_i) \cdot (\sum_{i=1}^{d} \alpha_i \overline{v}_i)$; we can expand it using the distributive property, and then we use the orthogonality of the eigenvectors to set $\overline{v}_i \cdot \overline{v}_j = 0$. The objective function value is $\sum_i \lambda_i \alpha_i^2$, where the different α_i^2 sum to 1. Clearly, the minimum and maximum possible values of this objective function are achieved by setting the weight α_i^2 of a single value of λ_i to 1, which corresponds to the minimum or maximum possible eigenvalue (depending on whether the optimization problem is posed in minimization or maximization form):

> The maximum value of the norm-constrained quadratic optimization problem is obtained by setting $\overline{x}$ to the largest eigenvector of A. The minimum value is obtained by setting $\overline{x}$ to the smallest eigenvector of A.

This problem can be generalized to finding a k-dimensional subspace. In other words, we want to find orthonormal vectors $\overline{x}_1 \ldots \overline{x}_k$, so that $\sum_i \overline{x}_i A \overline{x}_i$ is optimized:

$$\text{Optimize } \sum_{i=1}^{k} \overline{x}_i^T A \overline{x}_i$$

subject to:

$$\|\overline{x}_i\|^2 = 1 \ \ \forall i \in \{1 \ldots k\}$$

$$\overline{x}_1 \ldots \overline{x}_k \text{ are mutually orthogonal}$$

The optimal solution to this problem can be derived using a similar procedure. We provide an alternative solution with the use of Lagrangian relaxation in Section 6.6 of Chapter 6. Here, we simply state the optimal solution:

> The maximum value of the norm-constrained quadratic optimization problem is obtained by using the largest k eigenvectors of A. The minimum value is obtained by using the smallest k eigenvectors of A.

Intuitively, these results make geometric sense from the perspective of the anisotropic scaling caused by symmetric matrices like A. The matrix A distorts the space with scale factors corresponding to the eigenvalues along orthonormal directions corresponding to the eigenvectors. The objective function tries to either maximize or minimize the aggregate projections of the distorted vectors $A\overline{x}_i$ on the original vectors $\overline{x}_i$, which is sum of the dot products between $\overline{x}_i$ and $A\overline{x}_i$. By picking the *largest* k eigenvectors (scaling directions), this sum is maximized. On the other hand, by picking the smallest k directions, this sum is minimized.

3.5 Numerical Algorithms for Finding Eigenvectors

The simplest approach for finding eigenvectors of a $d \times d$ matrix A is to first find the d roots $\lambda_1 \ldots \lambda_d$ of the equation $\det(A - \lambda I) = 0$. Some of the roots might be repeated. In the next step, one has to solve linear systems of the form $(A - \lambda_j I)\overline{x} = 0$. This can be done using the

Gaussian elimination method (cf. Section 2.5.4 of Chapter 2). However, polynomial equation solvers are sometimes numerically unstable and have a tendency to show ill-conditioning in real-world settings. Finding the roots of a polynomial equation is numerically harder than finding eigenvalues of a matrix! In fact, one of the many ways in which high-degree polynomial equations are solved in engineering disciplines is to first construct a *companion matrix* of the polynomial, such that the matrix has the same characteristic polynomial, and then find its eigenvalues:

Problem 3.5.1 (Companion Matrix) *Consider the following matrix:*

$$A_2 = \begin{bmatrix} 0 & 1 \\ -c & -b \end{bmatrix}$$

Discuss why the roots of the polynomial equation $x^2 + bx + c = 0$ can be computed using the eigenvalues of this matrix. Also show that finding the eigenvalues of the following 3×3 matrix yields the roots of $x^3 + bx^2 + cx + d = 0$.

$$A_3 = \begin{bmatrix} 0 & 1 & 0 \\ 0 & 0 & 1 \\ -d & -c & -b \end{bmatrix}$$

Note that the matrix has a non-zero row and superdiagonal of 1s. Provide the general form of the $t \times t$ matrix A_t required for solving the polynomial equation $x^t + \sum_{i=0}^{t-1} a_i x^i = 0$.

In some cases, algorithms for finding eigenvalues also yield the eigenvectors as a byproduct, which is particularly convenient. In the following, we present alternatives both for finding eigenvalues and for finding eigenvectors.

3.5.1 The QR Method via Schur Decomposition

The QR algorithm uses the following two steps alternately in an iterative way:

1. Decompose the matrix $A = QR$ using the QR algorithm discussed in Section 2.7.2. Here, R is an upper-triangular matrix and Q is an orthogonal matrix.

2. Iterate by using $A \Leftarrow Q^T A Q$ and go to the previous step.

The matrix $Q^T A Q$ is similar to A, and therefore it has the same eigenvalues. A key result[2] is that applying the transformation $A \Leftarrow Q^T A Q$ repeatedly to A results in the upper-triangular matrix U of the Schur decomposition. In fact, if we keep track of the orthogonal matrices $Q_1 \ldots Q_s$ obtained using QR decomposition (in that order) and denote their product $Q_1 Q_2 \ldots Q_s$ by the single orthogonal matrix P, one can obtain the Schur decomposition of A in the following form:

$$A = PUP^T$$

The diagonal entries of this converged matrix U contain the eigenvalues. In general, the triangulization of a matrix is a natural way of finding its eigenvalues. After the eigenvalues $\lambda_1 \ldots \lambda_d$ have been found, the eigenvectors can be found by solving equations of the form $(A - \lambda_j I)\overline{x} = 0$ using the methods of Section 2.5.4 in Chapter 2. This approach is not fully optimized for computational speed, which can be improved by first transforming the matrix to *Hessenberg form*. The reader is referred to [52] for a detailed discussion.

---

[2]We do not provide a proof of this result here. Refer to [52].

3.5.2   The Power Method for Finding Dominant Eigenvectors

The power method finds the eigenvector with the largest *absolute* eigenvalue of a matrix, which is also referred to as its *dominant eigenvector* or *principal eigenvector*. One caveat is that it is possible for the principal eigenvalue of a matrix to be complex, in which case the power method might not work. The following discussion assumes that the matrix has real-valued eigenvectors/eigenvalues, which is the case in many real-world applications. Furthermore, we usually do not need all the eigenvectors, but only the top few eigenvectors. The power method is designed to find only the top eigenvector, although it can be used to find the top few eigenvectors with some modifications. Unlike the QR method, one can find eigenvectors and eigenvalues simultaneously, without the need to solve systems of equations after finding the eigenvalues. The power method is an iterative method, and the underlying iterations are also referred to as *von Mises* iterations.

Consider a $d \times d$ matrix A, which is diagonalizable with real eigenvalues. Since A is a diagonalizable matrix, multiplication with A results in anisotropic scaling. If we multiply any column vector $\overline{x} \in \mathcal{R}^d$ with A to create $A\overline{x}$, it will result in a linear distortion of $\overline{x}$, in which directions corresponding to larger (absolute) eigenvalues are stretched to a greater degree. As a result, the (acute) angle between $A\overline{x}$ and the largest eigenvector $\overline{v}$ will reduce from that between $\overline{x}$ and $\overline{v}$. If we keep repeating this process, the transformations will eventually result in a vector pointing in the direction of the largest (absolute) eigenvector. Therefore, the power method starts by first initializing the d components of the vector $\overline{x}$ to random values from a uniform distribution in $[-1, 1]$. Subsequently, the following von Mises iteration is repeated to convergence:

$$\overline{x} \Leftarrow \frac{A\overline{x}}{\|A\overline{x}\|}$$

Note that normalization of the vector in each iteration is essential to prevent overflow or underflow to arbitrarily large or small values. After convergence to the principal eigenvector $\overline{v}$, one can compute the corresponding eigenvalue as the ratio of $\overline{v}^T A \overline{v}$ to $\|\overline{v}\|^2$, which is referred to as the *Raleigh quotient*.

We now provide a formal justification. Consider a situation in which we represent the starting vector $\overline{x}$ as a linear combination of the basis of d eigenvectors $\overline{v}_1 \ldots \overline{v}_d$ with coefficients $\alpha_1 \ldots \alpha_d$:

$$\overline{x} = \sum_{i=1}^{d} \alpha_i \overline{v}_i \tag{3.41}$$

If the eigenvalue of $\overline{v}_i$ is λ_i, then multiplying with A^t has the following effect:

$$A^t \overline{x} = \sum_{i=1}^{t} \alpha_i A^t \overline{v}_i = \sum_{i=1}^{t} \alpha_i \lambda_i^t \overline{v}_i \propto \sum_{i=1}^{t} \alpha_i (-1)^t \frac{|\lambda_i|^t}{\sum_{j=1}^{t} |\lambda_j|^t} \overline{v}_i$$

When t becomes large, the quantity on the right-hand side will be dominated by the effect of the largest eigenvector. This is because the factor $|\lambda_1^t|$ increases the proportional weight of the first eigenvector, when λ_1 is the (strictly) largest eigenvalue. The fractional value $|\lambda_1^t|/\sum_{j=1}^{t} |\lambda_j^t|$ will converge to 1 for the largest (absolute) eigenvector and to 0 for all others. As a result, the normalized version of $A^t \overline{x}$ will point in the direction of the largest (absolute) eigenvector $\overline{v}_1$. Note that this proof does depend on the fact that λ_1 is strictly greater than the next eigenvalue, or else the convergence will not occur. Furthermore, if the top-2 eigenvalues are too similar, the convergence will be slow. However, large machine learning matrices (e.g., covariance matrices) are often such that the top few eigenvalues are quite different in magnitude, and most of the similar eigenvalues are at the bottom with values of 0. Furthermore, even when there are ties in the eigenvalues, the power method tends to find a vector that lies within the span of the tied eigenvectors.

Problem 3.5.2 (Inverse Power Iteration) *Let A be an invertible matrix. Discuss how you can use A^{-1} to discover the smallest eigenvector and eigenvalue of A in absolute magnitude.*

Finding the Top-k Eigenvectors for Symmetric Matrices

In most machine learning applications, one is looking not for the top eigenvector, but for the top-k eigenvectors. It is possible to use the power method to find the top-k eigenvectors. In symmetric matrices, the eigenvectors $\overline{v}_1 \ldots \overline{v}_d$, which define the columns of the basis matrix V, are orthonormal according to the following diagonalization:

$$A = V \Delta V^T \tag{3.42}$$

The above relationship can also be rearranged in terms of the column vectors of V and the eigenvalues $\lambda_1 \ldots \lambda_d$ of Δ:

$$A = V \Delta V^T = \sum_{i=1}^{d} \lambda_i [\overline{v}_i \overline{v}_i^T] \tag{3.43}$$

This result follows from the fact that any matrix product can be expressed as the sum of outer products (cf. Lemma 1.2.1 of Chapter 1). Applying Lemma 1.2.1 to the product of $(V\Delta)$ and V^T yields the above result. The decomposition implied by Equation 3.43 is referred to as a *spectral decomposition* of the matrix A. Each $\overline{v}_i \overline{v}_i^T$ is a rank-1 matrix of size $d \times d$, and λ_i is the weight of this matrix component. As discussed in Section 7.2.3 of Chapter 7, spectral decomposition can be applied to any type of matrix (and not just symmetric matrices) using an idea referred to as *singular value decomposition*.

Consider the case in which we have already found the top eigenvector λ_1 with eigenvalue $\overline{v}_1$. Then, one can remove the effect of the top eigenvalue by creating the following modified matrix:

$$A' = A - \lambda_1 \overline{v}_1 \overline{v}^T \tag{3.44}$$

As a result, the *second largest* eigenvalue of A becomes the dominant eigenvalue of A'. Therefore, by repeating the power iteration with A', one can now determine the second-largest eigenvector. The process can be repeated any number of times.

When the matrix A is sparse, one disadvantage of this method is that A' might not be sparse. Sparsity is a desirable feature of matrix representations, because of the space- and time-efficiency of sparse matrix operations. However, it is not necessary to represent the dense matrix A' explicitly. The matrix multiplication $A'\overline{x}$ for the power method can be accomplished using the following relationship:

$$A'\overline{x} = A\overline{x} - \lambda_1 \overline{v}_1 (\overline{v}_1^T \overline{x}) \tag{3.45}$$

It is important to note how we have bracketed the second term on the right-hand side. This avoids the explicit computation of a rank-1 matrix (which is dense), and it can be accomplished with simple dot product computation between $\overline{v}_1$ and $\overline{x}$. This is an example of the fact that the associativity property of matrix multiplication is often used to ensure the best efficiency of matrix multiplication. One can also generalize these ideas to finding the top-k eigenvectors by removing the effect of the top-r eigenvectors from A when finding the $(r+1)$th eigenvector.

Problem 3.5.3 (Generalization to Asymmetric Matrices) *The power method is designed to find the single largest eigenvector. The approach for finding the top-k eigenvectors makes the additional assumption of a symmetric matrix. Discuss where the assumption of a symmetric matrix was used in this section. Can you find a way to generalize the approach to arbitrary matrices assuming that the top-k eigenvalues are distinct?*

A hint for the above problem is that the left eigenvectors and right eigenvectors may not be the same in asymmetric matrices (as in symmetric matrices) and both are needed in order to subtract the effect of dominant eigenvectors.

Problem 3.5.4 (Finding Largest Eigenvectors) *The power method finds the top-k eigenvectors of largest **absolute** magnitude. In most applications, we also care about the sign of the eigenvector. In other words, an eigenvalue of +1 is greater than −2, when sign is considered. Show how you can modify the power method to find the top-k eigenvectors of a symmetric matrix when sign is considered.*

The key point in the above exercise is to translate the eigenvalues to nonnegative values by modifying the matrix using the ideas already discussed in this section.

3.6 Summary

Diagonalizable matrices represent a form of linear transformation, so that multiplication of a vector with such a matrix corresponds to anisotropic scaling of the vector in (possibly non-orthogonal) directions. Not all matrices are diagonalizable. Symmetric matrices are always diagonalizable, and they can be represented as scaling transformations in mutually orthogonal directions. When the scaling factors of symmetric matrices are nonnegative, they are referred to as positive semidefinite matrices. Such matrices frequently arise in different types of machine learning applications. Therefore, this chapter has placed a special emphasis on these types of matrices and their eigendecomposition properties. We also introduce a number of key optimization applications of such matrices, which sets the stage for more detailed discussions in later chapters.

3.7 Further Reading

The concepts of diagonalization are discussed in the books by Strang [122, 123], David Lay [77], Hoffman-Kunze [62], and Golub and Van Loan [52]. A discussion of numerical methods that combine linear algebra and optimization is provided in [99]. The field of convex optimization is studied in detail in [22].

3.8 Exercises

1. In Chapter 2, you learned that any $d \times d$ orthogonal matrix A can be decomposed into $O(d^2)$ Givens rotations and at most one elementary reflection. Discuss how the sign of the determinant of A determines whether or not a reflection is needed.

2. In Chapter 2, you learned that any $d \times d$ matrix A can be decomposed into at most $O(d)$ Householder reflections. Discuss the effect of the sign of the determinant on the number of Householder reflections.

3. Show that if a matrix A satisfies $A^2 = 4I$, then all eigenvalues of A are 2 and -2.

4. You are told that a 4×4 symmetric matrix has eigenvalues 4, 3, 2, and 2. You are given the values of eigenvectors belonging to the eigenvalues 4 and 3. Provide a procedure to reconstruct the entire matrix. [Hint: One eigenvalue is repeated and the matrix is symmetric.]

5. Suppose that A is a square $d \times d$ matrix. The matrix A' is obtained by multiplying the ith row of A with γ_i and dividing the ith column of A with γ_i for each i. How are the eigenvectors of A are related to those of A'? [Hint: Relate A and A' with matrix operators.]

6. For a 4×4 matrix A with the following list of eigenvalues obtained from the characteristic polynomial, state in each case whether the matrix is *guaranteed* to be diagonalizable, invertible, both, or neither: (a) $\{\lambda_1, \lambda_2, \lambda_3, \lambda_4\} = \{1, 3, 4, 9\}$ (b) $\{\lambda_1, \lambda_2, \lambda_3, \lambda_4\} = \{1, 3, 3, 9\}$ (c) $\{\lambda_1, \lambda_2, \lambda_3, \lambda_4\} = \{0, 3, 4, 9\}$ (d) $\{\lambda_1, \lambda_2, \lambda_3, \lambda_4\} = \{0, 3, 3, 9\}$ (e) $\{\lambda_1, \lambda_2, \lambda_3, \lambda_4\} = \{0, 0, 4, 9\}$.

7. Show that any real-valued matrix of odd dimension must have at least one real eigenvalue. Show the related fact that the determinant of a real-valued matrix without any real eigenvalues is always positive. Furthermore, show that a real-valued matrix of even dimension with a negative determinant must have at least two distinct real-valued eigenvalues. [Hint: Properties of polynomial roots.]

8. In the Jordan normal form $A = VUV^{-1}$, the upper triangular matrix U is in *block diagonal form*, where smaller upper-triangular matrices $U_1 \ldots U_r$ are arranged along the diagonal of U. What is the effect of applying a polynomial function $f(U)$ on the individual blocks $U_1 \ldots U_r$? Use this fact to provide a general proof of the Cayley-Hamilton theorem. [Hint: Strictly triangular matrices are nilpotent.]

9. Provide an example of a defective matrix whose square is diagonalizable. [Hint: Construct a singular matrix in Jordan normal form.]

10. Let A and B be $d \times d$ matrices. Show that the matrix $AB - BA$ can never be positive semidefinite unless it is the zero matrix. [Hint: Use properties of the trace.]

11. Can the square of a matrix that does not have real eigenvalues be diagonalizable with real eigenvalues? If no, provide a proof. If yes, provide an example.

12. If the matrices A, B, and AB are all symmetric, show that the matrices A and B must be simultaneously diagonalizable. [Hint: See Problem 1.2.7 in Chapter 1.]

13. Suppose that the $d \times d$ matrix S is symmetric, positive semidefinite matrix, and the matrix D is of size $n \times d$. Show that DSD^T must also be a symmetric, positive semidefinite matrix. Note that DSD^T is a matrix of inner products between rows of D, which is a generalization of the dot product matrix DD^T.

14. Let S be a positive semidefinite matrix, which can therefore be expressed in Gram matrix form as $S = B^T B$ (Lemma 3.3.14). Use this fact to show that a diagonal entry can never be negative. What does this imply for the convexity of quadratic functions?

15. Show that if a matrix P satisfies $P^2 = P$, then all its eigenvalues must be 1 or 0.

16. Show that a matrix A is always similar to its transpose A^T. [Hint: Show that if A is similar to U, then A^T is similar to U^T. Then show that a matrix U in Jordan normal form is similar to its transpose with the use of a permutation matrix.]

17. Let $\overline{x}$ be a right eigenvector (column vector) of square matrix A with eigenvalue λ_r. Let $\overline{y}$ be a left eigenvector (row vector) of A with eigenvalue $\lambda_l \neq \lambda_r$. Show that $\overline{x}$ and $\overline{y}^T$ are orthogonal. [Hint: The spectral theorem contains a special case of this result. Problem 3.3.6 is also a special case for diagonalizable matrices.]

18. True or False? (a) A matrix with all zero eigenvalues must be the zero matrix. (b) A symmetric matrix with all zero eigenvalues must be the zero matrix.

19. Show that if λ is a non-zero eigenvalue of AB, then it must also be a non-zero eigenvalue of BA. Why does this argument not work for zero eigenvalues? Furthermore, show that if either A or B is invertible, then AB and BA are similar.

20. Is the quadratic function $f(x_1, x_2, x_3) = 2x_1^2 + 3x_2^2 + 2x_3^2 - 3x_1x_2 - x_2x_3 - 2x_1x_3$ convex? How about the function $g(x_1, x_2, x_3) = 2x_1^2 - 3x_2^2 + 2x_3^2 - 3x_1x_2 - x_2x_3 - 2x_1x_3$? In each case, find the minimum of the objective function, subject to the constraint that the norm of $[x_1, x_2, x_3]^T$ is 1.

21. Consider the function $f(x_1, x_2) = x_1^2 + 3x_1x_2 + 6x_2^2$. Propose a linear transformation of the variables so that the function is separable in terms of the new variables. Use the separable form of the objective function to find an optimal solution.

22. Show that the difference between two similar, symmetric matrices must be indefinite, unless both matrices are the same. [Hint: Use properties of the trace.]

23. Show that an nth root of a $d \times d$ diagonalizable matrix can always be found, as long as we allow for complex roots. Provide a geometric interpretation of the resulting matrix in terms of its relationship to the original matrix in the case where the root is a real-valued matrix.

24. Generate the equation of an ellipsoid centered at $[1, -1, 1]^T$, and whose axes directions are the orthogonal vectors $[1, 1, 1]^T$, $[1, -2, 1]^T$, and $[1, 0, -1]^T$. The ellipsoid is stretched in these directions in the ratio $1 : 2 : 3$. The answer to this question is not unique, and it depends on the size of your ellipsoid. Use the matrix form of ellipsoids discussed in the chapter. [Be careful about the mapping of the stretching ratios to the eigenvalues of this matrix both in terms of magnitude and relative ordering.]

25. If A and B are symmetric matrices whose eigenvalues lie in $[\lambda_1, \lambda_2]$ and $[\gamma_1, \gamma_2]$, respectively, show that the eigenvalues of $A - B$ lie in $[\lambda_1 - \gamma_2, \lambda_2 - \gamma_1]$. [Think geometrically about the effect of the multiplication of a vector with $(A - B)$. Also think of the norm-constrained optimization problem of $\overline{x}^T C \overline{x}$ for C chosen appropriately.]

26. Nilpotent Matrix: Consider a non-zero, square matrix A satisfying $A^k = 0$ for some k. Such a matrix is referred to as nilpotent. Show that all eigenvalues are 0 and such a matrix is defective.

27. Show that A is diagonalizable in each case if (i) it satisfies $A^2 = A$, and (ii) it satisfies $A^2 = I$.

28. Elementary Row Addition Matrix Is Defective: Show that the $d \times d$ elementary row addition matrix with 1s on the diagonal and a single non-zero off-diagonal entry is not diagonalizable.

29. Symmetric and idempotent matrices: Show that any $n \times n$ matrix P satisfying $P^2 = P$ and $P = P^T$ can be expressed in the form QQ^T for some $n \times d$ matrix Q with orthogonal columns (and is hence an alternative definition of a projection matrix).

30. Diagonalizability and Nilpotency: Show that every square matrix can be expressed as the sum of a diagonalizable matrix and a nilpotent matrix (including zero matrices for either part).

31. Suppose you are given the Cholesky factorization LL^T of a positive-definite matrix A. Show how to compute the inverse of A using multiple applications of back substitution.

32. Rotation in 3-d with arbitrary axis: Suppose that the vector $[1, 2, -1]^T$ is the axis of a counter-clockwise rotation of θ degrees, just as $[1, 0, 0]^T$ is the axis of the counter-clockwise θ-rotation of a column vector with the Givens matrix:

$$R_{[1,0,0]} = \begin{bmatrix} 1 & 0 & 0 \\ 0 & \cos(\theta) & -\sin(\theta) \\ 0 & \sin(\theta) & \cos(\theta) \end{bmatrix}$$

Create a new orthogonal basis system of $\mathcal{R}^3$ that includes $[1, 2, -1]^T$. Now use the concept of similarity $R_{[1,2,-1]} = PR_{[1,0,0]}P^T$ to create a $60°$ rotation matrix M about the axis $[1, 2, -1]^T$. The main point is in knowing how to infer P from the aforementioned orthogonal basis system. Be careful of avoiding inadvertent reflections during the basis transformation by checking $\det(P)$. Now show how to recover the axis and angle of rotation from M using complex-valued diagonalization. [Hint: The eigenvalues are the same for similar matrices and the axis of rotation is an invariant direction.]

33. Show how you can use the Jordan normal form of a matrix to quickly identify its rank and its four fundamental subspaces.

34. Consider the following quadratic form:

$$f(x_1, x_2, x_3) = x_1^2 + 2x_2^2 + x_3^2 + a\,x_1 x_2 + x_2 x_3$$

Under what conditions on a is the function $f(x_1, x_2, x_3)$ convex?

35. Useful for Kernel Methods: Consider an $n \times n$ non-singular matrix $A = BB^T$, which is the left Gram matrix of $n \times n$ matrix B. Propose an algorithm that takes B as input and generates 100 different matrices, $B_1 \ldots B_{100}$, such that A is the left Gram matrix of each B_i. How many such matrices exist? Is it possible to obtain a B_i that is also symmetric like A? Is any B_i triangular? [Note: For an $n \times n$ similarity matrix A, the kth row of B_i is a multidimensional representation of the kth object.]

36. Let P be an $n \times n$ nonnegative stochastic transition matrix of probabilities, so that the probabilities in each row sum to 1. Find a right eigenvector with eigenvalue 1 by inspection. Prove that no eigenvalue can be larger than 1.

37. Suppose that $A = V\Delta V^{-1}$ is a diagonalizable matrix. Show that the matrix $\lim_{n \to \infty} (I + A/n)^n$ exists with finite entries. [This result holds for any square matrix, and the proof for the general case is a good challenge exercise.]

38. Eigenvalues are scaling factors along specific directions. Construct a 2×2 diagonalizable matrix A and 2-dimensional vector $\overline{x}$, so that each eigenvalue of A is less than 1 in absolute magnitude and the length of $A\overline{x}$ is larger than that of $\overline{x}$. Prove that any such matrix A cannot be symmetric. Explain both phenomena geometrically.

39. Mahalanobis distance: Let $C = D^T D/n$ be the covariance matrix of an $n \times d$ mean-centered data set. The squared Mahalanobis distance of the ith row $\overline{X}_i$ of D to the mean of the data set (which is the origin in this case) is given by the following:

$$\delta_i^2 = \overline{X}_i C^{-1} \overline{X}_i^T$$

Let $C = P\Delta P^T$ be the diagonalization of C, and each row vector $\overline{X}_i$ be transformed to $\overline{Z}_i = \overline{X}_i P$. Normalize each attribute of the transformed data matrix DP by dividing with its standard derivation to make its variance 1 along each dimension and to create the new rows $\overline{Z}_1' \ldots \overline{Z}_n'$. Show that the Mahalanobis distance δ_i is equal to $\|\overline{Z}_i'\|$.

40. Non-orthogonal diagonalization of symmetric matrix: Consider the following diagonalization of a symmetric matrix:

$$
\begin{bmatrix} 3 & 0 & 1 \\ 0 & 4 & 0 \\ 1 & 0 & 3 \end{bmatrix}
=
\begin{bmatrix} 1/\sqrt{2} & 0 & 1/\sqrt{2} \\ 0 & 1 & 0 \\ 1/\sqrt{2} & 0 & -1/\sqrt{2} \end{bmatrix}
\begin{bmatrix} 4 & 0 & 0 \\ 0 & 4 & 0 \\ 0 & 0 & 2 \end{bmatrix}
\begin{bmatrix} 1/\sqrt{2} & 0 & 1/\sqrt{2} \\ 0 & 1 & 0 \\ 1/\sqrt{2} & 0 & -1/\sqrt{2} \end{bmatrix}
$$

Find an alternative diagonalization $V \Delta V^{-1}$ in which at least some column pairs of V are not orthogonal. [Hint: Try modifying this diagonalization using tied eigenvectors.]

41. You have a 100000×100 sparse matrix D, and you want to compute the dominant eigenvector of its left Gram matrix DD^T. Unfortunately, DD^T is a non-sparse matrix of size 100000×100000, which causes computational problems. Show how you can implement the power method using only sparse matrix-vector multiplications.

42. Multiple choice: Suppose $\overline{x}_i^T A \overline{x}_i > 0$ for d vectors $\overline{x}_1 \ldots \overline{x}_d$ and $d \times d$ symmetric matrix A. Then, A is always positive definite if the different $\overline{x}_i$'s are (i) linearly independent, (ii) orthogonal, (iii) A-orthogonal, (iv) any of the above, or (v) none of the above? Justify your answer.

43. Convert the diagonalization in the statement of Exercise 40 into Gram matrix form $A = B^T B$ and then compute the Cholesky factorization $A = LL^T = R^T R$ using the QR decomposition $B = QR$.

Chapter 4

Optimization Basics: A Machine Learning View

"If you optimize everything, you will always be unhappy."–Donald Knuth

4.1   Introduction

Many machine learning models are often cast as continuous optimization problems in multiple variables. The simplest example of such a problem is least-squares regression, which is also viewed as a fundamental problem in linear algebra. This is because solving a (consistent) system of equations is a special case of least-squares regression. In least-squares regression, one finds the *best-fit solution* to a system of equations that may or may not be consistent, and the loss corresponds to the aggregate squared error of the best fit. The special case of a consistent system of equations yields a loss value of 0. Least-squares regression has a special place in linear algebra, optimization, and machine learning, because it serves as a foundational problem in all three disciplines. Least-squares regression historically preceded the classification problem in machine learning, and the optimization models for classification were often motivated as modifications of the least-squares regression model. The main difference between least-squares regression and classification is that the predicted target variable is numerical in the former, whereas it is discrete (typically binary) in the latter. Therefore, the optimization model for linear regression needs to be "repaired" in order to make it usable for discrete target variables. This chapter will make a special effort to show how least-squares regression is so foundational to machine learning.

   Most continuous optimization methods use differential calculus in one form or the other. Differential calculus is an old discipline, and it was independently invented by Isaac Newton and Gottfried Leibniz in the 17th century. The main idea of differential calculus is to provide a quantification of the *instantaneous* rate of change of an objective function with respect to each of the variables in its argument. Optimization methods based on differential calculus use the fact that the rate of change of an objective function at a particular set of values

C. C. Aggarwal, *Linear Algebra and Optimization for Machine Learning*,
https://doi.org/10.1007/978-3-030-40344-7_4

of the optimization variables provides hints on how to iteratively change the optimization variable(s) and bring them closer to an optimum solution. Such iterative algorithms are easy to implement on modern computers. Although computers had not been invented in the 17th century, Newton proposed several iterative methods to provide humans a systematic way to manually solve optimization problems (albeit with some rather tedious work). It was natural to adapt these methods later as computational algorithms, when computers were invented. This chapter will introduce the basics of optimization and the associated computational algorithms. Later chapters will expand on these ideas.

This chapter is organized as follows. The next section will discuss the basics of optimization. The notion of convexity is introduced in Section 4.3 because of its importance in machine learning. Important details of gradient descent are discussed in Section 4.4. There are several ways in which optimization problems are manifested in a different way in machine learning (than in traditional applications). This issue will be discussed in Section 4.5. Useful matrix calculus notations and identities are introduced in Section 4.6 for computing the derivatives of objective functions with respect to vectors. The least-squares regression problem is introduced in Section 4.7. The design of machine learning algorithms with discrete targets is presented in Section 4.8. Optimization models for multiway classification are discussed in Section 4.9. Coordinate descent methods are discussed in Section 4.10. A summary is given in Section 4.11.

4.2 The Basics of Optimization

An optimization problem has an *objective function* that is defined in terms of a set of variables, referred to as *optimization variables*. The goal of the optimization problem is to compute the values of the variables at which the objective function is either maximized or minimized. It is common to use a minimization form of the objective function in machine learning, and the corresponding objective function is often referred to as a *loss function*. Note that the term "loss function" often (semantically) refers to an objective function with certain types of properties quantifying a nonnegative "cost" associated with a particular configuration of variables. This term is used in the econometrics, statistics, and the machine learning communities, although the term "objective function" is a more general concept than the term "loss function." For example, a loss function is always associated with a minimization objective function, and it is often interpreted as a cost with a nonnegative value. Most objective functions in machine learning are multivariate loss functions over many variables. First, we will consider the simple case of optimization functions defined on a single variable.

4.2.1 Univariate Optimization

Consider a single-variable objective function $f(x)$ as follows:

$$f(x) = x^2 - 2x + 3 \tag{4.1}$$

This objective function is an upright parabola, which can also be expressed in the form $f(x) = (x - 1)^2 + 2$. The objective function is shown in Figure 4.2(a); it clearly takes on its minimum value at $x = 1$, where the nonnegative term $(x - 1)^2$ drops to 0. Note that at the minimum value, the rate of change of $f(x)$ with respect to x is zero, as the tangent to the

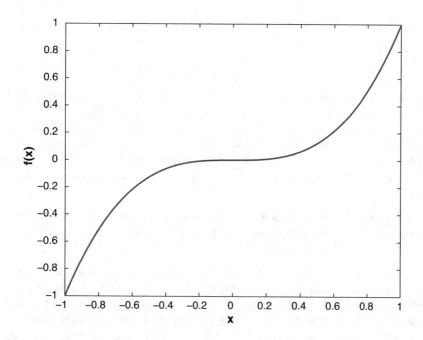

Figure 4.1: Example of 1-dimensional function $F(x) = x^3$

plot at that point is horizontal. One can also find this optimal value by computing the first derivative $f'(x)$ of the function $f(x)$ with respect to x and setting it to 0:

$$f'(x) = \frac{\mathrm{d}f(x)}{\mathrm{d}x} = 2x - 2 = 0 \tag{4.2}$$

Therefore, we obtain $x = 1$ as an optimum value. Intuitively, the function $f(x)$ changes at zero rate on slightly perturbing the value of x from $x = 1$, which suggests that it is an optimal point. However, this analysis alone is not sufficient to conclude that the point is a minimum. In order to understand this point, consider the *inverted* parabola, obtained by setting $g(x) = -f(x)$:

$$g(x) = -f(x) = -x^2 + 2x - 3 \tag{4.3}$$

Setting the derivative of $g(x)$ to 0 yields *exactly* the same solution of $x = 1$:

$$g'(x) = 2 - 2x = 0 \tag{4.4}$$

However, in this case the solution $x = 1$ is a maximum rather than a minimum. Furthermore, the point $x = 0$ is an *inflection point* or *saddle point* of the function $F(x) = x^3$ (cf. Figure 4.1), even though the derivative is 0 at $x = 0$. Such a point is neither a maximum nor a minimum.

All points for which the first derivative is zero are referred to as *critical points* of the optimization problem. A critical point might be a maximum, minimum, or saddle point. How does one distinguish between the different cases for critical points? One observation is that a function looks like an upright bowl at a minimum point, which implies that its first derivative increases at minima. In other words, the *second derivative* (i.e., derivative of the derivative) will be positive for minima (although there are a few exceptions to this rule).

For example, the second derivatives for the two quadratic functions $f(x)$ and $g(x)$ discussed above are as follows:

$$f''(x) = 2 > 0, \qquad g''(x) = -2 < 0$$

The case where the second derivative is zero is somewhat ambiguous, because such a point could be a minimum, maximum, or an inflection point. Such a critical point is referred to as *degenerate*. Therefore, for a single-variable optimization function $f(x)$ in minimization form, satisfying *both* $f'(x) = 0$ and $f''(x) > 0$ is sufficient to ensure that the point is a minimum with respect to its *immediate locality*. Such a point is referred to as a *local* minimum. This does not, however, mean that the point x is a *global* minimum across the entire range of values of x.

Lemma 4.2.1 (Optimality Conditions in Unconstrained Optimization) *A univariate function $f(x)$ is a minimum value at $x = x_0$ with respect to its immediate locality if it satisfies both $f'(x_0) = 0$ and $f''(x_0) > 0$.*

These conditions are referred to as *first-order* and *second-order* conditions for minimization. The above conditions are *sufficient* for a point to be minimum with respect to its *infinitesimal locality*, and they are "almost" *necessary* for the point to be a minimum with respect to its locality. We use the word "almost" in order to address the degenerate case where a point x_0 might satisfy $f'(x_0) = 0$ and $f''(x_0) = 0$. This type of setting is an ambiguous situation where the point x_0 might or might not be a minimum. As an example of this ambiguity, the functions $F(x) = x^3$ and $G(x) = x^4$ have zero first and second derivatives at $x = 0$, but only the latter is a minimum. One can understand the optimality condition of Lemma 4.2.1 by using a Taylor expansion of the function $f(x)$ within a small locality $x_0 + \Delta$ (cf. Section 1.5.1 of Chapter 1):

$$f(x_0 + \Delta) \approx f(x_0) + \underbrace{\Delta f'(x_0)}_{0} + \frac{\Delta^2}{2} f''(x_0)$$

Note that Δ might be either positive or negative, although Δ^2 will always be positive. The value of $|\Delta|$ is assumed to be extremely small, and successive terms rapidly drop off in magnitude. Therefore, it makes sense to keep only the first non-zero term in the above expansion in order to meaningfully compare $f(x_0)$ with $f(x_0 + \Delta)$. Since $f'(x_0)$ is zero, the first non-zero term is the second-order term containing $f''(x_0)$. Furthermore, since Δ^2 and $f''(x_0)$ are positive, it follows that $f(x_0 + \Delta) = f(x_0) + \epsilon$, where ϵ is some positive quantity. This means that $f(x_0)$ is less than $f(x_0 + \Delta)$ for any small value of Δ, whether it is positive or negative. In other words, x_0 is a minimum with respect to its immediate locality.

The Taylor expansion also provides insights as to why the degenerate case $f'(x_0) = f''(x_0) = 0$ is problematic. In the event that $f''(x)$ is zero, one would need to keep expanding the Taylor series until one reaches the first non-zero term. If the first non-zero term is positive, then one can show that $f(x_0 + \Delta) < f(x_0)$. An example of such a function is $f(x) = x^4$ at $x_0 = 0$. In such a case, x_0 is indeed a minimum with respect to its immediate locality. However, if the first non-zero term is negative or it depends on the sign of Δ, it could be a maximum or saddle point; an example is the inflection point of x^3 at the origin, which is shown in Figure 4.1.

Problem 4.2.1 *Consider the quadratic function $f(x) = ax^2 + bx + c$. Show that a point can be found at which $f(x)$ satisfies the optimality condition (for minimization) when $a > 0$. Show that the optimality condition (for maximization) is satisfied when $a < 0$.*

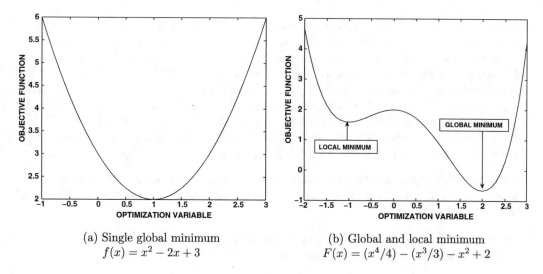

(a) Single global minimum
$f(x) = x^2 - 2x + 3$

(b) Global and local minimum
$F(x) = (x^4/4) - (x^3/3) - x^2 + 2$

Figure 4.2: Illustrations of local and global optima

A quadratic function is a rather simple case in which a single minimum or maximum exists, depending on the sign of the quadratic term. However, other functions have multiple turning points. For example, the function $\sin(x)$ is periodic, and has an infinite number of minima/maxima over $x \in (-\infty, +\infty)$. It is noteworthy that the optimality conditions of Lemma 4.2.1 only focus on defining a minimum in a local sense. In other words, the point is minimum with respect to its infinitesimal locality. A point that is a minimum only with respect to its immediate locality is referred to as a *local* minimum. Intuitively, the word "local" refers to the fact that the point is a minimum only within its neighborhood of (potentially) infinitesimal size. The minimum across the entire domain of values of the optimization variable is the *global* minimum. It is noteworthy that the conditions of Lemma 4.2.1 do not tell us with certainty whether or not a point is a global minimum. However, these conditions are sufficient for a point to be at least a local minimum and "almost" necessary to be a local minimum (i.e., necessary with the exception of the degenerate case discussed earlier with a zero second derivative).

Next, we will consider an objective function that has both local and global minima:

$$F(x) = (x^4/4) - (x^3/3) - x^2 + 2$$

This function is shown in Figure 4.2(b), and it has two possible minima. The minimum at $x = -1$ is a *local* minimum, and the minimum at $x = 2$ is a *global* minimum. Both the local and global minima are shown in Figure 4.2(b). On differentiating $F(x)$ with respect to x and setting it to zero, we obtain the following condition:

$$x^3 - x^2 - 2x = x(x+1)(x-2) = 0$$

The roots are $x \in \{-1, 0, 2\}$. The second derivative is $3x^2 - 2x - 2$, which is positive at -1 and 2 (minima), and negative at $x = 0$ (maximum). The value of the function at the two minima are as follows:

$$F(-1) = 1/4 + 1/3 - 1 + 2 = 19/12$$
$$F(2) = 4 - 8/3 - 4 + 2 = -2/3$$

Therefore, $x = 2$ is a *global* minimum, whereas $x = -1$ is a *local* minimum. It is noteworthy that $x = 0$ is a (local) maximum satisfying $F(0) = 2$. This local maximum appears as a small hill with a peak at $x = 0$ in Figure 4.2(b). Local optima pose a challenge for optimization problems, because there is often no way of knowing whether a solution satisfying the optimality conditions is the global optimum or not. Certain types of optimization functions, referred to as *convex* functions, are guaranteed to have a single global minimum. An example of a convex function is the univariate quadratic objective function of Figure 4.2(a). Before discussing convex functions, we will discuss the problem of reaching a solution that satisfies the conditions of Lemma 4.2.1 (and its generalization to multiple variables).

Problem 4.2.2 *Show that the function $F(x) = x^4 - 4x^3 - 2x^2 + 12x$ takes on minimum values at $x = -1$ and $x = 3$. Show that it takes on a maximum value at $x = 1$. Which of these are local optima?*

Problem 4.2.3 *Find the local and global optima of $F(x) = (x - 1)^2[(x - 3)^2 - 1]$. Which of these are maxima and which are minima?*

4.2.1.1 Why We Need Gradient Descent

Solving the equation $f'(x) = 0$ for x provides an *analytical* solution for a critical point. Unfortunately, it is not always possible to compute such analytical solutions in closed form. It is often difficult to exactly solve the equation $f'(x) = 0$ because this derivative might itself be a complex function of x. In other words, a *closed form solution* (like the example above) typically does not exist. For example, consider the following function that needs to be minimized:

$$f(x) = x^2 \cdot \log_e(x) - x \qquad (4.5)$$

Setting the first derivative of this function to 0 yields the following condition:

$$f'(x) = 2x \cdot \log_e(x) + x - 1 = 0$$

This equation is somewhat hard to solve, although iterative methods exist for solving it. By trial and error, one might get lucky and find out that $x = 1$ is indeed a solution to the first-order optimality condition because it satisfies $f'(1) = 2\log_e(1) + 1 - 1 = 0$. Furthermore, the second derivative $f''(x)$ can be shown to be positive at $x = 1$, and therefore this point is at least a local minimum. However, solving an equation like this numerically causes all types of numerical and computational challenges; these types of challenges increase when we move from univariate optimization to multivariate optimization.

A very popular approach for optimizing objective functions (irrespective of their functional form) is to use the method of *gradient descent*. In gradient descent, one starts at an initial point $x = x_0$ and successively updates x using the *steepest descent direction*:

$$x \Leftarrow x - \alpha f'(x)$$

Here, $\alpha > 0$ regulates the step size, and is also referred to as the *learning rate*. In the univariate case, the notion of "steepest" is hard to appreciate, as there are only two directions of movement (i.e., increase x or decrease x). One of these directions causes ascent, whereas the other causes descent. However, in multivariate problems, there can be an infinite number of possible directions of descent, and the generalization of the notion of univariate derivative leads to the steepest descent direction. The value of x changes in each iteration by $\delta x = -\alpha f'(x)$. Note that at infinitesimally small values of the learning rate $\alpha > 0$, the above updates will always reduce $f(x)$. This is because for very small α, we can use the first-order Taylor expansion to obtain the following:

$$f(x + \delta x) \approx f(x) + \delta x f'(x) = f(x) - \alpha[f'(x)]^2 < f(x) \qquad (4.6)$$

Using very small values of $\alpha > 0$ is not advisable because it will take a long time for the algorithm to converge. On the other hand, using large values of α could make the effect of the update unpredictable with respect to the computed gradient (as the first-order Taylor expansion is no longer a good approximation). After all, the gradient is only an instantaneous rate of change, and it does not apply over larger ranges. Therefore, large step-sizes could cause the solution to overshoot an optimal value, if the sign of the gradient changes over the length of the step. At extremely large values of the learning rate, it is even possible for the solution to *diverge*, where it moves at an increasing speed towards large absolute values, and typically terminates with a numerical overflow.

In the following, we will show two iterations of the gradient descent procedure for the function of Equation 4.5. Consider the case where we start at $x_0 = 2$, which is larger than the optimal value of $x = 1$. At this point, the value of $f'(x)$ can be shown to be $2\log_e(2) + 1 \approx 2.4$. If we use $\alpha = 0.2$, then the value of x gets updated from x_0 as follows:

$$x_1 \Leftarrow x_0 - 0.2 * 2.4 = 2 - 0.48 = 1.52$$

This new value of x is closer to the optimal solution. One can then recompute the derivative at $x_1 = 1.52$ and perform the update $x \Leftarrow 1.52 - 0.2 * f'(1.52)$. Performing this update again and again to construct the sequence $x_0, x_1, x_2 \ldots x_t$ will eventually converge to the optimal value of $x_t = 1$ for large values of t. Note that the choice of α does matter. For example, if we choose $\alpha = 0.8$, then it results in the following update:

$$x_1 \Leftarrow x_0 - \alpha f'(x_0) = 2 - 2.4 * 0.8 = 0.08$$

In this case, the solution has overshot the optimal value of $x = 1$, although it is still closer to the optimal solution than the initial point of $x_0 = 2$. The solution can still be shown to *converge* to an optimal value, but after a longer time. As we will see later, even this is not guaranteed in all cases.

4.2.1.2 Convergence of Gradient Descent

The execution of gradient-descent updates will generally result in a sequence of values $x_0, x_1 \ldots x_t$ of the optimization variable, which become successively closer to an optimum solution. As the value of x_t nears the optimum value, the derivative $f'(x_t)$ also tends to be closer and closer to zero (thereby satisfying the first-order optimality conditions of Lemma 4.2.1). In other words, the absolute step size will tend to reduce over the execution of the algorithm. As gradient descent nears an optimal solution, the objective function will also improve at a slower rate. This observation provides some natural ideas on making decisions regarding the termination of the algorithm (when the current solution is sufficiently close to an optimal value). The idea is to plot the current value of $f(x_t)$ with iteration index t as the algorithm progresses. A typical example of good progress during gradient descent is shown in Figure 4.3(a). The X-axis contains the iteration index, whereas the Y-axis contains the objective function value. The objective function value need not be monotonically decreasing over the course of the algorithm, but it will tend to show small noisy changes (without significant long-term direction) after some point. This situation can be treated as a good termination point for the algorithm. However, in some cases, the update steps can be shown to *diverge* from an optimal solution, if the step size is not chosen properly.

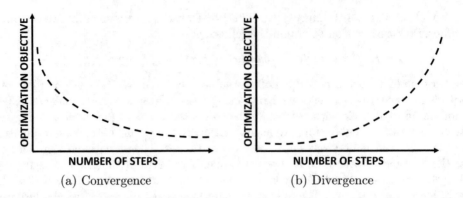

(a) Convergence   (b) Divergence

Figure 4.3: Typical behaviors of objective function during convergence and divergence

4.2.1.3   The Divergence Problem

Choosing a very large learning rate α can cause overshooting from the optimal solution, and even divergence in some cases. In order to understand this point, let us consider the quadratic function $f(x)$ of Figure 4.2(a), which takes on its optimal value at $x = 1$:

$$f(x) = x^2 - 2x + 3$$

Now imagine a situation where the starting point is $x_0 = 2$, and one chooses a large learning rate $\alpha = 10$. The derivative of $f(x) = 2x - 2$ evaluates to $f'(x_0) = f'(2) = 2$. Then, the update from the first step yields the following:

$$x_1 \Leftarrow x_0 - 10 * 2 = 2 - 20 = -18$$

Note that the new point x_1 is *much further away* from the optimal value of $x = 1$, which is caused by the overshooting problem. Even worse, the absolute gradient is very large at this point, and it evaluates to $f'(-18) = -38$. If we keep the learning rate fixed, it will cause the solution to move at an even faster rate in the opposite direction:

$$x_2 \Leftarrow x_1 - 10 * (-38) = -18 + 380 = 362$$

In this case, the solution has overshot back in the original direction but is even further away from the optimal solution. Further updates cause back-and-forth movements at increasingly large amplitudes:

$$x_3 \Leftarrow x_2 - 10 * 722 = 362 - 7220 = -6858, \quad x_4 \Leftarrow x_3 + 10 * 13718 = 130322$$

Note that each iteration flips the sign of the current solution and increases its magnitude by a factor of about 20. In other words, the solution moves away faster and faster from an optimal solution until it leads to a numerical overflow. An example of the behavior of the objective function during divergence is shown in Figure 4.3(b).

It is common to reduce the learning rate over the course of the algorithm, and one of the many purposes served by such an approach is to arrest divergence; however, in some cases, such an approach might not prevent divergence, especially if the initial learning rate is large. Therefore, when an analyst encounters a situation in gradient descent, where the size of the parameter vector seems to increase rapidly (and the optimization objective worsens), it is a tell-tale sign of divergence. The first adjustment should be to experiment with a lower initial learning rate. However, choosing a learning rate that is too small might lead to unnecessarily slow progress, which causes the entire procedure to take too much time. There is a considerable literature in finding the correct step size or adjusting it over the course of the algorithm. Some of these issues will be discussed in later sections.

4.2.2 Bivariate Optimization

The univariate optimization scenario is rather unrealistic, and most optimization problems in real-world settings have multiple variables. In order to understand the subtle differences between single-variable and multivariable optimization, we will first consider the case of an optimization function containing two variables. This setting is referred to as *bivariate optimization*, and it is helpful in bridging the gap in complexity from single-variable optimization to multivariate optimization. For ease in understanding, we will consider bivariate generalizations of the univariate optimization functions in Figure 4.2. We construct bivariate functions by adding two instances of the univariate function shown in Figure 4.2 as follows:

$$g(x, y) = f(x) + f(y) = x^2 + y^2 - 2x - 2y + 6$$
$$G(x, y) = g(x) + g(y) = ([x^4 + y^4]/4) - ([x^3 + y^3]/3) - x^2 - y^2 + 4$$

Note that these functions are simplified and have very special structure; they are *additively separable*. Additively separable functions are those in which univariate terms are added, and they do not interact with one another. In other words, an additively separable function might contain terms like $\sin(x^2)$ and $\sin(y^2)$, but not $\sin(xy)$. Nevertheless, these simplified polynomial functions are adequate for demonstrating the complexities associated with multivariable optimization. In fact, as discussed in Section 3.4.4 of Chapter 3, all quadratic functions can be represented in additively separable form (although this is not true for non-quadratic functions). The two bivariate functions $g(x, y)$ and $G(x, y)$ are shown in Figure 4.4(a) and (b), respectively. It is evident that the single-variable cross-sections of the objective functions in Figure 4.4(a) and (b) are similar to the 1-dimensional functions in Figure 4.2(a) and (b). The objective function of Figure 4.4(a) has a single global optimum (like the quadratic function of Figure 4.2(a) in one dimension). However, the objective function of Figure 4.4(b) has four minima, only one of which is global minimum at $[x, y] = [2, 2]$. Examples of local and global minima are annotated in Figure 4.4(b).

In this case, one can compute the *partial derivative* of the objective functions $g(x, y)$ and $G(x, y)$ (of Figure 4.2) in order to perform gradient descent. A partial derivative computes the derivative with respect to a particular variable, while treating other variables as constants. In fact, a "gradient" is naturally defined as a vector of partial derivatives. One can compute the gradient of the function $g(x, y)$ in Figure 4.4(a) as follows:

$$\nabla g(x, y) = \left[\frac{\partial g(x, y)}{\partial x}, \frac{\partial g(x, y)}{\partial y} \right]^T = \left[\begin{array}{c} 2x - 2 \\ 2y - 2 \end{array} \right]$$

The notation "∇" is added in front of a function to denote its gradient. This notation will be consistently used in the book, and we will occasionally add subscripts like $\nabla_{x,y} g(x, y)$ to clarify the choice of variables with respect to which the gradient is computed. In this case, the gradient is a column vector with two components, because we have two optimization variables x and y. Each component of the 2-dimensional vector is a partial derivative of the objective function with respect to one of the two variables. The simplest approach for

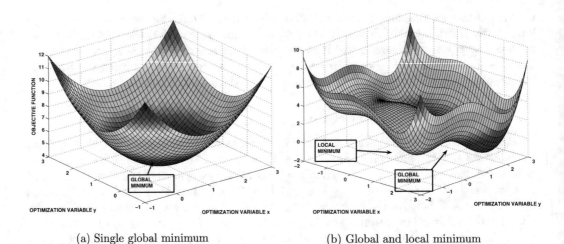

(a) Single global minimum
$$g(x, y) = x^2 + y^2 - 2x - 2y + 6$$

(b) Global and local minimum
$$G(x, y) = ([x^4 + y^4]/4)$$
$$-([x^3 + y^3]/3) - x^2 - y^2 + 4$$

Figure 4.4: Illustrations of local and global optima

solving the optimization problem is to set the gradient $\nabla g(x, y)$ to zero, which leads to the solution $[x, y] = [1, 1]$. We will discuss the second-order optimality conditions (to distinguish between maxima, minima, and inflection points) in Section 4.2.3.

The simple approach of setting the gradient of the objective function to zero might not always lead to a system of equations with a closed-form solution. The common solution is to use gradient-descent updates with respect to the optimization variables $[x, y]$ as follows:

$$\left[\begin{array}{c} x_{t+1} \\ y_{t+1} \end{array} \right] \Leftarrow \left[\begin{array}{c} x_t \\ y_t \end{array} \right] - \alpha \nabla g(x_t, y_t) = \left[\begin{array}{c} x_t \\ y_t \end{array} \right] - \alpha \left[\begin{array}{c} 2x_t - 2 \\ 2y_t - 2 \end{array} \right]$$

So far, we have only examined additively separable functions with simple structure. Now let us consider a somewhat more complicated function:

$$H(x, y) = x^2 - \sin(xy) + y^2 - 2x$$

In such a case, the term $\sin(xy)$ ensures that the function is not additively separable. In such a case, the gradient of the function can be shown to be the following:

$$\nabla H(x, y) = \left[\frac{\partial H(x, y)}{\partial x}, \frac{\partial H(x, y)}{\partial y} \right]^T = \left[\begin{array}{c} 2x - y\cos(xy) - 2 \\ 2y - x\cos(xy) \end{array} \right]$$

Although the partial derivative components are no longer expressed in terms of individual variables, gradient descent updates can be performed in a similar manner to the previous case.

As in the case of univariate optimization, the presence of local optima remains a consistent problem. For example, in the case of the function $G(x, y)$ shown in Figure 4.4(b), local optima are clearly visible. All critical points can be found by setting the gradient $\nabla G(x, y)$ to 0:

$$\nabla G(x, y) = \left[\begin{array}{c} x^3 - x^2 - 2x \\ y^3 - y^2 - 2y \end{array} \right] = \overline{0}$$

This optimization problem has an interesting structure, because any of the nine pairs $(x, y) \in \{-1, 0, 2\} \times \{-1, 0, 2\}$ satisfies the first order optimality conditions, and are therefore critical points. Among these, there is a single global minimum, three local minima, and a single local maximum at $(0, 0)$. The other four can be shown to be saddle points. The classification of points as minima, maxima, or saddle points can only be accomplished with the use of multivariate second-order conditions, which are direct generalizations of the univariate optimality conditions of Lemma 4.2.1. The discussion of second-order optimality conditions for the multivariate case is deferred to Section 4.2.3. Note the rapid proliferation of the number of possible critical points satisfying the optimality conditions when the optimization problem contains two variables instead of one. In general, when a multivariate problem is posed as sum of univariate functions, the number of local optima can proliferate exponentially fast with the number of optimization variables.

Problem 4.2.4 *Consider a univariate function $f(x)$, which has k values of x satisfying the optimality condition $f'(x) = 0$. Let $G(x, y) = f(x) + f(y)$ be a bivariate objective function. Show that there are k^2 pairs (x, y) satisfying $\nabla G(x, y) = \overline{0}$. How many tuples $[x_1, \ldots, x_d]^T$ would satisfy the first-order optimality condition for the d-dimensional function $H(x_1 \ldots x_d) = \sum_{i=1}^{d} f(x_i)$?*

In the case of the objective function of Figure 4.4(b), a single (local or global) optimum exists in each of the four quadrants. Furthermore, it can be shown that starting the gradient descent in a particular quadrant (at low learning rates) will converge to the single optimum in that quadrant because each quadrant contains its own local bowl. At higher learning rates, it is possible for the gradient descent to overshoot a local/global optimum and move to a different bowl (or even behave in an unpredictable way with numerical overflows). Therefore, the final resting point of gradient descent depends on (what would seem to be) small details of the computational procedure, such as the starting point or the learning rate. We will discuss many of these details in Section 4.4.

The function $g(x, y)$ of Figure 4.4(a) has a single global optimum and no local optima. In such cases, one is more likely to reach the global optimum, irrespective of where one starts the gradient-descent procedure. The better outcome in this case is a result of the structure of the optimization problem. Many optimization problems that are encountered in machine learning have the nice structure of Figure 4.4(a) (or something very close to it), as a result of which local optima cause fewer problems than would seem at first glance.

4.2.3   Multivariate Optimization

Most machine learning problems are defined on a large parameter space containing multiple optimization variables. The variables of the optimization problem are *parameters* that are used to create a *prediction function* of either observed or hidden attributes of the machine learning problem. For example, in a linear regression problem, the optimization variables $w_1, w_2 \ldots w_d$ are used to predict the dependent variable y from the independent variables $x_1 \ldots x_d$ as follows:

$$y = \sum_{i=1}^{d} w_i x_i$$

Starting from this section, we assume that only the notations $w_1 \ldots w_d$ represent optimization variables, whereas the other "variables" like x_i and y are really observed values from the data set at hand (which are constants from the optimization perspective). This notation is typical for machine learning problems. The objective functions often penalize differences in observed and predicted values of specific attributes, such as the variable y shown above. For example, if we have many observed tuples of the form $[x_1, x_2 \ldots x_d, y]$, one can sum up the values of $(y - \sum_{i=1}^{d} w_i x_i)^2$ over all the observed tuples. Such objective functions are often referred to as loss functions in machine learning parlance. Therefore, we will often substitute the term "objective function" with "loss function" in the remainder of this chapter. In this section, we will assume that the loss function $J(\overline{w})$ is a function of a vector of multiple optimization variables $\overline{w} = [w_1 \ldots w_d]^T$. Unlike the discussion in the preceding sections, we will use the notations $w_1 \ldots w_d$ for optimization variables, because the notations $\overline{X}$, x_i, $\overline{y}$, and y_i, will be reserved for the attributes in the data (whose values are observed). Although attributes are also sometimes referred to as "variables" (e.g., dependent and independent variables) in machine learning parlance, they are not variables from the perspective of the optimization problem. The values of the attributes are always fixed based on the observed data during training, and therefore appear among the (constant) coefficients of the optimization problem. Confusingly, these attributes (with constant observed values) are also referred to as "variables" in machine learning, because they are arguments of the prediction function that the machine learning algorithm is trying to model. The use of notations such as $\overline{X}$, x_i, $\overline{y}$, and y_i to denote attributes is a common practice in the machine learning community. Therefore, the subsequent discussion in this chapter will be consistent with this convention. The value of d corresponds to the number of optimization variables in the problem at hand, and the parameter vector $\overline{w} = [w_1 \ldots w_d]^T$ is assumed to be a column vector.

The computation of the gradient of an objective function of d variables is similar to the bivariate case discussed in the previous section. The main difference is that a d-dimensional vector of partial derivatives is computed instead of a 2-dimensional vector. The ith component of the d-dimensional gradient vector is the partial derivative of J with respect to the ith parameter w_i. The simplest approach to solve the optimization problem directly (without gradient descent) is to set the gradient vector to zero, which leads to the following set of d conditions:

$$\frac{\partial J(\overline{w})}{\partial w_i} = 0, \quad \forall i \in \{1 \ldots d\}$$

These conditions lead to a system of d equations, which can be solved to determine the parameters $w_1 \ldots w_d$. As in the case of univariate optimization, one would like to have a way to characterize whether a critical point (i.e., zero-gradient point) is a maximum, minimum, or inflection point. This brings us to the second-order condition. Recall that in single-variable optimization, the condition for $f(w)$ to be a minimum is $f''(w) > 0$. In multivariate optimization, this principle is generalized with the use of the *Hessian* matrix. Instead of a scalar second derivative, we have a $d \times d$ matrix of second-derivatives, which includes *pairwise* derivatives of J with respect to different pairs of variables. The Hessian of the loss function $J(\overline{w})$ with respect to the optimization variables $w_1 \ldots w_d$ is given by a $d \times d$ symmetric matrix H, in which the (i, j)th entry H_{ij} is defined as follows:

$$H_{ij} = \frac{\partial^2 J(\overline{w})}{\partial w_i \partial w_j} \tag{4.7}$$

Note that the (i, j)th entry of the Hessian is equal to the (j, i)th entry because partial derivatives are commutative according to *Schwarz's theorem*. The fact that the Hessian is a symmetric matrix is helpful in many computational algorithms that require eigendecomposition of the matrix.

The Hessian matrix is a direct generalization of the univariate second derivative $f''(w)$. For a univariate function, the Hessian is a 1×1 matrix containing $f''(w)$ as its only entry.

Strictly speaking, the Hessian is a *function* of $\overline{w}$, and should be denoted by $H(\overline{w})$, although we denote it by H for brevity. In the event that the function $J(\overline{w})$ is quadratic, the entries in the Hessian matrix do not depend on the parameter vector $\overline{w} = [w_1 \ldots w_d]^T$. This is similar to the univariate case, where the second derivative $f''(w)$ is a constant when the function $f(w)$ is quadratic. In general, however, the Hessian matrix depends on the value of the parameter vector $\overline{w}$ at which it is computed. For a parameter vector $\overline{w}$ at which the gradient is zero (i.e., critical point), one needs to test the Hessian matrix H in the same way we test $f''(w)$ in univariate functions. Just as $f''(w)$ needs to be positive for a point w to be a minimum, the Hessian matrix H needs to be positive-*definite* for a point to be guaranteed to be a minimum. In order to understand this point, we consider the second-order, multivariate Taylor expansion of $J(\overline{w})$ in the immediate locality of $\overline{w}_0$ along the direction $\overline{v}$ and small radius $\epsilon > 0$:

$$J(\overline{w}_0 + \epsilon\overline{v}) \approx J(\overline{w}_0) + \epsilon\,\overline{v}^T \underbrace{[\nabla J(\overline{w}_0)]}_{0} + \frac{\epsilon^2}{2}[\overline{v}^T H \overline{v}] \qquad (4.8)$$

The Hessian matrix H, which depends on the parameter vector, is computed at $\overline{w} = \overline{w}_0$. It is evident that the objective function $J(\overline{w}_0)$ will be less than $J(\overline{w}_0 + \epsilon\overline{v})$ when we have $\overline{v}^T H \overline{v} > 0$. If we can find even a single direction $\overline{v}$ where we have $\overline{v}^T H \overline{v} < 0$, then $\overline{w}$ is clearly not a minimum with respect to its immediate locality. A matrix H that satisfies $\overline{v}^T H \overline{v} > 0$ is positive definite (cf. Section 3.3.8). The notion of positive definiteness of the Hessian is the direct generalization of the second-derivative condition $f''(w) > 0$ for univariate functions. After all, the Hessian of a univariate function is a 1×1 matrix containing the second derivative. The single entry in this matrix needs to be positive for this 1×1 matrix to be positive-definite.

Assuming that the gradient is zero at critical point $\overline{w}$, we can summarize the following second-order optimality conditions:

1. If the Hessian is positive definite at $\overline{w} = [w_1 \ldots w_d]^T$, then $\overline{w}$ is a local minimum.

2. If the Hessian is negative definite at $\overline{w} = [w_1 \ldots w_d]^T$, then $\overline{w}$ is a local maximum.

3. If the Hessian is indefinite at $\overline{w}$, then $\overline{w}$ is a saddle point.

4. If the Hessian is positive- or negative **semi**-definite, then the test is inconclusive, because the point could either be a local optimum or a saddle point.

These conditions represent direct generalizations of univariate optimality conditions. It is helpful to examine what the saddle point for an indefinite Hessian matrix looks like. Consider the following optimization objective function $g(w_1, w_2) = w_1^2 - w_2^2$. The Hessian of this quadratic function is independent of the parameter vector $[w_1, w_2]^T$, and is defined as follows:

$$H = \begin{bmatrix} 2 & 0 \\ 0 & -2 \end{bmatrix}$$

This Hessian turns out to be a diagonal matrix, which is clearly indefinite because one of the two diagonal entries is negative. The point $[0, 0]$ is a critical point because the gradient is zero at that point. However, this point is a saddle point because of the indefinite nature of the Hessian matrix. This saddle point is illustrated in Figure 4.5.

Problem 4.2.5 *The gradient of the objective function $J(\overline{w})$ is 0 and the determinant of the Hessian is negative at $\overline{w} = \overline{w}_0$. Is $\overline{w}_0$ a minimum, maximum, or a saddle-point?*

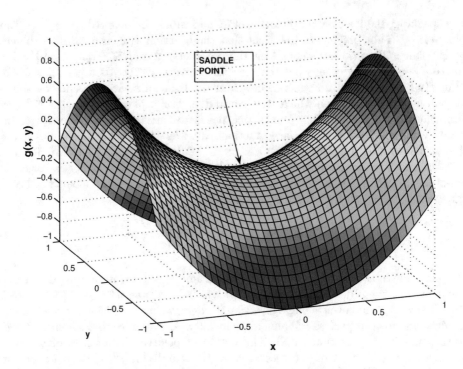

Figure 4.5: Re-visiting Figure 3.6: Illustration of saddle point created by indefinite Hessian

Setting the gradient of the objective function to 0 and then solving the resulting system of equations is usually computationally difficult. Therefore, gradient-descent is used. In other words, we use the following updates repeatedly with learning rate α:

$$[w_1 \ldots w_d]^T \Leftarrow [w_1 \ldots w_d]^T - \alpha \left[\frac{\partial J(\overline{w})}{\partial w_1} \ldots \frac{\partial J(\overline{w})}{\partial w_d} \right]^T \tag{4.9}$$

One can also write the above expression in terms of the gradient of the objective function with respect to $\overline{w}$:

$$\overline{w} \Leftarrow \overline{w} - \alpha \nabla J(\overline{w})$$

Here, $\nabla J(\overline{w})$ is a column vector containing the partial derivatives of $J(\overline{w})$ with respect to the different parameters in the column vector $\overline{w}$. Although the learning rate α is shown as a constant here, it usually varies over the course of the algorithm (cf. Section 4.4.2).

4.3   Convex Objective Functions

The presence of local minima creates uncertainty about the effectiveness of gradient-descent algorithms. Ideally, one would like to have an objective function without local minima. A specific type of objective function with this property is the class of *convex* functions. First, we need to define the concept of *convex sets*, as convex functions are defined only with domains that are convex.

Definition 4.3.1 (Convex Set) *A set S is convex, if for every pair of points $\overline{w}_1, \overline{w}_2 \in S$, the point $\lambda \overline{w}_1 + [1 - \lambda]\overline{w}_2$ must also be in S for all $\lambda \in (0,1)$.*

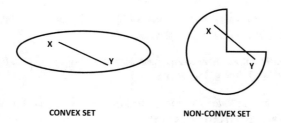

CONVEX SET **NON-CONVEX SET**

Figure 4.6: Examples of convex and non-convex sets

In other words, it is impossible to find a pair of points in the set such that any of the points on the straight line joining them do not lie in the set. A *closed convex set* is one in which the boundary points (i.e., limit points) of the set are included within the set, whereas an *open convex set* is one in which all points within the boundary are included but not the boundary itself. For example, in 1-dimensional space the set is $[-2, +2]$ is a closed convex set, whereas the set $(-2, +2)$ is an open convex set.

Examples of convex and non-convex sets are illustrated in Figure 4.6. A circle, an ellipse, a square, or a half-moon are all convex sets. However, a three-quarter circle is not a convex set because one can draw a line between the two points inside the set, so that a portion of the line lies outside the set (cf. Figure 4.6).

A convex function $F(\overline{w})$ is defined as a function with a convex domain that satisfies the following condition for any $\lambda \in (0, 1)$:

$$F(\lambda \overline{w}_1 + (1 - \lambda)\overline{w}_2) \leq \lambda F(\overline{w}_1) + (1 - \lambda)F(\overline{w}_2) \tag{4.10}$$

One can generalize the convexity condition to k points, as discussed in the practice problem below.

Problem 4.3.1 *For a convex function $F(\cdot)$, and k parameter vectors $\overline{w}_1 \ldots \overline{w}_k$, show that the following is true for any $\lambda_1 \ldots \lambda_k \geq 0$ and satisfying $\sum_i \lambda_i = 1$:*

$$F(\sum_{i=1}^{k} \lambda_i \overline{w}_i) \leq \sum_{i=1}^{k} \lambda_i F(\overline{w}_i)$$

The simplest example of a convex objective function is the class of quadratic functions in which the leading (quadratic) term has a nonnegative coefficient:

$$f(w) = a \cdot w^2 + b \cdot w + c$$

Here, a needs to be nonnegative for the function to be considered quadratic. The result can be shown by using the convexity condition above. All linear functions are always convex, because the convexity property holds with equality.

Lemma 4.3.1 *A linear function of the vector $\overline{w}$ is always convex.*

Convex functions have a number of useful properties that are leveraged in practical applications.

Lemma 4.3.2 *Convex functions obey the following properties:*

1. *The sum of convex functions is always convex.*

2. *The maximum of convex functions is convex.*

3. *The square of a nonnegative convex function is convex.*

4. *If $F(\cdot)$ is a convex function with a single argument and $G(\overline{w})$ is a linear function with a scalar output, then $F(G(\overline{w}))$ is convex.*

5. *If $F(\cdot)$ is a convex non-increasing function and $G(\overline{w})$ is a concave function with a scalar output, then $F(G(\overline{w}))$ is convex.*

6. *If $F(\cdot)$ is a convex non-decreasing function and $G(\overline{w})$ is a convex function with a scalar output, then $F(G(\overline{w}))$ is convex.*

We leave the detailed proofs of these results (which can be derived from Equation 4.10) as an exercise:

Problem 4.3.2 *Prove all the results of Lemma 4.3.2 using the definition of convexity.*

There are several natural combinations of convex functions that one might expect to be convex at first glance, but turn out to be non-convex on closer examination. The product of two convex functions is not necessarily convex. The functions $f(x) = x$ and $g(x) = x^2$ are convex functions, but their product $h(x) = f(x) \cdot g(x) = x^3$ is not convex (see Figure 4.1). Furthermore, the composition of two convex functions is not necessarily convex, and it might be indefinite or concave. As a specific example, consider the linear convex function $f(x) = -x$ and also the quadratic convex function $g(x) = x^2$. Then, we have $f(g(x)) = -x^2$, which is a concave function. The result on the composition of functions is important from the perspective of deep neural networks (cf. Chapter 11). *Even though the individual nodes of neural networks usually compute convex functions, the composition of the functions computed by successive nodes is often not convex.*

A nice property of convex functions is that a local minimum will also be a global minimum. If there are two "local" minima, then the above convexity condition ensures that the entire line joining them also has the same objective function value.

Problem 4.3.3 *Use the convexity condition to show that every local minimum in a convex function must also be a global minimum.*

The fact that every local minimum is a global minimum can also be characterized by using a geometric definition of convexity. This geometric definition, which is also referred to as the *first-derivative condition*, is that the entire convex function will always lie above a tangent to a convex function, as shown in Figure 4.7. This figure illustrates a 2-dimensional convex function, where the horizontal directions are arguments to the function (i.e., optimization variables), and the vertical direction is the objective function value. An important consequence of convexity is that one is often guaranteed to reach a global optimum if successful convergence occurs during the gradient-descent procedure.

The condition of Figure 4.7 can also be written algebraically using the gradient of the convex function at a given point $\overline{w}_0$. In fact, this condition provides an alternative definition of convexity. We summarize this condition below:

Lemma 4.3.3 (First-Derivative Characterization of Convexity) *A   differentiable function $F(\overline{w})$ is a convex function if and only if the following is true for any pair $\overline{w}_0$ and $\overline{w}$:*

$$F(\overline{w}) \geq F(\overline{w}_0) + [\nabla F(\overline{w}_0)] \cdot (\overline{w} - \overline{w}_0)$$

We omit a detailed proof of the lemma. Note that if the gradient of $F(\overline{w})$ is zero at $\overline{w} = \overline{w}_0$, it would imply that $F(\overline{w}) \geq F(\overline{w}_0)$ for any $\overline{w}$. In other words, $\overline{w}_0$ is a global minimum. Therefore, any critical point that satisfies the first-derivative condition is a global minimum.

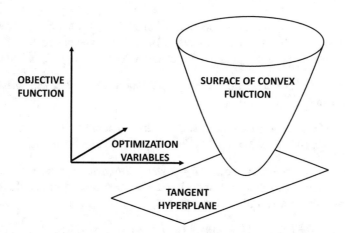

OBJECTIVE
FUNCTION

SURFACE OF CONVEX
FUNCTION

OPTIMIZATION
VARIABLES

TANGENT
HYPERPLANE

Figure 4.7: A convex function always lies entirely above any tangent to the surface. The example illustrates a 2-dimensional function, where the two horizontal axes are the optimization variables and the vertical axis is the objective function value

The main disadvantage of the first-derivative condition (with respect to the direct definition of convexity) is that it applies only to differentiable functions. Interestingly, there is a third characterization of convexity in terms of the second-derivative:

Lemma 4.3.4 (Second-Derivative Characterization of Convexity) *The twice differentiable function $F(\overline{w})$ is convex, if and only if it has a positive semidefinite Hessian at every value of the parameter $\overline{w}$ in the domain of $F(\cdot)$.*

The second derivative condition has the disadvantage of requiring the function $F(\overline{w})$ to be twice differentiable. Therefore, the following convexity definitions are equivalent for twice-differentiable functions defined over $\mathcal{R}^d$:

1. **Direct:** The convexity condition $F(\lambda \overline{w}_1 + [1 - \lambda]\overline{w}_2) \leq \lambda F(\overline{w}_1) + (1 - \lambda)F(\overline{w}_2)$ is satisfied for all $\overline{w}_1, \overline{w}_2$ and $\lambda \in (0, 1)$.

2. **First-derivative:** The first-derivative condition $F(\overline{w}) \geq F(\overline{w}_0) + [\nabla F(\overline{w}_0)] \cdot (\overline{w} - \overline{w}_0)$ is satisfied for all $\overline{w}$ and $\overline{w}_0$.

3. **Second-derivative:** The Hessian of $F(\overline{w})$ is positive semidefinite for all $\overline{w}$.

One can choose to use any of the above conditions as the definition of convexity, and then derive the other two as lemmas. However, the direct definition is slightly more general because it does not depend on differentiability, whereas the other definitions have the additional requirement of differentiability. For example, the function $F(\overline{w}) = \|\overline{w}\|_1$ is convex but only the first definition can be used because of its non-differentiability at any point where a component of $\overline{w}$ is 0. We refer the reader to [10, 15, 22] for detailed proofs of the equivalence of the various definitions in the differentiable case. It is often the case that a particular definition is easier to use than another when one is trying to prove the convexity of a specific function. Many machine learning objective functions are of the form $F(G(\overline{w}))$, where $G(\overline{w})$ is the linear function $\overline{w} \cdot \overline{X}^T$ for a row vector containing a d-dimensional data point $\overline{X}$, and $F(\cdot)$ is a univariate function. In such a case, one only needs to prove that the univariate function $F(\cdot)$ is convex, based on the final portion of Lemma 4.3.2. It is particularly easy to use the second-order condition $F''(\cdot) \geq 0$ for univariate functions. As a specific example, we provide a practice exercise for showing the convexity of the logarithmic logistic loss function. This function is useful for showing the convexity of logistic regression.

Problem 4.3.4 *Use the second derivative condition to show that the univariate function* $F(x) = log_e(1 + exp(-x))$ *is convex.*

Problem 4.3.5 *Use the second-derivative condition to show that if the univariate function* $F(x)$ *is convex, then the function* $G(x) = F(-x)$ *must be convex as well.*

A slightly stronger condition than convexity is *strict convexity* in which the convexity condition is modified to strict inequality. A strictly convex function $F(\overline{w})$ is defined as a function that satisfies the following condition for any $\lambda \in (0, 1)$:

$$F(\lambda \overline{w}_1 + (1 - \lambda)\overline{w}_2) < \lambda F(\overline{w}_1) + (1 - \lambda)F(\overline{w}_2)$$

For example, a bowl with a flat bottom is convex, but it is not strictly convex. A strictly convex function will have a unique global minimum. One can also adapt the first-order conditions to strictly convex functions. A function $F(\cdot)$ can be shown to be strictly convex if and only if the following condition holds for all $\overline{w}$ and $\overline{w}_0$:

$$F(\overline{w}) > F(\overline{w}_0) + [\nabla F(\overline{w}_0)] \cdot (\overline{w} - \overline{w}_0)$$

The second-derivative condition cannot, however, be fully generalized to strict convexity. If a function has a positive definite Hessian everywhere, then it is guaranteed to be strictly convex. However, the converse does not necessarily hold. For example, the function $f(x) = x^4$ is strictly convex, but its second derivative is 0 at $x = 0$. An important property of strictly convex functions is the following:

Lemma 4.3.5 *A strictly convex function can contain at most one critical point. If such a point exists, it will be the global minimum of the strictly convex function.*

The above property is easy to show by using either the direct definition or the first-order definition of strict convexity. One often constructs objective functions in machine learning by adding convex and strictly convex functions. In such cases, the sum of these functions is strictly convex.

Lemma 4.3.6 *The sum of a convex function and a strictly convex function is strictly convex.*

The proof of this lemma is not very different from that of the proof of Lemma 4.3.2 for the sum of two convex functions. Many objective functions in machine learning are convex, and they can often be made strictly convex by adding a strictly convex regularizer.

A special case of convex functions is that of quadratic convex functions, which can be directly expressed in terms of the positive semidefinite Hessian. Although the Hessian of a function depends on the value of the parameter vector at a specific point, it is a constant matrix in the case of quadratic functions. An example of a quadratic convex function $f(\overline{w})$ in terms of the constant Hessian matrix H is the following:

$$f(\overline{w}) = \frac{1}{2}[\overline{w} - \overline{b}]^T H[\overline{w} - \overline{b}] + c$$

Here, $\overline{b}$ is a d-dimensional column vector, and c is a scalar. The properties of such convex functions are discussed in Chapter 3. A convex objective function is an ideal setting for a gradient-descent algorithm; the approach will never get stuck in a local minimum. Although the objective functions in complex machine learning models (like neural networks) are not convex, they are often close to convex. As a result, gradient-descent methods work quite well in spite of the presence of local optima.

For any convex function $F(\overline{w})$, the region of the space bounded by $F(\overline{w}) \leq b$ for any constant b can be shown to be a convex set. This type of constraint is encountered often in optimization problems. Such problems are easier to solve because of the convexity of the space in which one wants to search for the parameter vector.

4.4 The Minutiae of Gradient Descent

An earlier section introduces gradient descent, which serves as the workhorse of much of optimization in machine learning. However, as the example in Section 4.2.1.3 shows, small details do matter; an improper choice of the learning rate can cause divergence of gradient descent, rather than convergence. This section discusses these important minutiae.

4.4.1 Checking Gradient Correctness with Finite Differences

Many machine learning algorithms use complex objective functions over millions of parameters. The gradients are computed either analytically and then hand-coded into the algorithm, or they are computed using automatic differentiation methods in applications like neural networks (cf. Chapter 11). In all these cases, analytical or coding errors remain a real possibility, which may or may not become obvious during execution. Knowing the reason for the poor performance of an algorithm is a critical step in deciding whether to simply debug the algorithm or to make fundamental design changes.

Consider a situation where we compute the gradient of the objective function $J(\overline{w}) = J(w_1 \ldots w_d)$. In the finite-difference method, we sample a few of the optimization parameters from $w_1 \ldots w_d$ and check their partial derivatives using the *finite-difference approximation*. The basic idea is to perturb an optimization parameter w_i by a small amount Δ and approximate the partial derivative with respect to w_i by using the difference between the perturbed value of the objective function and the original value:

$$\frac{\partial J(\overline{w})}{\partial w_i} \approx \frac{J(w_1 \ldots, w_i + \Delta, \ldots, w_d) - J(w_1, \ldots, w_i, \ldots, w_d)}{\Delta}$$

This way of estimating the gradient is referred to as a finite-difference approximation. As the name suggests, one would not obtain an exact value of the partial derivative in this way. However, in cases where the gradients are computed incorrectly, the value of the finite-difference approximation is often so wildly different from the analytical value that the error becomes self-evident. Typically, it suffices to check the partial derivatives of a small subset of the parameters in order to detect a systemic problem in gradient computation.

4.4.2 Learning Rate Decay and Bold Driver

A constant learning rate often poses a dilemma to the analyst. A lower learning rate used early on will cause the algorithm to take too long to reach anywhere close to an optimal solution. On the other hand, a large initial learning rate will allow the algorithm to come reasonably close to a good solution at first; however, the algorithm will then oscillate around the point for a very long time. Allowing the learning rate to decay over time can naturally achieve the desired learning-rate adjustment to avoid these challenges. Therefore, a decaying learning rate α_t is subscripted with the time-stamp t, and the update is as follows:

$$\overline{w} \Leftarrow \overline{w} - \alpha_t \nabla J$$

The time t is typically measured in terms of the number of cycles over all training points. The two most common decay functions are *exponential decay* and *inverse decay*. The learning rate α_t can be expressed in terms of the initial decay rate α_0 and time t as follows:

$$\alpha_t = \alpha_0 \exp(-k \cdot t) \quad \text{[Exponential Decay]}$$
$$\alpha_t = \frac{\alpha_0}{1 + k \cdot t} \quad \text{[Inverse Decay]}$$

The parameter k controls the rate of the decay. Another approach is to use step decay in which the learning rate is reduced by a particular factor every few steps of gradient descent.

Another popular approach for adjusting the learning rate is the *bold-driver algorithm*. In the bold-driver algorithm, the learning rate changes, depending on whether the objective function is improving or worsening. The learning rate is *increased* by factor of around 5% in each iteration as long as the steps improve the objective function. As soon as the objective function *worsens* because of a step, the step is *undone* and an attempt is made again with the learning rate reduced by a factor of around 50%. This process is continued to convergence. A tricky aspect of the bold-driver algorithm is that it does not work in some noisy settings of gradient descent, where the objective function is approximated by using samples of the data. An example of such a noisy setting is *stochastic gradient descent*, which is discussed later in this chapter. In such cases, it is important to test the objective function and adjust the learning rate after m steps, rather than a single step. The change in objective function can be measured more robustly across multiple steps, and all m steps must be undone when the objective function worsens over these steps.

4.4.3   Line Search

Line search directly uses the optimum step size in order to provide the best improvement. Although it is rarely used in vanilla gradient descent (because it is computationally expensive), it is helpful in some specialized variations of gradient descent. Some inexact variations (like the *Armijo rule*) can be used in vanilla gradient descent because of their efficiency.

Let $J(\overline{w})$ be the objective function being optimized and $\overline{g}_t$ be the descent direction at the beginning of the tth step with parameter vector $\overline{w}_t$. In the steepest-descent method, the direction $\overline{g}_t$ is the same as $-\nabla J(\overline{w}_t)$, although advanced methods (see next chapter) might use other descent directions. In the following, we will not assume that $\overline{g}_t$ is the steepest-descent direction in order to preserve generality of the exposition. Clearly, the parameter vector needs to be updated as follows:

$$\overline{w}_{t+1} \Leftarrow \overline{w}_t + \alpha_t \overline{g}_t$$

In line search, the learning rate α_t is chosen in each step, so as to minimize the value of the objective function at $\overline{w}_{t+1}$. The step-size α_t is computed as follows:

$$\alpha_t = \text{argmin}_\alpha J(\overline{w}_t + \alpha \overline{g}_t) \tag{4.11}$$

After performing the step, the gradient is computed at $\overline{w}_{t+1}$ for the next step. The gradient at $\overline{w}_{t+1}$ will be perpendicular to the search direction $\overline{g}_t$ or else α_t will not be optimal. This result can be shown by observing that if the gradient of the objective function at $\overline{w}_t + \alpha_t \overline{g}_t$ has a non-zero dot product with the current movement direction $\overline{g}_t$, then one can improve the objective function by moving an amount of either $+\delta$ or $-\delta$ along $\overline{g}_t$ from $\overline{w}_{t+1}$:

$$J(\overline{w}_t + \alpha_t \overline{g}_t \pm \delta \overline{g}_t) \approx J(\overline{w}_t + \alpha_t \overline{g}_t) \pm \delta \underbrace{\overline{g}_t^T [\nabla J(\overline{w}_t + \alpha_t \overline{g}_t)]}_{0} \quad \text{[Taylor Expansion]}$$

Therefore, we obtain the following:

$$\overline{g}_t^T [\nabla J(\overline{w}_t + \alpha_t \overline{g}_t)] = 0$$

We summarize the result below:

Lemma 4.4.1 *The gradient at the optimal point of a line search is always orthogonal to the current search direction.*

A natural question arises as to how the minimization of Equation 4.11 is performed. One important property of typical line-search settings is that the objective function $H(\alpha) = J(\overline{w}_t + \alpha \overline{g}_t)$, when expressed in terms of α is often a unimodal function. The main reason for this is that typical machine learning settings that use the line-search method use quadratic, convex approximations of the original objective function on which the search is done. Examples of such techniques include the *Newton method* and the *conjugate gradient method* (cf. Chapter 5).

The first step is to identify a range $[0, \alpha_{max}]$ in which to perform the search. This can be performed efficiently by evaluating the objective function value at geometrically increasing values of α (increasing every time by a factor of 2). Subsequently, it is possible to use a variety of methods to narrow the interval such as the *binary-search method*, the *golden-section search method*, and the *Armijo rule*. The first two of these methods and exact methods, and they leverage the unimodality of the objective function in terms of the step-size α. The Armijo rule is inexact, and it works even when $H(\alpha) = J(\overline{w}_t + \alpha \overline{g}_t)$ is multimodal/nonconvex in α. Therefore, the Armijo rule has broader use than exact line-search methods, especially as far as simple forms of gradient descent are concerned. In the following, we discuss these different methods.

4.4.3.1 Binary Search

We start by initializing the binary search interval for α to $[a, b] = [0, \alpha_{max}]$. In binary search over $[a, b]$, the interval is narrowed by evaluating the objective function at two closely spaced points near $(a + b)/2$. We evaluate the objective function at $(a + b)/2$ and $(a + b)/2 + \epsilon$, where ϵ is a numerically small value like 10^{-6}. In other words, we compute $H[(a+b)/2]$ and $H[(a+b)/2+\epsilon]$. This allows us to evaluate whether the function is increasing or decreasing at $(a+b)/2$ by determining which of the two evaluations is larger. If the function is increasing at $(a + b)/2$, the interval is narrowed to $[a, (a + b)/2 + \epsilon]$. Otherwise, it is narrowed to $[(a + b)/2, b]$. This process is repeated, until an interval is reached with the required level of accuracy.

4.4.3.2 Golden-Section Search

As in the case of binary search, we start by initializing $[a, b] = [0, \alpha_{max}]$. However, the process of narrowing the interval is different. The basic principle in golden-section search is to use the fact that if we pick any pair of middle samples m_1, m_2 for α in the interval $[a, b]$, where $a < m_1 < m_2 < b$, at least one of the intervals $[a, m_1]$ and $[m_2, b]$ can be dropped. In some cases, an even larger interval like $[a, m_2]$ and $[m_1, b]$ can be dropped. This is because the minimum value for a unimodal function must always lie in an adjacent interval to the choice of $\alpha \in \{a, m_1, m_2, b\}$ that yields the minimum value of $H(\alpha)$. When $\alpha = a$ yields the minimum value for $H(\alpha)$, we can exclude the interval $(m_1, b]$, and when $\alpha = b$ yields the minimum value for $H(\alpha)$, we can exclude the interval $[a, m_2)$. When $\alpha = m_1$ yields the minimum value, we can exclude the interval $(m_2, b]$, and when $\alpha = m_2$ yields the minimum value, we can exclude the interval $[a, m_1)$. The new bounds $[a, b]$ for the interval are reset based on these exclusions. At the end of the process, we are left with an interval containing either 0 or 1 evaluated point. If we have an interval containing no evaluated point, we first select a random point $\alpha = p$ in the (reset) interval $[a, b]$, and then another random point $\alpha = q$ in the larger of the two intervals $[a, p]$ and $[p, b]$. On the other hand, if we are left with an interval $[a, b]$ containing a single evaluated point $\alpha = p$, then we select $\alpha = q$ in the larger of the two intervals $[a, p]$ and $[p, b]$. This yields another set of four points over which we can apply golden-section search. This process is repeated until an interval is reached with the required level of accuracy.

4.4.3.3   Armijo Rule

The basic idea behind the Armijo rule is that the descent direction $\overline{g}_t$ at the starting point $\overline{w}_t$ (i.e., at $\alpha = 0$) often deteriorates in terms of rate of improvement of objective function as one moves further along this direction. The rate of improvement of the objective function along the search direction at the starting point is $|\overline{g}_t^T [\nabla F(\overline{w}_t)]|$. Therefore, the (typical) improvement of the objective function at a particular value of α can optimistically be expected to be $\alpha |\overline{g}_t^T [\nabla F(\overline{w}_t)]|$ for most[1] real-world objective functions. The Armijo rule is satisfied with a fraction $\mu \in (0, 0.5)$ of this improvement. A typical value of μ is around 0.25. In other words, we want to find the largest step-size α satisfying the following:

$$F(\overline{w}_t) - F(\overline{w}_t + \alpha \overline{g}_t) \geq \mu \alpha |\overline{g}_t^T [\nabla F(\overline{w}_t)]|$$

Note that for small enough values of α, the condition above will always be satisfied. In fact, one can show using the finite-difference approximation, that for infinitesimally small values of α, the condition above is satisfied at $\mu = 1$. However, we want a larger step size to ensure faster progress. What is the largest step-size one can use? We test successively *decreasing* values of α for the condition above, and stop the first time the condition above is satisfied. In backtracking line search, we start by testing $H(\alpha_{max})$, $H(\beta \alpha_{max}) \ldots H(\beta^r \alpha_{max})$, until the condition above is satisfied. At that point we use $\alpha = \beta^r \alpha_{max}$. Here, β is a parameter drawn from $(0, 1)$, and a typical value is 0.5.

When to Use Line Search

Although the line-search method can be shown to converge to at least a local optimum, it is expensive. This is the reason that it is rarely used in vanilla gradient descent. However, it is used in some specialized variations of gradient descent like *Newton's method* (cf. Section 5.4 of Chapter 5). Exact line search is required in some of these variations, whereas fast, inexact methods like the Armijo rule can be used in vanilla gradient descent. When exact line search is required, the number of steps is often relatively small, and the fewer number of steps more

---

[1]It is possible to construct pathological counter-examples where this is not true.

than compensate for the expensive nature of the individual steps. An important point with the use of line-search is that convergence is guaranteed, even if the resulting solution is a local optimum.

4.4.4 Initialization

The gradient-descent procedure always starts at an initial point, and successively improves the parameter vector at a particular learning rate. A critical question arises as to how the initialization point can be chosen. For some of the relatively simple problems in machine learning (like the ones discussed in this chapter), the vector components of the initialization point can be chosen as small random values from $[-1, +1]$. In case the parameters are constrained to be nonnegative, the vector components can be chosen from $[0, 1]$.

However, this simple way of initialization can sometimes cause problems for more complex algorithms. For example, in the case of neural networks, the parameters have complex dependencies on one another, and choosing good initialization points can be critical. In other cases, choosing improper magnitudes of the initial parameters can cause numerical overflows or underflows during the updates. It is sometimes effective to use some form of heuristic optimization for initialization. Such an approach already *pretrains* the algorithm to an initialization near an optimum point. The choice of the heuristic generally depends on the algorithm at hand. Some learning algorithms like neural networks have systematic ways of performing pretraining and choosing good initializations. In this chapter, we will give some examples of heuristic initializations.

4.5 Properties of Optimization in Machine Learning

The optimization problems in machine learning have some typical properties that are often not encountered in other generic optimization settings. This section will provide an overview of these specific quirks of optimization in machine learning.

4.5.1 Typical Objective Functions and Additive Separability

Most objective functions in machine learning penalize the deviation of a *predicted value* from an *observed value* in one form or another. For example, the objective function of least-squares regression is as follows:

$$J(\overline{w}) = \sum_{i=1}^{n} \|\overline{w} \cdot \overline{X}_i^T - y_i\|^2 \tag{4.12}$$

Here, $\overline{X}_i$ is a d-dimensional row vector containing the ith of n training points, $\overline{w}$ is a d-dimensional column vector of optimization variables, and y_i contains the real-valued observation of the ith training point. Note that this objective function represents an additively separable sum of squared differences between the *predicted values* $\hat{y}_i = \overline{w} \cdot \overline{X}_i^T$ and the *observed values* y_i in the actual data.

Another form of penalization is the negative *log-likelihood objective function*. This form of the objective function uses the probability that the model's prediction of a dependent variable matches the observed value in the data. Clearly, higher values of the probability are desirable, and therefore the model should learn parameters that maximize these probabilities (or *likelihoods*). For example, such a model might output the probability of each class in a binary classification setting, and it is desired to maximize the probability of the true (observed) class. For the ith training point, this probability is denoted by $P(\overline{X}_i, y_i, \overline{w})$, which depends on the parameter vector $\overline{w}$ and training pair $(\overline{X}_i, y_i)$. The probability of correct prediction over all training points is given by the products of probabilities $P(\overline{X}_i, y_i, \overline{w})$ over all $(\overline{X}_i, y_i)$. The negative logarithm is applied to this product to convert the maximization problem into a minimization problem (while addressing numerical underflow issues caused by repeated multiplication):

$$J(\overline{w}) = -\log_e \left[\prod_{i=1}^{n} P(\overline{X}_i, y_i, \overline{w}) \right] = -\sum_{i=1}^{n} \log_e \left[P(\overline{X}_i, y_i, \overline{w}) \right] \qquad (4.13)$$

Using the logarithm also makes the objective function appear as an *additively separable sum* over the training points.

As evident from the aforementioned examples, many machine learning problems use additively separable data-centric objective functions, whether squared loss or log-likelihood loss is used. This means that each individual data point creates a small (additive) component of the objective function. In each case, the objective function contains n additively separable terms, and each point-specific error [such as $J_i = (y_i - \overline{w} \cdot \overline{X}_i^T)^2$ in least-squares regression] can be viewed as a point-specific loss. Therefore, the overall objective function can be expressed as the sum of these point-specific losses:

$$J(\overline{w}) = \sum_{i=1}^{n} J_i(\overline{w}) \qquad (4.14)$$

This type of linear separability is useful, because it enables the use of fast optimization methods like *stochastic gradient descent* and *mini-batch stochastic gradient descent*, where one can replace the objective function with a sampled approximation.

4.5.2   Stochastic Gradient Descent

The linear and additive nature of the objective functions in machine learning, enables the use of techniques referred to as *stochastic gradient descent*. Stochastic gradient descent is particularly useful in the case in which the data sets are very large and one can often estimate good descent directions using modest samples of the data. Consider a sample S of the n data points $\overline{X}_1 \ldots \overline{X}_n$, where S contains the indices of the relevant data points from $\{1 \ldots n\}$. The set S of data points is referred to as a *mini-batch*. One can set up a sample-centric objective function $J(S)$ as follows:

$$J(S) = \frac{1}{2} \sum_{i \in S} (y_i - \overline{w} \cdot \overline{X}_i^T)^2 \qquad (4.15)$$

The key idea in mini-batch stochastic gradient descent is that *the gradient of $J(S)$ with respect to the parameter vector $\overline{w}$ is an excellent approximation of the gradient of the full objective function J.* Therefore, the gradient-descent update of Equation 4.9 is modified to mini-batch stochastic gradient descent as follows:

$$[w_1 \ldots w_d]^T \Leftarrow [w_1 \ldots w_d]^T - \alpha \left[\frac{\partial J(S)}{\partial w_1} \ldots \frac{\partial J(S)}{\partial w_d} \right]^T \qquad (4.16)$$

This approach is referred to as *mini-batch* stochastic gradient descent. Note that computing the gradient of $J(S)$ is far less computationally intensive compared to computing the gradient of the full objective function. A special case of mini-batch stochastic gradient descent is one in which the set S contains a single randomly chosen data point. This approach is referred to as stochastic gradient descent. The use of stochastic gradient descent is rare, and one tends to use the mini-batch method more often. Typical mini-batch sizes are powers of 2, such as 64, 128, 256, and so on. The reason for this is purely practical rather than mathematical; using powers of 2 for mini-batch sizes often results in the most efficient use of resources such as Graphics Processor Units (GPUs).

Stochastic gradient-descent methods typically cycle through the full data set, rather than simply sampling the data points at random. In other words, the data points are permuted in some random order and blocks of points are drawn from this ordering. Therefore, all other points are processed before arriving at a data point again. Each cycle of the mini-batch stochastic gradient descent procedure is referred to as an *epoch*. In the case where the mini-batch size is 1, an epoch will contain n updates, where n is the training data size. In the case where the mini-batch size is k, an epoch will contain $\lceil n/k \rceil$ updates. An epoch essentially means that every point in the training data set has been seen exactly once.

Stochastic gradient-descent methods have much lower memory requirements than pure gradient-descent, because one is processing only a small sample of the data in each step. Although each update is more noisy, the sampled gradient can be computed much faster. Therefore, even though more updates are required, the overall process is much faster. Why does stochastic gradient descent work so well in machine learning? At its core, mini-batch methods are random sampling methods. One is trying to estimate the gradient of a loss function using a random subset of the data. At the very beginning of the gradient-descent, the parameter vector $\overline{w}$ is grossly incorrect. Therefore, using only a small subset of the data is often sufficient to estimate the direction of descent very well, and the updates of mini-batch stochastic gradient descent are almost as good as those obtained using the full data (but with a tiny fraction of the computational effort). This is what contributes to the significant improvement in running time. When the parameter vector $\overline{w}$ nears the optimal value during descent, the effect of sampling error is more significant. Interestingly, it turns out that this type of error is actually *beneficial* in machine learning applications because of an effect referred to as *regularization*! The reason has to do with the subtle differences between how optimization is used traditionally as opposed to how it is used in machine learning applications. This will be the subject of the discussion in the next section.

4.5.3 How Optimization in Machine Learning Is Different

There are some subtle differences in how optimization is used in machine learning from the way it is used in traditional optimization. An important difference is that traditional optimization focuses on learning the parameters so as to optimize the objective function as much as possible. However, in machine learning, there is a differentiation between the *training data* and the (roughly similar) unseen *test data*. For example, an entrepreneur may build an optimization model based on a history of how the *independent attributes* (like forecasting indicators) relate to the *dependent variable* (like actual sales) by minimizing the squared error of prediction of the dependent variable. The assumption is that the entrepreneur is using this model to make future predictions that are not yet known, and therefore the model can only be evaluated in retrospect on new data. Predicting the training data accurately does not always help one predict unseen test data more accurately. The general rule is that the optimized model will almost always predict the dependent variable of the training data more accurately than that of the test data (since it was directly used in modeling). This difference results in some critical design choices for optimization algorithms.

Consider the example of linear regression, where one will often have training examples $(\overline{X}_1, y_1) \ldots (\overline{X}_n, y_n)$ and a separate set of test examples $(\overline{Z}_1, y_1') \ldots (\overline{Z}_t, y_t')$. The labels of the test examples are unavailable in real-world applications at the time they are predicted.

In practice, they often become available only in *retrospect*, when the true accuracy of the machine learning algorithm can be computed. Therefore, the labels of the test examples cannot be made available during training. *In machine learning, one only cares about accuracy on the unseen test examples rather than training examples.* It is possible for excellently designed optimization methods to perform very well on the training data, but have disastrously poor results on the test data. This separation between training and test data is also respected during benchmarking of machine learning algorithms by creating simulated training and test data sets from a single labeled data set. In order to achieve this goal, one simply hides a part of the labeled data, and refers to the available part as the training data and the remainder as the test data. After building the model on the training data, one evaluates the performance of the model *on the test data, which was never seen during the training phase.* This is a key difference from traditional optimization, because the model is constructed using a particular data set; yet, a different (but similar) data set is used to evaluate performance of the optimization algorithm. This difference is crucial because models that perform very well on the training data might not perform very well on the test data. In other words, the model needs to *generalize* well to unseen test data. When a model performs very well on the training data, but does not perform very well on the unseen test data, the phenomenon is referred to as *overfitting*.

In order to understand this point, consider a case where one has a 4-dimensional data set of individuals, in which the four attributes x_1, x_2, x_3, and x_4 correspond to arm span, number of freckles, length of hair, and the length of nails. The arm span is defined as the maximum distance between fingertips when an individual holds their arms out wide. The target attribute is the height of the individual. The arm span is known to be almost equal to the height of an individual (with minor variations across races, genders, and individuals), although the goal of the machine learning application is to *infer* this fact in a data-driven manner. The predicted height of the individual is modeled by the linear function $\hat{y} = w_1 x_1 + w_2 x_2 + w_3 x_3 + w_4 x_4 + w_5$ for the purposes of prediction. The *best-fit* coefficients $w_1 \ldots w_5$ can be learned in a data-driven manner by minimizing the squared loss between predicted $\hat{y}$ and observed y. One would expect that the height of an individual is highly correlated with their arm span, but the number of freckles and lengths of hair/nails are not similarly correlated. As a result, one would typically expect $w_1 x_1$ to make most of the contribution to the prediction, and the other three attributes would contribute very little (or noise). If the number of training examples is large, one would typically learn values of w_i that show this type of behavior. However, a different situation arises, if the number of training examples is small. For a problem with five parameters $w_1 \ldots w_5$, one needs at least 5 training examples to avoid a situation where an infinite number of solutions to the parameter vector exist (typically with zero error *on the training data*). This is because a system of equations of the form $y = w_1 x_1 + w_2 x_2 + w_3 x_3 + w_4 x_4 + w_5$ has an infinite number of equally good best-fit solutions if there are fewer equations than the number of variables. In fact, one can often find at least one solution in which w_1 is 0, and the squared error $(y - \sum_{i=1}^{4} w_i x_i - w_5)^2$ takes on its lowest possible value of zero on the training data. In spite of this fact, the error in the test data will typically be very high. Consider an example of a training set containing the following three data points:

Arm Span (inches)	Freckles (number)	Hair Length (inches)	Nail Length (inches)	Height (inches)
61	2	3	0.1	59
40	0	4	0.5	40
68	0	10	1.0	70

In this case, setting w_1 to 1 and all other coefficients to 0 is the "correct" solution, based on what is likely to happen *over an infinite number of training examples*. Note that this solution does not provide zero training error on this specific training data set, because there are always empirical variations across individuals. If we had an large number of examples (unlike the case of this table), it would also be possible for a model to learn this behavior well with a loss function that penalizes only the squared errors of predictions. However, with only three training examples, many other solutions exist that have zero training error. For example, setting $w_1 = 0$, $w_2 = 7$, $w_3 = 5$, $w_4 = 0$, and $w_5 = 20$ provides zero error on the training data. Here, the arm span and the nail length are not used at all. At the same time, setting $w_1 = 0$, $w_2 = 21.5$, $w_3 = 0$, $w_4 = 60$, and $w_5 = 10$ also yields zero error on the training data. This solution does not use the arm span or the hair length. Furthermore, any convex combination of these coefficients also provides zero error on the training data. Therefore, an infinite number of solutions that use irrelevant attributes provide better training error than the natural and intuitive solution that uses arm span. This is primarily because of overfitting to the specific training data at hand; this solution will generalize poorly to unseen test data.

All machine learning applications are used on unseen test data in real settings; therefore, it is unacceptable to have models that perform well on training data but perform poorly on test data. *Poor generalization is a result of models adapting to the quirks and random nuances of a specific training data set; it is likely to occur when the training data is small.* When the number of training instances is fewer than the number of features, an infinite number of equally "good" solutions exist. In such cases, poor generalization is almost inevitable unless steps are taken to avoid this problem. Therefore, there are a number of special properties of optimization in machine learning:

1. In traditional optimization, one optimizes the parameters as much as possible to improve the objective function. However, in machine learning, optimizing the parameter vector beyond a certain point often leads to overfitting. One approach is to hide a portion of the labeled data (which is referred to as the *held-out data*), perform the optimization, and always calculate the *out-of-sample accuracy* on this held-out data. Towards the end of the optimization process, the accuracy on the out-of-sample data begins to rise (even though the loss on the training data might continue to reduce). At this point, the learning is terminated. Therefore, the criterion for termination is different from that in traditional optimization.

2. While stochastic gradient-descent methods have lower accuracy than gradient-descent methods on training data (because of a sampling approximation), they often perform comparably (or even better) on the test data. This is because the random sampling of training instances during optimization reduces overfitting.

3. The objective function is sometimes modified by penalizing the squared norms of weight vectors. While the unmodified objective function is the most direct surrogate for the performance on the training data, the penalized objective function performs better on the out-of-sample test data. Concise parameter vectors with smaller squared norms are less prone to overfitting. This approach is referred to as *regularization*.

These differences between traditional optimization and machine learning are important because they affect the design of virtually every optimization procedure in machine learning.

4.5.4   Tuning Hyperparameters

As we have already seen, the learning process requires us to specify a number of hyperparameters such as the learning rate, the weight of regularization, and so on. The term "hyperparameter" is used to specifically refer to the parameters regulating the design of the model (like learning rate and regularization), and they are different from the more fundamental parameters such as the weights of the linear regression model. Machine learning always uses a two-tiered organization of parameters in the model, in which primary model parameters like weights are optimized with computational learning algorithms (e.g., stochastic gradient descent) only after fixing the hyperparameters either manually or with the use of a *tuning* phase. Here, it is important to note that the hyperparameters should not be tuned using the same data used for gradient descent. Rather, a portion of the data is held out as *validation data*, and the performance of the model is tested on the validation set with various choices of hyperparameters. This type of approach ensures that the tuning process does not overfit to the training data set.

The main challenge in hyperparameter optimization is that different combinations of hyperparameters need to be tested for their performance. The most well-known technique is *grid search*, in which all combinations of selected values of the hyperparameters are tested in order to determine the optimal choice. One issue with this procedure is that the number of hyperparameters might be large, and the number of points in the grid increases *exponentially* with the number of hyperparameters. For example, if we have 5 hyperparameters, and we test 10 values for each hyperparameter, the training procedure needs to be executed $10^5 = 100000$ times to test its accuracy. Therefore, a commonly used trick is to first work with coarse grids. Later, when one narrows down to a particular range of interest, finer grids are used. One must be careful when the optimal hyperparameter selected is at the edge of a grid range, because one would need to test beyond the range to see if better values exist.

The testing approach may at times be too expensive even with the coarse-to-fine-grained process. In some cases, it makes sense to randomly sample the hyperparameters uniformly within the grid range [14]. As in the case of grid ranges, one can perform multi-resolution sampling, where one first samples in the full grid range. One then creates a new set of grid ranges that are geometrically smaller than the previous grid ranges and centered around the optimal parameters from the previously explored samples. Sampling is repeated on this smaller box and the entire process is iteratively repeated multiple times to refine the parameters.

Another key point about sampling many types of hyperparameters is that the *logarithms* of the hyperparameters are sampled uniformly rather than the hyperparameters themselves. Two examples of such parameters include the regularization rate and the learning rate. For example, instead of sampling the learning rate α between 0.1 and 0.001, we first sample $\log_{10}(\alpha)$ uniformly between -1 and -3, and then exponentiate it as a power of 10. It is more common to search for hyperparameters in the logarithmic space, although there are some hyperparameters that should be searched for on a uniform scale.

4.5.5   The Importance of Feature Preprocessing

Vastly varying sensitivities of the loss function to different parameters tend to hurt the learning, and this aspect is controlled by the scale of the features. Consider a model in which a person's wealth is modeled as a linear function of the age x_1 (in the range $[0, 100]$), and the number of years of college education x_2 (in the range $[0, 10]$) as follows:

$$y = w_1 x_1^2 + w_2 x_2^2 \tag{4.17}$$

In such a case, the partial derivative $\frac{\partial y}{\partial w_1} = x_1^2$ and $\frac{\partial y}{\partial w_2} = x_2^2$ will show up as multiplicative terms in the components of the error gradient with respect to w_1 and w_2, respectively. Since x_1^2 is usually much larger than x_2^2 (and often by a factor of 100), the components of the error gradient with respect to w_1 will typically be much greater in magnitude than those with respect to w_2. Often, small steps along w_2 will lead to large steps along w_1 (and therefore an overshooting of the optimal value along w_1). Note that the sign of the gradient component along the w_1 direction will often keep flipping in successive steps to compensate for the overshooting along the w_1 direction after large steps. In practice, this leads to a back-and-forth "bouncing" behavior along the w_1 direction and tiny (but consistent) progress along the w_2 direction. As a result, convergence will be very slow. This type of behavior is discussed in greater detail in the next chapter. Therefore, it is often helpful to have features with similar variance. There are two forms of feature preprocessing used in machine learning algorithms:

1. *Mean-centering:* In many models, it can be useful to mean-center the data in order to remove certain types of bias effects. Many algorithms in traditional machine learning (such as principal component analysis) also work with the assumption of mean-centered data. In such cases, a vector of column-wise means is subtracted from each data point.

2. *Feature normalization:* A common type of normalization is to divide each feature value by its standard deviation. When this type of feature scaling is combined with mean-centering, the data is said to have been *standardized.* The basic idea is that each feature is presumed to have been drawn from a *standard* normal distribution with zero mean and unit variance.

   Min-max normalization is useful when the data needs to be scaled in the range $(0, 1)$. Let min_j and max_j be the minimum and maximum values of the jth attribute. Then, each feature value x_{ij} for the jth dimension of the ith point is scaled by min-max normalization as follows:

   $$x_{ij} \Leftarrow \frac{x_{ij} - min_j}{max_j - min_j} \tag{4.18}$$

Feature normalization avoids ill-conditioning and ensures much smoother convergence of gradient-descent methods.

4.6 Computing Derivatives with Respect to Vectors

In typical optimization models encountered in machine learning, one is differentiating scalar objective functions (or even vectored quantities) with respect to vectors of parameters. This is because the loss function $J(\overline{w})$ is often a function of a vector of parameters $\overline{w}$. Rather than having to write out large numbers of partial derivatives with respect to each component of the vector, it is often convenient to represent such derivatives in *matrix calculus* notation. In the matrix calculus notation, one can compute a derivative of a scalar, vector, or matrix with respect to another scalar, vector, or matrix. The result might be a scalar, vector, matrix, or tensor; the final result can often be compactly expressed in terms of the vectors/matrices in the partial derivative (and therefore one does not have to tediously compute them in elementwise form). In this book, we will restrict ourselves to computing the derivatives of scalars/vectors with respect to other scalars/vectors. Occasionally, we will consider derivatives of scalars with respect to matrices. The result is always a scalar, vector, or matrix. Being able to differentiate blocks of variables with respect to other blocks is useful from the perspective of brevity and quick computation. Although the field of matrix calculus is very broad, we will focus on a few important identities, which are useful for addressing the vast majority of machine learning problems one is likely to encounter in practice.

4.6.1   Matrix Calculus Notation

The simplest (and most common) example of matrix calculus notation arises during the computation of gradients. For example, consider the gradient-descent update for multivariate optimization problems, as discussed in the previous section:

$$\overline{w} \Leftarrow \overline{w} - \alpha \nabla J$$

An equivalent notation for the gradient ∇J is the matrix-calculus notation $\frac{\partial J(\overline{w})}{\partial \overline{w}}$. This notation is a scalar-to-vector derivative, which always returns a vector. Therefore, we have the following:

$$\nabla J = \frac{\partial J(\overline{w})}{\partial \overline{w}} = \left[\frac{\partial J(\overline{w})}{\partial w_1} \cdots \frac{\partial J(\overline{w})}{\partial w_d} \right]^T$$

Here, it is important to note that there is some convention-centric ambiguity in the treatments of matrix calculus by various communities as to whether the derivative of a scalar with respect to a column vector is a row vector or whether it is a column vector. Throughout this book, we use the convention that the derivative of a scalar with respect to a column vector is also a column vector. This convention is referred to as the *denominator layout* (although the numerator layout is more common in which the derivative is a row vector). We use the denominator layout because it frees us from the notational clutter of always having to transpose a row vector into a column vector in order to perform gradient descent updates on $\overline{w}$ (which are extremely common in machine learning). Indeed, the choice of using the numerator layout and denominator layout in different communities is often regulated by these types of notational conveniences. Therefore, we can directly write the update in matrix calculus notation as follows:

$$\overline{w} \Leftarrow \overline{w} - \alpha \left[\frac{\partial J(\overline{w})}{\partial \overline{w}} \right]$$

The matrix calculus notation also allows derivatives of vectors with respect to vectors. Such a derivative results in a matrix, referred to as the *Jacobian*. Jacobians arise frequently when computing the gradients of recursively nested multivariate functions; a specific example is the case of multilayer neural networks (cf. Chapter 11). For example, the derivative of an m-dimensional column vector $\overline{h} = [h_1, \ldots, h_m]^T$ with respect to a d-dimensional column vector $\overline{w} = [w_1, \ldots, w_d]^T$ is a $d \times m$ matrix in the denominator layout. The (i, j)th entry of this matrix is the derivative of h_j with respect to w_i:

$$\left[\frac{\partial \overline{h}}{\partial \overline{w}} \right]_{ij} = \frac{\partial h_j}{\partial w_i} \tag{4.19}$$

The (i, j)th element of the Jacobian is always $\frac{\partial h_i}{\partial w_j}$, and therefore it is the transpose of the matrix $\frac{\partial \overline{h}}{\partial \overline{w}}$ shown in Equation 4.19.

Another useful derivative that arises frequently in different types of matrix factorization is the derivative of a scalar objective function J with respect to an $m \times n$ matrix W. In the denominator layout, the result inherits the shape of the matrix in the denominator. The (i, j)th entry of the derivative is simply the derivative of J with respect to the (i, j)th entry in W.

$$\left[\frac{\partial J}{\partial W}\right]_{ij} = \frac{\partial J}{\partial W_{ij}} \tag{4.20}$$

A review of matrix calculus notations and conventions is provided in Table 4.1.

4.6.2 Useful Matrix Calculus Identities

In this section, we will introduce a number of matrix calculus identities that are used frequently in machine learning. A common expression that arises commonly in machine learning is of the following form:

$$F(\overline{w}) = \overline{w}^T A \overline{w} \tag{4.21}$$

Here, A is a $d \times d$ symmetric matrix of constant values and $\overline{w}$ is a d-dimensional column vector of optimization variables. Note that this type of objective function occurs in virtually every convex quadratic loss function like least-squares regression and in the (dual) support-vector machine. In such a case, the gradient $\nabla F(\overline{w})$ can be written as follows:

$$\nabla F(\overline{w}) = \frac{\partial F(\overline{w})}{\partial \overline{w}} = 2A\overline{w} \tag{4.22}$$

The algebraic similarity of the derivative to the scalar case is quite noticeable. The reader is encouraged to work out each element-wise partial derivative and verify that the above expression is indeed correct. Note that $\nabla F(\overline{w})$ is a column vector.

Another common objective function $G(\overline{w})$ in machine learning is the following:

$$G(\overline{w}) = \overline{b}^T B \overline{w} = \overline{w}^T B^T \overline{b} \tag{4.23}$$

Here, B is an $n \times d$ matrix of constant values and $\overline{w}$ is a d-dimensional column vector of optimization variables. Furthermore, $\overline{b}$ is an n-dimensional constant vector that does not depend on $\overline{w}$. Therefore, this is a linear function in $\overline{w}$ and all components of the gradient are constants. The values $\overline{b}^T B \overline{w}$ and $\overline{w}^T B^T \overline{b}$ are the same because the transposition of a scalar is the same scalar. In such cases, the gradient of $G(\overline{w})$ is computed as follows:

$$\nabla G(\overline{w}) = \frac{\partial G(\overline{w})}{\partial \overline{w}} = B^T \overline{b} \tag{4.24}$$

In this case, every component of the gradient is a constant. We leave the proofs of these results as a practice exercise:

Problem 4.6.1 *Let $A = [a_{ij}]$ be a symmetric $d \times d$ matrix of constant values, $B = [b_{ij}]$ be an $n \times d$ matrix of constant values, $\overline{w}$ be a d-dimensional column vector of optimization variables, and $\overline{b}$ be an n-dimensional column vector of constants. Let $F(\overline{w}) = \overline{w}^T A \overline{w}$ and let $G(\overline{w}) = \overline{b}^T B \overline{w}$. Show using component-wise partial derivatives that $\nabla F(\overline{w}) = 2A\overline{w}$ and $\nabla G(\overline{w}) = B^T \overline{b}$.*

The above practice exercise would require one to expand each expression in terms of the scalar values in the matrices and vectors. One can then appreciate the compactness of the matrix calculus approach for quick computation. We provide a list of the commonly used identities in Table 4.2. Many of these identities are useful in machine learning models.

Table 4.1: Matrix calculus operations in numerator and denominator layouts

Derivative of:	with respect to:	Output size	ith or (i,j)th element
Scalar J	Scalar x	Scalar	$\frac{\partial J}{\partial x}$
Column vector $\overline{h}$ in m dimensions	Scalar x	Column vector in m dimensions	$\left[\frac{\partial \overline{h}}{\partial x}\right]_i = \frac{\partial h_i}{\partial x}$
Scalar J	Column vector $\overline{w}$ in d dimensions	Row vector in d dimensions	$\left[\frac{\partial J}{\partial \overline{w}}\right]_i = \frac{\partial J}{\partial w_i}$
Column vector $\overline{h}$ in m dimensions	Column vector $\overline{w}$ in d dimensions	$m \times d$ matrix	$\left[\frac{\partial \overline{h}}{\partial \overline{w}}\right]_{ij} = \frac{\partial h_i}{\partial w_j}$
Scalar J	$m \times n$ matrix W	$n \times m$ matrix	$\left[\frac{\partial J}{\partial W}\right]_{ij} = \frac{\partial J}{\partial W_{ji}}$

(a) Numerator layout

Derivative of:	with respect to:	Output size	ith or (i,j)th element
Scalar J	Scalar x	Scalar	$\frac{\partial J}{\partial x}$
Column vector $\overline{h}$ in m dimensions	Scalar x	Row vector in m dimensions	$\left[\frac{\partial \overline{h}}{\partial x}\right]_i = \frac{\partial h_i}{\partial x}$
Scalar J	Column vector $\overline{w}$ in d dimensions	Column vector in d dimensions	$\left[\frac{\partial J}{\partial \overline{w}}\right]_i = \frac{\partial J}{\partial w_i}$
Column vector $\overline{h}$ in m dimensions	Column vector $\overline{w}$ in d dimensions	$d \times m$ matrix	$\left[\frac{\partial \overline{h}}{\partial \overline{w}}\right]_{ij} = \frac{\partial h_j}{\partial w_i}$
Scalar J	$m \times n$ matrix W	$m \times n$ matrix	$\left[\frac{\partial J}{\partial W}\right]_{ij} = \frac{\partial J}{\partial W_{ij}}$

(b) Denominator layout

Table 4.2: List of common matrix calculus identities in denominator layout. A is a constant $d \times d$ matrix, B is a constant $n \times d$ matrix, and $\bar{b}$ is a constant n-dimensional vector independent of the parameter vector $\overline{w}$. C is a $k \times d$ matrix

	Objective J	Derivative of J with respect to $\overline{w}$
(i)	$\overline{w}^T A \overline{w}$	$2A\overline{w}$ (symmetric A)
		$(A + A^T)\overline{w}$ (asymmetric A)
(ii)	$\bar{b}^T B \overline{w}$ or $\overline{w}^T B^T \bar{b}$	$B^T \bar{b}$
(iii)	$\|B\overline{w} + \bar{b}\|^2$	$2B^T(B\overline{w} + \bar{b})$
(iv)	$f(g(\overline{w}))$	$f'(g(\overline{w}))\nabla_w g(\overline{w})$
	[$g(\overline{w})$ is scalar: example below]	
(v)	$f(\overline{w} \cdot \overline{a})$	$f'(\overline{w} \cdot \overline{a})\overline{a}$
	[Example $g(\overline{w}) = \overline{w} \cdot \overline{a}$ of above]	

(a) Scalar-to-vector derivatives

	Vector $\overline{h}$	Derivative of $\overline{h}$ with respect to $\overline{w}$
(i)	$\overline{h} = C\overline{w}$	C^T
(ii)	$\overline{h} = F(\overline{w})$ [$F(\cdot)$ is elementwise function]	Diagonal matrix with (i,i)th entry containing partial derivative of ith component of $F(\overline{w})$ w.r.t. w_i
(iii)	Product-of-variables identity $\overline{h} = f_s(\overline{w})\overline{x}$ [$f_s(\overline{w})$ is vector-to-scalar function]	$\frac{\partial f_s(\overline{w})}{\partial \overline{w}}\overline{x}^T + f_s(\overline{w})\frac{\partial \overline{x}}{\partial \overline{w}}$

(b) Vector-to-vector derivatives

Since it is common to compute the gradient with respect to a column vector of parameters, all these identities represent the derivatives with respect to a column vector. Note that Table 4.2(b) represent some simple vector-to-vector derivatives, which always lead to the transpose of the Jacobian. Beyond these commonly used identities, a full treatment of matrix calculus is beyond the scope of the book, although interested readers are referred to [20].

4.6.2.1 Application: Unconstrained Quadratic Programming

In *quadratic programming*, the objective function contains a quadratic term of the form $\overline{w}^T A \overline{w}$, a linear term $\bar{b}^T \overline{w}$, and a constant. An unconstrained quadratic program has the following form:

$$\text{Minimize}_{\overline{w}} \, \frac{1}{2}\overline{w}^T A \overline{w} + \bar{b}^T \overline{w} + c$$

Here, we assume that A is a *positive definite* $d \times d$ matrix, $\bar{b}$ is a d-dimensional column vector, c is a scalar constant, and the optimization variables are contained in the d-dimensional column vector $\overline{w}$. An unconstrained quadratic program is a direct generalization of 1-dimensional quadratic functions like $\frac{1}{2}ax^2 + bx + c$. Note that a minimum exists at $x = -b/a$ for 1-dimensional quadratic functions when $a > 0$, and a minimum exists for multidimensional quadratic functions when A is positive definite.

The two terms in the objective function can be differentiated with respect to $\overline{w}$ by using the identities (i) and (ii) in Table 4.2(a). Since the matrix A is positive definite, it follows that the Hessian A is positive definite irrespective of the value of $\overline{w}$. Therefore, the objective function is strictly convex, and setting the gradient to zero is a necessary and sufficient condition for minimization of the objective function. Using the identities (i) and (ii) of Table 4.2(a), we obtain the following optimality condition:

$$A\overline{w} + \overline{b} = \overline{0}$$

Therefore, we obtain the solution $\overline{w} = -A^{-1}\overline{b}$. Note that this is a direct generalization of the solution for the 1-dimensional quadratic function. In the event that A is singular, a solution is not guaranteed even when A is positive semidefinite. For example, when A is the zero matrix, the objective function becomes linear with no minimum. When A is positive semidefinite, it can be shown that a minimum exists if and only if $\overline{b}$ lies in the column space of A (see Exercise 8).

4.6.2.2 Application: Derivative of Squared Norm

A special case of unconstrained quadratic programming is the norm of a vector that is itself a linear function of another vector (with an additional constant offset). Such a problem arises in least-squares regression, which is known to have a closed form solution (cf. Section 4.7) like the quadratic program of the previous section. This particular objective function has the following form:

$$J(\overline{w}) = \|B\overline{w} + \overline{b}\|^2$$
$$= \overline{w}^T B^T B\overline{w} + 2\overline{b}^T B\overline{w} + \overline{b}^T \overline{b}$$

Here, B is an $n \times d$ data matrix, $\overline{w}$ is a d-dimensional vector, and $\overline{b}$ is an n-dimensional vector. This form of the objective function arises frequently in least-squares-regression, where B is set to the observed data matrix D, and the constant vector $\overline{b}$ is set to the negative of the response vector $\overline{y}$. One needs to compute the gradient with respect to $\overline{w}$ in order to perform the updates.

We have expanded the squared norm in terms of matrix vector products above. The individual terms are of the same form as the results (i) and (ii) of Table 4.2(a). In such a case, we can compute the derivative of the squared norm with respect to $\overline{w}$ by substituting for the scalar-to-vector derivatives in results (i) and (ii) Table 4.2(a). Therefore, we obtain the following results:

$$\frac{\partial J(\overline{w})}{\partial \overline{w}} = 2B^T B\overline{w} + 2B^T \overline{b} \tag{4.25}$$
$$= 2B^T (B\overline{w} + \overline{b}) \tag{4.26}$$

This form of the gradient is used often in least-squares regression. Setting this gradient to zero yields the closed-form solution to least-squares regression (cf. Section 4.7).

4.6.3   The Chain Rule of Calculus for Vectored Derivatives

The chain rule of calculus is extremely useful for differentiating compositions of functions. In the univariate case with scalars, the rule is quite simple. For example, consider the case where the scalar objective J is a function of the scalar w as follows:

$$J = f(g(h(w))) \tag{4.27}$$

All of $f(\cdot)$, $g(\cdot)$, and $h(\cdot)$ are assumed to be scalar functions. In such a case, the derivative of J with respect to the scalar w is simply $f'(g(h(w)))g'(h(w))h'(w)$. This rule is referred to as the univariate chain rule of differential calculus. Note that the order of multiplication does not matter because scalar multiplication is commutative.

Similarly, consider the case where you have the following functions, where one of the functions is a vector-to-scalar function:

$$J = f(g_1(w), g_2(w), \ldots, g_k(w))$$

In such a case, the *multivariate chain rule* states that one can compute the derivative of J with respect to w as the sum of the products of the partial derivatives using all arguments of the function:

$$\frac{\partial J}{\partial w} = \sum_{i=1}^{k} \left[\frac{\partial J}{\partial g_i(w)} \right] \left[\frac{\partial g_i(w)}{\partial w} \right]$$

One can generalize *both* of the above results into a single form by considering the case where the functions are vector-to-vector functions. Note that vector-to-vector derivatives are matrices, and therefore we will be multiplying matrices together instead of scalars. Surprisingly, very large classes of machine learning algorithms perform the repeated composition of only two types of functions, which are shown in Table 4.2(b). *Unlike the case of the scalar chain rule, the order of multiplication is important when dealing with matrices and vectors.* In a composition function, the derivative of the argument (inner level variable) is always pre-multiplied with the derivative of the function (outer level variable). In many cases, the order of multiplication is self-evident because of the size constraints associated with matrix multiplication. We formally define the vectored chain rule as follows:

Theorem 4.6.1 (Vectored Chain Rule) *Consider a composition function of the following form:*

$$\overline{o} = F_k(F_{k-1}(\ldots F_1(\overline{x})))$$

Assume that each $F_i(\cdot)$ takes as input an n_i-dimensional column vector and outputs an n_{i+1}-dimensional column vector. Therefore, the input $\overline{x}$ is an n_1-dimensional vector and the final output $\overline{o}$ is an n_{k+1}-dimensional vector. For brevity, denote the vector output of $F_i(\cdot)$ by $\overline{h}_i$. Then, the vectored chain rule asserts the following:

$$\underbrace{\left[\frac{\partial \overline{o}}{\partial \overline{x}} \right]}_{n_1 \times n_{k+1}} = \underbrace{\left[\frac{\partial \overline{h}_1}{\partial \overline{x}} \right]}_{n_1 \times n_2} \underbrace{\left[\frac{\partial \overline{h}_2}{\partial \overline{h}_1} \right]}_{n_2 \times n_3} \cdots \underbrace{\left[\frac{\partial \overline{h}_{k-1}}{\partial \overline{h}_{k-2}} \right]}_{n_{k-1} \times n_k} \underbrace{\left[\frac{\partial \overline{o}}{\partial \overline{h}_{k-1}} \right]}_{n_k \times n_{k+1}}$$

It is easy to see that the size constraints of matrix multiplication are respected in this case.

4.6.3.1 Useful Examples of Vectored Derivatives

In the following, we provide some examples of vectored derivatives that are used frequently in machine learning. Consider the case where the function $g(\cdot)$ has a d-dimensional vector argument and its output is scalar. Furthermore, the function $f(\cdot)$ is a scalar-to-scalar function.

$$J = f(g(\overline{w}))$$

In such a case, we can apply the vectored chain rule to obtain the following:

$$\nabla J = \frac{\partial J}{\partial \overline{w}} = \nabla g(\overline{w}) \underbrace{f'(g(\overline{w}))}_{\text{scalar}} \tag{4.28}$$

In this case, the order of multiplication does not matter, because one of the factors in the product is a scalar. Note that this result is used frequently in machine learning, because many loss-functions in machine learning are computed by applying a scalar function $f(\cdot)$ to the dot product of $\overline{w}$ with a training point $\overline{a}$. In other words, we have $g(\overline{w}) = \overline{w} \cdot \overline{a}$. Note that $\overline{w} \cdot \overline{a}$ can be written as $\overline{w}^T (I) \overline{a}$, where I represents the identity matrix. This is in the form of one of the matrix identities of Table 4.2(a) [see identity (ii)]. In such a case, one can use the chain rule to obtain the following:

$$\frac{\partial J}{\partial \overline{w}} = \underbrace{[f'(g(\overline{w}))]}_{\text{scalar}} \overline{a} \qquad (4.29)$$

This result is extremely useful, and it can be used for computing the derivatives of many loss functions like least-squares regression, SVMs, and logistic regression. The vector $\overline{a}$ is simply replaced with the vector of the training point at hand. The function $f(\cdot)$ defines the specific form of the loss function for the model at hand. We have listed these identities as results (iv) and (v) of Table 4.2(a).

Table 4.2(b) contains a number of useful derivatives of vector-to-vector functions. The first is the linear transformation $\overline{h} = C\overline{w}$, where C is a matrix that does not depend on the parameter vector $\overline{w}$. The corresponding vector-to-vector derivative of $\overline{h}$ with respect to $\overline{w}$ is C^T [see identity (i) of Table 4.2(b)]. This type of transformation is used commonly in *linear layers of feed-forward neural networks*. Another common vector-to-vector function is the element-wise function $F(\overline{w})$, which is also used in neural networks (in the form of *activation functions*). In this case, the corresponding derivative is a diagonal matrix containing the element-wise derivatives as shown in the second identity of Table 4.2(b).

Finally, we consider a generalization of the *product identity* in differential calculus. Instead of differentiating the product of two scalar variables, we consider the product of a scalar and a vector variable. Consider the relationship $\overline{h} = f_s(\overline{w})\overline{x}$, which is the product of a vector and a scalar. Here, $f_s(\cdot)$ is a vector-to-scalar function and $\overline{x}$ is a column vector that depends on $\overline{w}$. In such a case, the derivative of $\overline{h}$ with respect to $\overline{w}$ is the matrix $\frac{\partial f_s(\overline{w})}{\partial \overline{w}}\overline{x}^T + f_s(\overline{w})\frac{\partial \overline{x}}{\partial \overline{w}}$ [see identity (iii) of Table 4.2(b)]. Note that the first term is the outer product of the two vectors $\frac{\partial f_s(\overline{w})}{\partial \overline{w}}$ and $\overline{x}$, whereas the second term is a scalar multiple of a vector-to-vector derivative.

4.7   Linear Regression: Optimization with Numerical Targets

Linear regression is also referred to as least-squares regression, because it is usually paired with a least-squares objective function. Least-squares regression was introduced briefly in Section 2.8 of Chapter 2 in order to provide an optimization-centric view of solving systems of equations. A more natural application of least-squares regression is to model the dependence of a target variable on the feature variables. We have n pairs of observations $(\overline{X}_i, y_i)$ for $i \in \{1 \ldots n\}$. The target y_i is predicted using $\hat{y}_i \approx \overline{W} \cdot \overline{X}_i^T$. The circumflex on top of $\hat{y}_i$ indicates that it is a predicted value. Here, $\overline{W} = [w_1 \ldots w_d]^T$ is a d-dimensional column vector of optimization parameters.

Each vector $\overline{X}_i$ is referred to as the set of independent variables or *regressors*, whereas the variable y_i is referred to as the target variable, response variable, or *regressand*. Each $\overline{X}_i$ is a row vector, because it is common for data points to be represented as rows of data matrices in machine learning. Therefore, the row vector $\overline{X}_i$ needs to be transposed before performing a dot product with the column vector $\overline{W}$. The vector $\overline{W}$ needs to be learned in a data driven manner, so that $\hat{y}_i = \overline{W} \cdot \overline{X}_i^T$ is as close to each y_i as possible. Therefore, we compute the loss $(y_i - \overline{W} \cdot \overline{X}_i^T)^2$ for each training data point, and then add up this losses over all points in order to create the objective function:

$$J = \frac{1}{2} \sum_{i=1}^{n} (y_i - \overline{W} \cdot \overline{X}_i^T)^2 \tag{4.30}$$

Once the vector $\overline{W}$ has been learned from the training data by optimizing the aforementioned objective function, the numerical value of the target variable of an unseen test instance $\overline{Z}$ (which is a d-dimensional row vector) can be predicted as $\overline{W} \cdot \overline{Z}^T$.

It is particularly convenient to write this objective function in terms of an $n \times d$ data matrix. The $n \times d$ data matrix D is created by stacking up the n rows $\overline{X}_1 \ldots \overline{X}_n$. Similarly, $\overline{y}$ is an n-dimensional column vector of response variables for which the ith entry is y_i. Note that $D\overline{W}$ is an n-dimensional column vector of *predictions* which should ideally equal the observed vector $\overline{y}$. Therefore, the vector of errors is given by $(D\overline{W} - \overline{y})$, and the squared norm of the error vector is the loss function. Therefore, the minimization loss function of least-squares regression may be written as follows:

$$J = \frac{1}{2} \|D\overline{W} - \overline{y}\|^2 = \frac{1}{2} [D\overline{W} - \overline{y}]^T [D\overline{W} - \overline{y}] \tag{4.31}$$

One can expand the above expression as follows:

$$J = \frac{1}{2} \overline{W}^T D^T D \overline{W} - \frac{1}{2} \overline{W}^T D^T \overline{y} - \frac{1}{2} \overline{y}^T D \overline{W} + \frac{1}{2} \overline{y}^T \overline{y} \tag{4.32}$$

It is easy to see that the above expression is convex, because $D^T D$ is the positive semidefinite Hessian in the quadratic term. This means that if we find a value of the vector $\overline{W}$ at which the gradient is zero (i.e., a critical point), it will be a global minimum of the objective function.

In order to compute the gradient of J with respect to $\overline{W}$, one can directly use the squared-norm result of Section 4.6.2.2 to yield the following:

$$\nabla J = D^T D \overline{W} - D^T \overline{y} \tag{4.33}$$

Setting the gradient to zero yields the following condition:

$$D^T D \overline{W} = D^T \overline{y} \tag{4.34}$$

Pre-multiplying both sides with $(D^T D)^{-1}$, one obtains the following:

$$\overline{W} = (D^T D)^{-1} D^T \overline{y} \tag{4.35}$$

Note that this formula is identical to the use of the left-inverse of D for solving a system of equations (cf. Section 2.8 of Chapter 2), and the derivation of Section 2.8 uses the normal equation rather than calculus. The problem of solving a system of equations is a special case of least-squares regression. When the system of equations has a feasible solution, the optimal solution has zero loss on the training data. In the case that the system is inconsistent, we obtain the best-fit solution.

How can one compute $\overline{W}$ efficiently, when $D^T D$ is invertible? This can be achieved via QR decomposition of matrix D as $D = QR$ (see end of Section 2.8.2), where Q is an $n \times d$ matrix with orthonormal columns and R is a $d \times d$ upper-triangular matrix. One can simply substitute $D = QR$ in Equation 4.34, and use $Q^T Q = I_d$ to obtain the following:

$$R^T R \overline{W} = R^T Q^T \overline{y} \qquad (4.36)$$

Multiplying both sides with $(R^T)^{-1}$, one obtains $R\overline{W} = Q^T \overline{y}$. This triangular system of equations can be solved efficiently using back-substitution.

The above solution assumes that the matrix $D^T D$ is invertible. However, in cases where the number of data points is small, the matrix $D^T D$ might not be invertible. In such cases, infinitely many solutions exist to this system of equations, which will overfit the training data; such methods will not generalize easily to unseen test data. In such cases, regularization is important.

4.7.1   Tikhonov Regularization

The closed-form solution to the problem does not work in under-determined cases, where the number of optimization variables is greater than the number of points. One possible solution is to reduce the number of variables in the data by posing the problem as a constrained optimization problem. In other words, we could try to optimize the same loss function while posing the hard constraint that at most k values of w_i are non-zero. However, such a constrained optimization problem is hard to solve. A softer solution is to impose a small penalty on the absolute value of each w_i in order to discourage non-zero values of w_i. Therefore, the resulting loss function is as follows:

$$J = \frac{1}{2}\|D\overline{W} - \overline{y}\|^2 + \frac{\lambda}{2}\|\overline{W}\|^2 \qquad (4.37)$$

Here, $\lambda > 0$ is the regularization parameter. By adding the squared norm penalty, we are encouraging each w_i to be small in magnitude, unless it is absolutely essential for learning. Note that the addition of the strictly convex term $\lambda\|\overline{W}\|^2$ to the convex least-squares regression loss function makes the regularized objective function *strictly* convex (see Lemma 4.3.6 on addition of convex and strictly convex functions). A strictly convex objective function has a unique optimal solution.

In order to solve the optimization problem, one can set the gradient of J to 0. The gradient of the added term $\lambda\|\overline{W}\|^2/2$ is $\lambda\overline{W}$, based on the discussion in Section 4.6.2.2. On setting the gradient of J to 0, we obtain the following modified condition:

$$(D^T D + \lambda I)\overline{W} = D^T \overline{y} \qquad (4.38)$$

Pre-multiplying both sides with $(D^T D + \lambda I)^{-1}$, one obtains the following:

$$\overline{W} = (D^T D + \lambda I)^{-1} D^T \overline{y} \qquad (4.39)$$

Here, it is important to note that $(D^T D + \lambda I)$ is always invertible for $\lambda > 0$, since the matrix is positive definite (see Problem 2.4.2 of Chapter 2). The resulting solution is regularized, and it generalizes much better to out-of-sample data. Because of the *push-through identity* (see Problem 1.2.13), the solution can also be written in the following alternative form:

$$\overline{W} = D^T (DD^T + \lambda I)^{-1} \overline{y} \qquad (4.40)$$

4.7.1.1 Pseudoinverse and Connections to Regularization

A special case of Tikhonov regularization is the *Moore-Penrose pseudoinverse*, which is introduced in Section 2.8.1 of Chapter 2. The Moore-Penrose pseudoinverse D^+ of the matrix D is the limiting case of Tikhonov regularization in which $\lambda > 0$ is infinitesimally small:

$$D^+ = \lim_{\lambda \to 0+} (D^T D + \lambda I)^{-1} D^T = \lim_{\lambda \to 0+} D^T (D D^T + \lambda I)^{-1} \qquad (4.41)$$

Therefore, one can simply write the solution $\overline{W}$ in terms of the Moore-Penrose pseudoinverse as $\overline{W} = D^+ \overline{y}$.

4.7.2 Stochastic Gradient Descent

In machine learning, it is rare to obtain a closed-form solution like Equation 4.39. In most cases, one uses (stochastic) gradient-descent updates of the following form:

$$\overline{W} \Leftarrow \overline{W} - \alpha \nabla J \qquad (4.42)$$

One advantage of (stochastic) gradient descent is that it is an efficient solution both in terms of memory requirements and computational efficiency. In the case of least-squares regression, the update of Equation 4.42 can be instantiated as follows:

$$\overline{W} \Leftarrow \overline{W}(1 - \alpha\lambda) - \alpha D^T \underbrace{(D\overline{W} - \overline{y})}_{\text{Error vector } \overline{e}} \qquad (4.43)$$

Here, $\alpha > 0$ is the learning rate. In order to implement the approach efficiently, one first computes the n-dimensional error vector $\overline{e} = (D\overline{W} - \overline{y})$, which is marked in the above equation. Subsequently, the d-dimensional vector $D^T \overline{e}$ is computed for the update. Such an approach only requires matrix-vector multiplication, rather than requiring the materialization of the potentially large matrix $D^T D$.

One can also perform mini-batch stochastic gradient descent by selecting a subset of examples (rows) from the data matrix D. Let S be a set of training examples in the current mini-batch, where each example in S contains the feature-target pair in the form $(\overline{X}_i, y_i)$. Then, the gradient-descent update can be modified to the mini-batch update as follows:

$$\overline{W} \Leftarrow \overline{W}(1 - \alpha\lambda) - \alpha \sum_{(\overline{X}_i, y_i) \in S} \overline{X}_i^T \underbrace{(\overline{W} \cdot \overline{X}_i^T - y_i)}_{\text{Error value}} \qquad (4.44)$$

Note that Equation 4.44 can be derived directly from Equation 4.43 by simply assuming that only the (smaller) matrix corresponding to the mini-batch is available at the time of the update.

4.7.3 The Use of Bias

It is common in machine learning to introduce an additional bias variable to account for unexplained constant effects in the targets. For example, consider the case in which the target variable is the temperature in a tropical city in Fahrenheit and the two feature variables respectively correspond to the number of days since the beginning of the year, and the number of minutes since midnight. The modeling $y_i = \overline{W} \cdot \overline{X}_i^T$ is bound to lead to large errors because of unexplained constant effects. For example, when both feature variables are 0, it corresponds to the New Year's eve. The temperature in a tropical city is bound to be much higher than 0 on New Year's eve. However, the modeling $y_i = \overline{W} \cdot \overline{X}_i^T$ will always yield 0 as a predicted value. This problem can be avoided with the use of a bias variable b, so that the new model is $y_i = \overline{W} \cdot \overline{X}_i^T + b$. The bias variable absorbs the additional constant effects (i.e., bias specific to the city at hand) and it needs to be learned like the other parameters in $\overline{W}$. In such a case, it can be shown that the gradient-descent updates of Equation 4.44 are modified as follows:

$$\overline{W} \Leftarrow \overline{W}(1 - \alpha\lambda) - \alpha \sum_{(\overline{X}_i, y_i) \in S} \overline{X}_i^T \underbrace{(\overline{W} \cdot \overline{X}_i^T + b - y_i)}_{\text{Error value}}$$

$$b \Leftarrow b(1 - \alpha\lambda) - \alpha \sum_{(\overline{X}_i, y_i) \in S} \underbrace{(\overline{W} \cdot \overline{X}_i^T + b - y_i)}_{\text{Error value}}$$

It turns out that it is possible to achieve *exactly* the same effect as the above updates without changing the original (i.e., bias-free) model. The trick is to add an additional dimension to the training and test data with a constant value of 1. Therefore, one would have an additional $(d+1)$th parameter w_{d+1} in vector $\overline{W}$, and the target variable for $\overline{X} = [x_1 \ldots x_d]$ is predicted as follows:

$$\hat{y} = [\sum_{i=1}^{d} w_i x_i] + w_{i+1}(1)$$

It is not difficult to see that this is exactly the same prediction function as the one with bias. The coefficient w_{d+1} of this additional dimension is the bias variable b. Since the bias variable can be incorporated with a feature engineering trick, it will largely be omitted in most of the machine learning applications in this book. However, as a practical matter, it is very important to use the bias (in some form) in order to avoid undesirable constant effects.

4.7.3.1 Heuristic Initialization

Choosing a good initialization can sometimes be helpful in speeding up the updates. Consider a linear regression problem with an $n \times d$ data matrix D. In most cases, the number of training examples n is much greater than the number of features d. A simple approach for heuristic initialization is to select d randomly chosen training points and solve the $d \times d$ system of equations using any of the methods discussed in Chapter 2. Solving a system of linear equations is a special case of linear regression, and it is also much simpler. This provides a good initial starting point for the weight vector.

Problem 4.7.1 (Matrix Least-Squares) *Consider an $n \times d$ tall data matrix D and $n \times k$* **matrix** Y *of numerical targets. You want to find the $d \times k$ weight matrix W so that $\|DW - Y\|_F^2$ is as small as possible. Show that the optimal weight matrix is $W = (D^T D)^{-1} D^T Y$, assuming that D has linearly independent columns. Show that the left-inverse of a tall matrix D is the best least-squares solution to the matrix R satisfying the right-inverse relationship $DR \approx I_n$, and the resulting approximation of I_n is a projection matrix.*

4.8 Optimization Models for Binary Targets

Least-squares regression learns how to relate numerical feature variables (independent variables or regressor) to a numerical target (i.e., dependent variable or regressand). In many applications, the targets are discrete rather than real-valued. An example of such a target is the color such as $\{Blue, Green, Red\}$. Note that there is no natural ordering between these targets, which is different from the case of numerical targets unless the target variable is binary.

A special case of discrete targets is the case in which the target variable y is binary and drawn from $\{-1, +1\}$. The instances with label $+1$ are referred to as *positive class instances*, and those with label -1 are referred to as *negative class instances*. For example, the feature variables in a cancer detection application might correspond to patient clinical measurements, and the class variable can be an indicator of whether or not the patient has cancer. In the binary-class case, we *can* impose an ordering between the two possible target values. In other words, we can pretend that the targets are numeric, and simply perform linear regression. This method is referred to as *least-squares classification*, which is discussed in the next section. Treating discrete targets as numerical values does have its disadvantages. Therefore, many alternative loss functions have been proposed for discrete (binary) data that avoid these disadvantages. Examples include the support vector machine and logistic regression. In the following, we will provide an overview of these models and their relationships with one another. While discussing these relationships, it will become evident that the ancient problem of least-squares regression serves as the parent model and the motivating force to all these (relatively recent) models for discrete-valued targets.

4.8.1 Least-Squares Classification: Regression on Binary Targets

In least-squares classification, linear regression is directly applied to binary targets. The $n \times d$ data matrix D still contains numerical values, and its rows $\overline{X}_1 \ldots \overline{X}_n$ are d-dimensional row vectors. However, the n-dimensional target vector $\overline{y} = [y_1 \ldots y_n]^T$ will only contain binary values drawn from -1 or $+1$. In least-squares classification, we pretend that the binary targets are real-valued. Therefore, we model each target as $y_i \approx \overline{W} \cdot \overline{X}_i^T$, where $\overline{W} = [w_1, \ldots, w_d]^T$ is a column vector containing the weights. We set up the same squared loss function as least-squares regression by treating binary targets as special cases of numerical targets. This results in the same closed-form solution for $\overline{W}$:

$$\overline{W} = (D^T D + \lambda I)^{-1} \overline{y} \tag{4.45}$$

Even though $\overline{W} \cdot \overline{X}_i^T$ yields a real-valued prediction for instance $\overline{X}_i$ (like regression), it makes more sense to view the hyperplane $\overline{W} \cdot \overline{X}^T = 0$ as a *separator* or *modeled decision boundary*, where any instance $\overline{X}_i$ with label $+1$ will satisfy $\overline{W} \cdot \overline{X}_i^T > 0$, and any instance with label -1 will satisfy $\overline{W} \cdot \overline{X}_i^T < 0$. Because of the way in which the model has been trained, most *training* points will align themselves on the two sides of the separator, so that the sign of the training label y_i matches the sign of $\overline{W} \cdot \overline{X}_i^T$. An example of a two-class data set in two dimensions is illustrated in Figure 4.8 in which the two classes are denoted by '+' and '*', respectively. In this case, it is evident that the value of $\overline{W} \cdot \overline{X}_i^T = 0$ is true only for points on the separator. The training points on the two sides of the separator satisfy either $\overline{W} \cdot \overline{X}_i^T < 0$ or $\overline{W} \cdot \overline{X}_i^T > 0$. The separator $\overline{W} \cdot \overline{X}^T = 0$ between the two classes is the modeled decision boundary. Note that some data distributions might not have the kind of neat separability as shown in Figure 4.8. In such cases, one either needs to live with errors or use feature transformation techniques to create linear separability. These techniques (such as *kernel methods*) are discussed in Chapter 9.

Once the weight vector $\overline{W}$ has been learned in the training phase, the classification is performed on an unseen test instance $\overline{Z}$. Since the test instance $\overline{Z}$ is a row vector, whereas

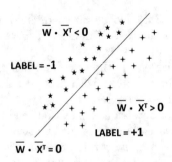

Figure 4.8: An example of linear separation between two classes

$\overline{W}$ is a column vector, the test instance needs to be transposed before computing the dot product between $\overline{W}$ and $\overline{Z}^T$. This dot product yields a real-valued prediction, which is converted to a binary prediction with the use of sign function:

$$\hat{y} = \text{sign}\{\overline{W} \cdot \overline{Z}^T\} \tag{4.46}$$

In effect, the model learns a linear hyperplane $\overline{W} \cdot \overline{X}^T = 0$ separating the positive and negative classes. All test instances for which $\overline{W} \cdot \overline{Z}^T > 0$ are predicted to belong to the positive class, and all instances for which $\overline{W} \cdot \overline{Z}^T < 0$ are predicted to belong to the negative class.

As in the case of real-valued targets, one can also use mini-batch stochastic gradient-descent for regression on binary targets. Let S be a mini-batch of pairs $(\overline{X}_i, y_i)$ of feature variables and targets. Each $\overline{X}_i$ is a row of the data matrix D and y_i is a target value drawn from $\{-1, +1\}$. Then, the mini-batch update for least-squares classification is identical to that of least-squares regression:

$$\overline{W} \Leftarrow \overline{W}(1 - \alpha\lambda) - \alpha \sum_{(\overline{X}_i, y_i) \in S} \overline{X}_i^T (\overline{W} \cdot \overline{X}_i^T - y_i) \tag{4.47}$$

Here, $\alpha > 0$ is the learning rate, and $\lambda > 0$ is the regularization parameter. Note that this update is *identical* to that in Equation 4.44. However, since each target y_i is drawn from $\{-1, +1\}$, an alternative approach also exists for writing the targets by using the fact that $y_i^2 = 1$. This alternative form of the update is as follows:

$$\overline{W} \Leftarrow \overline{W}(1 - \alpha\lambda) - \alpha \sum_{(\overline{X}_i, y_i) \in S} \underbrace{y_i^2}_{1} \overline{X}_i^T (\overline{W} \cdot \overline{X}_i^T - y_i)$$

$$= \overline{W}(1 - \alpha\lambda) - \alpha \sum_{(\overline{X}_i, y_i) \in S} y_i \overline{X}_i^T (y_i[\overline{W} \cdot \overline{X}_i^T] - y_i^2)$$

Setting $y_i^2 = 1$, we obtain the following:

$$\overline{W} \Leftarrow \overline{W}(1 - \alpha\lambda) + \alpha \sum_{(\overline{X}_i, y_i) \in S} y_i \overline{X}_i^T (1 - y_i[\overline{W} \cdot \overline{X}_i^T]) \tag{4.48}$$

This form of the update is more convenient because it is more closely related to updates of other classification models discussed later in this chapter. Examples of these models are the *support vector machine* and *logistic regression*. The loss function can also be converted to a more convenient representation for binary targets drawn from $\{-1, +1\}$.

Alternative Representation of Loss Function

The alternative form of the aforementioned updates can also be derived from an alternative form of the loss function. The loss function of (regularized) least-squares classification can be written as follows:

$$J = \frac{1}{2} \sum_{i=1}^{n} (y_i - \overline{W} \cdot \overline{X}_i^T)^2 + \frac{\lambda}{2} \|\overline{W}\|^2 \tag{4.49}$$

Using the fact that $y_i^2 = 1$ for binary targets, we can modify the objective function as follows:

$$J = \frac{1}{2} \sum_{i=1}^{n} y_i^2 (y_i - \overline{W} \cdot \overline{X}_i^T)^2 + \frac{\lambda}{2} \|\overline{W}\|^2$$

$$= \frac{1}{2} \sum_{i=1}^{n} (y_i^2 - y_i[\overline{W} \cdot \overline{X}_i^T])^2 + \frac{\lambda}{2} \|\overline{W}\|^2$$

Setting $y_i^2 = 1$, we obtain the following loss function:

$$J = \frac{1}{2} \sum_{i=1}^{n} (1 - y_i[\overline{W} \cdot \overline{X}_i^T])^2 + \frac{\lambda}{2} \|\overline{W}\|^2 \tag{4.50}$$

Differentiating this loss function directly leads to Equation 4.48. However, it is important to note that the loss function/updates of least-squares classification are identical to the loss function/updates of least-squares regression, even though one might use the binary nature of the targets in the former case in order to make them *look* superficially different.

The updates of least-squares classification are also referred to as Widrow-Hoff updates [132]. The rule was proposed in the context of neural network learning, and it was the second major neural learning algorithm proposed after the perceptron [109]. Interestingly, the neural models were proposed independently of the classical literature on least-squares regression; yet, the updates turn out to be identical.

Heuristic Initialization

A good way to perform heuristic initialization is to determine the mean $\overline{\mu}_0$ and $\overline{\mu}_1$ of the points belonging to the negative and positive classes, respectively. The difference between the two means is $\overline{w}_0 = \overline{\mu}_1^T - \overline{\mu}_0^T$ is a d-dimensional column vector, which satisfies $\overline{w}_0 \cdot \overline{\mu}_1^T \geq \overline{w}_0 \cdot \overline{\mu}_0^T$. The choice $\overline{W} = \overline{w}_0$ is a good starting point, because positive-class instances will have larger dot products with $\overline{w}_0$ than will negative-class instances (on the average). In many real applications, the classes are roughly separable with a linear hyperplane, and the normal hyperplane to the line joining the class centroids provides a good initial separator.

4.8.1.1 Why Least-Squares Classification Loss Needs Repair

The least-squares classification model has an important weakness, which is revealed when one examines its loss function:

$$J = \frac{1}{2} \sum_{i=1}^{n} (1 - y_i[\overline{W} \cdot \overline{X}_i^T])^2 + \frac{\lambda}{2} \|\overline{W}\|^2$$

Now consider a positive class instance for which $\overline{W} \cdot \overline{X}_i^T = 100$ is highly positive. This is obviously an desirable situation at least from a predictive point of view because the training instance is being classified on the correct side of the linear separator between the two classes in a positive way. However, the loss function in the training model treats this prediction as a large loss contribution of $(1 - y_i[\overline{W} \cdot \overline{X}_i^T])^2 = (1 - (1)(100))^2 = 99^2 = 9801$. Therefore, a large gradient descent update will be performed for a training instance that is located at a large distance from the hyperplane $\overline{W} \cdot \overline{X}^T = 0$ on the correct side. Such a situation is undesirable because it tends to confuse least-squares classification; the updates from these points on the correct side of the hyperplane $\overline{W} \cdot \overline{X}^T = 0$ tend to push the hyperplane in the same direction as some of the incorrectly classified points. In order to address this issue, many machine learning algorithms treat such points in a more nuanced way. These nuances will be discussed in the following sections.

4.8.2 The Support Vector Machine

As in the case of the least-squares classification model, we assume that we have n training pairs of the form $(\overline{X}_i, y_i)$ for $i \in \{1 \ldots n\}$. Each $\overline{X}_i$ is a d-dimensional row vector, and each $y_i \in \{-1, +1\}$ is the label. We would like to find a d-dimensional column vector $\overline{W}$ so that the sign of $\overline{W} \cdot \overline{X}_i^T$ yields the class label.

The support vector machine (SVM) treats *well-separated points* in the loss function in a more careful way by not penalizing them at all. What is a well separated point? Note that a point is correctly classified by the least-squares classification model when $y_i[\overline{W} \cdot \overline{X}_i^T] > 0$. In other words, y_i has the same sign as $\overline{W} \cdot \overline{X}_i^T$. Furthermore, the point is well-separated when $y_i[\overline{W} \cdot \overline{X}_i^T] > 1$. Therefore, the loss function of least-squares classification can be modified by setting the loss to 0, when this condition is satisfied. This can be achieved by modifying the least-squares loss to SVM loss as follows:

$$J = \frac{1}{2} \sum_{i=1}^{n} \max \left\{ 0, \left(1 - y_i[\overline{W} \cdot \overline{X}_i^T] \right) \right\}^2 + \frac{\lambda}{2} \|\overline{W}\|^2 \quad [L_2\text{-loss SVM}]$$

Note that the *only* difference from the least-squares classification model is the use of the maximization term in order to set the loss of well-separated points to 0. Once the vector $\overline{W}$ has been learned, the classification process for an unseen test instance is the same in the SVM as it is in the case of least-squares classification. For an unseen test instance $\overline{Z}$, the sign of $\overline{W} \cdot \overline{Z}^T$ yields the class label.

A more common form of the SVM loss is the *hinge-loss*. The hinge-loss is the L_1-version of the (squared) loss above:

$$J = \sum_{i=1}^{n} \max\{0, (1 - y_i[\overline{W} \cdot \overline{X}_i^T])\} + \frac{\lambda}{2} \|\overline{W}\|^2 \quad [\text{Hinge-loss SVM}] \tag{4.51}$$

Both forms of these objective functions can be shown to be convex.

Lemma 4.8.1 *Both the L_2-Loss SVM and the hinge loss are convex in the parameter vector $\overline{W}$. Furthermore, these functions are strictly convex when the regularization term is included.*

Proof: The proof of the above lemmas follow from the properties enumerated in Lemma 4.3.2. The point-specific hinge-loss is obtained by taking the maximum of two convex functions (one of which is linear and the other is a constant). Therefore, it is a convex function as well. The L_2-loss SVM squares the nonnegative hinge loss. Since the square of a nonnegative convex function is convex (according to Lemma 4.3.2), it follows that the point-specific L_2-loss is convex. The sum of the point-specific losses (convex functions) is convex according to Lemma 4.3.2. Therefore, the unregularized loss is convex.

Regularized Loss: We have already shown earlier in Section 4.7.1 that the L_2-regularization term is strictly convex. Since the sum of a convex and a strictly convex function is strictly convex according to Lemma 4.3.6, both objective functions (including the regularization term) are strictly convex. ∎

Therefore, one can find the *global* optimum of an SVM by using gradient descent.

4.8.2.1 Computing Gradients

The objective functions for the L_1-loss (hinge loss) and L_2-loss SVM are both in the form $J = \sum_i J_i + \Omega(\overline{W})$, where J_i is a point-specific loss and $\Omega(\overline{W}) = \lambda \|\overline{W}\|^2 / 2$ is the regularization term. The gradient of the latter term is $\lambda \overline{W}$. The main challenge is in computing the gradient of the point-specific loss J_i. Here, the key point is that the point-specific loss of both the L_1-loss (hinge loss) and L_2-loss can be expressed in the form of identity (v) of Table 4.2(a) for an appropriately chosen function $f(\cdot)$:

$$J_i = f_i(\overline{W} \cdot \overline{X}_i^T)$$

Here, the function $f_i(\cdot)$ is defined for the hinge-loss and L_2-loss SVMs as follows:

$$f_i(z) = \begin{cases} \max\{0, 1 - y_i z\} & [\text{Hinge Loss}] \\ \frac{1}{2}\max\{0, 1 - y_i z\}^2 & [L_2\text{-Loss}] \end{cases}$$

Therefore, according to Table 4.2(a) (also see Equation 4.29), the gradient of J_i with respect to $\overline{W}$ is the following:

$$\frac{\partial J_i}{\partial \overline{W}} = \overline{X}_i^T f_i'(\overline{W} \cdot \overline{X}_i^T) \tag{4.52}$$

The derivatives for the L_1-loss and the L_2-loss SVMs depend on the corresponding derivatives of $f_i(z)$, as they are defined in the two cases:

$$f_i'(z) = \begin{cases} -y_i I([1 - y_i z] > 0) & [\text{Hinge Loss}] \\ -y_i \max\{0, 1 - y_i z\} & [L_2\text{-Loss}] \end{cases}$$

Here, $I(\cdot)$ is an indicator function, which takes on the value of 1 when the condition inside it is true, and 0, otherwise. Therefore, by plugging in the value of $f'(z)$ in Equation 4.52, one obtains the following loss derivatives in the two cases:

$$\frac{\partial J_i}{\partial \overline{W}} = \begin{cases} -y_i \overline{X}_i^T I([1 - y_i(\overline{W} \cdot \overline{X}_i^T)] > 0) & [\text{Hinge Loss}] \\ -y_i \overline{X}_i^T \max\{0, 1 - y_i(\overline{W} \cdot \overline{X}_i^T)\} & [L_2\text{-Loss}] \end{cases}$$

These point-wise loss derivatives can be used to derive the stochastic gradient-descent updates.

4.8.2.2   Stochastic Gradient Descent

For the greatest generality, we will use mini-batch stochastic gradient descent in which a set S of training instances contains feature-label pairs of the form $(\overline{X}_i, y_i)$. For the hinge-loss SVM, we first determine the set $S^+ \subseteq S$ of training instances in which $y_i[\overline{W} \cdot \overline{X}_i^T] < 1$.

$$S^+ = \{(\overline{X}_i, y_i) : (\overline{X}_i, y_i) \in S, y_i[\overline{W} \cdot \overline{X}_i^T] < 1\} \tag{4.53}$$

The subset of instances in S^+ correspond to those for which the indicator function $I(\cdot)$ of the previous section takes on the value of 1. These instances are of two types; those corresponding to $y_i[\overline{W} \cdot \overline{X}_i^T] < 0$ are misclassified instances on the wrong side of the decision boundary, whereas the remaining instances corresponding to $y_i[\overline{W} \cdot \overline{X}_i^T] \in (0, 1)$ lie on the correct side of the decision boundary, but they are uncomfortably close to the decision boundary. Both these types of instances trigger updates in the SVM. In other words, the well-separated points do not play a role in the update. By using the gradient of the loss function, the updates in the L_1-loss SVM can be shown to be the following:

$$\overline{W} \Leftarrow \overline{W}(1 - \alpha\lambda) + \sum_{(\overline{X}_i, y_i) \in S^+} \alpha y_i \overline{X}_i^T \tag{4.54}$$

This algorithm is referred to as the primal support vector machine algorithm. The hinge-loss update seems somewhat different from the update for least-squares classification. The primary reason for this is that the least-squares classification model uses a squared loss function, whereas the hinge-loss is a piece-wise linear function. The similarity with the updates of least-squares classification becomes more obvious when one compares the updates of least-squares classification with those of the SVM with L_2-loss. The updates of the SVM with L_2-loss are as follows:

$$\overline{W} \Leftarrow \overline{W}(1 - \alpha\lambda) + \alpha \sum_{(\overline{X}_i, y_i) \in S} y_i \overline{X}_i^T (\max\{1 - y_i[\overline{W} \cdot \overline{X}_i^T], 0\}) \tag{4.55}$$

In this case, it is evident that the updates of the L_2-SVM are different from those of least-squares classification (cf. Equation 4.48) only in terms of the treatment of well-separated points; *identical updates are made for misclassified points and those near the decision boundary, whereas no updates are made for well-separated points on the correct side of the decision boundary.* This difference in the nature of the updates fully explains the difference between the L_2-SVM and least-squares classification. It is noteworthy that the loss function of the L_2-SVM was proposed [60] by Hinton much earlier than the Cortes and Vapnik [30] work on the hinge-loss SVM. Interestingly, Hinton proposed the L_2-loss as a way to repair the Widrow-Hoff loss (i.e., least-squares classification loss), which makes a lot of sense from an intuitive point of view. Hinton's work remained unnoticed by the community of researchers working on SVMs during the early years. However, the approach was eventually rediscovered in the recent focus on deep learning, where many of the early works were revisited.

4.8.3   Logistic Regression

We use the same notations as earlier sections by assuming that we have n training pairs of the form $(\overline{X}_i, y_i)$ for $i \in \{1 \ldots n\}$. Each $\overline{X}_i$ is a d-dimensional row vector, and each $y_i \in \{-1, +1\}$ is the label. We would like to find a d-dimensional column vector $\overline{W}$ so that the sign of $\overline{W} \cdot \overline{X}_i^T$ yields the class label of $\overline{X}_i$.

Logistic regression uses a loss function, which has a very similar shape to the hinge-loss SVM. However, the hinge-loss is piecewise linear, whereas logistic regression is a smooth loss function. Logistic regression has a probabilistic interpretation in terms of the log-likelihood loss of a data point. The loss function of logistic regression is formulated as follows:

$$J = \sum_{i=1}^{n} \underbrace{\log(1 + \exp(-y_i[\overline{W} \cdot \overline{X}_i^T]))}_{J_i} + \frac{\lambda}{2}\|\overline{W}\|^2 \quad \text{[Logistic Regression]} \tag{4.56}$$

All logarithms in this section are natural logarithms. When $\overline{W} \cdot \overline{X}_i^T$ is large in absolute magnitude and has the same sign as y_i, the point-specific loss J_i is close to $\log(1+\exp(-\infty)) = 0$. On the other hand, the loss is larger than $\log(1 + \exp(0)) = \log(2)$ when the signs of y_i and $\overline{W} \cdot \overline{X}_i^T$ disagree. For cases in which the signs disagree, the loss increases almost linearly with $\overline{W} \cdot \overline{X}_i^T$, as the magnitude of $\overline{W} \cdot \overline{X}_i^T$ becomes increasingly large. This is because of the following relationship:

$$\lim_{z \to -\infty} \frac{\log(1 + \exp(-z))}{-z} = \lim_{z \to -\infty} \frac{\exp(-z)}{1 + \exp(-z)} = \lim_{z \to -\infty} \frac{1}{1 + \exp(z)} = 1$$

The above limit is computed using *L'Hopital's rule*, which differentiates the numerator and denominator of a limit to evaluate it. Note that the hinge loss of an SVM is always $(1 - z)$ for $z = y_i \overline{W} \cdot \overline{X}_i^T < 1$. One can show that the logistic loss differs from the hinge loss by a constant offset of 1 for grossly misclassified instances:

Problem 4.8.1 *Show that* $\lim_{z \to -\infty} \underbrace{(1 - z)}_{SVM} - \underbrace{\log(1 + \exp(-z))}_{Logistic} = 1.$

Since constant offsets do not affect gradient descent, logistic loss and hinge loss treat grossly misclassified training instances in a similar way. However, unlike the hinge loss, all instances have non-zero logistic losses. Like SVMs, the loss function of logistic regression is convex:

Lemma 4.8.2 *The loss function of logistic regression is a convex function. Adding the regularization term makes the loss function strictly convex.*

Proof: This result can be shown by using the fact that the point-wise loss is of the form $\log[1+\exp(G(\overline{X}))]$, where $G(\overline{X}_i)$ is the linear function $G(\overline{X}_i) = -y_i(\overline{W} \cdot \overline{X}_i^T)$. Furthermore, the function $\log[1 + \exp(-z)]$ is convex (see Problem 4.3.4). Then, by using Lemma 4.3.2 on the composition of convex and linear functions, it is evident that each point-specific loss is convex. Adding all the point-specific losses also results in a convex function because of the first part of the same lemma. Furthermore, adding the regularization term makes the function strictly convex according to Lemma 4.3.6, because the regularization term is strictly convex. ∎

It is, in fact, possible to show that logistic regression is strictly convex even without regularization. We leave the proof of this result as an exercise.

Problem 4.8.2 *Show that the loss function in logistic regression is strictly convex even without regularization.*

4.8.3.1   Computing Gradients

Since the logistic regression loss function is strictly convex, it means that one can reach a global optimum with stochastic gradient-descent methods. As in the case of SVMs, the objective function for logistic regression is in the form $J = \sum_i J_i + \Omega(\overline{W})$, where J_i is a point-specific loss and $\Omega(\overline{W}) = \lambda\|\overline{W}\|^2/2$ is the regularization term. The gradient of the regularization term is $\lambda\overline{W}$. We also need to compute the gradient of the point-specific loss J_i. The logistic loss can be expressed in the form of identity (v) of Table 4.2(a) for an appropriately chosen function $f(\cdot)$:

$$J_i = f_i(\overline{W} \cdot \overline{X}_i^T)$$

Here, the function $f_i(\cdot)$ is defined as follows for constant y_i:

$$f_i(z) = \log(1 + \exp(-y_i z))$$

Therefore, according to Table 4.2(a) (see also Equation 4.29), the gradient of J_i with respect to $\overline{W}$ is the following:

$$\frac{\partial J_i}{\partial \overline{W}} = \overline{X}_i^T f_i'(\overline{W} \cdot \overline{X}_i^T) \tag{4.57}$$

The corresponding derivative is as follows:

$$f_i'(z) = \frac{-y_i \exp(-y_i z)}{1 + \exp(-y_i z)} = \frac{-y_i}{1 + \exp(y_i z)}$$

Therefore, by plugging in the value of $f_i'(z)$ in Equation 4.57 after setting $z = \overline{W} \cdot \overline{X}_i^T$, one obtains the following loss derivative:

$$\frac{\partial J_i}{\partial \overline{W}} = \frac{-y_i \overline{X}_i^T}{(1 + \exp(y_i[\overline{W} \cdot \overline{X}_i^T]))}$$

These point-wise loss derivatives can be used to derive the stochastic gradient-descent updates.

4.8.3.2   Stochastic Gradient Descent

Given a mini-batch of S of feature-target pairs $(\overline{X}_i, y_i)$, one can define an objective function $J(S)$, which uses the loss of only the training instances in S. The regularization term remains unchanged, as one can simply re-scale the regularization parameter by $|S|/n$. It is relatively easy to compute the gradient $\nabla J(S)$ based on mini-batch S as follows:

$$\nabla J(S) = \lambda\overline{W} - \sum_{(\overline{X}_i, y_i) \in S} \frac{y_i \overline{X}_i^T}{(1 + \exp(y_i[\overline{W} \cdot \overline{X}_i^T]))} \tag{4.58}$$

Therefore, the mini-batch stochastic gradient-descent method can be implemented as follows:

$$\overline{W} \Leftarrow \overline{W}(1 - \alpha\lambda) + \sum_{(\overline{X}_i, y_i) \in S} \frac{\alpha y_i \overline{X}_i^T}{(1 + \exp(y_i[\overline{W} \cdot \overline{X}_i^T]))} \tag{4.59}$$

Logistic regression makes similar updates as the hinge-loss SVM. The main difference is in terms of the treatment of well-separated points, where SVM does not make any updates and logistic regression makes (small) updates.

4.8.4 How Linear Regression Is a Parent Problem in Machine Learning

Many binary classification models use loss functions that are modifications of the least-squares regression loss function in order to handle binary target variables. The most extreme example of this inheritance is least-squares classification, where one directly uses the regression loss function by pretending that the labels from $\{-1, +1\}$ are numerical values. As discussed in Section 4.8.1.1, this direct inheritance of the regression loss function has undesirable consequences for binary data. In least-squares classification, the value of the loss first decreases as $\overline{W} \cdot \overline{X}^T$ increases as long as $\overline{W} \cdot \overline{X}^T \leq 1$; however, this loss increases for the same positive instance when $\overline{W} \cdot \overline{X}^T$ increases beyond 1. This is counter-intuitive behavior because one should not expect the loss to increase with increasingly correct classification of a point. After all, the sign of the predicted class label does not change with increasing positive values of $\overline{W} \cdot \overline{X}^T$. This situation is caused by the fact that least-squares classification is a blind application of linear regression to the classification problem, and it does not bother to make adjustments for the discrete nature of the class variable. In support-vector machines, increasing distance in the correct direction from the decision boundary beyond the point where $\overline{W} \cdot \overline{X}^T = 1$ is neither rewarded nor penalized, because the loss function is $\max\{1 - \overline{W} \cdot \overline{X}^T, 0\}$ (for positive class instances). This point is referred to as the *margin boundary* in support vector machines. In logistic regression, increasing distance of a training point $\overline{X}$ from the hyperplane $\overline{W} \cdot \overline{X}^T = 0$ on the correct side is slightly rewarded.

To show the differences among least-squares classification, SVM, and logistic regression, we have shown their loss at varying values of $\overline{W} \cdot \overline{X}^T$ of a **positive** training point $\overline{X}$ with label $y = +1$ [cf. Figure 4.9(a)]. Therefore, positive and increasing $\overline{W} \cdot \overline{X}^T$ is desirable for correct predictions. The loss functions of logistic regression and the support vector machine look strikingly similar, except that the former is a smooth function, and the SVM sharply bottoms at zero loss beyond $\overline{W} \cdot \overline{X}^T \geq 1$. This similarity in loss functions is also reflected

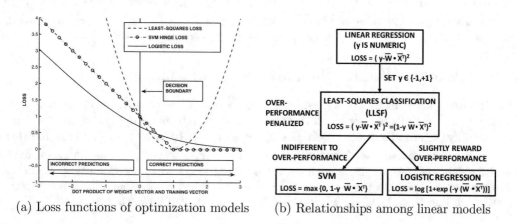

(a) Loss functions of optimization models   (b) Relationships among linear models

Figure 4.9: (a) The loss for a training instance $\overline{X}$ belonging to the **positive class** at varying values of $\overline{W} \cdot \overline{X}^T$. Logistic regression can be viewed as a smooth variant of SVM hinge loss. Least-squares classification is the only case in which the loss *increases* with increasingly correct classification in some regions. (b) All linear models in classification derive their motivation from the parent problem of linear regression in the real-world experiences of machine learning practitioners who often find that the two models seem to provide similar results. The least-squares classification model provides the only loss function where increasing the magnitude of $\overline{W} \cdot \overline{X}^T$ increases the loss for correctly classified instances. The semantic relationships among different loss functions are illustrated in Figure 4.9(b). It is evident that all the binary classification models inherit the basic structure of their loss functions from least-squares regression (while making adjustments for the binary nature of the target variable).

These relationships among their loss functions are also reflected as relationships among their updates in gradient descent. The updates for all three models can be expressed in a unified way in terms of a model-specific *mistake function* $\delta(\overline{X}_i, y_i)$ for the training pair $(\overline{X}_i, y_i)$ at hand. In particular, it can be shown that the stochastic gradient-descent updates of all the above algorithms are of the following form:

$$\overline{W} \Leftarrow \overline{W}(1 - \alpha\lambda) + \alpha y_i [\delta(\overline{X}_i, y_i)] \overline{X}_i^T \tag{4.60}$$

The mistake function $\delta(\overline{X}_i, y_i)$ is $(y_i - \overline{W} \cdot \overline{X}_i^T)$ for least-squares regression and classification, an indicator variable for SVMs, and a probability value for logistic regression.

4.9   Optimization Models for the MultiClass Setting

In multi-class classification, the discrete labels are no longer binary. Rather, they are drawn from a set of k *unordered* possibilities, whose indices are $\{1, \ldots, k\}$. For example, the color of an object could be a label, and there is no ordering between the values of the targets. This lack of ordering of target attributes requires further algorithmic modifications.

Each training instance $(\overline{X}_i, c(i))$ contains a d-dimensional feature vector $\overline{X}_i$ (which is a row vector) and the index $c(i) \in \{1 \ldots k\}$ of its observed class. We would like to find k different column vectors $\overline{W}_1 \ldots \overline{W}_k$ simultaneously so that the value of $\overline{W}_{c(i)} \cdot \overline{X}_i^T$ is greater than $\overline{W}_r \cdot \overline{X}_i^T$ for each $r \neq c(i)$. In other words, the training instance $\overline{X}_i$ is predicted to the class r with the largest value of $\overline{W}_r \cdot \overline{X}_i^T$. After training, the test instances are predicted to the class with the largest dot product with the weight vector.

4.9.1   Weston-Watkins Support Vector Machine

For the ith training instance, $\overline{X}_i$, we would like $\overline{W}_{c(i)} \cdot \overline{X}_i^T - \overline{W}_j \cdot \overline{X}_i^T$ to be greater than 0 (for each $j \neq c(i)$). In keeping with the notion of margin in a support vector machine, we not only penalize incorrect classification, but also "barely correct" predictions. In other words, we would like to penalize cases in which $\overline{W}_{c(i)} \cdot \overline{X}_i^T - \overline{W}_j \cdot \overline{X}_i^T$ is less than some fixed positive value of the margin. This margin value can be set to 1, because using any other value a simply scales up the parameters by the same factor a. In other words, our "ideal" setting with zero loss is one in which the following is satisfied for each $j \neq c(i)$:

$$\overline{W}_{c(i)} \cdot \overline{X}_i^T - \overline{W}_j \cdot \overline{X}_i^T \geq 1 \tag{4.61}$$

Therefore, one can set up a loss value J_i for the ith training instance as follows:

$$J_i = \sum_{j:j \neq c(i)} \max(\overline{W}_j \cdot \overline{X}_i^T - \overline{W}_{c(i)} \cdot \overline{X}_i^T + 1, 0) \tag{4.62}$$

It is not difficult to see the similarity between this loss function and that of the binary SVM. The overall objective function can be computed by adding the losses over the different training instances, and also adding a regularization term $\Omega(\overline{W}_1 \ldots \overline{W}_k) = \lambda \sum_r \|\overline{W}_r\|^2/2$:

$$J = \sum_{i=1}^{n} \sum_{j:j \neq c(i)} \max(\overline{W}_j \cdot \overline{X}_i^T - \overline{W}_{c(i)} \cdot \overline{X}_i^T + 1, 0) + \frac{\lambda}{2} \sum_{r=1}^{k} \|\overline{W}_r\|^2$$

The fact that the Weston-Watkins loss function is convex has a proof that is very similar to the binary case. One needs to show that each additive term of J_i is convex in terms of the parameter vector; after all, this additive term is the composition of a linear and a maximization function. This can be used to show that J_i is convex as well. We leave this proof as an exercise for the reader:

Problem 4.9.1 *The Weston-Watkins loss function is convex in terms of its parameters.*

As in the case of the previous models, one can learn the weight vectors with the use of gradient descent.

4.9.1.1 Computing Gradients

The main point in computing gradients is the vector derivative of J_i with respect to $\overline{W}_r$. The above gradient is computed using the chain rule, while recognizing that J_i contains additive terms of the form $\max\{v_{ji}, 0\}$, where v_{ji} is defined as follows:

$$v_{ji} = \overline{W}_j \cdot \overline{X}_i^T - \overline{W}_{c(i)} \cdot \overline{X}_i^T + 1$$

Furthermore, the derivative of J_i can be written with respect to $\overline{W}_r$ by using the multivariate chain rule as follows:

$$\frac{\partial J_i}{\partial \overline{W}_r} = \sum_{j=1}^{k} \frac{\partial J_i}{\partial v_{ji}} \underbrace{\frac{\partial v_{ji}}{\overline{W}_r}}_{\delta(j,\overline{X}_i)} \tag{4.63}$$

The partial derivative of $J_i = \sum_r \max\{v_{ri}, 0\}$ with respect to v_{ji} is equal to the partial derivative of $\max\{v_{ji}, 0\}$ with respect to v_{ji}. The partial derivative of the function $\max\{v_{ji}, 0\}$ with respect to v_{ji} is 1 for positive v_{ji}, and 0, otherwise. We denote this value by $\delta(j, \overline{X}_i)$. In other words, the binary value $\delta(j, \overline{X}_i)$ is 1, when $\overline{W}_{c(i)} \cdot \overline{X}_i^T < \overline{W}_j \cdot \overline{X}_i^T + 1$, and therefore the correct class is not preferred with respect to class j with sufficient margin.

The right-hand side of Equation 4.63 requires us to compute the derivative of $v_{ji} = \overline{W}_j \cdot \overline{X}_i^T - \overline{W}_{c(i)} \cdot \overline{X}_i^T + 1$ with respect to $\overline{W}_r$. This is an easy derivative to compute because of its linearity, as long as we are careful to track which weight vectors $\overline{W}_r$ appear with positive signs in v_{ji}. In the case when $r \neq c(i)$ (separator for wrong class), the derivative of v_{ji} with respect to $\overline{W}_r$ is $\overline{X}_i^T$ when $j = r$, and 0, otherwise. In the case when $r = c(i)$, the derivative is $-\overline{X}_i^T$ when $j \neq r$, and 0, otherwise. On substituting these values, one obtains the gradient of J_i with respect to $\overline{W}_r$ as follows:

$$\frac{\partial J_i}{\partial \overline{W}_r} = \begin{cases} \delta(r, \overline{X}_i)\overline{X}_i^T & r \neq c(i) \\ -\sum_{j \neq r} \delta(j, \overline{X}_i)\overline{X}_i^T & r = c(i) \end{cases}$$

One can obtain the gradient of J with respect to $\overline{W}_r$ by summing up the contributions of the different J_i and the regularization component of $\lambda \overline{W}_r$. Therefore, the updates for stochastic gradient descent are as follows:

$$\overline{W}_r \Leftarrow \overline{W}_r(1 - \alpha\lambda) - \alpha\frac{\partial J_i}{\partial \overline{W}_r} \quad \forall r \in \{1 \ldots k\}$$

$$= \overline{W}_r(1 - \alpha\lambda) - \alpha \begin{cases} \delta(r, \overline{X}_i)\overline{X}_i^T & r \neq c(i) \\ -\sum_{j \neq r} \delta(j, \overline{X}_i)\overline{X}_i^T & r = c(i) \end{cases} \quad \forall r \in \{1 \ldots k\}$$

An important special case is one in which there are only two classes. In such a case, it can be shown that the resulting updates of the separator belonging to the positive class will be identical to those in the hinge-loss SVM. Furthermore, the relationship $\overline{W}_1 = -\overline{W}_2$ will always be maintained, assuming that the parameters are initialized in this way. This is because the update to each separator will be the negative of the update to the other separator. We leave the proof of this result as a practice exercise.

Problem 4.9.2 *Show that the Weston-Watkins SVM defaults to the binary hinge-loss SVM in the special case of two classes.*

One observation from the relationship $\overline{W}_1 = -\overline{W}_2$ in the binary case is that there is a slight redundancy in the number of parameters of the multiclass SVM. This is because we really need $(k - 1)$ separators in order to model k classes, and one separator is redundant. However, since the update of the kth separator is always exactly defined by the updates of the other $(k - 1)$ separators, this redundancy does not make a difference.

Problem 4.9.3 *Propose a natural L_2-loss function for the multclass SVM. Derive the gradient and the details of stochastic gradient descent in this case.*

4.9.2   Multinomial Logistic Regression

Multinomial logistic regression is a generalization of logistic regression to multiple classes. As in the case of the Weston-Watkins SVM, each training instance $(\overline{X}_i, c(i))$ contains a d-dimensional feature vector $\overline{X}_i$ (which is a row vector) and the index $c(i) \in \{1 \ldots k\}$ of its observed class. Furthermore, similar to the Weston-Watkins SVM, k different separators are learned whose parameter vectors are $\overline{W}_1 \ldots \overline{W}_k$. The prediction rule for test instances is also the same as the Weston-Watkins SVM, since the class j with the largest dot product $\overline{W}_j \cdot \overline{Z}^T$ is predicted as the class of test instance $\overline{Z}$. Multinomial logistic regression models the *probability* of a point belonging to the rth class. The probability of training point $\overline{X}_i$ belonging to class r is given by applying the *softmax function* to $\overline{W}_1 \cdot \overline{X}_i^T \ldots \overline{W}_k \cdot \overline{X}_i^T$:

$$P(r|\overline{X}_i) = \frac{\exp(\overline{W}_r \cdot \overline{X}_i^T)}{\sum_{j=1}^k \exp(\overline{W}_j \cdot \overline{X}_i^T)} \tag{4.64}$$

It is easy to verify that the probability of $\overline{X}_i$ belonging to the rth class increases exponentially with increasing dot product between $\overline{W}_r$ and $\overline{X}_i^T$.

The goal in learning $\overline{W}_1 \ldots \overline{W}_k$ is to ensure that the aforementioned probability is high for the class $c(i)$ for (each) instance $\overline{X}_i$. This is achieved by using the *cross-entropy loss*, which is the negative logarithm of the probability of the instance $\overline{X}_i$ belonging to the correct class $c(i)$:

$$J = -\sum_{i=1}^{n} \underbrace{\log[P(c(i)|\overline{X}_i)]}_{J_i} + \frac{\lambda}{2}\sum_{r=1}^{k}\|\overline{W}_r\|^2$$

It is relatively easy to show that each $J_i = -\log[P(c(i)|\overline{X}_i)]$ is convex using an approach similar to the case of binary logistic regression.

4.9.2.1 Computing Gradients

We would like the determine the gradient of J with respect to each $\overline{W}_r$. We can decompose this gradient into the sum of the gradients of $J_i = -\log[P(c(i)|\overline{X}_i)]$ (along with the gradient of the regularization term). We denote this quantity by $\frac{\partial J_i}{\partial \overline{W}_r}$. Let v_{ji} denote the quantity $\overline{W}_j \cdot \overline{X}_i^T$. Then, the value of $\frac{\partial J_i}{\partial \overline{W}_r}$ is computed using the chain rule as follows:

$$\frac{\partial J_i}{\partial \overline{W}_r} = \sum_j \left(\frac{\partial J_i}{\partial v_{ji}}\right)\frac{\partial v_{ji}}{\partial \overline{W}_r} = \frac{\partial J_i}{\partial v_{ri}}\underbrace{\frac{\partial v_{ri}}{\partial \overline{W}_r}}_{\overline{X}_i^T} = \overline{X}_i^T\frac{\partial J_i}{\partial v_{ri}} \tag{4.65}$$

In the above simplification, we used the fact that v_{ji} has a zero gradient with respect to $\overline{W}_r$ for $j \neq r$, and therefore all terms in the summation except for the case of $j = r$ drop out to 0. We still need to compute the partial derivative of J_i with respect to v_{ri}. First, we express J_i directly as a function of $v_{1i}, v_{2i}, \ldots, v_{ki}$ as follows:

$$J_i = -\log[P(c(i)|\overline{X}_i)] = -\overline{W}_{c(i)} \cdot \overline{X}_i^T + \log[\sum_{j=1}^{k}\exp(\overline{W}_j \cdot \overline{X}_i^T)] \quad \text{[Using Equation 4.64]}$$

$$= -v_{c(i),i} + \log[\sum_{j=1}^{k}\exp(v_{ji})]$$

Therefore, we can compute the partial derivative of J_i with respect to v_{ri} as follows:

$$\frac{\partial J_i}{\partial v_{ri}} = \begin{cases} -\left(1 - \frac{\exp(v_{ri})}{\sum_{j=1}^{k}\exp(v_{ji})}\right) & \text{if } r = c(i) \\ \left(\frac{\exp(v_{ri})}{\sum_{j=1}^{k}\exp(v_{ji})}\right) & \text{if } r \neq c(i) \end{cases}$$

$$= \begin{cases} -(1 - P(r|\overline{X}_i)) & \text{if } r = c(i) \\ P(r|\overline{X}_i) & \text{if } r \neq c(i) \end{cases}$$

By substituting the value of the partial derivative $\frac{\partial J_i}{\partial v_{ri}}$ in Equation 4.65, we obtain the following:

$$\frac{\partial J_i}{\partial \overline{W}_r} = \begin{cases} -\overline{X}_i^T(1 - P(r|\overline{X}_i)) & \text{if } r = c(i) \\ \overline{X}_i^T P(r|\overline{X}_i) & \text{if } r \neq c(i) \end{cases} \tag{4.66}$$

4.9.2.2 Stochastic Gradient Descent

One can then use this point-specific gradient to compute the stochastic gradient descent updates:

$$\overline{W}_r \Leftarrow \overline{W}_r(1 - \alpha\lambda) + \alpha \begin{cases} \overline{X}_i^T(1 - P(r|\overline{X}_i)) & \text{if } r = c(i) \\ -\overline{X}_i^T P(r|\overline{X}_i) & \text{if } r \neq c(i) \end{cases} \quad \forall r \in \{1 \ldots k\} \qquad (4.67)$$

The probabilities in the above update can be substituted using Equation 4.64. It is noteworthy that the updates use the *probabilities of mistakes* in order to change each separator. In comparison, methods like least-squares regression use the *magnitudes of mistakes* in the updates. This difference is natural, because the softmax method is a probabilistic model. The above stochastic gradient descent is proposed for a mini-batch size of 1. We leave the derivation for a mini-batch S as an exercise for the reader.

Problem 4.9.4 *The text provides the derivation of stochastic gradient descent in multinomial logistic regression for a mini-batch size of 1. Provide a derivation of the update of each separator $\overline{W}_r$ for a mini-batch S containing pairs of the form $(\overline{X}, c)$ as follows:*

$$\overline{W}_r \Leftarrow \overline{W}_r(1 - \alpha\lambda) + \alpha \sum_{(\overline{X},c)\in S, r=c} \overline{X}^T \cdot (1 - P(r|\overline{X})) - \alpha \sum_{(\overline{X},c)\in S, r\neq c} \overline{X}^T \cdot P(r|\overline{X}) \quad (4.68)$$

Just as the Weston-Watkins SVM defaults to the hinge-loss SVM for the two-class case, multinomial logistic regression defaults to logistic regression in the special case of two classes. We leave the proof of this result as an exercise.

Problem 4.9.5 *Show that multinomial logistic regression defaults to binary logistic regression in the special case of two classes.*

4.10   Coordinate Descent

Coordinate descent is a method that optimizes the objective function one variable at a time. Therefore, if we have an objective function $J(\overline{w})$, which is a function of d-dimensional vector variables, we can try to optimize a single variable w_i from the vector $\overline{w}$, while holding all the other parameters fixed. This corresponds to the following optimization problem:

$$\overline{w} = \text{argmin}_{[w_i \text{ varies only}]} \; J(\overline{w}) \qquad [\text{All parameters except } w_i \text{ are fixed}]$$

Note that this is a single-variable optimization problem, which is usually much simpler to solve. In some cases, one might need to use line-search to determine w_i, when a closed form of the solution is not available. If one cycles through all the variables, and no improvement occurs, convergence has occurred. In the event that the optimized function is convex and differentiable in minimization form, the solution at convergence will be the optimal one. For non-convex functions, optimality is certainly not guaranteed, as the system can get stuck at a local minimum. Even for functions that are convex but non-differentiable, it is possible for coordinate descent to reach a suboptimal solution. An important point about coordinate descent is that it implicitly uses more than first-order gradient information; after all, it finds an optimal solution with respect to the variable it is optimizing. As a result, convergence can sometimes be faster with coordinate descent, as compared to stochastic

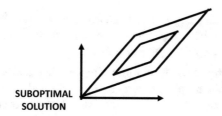

SUBOPTIMAL
SOLUTION

Figure 4.10: The contour plot of a non-differentiable function is shown. The center of the parallelogram-like contour plot is the optimum. Note that the axis-parallel moves can only worsen the objective function from acute-angled positions

gradient descent. Another important point about coordinate descent is that convergence is usually guaranteed, even if the resulting solution is a local optimum.

There are two main problems with coordinate descent. First, it is inherently sequential in nature. The approach optimizes one variable at a time, and therefore it would need to have optimized with respect to one variable in order to perform the next optimization step. Therefore, the parallelization of coordinate descent is always a challenge. Second, it can get stuck at suboptimal points (local minima). Even though the convergence to a local minimum is guaranteed, the use of a single variable can sometimes be myopic. This type of problem could occur even for convex functions, if the function is not differentiable. For example, consider the following function:

$$f(x, y) = |x + y| + 2|x - y| \qquad (4.69)$$

This objective function is convex but not differentiable. The optimal point of this function is $(0, 0)$. However, if coordinate descent reaches the point $(1, 1)$, it will cycle through both variables without improving the solution. The problem is that *no path exists to the optimal solution using axis-parallel directions.* Such a situation can occur with non-differentiable functions having pointed contour plots; if one ends up at one of the corners of the contour plot, there might not be a suitable axis-parallel direction of movement in order to improve the objective function. An example of such a scenario is illustrated in Figure 4.10. Such a situation can never arise in a differentiable function, where at least one axis-parallel direction will always improve the objective function.

A natural question that arises is to characterize the conditions under which coordinate descent is well behaved in non-differentiable function optimization. One observation is that even though the function $f(x, y)$ of Equation 4.69 is convex, its additive components are not separable in terms of the individual variables. In general, a sufficient condition for coordinate descent to reach a global optimum solution is that *the additive components of the non-differentiable portion of the multivariate function need to be expressed in terms of individual variables, and each of them must be convex.* We summarize a general version of the above result:

Lemma 4.10.1 *Consider a multivariate function $F(\overline{w})$ that can be expressed in the following form:*

$$F(\overline{w}) = G(\overline{w}) + \sum_{i=1}^{d} H_i(w_i)$$

The function $G(\overline{w})$ is a convex and differentiable function, whereas each $H_i(w_i)$ is a convex, univariate function of w_i, which might be non-differentiable. Then, coordinate descent will converge to a global optimum of the function $F(\overline{w})$.

An example of a non-differentiable function $H_i(w_i)$, which is also convex, is $H_i(w_i) = |w_i|$. This function is used for L_1-regularization. In fact, we will discuss the use of coordinate descent for L_1-regularized regression in Section 5.8.1.2 of Chapter 5.

The issue of additive separability is important, and it is sometimes helpful to perform a variable transformation, so that the non-differentiable part is additively separable. For example, consider a generalization of the objective function of Equation 4.69:

$$f(x,y) = g(x,y) + |x + y| + 2|x - y| \tag{4.70}$$

Assume that $g(x,y)$ is differentiable. Now, we make the following variable transformations $u = x + y$ and $v = x - y$. Then, one can rewrite the objective function after the variable transformation as $f([u + v]/2, [u - v]/2)$. In other words, we always substitute $[u + v]/2$ everywhere for x and $[u - v]/2$ everywhere for y to obtain the following:

$$F(u,v) = g([u + v]/2, [u - v]/2) + |u| + 2|v| \tag{4.71}$$

Each of the non-differentiable components is a convex function. Now, one can perform coordinate descent with respect to u and v without any problem. The main point of this trick is that the variable transformation changes the directions of movement, so that a path to the optimum solution exists.

Interestingly, even though non-differentiable functions cause problems for coordinate descent, such functions (and even discrete optimization problems) are often better solved by coordinate descent than gradient descent. This is because coordinate descent often enables the decomposition of a complex problem into smaller subproblems. As a specific example of this decomposition, we will show how the well-known k-means algorithm is an example of coordinate descent, when applied to a potentially difficult *mixed integer program* (cf. Section 4.10.3).

4.10.1   Linear Regression with Coordinate Descent

Consider an $n \times d$ data matrix D (with rows containing training instances), an n-dimensional column vector $\overline{y}$ of response variables, and a d-dimensional column vector $\overline{W} = [w_1 \ldots w_d]^T$ of parameters. We revisit the linear-regression objective function of Equation 4.31 as follows:

$$J = \frac{1}{2}\|D\overline{W} - \overline{y}\|^2 \tag{4.72}$$

The corresponding gradient with respect to *all* variables is used in straightforward gradient-descent methods (cf. Equation 4.33):

$$\nabla J = D^T(D\overline{W} - \overline{y}) \tag{4.73}$$

Coordinate descent optimizes the objective with respect to only a *single* variable at a time. In order to optimize with respect to w_i, we need to pick out the ith component of ∇J and set it to zero. Let $\overline{d}_i$ be the ith column of D. Furthermore, let $\overline{r}$ denote the n-dimensional residual vector $\overline{y} - D\overline{W}$. Then, we obtain the following condition:

$$\overline{d}_i^T(D\overline{W} - \overline{y}) = 0$$
$$\overline{d}_i^T(\overline{r}) = 0$$
$$\overline{d}_i^T\overline{r} + w_i\overline{d}_i^T\overline{d}_i = w_i\overline{d}_i^T\overline{d}_i$$

Note that the left-hand side is free of w_i because the two terms involving w_i cancel each other out. This is because the term $\overline{d}_i^T \overline{r}$ contributes $-w_i \overline{d}_i^T \overline{d}_i$, which cancels with $w_i \overline{d}_i^T \overline{d}_i$. Because of the fact that one of the sides does not depend on w_i, we obtain an update that yields the optimal value of w_i in a *single* iteration:

$$\overline{w}_i \Leftarrow \overline{w}_i + \frac{\overline{d}_i^T \overline{r}}{\|\overline{d}_i\|^2} \tag{4.74}$$

In the above update, we have used the fact that $\overline{d}_i^T \overline{d}_i$ is the same as the squared norm of $\overline{d}_i$. It is common to standardize each column of the data matrix to zero mean and unit variance. In such a case, the value of $\|\overline{d}_i\|^2$ will be 1, and the update further simplifies to the following:

$$\overline{w}_i \Leftarrow \overline{w}_i + \overline{d}_i^T \overline{r} \tag{4.75}$$

This update is extremely efficient. One full cycle of coordinate descent through all the variables requires asymptotically similar time as one full cycle of stochastic gradient descent through all the points. However, the number of cycles required by coordinate descent tends to be smaller than that in least-squares regression. Therefore, the coordinate-descent approach is more efficient. One can also derive a form of coordinate descent for regularized least-squares regression. We leave this problem as a practice exercise.

Problem 4.10.1 *Show that if Tikhonov regularization is used with parameter λ on least-squares regression, then the update of Equation 4.74 needs to be modified to the following:*

$$w_i \Leftarrow \frac{w_i \|\overline{d}_i\|^2 + \overline{d}_i^T \overline{r}}{\|\overline{d}_i\|^2 + \lambda}$$

The simplification of optimization subproblems that are inherent in solving for one variable at a time (while keeping others fixed) is very significant in coordinate descent.

4.10.2 Block Coordinate Descent

Block coordinate descent generalizes coordinate descent by optimizing a *block* of variables at a time, rather than a single variable. Although each step in block coordinate descent is more expensive, fewer steps are required. An example of block coordinate descent is the *alternating least-squares method*, which is often used in matrix factorization (cf Section 8.3.2.3 of Chapter 8). Block coordinate descent is often used in multi-convex problems where the objective function is non-convex, but each block of variables can be used to create a convex subproblem. Alternatively, each block admits to easy optimization, even when some of the variables are discrete. It is sometimes also easy to handle constrained optimization problems with coordinate descent, because the constraints tend to simplify themselves, when one is considering only a few carefully chosen variables. A specific example of this type of setting is the k-means algorithm.

4.10.3 K-Means as Block Coordinate Descent

The k-means algorithm is a good example of how choosing specific blocks of variables carefully allows good alternating minimization over different blocks of variables. One often views k-means as a simple heuristic method, although the reality is that it is fundamentally rooted in important ideas from coordinate descent.

It is assumed that there are a total of n data points denoted by the d-dimensional row vectors $\overline{X}_1 \ldots \overline{X}_n$. The k-means algorithms creates k *prototypes*, which are denoted by $\overline{z}_1 \ldots \overline{z}_k$, so that the sum of squared distances of the data points from their nearest prototypes is as small as possible. Let y_{ij} be a 0-1 indicator of whether point i gets assigned to cluster j. Each point gets assigned to only a single cluster, and therefore we have $\sum_j y_{ij} = 1$. One can therefore, formulate the k-means problem as a *mixed integer program* over the *real-valued* d-dimensional prototype row vectors $\overline{z}_1 \ldots \overline{z}_k$ and the matrix $Y = [y_{ij}]_{n \times k}$ of *discrete* assignment variables:

$$\text{Minimize} \sum_{j=1}^{k} \underbrace{\sum_{i=1}^{n} y_{ij} \|\overline{X}_i - \overline{z}_j\|^2}_{O_j}$$

subject to:

$$\sum_{j=1}^{k} y_{ij} = 1$$

$$y_{ij} \in \{0, 1\}$$

This is a mixed integer program, and such optimization problems are known to be very hard to solve in general. However, in this case, carefully choosing the blocks of variables is essential. Choosing the blocks of variables carefully also trivializes the underlying constraints. In this particular case, the variables are divided into two blocks corresponding to the $k \times d$ prototype variables in the vectors $\overline{z}_1 \ldots \overline{z}_k$ and the $n \times k$ assignment variables $Y = [y_{ij}]$. We alternately minimize over these two blocks of variables, because it provides the best possible decomposition of the problem into smaller subproblems. Note that if the prototype variables are fixed, the resulting assignment problem becomes trivial and one assigns each point to the nearest prototype. On the other hand, if the cluster assignments are fixed, then the objective function can be decomposed into separate objective functions over different clusters. The portion of the objective function O_j contributed by the jth cluster is shown by an underbrace in the optimization formulation above. For each cluster, the relevant optimal solution $\overline{z}_j$ is the mean of the points assigned to that cluster. This result can be shown by setting the gradient of the objective function O_j with respect to each $\overline{z}_j$ to 0:

$$\frac{\partial O_j}{\partial z_j} = 2 \sum_{i=1}^{n} y_{ij}(\overline{X}_i - \overline{z}_j) = 0 \quad \forall j \in \{1 \ldots k\} \tag{4.76}$$

The points that do not belong to cluster j drop out in the above condition because $y_{ij} = 0$ for such points. As a result, $\overline{z}_j$ is simply the mean of the points in its cluster. Therefore, we need to alternative assign points to their closest prototypes, and set the prototypes to the centroids of the clusters defined by the assignment; these are exactly the steps of the well-known k-means algorithm. The centroid computation is a continuous optimization step, whereas cluster assignment is a discrete optimization step (which is greatly simplified by the decomposition approach of coordinate descent).

4.11   Summary

This chapter introduces the basic optimization models in machine learning. We discussed the conditions for optimality, as well as the cases in which a global optimum is guaranteed. Optimization problems in machine learning often have objective functions which can be separated into components across individual data points. This property enables the use of efficient sampling methods like stochastic gradient descent. Optimization models in machine learning are significantly different from traditional optimization in terms of the need to maximize performance on out-of-sample data rather than on the original optimization problem defined on the training data. Several examples of optimization in machine learning, such as linear regression, support vector machine, and logistic regression were discussed. Generalizations to multiclass models were also discussed. An alternative to stochastic gradient descent is coordinate descent, which can be more efficient in some situations.

4.12   Further Reading

Optimization is a field that has applications in many disciplines, and several books with a generic focus may be found in [10, 15, 16, 22, 99]. The work in [22] is particularly notable in providing a detailed exposition on convex optimization. Some of the books on linear algebra [130] are numerically focused, and provide several details of linear optimization algorithms. Methods for numerical optimization are also discussed in [52]. Some of the basic linear algebra books [122, 123] discuss the basics of optimization. A detailed discussion of linear regression methods may be found in linear algebra, optimization, and machine learning books. We recommend the The available machine learning books [1–4, 18, 19, 39, 46, 53, 56, 85, 94, 95] cover various machine learning applications.

Least-squares regression and classification dates back to the Widrow-Hoff algorithm [132] and Tikhonov-Arsenin's seminal work [127]. A detailed discussion of regression analysis may be found in [36]. The Fisher discriminant was proposed by Ronald Fisher [45] in 1936, and it turns out to be a special case of least-squares regression in which the binary response variable is used as the regressand [18]. The support-vector machine is generally credited to Cortes and Vapnik [30], although the primal method for L_2-loss SVMs was proposed several years earlier by Hinton [60]. This approach repairs the loss function in least-squares classification by keeping only one-half of the quadratic loss curve and setting the remaining to zero to create a smooth version of hinge loss (try this on Figure 4.9(a)). The specific significance of this contribution was lost within the broader literature on neural networks. A number of practical implementations of LIBSVM are available in [27] and those of linear classifiers are available in LIBLINEAR [44]. Detailed discussions of SVMs are provided in [31]. Discussions of numerical optimization techniques for logistic regression are provided in [93]. Coordinate descent is discussed in Hastie *et al.* [56], and more recently in [134].

4.13   Exercises

1. Find the saddle points, minima, and the maxima of the following functions:

    (a) $F(x) = x^2 - 2x + 2$
    (b) $F(x, y) = x^2 - 2x - y^2$

2. Suppose that $\overline{y}$ is a d-dimensional vector with very small norm $\epsilon = \|\overline{y}\|_2$. Consider a continuous and differentiable objective function $J(\overline{w})$ with zero gradient and Hessian H at $\overline{w} = \overline{w}_0$. Show that $\overline{y}^T H \overline{y}$ is approximately equal to twice the change in $J(\overline{w})$ by perturbing $\overline{w} = \overline{w}_0$ by ϵ in direction $\overline{y}/\|\overline{y}\|$.

3. Suppose that an optimization function $J(\overline{w})$ has a gradient of 0 at $\overline{w} = \overline{w}_0$. Furthermore, the Hessian of $J(\overline{w})$ at $w = \overline{w}_0$ has both positive and negative eigenvalues. Show

how you would use the Hessian to (i) find a vector direction along which infinitesimal movements in either direction from $\overline{w}_0$ decrease $J(\overline{w})$; (ii) find a vector direction along which infinitesimal movements in either direction from $\overline{w}_0$ increase $J(\overline{w})$. Is $\overline{w}_0$ a maximum, minimum, or saddle-point?

4. We know that the maximum of two convex functions is a convex function. Is the minimum of two convex functions convex? Is the intersection of two convex sets convex? If the union of two convex sets convex? Justify your answer in each case.

5. Either prove each statement or give a counterexample: (i) If $f(x)$ and $g(x)$ are convex, then $F(x, y) = f(x) + g(y)$ is convex. (ii) If $f(x)$ and $g(x)$ are convex, then $F(x, y) = f(x) \cdot g(y)$ is convex.

6. **Hinge-loss without margin:** Suppose that we modified the hinge-loss on page 184 by removing the constant value within the maximization function as follows:

$$J = \sum_{i=1}^{n} \max\{0, (-y_i[\overline{W} \cdot \overline{X}_i^T])\} + \frac{\lambda}{2}\|\overline{W}\|^2$$

This loss function is referred to as the *perceptron criterion*. Derive the stochastic gradient descent updates for this loss function.

7. Compare the perceptron criterion of the previous exercise to the hinge-loss in terms of its sensitivity to the magnitude of $\overline{W}$. State one non-informative weight vector $\overline{W}$, which will always be an optimal solution to the optimization problem of the previous exercise. Use this observation to explain why a perceptron (without suitable modifications) can sometimes provide much poorer solutions with an SVM when the points of the two classes cannot be separated by a linear hyperplane.

8. Consider an unconstrained quadratic program of the form $\overline{w}^T A \overline{w} + \overline{b}^T \overline{w} + c$, where $\overline{w}$ is a d-dimensional vector of optimization variables, and the $d \times d$ matrix A is positive semidefinite. The constant vector $\overline{b}$ is d-dimensional. Show that a global minimum exists for this quadratic program if and only if $\overline{b}$ lies in the column space of A.

9. The text of the book discusses a stochastic gradient descent update of the Weston-Watkins SVM, but not a mini-batch update. Consider a setting in which the mini-batch S contains training pairs of the form $(\overline{X}, c)$, where each $c \in \{1, \ldots, k\}$ is the categorical class label. Show that the stochastic gradient-descent step for each separator $\overline{W}_r$ at learning rate α:

$$\overline{W}_r \Leftarrow \overline{W}_r(1 - \alpha\lambda) + \alpha \sum_{(\overline{X},c)\in S, r=c} \overline{X}^T [\sum_{j\neq r} \delta(j, \overline{X})] - \alpha \sum_{(\overline{X},c)\in S, r\neq c} \overline{X}^T [\delta(r, \overline{X})] \quad (4.77)$$

Here, $\overline{W}_r$ is defined in the same way as the text of the chapter.

10. Consider the following function $f(x, y) = x^2 + 2y^2 + axy$. For what values of a (if any) is the function $f(x, y)$ concave, convex, and indefinite?

11. Consider the bivariate function $f(x, y) = x^3/6 + x^2/2 + y^2/2 + xy$. Define a domain of values of the function, at which it is convex.

12. Consider the L_1-loss function for binary classification, where for feature-class pair $(\overline{X}_i, y_i)$ and d-dimensional parameter vector $\overline{W}$, the point-specific loss for the ith instance is defined as follows:

$$L_i = \|y_i - \overline{W} \cdot \overline{X}_i^T\|_1$$

Here, we have $y_i \in \{-1, +1\}$, and $\overline{X}_i$ is a d-dimensional row vector of features. The norm used above is the L_1-norm instead of the L_2-norm of least-squares classification. Discuss why the loss function can be written as follows for $y_i \in \{-1, +1\}$:

$$L_i = \|1 - y_i \overline{W} \cdot \overline{X}_i^T\|_1$$

Show that the stochastic gradient descent update is as follows:

$$\overline{W} \Leftarrow \overline{W}(1 - \alpha\lambda) + \alpha y_i \overline{X}_i^T \operatorname{sign}(1 - y_i \overline{W} \cdot \overline{X}_i^T)$$

Here, λ is the regularization parameter, and α is the learning rate. Compare this update with the hinge-loss update for SVMs.

13. Let $\overline{x}$ be an n_1-dimensional vector, and W be an $n_2 \times n_1$-dimensional matrix. Show how to use the vector-to-vector chain rule to compute the vector derivative of $(W\overline{x}) \odot (W\overline{x})$ with respect to $\overline{x}$. Is the resulting vector derivative a scalar, vector, or matrix? Now repeat this exercise for $F((W\overline{x}) \odot (W\overline{x}))$, where $F(\cdot)$ is a function summing the elements of its argument into a scalar.

14. Let $\overline{x}$ be an n_1-dimensional vector, and W be an $n_2 \times n_1$-dimensional matrix. Show how to use the vector-to-vector chain rule to compute the vector derivative of $W(\overline{x} \odot \overline{x} \odot \overline{x})$ with respect to $\overline{x}$. Is the resulting vector derivative a scalar, vector, or matrix? Now repeat this exercise for $G(W(\overline{x} \odot \overline{x} \odot \overline{x}) - \overline{y})$, where $\overline{y}$ is a constant vector in n_2-dimensions, and $G(\cdot)$ is a function summing the *absolute value of the* elements of its argument into a scalar.

15. Show that if scalar L can be expressed as $L = f(W\overline{x})$ for $m \times d$ matrix W and d-dimensional vector $\overline{x}$, then $\frac{\partial L}{\partial W}$ will always be a rank-1 matrix or a zero matrix irrespective of the choice of function $f(\cdot)$. [This type of derivative is encountered frequently in neural networks.]

16. **Incremental linear regression with added points:** Suppose that you have a data matrix D and target vector $\overline{y}$ in linear regression. You have done all the hard work to invert $(D^T D)$ and then compute the closed-form solution $\overline{W} = (D^T D)^{-1} D^T \overline{y}$. Now you are given an additional training point $(\overline{X}, y)$, and are asked to compute the updated parameter vector $\overline{W}$. Show how you can do this efficiently without having to invert a matrix from scratch. Use this result to provide an efficient strategy for incremental linear regression. [Hint: Matrix inversion lemma.]

17. **Incremental linear regression with added features:** Suppose that you have a data set with a fixed number of points, but with an ever-increasing number of dimensions (as data scientists make an ever-increasing number of measurements and surveys). Provide an efficient strategy for incremental linear regression with regularization. [Hint: There are multiple ways to express the closed-form solution in linear regression because of the push-through identity of Problem 1.2.13.]

18. **Frobenius norm to matrix derivative:** Let A be an $n \times d$ constant matrix and V be a $d \times k$ matrix of parameters. Let $\overline{v}_i$ be the ith row of V and $\overline{V}_j$ be the jth column of V. Let J be a scalar function of the entries of V. Show the following:

    (a) Discuss the relationship between $\frac{\partial J}{\partial V}$ and each of $\frac{\partial J}{\partial \overline{v}_i}$ and $\frac{\partial J}{\partial \overline{V}_j}$. This relationship enables the use of scalar-to-vector identities in the chapter for scalar-to-matrix derivatives.

    (b) Let $J = \|V\|_F^2$. Show that $\frac{\partial J}{\partial V} = 2V$. You may find it helpful to express the Frobenius norm as the sum of vector norms and then use scalar-to-vector identities.

    (c) Let $J = \|AV\|_F^2$. Express J using vector norms and the columns of V. Show that $\frac{\partial J}{\partial V} = 2A^T A V$ by using the scalar-to-vector identities discussed in the chapter. Now show that the derivative of $J = \|AV + B\|^2$ is $2A^T(AV + B)$, where B is an $n \times k$ matrix. What you just derived is gradient descent in matrix factorization.

19. Consider an additively separable multivariate function of the form $J(w_1, w_2, \ldots w_{100}) = \sum_{i=1}^{100} J_i(w_i)$. Each $J_i(w_i)$ is a univariate function, which has one global optimum and one local optimum. Discuss why the chances of coordinate descent to reach the global optimum with a randomly chosen starting point are likely to be extremely low.

20. Propose a computational procedure to use single-variable coordinate descent in order to solve the L_2-loss SVM. You may use line search for each univariate problem. Implement the procedure in a programming language of your choice.

21. Consider a bivariate quadratic loss function of the following form:

$$f(x, y) = a x^2 + b y^2 + 2c xy + d x + e y + f$$

    Show that $f(x, y)$ is convex if and only if a and b are non-negative, and c is at most equal to the geometric mean of a and b in absolute magnitude.

22. Show that the functions $f(\overline{x}) = \sqrt{\langle \overline{x}, \overline{x} \rangle}$ and $g(\overline{x}) = \langle \overline{x}, \overline{x} \rangle$ are both convex. With regard to inner products, you are allowed to use only the basic axioms, and the Cauchy-Schwarz/triangle inequality.

23. **Two-sided matrix least-squares:** Let A be an $n \times m$ matrix and B be a $k \times d$ matrix. You want to find the $m \times k$ matrix X so that $J = \|C - AXB\|_F^2$ is minimized, where C is a known $n \times d$ matrix. Derive the derivative of J with respect to X and the optimality conditions. Show that *one possible solution* to the optimality conditions is $X = A^+ C B^+$, where A^+ and B^+ represent the Moore-Penrose pseudo-inverses of A and B, respectively. [Hint: Compute the scalar derivatives with respect to individual elements of X and then convert to matrix calculus form. Also see Exercises 47–51 of Chapter 2.]

24. Suppose that you replace the sum-of-squared-Euclidean objective with a sum-of-Manhattan objective for the k-means algorithm (pp. 198). Show that block coordinate descent results in the k-medians clustering algorithm, where the each dimension of the "centroid" representative is chosen as the median of the cluster along that dimension and assignment of points to representatives is done using the Manhattan distance instead of Euclidean distance. [Interesting fact: Many other representative-based clustering variants like k-modes and k-medoids are coordinate descent algorithms.]

25. Consider the cubic polynomial objective function $f(x) = ax^3 + bx^2 + cx + d$. Under what conditions does this objective function not have a critical point? Under what conditions is it strictly increasing in $[-\infty, +\infty]$?

26. Consider the cubic polynomial objective function $f(x) = ax^3 + bx^2 + cx + d$. Under what conditions does this objective have exactly one critical point? What kind of critical point is it? Give an example of such an objective function.

27. Let $f(x)$ be a univariate polynomial of degree n. What is the maximum number of critical points of this polynomial? What is the maximum number of minima, maxima, and saddle points?

28. What is the maximum number of critical points of a multivariate polynomial of degree n in d dimensions? Give an example of a polynomial where this maximum is met.

29. Suppose that $\overline{h}$ and $\overline{x}$ are column vectors, and W_1, W_2, and W_3 are matrices satisfying $\overline{h} = W_1 W_2 \overline{x} - W_2^2 W_3 \overline{x} + W_1 W_2 W_3 \overline{x}$. Derive an expression for $\frac{\partial \overline{h}}{\partial \overline{x}}$.

30. Consider a situation in which $\overline{h}_i = W_i W_{i-1} \overline{h}_{i-1}$, for $i \in \{1 \ldots n\}$. Here, each W_i is a matrix and each $\overline{h}_i$ is a vector. Use the vector-centric chain rule to derive an expression for $\frac{\partial \overline{h}_i}{\partial \overline{h}_0}$.

Chapter 5

Advanced Optimization Solutions

"The journey of a thousand miles begins with one step." –Lao Tzu

5.1 Introduction

The previous chapter introduced several basic algorithms for gradient descent. However, these algorithms do not always work well because of the following reasons:

- *Flat regions and local optima:* The objective functions of machine learning algorithms might have local optima and flat regions in the loss surface. As a result, the learning process might be too slow or arrive at a poor solution.

- *Differential curvature:* The directions of gradient descent are only *instantaneous* directions of best movement, which usually change over steps of finite length. Therefore, a steepest direction of descent no longer remains the steepest direction, after one makes a finite step in that direction. If the step is too large, the different components of the gradient might flip signs, and the objective function might worsen. A direction is said to show high *curvature*, if the gradient changes rapidly in that direction. Clearly, directions of high curvature cause uncertainty in the outcomes of gradient descent.

- *Non-differentiable objective functions:* Some objective functions are non-differentiable, which causes problems for gradient descent. If differentiability is violated at a relatively small number of points and the loss function is informative for the large part, one can use gradient descent with minor modifications. More challenging cases arise when the objective functions have steep cliffs or flat surfaces in large regions of the space, and the gradients are not informative at all.

The simplest approach to address both flat regions and differential curvature is to adjust the gradients in some way to account for poor convergence. These methods *implicitly* use the curvature to adjust the gradients of the objective function with respect to different parameters. Examples of such techniques include the pairing of vanilla gradient-descent methods with computational algorithms like the *momentum method*, *RMSProp*, or *Adam*.

Another class of methods uses second-order derivatives to *explicitly* measure the curvature; after all, a second derivative is the rate of *change* in gradient, which is a direct measure of the unpredictability of using a constant gradient direction over a finite step. The second-derivative matrix, also referred to as the *Hessian*, contains a wealth of information about directions along which the greatest curvature occurs. Therefore, the Hessian is used by many second-order techniques like the *Newton method* in order to adjust the directions of movement by using a trade-off between the steepness of the descent and the curvature along a direction.

Finally, we discuss the problem of non-differentiable objective functions. Consider the L_1-loss function, which is non-differentiable at some points in the parameter space:

$$f(x_1, x_2) = |x_1| + |x_2|$$

The point $(x_1, x_2) = (0, 0)$ is a non-differentiable point of the optimization. This type of setting can be addressed easily by having special rules for the small number of non-differentiable points in the space. However, in some cases, non-informative loss surfaces contain only flat regions and vertical cliffs. For example, trying to directly optimize a ranking-based objective function will cause non-differentiability in large regions of the space. Consider the following objective function containing training points $\overline{X}_1 \ldots \overline{X}_n$, of which a subset S belong to a *positive class* (e.g., fraud instances versus normal instances):

$$J(\overline{W}) = \sum_{i \in S} \text{Rank}(\overline{W} \cdot \overline{X}_i)$$

Here, the function "Rank" simply computes a value from 1 through n, based on sorting the values of $\overline{W} \cdot \overline{X}_i$ over the n training points and returning the rank of each $\overline{X}_i$. Minimizing the function $J(\overline{W})$ tries to set $\overline{W}$ to ensure that positive examples are always ranked before negative examples. This kind of objective function will contain only flat surfaces and vertical cliffs with respect to $\overline{W}$, because the ranks can suddenly change at specific values of the parameter vector $\overline{W}$. In most regions, the ranks will not change on perturbing $\overline{W}$ slightly, and therefore $J(\overline{W})$ will have a zero gradient in most regions. This type of setting can cause serious problems for gradient descent because the gradients are not informative at all. In such cases, more complex methods like the proximal gradient method need to be used. This chapter will discuss several such options.

This chapter is organized as follows. The next section will discuss the challenges associated with optimization of differentiable functions. Methods that modify the first-order derivative of the loss function to account for curvature are discussed in Section 5.3. The Newton method is introduced in Section 5.4. Applications of the Newton method to machine learning are discussed in Section 5.5. The challenges associated with the Newton method are discussed in Section 5.6. Computationally efficient approximations of the Newton method are discussed in Section 5.7. The optimization of non-differentiable functions is discussed in Section 5.8. A summary is given in Section 5.9.

5.2   Challenges in Gradient-Based Optimization

In this section, we will discuss the two main problems associated with gradient-based optimization. The first problem has to do with flat regions and local optima, whereas the second problem has to do with the different levels of curvature in different directions. Understanding these problems is one of the keys in designing good solutions for them. Therefore, this section will discuss these issues in detail.

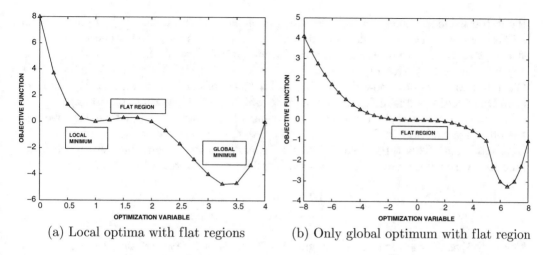

(a) Local optima with flat regions   (b) Only global optimum with flat region

Figure 5.1: Illustrations of local optima and flat regions

5.2.1 Local Optima and Flat Regions

The previous chapter discussed several optimization models that correspond to convex functions, which have a single global optimum and no local optima. However, more complex machine learning settings like neural networks are typically not convex, and they might have multiple local optima. Such local optima create challenges for gradient descent.

Consider the following 1-dimensional function:

$$F(x) = (x - 1)^2[(x - 3)^2 - 1]$$

Computing the derivative and setting it to zero yields the following condition:

$$F'(x) = 2(x - 1)[(x - 1)(x - 3) + (x - 3)^2 - 1] = 0$$

The solutions to this equation are $x = 1, \frac{5}{2} - \frac{\sqrt{3}}{2} = 1.634, \frac{5}{2} + \frac{\sqrt{3}}{2} = 3.366$. From the second-derivative conditions, it is possible to show that the first and third roots are minima with $F''(x) > 0$, whereas the second root is a maximum with $F''(x) < 0$. When the function $F(x)$ is evaluated at these points, we obtain $F(1) = 0$, $F(1.634) = 0.348$, and $F(3.366) = -4.848$. The plot of this function is shown in Figure 5.1(a). It is evident that the first of the optima is a local minimum, whereas the second is a local maximum. The last point $x = 3.366$ is the *global* minimum we are looking for.

In this case, we were able to solve for both the potential minima by using the optimality condition, and then plug in these values to determine which of them is the global minimum. But what happens when we try to use gradient descent? The problem is that if we start the gradient descent from any point less than 1.634, one will arrive at a local minimum. Furthermore, one might never arrive at a global minimum (if we always choose the wrong starting point in multiple runs), and there would be no way of knowing that a better minimum exists. This problem becomes even more severe when there are multiple dimensions, and the number of local minima proliferate. We point the reader to Problem 4.2.4 of the previous chapter as an example of how local minima proliferate exponentially fast with increasing dimensionality. It is relatively easy to show that if we have d univariate functions (in different variables $x_1 \ldots x_d$), so that the ith function has k_i local/global minima, then the d-dimensional function created by the sum of these functions has $\prod_{i=1}^{d} k_i$ local/global minima. For example, a 10-dimensional function, which is a sum of 10 instances of the function represented in Equation 5.2.1 (over different variables) would have $2^{10} = 1024$ minima obtained by setting each of the 10 dimensions to any one of the values from $\{1, 3.366\}$. Clearly, if one does not know the number and location of the local minima, it is hard to be confident about the optimality of the point to which gradient descent converges.

Another problem is the presence of flat regions in the objective function. For example, the objective function in Figure 5.1(a) has a flat region between a local minimum and a local maximum. This type of situation is quite common and is possible even in objective functions where there are no local optima. Consider the following objective function:

$$F(x) = \begin{cases} -(x/5)^3 & \text{if } x \leq 5 \\ x^2 - 13x + 39 & \text{if } x > 5 \end{cases}$$

The objective function is shown in Figure 5.1(b). This objective function has a flat region in the range $[-1, +1]$, where the absolute value of the gradient is less than 0.1. On the other hand, the gradient increases rapidly for values of $x > 5$. Why are flat regions problematic? The main issue is that the speed of descent depends on the magnitude of the gradient (if the learning rate is fixed). In such cases, the optimization procedure will take a long time to cross flat regions of the space. This will make the optimization process excruciatingly slow. As we will see later, techniques like momentum methods use analogies from physics in order to inherit the rate of descent from previous steps as a type of momentum. The basic idea is that if you roll a marble down a hill, it gathers speed as it rolls down, and it is often able to navigate local potholes and flat regions better because of its momentum. We will discuss this principle in more detail in Section 5.3.1.

5.2.2 Differential Curvature

In multidimensional settings, the components of the gradients may have very different magnitudes, which causes problems for gradient-descent methods. For example, neural networks often have large differences in the magnitudes of the partial derivatives with respect to parameters of different layers; this phenomenon is popularly referred to as the *vanishing and exploding gradient problem*. Minor manifestations of this problem occur even in simple cases like convex and quadratic objective functions. Therefore, we will start by studying these simple cases, because they provide excellent insight into the source of the problem and possible solutions.

Consider the simplest possible case of a convex, quadratic objective function with a bowl-like shape and a single global minimum. Two such bivariate loss functions are illustrated in Figure 5.2. In this figure, the contour plots of the loss function are shown, in which each line corresponds to points in the XY-plane where the loss function has the same value. The direction of steepest descent is always perpendicular to this line. The first loss function is of the form $L = x^2 + y^2$, which takes the shape of a perfectly circular bowl, if one were to view the height as the objective function value. This loss function treats x and y in a symmetric way. The second loss function is of the form $L = x^2 + 4y^2$, which is an elliptical bowl. Note that this loss function is more sensitive to changes in the value of y as compared to changes in the value of x, although the specific sensitivity depends on the position of the data point. In other words, the *second-order* derivatives $\frac{\partial^2 L}{\partial x^2}$ and $\frac{\partial^2 L}{\partial y^2}$ are different in the case of the loss $L = x^2 + 4y^2$. A high second-order derivative is also referred to as high curvature, because it affects how quickly the gradient changes. This is important from the perspective

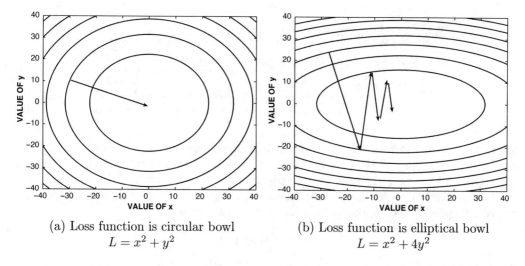

(a) Loss function is circular bowl
$$L = x^2 + y^2$$

(b) Loss function is elliptical bowl
$$L = x^2 + 4y^2$$

Figure 5.2: The effect of the shape of the loss function on steepest-gradient descent

of gradient descent because it tells us that some directions have more consistent gradients that do not change rapidly. Consistent gradients are more desirable from the perspective of making gradient-descent steps of larger sizes.

In the case of the circular bowl of Figure 5.2(a), the gradient points directly at the optimum solution, and one can reach the optimum in a single step, as long as the correct step-size is used. This is not quite the case in the loss function of Figure 5.2(b), in which the gradients are often more significant in the y-direction as compared to the x-direction. Furthermore, the gradient never points to the optimal solution, as a result of which many course corrections are needed over the descent. A salient observation is that the steps along the y-direction are large, but subsequent steps undo the effect of previous steps. On the other hand, the progress along the x-direction is consistent but tiny. In other words, the long-term progress along each direction is very limited; therefore, it is possible to get into situations where very little progress is made even after training for a long time.

The above example represents a very simple quadratic, convex, and additively separable function, which represents a straightforward scenario compared to any real-world setting in machine learning. In fact, *with very few exceptions, the path of steepest descent in most objective functions is only an instantaneous direction of best movement, and is not the correct direction of descent in the longer term.* In other words, small steps with "course corrections" are always needed; the only way to reach the optimum with steepest-descent updates is by using an *extremely large number of tiny updates and course corrections*, which is obviously very inefficient. At first glance, this might seem almost ominous, but it turns out that there are numerous solutions of varying complexity to address these issues. The simplest example is feature normalization.

5.2.2.1 Revisiting Feature Normalization

As discussed in Chapter 4, it is common to standardize features before applying gradient descent. An important reason for scaling the features is to ensure better performance of gradient descent. In order to understand this point, we will use an example. Consider a (hypothetical) data set containing information about the classical guns-butter trade-off in

Table 5.1: A hypothetical data set of guns, butter, and happiness

Guns (number per capita)	Butter (ounces per capita)	Happiness (index)
0.1	25	7
0.8	10	1
0.4	10	4

the expenditure of various nations, together with the happiness index. The goal is to predict the happiness index y of the nation as a function of the guns per capita x_1 and the ounces per capita of butter x_2. An example of a toy data set of three points is shown in Table 5.1. A linear regression model uses the coefficient w_1 for guns and the coefficient w_2 for butter in order to predict the happiness index from guns and butter:

$$y = w_1 x_1 + w_2 x_2$$

Then, one can model the least-squares objective function as follows:

$$J = (0.1w_1 + 25w_2 - 7)^2 + (0.8w_1 + 10w_2 - 1)^2 + (0.4w_1 + 10w_2 - 4)^2$$
$$= 0.81w_1^2 + 825w_2^2 + 29w_1 w_2 - 6.2w_1 - 450w_2 + 66$$

Note that this objective function is far more sensitive to w_2 as compared to w_1. This is caused by the fact that the butter feature has a much larger variance than the gun feature, which shows up in the coefficients of the objective function. As a result, the gradient will often bounce along the w_2 direction, while making tiny progress along the w_1 direction. However, if we standardize each column in Table 5.1 to zero mean and unit variance, the coefficients of w_1^2 and w_2^2 will become much more similar. As a result, the bouncing behavior of gradient descent is reduced. In this particular case, the interaction terms of the form $w_1 w_2$ will cause the ellipse to be oriented at an angle to the original axes. This causes additional challenges in terms of bouncing of gradient descent along directions that are not parallel to the original axes. Such interaction terms can be addressed by a procedure called *whitening*, and it is an application of the method of principal component analysis (cf. Section 7.4.6 of Chapter 7).

5.2.3   Examples of Difficult Topologies: Cliffs and Valleys

It is helpful to examine a number of specific manifestations of high-curvature topologies in loss surfaces. Two examples of high-curvature surfaces are cliffs and valleys. An example of a cliff is shown in Figure 5.3. In this case, there is a gently sloping surface that rapidly changes into a cliff. However, if one computed only the first-order partial derivative with respect to the variable x shown in the figure, one would only see a gentle slope. As a result, a modest learning rate might cause very slow progress in gently sloping regions, whereas the same learning rate can suddenly cause overshooting to a point far from the optimal solution in steep regions. This problem is caused by the nature of the curvature (i.e., changing gradient), where the first-order gradient does not contain the information needed to control the size of the update. As we will see later, several computational solutions directly or indirectly make use of second-order derivatives in order to account for the curvature. Cliffs are not desirable because they manifest a certain level of instability in the loss function. This implies that a small change in some of the weights can suddenly change the local topology so drastically that continuous optimization algorithms (like gradient descent) have a hard time.

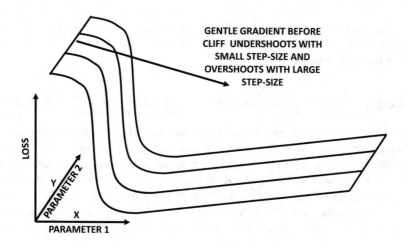

Figure 5.3: An example of a cliff in the loss surface

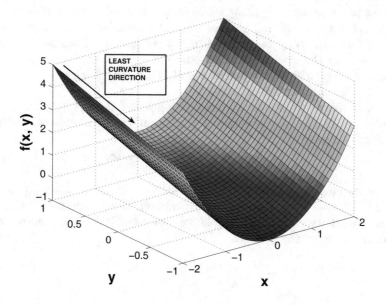

Figure 5.4: The curvature effect in valleys

The specific effect of curvature is particularly evident when one encounters loss functions in the shape of sloping or winding valleys. An example of a sloping valley is shown in Figure 5.4. A valley is a dangerous topography for a gradient-descent method, particularly if the bottom of the valley has a steep and rapidly changing surface (which creates a narrow valley). In narrow valleys, the gradient-descent method will bounce violently along the steep sides of the valley without making much progress in the gently sloping direction, where the greatest *long-term* gains are present. As we will see later in this chapter, many computational methods magnify the components of the gradient along consistent directions of movement (to discourage back-and-forth bouncing). In some cases, the steepest descent directions are modified using such ad hoc methods, whereas in others, the curvature is explicitly used with the help of second-order derivatives. The first of these methods will be the topic of discussion in the next section.

5.3   Adjusting First-Order Derivatives for Descent

In this section, we will study computational methods that modify first-order derivatives. Implicitly, these methods do use second-order information by taking the curvature into account while modifying the components of the gradient. Many of these methods use different learning rates for different parameters. The idea is that parameters with large partial derivatives are often oscillating and zigzagging, whereas parameters with small partial derivatives tend to be more consistent but move in the same direction. These methods are also more popular than second-order methods, because they are computationally efficient to implement.

5.3.1   Momentum-Based Learning

Momentum-based methods address the issues of local optima, flat regions, and curvature-centric zigzagging by recognizing that *emphasizing medium-term to long-term directions of consistent movement* is beneficial, because they de-emphasize local distortions in the loss topology. Consequently, an aggregated measure of the feedback from previous steps is used in order to speed up the gradient-descent procedure. As an analogy, a marble that rolls down a sloped surface with many potholes and other distortions is often able to use its momentum to overcome such minor obstacles.

Consider a setting in which one is performing gradient-descent with respect to the parameter vector $\overline{W}$. The normal updates for gradient-descent with respect to the objective function J are as follows:

$$\overline{V} \Leftarrow -\alpha \frac{\partial J}{\partial \overline{W}}; \quad \overline{W} \Leftarrow \overline{W} + \overline{V}$$

Here, α is the learning rate. We are using the matrix calculus notation $\frac{\partial J}{\partial \overline{W}}$ in lieu of ∇J. As discussed in Chapter 4, we are using the convention that the derivative of a scalar with respect to a column vector is a column vector (see page 170), which corresponds to the denominator layout in matrix calculus:

$$\nabla J = \frac{\partial J}{\partial \overline{W}} = \left[\frac{\partial J}{\partial w_1} \cdots \frac{\partial J}{\partial w_d} \right]^T$$

In momentum-based descent, the vector $\overline{V}$ inherits a fraction β of the velocity from its previous step in addition to the current gradient, where $\beta \in (0, 1)$ is the momentum parameter:

$$\overline{V} \Leftarrow \beta \overline{V} - \alpha \frac{\partial J}{\partial \overline{W}}; \quad \overline{W} \Leftarrow \overline{W} + \overline{V}$$

Setting $\beta = 0$ specializes to straightforward gradient descent. Larger values of $\beta \in (0, 1)$ help the approach pick up a consistent velocity $\overline{V}$ in the correct direction. The parameter β is also referred to as the *momentum parameter* or the *friction parameter*. The word "friction" is derived from the fact that small values of β act as "brakes," much like friction.

Momentum helps the gradient descent process in navigating flat regions and local optima, such as the ones shown in Figure 5.1. A good analogy for momentum-based methods is to visualize them in a similar way as a marble rolls down a bowl. As the marble picks up

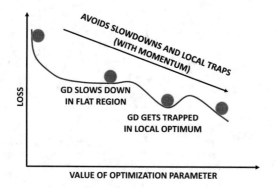

Figure 5.5: Effect of momentum in navigating complex loss surfaces. The annotation "GD" indicates pure gradient descent without momentum. Momentum helps the optimization process retain speed in flat regions of the loss surface and avoid local optima

speed, it will be able to navigate flat regions of the surface quickly and escape form local potholes in the bowl. This is because the gathered momentum helps it escape potholes. Figure 5.5, which shows a marble rolling down a complex loss surface (picking up speed as it rolls down), illustrates this concept. The use of momentum will often cause the solution to slightly overshoot in the direction where velocity is picked up, just as a marble will overshoot when it is allowed to roll down a bowl. However, with the appropriate choice of β, it will still perform better than a situation in which momentum is not used. The momentum-based method will generally perform better because the marble gains speed as it rolls down the bowl; the quicker arrival at the optimal solution more than compensates for the overshooting of the target. Overshooting is desirable to the extent that it helps avoid local optima. The parameter β controls the amount of friction that the marble encounters while rolling down the loss surface. While increased values of β help in avoiding local optima, it might also increase oscillation at the end. In this sense, the momentum-based method has a neat interpretation in terms of the physics of a marble rolling down a complex loss surface. Setting $\beta > 1$ can cause instability and divergence, because gradient descent can pick up speed in an uncontrolled way.

In addition, momentum-based methods help in reducing the undesirable effects of curvature in the loss surface of the objective function. Momentum-based techniques recognize that zigzagging is a result of highly contradictory steps that cancel out one another and reduce the *effective* size of the steps in the correct (long-term) direction. An example of this scenario is illustrated in Figure 5.2(b). Simply attempting to increase the size of the step in order to obtain greater movement in the correct direction might actually move the current solution even further away from the optimum solution. In this point of view, it makes a lot more sense to move in an "averaged" direction of the last few steps, so that the zigzagging is smoothed out. This type of averaging is achieved by using the momentum from the previous steps. Oscillating directions do not contribute consistent velocity to the update.

With momentum-based descent, the learning is accelerated, because one is generally moving in a direction that often points closer to the optimal solution and the useless "sideways" oscillations are muted. The basic idea is to give greater preference to *consistent* directions over multiple steps, which have greater importance in the descent. This allows the use of larger steps in the correct direction without causing overflows or "explosions" in the sideways direction. As a result, learning is accelerated. An example of the use of

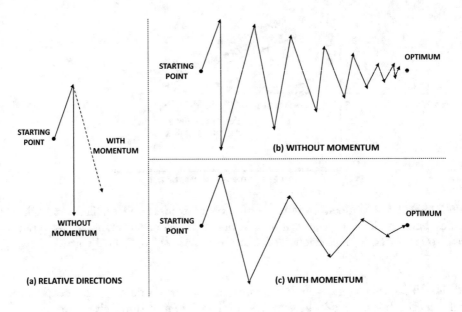

Figure 5.6: Effect of momentum in smoothing zigzag updates

momentum is illustrated in Figure 5.6. It is evident from Figure 5.6(a) that momentum increases the relative component of the gradient in the correct direction. The corresponding effects on the updates are illustrated in Figure 5.6(b) and (c). It is evident that momentum-based updates can reach the optimal solution in fewer updates. One can also understand this concept by visualizing the movement of a marble down the valley of Figure 5.4. As the marble gains speed down the gently sloping valley, the effects of bouncing along the sides of the valley will be muted over time.

5.3.2   AdaGrad

In the AdaGrad algorithm [38], one keeps track of the aggregated squared magnitude of the partial derivative with respect to each parameter over the course of the algorithm. The square-root of this value is *proportional* to the root-mean-squared slope for that parameter (although the absolute value will increase with the number of epochs because of successive aggregation).

Let A_i be the aggregate value for the ith parameter. Therefore, in each iteration, the following update is performed with respect to the objective function J:

$$A_i \Leftarrow A_i + \left(\frac{\partial J}{\partial w_i}\right)^2; \quad \forall i \tag{5.1}$$

The update for the ith parameter w_i is as follows:

$$w_i \Leftarrow w_i - \frac{\alpha}{\sqrt{A_i}}\left(\frac{\partial J}{\partial w_i}\right); \quad \forall i$$

If desired, one can use $\sqrt{A_i + \epsilon}$ in the denominator instead of $\sqrt{A_i}$ to avoid ill-conditioning. Here, ϵ is a small positive value such as 10^{-8}.

Scaling the derivative inversely with $\sqrt{A_i}$ is a kind of "signal-to-noise" normalization because A_i only measures the historical magnitude of the gradient rather than its sign; it encourages faster *relative* movements along gently sloping directions with consistent sign of the gradient. If the gradient component along the ith direction keeps wildly fluctuating between $+100$ and -100, this type of magnitude-centric normalization will penalize that component far more than another gradient component that consistently takes on the value in the vicinity of 0.1 (but with a consistent sign). For example, in Figure 5.6, the movements along the oscillating direction will be de-emphasized, and the movement along the consistent direction will be emphasized. However, absolute movements along all components will tend to slow down over time, which is the main problem with the approach. The slowing down is caused by the fact that A_i is the *aggregate* value of the entire history of partial derivatives. This will lead to diminishing values of the scaled derivative. As a result, the progress of AdaGrad might prematurely become too slow, and it will eventually (almost) stop making progress. Another problem is that the aggregate scaling factors depend on ancient history, which can eventually become stale. It turns out that the exponential averaging of RMSProp can address both issues.

5.3.3 RMSProp

The RMSProp algorithm [61] uses a similar motivation as AdaGrad for performing the "signal-to-noise" normalization with the absolute magnitude $\sqrt{A_i}$ of the gradients. However, instead of simply adding the squared gradients to estimate A_i, it uses *exponential averaging*. Since one uses *averaging* to normalize rather than *aggregate* values, the progress is not slowed prematurely by a constantly increasing scaling factor A_i. The basic idea is to use a decay factor $\rho \in (0,1)$, and weight the squared partial derivatives occurring t updates ago by ρ^t. Note that this can be easily achieved by multiplying the current squared aggregate (i.e., *running* estimate) by ρ and then adding $(1 - \rho)$ times the current (squared) partial derivative. The running estimate is initialized to 0. This causes some (undesirable) bias in early iterations, which disappears over the longer term. Therefore, if A_i is the exponentially averaged value of the ith parameter w_i, we have the following way of updating A_i:

$$A_i \Leftarrow \rho A_i + (1 - \rho)\left(\frac{\partial J}{\partial w_i}\right)^2 ; \quad \forall i \tag{5.2}$$

The square-root of this value for each parameter is used to normalize its gradient. Then, the following update is used for (global) learning rate α:

$$w_i \Leftarrow w_i - \frac{\alpha}{\sqrt{A_i}}\left(\frac{\partial J}{\partial w_i}\right) ; \quad \forall i$$

If desired, one can use $\sqrt{A_i + \epsilon}$ in the denominator instead of $\sqrt{A_i}$ to avoid ill-conditioning. Here, ϵ is a small positive value such as 10^{-8}. Another advantage of RMSProp over AdaGrad is that the importance of ancient (i.e., stale) gradients decays exponentially with time. The drawback of RMSProp is that the running estimate A_i of the second-order moment is biased in early iterations because it is initialized to 0.

5.3.4 Adam

The Adam algorithm uses a similar "signal-to-noise" normalization as AdaGrad and RMSProp; however, it also incorporates momentum into the update. In addition, it directly addresses the initialization bias inherent in the exponential smoothing of pure RMSProp.

As in the case of RMSProp, let A_i be the exponentially averaged value of the ith parameter w_i. This value is updated in the same way as RMSProp with the decay parameter $\rho \in (0, 1)$:

$$A_i \Leftarrow \rho A_i + (1 - \rho)\left(\frac{\partial J}{\partial w_i}\right)^2; \quad \forall i \tag{5.3}$$

At the same time, an exponentially smoothed value of the gradient is maintained for which the ith component is denoted by F_i. This smoothing is performed with a different decay parameter ρ_f:

$$F_i \Leftarrow \rho_f F_i + (1 - \rho_f)\left(\frac{\partial J}{\partial w_i}\right); \quad \forall i \tag{5.4}$$

This type of exponentially smoothing of the gradient with ρ_f is a variation of the momentum method discussed in Section 5.3.1 (which is parameterized by a friction parameter β instead of ρ_f). Then, the following update is used at learning rate α_t in the tth iteration:

$$w_i \Leftarrow w_i - \frac{\alpha_t}{\sqrt{A_i}}F_i; \quad \forall i$$

There are two key differences from the RMSProp algorithm. First, the gradient is replaced with its exponentially smoothed value in order to incorporate momentum. Second, the learning rate α_t now depends on the iteration index t, and is defined as follows:

$$\alpha_t = \alpha \underbrace{\left(\frac{\sqrt{1 - \rho^t}}{1 - \rho_f^t}\right)}_{\text{Adjust Bias}} \tag{5.5}$$

Technically, the adjustment to the learning rate is actually a bias correction factor that is applied to account for the unrealistic initialization of the two exponential smoothing mechanisms, and it is particularly important in early iterations. Both F_i and A_i are initialized to 0, which causes bias in early iterations. The two quantities are affected differently by the bias, which accounts for the ratio in Equation 5.5. It is noteworthy that each of ρ^t and ρ_f^t converge to 0 for large t because $\rho, \rho_f \in (0, 1)$. As a result, the initialization bias correction factor of Equation 5.5 converges to 1, and α_t converges to α. The default suggested values of ρ_f and ρ are 0.9 and 0.999, respectively, according to the original Adam paper [72]. Refer to [72] for details of other criteria (such as parameter sparsity) used for selecting ρ and ρ_f. Like other methods, Adam uses $\sqrt{A_i + \epsilon}$ (instead of $\sqrt{A_i}$) in the denominator of the update for better conditioning. The Adam algorithm is extremely popular because it incorporates most of the advantages of other algorithms, and often performs competitively with respect to the best of the other methods [72].

5.4   The Newton Method

The use of second-order derivatives has found a modest level of renewed popularity in recent years. Such methods can partially alleviate some of the problems caused by the high curvature of the loss function. This is because second-order derivatives encode the rate of change of the gradient in each direction, which is a more formal description of the concept of curvature. The Newton method uses a trade-off between the first- and second-order derivatives in order to descend in directions that are sufficiently steep and also do not have drastically changing gradients. Such directions allow the use of fewer steps with better individual loss improvements. In the special case of quadratic loss functions, the Newton method requires a single step.

5.4.1 The Basic Form of the Newton Method

Consider the parameter vector $\overline{W} = [w_1 \ldots w_d]^T$ for which the second-order derivatives of the objective function $J(\overline{W})$ are of the following form:

$$H_{ij} = \frac{\partial^2 J(\overline{W})}{\partial w_i \partial w_j}$$

Note that the partial derivatives use all pairwise parameters in the denominator. Therefore, for a neural network with d parameters, we have a $d \times d$ *Hessian matrix* H, for which the (i, j)th entry is H_{ij}.

The Hessian can also be defined as the *Jacobian* of the gradient with respect to the weight vector. As discussed in Chapter 4, a Jacobian is a vector-to-vector derivative in matrix calculus, and therefore the result is a matrix. The derivative of an m-dimensional column vector with respect to an d-dimensional column vector is a $d \times m$ matrix in the denominator layout of matrix calculus, whereas it is an $m \times d$ matrix in the numerator layout (see page 170). The Jacobian is an $m \times d$ matrix, and therefore conforms to the numerator layout. In this book, we are consistently using the denominator layout, and therefore, the Jacobian of the m-dimensional vector $\overline{h}$ with respect to the d-dimensional vector $\overline{w}$ is defined as the *transpose* of the vector-to-vector derivative:

$$\text{Jacobian}(\overline{h}, \overline{w}) = \left[\frac{\partial \overline{h}}{\partial \overline{w}}\right]^T = \left[\frac{\partial h_i}{\partial w_j}\right]_{m \times d \text{ matrix}} \tag{5.6}$$

However, the transposition does not really matter in the case of the Hessian, which is symmetric. Therefore, the Hessian can also be defined as follows:

$$H = \left[\frac{\partial \nabla J(\overline{W})}{\partial \overline{W}}\right]^T = \frac{\partial \nabla J(\overline{W})}{\partial \overline{W}} \tag{5.7}$$

The Hessian can be viewed as the natural generalization of the second derivative to multivariate data. Like the univariate Taylor series expansion of the second derivative, it can be used for the multivariate Taylor-series expansion by replacing the scalar second derivative with the Hessian. Recall that the (second-order) Taylor-series expansion of a univariate function $f(w)$ about the scalar w_0 may be defined as follows (cf. Section 1.5.1 of Chapter 1):

$$f(w) \approx f(w_0) + (w - w_0)f'(w_0) + \frac{(w - w_0)^2}{2} f''(w_0) \tag{5.8}$$

It is noteworthy that the Taylor approximation is accurate when $|w - w_0|$ is small, and it starts losing its accuracy for non-quadratic functions as $|w - w_0|$ increases (as the contribution of the higher-order terms increases as well). One can also write a quadratic approximation of the *multivariate* loss function $J(\overline{W})$ in the vicinity of parameter vector $\overline{W}_0$ by using the following Taylor expansion:

$$J(\overline{W}) \approx J(\overline{W}_0) + [\overline{W} - \overline{W}_0]^T [\nabla J(\overline{W}_0)] + \frac{1}{2}[\overline{W} - \overline{W}_0]^T H [\overline{W} - \overline{W}_0] \tag{5.9}$$

As in the case of the univariate expansion, the accuracy of this approximation falls off with increasing value of $\|\overline{W} - \overline{W}_0\|$, which is the Euclidean distance between $\overline{W}$ and $\overline{W}_0$. Note that the Hessian H is computed at $\overline{W}_0$. Here, the parameter vectors $\overline{W}$ and $\overline{W}_0$ are d-dimensional column vectors. This is a quadratic approximation, and one can simply set the gradient to 0, which results in the following optimality condition for the quadratic approximation:

$$\nabla J(\overline{W}) = \overline{0}, \qquad\qquad\qquad \text{[Gradient of Loss Function]}$$
$$\nabla J(\overline{W}_0) + H[\overline{W} - \overline{W}_0] = \overline{0}, \qquad \text{[Gradient of Taylor approximation]}$$

The optimality condition above only finds a critical point, and the convexity of the function is important to ensure that this critical point is a minimum. One can rearrange the above optimality condition to obtain the following Newton update:

$$\overline{W}^* \Leftarrow \overline{W}_0 - H^{-1}[\nabla J(\overline{W}_0)] \qquad\qquad (5.10)$$

One interesting characteristic of this update is that it is directly obtained from an optimality condition, and therefore there is no learning rate. In other words, this update is approximating the loss function with a quadratic bowl and moving *exactly* to the bottom of the bowl *in a single step*; the learning rate is already incorporated implicitly. Recall from Figure 5.2 that first-order methods bounce along directions of high curvature. Of course, the bottom of the quadratic approximation is not the bottom of the true loss function, and therefore multiple Newton updates will be needed. Therefore, the basic Newton method for non-quadratic functions initializes $\overline{W}$ to an initial point $\overline{W}_0$, performs the updates as follows:

1. Compute the gradient $\nabla J(\overline{W})$ and the Hessian H at the current parameter vector $\overline{W}$.

2. Perform the Newton update:

$$\overline{W} \Leftarrow \overline{W} - H^{-1}[\nabla J(\overline{W})]$$

3. If convergence has not occurred, go back to step 1.

Although the algorithm above is iterative, the Newton method requires only a single step for the special case of quadratic functions. The main difference of Equation 5.10 from the update of steepest-gradient descent is pre-multiplication of the steepest direction (which is $[\nabla J(\overline{W}_0)]$) with the inverse of the Hessian. This multiplication with the inverse Hessian plays a key role in changing the direction of the steepest-gradient descent, so that one can take larger steps in that direction (resulting in better improvement of the objective function) even if the *instantaneous* rate of change in that direction is not as large as the steepest-descent direction. This is because the Hessian encodes how fast the gradient is changing in each direction. Changing gradients are bad for larger updates because one might inadvertently worsen the objective function, if the signs of many components of the gradient change during the step. It is profitable to move in directions where the ratio of the gradient to the rate of change of the gradient is large, so that one can take larger steps while being confident that the movement is not causing unexpected changes because of the changed gradient. Pre-multiplication with the inverse of the Hessian achieves this goal. The effect of the pre-multiplication of the steepest-descent direction with the inverse Hessian is shown in Figure 5.7. It is helpful to reconcile this figure with the example of the quadratic bowl in Figure 5.2. In a sense, pre-multiplication with the inverse Hessian biases

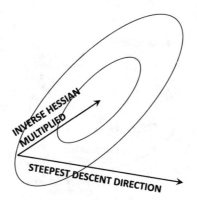

Figure 5.7: The effect of pre-multiplication of steepest-descent direction with the inverse Hessian

the learning steps towards low-curvature directions. This situation also arises in valleys like the ones shown in Figure 5.4. Multiplication with the inverse Hessian will tend to favor the gently sloping (but low curvature) direction, which is a better direction of long-term movement. Furthermore, if the Hessian is negative semi-definite at a particular point (rather than positive semi-definite), the Newton method might move in the wrong direction towards a maximum (rather than a minimum). Unlike gradient descent, the Newton method only finds critical points rather than minima.

5.4.2 Importance of Line Search for Non-quadratic Functions

It is noteworthy that the update for a non-quadratic function can be somewhat unpredictable because one moves to the bottom of a *local quadratic approximation* caused by the Taylor expansion. This local quadratic approximation can sometimes be very poor as one moves further away from the point of the Taylor approximation. Therefore, it is possible for a Newton step to worsen the quality of the objective function if one simply moves to the bottom of the local quadratic approximation. In order to understand this point, we will consider the simple case of a univariate function in Figure 5.8, where both the original function and its quadratic approximation are shown. Both the starting and ending points of a Newton step are shown, and the objective function value of the ending point differs considerably between the true function and the quadratic approximation (although the starting points are the same). As a result, the Newton step actually *worsens* the objective function value. One can view this situation in an analogous way to the problems faced by gradient descent; while gradient-descent faces problems even in quadratic functions (in terms of bouncing behavior), a "quadratically-savvy" method like the Newton technique faces problems in the case of higher-order functions.

This problem can be alleviated by exact or approximate line search, as discussed in Section 4.4.3 of Chapter 4. Line search adjusts the size of the step, so as to terminate at a better point in terms of the *true* objective function value. For example, when line search is used for the objective function in Figure 5.8, the size of the step is much smaller. It also has a much lower value of the (true) objective function. Note that line search could result in either smaller or larger steps than those computed by the vanilla Newton method.

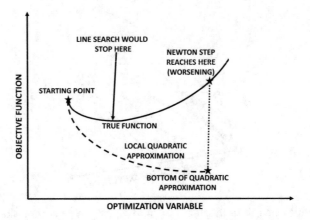

Figure 5.8: A Newton step can worsen the objective function in large steps for non-quadratic functions, because the quadratic approximation increasingly deviates from the true function. A line search can ameliorate the worsening

5.4.3   Example: Newton Method in the Quadratic Bowl

We will revisit how the Newton method behaves in the quadratic bowl of Figure 5.2. Consider the following elliptical objective function, which is the same as the one discussed in Figure 5.2(b):

$$J(w_1, w_2) = w_1^2 + 4w_2^2$$

This is a very simple convex quadratic, whose optimal point is the origin. Applying straightforward gradient descent starting at any point like $[w_1, w_2] = [1, 1]$ will result in the type of bouncing behavior shown in Figure 5.2(b). On the other hand, consider the Newton method, starting at the point $[w_1, w_2] = [1, 1]$. The gradient may be computed as $\nabla J(1, 1) = [2w_1, 8w_2]^T = [2, 8]^T$. Furthermore, the Hessian of this function is a constant that is independent of $[w_1, w_2]^T$:

$$H = \begin{bmatrix} 2 & 0 \\ 0 & 8 \end{bmatrix}$$

Applying the Newton update results in the following:

$$\begin{bmatrix} w_1 \\ w_2 \end{bmatrix} \Leftarrow \begin{bmatrix} 1 \\ 1 \end{bmatrix} - \begin{bmatrix} 2 & 0 \\ 0 & 8 \end{bmatrix}^{-1} \begin{bmatrix} 2 \\ 8 \end{bmatrix} = \begin{bmatrix} 0 \\ 0 \end{bmatrix}$$

In other words, a single step suffices to reach the optimum point of this quadratic function. This is because the second-order Taylor "approximation" of a quadratic function is exact, and the Newton method solves this approximation in each iteration. Of course, real-world functions are not quadratic, and therefore multiple steps are typically needed.

5.4.4   Example: Newton Method in a Non-quadratic Function

In this section, we will modify the objective function of the previous section to make it non-quadratic. The corresponding function is as follows:

$$J(w_1, w_2) = w_1^2 + 4w_2^2 - \cos(w_1 + w_2)$$

It is assumed that w_1 and w_2 are expressed[1] in radians. Note that the optimum of this objective function is still $[w_1, w_2] = [0, 0]$, since the value of $J(0, 0)$ is -1 at this point, where each additive term of the above expression takes on its minimum value. We will again start at $[w_1, w_2] = [1, 1]$, and show that one iteration no longer suffices in this case. In this case, we can show that the gradients and Hessian are as follows:

$$\nabla J(1, 1) = \begin{bmatrix} 2 + \sin(2) \\ 8 + \sin(2) \end{bmatrix} = \begin{bmatrix} 2.91 \\ 8.91 \end{bmatrix}$$

$$H = \begin{bmatrix} 2 + \cos(2) & \cos(2) \\ \cos(2) & 8 + \cos(2) \end{bmatrix} = \begin{bmatrix} 1.584 & -0.416 \\ -0.416 & 7.584 \end{bmatrix}$$

The inverse of the Hessian is as follows:

$$H^{-1} = \begin{bmatrix} 0.64 & 0.035 \\ 0.035 & 0.134 \end{bmatrix}$$

Therefore, we obtain the following Newton update:

$$\begin{bmatrix} w_1 \\ w_2 \end{bmatrix} \Leftarrow \begin{bmatrix} 1 \\ 1 \end{bmatrix} - \begin{bmatrix} 0.64 & 0.035 \\ 0.035 & 0.134 \end{bmatrix} \begin{bmatrix} 2.91 \\ 8.91 \end{bmatrix} = \begin{bmatrix} 1 \\ 1 \end{bmatrix} - \begin{bmatrix} 2.1745 \\ 1.296 \end{bmatrix} = \begin{bmatrix} -1.1745 \\ -0.2958 \end{bmatrix}$$

Note that we do reach closer to an optimal solution, although we certainly do not reach the optimum point. This is because the objective function is not quadratic in this case, and one is only reaching the bottom of the *approximate* quadratic bowl of the objective function. However, Newton's method does find a better point in terms of the true objective function value. The approximate nature of the Hessian is why one must use either exact or approximate line search to control the step size. Note that if we used a step-size of 0.6 instead of the default value of 1, one would obtain the following solution:

$$\begin{bmatrix} w_1 \\ w_2 \end{bmatrix} \Leftarrow \begin{bmatrix} 1 \\ 1 \end{bmatrix} - 0.6 \begin{bmatrix} 2.1745 \\ 1.296 \end{bmatrix} = \begin{bmatrix} -0.30 \\ 0.22 \end{bmatrix}$$

Although this is only a very rough approximation to the optimal step size, it still reaches much closer to the true optimal value of $[w_1, w_2] = [0, 0]$. It is also relatively easy to show that this set of parameters yields a much better objective function value. This step would need to be repeated in order to reach closer and closer to an optimal solution.

5.5 Newton Methods in Machine Learning

In this section, we will provide some examples of the use of the Newton method for machine learning.

5.5.1 Newton Method for Linear Regression

We will start with the linear-regression loss function. Even though linear regression is relatively easy to solve with first-order methods, the approach is instructive because it allows us to relate the Newton method to the most straightforward closed-form solution of linear regression (cf. Section 4.7 of Chapter 4). The objective function of linear regression for an

---

[1]This ensures simplicity, as all calculus operations assume that angles are expressed in radians.

$n \times d$ data matrix D, n-dimensional column vector of target variables $\overline{y}$, and d-dimensional column vector $\overline{W}$ of parameters, is as follows:

$$J(\overline{W}) = \frac{1}{2}\|D\overline{W} - \overline{y}\|^2 = \frac{1}{2}[D\overline{W} - \overline{y}]^T[D\overline{W} - \overline{y}] \tag{5.11}$$

The Newton method requires us to compute both the gradient and the Hessian. We will start by computing the gradient, and then compute the Jacobian of the gradient in order to compute the Hessian. The loss function can be expanded as $\overline{W}^T D^T D\overline{W}/2 - \overline{y}^T D\overline{W} + \overline{y}^T\overline{y}/2$. We can use identities (i) and (ii) from Table 4.2(a) of Chapter 4 to compute the gradients of the individual terms. Therefore, we obtain the gradient of the loss function as follows:

$$\nabla J(\overline{W}) = D^T D\overline{W} - D^T\overline{y} \tag{5.12}$$

The Hessian is obtained by computing the Jacobian of this gradient. The second term of the gradient is a constant and therefore further differentiating it will yield 0; we need only differentiate the first term. On computing the vector-to-vector derivative of the first term of the gradient with respect to $\overline{W}$, we obtain the fact that the Hessian is $D^T D$. This observation can be verified directly using the matrix calculus identity (i) of Table 4.2(b) in Chapter 4. We summarize this observation as follows:

Observation 5.5.1 (Hessian of Squared Loss) *Let $J(\overline{W}) = \frac{1}{2}\|D\overline{W} - \overline{y}\|^2$ be the loss function of linear regression for an $n \times d$ data matrix D, a d-dimensional column vector $\overline{W}$ of coefficients and n-dimensional column vector $\overline{y}$ of targets. Then, the Hessian of the loss function is given by $D^T D$.*

It is also helpful to view the Hessian as the sum of point-specific Hessians, since the Hessian of any linearly additive function is the sum of the Hessians of the individual terms:

Observation 5.5.2 (Point-Specific Hessian of Squared Loss) *Let $J_i = \frac{1}{2}(\overline{W} \cdot \overline{X}_i - y_i)^2$ be the loss function of linear regression for a single training pair $(\overline{X}_i, y_i)$. Then, the point specific Hessian of the squared loss of J_i is given by the outer-product $\overline{X}_i^T \overline{X}_i$.*

Note that $D^T D$ is simply the sum over all $\overline{X}_i^T \overline{X}_i$, since any matrix multiplication can be decomposed into the sum of outer-products (Lemma 1.2.1 of Chapter 1):

$$D^T D = \sum_{i=1}^{n} \overline{X}_i^T \overline{X}_i$$

This is consistent with the fact that Hessian of the full data-specific loss function is the sum of the point-specific Hessians.

One can now combine the Hessian and gradient to obtain the Newton update. A neat result is that the Newton update for least-squares regression and classification simplifies to the closed-form solution of linear regression result discussed in Chapter 4. Given the current vector $\overline{W}$, the Newton update is as follows (based on Equation 5.10):

$$\overline{W} \Leftarrow \overline{W} - H^{-1}[\nabla J(\overline{W})] = \overline{W} - (D^T D)^{-1}[D^T D\overline{W} - D^T\overline{y}]$$
$$= \underbrace{\overline{W} - \overline{W}}_{0} + (D^T D)^{-1}D^T\overline{y} = (D^T D)^{-1}D^T\overline{y}$$

Note that the right-hand side is free of $\overline{W}$, and therefore we need a single "update" step in closed form. This solution is identical to Equation 4.39 of Chapter 4! This equivalence is not surprising. The closed-form solution of Chapter 4 is obtained by setting the gradient of the loss function to 0. The Newton method also sets the gradient of the loss function to 0 after representing it using a second-order Taylor expansion (which is exact for quadratic functions).

Problem 5.5.1 *Derive the Newton update for least-squares regression, when Tikhonov regularization with parameter $\lambda > 0$ is used. Show that the final solution is $\overline{W}^* = (D^T D + \lambda I)^{-1} D^T \overline{y}$, which is the same regularized solution derived in Chapter 4.*

5.5.2 Newton Method for Support-Vector Machines

Next, we will discuss the case of the support vector machine with binary class variables $\overline{y} = [y_1, \ldots, y_n]^T$, where each $y_i \in \{-1, +1\}$. All other notations, such as D, $\overline{W}$, and $\overline{X}_i$ are the same as those of the previous section. The use of the hinge-loss is not common with the Newton method because of its non-differentiability at specific points. Although the non-differentiability does not cause too many problems for straightforward gradient descent (see Section 4.8.2 of Chapter 4), it becomes a bigger problem when dealing with second-order methods. Although one can create a differentiable *Huber loss* approximation [28], we will only discuss the L_2-SVM here. One can write its objective function in terms of the rows of matrix D, which are $\overline{X}_1 \ldots \overline{X}_n$, and the elements of $\overline{y}$, which are $y_1 \ldots y_n$:

$$J(\overline{W}) = \frac{1}{2} \sum_{i=1}^{n} \max \left\{ 0, \left(1 - y_i [\overline{W} \cdot \overline{X}_i^T] \right) \right\}^2 \quad [L_2\text{-loss SVM}]$$

We have omitted the regularization term for simplicity. This loss can be decomposed as $J(\overline{W}) = \sum_i J_i$, where J_i is the point-specific loss. The point-specific loss for the ith point can be expressed in a form corresponding to identity (v) of Table 4.2(a) in Chapter 4:

$$J_i = f_i(\overline{W} \cdot \overline{X}_i^T) = \frac{1}{2} \max \left\{ 0, \left(1 - y_i [\overline{W} \cdot \overline{X}_i^T] \right) \right\}^2$$

Note the use of the function $f_i(\cdot)$ in the above expression, which is defined for L_2-loss SVMs as follows:

$$f_i(z) = \frac{1}{2} \max\{0, 1 - y_i z\}^2$$

This function will eventually need to be differentiated during gradient descent:

$$\frac{\partial f_i(z)}{\partial z} = f_i'(z) = -y_i \max\{0, 1 - y_i z\}$$

Therefore, we have $J_i = f_i(z_i)$, where $z_i = \overline{W} \cdot \overline{X}_i^T$. The derivative of $J_i = f_i(z_i)$ with respect to $\overline{W}$ is computed using the chain rule:

$$\frac{\partial J_i}{\partial \overline{W}} = \frac{\partial f_i(z_i)}{\partial \overline{W}} = \frac{\partial f_i(z_i)}{\partial z_i} \underbrace{\frac{\partial z_i}{\partial \overline{W}}}_{\overline{X}_i^T} = -y_i \max\{0, 1 - y_i (\overline{W} \cdot \overline{X}_i)\} \overline{X}_i^T \quad (5.13)$$

Note that this derivative is in the same form as identity (v) of Table 4.2(a). In order to compare the gradients of least-squares classification and the L_2-SVM, we restate them next to each other:

$$\frac{\partial J_i}{\partial \overline{W}} = -y_i(1 - y_i(\overline{W} \cdot \overline{X}_i^T))\overline{X}_i^T \qquad \text{[Least-Squares Classification]}$$

$$\frac{\partial J_i}{\partial \overline{W}} = -y_i \max\{0, 1 - y_i(\overline{W} \cdot \overline{X}_i^T)\}\overline{X}_i^T \quad [L_2\text{-SVM}]$$

The least-squares classification and the L_2-SVM have a similar gradient, except that the contributions of instances that are correctly classified in a confident way (i.e., instances satisfying $y_i(\overline{W} \cdot \overline{X}_i^T) \geq 1$) are not included in the SVM. One can use $y_i^2 = 1$ to rewrite the gradient of the L_2-SVM in terms of the indicator function as follows:

$$\frac{\partial J_i}{\partial \overline{W}} = \underbrace{(\overline{W} \cdot \overline{X}_i^T - y_i)I([1 - y_i(\overline{W} \cdot \overline{X}_i^T)] > 0)}_{\text{scalar}} \underbrace{\overline{X}_i^T}_{\text{vector}} \quad [L_2\text{-SVM}]$$

The binary indicator function $I(\cdot)$ takes on the value of 1 when the condition inside it is satisfied. Therefore, the overall gradient of $J(\overline{W})$ with respect to $\overline{W}$ can be written as follows:

$$\nabla J(\overline{W}) = \sum_{i=1}^{n} \frac{\partial J_i}{\partial \overline{W}} = \sum_{i=1}^{n} \underbrace{(\overline{W} \cdot \overline{X}_i^T - y_i)I([1 - y_i(\overline{W} \cdot \overline{X}_i^T)] > 0)}_{\text{scalar}} \underbrace{\overline{X}_i^T}_{\text{vector}}$$

$$= D^T \Delta_w (D\overline{W} - \overline{y})$$

Here, Δ_w is an $n \times n$ diagonal matrix in which the (i, i)th entry contains the indicator function $I([1 - y_i(\overline{W} \cdot \overline{X}_i^T)] > 0)$ for the ith training instance.

Next, we focus on the computation of the Hessian. We would first like to compute the Jacobian of the point-specific gradient $\frac{\partial J_i}{\partial \overline{W}}$ in order to compute the point-specific Hessian, and then add up the point-specific Hessians. In important point is that the gradient is the product of a scalar $s = -y_i \max\{0, 1 - y_i(\overline{W} \cdot \overline{X}_i)\}$ (dependent on $\overline{W}$) and the vector $\overline{X}_i^T$ (independent of $\overline{W}$). This fact simplifies the computation of the *point-specific* Hessian H_i (i.e., transposed vector derivative of the gradient), using the product-of-variables identity in Table 4.2(b):

$$H_i = \overline{X}_i^T \left[\frac{\partial s}{\partial \overline{W}}\right]^T = \overline{X}_i^T \left[y_i^2 I([1 - y_i(\overline{W} \cdot \overline{X}_i^T)] > 0)\overline{X}_i\right]$$

$$= I([1 - y_i(\overline{W} \cdot \overline{X}_i^T)] > 0)[\overline{X}_i^T \overline{X}_i] \quad \text{[Setting } y_i^2 = 1\text{]}$$

The overall Hessian H is the sum of the point-specific Hessians:

$$H = \sum_{i=1}^{n} H_i = \sum_{i=1}^{n} \underbrace{I([1 - y_i(\overline{W} \cdot \overline{X}_i^T)] > 0)}_{\text{Binary Indicator}} \underbrace{[\overline{X}_i^T \overline{X}_i]}_{\text{Outer Prod.}}$$

How is the Hessian of the L_2-SVM different from that in least-squares classification? Note that the Hessian of least-squares classification can be written as the sum of outer products $\sum_i[\overline{X}_i^T \overline{X}_i]$ of the individual points. The Hessian of the L_2-SVM also sums the outer products, except that it uses an indicator function to drop out the points that meet the margin condition (of being classified correctly with sufficient margin). Such points do not contribute to the Hessian. Therefore, one can write the Hessian of the L_2-SVM loss as follows:

$$H = D^T \Delta_w D$$

Here, Δ_w is the same $n \times n$ binary diagonal matrix Δ_w that is used in the expression for the gradient. The value of Δ_w will change over time during learning, as different training instances move in and out of correct classification and therefore contribute in varying ways to Δ_w. The key point is that rows drop in and out in terms of their contributions to the gradient and the Hessian, as $\overline{W}$ changes. This is the reason that we have subscripted Δ with w to indicate that it depends on the parameter vector.

Therefore, at any given value of the parameter vector, the Newton update of the L_2-loss SVM is as follows:

$$\overline{W} \Leftarrow \overline{W} - H^{-1}[\nabla J(\overline{W})] = \overline{W} - (D^T \Delta_w D)^{-1}[D^T \Delta_w (D\overline{W} - \overline{y})]$$
$$= \underbrace{\overline{W} - \overline{W}}_{0} + (D^T \Delta_w D)^{-1} D^T \Delta_w \overline{y} = (D^T \Delta_w D)^{-1} D^T \Delta_w \overline{y}$$

This form is almost identical to least-squares classification, except that we are dropping the instances that are correctly classified in a strong way. At first glance, it might seem that the L_2-SVM also requires a single iteration like least-squares regression, because the vector $\overline{W}$ has disappeared on the right-hand side. However, this does not mean that the right-hand side is independent of $\overline{W}$. The matrix Δ_w *does* depend on the weight vector, and will change once $\overline{W}$ is updated. Therefore, one must recompute Δ_w in each iteration and repeat the above step to convergence.

The second point is that *line search becomes important in each update of the L_2-SVM, as we are no longer dealing with a quadratic function.* Therefore, we can add line search to compute the learning rate α_t in the tth iteration. This results in the following update:

$$\overline{W} \Leftarrow \overline{W} - \alpha_t (D^T \Delta_w D)^{-1}[D^T \Delta_w D\overline{W} - D_w^T \Delta_w \overline{y}]$$
$$= \overline{W}(1 - \alpha_t) + \alpha_t (D^T \Delta_w D)^{-1} D^T \overline{y}$$

Note that it is possible for line search to obtain a value of $\alpha_t > 1$, and therefore the coefficient $(1 - \alpha_t)$ of the first term can be negative. One can also derive a form of the update for the regularized SVM. We leave this problem as a practice exercise.

Problem 5.5.2 *Derive the Newton update without line-search for the L_2-SVM, when Tikhonov regularization with parameter $\lambda > 0$ is used. Show that the* **iterative update** *of the Newton method is* $\overline{W} \Leftarrow (D^T \Delta_w D + \lambda I)^{-1} D^T \Delta_w \overline{y}$. *All notations are the same as those used for the L_2-SVM in this section.*

It is noteworthy that the Newton's update uses the quadratic Taylor expansion of the non-quadratic objective function of the L_2-SVM; the second-order Taylor expansion is, therefore, only an approximation. On the other hand, least-squares regression already has a quadratic objective function, and its second-order Taylor approximation is exact. This point of view is critical in understanding why certain objective functions like least-squares regression require a single Newton update, whereas others like the SVM do not.

Problem 5.5.3 *Discuss why the Hessian is more likely to become singular towards the end of learning in the Newton method for the L_2-SVM. How would you address the problem caused by the non-invertibility of the Hessian? Also discuss the importance of line search in these cases.*

5.5.3 Newton Method for Logistic Regression

We revisit logistic regression (cf. Section 4.8.3 of Chapter 4) with training pairs $(\overline{X}_i, y_i)$. Here, each $\overline{X}_i$ is a d-dimensional row vector and $y_i \in \{-1, +1\}$. There are a total of n

training pairs, and therefore stacking up all the d-dimensional rows results in an $n \times d$ matrix D. The resulting loss function (cf. Section 4.8.3) is as follows:

$$J(\overline{W}) = \sum_{i=1}^{n} \log(1 + \exp(-y_i[\overline{W} \cdot \overline{X}_i^T]))$$

We start by defining a function for logistic loss in order to enable the (eventual) use of the chain rule:

$$f_i(z) = \log(1 + \exp(-y_i z)) \tag{5.14}$$

When z_i is set to $\overline{W} \cdot \overline{X}_i^T$, the function $f_i(z_i)$ contains the loss for the ith training point. The derivative of $f_i(z_i)$ is as follows:

$$\frac{\partial f_i(z_i)}{\partial z_i} = -y_i \frac{\exp(-y_i z_i)}{1 + \exp(-y_i z_i)} = -y_i \underbrace{\frac{1}{1 + \exp(y_i z_i)}}_{p_i}$$

The quantity $p_i = 1/(1 + \exp(y_i z_i))$ in the above expression is always interpreted as the probability of the model to make[2] a mistake, when $z_i = \overline{W} \cdot \overline{X}_i^T$. Therefore, one can express the derivative of $f_i(z_i)$ as follows:

$$\frac{\partial f_i(z_i)}{\partial z_i} = -y_i p_i$$

With this machinery and notations, one can write the objective function of logistic regression in terms of the individual losses:

$$J(\overline{W}) = \sum_{i=1}^{n} f_i(\overline{W} \cdot \overline{X}_i^T) = \sum_{i=1}^{n} f_i(z_i)$$

Then, one can compute the gradient of the loss function using the chain rule as follows:

$$\nabla J(\overline{W}) = \sum_{i=1}^{n} \underbrace{\frac{\partial f_i(z_i)}{\partial z_i}}_{-y_i p_i} \underbrace{\frac{\partial z_i}{\partial \overline{W}}}_{\overline{X}_i^T} = -\sum_{i=1}^{n} y_i p_i \overline{X}_i^T \tag{5.15}$$

The derivative of $z_i = \overline{W} \cdot \overline{X}_i^T$ with respect to $\overline{W}$ is based on identity (v) of Table 4.2(a). To represent the gradient compactly using matrices, one can introduce an $n \times n$ diagonal matrix Δ_w^p, in which the ith diagonal entry contains the probability p_i:

$$\nabla J(\overline{W}) = -D^T \Delta_w^p \overline{y} \tag{5.16}$$

One can view Δ_w^p as a soft version of the binary matrix Δ_w used for the L_2-SVM. Therefore, we have added the superscript p to the matrix Δ_w^p in order to indicate that it is a probabilistic matrix.

The Hessian is given by the Jacobian of the gradient:

$$H = \left[\frac{\partial \nabla J(\overline{W})}{\partial \overline{W}}\right]^T = -\sum_{i=1}^{n} \left[\frac{\partial[y_i p_i \overline{X}_i^T]}{\partial \overline{W}}\right]^T = -\sum_{i=1}^{n} y_i \left[\frac{\partial[p_i \overline{X}_i^T]}{\partial \overline{W}}\right]^T \tag{5.17}$$

---

[2]This conclusion follows from the modeling assumption in logistic regression that the probability of a correct prediction is $p_i' = 1/(1 + \exp(-y_i z_i))$. It can be easily shown that $p_i + p_i' = 1$.

The vector $\overline{X}_i$ is independent of $\overline{W}$, whereas p_i is a scalar that depends on $\overline{W}$. In the denominator layout, the derivative of the column vector $p_i \overline{X}_i^T$ with respect to the column vector $\overline{W}$ is the matrix $\frac{\partial p_i}{\partial \overline{W}} \overline{X}_i$ based on identity (iii) of Table 4.2(b). Therefore, the Hessian can be written in matrix calculus notation as $H = -\sum_i y_i \left[\frac{\partial p_i}{\partial \overline{W}} \overline{X}_i \right]^T$. The gradient of p_i with respect to $\overline{W}$ can be computed using the chain rule with respect to intermediate variable $z_i = \overline{W} \cdot \overline{X}_i^T$ as follows:

$$\frac{\partial p_i}{\partial \overline{W}} = \frac{\partial p_i}{\partial z_i} \frac{\partial z_i}{\partial \overline{W}} = \frac{\partial p_i}{\partial z_i} \overline{X}_i^T = -\frac{y_i \exp(y_i z_i)}{(1 + \exp(y_i z_i))^2} \overline{X}_i^T = -y_i p_i (1 - p_i) \overline{X}_i^T \qquad (5.18)$$

Substituting the gradient of p_i from Equation 5.18 in the expression $H = -\sum_i y_i \left[\frac{\partial p_i}{\partial \overline{W}} \overline{X}_i \right]^T$, we obtain the following:

$$H = \sum_i \underbrace{y_i^2}_{=1} p_i (1 - p_i) \overline{X}_i^T \overline{X}_i \qquad (5.19)$$

Now observe that this form is the weighted sum of matrices, where each matrix is the outer-product between a vector and itself. This form is also used in the spectral decomposition of matrices (cf. Equation 3.43 of Chapter 3), in which the weighting is handled by a diagonal matrix. Consequently, we can convert the Hessian to a form using the data matrix D as follows:

$$H = D^T \Lambda_w^u D \qquad (5.20)$$

Here, Λ_w^u is a diagonal matrix of *uncertainties* in which the ith diagonal entry is simply $p_i(1 - p_i)$, where p_i is the probability of making a mistake on the ith training instance with weight vector $\overline{W}$. When a point is classified with probability close to 0 or 1, the value of p_i will always be closer to 0. On the other hand, if the model is unsure about the class label of p_i, its probability will be high. Note that Λ_w^u depends on the value of the parameter vector, and we have added the notations w, u to it in order to emphasize that it is an uncertainty matrix that depends on the parameter vector. It is helpful to note that the Hessian of logistic regression is similar in form to the Hessian $D^T D$ in the "parent problem" of linear regression and the Hessian $D^T \Delta_w D$ in the L_2-SVM. The L_2-SVM explicitly drops rows that are *correctly* classified in a confident way, whereas logistic regression gives each row a soft weight depending on the level of *uncertainty* (rather than correctness) in classification.

One can now derive an expression for the Newton update for logistic regression by plugging in the expressions for the Hessian and the gradient. At any given value of the parameter vector $\overline{W}$, the update is as follows:

$$\overline{W} \Leftarrow \overline{W} + (D^T \Lambda_w^u D)^{-1} D^T \Delta_w^p \overline{y}$$

This iterative update needs to be executed to convergence. Note that Δ_w^p simply weights each class label from $\{-1, +1\}$ by the probability of making a mistake for that training instance. Therefore, instances with larger mistake probabilities are emphasized in the update. This is also an important difference from the L_2-SVM where only incorrect or marginally classified instances are used, and other "confidently correct" instances are discarded. Furthermore, the update of logistic regression uses the "uncertainty weight" in the matrix Λ_w^u. Finally, it is common to use line search in conjunction with learning rate α in order to modify the aforementioned update to the following:

$$\overline{W} \Leftarrow \overline{W} + \alpha (D^T \Lambda_w^u D)^{-1} D^T \Delta_w^p \overline{y}$$

Problem 5.5.4 *Derive the Newton update for logistic regression, when Tikhonov regularization with parameter λ is used. Show that the update is modified to the following:*

$$\overline{W} \Leftarrow \overline{W} + \alpha(D^T \Lambda_w^u D + \lambda I)^{-1}\{[D^T \Delta_w^p \overline{y}] - \lambda \overline{W}\}$$

The notations here are the same as those in the discussion of this section.

5.5.4   Connections Among Different Models and Unified Framework

The Newton update for the different models, corresponding to least-squares regression, the L_2-SVM, and logistic regression are closely related. This is not particularly surprising, since their loss functions are closely related (cf. Figure 4.9 of Chapter 4). In the following table, we list all the updates for the various Newton Methods, so that they can be compared:

Method	Update (no line search)	Update (with line search)
Linear regression and classification	$\overline{W} = (D^T D)^{-1} D^T \overline{y}$ (single step: no iterations)	Line search not needed (single step: no iterations)
L_2-SVM	$\overline{W} \Leftarrow (D^T \Delta_w D)^{-1} D^T \Delta_w \overline{y}$ (Δ_w is binary diagonal matrix) (Δ_w excludes selected points)	$\overline{W} \Leftarrow (1 - \alpha_t)\overline{W} + \alpha_t(D^T \Delta_w D)^{-1} D^T \Delta_w \overline{y}$ (Δ_w is binary diagonal matrix) (Δ_w excludes selected points)
Logistic regression	$\overline{W} \Leftarrow \overline{W} + (D^T \Lambda_w^u D)^{-1} D^T \Delta_w^p \overline{y}$ (Λ_w^u, Δ_w^p are soft diagonal matrices) (Matrices use soft weights)	$\overline{W} \Leftarrow \overline{W} + \alpha_t(D^T \Lambda_w^u D)^{-1} D^T \Lambda_w^p \overline{y}$ (Λ_w^u, Δ_w^p are soft diagonal matrices) (Matrices use soft weights)

It is evident that all the updates are very similar. One can explain these differences in terms of the similarities and differences of the loss functions. For example, when the L_2-SVM is compared to least-squares classification, it is primarily different in terms of assuming zero loss for points that are classified correctly in a sufficiently "confident" way (i.e., meet the margin requirement). Similarly, when we compare the Hessian and the gradient used in the case of the L_2-SVM to that used in least-squares classification, a *binary* diagonal matrix Δ_w is used to remove the effect of these correctly classified points (whereas least-squares classification includes these points as well). The impact of changing the loss function is more complex in the case of logistic regression; points that are correctly classified with high probability are de-emphasized in the gradient, and points that the model is certain about (whether correct or incorrect) are de-emphasized in the Hessian. Furthermore, unlike the L_2-SVM, logistic regression uses soft weighting rather than hard weighting. All these connections are naturally related to the connections among their loss functions (cf. Figure 4.9 of Chapter 4). The logistic regression update is considered a soft and iterative version of the closed-form solution to least-squares regression — as a result, the Newton method for logistic regression is sometimes also referred to as the *iteratively re-weighted least-squares algorithm.*

One can also understand all these updates in the context of a unified framework, where the regularized loss function for many machine learning models can be expressed as follows:

$$J = \sum_{i=1}^{n} f_i(\overline{W} \cdot \overline{X}_i^T) + \frac{\lambda}{2}\|\overline{W}\|^2$$

Note that each $f_i(\cdot)$ also uses the observed value y_i to compute the loss, and can also be written as $L(y_i, \overline{W} \cdot \overline{X}_i^T)$. All the updates can be written in a single unified form as discussed in the result below:

Lemma 5.5.1 (Unified Newton Update for Machine Learning) *Let the objective function for a machine learning problem with d-dimensional parameter vector $\overline{W}$, and $n \times d$ data matrix D containing rows (feature vectors) $\overline{X}_1 \ldots \overline{X}_n$ be as follows:*

$$J = \sum_{i=1}^{n} L(y_i, \overline{W} \cdot \overline{X}_i^T) + \frac{\lambda}{2} \|\overline{W}\|^2$$

Here, $\overline{y} = [y_1 \ldots y_n]^T$ is the observed dependent variable parameter vector for matrix D. Then, the regularized Newton update can be written in the following form:

$$\overline{W} \Leftarrow \overline{W} - \alpha (D^T \Delta_2 D + \lambda I)^{-1} (D^T \Delta_1 \overline{1} + \lambda \overline{W})$$

Here Δ_2 is an $n \times n$ diagonal matrix whose diagonal entries contain the second derivative $L''(y_i, z_i)$ [with respect to $z_i = \overline{W} \cdot \overline{X}_i^T$] evaluated at each $(\overline{X}_i, y_i)$, and Δ_1 is an $n \times n$ diagonal matrix whose diagonal entries contain the corresponding first derivative $L'(y_i, z_i)$ evaluated at each $(\overline{X}_i, y_i)$.

We leave the proof of this lemma as an exercise for the reader (see Exercise 14).

5.6 Newton Method: Challenges and Solutions

Although the Newton method avoids many of the problems associated with gradient descent, it comes with its own set of challenges, which will be studied in this section.

5.6.1 Singular and Indefinite Hessian

Newton's method is inherently designed for convex quadratic functions with positive-definite Hessians. The Hessian can sometimes be singular or indefinite. For example, in the case of the (unregularized) L_2-SVM, the Hessian is the (signed) sum of outer products $\overline{X}_i \overline{X}_i^T$ of points that are marginally correct or incorrect in terms of prediction. Each of these point-specific Hessians is a rank-1 matrix. We need at least d of them in order to create a $d \times d$ Hessian of full rank d (cf. Lemma 2.6.2 of Chapter 2). This might not occur near convergence.

When the Hessian is not invertible, one can either add λI to the Hessian (for regularization) or work with the pseudoinverse of the Hessian. Regularization can also convert an indefinite Hessian to a positive definite matrix by using a large enough value of λ. In particular, choosing λ to be slightly greater than the absolute value of the most negative eigenvalue (of the Hessian) will result in a positive definite Hessian. It is noteworthy that ill-conditioning problems continue to arise even with regularization (cf. Sections 2.9 and 7.4.4.1), when the Hessian is nearly singular.

5.6.2 The Saddle-Point Problem

So far, we have looked at the performance of the Newton method with convex functions. Non-convex functions bring other types of challenges such as *saddle points*. Saddle points occur when the Hessian of the loss function is indefinite. A saddle point is a stationary point

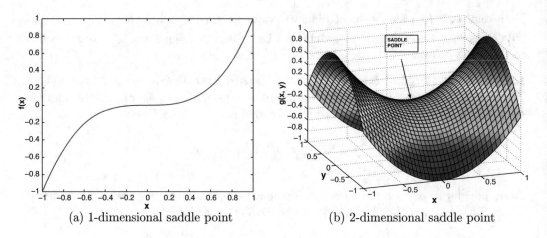

(a) 1-dimensional saddle point   (b) 2-dimensional saddle point

Figure 5.9: Illustration of saddle points

(i.e., a critical point) of a gradient-descent method because its gradient is zero, but it is not a minimum (or maximum). A saddle point is an *inflection point*, which appears to be either a minimum or a maximum depending on which direction we approach it from. Therefore, the quadratic approximation of the Newton method will result in vastly different shapes depending on the precise location of current parameter vector with respect to a nearby saddle point. A 1-dimensional function with a saddle point is the following:

$$f(x) = x^3$$

This function is shown in Figure 5.9(a), and it has an inflection point at $x = 0$. Note that a quadratic approximation at $x > 0$ will look like an upright bowl, whereas a quadratic approximation at $x < 0$ will look like an inverted bowl. The second-order Taylor approximations at $x = 1$ and $x = -1$ are as follows:

$$F(x) = 1 + 3(x - 1) + \frac{6(x - 1)^2}{2} = 3x^2 - 3x + 1 \ \ [\text{At } x = 1]$$

$$G(x) = -1 + 3(x + 1) - \frac{6(x + 1)^2}{2} = -3x^2 - 3x - 1 \ \ [\text{At } x = -1]$$

It is not difficult to verify that one of these functions is an upright bowl (convex function) with a minimum and no maximum, whereas another is an inverted bowl (concave function) with a maximum and no minimum. Therefore, the Newton optimization will behave in an unpredictable way, depending on the current value of the parameter vector. Furthermore, even if one reaches $x = 0$ in the optimization process, both the second derivative and the first derivative will be zero. Therefore, a Newton update will take the $0/0$ form and become indefinite. Such a point is a degenerate point from the perspective of numerical optimization. In general, a degenerate critical point is one where the Hessian is singular (along with the first-order condition that the gradient is zero). The problem is complicated by the fact that a degenerate critical point can be either a true optimum or a saddle point. For example, the function $h(x) = x^4$ has a degenerate critical point at $x = 0$ in which both first-order and second-order derivatives are 0. However, the point $x = 0$ is a true minimum.

It is also instructive to examine the case of a saddle point in a multivariate function, where the Hessian is not singular. An example of a 2-dimensional function with a saddle point is as follows:

$$g(x, y) = x^2 - y^2$$

This function is shown in Figure 5.9(b). The saddle point is $(0, 0)$. The Hessian of this function is as follows:

$$H = \begin{bmatrix} 2 & 0 \\ 0 & -2 \end{bmatrix}$$

It is easy to see that the shape of this function resembles a riding saddle. In this case, approaching from the x direction or from the y direction will result in very different quadratic approximations. In one case, the function will appear to be a minimum, and in another case, the function will appear to be a maximum. Furthermore, the saddle point $[0, 0]$ will be a stationary point from the perspective of a Newton update, even though it is not an extremum. Saddle points occur frequently in regions between two hills of the loss function, and they present a problematic topography for the Newton method. Interestingly, straightforward gradient-descent methods are often able to escape from saddle points [54], because they are simply not attracted by such points. On the other hand, Newton's method is indiscriminately attracted to all critical points (such as maxima or saddle points). High-dimensional objective functions seem to contain a large number of saddle points compared to true optima (see Exercise 14). The Newton method does not always perform better than gradient descent, and the specific topography of a particular loss function may have an important role to play. The Newton method is needed for loss functions with complex curvatures, but without too many saddle points. Note that the pairing of computational algorithms (like Adam) with gradient-descent methods already changes the steepest direction in a way that incorporates several advantages of second-order methods in an implicit way. Therefore, real-world practitioners often prefer gradient-descent methods in combination with computational algorithms like Adam. Recently, some methods have been proposed [32] to address saddle points in second-order methods.

5.6.3 Convergence Problems and Solutions with Non-quadratic Functions

The first-order gradient-descent method works well with the SVM and logistic regression, because these are convex functions. In such cases, gradient descent is almost always guaranteed to converge to an optimum, as long as step-sizes are chosen appropriately. However, a surprising fact is that the (more sophisticated) Newton method is not guaranteed to converge to an optimal solution. Furthermore, one is not even guaranteed to improve the objective function value with a given update, if one uses the most basic form of the Newton method.

Here, it is important to understand that the Newton method uses a local Taylor approximation at the current parameter vector $\overline{w}$ to compute both the gradient and the Hessian; if the quadratic approximation deteriorates rapidly with increasing distance from the parameter vector $\overline{W}$, the results can be uncertain. Just as first-order gradient descent uses the instantaneous direction of steepest descent as an approximation, the second-order method uses a local Taylor approximation which is correct only over an infinitesimal region of the space. As one makes steps of larger size, the effect of the step can be uncertain.

In order to understand this point, let us examine a simple 1-dimensional classification problem in which the feature-label pairs are $(1, 1)$, $(2, 1)$, and $(3, -1)$. We have a single parameter w that needs to be learned. The objective function of least-squares classification is as follows:

$$J = (1 - w)^2 + (1 - 2w)^2 + (1 + 3w)^2$$

This is a quadratic objective function, and the individual losses are the three terms of the above expression. The aggregate loss can also be written as $J = 14w^2 + 3$. Therefore, the loss functions of the three individual points and the aggregate loss are both quadratic. This is the reason that the Newton method converges to the optimal solution in a single step in least-squares classification/regression; the Taylor "approximation" is exact.

Let us now examine, how this objective function would be modified by the L_2-SVM:

$$J = \max\{(1 - w), 0\}^2 + \max\{(1 - 2w), 0\}^2 + \max\{(1 + 3w), 0\}^2$$

This objective function is no longer quadratic because of the use of the maximization function within the loss. As a result, the Taylor approximation is no longer exact, and a finite step will lead to a point where the Taylor approximation deteriorates. Note that different points contribute non-zero values at different values of w. Therefore, for any Newton step of finite size, points may drop off or add into the loss, which can cause unexpected results. For example, as one reaches near an optimal solution many misclassified training points may be the result of noise and errors in the training data. In this situation, the Newton method will define the update of the weight vector based on such unreliable training points. This is one of the reasons that line search in important in the Newton method. Another solution is to use the *trust region method*.

5.6.3.1 Trust Region Method

The trust-region method can be viewed as a complementary approach to line-search; whereas line-search selects the step-size after choosing the direction, a trust-region method selects the direction after choosing a step-size (trust region), which is incorporated within the optimization formulation for selecting the direction of movement. Let $\overline{W} = \overline{a}_t$ be the value of the parameter vector at the tth iteration of optimizing the objective function $J(\overline{W})$. Similarly, let H_t be the Hessian of the loss function, when evaluated at $\overline{a}_t$. Then, the trust-region method solves the following subproblem using an important quantity $\delta_t > 0$ that controls the trust-region size:

$$\text{Minimize } F(\overline{W}) = J(\overline{a}_t) + (\overline{W} - \overline{a}_t)^T [\nabla J(\overline{a}_t)] + \frac{1}{2}(\overline{W} - \overline{a}_t)^T H_t (\overline{W} - \overline{a}_t)$$

$$\text{subject to:}$$

$$\|\overline{W} - \overline{a}_t\| \leq \delta_t$$

The objective function $F(\overline{W})$ contains the second-order Taylor approximation of the true objective function $J(\overline{W})$ in the locality of the current parameter vector $\overline{a}_t$. Note that this approach is also working with the approximate quadratic bowl like the Newton method, except that it does not move to the bottom of the quadratic bowl. Rather, one uses the trust radius δ_t to restrict the amount of movement as a constraint. This type of restriction is referred to as the *trust constraint*. The key point here is that the direction of best movement is also affected by regulating the maximum step-size, which makes it complementary to line-search methods. For example, if the maximum step-size δ_t is chosen to be very small, then the direction of movement will be very similar to a vanilla gradient-descent method, rather than the inverse-Hessian biased Newton method. The basic idea is that the Taylor approximation becomes less and less reliable with increasing distance from the point of expansion, and therefore one needs to restrict the radius in order to obtain better improvements. The broad process of solving such convex optimization problems with constraints is provided in Chapter 6, and a specific method for solving this type of optimization problem is provided in Section 6.5.1.

A key point is in terms of how the radius δ_t should be selected. The radius δ_t is either increased or decreased, by comparing the improvement $F(\overline{a}_t) - F(\overline{a}_{t+1})$ of the Taylor approximation $F(\overline{W})$ to the improvement $J(\overline{a}_t) - J(\overline{a}_{t+1})$ of the true objective function:

$$I_t = \frac{J(\overline{a}_t) - J(\overline{a}_{t+1})}{F(\overline{a}_t) - F(\overline{a}_{t+1})} \quad \text{[Improvement Ratio]}$$

Intuitively, we would like the true objective function to improve as much as possible, and not just the Taylor approximation. The value of the improvement ratio I_t is usually less than 1, as one is optimizing the Taylor approximation rather than the true objective function. For example, choosing extremely small values of δ_t will lead to improvement ratios near 1, but it is not helpful in terms of making sufficient progress.

Therefore, the change in δ_t from iteration to iteration is accomplished by using the improvement ratio as a hint about whether it is too conservative or too liberal. Similarly, the trust constraint $\|\overline{W} - \overline{a}_t\| \leq \delta_t$ needs to be satisfied tightly by the optimization solution $\overline{W} = \overline{a}_{t+1}$ in order to increase the size of the trust region in the next iteration. If the improvement ratio is too small (say, less than 0.25), then the trust radius δ_t needs to be reduced by a factor of 2 in the next iteration. If the ratio is too large (say, greater than 0.75) and a full step of δ_t was used in the current iteration (i.e., tightly satisfied trust constraint), the trust radius δ_t needs to be increased. Otherwise, the trust radius does not change. Furthermore, if the improvement ratio is smaller than a critical point (say, negative), then the current step is not accepted, and we set $\overline{a}_{t+1} = \overline{a}_t$ and the optimization problem is solved again with a smaller step size. This process is repeated to convergence. An example of the implementation of logistic regression with a trust-region method is given in [80].

5.7 Computationally Efficient Variations of Newton Method

The Newton method requires fewer iterations than vanilla gradient descent, but each iteration is more expensive. The main challenge arises in the inversion of the Hessian. When the number of parameters is large, the Hessian is too large to store or compute explicitly. This situation arises commonly in domains such as neural network optimization. It is not uncommon to have neural networks with millions of parameters. Trying to compute the inverse of a $10^6 \times 10^6$ Hessian matrix is impractical. Therefore, many approximations and variations of the Newton method have been developed. All these methods borrow the quadratic-approximation principles of the Newton method, but are able to implement these methods more efficiently. Examples of such methods include the method of *conjugate gradients* [19, 59, 86, 87] and quasi-Newton methods that approximate the Hessian. The method of conjugate gradients does not materialize even an approximation of the Hessian, but it tries to express the Newton step as a sequence of d simpler steps, where d is the dimensionality of the data. The d directions of these steps are referred to as *conjugate directions*, which is how this method derives its name. Since the Hessian is never explicitly computed, this technique is also referred to as *Hessian-free optimization*.

5.7.1 Conjugate Gradient Method

The *conjugate gradient method* [59] requires d steps to reach the optimal solution of a quadratic loss function (instead of a single Newton step). The basic idea is that any quadratic function can be transformed to a sum of additively separable univariate functions by using an

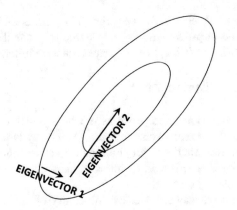
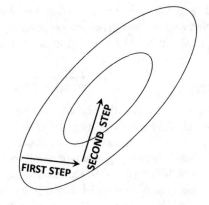

(a) Eigenvectors of Hessian   (b) Arbitrary conjugate pair
Mutually Orthogonal: $\overline{q}_i^{\mathrm{T}} \overline{q}_j = 0$ Non-orthogonal: $\overline{q}_i^{\mathrm{T}} H \overline{q}_j = 0$

Figure 5.10: The eigenvectors of the Hessian of a quadratic function represent the orthogonal axes of the quadratic ellipsoid and are also mutually orthogonal. The eigenvectors of the Hessian are orthogonal conjugate directions. The generalized definition of conjugacy may result in non-orthogonal directions

appropriate basis transformation of variables (cf. Section 3.4.4 of Chapter 3). These variables represent directions in the data that do not interact with one another. Such noninteracting directions are extremely convenient for optimization because they can be independently optimized with line search. Since it is possible to find such directions only for quadratic loss functions, we will first discuss the conjugate gradient method under the assumption that the objective function $J(\overline{W})$ is quadratic. Later, we will discuss the generalization to non-quadratic functions.

A quadratic and convex loss function $J(\overline{W})$ has an ellipsoidal contour plot of the type shown in Figure 5.10, and has a constant Hessian over all regions of the optimization space. The orthonormal eigenvectors $\overline{q}_0 \ldots \overline{q}_{d-1}$ of the symmetric Hessian represent the axes directions of the ellipsoidal contour plot. One can rewrite the loss function in a new coordinate space defined by the eigenvectors as the basis vectors (cf. Section 3.4.4 of Chapter 3) *to create an additively separable sum of univariate quadratic functions in the different variables.* This is because the new coordinate system creates a basis-aligned ellipse, which does not have interacting quadratic terms of the type $x_i x_j$. Therefore, each transformed variable can be optimized independently of the others. Alternatively, one can work with the original variables (without transformation), and simply perform line search along each eigenvector of the Hessian to select the step size. The nature of the movement is illustrated in Figure 5.10(a). Note that movement along the jth eigenvector does not disturb the work done along other eigenvectors, and therefore d steps are sufficient to reach the optimal solution in quadratic loss functions.

Although it is impractical to compute the eigenvectors of the Hessian, there are other efficiently computable directions satisfying similar properties; this key property is referred to as *mutual conjugacy* of vectors. Note that two eigenvectors $\overline{q}_i$ and $\overline{q}_j$ of the Hessian satisfy $\overline{q}_i^T \overline{q}_j = 0$ because of orthogonality of the eigenvectors of a symmetric matrix. Furthermore, since $\overline{q}_j$ is an eigenvector of H, we have $H\overline{q}_j = \lambda_j \overline{q}_j$ for some scalar eigenvalue λ_j. Multiplying both sides with $\overline{q}_i^T$, we can easily show that the eigenvectors of the Hessian satisfy

$\bar{q}_i^T H \bar{q}_j = 0$ in pairwise fashion. The condition $\bar{q}_i^T H \bar{q}_j = 0$ is referred to as H-orthogonality in linear algebra, and is also referred to as the *mutual conjugacy condition* in optimization. It is this mutual conjugacy condition that results in linearly separated variables. However, the eigenvectors are not the only set of mutually conjugate conditions. Just as there are an infinite number of orthonormal basis sets, there are an infinite number of H-orthogonal basis sets in d-dimensional space. In fact, the expression $\langle \bar{q}_i, \bar{q}_j \rangle = \bar{q}_i^T H \bar{q}_j$ is a generalized form of the dot product, referred to as the *inner product*, which has particular significance to quadratic optimization with an elliptical Hessian. If we re-write the quadratic loss function in terms of coordinates in *any* axis system of H-orthogonal directions, the objective function will contain a sum of univariate quadratic functions in terms of the transformed variables. In order to understand why this is the case, let us construct the $d \times d$ matrix $Q = [\bar{q}_0 \ldots \bar{q}_{d-1}]$, whose columns contain H-orthogonal directions. Therefore $\Delta = Q^T H Q$ is diagonal by definition of H-orthogonality. Now note that a quadratic objective function with Hessian H is always of the form $J(\overline{W}) = \overline{W}^T H \overline{W}/2 + \bar{b}^T \overline{W} + c$. Here, $\bar{b}$ is a d-dimensional vector and c is a scalar. This same quadratic function can be expressed in terms of the transformed variables $\overline{W}'$ satisfying $\overline{W} = Q\overline{W}'$ as follows:

$$J\left(Q\overline{W}'\right) = \overline{W}'^T \left[Q^T H Q\right] \overline{W}'/2 + \bar{b}^T Q\overline{W}' + c$$
$$= \overline{W}'^T \Delta \overline{W}'/2 + \bar{b}^T Q\overline{W}' + c$$

Note that the second-order term in the above objective function uses the diagonal matrix Δ, where $\overline{W}'$ contains the coordinates of the parameter vector in the basis corresponding to the conjugate directions. Of course, we do not need to be explicit about performing a basis transformation into an additively separable objective function. Rather, one can separately optimize along each of these d H-orthogonal directions (in terms of the original variables) to solve the quadratic optimization problem in d steps. Each of these optimization steps can be performed using line search along an H-orthogonal direction. Hessian eigenvectors represent a rather special set of H-orthogonal directions that are also orthogonal; conjugate directions other than Hessian eigenvectors, such as those shown in Figure 5.10(b), are not mutually orthogonal. Therefore, conjugate gradient descent optimizes a quadratic objective function by *implicitly* transforming the loss function into a *non-orthogonal* basis with an additively separable representation of the objective function in which each additive term is a univariate quadratic. One can state this observation as follows:

Observation 5.7.1 (Properties of H-Orthogonal Directions) *Let H be the Hessian of a quadratic objective function. If any set of d H-orthogonal directions are selected for movement, then one is implicitly moving along separable variables in a transformed representation of the function. Therefore, at most d steps are required for quadratic optimization.*

The independent optimization along each non-interacting direction (with line search) ensures that the component of the gradient along each conjugate direction will be 0. Strictly convex loss functions have linearly independent conjugate directions (see Exercise 9). In other words, the final gradient will have zero dot product with d linearly independent directions; this is possible only when the final gradient is the zero vector (see Exercise 10), which implies optimality for a convex function. In fact, one can often reach a near-optimal solution in far fewer than d updates.

How can one identify conjugate directions? The simplest approach is to use generalized Gram-Schmidt orthogonalization on the Hessian of the quadratic function in order to generate H-orthogonal directions (cf. Problem 2.7.1 of Chapter 2 and Exercise 11 of this chapter). Such an orthogonalization is easy to achieve using arbitrary vectors as starting points. However, this process can still be quite expensive because each direction $\overline{q}_t$ needs to use *all* the previous directions $\overline{q}_0 \ldots \overline{q}_{t-1}$ for iterative generation in the Gram-Schmidt method. Since each direction is a d-dimensional vector, and there are $O(d)$ such directions towards the end of the process, it follows that each step will require $O(d^2)$ time. Is there a way to do this using only the previous direction in order to reduce this time from $O(d^2)$ to $O(d)$? Surprisingly, only the most recent conjugate direction is needed to generate the next direction [99, 114], when steepest descent directions are used for iterative generation. In other words, one should not use Gram-Schmidt orthogonalization with arbitrary vectors, but should use steepest descent directions as the raw vectors to be orthogonalized. This choice makes all the difference in ensuring a more efficient form of orthogonalization. This is not an obvious result (see Exercise 12). The direction $\overline{q}_{t+1}$ is, therefore, defined iteratively as a linear combination of *only* the previous conjugate direction $\overline{q}_t$ and the current steepest descent direction $\nabla J(\overline{W}_{t+1})$ with combination parameter β_t:

$$\overline{q}_{t+1} = -\nabla J(\overline{W}_{t+1}) + \beta_t \overline{q}_t \tag{5.21}$$

Premultiplying both sides with $\overline{q}_t^T H$ and using the conjugacy condition to set the left-hand side to 0, one can solve for β_t:

$$\beta_t = \frac{\overline{q}_t^T H [\nabla J(\overline{W}_{t+1})]}{\overline{q}_t^T H \overline{q}_t} \tag{5.22}$$

This leads to an iterative update process, which initializes $\overline{q}_0 = -\nabla J(\overline{W}_0)$, and computes $\overline{q}_{t+1}$ iteratively for $t = 0, 1, 2, \ldots T$:

1. Update $\overline{W}_{t+1} \Leftarrow \overline{W}_t + \alpha_t \overline{q}_t$. Here, the step size α_t is computed using line search to minimize the loss function.

2. Set $\overline{q}_{t+1} = -\nabla J(\overline{W}_{t+1}) + \left(\frac{\overline{q}_t^T H [\nabla J(\overline{W}_{t+1})]}{\overline{q}_t^T H \overline{q}_t} \right) \overline{q}_t$. Increment t by 1.

It can be shown [99, 114] that $\overline{q}_{t+1}$ satisfies conjugacy with respect to *all* previous $\overline{q}_i$. A systematic road-map of this proof is provided in Exercise 12.

The conjugate-gradient method is also referred to as Hessian-free optimization. However, the above updates do not *seem* to be Hessian-free, because the matrix H is included in the above updates. However, the underlying computations only need the *projection* of the Hessian along particular directions; we will see that these can be computed indirectly using the method of finite differences without explicitly computing the individual elements of the Hessian. Let $\overline{v}$ be the vector direction for which the projection $H\overline{v}$ needs to be computed. The method of finite differences computes the loss gradient at the current parameter vector $\overline{W}$ and at $\overline{W} + \delta \overline{v}$ for some small value of δ in order to perform the approximation:

$$H\overline{v} \approx \frac{\nabla J(\overline{W} + \delta \overline{v}) - \nabla J(\overline{W})}{\delta} \propto \nabla J(\overline{W} + \delta \overline{v}) - \nabla J(\overline{W}) \tag{5.23}$$

The right-hand side is free of the Hessian. The condition is exact for quadratic functions. Other alternatives for Hessian-free updates are discussed in [19].

So far, we have discussed the simplified case of quadratic loss functions, in which the Hessian is a constant matrix (i.e., independent of the current parameter vector). However, most loss functions in machine learning are not quadratic and, therefore, the Hessian matrix is dependent on the current value of the parameter vector $\overline{W}_t$. This leads to several choices in terms of how one can create a modified algorithm for non-quadratic functions. Do we first create a quadratic approximation at a point and then solve it for a few iterations with the Hessian (quadratic approximation) fixed at that point, or do we change the Hessian every iteration along with the change in parameter vector? The former is referred to as the *linear conjugate gradient method*, whereas the latter is referred to as the *nonlinear conjugate gradient method*.

In the nonlinear conjugate gradient method, the mutual conjugacy (i.e., H-orthogonality) of the directions will deteriorate over time, as the Hessian changes from one step to the next. This can have an unpredictable effect on the overall progress from one step to the next. Furthermore, the computation of conjugate directions needs to be restarted every few steps, as the mutual conjugacy deteriorates. If the deterioration occurs too fast, the restarts occur very frequently, and one does not gain much from conjugacy. On the other hand, each quadratic approximation in the linear conjugate gradient method can be solved exactly, and will typically be (almost) solved in much fewer than d iterations. Therefore, one can make similar progress to the Newton method in each iteration. As long as the quadratic approximation is of high quality, the required number of approximations is often not too large. The nonlinear conjugate gradient method has been extensively used in traditional machine learning from a historical perspective [19], although recent work [86, 87] has advocated the use of linear conjugate methods. Experimental results in [86, 87] suggest that linear conjugate gradient methods have some advantages.

5.7.2   Quasi-Newton Methods and BFGS

The acronym BFGS stands for the Broyden–Fletcher–Goldfarb–Shanno algorithm, and it is derived as an approximation of the Newton method. Let us revisit the updates of the Newton method. A typical update of the Newton method is as follows:

$$\overline{W}^* \Leftarrow \overline{W}_0 - H^{-1}[\nabla J(\overline{W}_0)] \tag{5.24}$$

In quasi-Newton methods, a sequence of approximations of the inverse Hessian matrix are used in various steps. Let the approximation of the inverse Hessian matrix in the tth step be denoted by G_t. In the very first iteration, the value of G_t is initialized to the identity matrix, which amounts to moving along the steepest-descent direction. This matrix is continuously updated from G_t to G_{t+1} with low-rank updates (derived from the matrix inversion lemma of Chapter 1). A direct restatement of the Newton update in terms of the inverse Hessian $G_t \approx H_t^{-1}$ is as follows:

$$\overline{W}_{t+1} \Leftarrow \overline{W}_t - G_t[\nabla J(\overline{W}_t)] \tag{5.25}$$

The above update can be improved with an optimized learning rate α_t for non-quadratic loss functions working with (inverse) Hessian approximations like G_t:

$$\overline{W}_{t+1} \Leftarrow \overline{W}_t - \alpha_t G_t[\nabla J(\overline{W}_t)] \tag{5.26}$$

The optimized learning rate α_t is identified with line search. The line search does not need to be performed exactly (like the conjugate gradient method), because maintenance of conjugacy is no longer critical. Nevertheless, approximate conjugacy of the early set of directions is maintained by the method when starting with the identity matrix. One can (optionally) reset G_t to the identity matrix every d iterations (although this is rarely done).

It remains to be discussed how the matrix G_{t+1} is approximated from G_t. For this purpose, the *quasi-Newton condition*, also referred to as the *secant condition*, is needed:

$$\underbrace{\overline{W}_{t+1} - \overline{W}_t}_{\text{Parameter Change}} = G_{t+1} \underbrace{[\nabla J(\overline{W}_{t+1}) - \nabla J(\overline{W}_t)]}_{\text{First derivative change}} \tag{5.27}$$

The above formula is simply a finite-difference approximation. Intuitively, multiplication of the second-derivative matrix (i.e., Hessian) with the parameter change (vector) approximately provides the gradient change. Therefore, multiplication of the inverse Hessian approximation G_{t+1} with the gradient change provides the parameter change. The goal is to find a symmetric matrix G_{t+1} satisfying Equation 5.27, but it represents an underdetermined system of equations with an infinite number of solutions. Among these, BFGS chooses the closest symmetric G_{t+1} to the current G_t, and achieves this goal by posing a minimization objective function $\|G_{t+1} - G_t\|_w$ in the form of a *weighted Frobenius norm*. In other words, we want to find G_{t+1} satisfying the following:

$$\text{Minimize}_{[G_{t+1}]} \|G_{t+1} - G_t\|_w$$
$$\text{subject to:}$$
$$\overline{W}_{t+1} - \overline{W}_t = G_{t+1}[\nabla J(\overline{W}_{t+1}) - \nabla J(\overline{W}_t)]$$
$$G_{t+1}^T = G_{t+1}$$

The subscript of the norm is annotated by "w" to indicate that it is a weighted[3] form of the norm. This weight is an "averaged" form of the Hessian, and we refer the reader to [99] for details of how the averaging is done. Note that one is not constrained to using the weighted Frobenius norm, and different variations of how the norm is constructed lead to different variations of the quasi-Newton method. For example, one can pose the same objective function and secant condition in terms of the Hessian rather than the inverse Hessian, and the resulting method is referred to as the Davidson–Fletcher–Powell (DFP) method. In the following, we will stick to the use of the inverse Hessian, which is the BFGS method.

Since the weighted norm uses the Frobenius matrix norm (along with a weight matrix) the above is a quadratic optimization problem with linear constraints. Such constrained optimization problems are discussed in Chapter 6. In general, when there are linear equality constraints paired with a quadratic objective function, the structure of the optimization problem is quite simple, and closed-form solutions can sometimes be found. This is because the equality constraints can often be eliminated along with corresponding variables (using methods like Gaussian elimination), and an unconstrained, quadratic optimization problem can be defined in terms of the remaining variables. These problems sometimes turn out to have closed-form solutions like least-squared regression. In this case, the closed-form solution to the above optimization problem is as follows:

$$G_{t+1} \Leftarrow (I - \Delta_t \overline{q}_t \overline{v}_t^T)G_t(I - \Delta_t \overline{v}_t \overline{q}_t^T) + \Delta_t \overline{q}_t \overline{q}_t^T \tag{5.28}$$

Here, the (column) vectors $\overline{q}_t$ and $\overline{v}_t$ represent the parameter change and the gradient change; the scalar $\Delta_t = 1/(\overline{q}_t^T \overline{v}_t)$ is the inverse of the dot product of these two vectors.

$$\overline{q}_t = \overline{W}_{t+1} - \overline{W}_t; \quad \overline{v}_t = \nabla L(\overline{W}_{t+1}) - \nabla L(\overline{W}_t)$$

---

[3]The form of the objective function is $\|A^{1/2}(G_{t+1} - G_t)A^{1/2}\|_F$ norm, where A is an averaged version of the Hessian matrix over various lengths of the step. We refer the reader to [99] for details.

The update in Equation 5.28 can be made more space efficient by expanding it, so that fewer temporary matrices need to be maintained. Interested readers are referred to [83, 99, 104] for implementation details and derivation of these updates.

Even though BFGS benefits from approximating the inverse Hessian, it does need to carry over a matrix G_t of size $O(d^2)$ from one iteration to the next. The *limited memory BFGS* (L-BFGS) reduces the memory requirement drastically from $O(d^2)$ to $O(d)$ by not carrying over the matrix G_t from the previous iteration. In the most basic version of the L-BFGS method, the matrix G_t is replaced with the identity matrix in Equation 5.28 in order to derive G_{t+1}. A more refined choice is to store the $m \approx 30$ most recent vectors $\overline{q}_t$ and $\overline{v}_t$. Then, L-BFGS is equivalent to initializing G_{t-m+1} to the identity matrix and recursively applying Equation 5.28 m times to derive G_{t+1}. In practice, the implementation is optimized to directly compute the direction of movement from the vectors without explicitly storing large intermediate matrices from G_{t-m+1} to G_t.

5.8 Non-differentiable Optimization Functions

Several optimization functions in machine learning are non-differentiable. A mild example is the case in which an L_1-loss or L_1-regularization is used. A key point is that any type of L_1-norm of the vector $\overline{v} = [v_1, \ldots, v_d]$ uses the modulus $|v_i|$ of each of the vector components in the norm $\sum_{i=1}^{d} |v_i|$. The derivative of $|v_i|$ is non-differentiable at $v_i = 0$. Furthermore, any type of L_1-loss is non-differentiable. For example, the hinge loss of the support vector machine is non-differentiable.

A more severe form of non-differentiability is one in which one is trying optimize an inherently discrete objective function such as a *ranking objective function*. In many rare-class settings of classification, one of the labels is far less frequent compared to the others. For example, in a labeled database of intrusion records, the intrusion records are likely to less frequent compared to the normal records. In such cases, the objective function is often defined based on a function of the ranking of *instances with respect to their propensity to belong to the rare class*. For example, one might minimize the sum of (algorithm-determined) ranks of instances that truly belong to the rare class (based on ground-truth information). Note that this is a non-differentiable function because significant changes in the parameter vector might sometimes not affect the algorithmic ranking at all, and at other times infinitesimal changes in parameters might drastically affect the ranking. This results in a loss function with vertical walls and flat regions. As a specific example, consider a 1-dimensional example, in which the points are ranked according to decreasing value of $w \cdot x$, where x is the 1-dimensional feature value and w is the scalar parameter. The four training-label pairs are $(1, +1)$, $(2, +1)$, $(-1, -1)$, and $(-2, -1)$. Ideally, we would like to choose values of w so that all positive examples are ranked above the negative examples. In this simple problem, choosing any value $w > 0$ provides an ideal ranking in which the two positive examples have ranks of 1 and 2. Therefore, the sum of the ranks of positive instances is 3. Choosing $w < 0$ provides the worst-possible ranking in which the two positive instances have ranks of 3 and 4 (with a sum of 7). Choosing $w = 0$ leads to a tied rank of 2.5 for all training instances, and the sum of the ranks is 5. The objective function corresponding to the sum of the ranks (of only the positive instances) is shown in Figure 5.11. The problem with this staircase-like objective function is that it is not really informative anywhere from the perspective of gradient descent. Although the loss function is differentiable almost everywhere except for a single point, the zero gradient at all points provides no clues about the best direction of descent.

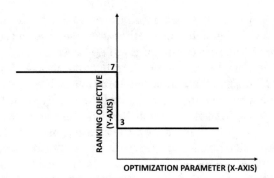

Figure 5.11: An example of a non-differentiable optimization problem caused by a ranking objective function

These types of non-differentiability are often addressed by either making fundamental changes to the underlying optimization algorithms, or by changing the loss function in order to make it smooth. After all, the loss functions of machine learning algorithms are almost always smooth approximations to discrete objective functions (like classification accuracy). In the following, we will provide an overview of the different types of methods used to handle non-differentiability in machine learning.

5.8.1   The Subgradient Method

The subgradient method is designed to work for convex minimization problems, where the gradient is informative at most points except for a few specific points where the objective function is non-differentiable. In such cases, subgradient mainly serves the purpose of bringing the optimization problem out of its non-differentiable "rut." Since the function is differentiable at most other points, it does not face many challenges in terms of optimization, once it gets out of this non-differentiable rut.

The main issue with non-differentiability is that the *one-sided derivatives* are different. For example, $|x|$ has a right-derivative of $+1$ and a left-derivative of -1. A subgradient corresponds to the interval $[-1, +1]$. The presence of the zero vector among the subgradients is an optimality condition for the subgradient method. In Figure 5.12(a), one possible subgradient of a 1-dimensional function is illustrated. Intuitively, the subgradient always lies "below" the loss function, as shown in Figure 5.12(a). Note that there are many possible subgradients in this case because one can construct the line below the loss function in many possible ways. For the d-dimensional function corresponding to the L_1-norm $\|\overline{w}\|_1$ of $\overline{w}$, one can select any d-dimensional vector for which each component is sampled uniformly at random from $(-1, 1)$ to create a subgradient. In Figure 5.12(a), we have shown an example of a subgradient for a 1-dimensional function. Note that one can draw many possible "tangents" at non-differentiable points for convex functions, which are (more precisely) referred to as *subtangents* at non-differentiable points. Each of these subtangents corresponds to a subgradient. For multidimensional functions, the subgradient is defined by any hyperplane lying fully below the loss function, as shown in Figure 5.12(b). For differentiable functions, we can draw only one tangent hyperplane. However, non-differentiable functions allow the construction of an infinite number of possibilities.

A subgradient of a function $J(\overline{w})$ at point $\overline{w}_0$ is formally defined as follows:

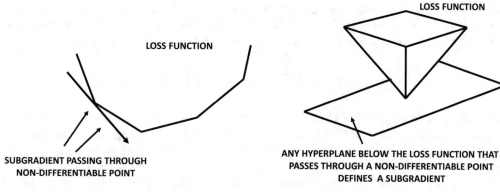

LOSS FUNCTION

LOSS FUNCTION

(a) Subgradient in one dimension   (b) Subgradient in two dimensions

Figure 5.12: Subgradients in one and two dimensions. Any vector residing on the hyperplane, which originates at the contact point between the loss function and the hyperplane, is a subgradient. The vertical direction is the loss function value in each case

Definition 5.8.1 (Subgradient) *Let $J(\overline{w})$ be a multivariate, convex loss function in d dimensions. The subgradient at point $\overline{w}_0$ is a d-dimensional vector $\overline{v}$ that satisfies the following for any $\overline{w}$:*

$$J(\overline{w}) \geq J(\overline{w}_0) + \overline{v} \cdot (\overline{w} - \overline{w}_0)$$

Note that the notion of subgradient is primarily used in a *convex* function rather than an arbitrary function (as in conventional gradients). Although it is possible to also apply the above definition for nonconvex functions, the definition loses its usefulness in those cases. The subgradient is not unique unless the function is differentiable at that point. At differentiable points, the subgradient is simply the gradient. It can be shown that any convex combination of subgradients is a subgradient.

Problem 5.8.1 *Show using Definition 5.8.1 that if $\overline{v}_1$ and $\overline{v}_2$ are subgradients of $J(\overline{w})$ at $\overline{w} = \overline{w}_0$, then $\lambda\overline{v}_1 + (1 - \lambda)\overline{v}_2$ is also a subgradient of $J(\overline{w})$ for any $\lambda \in (0,1)$.*

The above practice problem shows that the set of subgradients is a convex closed set. Furthermore, if the zero vector is a subgradient at $\overline{w}_0$, then Definition 5.8.1 implies that we have $J(\overline{w}) \geq J(\overline{w}_0)$ for all $\overline{w}$. In other words, $\overline{w}_0$ is an optimal solution. In the following, we mention some key properties of subgradients:

1. The conventional gradient at a differentiable point is its unique subgradient.

2. For convex functions, the optimality condition for a particular value of the optimization variables $\overline{w}_0$ is that the set of subgradients at $\overline{w}_0$ must include the zero vector.

3. At any point $\overline{w}_0$, the sum of any subgradient of $J_1(\overline{w}_0)$ and any subgradient of $J_2(\overline{w}_0)$ is a subgradient of $(J_1 + J_2)(\overline{w}_0)$. In other words, we can decompose the subgradient of a separably additive function into its constituent subgradients. This property is relevant to loss functions of various machine learning algorithms that add up loss contributions of individual training points.

While it might not be immediately obvious, we have already used the subgradient method (implicitly) in the hinge-loss SVM in Chapter 4. We repeat the objective function of the hinge-loss SVM here (cf. page 184), which is based on the training pairs $(\overline{X}_i, y_i)$:

$$J = \sum_{i=1}^{n} \max\{0, (1 - y_i[\overline{W}^T \cdot \overline{X}_i])\} + \frac{\lambda}{2}\|\overline{W}\|^2 \quad \text{[Hinge-loss SVM]}$$

As evident from Figure 4.9 of Chapter 4, the use of the maximization function causes non-differentiability at the sharp "hinge" of the hinge-loss function; these are values of $\overline{W}$ where the second argument of the max-function is 0 for any training point. So what happens at these points? The update of the SVM uses only those training points where the second argument is *not* zero. Therefore, at the non-differentiable points, the gradient is simply set to 0, which is a valid subgradient. Therefore, the primal updates of the hinge-loss SVM implicitly use the subgradient method, although the use is straightforward and natural. In this case, the subgradient does not point in a direction of instantaneous movement that *worsens* the objective function (for infinitesimal steps). This is not the case for more aggressive uses of the subgradient method.

5.8.1.1   Application: L_1-Regularization

A more aggressive use of the subgradient method appears in least-squares regression with L_1-regularization.

$$\text{Minimize } J = \underbrace{\frac{1}{2}\|D\overline{W} - \overline{y}\|^2}_{\text{Prediction Error}} + \underbrace{\lambda \sum_{j=1}^{d} |w_j|}_{L_1\text{-Regularization}}$$

Here D is an $n \times d$ data matrix whose rows contain the training instances, and $\overline{y}$ is an n-dimensional column vector containing the target variables. The column vector $\overline{W}$ contains the coefficients. Note that the regularization term now uses the L_1-norm of the coefficient vector rather than the L_2-norm. The function J is non-differentiable for any $\overline{W}$ in which even a single component w_j is 0. Specifically, if w_j is infinitesimally larger than 0, then the partial derivative of $|w_j|$ is $+1$, whereas if w_j is infinitesimally smaller than 0, then the partial derivative of $|w_j|$ is -1. In these methods, the partial derivative of w_j at 0 is selected randomly from $[-1, +1]$, whereas the derivative at values different from 0 is computed in the same way as the gradient. Let the subgradient of w_j be denoted by s_j. Then, for step-size $\alpha > 0$, the update is as follows:

$$\overline{W} \Leftarrow \overline{W} - \alpha\lambda[s_1, s_2, \ldots, s_d]^T - \alpha D^T \underbrace{(D\overline{W} - \overline{y})}_{\text{Error}}$$

Here, each s_j is the subgradient of w_j and is defined as follows:

$$s_j = \begin{cases} -1 & w_j < 0 \\ +1 & w_j > 0 \\ \text{Sample from } [-1, +1] & w_j = 0 \end{cases} \quad (5.29)$$

In this particular case, movement along the subgradient might worsen the objective function value because of the random choice of s_j from $[-1, +1]$. Therefore, one always maintains the best possible value of $\overline{W}_{best}$ that was obtained in any iteration. At the beginning of the process, both $\overline{W}$ and $\overline{W}_{best}$ are initialized to the same random vector. After each update of $\overline{W}$, the objective function value is evaluated with respect to $\overline{W}$, and $\overline{W}_{best}$, and is set to the recently updated $\overline{W}$ if the objective function value provided by $\overline{W}$ is better than that obtained by the stored value of $\overline{W}_{best}$. At the end of the process, the vector $\overline{W}_{best}$ is returned by the algorithm as the final solution. Note that $s_j = 0$ is also a subgradient at $w_j = 0$, and it is a choice that is sometimes used.

5.8.1.2 Combining Subgradients with Coordinate Descent

The subgradient method can also be combined with coordinate descent (cf. Section 4.10 of Chapter 4) by applying the subgradient optimality condition to the coordinate being learned. The learning problem is often greatly simplified in coordinate descent because only one variable is optimized at a time. As in all coordinate descent methods, one cycles through all the variables one by one in order to perform the optimization.

We provide an example of the use of coordinate descent in linear regression. As in the previous section, let D be an $n \times d$ data matrix with rows containing training instances, and $\overline{y}$ be an n-dimensional column vector of response variables. The d-dimensional column vector of parameters is denoted by $\overline{W} = [w_1 \ldots w_d]^T$. The objective function of least-squares regression with L_1-regularization is repeated below:

$$\text{Minimize } J = \underbrace{\frac{1}{2}\|D\overline{W} - \overline{y}\|^2}_{\text{Prediction Error}} + \underbrace{\lambda \sum_{j=1}^{d} |w_j|}_{L_1\text{-Regularization}}$$

As discussed in Section 4.10 of Chapter 4, coordinate descent can sometimes get stuck for non-differentiable functions. However, a sufficient condition for coordinate descent to work for convex loss functions is that the non-differentiable portion can be decomposed into separable univariate functions (cf. Lemma 4.10.1 of Chapter 4). In this case, the regularization term is clearly a sum of separable and convex functions. Therefore, one can use coordinate descent without getting stuck at a local optimum. The subgradient with respect to *all* the variables is as follows:

$$\nabla J = D^T(D\overline{W} - \overline{y}) + \lambda[s_1, s_2, \ldots s_d]^T \tag{5.30}$$

Here, each s_i is a subgradient drawn from $[-1, +1]$. Since we are optimizing with respect to only the ith variable, we only need to set the ith component of ∇J to zero. Let $\overline{d}_i$ be the ith column of D. Furthermore, let $\overline{r}$ denote the n-dimensional residual vector $\overline{y} - D\overline{W}$. One can then write the optimality condition for the ith component in terms of these variables as follows:

$$\overline{d}_i^T(\overline{y} - D\overline{W}) - \lambda s_j = 0$$
$$\overline{d}_i^T \overline{r} - \lambda s_j = 0$$
$$\overline{d}_i^T \overline{r} + w_i \overline{d}_i^T \overline{d}_i - \lambda s_j = w_i \overline{d}_i^T \overline{d}_i$$

The left-hand side is free of w_i because the term $\overline{d}_i^T \overline{r}$ contributes $-w_i \overline{d}_i^T \overline{d}_i$, which cancels with $w_i \overline{d}_i^T \overline{d}_i$. Therefore, we obtain the coordinate update for w_i:

$$\overline{w}_i \Leftarrow \overline{w}_i + \frac{\overline{d}_i^T \overline{r} - \lambda s_i}{\|\overline{d}_i\|^2} \tag{5.31}$$

The value of the subgradient s_i is defined in the same way as in the previous section. The main problem is that each s_i could be chosen to be any value between -1 and $+1$ when the updated value of w_i is close enough to 0; only one of these values will arrive at the optimal solution. How can one determine the exact value of s_i that optimizes the objective function in such cases? This is achieved by the use of *soft thresholding* of such "close enough" values of w_i to 0. Soft thresholding of w_i automatically sets the value of s_i to an appropriate intermediate value between -1 and $+1$. Therefore, the value of each w_i is set as follows:

$$w_i \Leftarrow \begin{cases} 0, & -\frac{\lambda}{\|\overline{d}_i\|^2} \leq \overline{w}_i + \frac{\overline{d}_i^T \overline{r}}{\|\overline{d}_i\|^2} \leq \frac{\lambda}{\|\overline{d}_i\|^2} \\ \overline{w}_i + \frac{\overline{d}_i^T \overline{r} - \lambda \operatorname{sign}(w_i)}{\|\overline{d}_i\|^2}, & \text{otherwise} \end{cases} \tag{5.32}$$

As in any form of coordinate-descent, one cycles through the variables one by one until convergence is reached. The *elastic-net* combines both L_1- and L_2-regularization, and we leave the derivation of the resulting updates as a practice problem.

Problem 5.8.2 (Elastic-Net Regression) *Consider the problem of elastic-net regression with the following objective function:*

$$Minimize \; J = \frac{1}{2} \|D\overline{W} - \overline{y}\|^2 + \lambda_1 \sum_{j=1}^{d} |w_j| + \frac{\lambda_2}{2} \sum_{j=1}^{d} w_j^2$$

Show that the updates of coordinate decent can be expressed as follows:

$$w_i \Leftarrow \begin{cases} 0, & -\frac{\lambda_1}{\|\overline{d}_i\|^2 + \lambda_2} \leq \frac{w_i \|\overline{d}_i\|^2 + \overline{d}_i^T \overline{r}}{\|\overline{d}_i\|^2 + \lambda_2} \leq \frac{\lambda_1}{\|\overline{d}_i\|^2 + \lambda_2} \\ \frac{w_i \|\overline{d}_i\|^2 + \overline{d}_i^T \overline{r} - \lambda_1 \, sign(w_i)}{\|\overline{d}_i\|^2 + \lambda_2}, & \text{otherwise} \end{cases}$$

The main challenge in coordinate descent is to avoid getting stuck in a local optimum because of non-differentiability (see Figure 4.10 of Chapter 4 for an example). In many cases, one can use variable transformations to convert the objective function to a well-behaved form (cf. Lemma 4.10.1) in which convergence to a global optimum is guaranteed. An example is the *graphical lasso* [48], which implicitly uses variable transformations.

5.8.2 Proximal Gradient Method

The proximal gradient method is particularly useful when the optimization function $J(\overline{W})$ can be broken up into two parts $G(\overline{W})$ and $H(\overline{W})$, one of which is differentiable, and the other is not:

$$J(\overline{W}) = G(\overline{W}) + H(\overline{W})$$

In this form, the portion $G(\overline{W})$ is assumed to be differentiable, whereas $H(\overline{W})$ is not. Both functions are assumed to be convex. The proximal gradient method uses an iterative approach, in which each iteration taking a *gradient step* on $G(\cdot)$ and a *proximal step* on $H(\cdot)$. The proximal step is essentially a minimum value of $H(\cdot)$ in the *locality* of the current value of the parameter vector $\overline{W} = \overline{w}$. This type of minimum in a local region around $\overline{w}$ may be discovered by adding a quadratic penalty to $H(\overline{w})$ depending on how far one ventures from the current value of the parameter vector. Here, a key point is to define the *proximal operator* for the function $H(\cdot)$. The proximal operator $\mathcal{P}$ is defined with the use of a step-size parameter α as follows:

$$\mathcal{P}_{H,\alpha}(\overline{w}) = \operatorname{argmin}_{\overline{u}} \left\{ \alpha H(\overline{u}) + \frac{1}{2} \|\overline{u} - \overline{w}\|^2 \right\} \tag{5.33}$$

In other words, we are trying to minimize the function $H(\cdot)$ in the proximity of $\overline{w}$ by adding a quadratic penalty term to penalize distance from $\overline{w}$. Therefore, the proximity operator will try to find a "better" $\overline{u}$ than $\overline{w}$, but only in the proximity of $\overline{w}$ because distance from $\overline{w}$ is quadratically penalized. Now let us examine what happens with a few examples:

- When $H(\overline{w})$ is set to be a constant, the $\mathcal{P}_{H,\alpha}(\overline{w}) = \overline{w}$. This is because one cannot improve $\overline{w}$ any further from its current argument, and the quadratic penalty encourages staying at the current point.

- When $H(\overline{w})$ is differentiable, then the proximity operator makes an approximate gradient-descent move at step size α. One can derive this result by setting the gradient of the expression inside the argmin of Equation 5.33 to 0:

$$\overline{u} = \overline{w} - \alpha \frac{\partial H(\overline{u})}{\partial \overline{u}} \qquad (5.34)$$

Note that this step is similar to gradient-descent except that the gradient of $H(\cdot)$ is computed at $\overline{u}$ rather than $\overline{w}$. However, the quadratic penalization ensures that the step-size is relatively small, and the computation of the gradient of $H(\overline{u})$ happens only in the proximity of $\overline{w}$. This is a key motivational point. The proximity operator makes sensible moves when $H(\cdot)$ is differentiable. However, it works for non-differentiable functions as well.

Armed with this definition of the proximal operator, one can then write the proximal gradient algorithm in terms of repeating the following two iterative steps as follows:

1. Make a standard gradient-descent step on the differentiable function $G(\cdot)$ with step-size α:

$$\overline{w} \Leftarrow \overline{w} - \alpha \left[\frac{\partial G(\overline{w})}{\partial \overline{w}} \right]$$

2. Make a proximal descent step on the non-differentiable function $H(\cdot)$ with step-size α:

$$\overline{w} \Leftarrow \mathcal{P}_{H,\alpha}(\overline{w}) = \operatorname{argmin}_{\overline{u}} \left\{ \alpha H(\overline{u}) + \frac{1}{2} \|\overline{u} - \overline{w}\|^2 \right\}$$

Note that if the function $H(\cdot)$ is differentiable, then the approach roughly simplifies to alternate gradient descent on $G(\cdot)$ and $H(\cdot)$.

Another key point is in terms of how hard it is to compute the proximal operator. The approach is only used for problems with "simple" proximal operators that are easy to compute; furthermore, the underlying functions have a small number of non-differentiable points. A typical example of such a non-differentiable function is the L_1-norm of a vector. For this reason, the proximal method is less general than the subgradient method; however, when it works, it provides better performance.

5.8.2.1 Application: Alternative for L_1-Regularized Regression

In the previous section, we introduced a subgradient method for least-squares regression with L_1-regularization. In this section, we discuss an alternative based on the proximal gradient method. We rewrite the objective function of least-squares regression and separate it out into the differentiable and non-differentiable parts as follows:

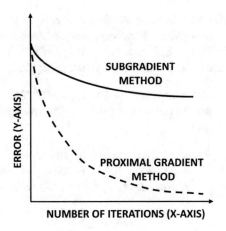

Figure 5.13: An illustrative comparison of the subgradient and the proximal gradient method in terms of typical behavior

$$\text{Minimize } J = \underbrace{\frac{1}{2}\|D\overline{W} - \overline{y}\|^2}_{G(\overline{W})} + \underbrace{\lambda \sum_{j=1}^{d} |w_j|}_{H(\overline{W})}$$

A key point is the definition of the proximal operator on the function $H(\overline{W})$, which the L_1-norm of $\overline{W}$. The proximal operator for $H(\overline{w})$ with step-size α is as follows:

$$[\mathcal{P}_{H,\alpha}]_j = \begin{cases} w_j + \alpha\lambda & w_j < -\alpha\lambda \\ 0 & -\alpha\lambda \le w_j \le \alpha\lambda \\ w_j - \alpha\lambda & w_j > \alpha\lambda \end{cases} \tag{5.35}$$

Note that the proximity operator essentially shrinks each w_j by exactly $\alpha\lambda$ as long as it is far away from the non-differentiable point. However, if it is close enough to the non-differentiable point then it simply moves to 0. This is the main difference from the subgradient method, which always updates by exactly $\alpha\lambda$ in either direction at all differentiable points, and updates by a random sample from $[-\alpha\lambda, \alpha\lambda]$ at the non-differentiable point. As a result, the subgradient method is more likely to oscillate around non-differentiable points as compared to the proximal gradient method. An illustrative comparison of the "typical" convergence behavior of the subgradient and proximal gradient method is shown in Figure 5.13. In most cases, the proximal gradient method performance *significantly* faster than the subgradient method. The faster convergence is because of the thresholding approach used in the neighborhood of non-differentiable points. This approach is referred to as the *iterative soft thresholding algorithm*, or *ISTA* in short.

5.8.3   Designing Surrogate Loss Functions for Combinatorial Optimization

Some problems like optimizing the ranking of a set of training instances are inherently combinatorial in nature, which do not provide informative loss surfaces in most regions of the space. For example, as shown in Figure 5.11, the sum of the ranks of positive class instances results in a highly non-informative function for the purposes of optimization. This function is not only non-differentiable at several points, but its staircase-like nature makes the gradient zero at all differentiable points. In other words, a gradient descent procedure would not know which direction to proceed. This type of problem does not occur with objective functions like the L_1-norm (which enables the use of a subgradient method). In such cases, it makes sense to design a surrogate loss function for the optimization problem at hand. This approach is inherently not a new one; *almost all objective functions for classification are surrogate loss functions anyway.* Strictly speaking, a classification problem should be directly optimizing the classification accuracy with respect to the parameter $\overline{W}$. However, the classification accuracy is another staircase-like function. Therefore, all the models we have seen so far use some form of surrogate loss, such as the least-squares (classification) loss, the hinge loss, and the logistic loss. Extending such methods to ranking problems is therefore not a fundamental innovation at least from a methodological point of view. However, the solutions to ranking objective functions have their own unique characteristics. In the following, we examine some surrogate objective functions designed for the ranking problem for classification.

Most classification objective functions are designed to penalize accuracy of classification by using some surrogate loss, such as the hinge-loss (which is a one-sided penalty from the target values of $+1$ and -1). Ranking-based objective functions are based on exactly the same principle. The only difference is that we penalize the deviation from an ideal ranking with a surrogate loss function. Two examples of such loss functions correspond to the *pairwise* and the *listwise* approaches. In the following, we discuss a simple pairwise approach for defining the loss function.

5.8.3.1 Application: Ranking Support Vector Machine

We will now formalize the optimization model for the ranking SVM. First, the training data is converted into pair-wise examples. For example, in the rare-class ranking problem, one would create pairs of positive and negative class instances, and always rank the positive class above the negative class. The training data $\mathcal{D}_R$ contains the following set of ranked pairs:

$$\mathcal{D}_R = \{(\overline{X}_i, \overline{X}_j) : \overline{X}_i \text{ should be ranked above } \overline{X}_j\}$$

For each such pair in the ranking support vector machine, the goal is learn a d-dimensional weight vector $\overline{W}$, so that $\overline{W} \cdot \overline{X}_i^T > \overline{W} \cdot \overline{X}_j^T$ when $\overline{X}_i$ is ranked above $\overline{X}_j$. Therefore, given an unseen set of test instances $\overline{Z}_1 \ldots \overline{Z}_t$, we can compute each $\overline{W} \cdot \overline{Z}_i^T$, and rank the test instances on the basis of this value.

In the traditional support vector machine, we always impose a *margin requirement* by penalizing points that are uncomfortably close to the decision boundary. Correspondingly, in the ranking SVM, we penalize pairs where the difference between $\overline{W} \cdot \overline{X}_i^T$ and $\overline{W} \cdot \overline{X}_j^T$ is not sufficiently large. Therefore, we would like to impose the following stronger requirement:

$$\overline{W} \cdot (\overline{X}_i - \overline{X}_j)^T > 1$$

Any violations of this condition are penalized by $1 - \overline{W} \cdot (\overline{X}_i - \overline{X}_j)^T$ in the objective function. Therefore, one can formulate the problem as follows:

$$\text{Minimize } J = \sum_{(\overline{X}_i, \overline{X}_j) \in \mathcal{D}_R} \max\{0, [1 - (\overline{W} \cdot [\overline{X}_i - \overline{X}_j]^T)]\} + \frac{\lambda}{2}\|\overline{W}\|^2$$

Here, $\lambda > 0$ is the regularization parameter. Note that one can replace each pair $(\overline{X}_i, \overline{X}_j)$ with the new set of features $\overline{X}_i - \overline{X}_j$. In other words, each $\overline{U}_p$ is of the form $\overline{U}_p = \overline{X}_i - \overline{X}_j$ for a ranked pair $(\overline{X}_i, \overline{X}_j)$ in the training data. Then, the ranking SVM formulates the following optimization problem for the t different pairs in the training data with corresponding features $\overline{U}_1 \ldots \overline{U}_t$:

$$\text{Minimize } J = \sum_{i=1}^{t} \max\{0, [1 - \overline{W} \cdot \overline{U}_i^T]\} + \frac{\lambda}{2}\|\overline{W}\|^2$$

Note that the only difference from a traditional support-vector machine is that the class variable y_i is missing in this optimization formulation. However, this change is extremely easy to incorporate in all the optimization techniques discussed in Section 4.8.2 of Chapter 4. In each case, the class variable y_i is replaced by 1 in the corresponding gradient-descent steps of various methods discussed in Section 4.8.2.

5.8.4   Dynamic Programming for Optimizing Sequential Decisions

Dynamic programming is an approach that is used for optimizing sequential decisions, and the most well-known machine learning application of this approach occurs in *reinforcement learning* [6]. The most general form of reinforcement learning optimizes an objective function $J(a_1 \ldots a_m)$, where $a_1 \ldots a_m$ is a sequence of actions or decisions. For example, finding a shortest path or a longest path from one point to another in a directed acyclic graph requires a sequence of decisions as to which node to select in the next step. Similarly, a two-player game like tic-tac-toe also requires a sequence of decisions about moves to be made in the game, although alternate decisions are made by opponents, and have opposite goals. This principle is used for game learning strategies in reinforcement learning. Another example is that of finding the edit distance between two strings, which requires a sequence of decisions of which edits to make. In all these cases, one has a sequence of decisions $a_1 \ldots a_m$ to make, and after making a decision, one is left with a smaller subproblem to solve. For example, if one has to choose the shortest path from source to sink in a graph, then after choosing the first outgoing node i from the source, one still has to compute the shortest path from i to the sink. In other words, dynamic programming breaks up a larger problems into smaller problems, *each of which would need to be optimally solved*. Dynamic programming works precisely in those scenarios that have the all-important *optimal substructure property*:

Property 5.8.1 (Optimal Substructure Property) *Dynamic programming works in those optimization settings, where a larger problem can be broken down into smaller subproblems of an identical nature. In other words, every optimal solution to the larger problem must also contain optimal solutions to the smaller subproblems.*

Here, the key point is that even though the number of solutions is extremely large, the optimal substructure property allows us to consider only a small subset of them. For example, the number of paths from the source to sink in a graph may be exponentially large, but one can easily compute all shortest paths containing at most 2 nodes from the source to all nodes. Because the optimal substructure property, these paths can be extended to paths containing at most 3 nodes in linear time. This process can be repeated for an increasing number of nodes, until the number of nodes in the graph is reached. One generally implements dynamic programming via an iterative table-filling approach where smaller subproblems are solved first and their solutions are saved. Larger problems are then solved as a function of the known solutions of the smaller problems using the optimal substructure property. In order to elucidate this point, we will use the example of optimizing the number of operations in chain matrix multiplication.

5.8.4.1 Application: Fast Matrix Multiplication

Consider the problem of multiplying the matrices A_1, A_2, A_3, A_4, and A_5 in that order. Because of the associative property of matrix multiplication, one can group the multiplications in a variety of ways without changing the result (as long as the sequential order of matrices is not changed). For example, one can group the multiplication as $[(A_1A_2)(A_3A_4)](A_5)$, or one can group the multiplication as $[(A_1)(A_2A_3)](A_4A_5)$. Consider the case where each A_i for odd i is a 1×1000 matrix, and each A_i for even i is a 1000×1 matrix. In such a case, the first grouping will require only about 3000 scalar multiplications to yield the final result of size 1×1000. All intermediate results will be compact scalars. On the other hand, the second grouping will create large intermediate matrices of size 1000×1000, the computation of which will require a million scalar multiplications. Clearly, the way in which the nesting is done is critical to the efficiency of matrix multiplication.

The decision problem in this case is to choose the top level grouping, since the subproblems are identical and can be solved in a similar way. For example, the top-level grouping in the first case is $[A_1A_2A_3A_4](A_5)$, and the top-level grouping in the second case above is $[A_1A_2A_3](A_4A_5)$. There are only four possible top-level groupings, and one needs to compute the number of operations in each case and choose the best among them. For each grouping, the smaller subproblems like $[A_1A_2A_3]$ and (A_4A_5) also need to be solved optimally. The complexity of multiplying the two intermediate matrices like $A_1A_2A_3$ and A_4A_5 of size $p \times q$ and $q \times r$, respectively, is pqr. This overhead is added to the complexity of the two subproblems to yield the complexity of that grouping.

Consider the matrices $A_1A_2 \ldots A_m$, where the matrix A_i is of size $n_i \times n_{i+1}$, and the optimal number of operations required for multiplying matrices i through j is $N[i, j]$. This leads to the following *dynamic programming recursion* for computing $N[1, m]$:

$$N[i, j] = \min_{k \in [i+1, j]} \{N[i, k-1] + N[k, j] + n_i n_k n_j\} \tag{5.36}$$

Note that the values on the right-hand side are computed earlier than the ones on the left using *iterative table filling*, where we compute all $N[i, j]$ in cases where $(j - i)$ is 1, 2, and so on in that order till $j - i$ is $(m - 1)$. There are at most $O(m^2)$ slots in the table to fill, and each slot computation needs the evaluation of the right-hand side of Equation 5.36. This evaluation requires a minimization over at most $(m - 1)$ possibilities, each of which requires two table lookups of the evaluations of smaller subproblems. Therefore, each evaluation of Equation 5.36 requires $O(m)$ time, and the overall complexity is $O(m^3)$. One can summarize this algorithm as follows:

Initialize $N[i, i] = 0$ and Split$[i, i] = -1$ for all i;
for $\delta = 1$ to $m - 1$ **do**
  for $i = 1$ to $m - \delta$ **do**
  $N[i, i + \delta] = \min_{k \in [i+1, i+\delta]} \{N[i, k-1] + N[k, i+\delta] + n_i n_k n_{i+\delta}\}$;
  Split$[i, i + \delta] = \text{argmin}_{k \in [i+1, i+\delta]} \{N[i, k-1] + N[k, i+\delta] + n_i n_k n_{i+\delta}\}$;
  endfor;
endfor

One also needs to keep track of the optimal split position for each pair $[i, j]$ in a separate table Split$[i, j]$ in order to reconstruct the nesting. For example, one will first access $k = \text{Split}(1, m)$ in order to divide the matrix into two groups $A_1 \ldots A_{k-1}$ and $A_k \ldots A_m$.

Subsequently Split$[1, k - 1]$ and Split$[k, m]$ will be accessed again to find the top-level nesting for the individual subproblems. This process will be repeated until we reach singleton matrices.

The word "dynamic programming" is used in settings beyond pure optimization. Many types of iterative table filling that achieve polynomial complexity by avoiding repeated operations are considered dynamic programming (even when no optimization occurs). For example, the backpropagation algorithm (cf. Chapter 11) uses the summation operation in the dynamic-programming recursion, but it is still considered dynamic programming. One can easily change the shortest-path algorithm between a source-sink pair to an algorithm for finding the number of paths between a source-sink pair (in a graph without cycles) with a small change to the form of the key table-filling step. Instead of computing the shortest path using each incident node i on source node s, one can compute the sum of the paths from each incident node i (on the source) to the sink. The key point is that an *additive* version of the substructure property holds, where the number of paths from source to sink is to equal to the sum of the number of paths from node i (incident on source) to sink. However, this is not an optimization problem. Therefore, the dynamic programming principle can also be viewed as a general computer programming paradigm that works in problem settings beyond optimization by exploiting any version of the substructure property — in general, the substructure property needs to be able to compute the statistics of superstructures from those of substructures via bottom-up table filling.

5.9 Summary

This chapter introduces a number of advanced methods for optimization, when simpler methods for gradient descent are not very effective. The simplest approach is to modify gradient descent methods, and incorporate several ideas from second-order methods into the descent process. The second approach is to directly use second-order methods such as the Newton technique. While the Newton technique can solve quadratic optimization problems in a single step, it can be used to solve non-quadratic problems with the use of local quadratic approximations. Several variations of the Newton method, such as the conjugate gradient method and the quasi-Newton method, can be used to make it computationally efficient. Finally, non-differentiable optimization problems present significant challenges in various machine learning settings. The simplest approach is to change the loss function to a differentiable surrogate. Other solutions include the use of the subgradient and the proximal gradient methods.

5.10 Further Reading

A discussion of momentum methods in gradient descent is provided in [106]. Nesterov's algorithm for gradient descent may be found in [97]. The delta-bar-delta method was proposed by [67]. The AdaGrad algorithm was proposed in [38]. The RMSProp algorithm is discussed in [61]. Another adaptive algorithm using stochastic gradient descent, which is *AdaDelta*, is discussed in [139]. This algorithms shares some similarities with second-order methods, and in particular to the method in [111]. The Adam algorithm, which is a further enhancement along this line of ideas, is discussed in [72]. The strategy of *Polyak averaging* is discussed in [105].

A description of several second-order gradient optimization methods (such as the Newton method) is provided in [19, 66, 83]. The implementation of the SVM approach with the

Newton method is presented in [28] and an implementation of logistic regression is presented in [80]. Discussions of various numerical optimization techniques for logistic regression (including the Newton method) are provided in [93]. The basic principles of the conjugate gradient method have been described in several classical books and papers [19, 59, 114], and the work in [86, 87] discusses applications to neural networks. The work in [89] leverages a Kronecker-factored curvature matrix for fast gradient descent. Another way of approximating the Newton method is the quasi-Newton method [78, 83], with the simplest approximation being a diagonal Hessian [13]. The acronym BFGS stands for the Broyden-Fletcher-Goldfarb-Shanno algorithm. A variant known as limited memory BFGS or L-BFGS [78, 83] does not require as much memory. Another popular second-order method is the Levenberg–Marquardt algorithm. Overviews of the approach may be found in [51, 83].

Methods for non-differentiable optimization are discussed in [96, 116]. The use of coordinate descent for L_1-regularized regression is discussed in [135]. Another variant, referred to as the *graphical lasso*, is discussed in [48]. These include discussions of the subgradient and the proximal gradient methods. A specific overview of proximal algorithms may be found in [100]. An in-depth discussion of methods for handling L_1-regularization is presented in [57]. A fast version of the iterative shrinkage thresholding algorithm is presented in [12]. Algorithms for learning to rank are presented in [81].

5.11   Exercises

1. Consider the loss function $L = x^2 + y^{10}$. Implement a simple steepest-descent algorithm to plot the coordinates as they vary from the initialization point to the optimal value of 0. Consider two different initialization points of $(0.5, 0.5)$ and $(2, 2)$ and plot the trajectories in the two cases at a constant learning rate. What do you observe about the behavior of the algorithm in the two cases?

2. As shown in this chapter with examples like Figure 5.2, the number of steps taken by gradient descent is very sensitive to the scaling of the variables. In this exercise, we will show that the Newton method is completely insensitive to the scaling of the variables. Let $\overline{x}$ be the set of optimization variables for a particular optimization problem (OP). Suppose we transform $\overline{x}$ to $\overline{y}$ by the linear scaling $\overline{y} = B\overline{x}$ with invertible matrix B, and pose the same optimization problem in terms of $\overline{y}$. The objective function might be non-quadratic. Show that the sequences $\overline{x}_0, \overline{x}_1 \ldots \overline{x}_r$ and $\overline{y}_0, \overline{y}_1 \ldots \overline{y}_r$ obtained by iteratively applying Newton's method will be related as follows:

$$\overline{y}_k = B\overline{x}_k \quad \forall k \in \{1 \ldots r\}$$

[As a side note, the preprocessing and scaling of features is extremely common in machine learning, which also affects the scaling of the optimization variables.]

3. Write down the second-order Taylor expansion of each of the following functions about $x = 0$: (a) x^2; (b) x^3; (c) x^4; (d) $\cos(x)$.

4. Suppose that you have the quadratic function $f(x) = ax^2 + bx + c$ with $a > 0$. It is well known that this quadratic function takes on its minimum value at $x = -b/2a$. Show that a single Newton step starting at any point $x = x_0$ will always lead to $x = -b/2a$ irrespective of the starting point x_0.

5. Consider the objective function $f(x) = [x(x-2)]^2 + x^2$. Write the Newton update for this objective function starting at $x = 1$.

6. Consider the objective function $f(x) = \sum_{i=1}^{4} x^i$. Write the Newton update starting at $x = 1$.

7. Is it possible for a Newton update to reach a maximum rather than a minimum? Justify your answer. In what types of functions is the Newton method guaranteed to reach a maximum rather than a minimum?

8. Consider the objective function $f(x) = \sin(x) - \cos(x)$, where the angle x is measured in radians. Write the Newton update starting at $x = \pi/8$.

9. The Hessian H of a strongly convex quadratic function always satisfies $\overline{x}^T H \overline{x} > 0$ for any non-zero vector $\overline{x}$. For such problems, show that all conjugate directions are linearly independent.

10. Show that if the dot product of a d-dimensional vector $\overline{v}$ with d linearly independent vectors is 0, then $\overline{v}$ must be the zero vector.

11. The chapter uses steepest descent directions to iteratively generate conjugate directions. Suppose we pick d *arbitrary* directions $\overline{v}_0 \ldots \overline{v}_{d-1}$ that are linearly independent. Show that (with appropriate choice of β_{ti}) we can start with $\overline{q}_0 = \overline{v}_0$ and generate successive conjugate directions in the following form:

$$\overline{q}_{t+1} = \overline{v}_{t+1} + \sum_{i=0}^{t} \beta_{ti} \overline{q}_i$$

Discuss why this approach is more expensive than the one discussed in the chapter.

12. The definition of β_t in Section 5.7.1 ensures that $\overline{q}_t$ is conjugate to $\overline{q}_{t+1}$. This exercise systematically shows that *any* direction $\overline{q}_i$ for $i \leq t$ satisfies $\overline{q}_i^T H \overline{q}_{t+1} = 0$.
[Hint: Prove (b), (c), and (d) *jointly* with induction on t while staring at (a).]

   (a) Recall from Equation 5.23 that $H\overline{q}_i = [\nabla J(\overline{W}_{i+1}) - \nabla J(\overline{W}_i)]/\delta_i$ for quadratic loss functions, where δ_i depends on ith step-size. Combine this condition with Equation 5.21 to show the following for all $i \leq t$:
   $$\delta_i[\overline{q}_i^T H \overline{q}_{t+1}] = -[\nabla J(\overline{W}_{i+1}) - \nabla J(\overline{W}_i)]^T [\nabla J(\overline{W}_{t+1})] + \delta_i \beta_t (\overline{q}_i^T H \overline{q}_t)$$
   Also show that $[\nabla J(\overline{W}_{t+1}) - \nabla J(\overline{W}_t)] \cdot \overline{q}_i = \delta_t \overline{q}_i^T H \overline{q}_t$.

   (b) Show that $\nabla J(\overline{W}_{t+1})$ is orthogonal to each $\overline{q}_i$ for $i \leq t$.

   (c) Show that the loss gradients at $\overline{W}_0 \ldots \overline{W}_{t+1}$ are mutually orthogonal.

   (d) Show that $\overline{q}_i^T H \overline{q}_{t+1} = 0$ for $i \leq t$. [The case for $i = t$ is trivial.]

13. Consider the use of the Newton method for a regularized L_2-loss SVM, and a wide data matrix D. Discuss how you can make the update in the chapter text more efficient by inverting a smaller matrix. [Hint: Use the push-through identity of Problem 1.2.13 by defining $D_w = \sqrt{\Delta_w}D$. The notations are the same as in the text.]

14. **Saddle points proliferate in high dimensions:** Consider the univariate function $f(x) = x^3 - 3x$, and its natural multivariate extension:

$$F(x_1 \ldots x_d) = \sum_{i=1}^{d} f(x_i)$$

Show that this function has one minimum, one maximum, and $2^d - 2$ saddle points. Argue why high-dimensional functions have proliferating saddle points.

15. Give a proof of the unified Newton update for machine learning in Lemma 5.5.1.

16. Preparing for backpropagation: Consider a directed-acyclic graph G (i.e., graph without cycles) with source node s and sink t. Each edge is associated with a length and a multiplier. The length of a path from s to t is equal to the *sum* of the edge lengths on the path and the multiplier of the path is the *product* of the corresponding edge multipliers. Devise dynamic programming algorithms to find (i) the longest path from s to t, (ii) the shortest path from s to t, (iii) the average path length from s to t, and (iv) the sum of the path-multipliers of all paths from s to t. [Part (iv) is the core idea behind the backpropagation algorithm.]

17. Give an example of a univariate cubic objective function along with two possible starting points for Newton's method, which terminate in maxima and minima, respectively.

18. Linear regression with L_1-loss minimizes $\|D\overline{W} - \overline{y}\|_1$ for data matrix D and target vector $\overline{y}$. Discuss why the Newton method cannot be used in this case.

Chapter 6

Constrained Optimization and Duality

"Virtuous people often revenge themselves for the constraints to which they submit by the boredom that they inspire." – Confucius

6.1 Introduction

In many machine learning settings, such as *nonnegative regression* and *box regression*, the optimization variables are constrained. Therefore, one needs to find an optimal solution only over the region of the optimization space that satisfies these constraints. This region is referred to as the *feasible region* in optimization parlance. The straightforward use of a gradient-descent procedure does not work, because an unconstrained step might move the optimization variables outside the feasible region of the optimization problem. In general, there are two approaches to addressing optimization constraints:

1. *Primal approach:* In the primal approach, one attempts to modify gradient descent so as to stay within the feasible regions of the space. Many of the methods discussed in the previous chapters, such as gradient descent, coordinate descent, and Newton's method, can be modified to stay within feasible regions of the space.

2. *Dual approach:* The dual approach uses Lagrangian relaxation in order to create a new *dual* problem in which primal constraints are converted into dual variables. In many cases, the structure of the dual problem is simpler to solve. However, the dual problem is often constrained as well, and might require similar optimization methods (to the primal methods above) that can work with constraints.

This chapter discusses both primal and the dual methods for constrained optimization. Some techniques like *penalty methods* incorporate aspects of both primal and dual methods.

The complexity of an optimization problem depends on the structure of its constraints. Luckily, many machine learning applications involve two simple types of constraints:

1. *Linear and convex constraints:* Linear constraints are of the form $F(\overline{w}) \leq b$ or of the form $G(\overline{w}) = c$, where $F(\overline{w})$ and $G(\overline{w})$ are linear functions. A more general type of constraint is the convex constraint of the form $H(\overline{w}) \leq d$, where $H(\overline{w})$ is convex.

© Springer Nature Switzerland AG 2020

C. C. Aggarwal, *Linear Algebra and Optimization for Machine Learning*,

https://doi.org/10.1007/978-3-030-40344-7_6

2. *Norm constraints:* Many machine learning problems are norm constrained, where we wish to minimize or maximize $F(\overline{w})$ subject to the constraint that $\|\overline{w}\|^2 = 1$. This problem arises in spectral clustering and principal component analysis.

This chapter is organized as follows. The next section will introduce constrained methods for (primal) gradient descent. Methods for coordinate descent are discussed in Section 6.3. The approach of Lagrangian relaxation is introduced in Section 6.4. Penalty methods are discussed in Section 6.5. Methods for norm-constrained optimization are discussed in Section 6.6. A discussion of the relative advantages of primal and dual methods is provided in Section 6.7. A summary is given in Section 6.8.

6.2   Primal Gradient Descent Methods

The projected gradient-descent method is also referred to as the *feasible direction method.* Such methods either make steps along a projection of the gradient-descent direction (that retains feasibility), or they immediately "repair" a movement outside the feasible space to a feasible solution. In its most basic form, an unconstrained steepest-descent update is first performed. However, such an update might move the current optimization variables outside the feasible space. At this point, one projects the parameter vector to the closest point in the feasible space. A key point is that this sequence of two steps works well, as long as the optimization problem has the following convex structure:

Definition 6.2.1 (Convex Objective Function Over a Convex Set) *The problem of minimizing a convex objective function over a convex set is defined as follows:*

$$\text{Minimize } F(\overline{w})$$
$$\text{subject to:}$$
$$\overline{w} \in C$$

Here, $F(\overline{w})$ is a convex function, and C is a convex set.

The above definition is the most general form of this type of optimization problem. However, there are many special cases of the set C that arise commonly in machine learning:

1. **Linear constraints:** The set C is the intersection of linear constraints of the form $f_i(\overline{w}) \leq 0$ or of the form $f_i(\overline{w}) = 0$. The value of i can range from $1 \ldots m$. Here, each $f_i(\overline{w})$ is a linear function. Note that an equality constraint can be expressed as the intersection of two linear inequality constraints $f_i(\overline{w}) \leq 0$ and $-f_i(\overline{w}) \leq 0$. Therefore, inequality constraints are more general than equality constraints, although inequality constraints create a more challenging optimization problem.

2. **Convex constraints:** The set C is the intersection of convex constraints of the form $f_i(\overline{w}) \leq 0$, where $i \in \{1 \ldots m\}$. Here, each $f_i(\overline{w})$ is a convex function (including the possibility of a linear function).

We will present a general algorithm for the feasible direction method, and then present its simplifications in special cases. For the general optimization formulation of Definition 6.2.1, the feasible direction method repeats the following two steps iteratively:

1. At the current parameter vector $\overline{w}$ perform the following steepest-descent update:

$$\overline{w} \Leftarrow \overline{w} - \alpha \nabla F(\overline{w})$$

Here, $\alpha > 0$ is the step-size. This step might move $\overline{w}$ outside the feasible set C.

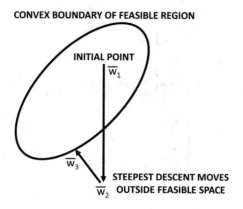

Figure 6.1: The projected gradient-descent method. Steepest descent first moves outside the feasible region and then projects back to nearest point inside feasible region

2. Project $\overline{w}$ onto its nearest point in the set C. This projection can be expressed as an optimization problem of the following form:

$$\overline{w} \Leftarrow \operatorname{argmin}_{\overline{v} \in C} \|\overline{w} - \overline{v}\|^2$$

This step is required only when the first step moves $\overline{w}$ outside the feasible region.

These two steps are iterated to convergence. When the set C is convex and the objective function $F(\overline{w})$ is convex, this approach can be shown to converge to an optimal solution. Note that the second step is itself an optimization problem, albeit with a simpler structure. The projected gradient descent method is pictorially illustrated in Figure 6.1.

6.2.1 Linear Equality Constraints

Certain types of optimization problems with linear constraints arise frequently in machine learning. A common example is that of *quadratic programming*, in which the objective function contains quadratic and linear terms of the form $\overline{w}^T Q \overline{w} + \overline{c}^T \overline{w}$ and the constraints are linear. Here, $\overline{w}$ is a d-dimensional parameter vector, $\overline{c}$ is a d-dimensional column vector, and Q is a $d \times d$ matrix. When the objective function is linear, the resulting formulation is referred to as *linear programming*.

Linear *equality* constraints can be considered almost equivalent to the unconstrained version of the problem, because one can eliminate the variables of an equality-constrained problem in order to create an unconstrained objective function. This type of elimination cannot be achieved in inequality constrained problems. In general, equality constraints in optimization problems are simpler to address than are inequality constraints (whether the constraints are linear or not). This is because equality constraints always allow the possibility of eliminating some subsets of the variables and constraints.

Observation 6.2.1 *One can use Gaussian elimination to convert a linear-equality-constrained optimization problem into an unconstrained form by eliminating a subset of the variables and constraints from the optimization problem.*

In order to understand this point, consider the case in which we wish to minimize the objective function $x^2 + y^2$ subject to the constraint $x + y = 1$. In this case, we substitute

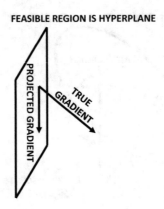

Figure 6.2: Projected gradient descent with different types of linear constraints

$y = 1 - x$, and drop both y and the constraint to create the following unconstrained objective function:

$$J = x^2 + (1 - x)^2$$

It is easy to verify that the optimal value of x is $1/2$. When we have a larger number of constraints, it is necessary to use row reduction in order to create row echelon form. Subsequently, one can express the variables for which leading non-zero entries exist in the row-reduced form of A in terms of all the remaining *free* variables (for which leading non-zero entries do not exist). As a result, an unconstrained objective function can be expressed only in terms of the free variables. An example of this type of elimination is shown in Section 2.5.4 of Chapter 2. Subsequently, one can use simple gradient descent on the unconstrained objective in order to solve the optimization problem.

In spite of the possibility of eliminating a subset of the variables (and the constraints) using Gaussian elimination, one can also use projected gradient descent with equality constraints. An example of a 2-dimensional hyperplane space in three dimensions is shown in Figure 6.2. Note that one need not separate out the two iterative steps of steepest direction movement and projection in this special case. Rather, the gradient can be directly projected onto the linear hyperplane in order to perform the descent. The corresponding projection of the steepest-descent direction on the 2-dimensional hyperplane is illustrated in Figure 6.2.

It is helpful to work out what the steepest-descent direction means in algebraic terms. Consider a situation where one is minimizing $F(\overline{w})$ subject to the constraint system $A\overline{w} = \overline{b}$. Here, $\overline{w}$ is a d-dimensional column vector, and A is an $m \times d$ matrix with $m \leq d$. Therefore, the vector $\overline{b}$ is m-dimensional. Note that it is important for $m \leq d$, or else the set of constraints might be infeasible. For simplicity, we will assume that *the rows of A are linearly independent*.

Consider the situation where the current parameter vector $\overline{w} = \overline{w}_t$. Assume that $\overline{w}_t$ is already feasible and therefore it satisfies the constraints $A\overline{w}_t = \overline{b}$ of the optimization problem. Then, the current steepest-descent direction is given by $\overline{g}_t = \nabla F(\overline{w}_t)$. Note that if $A\overline{g}_t \neq \overline{0}$, then the point $\overline{w}_t - \alpha \overline{g}_t$ will no longer be feasible. This is because we will have $A[\overline{w}_t - \alpha \overline{g}_t] = \overline{b} - \alpha A \overline{g}_t \neq \overline{b}$. This situation is shown in Figure 6.2, where the steepest-descent direction moves off the feasible hyperplane.

Therefore, in order for the steepest-descent step to stay feasible, the vector $\overline{g}_t$ needs to be projected onto the hyperplane $A\overline{w} = \overline{0}$, so that the projected vector $\overline{g}_t'$ satisfies $A\overline{g}_t' = 0$. In other words, projected steepest descent needs to project $\overline{g}_t$ onto the right null space of A.

This is achieved by expressing $\overline{g}_t = \overline{g}_\| + \overline{g}_\perp$ in terms of the portion $\overline{g}_\|$ lying in the subspace corresponding to the rows of A and the portion $\overline{g}_\perp$ in its orthogonal complementary subspace (cf. Definition 2.3.10 of Chapter 2). Note that it is the portion $\overline{g}_\perp$ that lies on $A\overline{w} = \overline{0}$. An example of a projected vector $\overline{g}_t' = \overline{g}_\perp$ is shown in Figure 6.2. Note that the notation $\perp$ refers to the fact that the vector $\overline{g}_\perp$ is perpendicular to the subspace defined by the rows of A, even though such a vector is actually *parallel* to the hyperplane $A\overline{w} = \overline{0}$. We mention this point because the reader might find it confusing to see a vector parallel to the hyperplane being annotated by "$\perp$." Here, it is important to note that even though the vector is parallel to the hyperplane $A\overline{w} = \overline{b}$, it needs to lie in the orthogonal complementary subspace of the rows of A to do so. In general, the coordinates $\overline{w}$ of all points on the hyperplane $A\overline{w} = \overline{0}$ form a vector space orthogonal to the rows of A. Therefore, the notation "$\perp$" refer to the *linear algebra* concept of orthogonal complementary subspace, rather than the more intuitive or geometric concept of being parallel to a hyperplane. Therefore, we need to subtract the component $\overline{g}_\|$ from $\overline{g}_t$ that lies in the span of the rows of A. The simplest approach is to use the row-wise[1] *projection matrix* discussed in Equation 2.17 of Chapter 2, although this result assumes that the rows of A are linearly independent (i.e., no redundant constraints). In other words, one can simply express $\overline{g}_t'$ in closed form as follows:

$$\overline{g}_t' = \overline{g}_t - \overline{g}_\| = [I - A^T(AA^T)^{-1}A]\overline{g}_t \tag{6.1}$$

In cases when the rows of A are not linearly independent, the computation of $\overline{g}_t' = \overline{g}_\perp$ can also be achieved easily by Gram-Schmidt orthogonalization (cf. Section 2.7.1 of Chapter 2) of the m rows of A to create $r < m$ orthonormal vectors $\overline{v}_1 \ldots \overline{v}_r$. Then, $\overline{g}_\perp$ can be computed as follows:

$$\overline{g}_\| = \sum_{i=1}^{r} [\overline{g}_t \cdot \overline{v}_i]\,\overline{v}_i$$

$$\overline{g}_\perp = \overline{g}_t - \overline{g}_\|$$

Subsequently, the iterative projected gradient descent steps can be written as follows:

1. Compute $\overline{g}_t = \nabla F(\overline{w}_t)$ and compute $\overline{g}_\perp$ from $\overline{g}_t$ as discussed above.

2. Update $\overline{w}_{t+1} \Leftarrow \overline{w}_t - \alpha \overline{g}_\perp$ and increment t by 1.

The above two steps are repeated to convergence. The procedure can be initialized with any feasible value of the vector $\overline{w} = \overline{w}_0$. The initial feasible value can be found by solving the system of equations $A\overline{w} = \overline{b}$ using any of the methods discussed in Chapter 2.

Problem 6.2.1 *Suppose that you use line search to determine the step-size α in each iteration for projected gradient descent in convex functions and linear equality constraints. Show that successive directions of projected descent are always orthogonal to one another.*

6.2.1.1 Convex Quadratic Program with Equality Constraints

We have already addressed the problem of unconstrained quadratic programming in Section 4.6.2.1 of Chapter 4. In this section, we will discuss quadratic programming with equality constraints. The quadratic programming problem is defined as follows:

---

[1]The default definition of projection matrix (cf. Equation 2.17) always projects in the span of the *columns* of A, which is a *column-wise* projection matrix. Here, we project in the span of the *rows* of A, and therefore the formula of Equation 2.17 has been modified by transposing A.

$$\text{Minimize } J(\overline{w}) = \frac{1}{2}\overline{w}^T Q \overline{w} + \overline{p}^T \overline{w} + q$$

$$\text{subject to:}$$

$$A\overline{w} = \overline{b}$$

Here, Q is a $d \times d$ positive definite matrix, $\overline{p}$ and $\overline{w}$ are d-dimensional column vectors, and q is a scalar. This objective function is strictly convex, since it has a positive-definite Hessian Q everywhere. For simplicity in discussion, we assume that the matrix A has linearly independent rows. Therefore, A is an $m \times d$ matrix with $m \leq d$, and the vector $\overline{b}$ is m-dimensional.

We already know from Section 4.6.2.1 that unconstrained quadratic programs with positive definite Hessians have closed-form solutions. Since equality constraints can always be eliminated with the Gaussian method, it stands to reason that one should be able to find a closed-form solution in this case as well. After all, the projection of a strictly convex function on a linear hyperplane $A\overline{w} = \overline{b}$ will continue to be strictly convex as well, and therefore we should be able to find a closed form solution in this case. However, to achieve this goal, we need to use a *variable transformation* so that the objective function contains linearly separable variables (cf. Section 3.4.4 of Chapter 3). This process is similar to that of converting a univariate quadratic function into vertex form. First we express $Q = P \Delta P^T$, where Δ is a diagonal matrix with strictly positive entries. Therefore, both the matrix $\sqrt{\Delta}$ and $\Delta^{-1/2}$ can be defined. The objective function can be rewritten as follows:

$$J(\overline{w}) = \frac{1}{2}\overline{w}^T Q \overline{w} + \overline{p}^T \overline{w} + q$$

$$= \frac{1}{2}\overline{w}^T [P\Delta P^T]\overline{w} + \overline{p}^T \overline{w} + q$$

$$= \frac{1}{2}\|\sqrt{\Delta}P^T\overline{w} + \Delta^{-1/2}P^T\overline{p}\|^2 + [q - \frac{1}{2}\overline{p}^T \underbrace{[P\Delta^{-1}P^T]}_{Q^{-1}}\overline{p}]$$

Note that the modified constant term is defined by $q' = q - \frac{1}{2}\overline{p}^T[P\Delta^{-1}P^T]\overline{p}$. In order to solve the problem, we make the following variable transformation:

$$\overline{w}' = \sqrt{\Delta}P^T\overline{w} + \Delta^{-1/2}P^T\overline{p} \qquad (6.2)$$

This variable transformation is invertible, since we can express $\overline{w}$ in terms of $\overline{w}'$ as well by left-multiplying both sides with $P\Delta^{-1/2}$:

$$P\Delta^{-1/2}\overline{w}' = \overline{w} + P\Delta^{-1}P^T\overline{p}$$

$$= \overline{w} + Q^{-1}\overline{p}$$

In other words, $\overline{w}$ can be expressed in terms of $\overline{w}'$ as follows:

$$\overline{w} = P\Delta^{-1/2}\overline{w}' - Q^{-1}\overline{p} \qquad (6.3)$$

The linear constraints $A\overline{w} = \overline{b}$ can be expressed in terms of the new variables $\overline{w}'$ as follows:

$$A\overline{w} = \overline{b}$$

$$A[P\Delta^{-1/2}\overline{w}' - Q^{-1}\overline{p}] = \overline{b}$$

$$\underbrace{[AP\Delta^{-1/2}]}_{A'}\overline{w}' = \underbrace{\overline{b} + AQ^{-1}\overline{p}}_{\overline{b}'}$$

Therefore, we again obtain linear constraints with new matrices/vectors A' and $\vec{b}'$. In other words, the optimization problem can be expressed in the following form:

$$\text{Minimize } J(\overline{w}') = \frac{1}{2}\|\overline{w}'\|^2 + q'$$

$$\text{subject to:}$$

$$A'\overline{w}' = \vec{b}'$$

Note that the rows of A' are linearly independent like those of A because A' is obtained by multiplying A with square matrices of full rank. This is exactly the optimization problem discussed in Section 2.8 of Chapter 2, where the right-inverse of A' can be used to find a solution for $\overline{w}'$:

$$\overline{w}' = A'^T(A'A'^T)^{-1}\vec{b}' \tag{6.4}$$

What does this mean in terms of the original coefficients and optimization variables? By substituting $A' = AP\Delta^{-1/2}$, it can be shown that $A'A'^T = A(P\Delta^{-1}P^T)A^T = AQ^{-1}A^T$. One can therefore obtain $\overline{w}$ in terms of the original coefficients:

$$\begin{aligned}
\overline{w} &= P\Delta^{-1/2}\overline{w}' - Q^{-1}\overline{p} \\
&= P\Delta^{-1/2}[\Delta^{-1/2}P^TA^T(AQ^{-1}A^T)^{-1}\vec{b}'] - Q^{-1}\overline{p} \\
&= Q^{-1}A^T[AQ^{-1}A^T]^{-1}\vec{b}' - Q^{-1}\overline{p} \\
&= Q^{-1}\{A^T[AQ^{-1}A^T]^{-1}[\overline{b} + AQ^{-1}\overline{p}] - \overline{p}\}
\end{aligned}$$

One can also express this solution in the following form:

$$\overline{w} = -Q^{-1}\overline{p} + \underbrace{Q^{-1}A^T[AQ^{-1}A^T]^{-1}[\overline{b} + AQ^{-1}\overline{p}]}_{\text{Adjustment caused by constraints}} \tag{6.5}$$

As discussed in Section 4.6.2.1 of Chapter 4 (with different notations), the solution to the unconstrained version of the problem is $-Q^{-1}\overline{p}$. This is the same as the first part of the above solution. The second part of the above solution is the adjustment caused by the equality constraints. It is noteworthy that the adjustment contains $\overline{b} - A[\overline{z}]$ as a factor, where $\overline{z} = -Q^{-1}\overline{p}$ is the solution to the unconstrained problem. In other words, the adjustment from the unconstrained solution also depends directly on how far the unconstrained solution is from feasibility.

6.2.1.2 Application: Linear Regression with Equality Constraints

The fact that one can find a closed-form solution to the problem of convex quadratic programming with equality constraints implies that one can also solve the problem of least-squares regression with equality constraints. After all, the objective function of linear regression is a convex quadratic function as well. Consider an $n \times d$ data matrix D containing the feature variables, and an n-dimensional response vector $\overline{y}$. Assume that we have some domain-specific insight about the data because of which the d-dimensional coefficient vector $\overline{w}$ is subject to the linear system of constraints $A\overline{w} = \overline{b}$. Here, A is an $m \times d$ matrix with

$m \leq d$ and $\overline{b}$ is an m-dimensional vector. In such a case, the optimization problem may be expressed as follows:

$$\text{Minimize } J(\overline{w}) = \frac{1}{2}\|D\overline{w} - \overline{y}\|^2 + \frac{\lambda}{2}\|\overline{w}\|^2$$
$$\text{subject to:}$$
$$A\overline{w} = \overline{b}$$

This objective function is *exactly* in the same form as the convex quadratic program of Section 6.2.1.1. This implies that we can use the closed-form solution of Equation 6.5. The key point is to able to transform the problem to the same form. We leave this transformation as an exercise.

Problem 6.2.2 *Show that one can express the solution to equality-constrained linear regression in the same form as the solution to the quadratic optimization formulation of Section 6.2.1.1 by using $Q = D^T D + \lambda I$ and $\overline{p} = \overline{D}^T \overline{y}$ in Equation 6.5.*

6.2.1.3   Application: Newton Method with Equality Constraints

One can adapt the Newton method to any convex function with linear equality constraints (even if the objective function is not quadratic). The overall idea is the same as that discussed in Chapter 5. Consider the case where we are trying to minimize the *arbitrary* convex function $J(\overline{w})$ subject to the equality constraints $A\overline{w} = \overline{b}$. Here, A is an $m \times d$ matrix, and $\overline{w}$ is a d-dimensional vector of optimization variables. The Newton method first initializes $\overline{w} = \overline{w}_0$ to a feasible point on the hyperplane $A\overline{w} = \overline{b}$. Then, we start with $t = 0$ and perform the following steps iteratively:

1. Compute the second-order Taylor approximation of the function $J(\overline{w})$ centered at $\overline{w} = \overline{w}_t$ (cf. Section 1.5.1 of Chapter 1).

2. Compute $\overline{w}_{t+1}$ using Equation 6.5 on the Taylor approximation.

3. Increment t by 1 and go to step 1.

Note that the second-order Taylor approximation can always be expressed in the form of Equation 6.5, and therefore its closed-form solution can be plugged in directly. This iterative approach can converge to the optimal solution in fewer steps than gradient descent.

6.2.2   Linear Inequality Constraints

Linear inequality constraints are much harder to address than linear equality constraints. This is because one can no longer use Gaussian elimination to get rid of sets of variables and constraints simultaneously. Inequality constraints are handled by formulating the *conditional gradient optimization problem*. Consider the case where $\overline{w}_t$ is the current value of the parameter vector, and one wishes to move to a new value $\overline{w}_{t+1}$ that reduces the objective function as much as possible while satisfying the feasibility constraints. This value of $\overline{w}_{t+1}$ is approximately obtained by using an objective function based on the first-order Taylor expansion:

$$\overline{w}_{t+1} = \text{argmin}_{\overline{w}} \underbrace{F(\overline{w}_t) + [\nabla F(\overline{w}_t)] \cdot [\overline{w} - \overline{w}_t]}_{\text{First-order Taylor expansion}}$$

subject to:

$$A\overline{w} \le \overline{b}$$

Here, it is important to note that we are solving one optimization problem as a subproblem of another; clearly, the subproblem has to be simple for the approach to make sense. As it turns out, this subproblem is indeed much easier than the original problem because it is a *linear programming problem*; it has a linear objective function and linear constraints. Such problems can be solved efficiently with off-the-shelf solvers, and we refer the reader to [16] for an introduction to linear optimization. Therefore, the conditional gradient method simply solves the above optimization problem repeatedly to convergence.

The main issue with the above optimization problem is that minimizing the objective function does not necessarily lead to the optimum point, as we are using the *instantaneous* gradient at $\overline{w}_t$ in order to determine $\overline{w}_{t+1}$. Obviously, the gradient will change as we move from $\overline{w}_t$ to $\overline{w}_{t+1}$, and the objective function might even start worsening as one approaches $\overline{w}_{t+1}$. This problem can be partially addressed as follows. We first solve the above optimization problem to find a *tentative* value of $\overline{w}_{t+1}$. At this point, we only obtain a *direction* of movement $\overline{q}_t = \overline{w}_{t+1} - \overline{w}_t$. Subsequently, the update is modified to $\overline{w}_t + \alpha_t \overline{q}_t$, where α_t is selected using line search. However, in this case, α_t would need to selected to ensure both feasibility and an optimum solution.

6.2.2.1 The Special Case of Box Constraints

Box constraints arise frequently in machine learning, and they represent a special case of linear constraints of the form $A\overline{w} \le \overline{b}$. All box constraints are of the form $l_i \le w_i \le u_i$. Therefore, the feasible region is a hypercube in d-dimensional space (although the box might be an open set when $l_i = -\infty$ or $u_i = \infty$). Box constraints are relatively easy to handle because of the ease in projecting an infeasible solution to its nearest point on a box. We show examples of violation of box constraints in Figure 6.3. In each case, the closest point on the box is obtained by simply setting the variable values of the violated constraints to the bounds that are violated. Figure 6.3 shows two cases. In the first case, only one constraint is violated by a gradient-descent step, whereas in the second case, two constraints are violated by a gradient descent step. In each case, the variable values for the violated constraints are set to their bounds by the projection step. Therefore, the computational algorithm is as follows:

1. Perform the gradient-descent step $\overline{w} \Leftarrow \overline{w} - \alpha [\nabla F(\overline{w})]$.

2. Find the components in $\overline{w}$ for which the interval bounds (box constraints) are violated, and set the component value to the end-point of the interval that is violated.

The above two steps are applied iteratively to convergence. One must take care to select the initialization points within the feasible box.

Problem 6.2.3 (Linear Regression with Box Constraints) *The linear regression problem optimizes the following objective function:*

$$J = \|D\overline{w} - \overline{y}\|^2$$

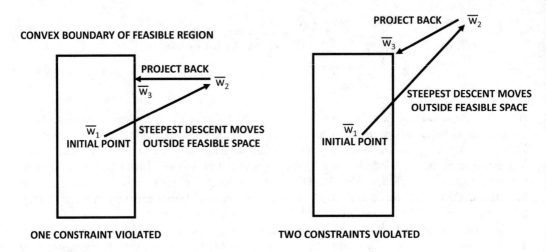

Figure 6.3: Violation of box constraints

Here, D is an n × d data matrix, $\overline{w} = [w_1 \dots w_d]^T$ contains the optimization variables, and $\overline{y}$ is an n-dimensional column vector of response variables. Suppose that we add box constraints of the form $l_i \leq w_i \leq u_i$ for each optimization variable w_i. Discuss how you will apply projected gradient descent in this case.

The dual problem for support vector machines is also a convex optimization problem with box constraints. This problem is discussed in Section 6.4.4.1.

Problem 6.2.4 *Consider the problem in which you want to use the L_2-loss SVM as the objective function (see page 184). However, you have the additional domain-specific knowledge that all coefficients are nonnegative (possibly because of known positive correlations between features and class label). Discuss how you would solve the L_2-SVM optimization problem.*

6.2.2.2   General Conditions for Projected Gradient Descent to Work

Box constraints represent a simple case in which it is relatively easy to find the nearest points of projection by using the violated constraints. All that one needs to do is to set all the variables to their nearest feasible points on the box. In the case of box constraints, this amounts to finding the closest point to the current point, so that all the violated box constraints are satisfied at equality.

The success of this approach in the case of box constraints leads one to wonder whether one can apply this approach to the general case. In other words, consider a problem in which we are minimizing $J(\overline{w})$ subject to $A\overline{w} \leq \overline{b}$. Consider a situation where we have used unconstrained gradient descent to move from a feasible point $\overline{w}_t$ to a (possibly infeasible) point $\overline{w}_{t+1}$. Suppose that the subset of the violated constraints is $A_v \overline{w} \leq \overline{b}_v$, where A_v and $\overline{b}_v$ are respectively obtained by extracting the corresponding rows from A and $\overline{b}$. As in the case of box constraints, can we simply find the closest point $\overline{w}$ to $\overline{w}_{t+1}$ that satisfies the violated constraints $A_v \overline{w} = \overline{b}_v$ at equality? Unfortunately, this is not the case when the rows of A_v are *linearly dependent*. Box constraints have the property that all the violated hyperplanes are mutually orthogonal; therefore, this situation does not arise. Furthermore, this problem also does not arise in linear equality constraints where one can use Gaussian elimination to

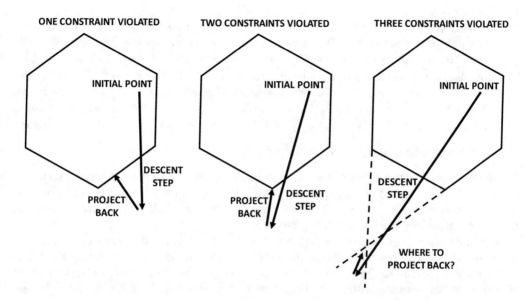

Figure 6.4: Linearly dependent violations can create challenges

remove linearly dependent constraints. Unfortunately, such elimination methods cannot be used when working with linear inequality constraints.

In order to understand why linearly dependent inequality constraints can create challenges, we will use an example of a convex region in 2-dimensional space, which is created by six inequality constraints. This convex region is illustrated in Figure 6.4. Note that any set of three constraints are linearly dependent, when working in 2-dimensional space. As shown in Figure 6.4, it is possible to project back to the closest point on the intersection of the violated constraints when these constraints are linearly independent. This situation corresponds to the left and the middle example of Figure 6.4. However, at the right side of Figure 6.4, we have a case where three constraints are violated, and these constraints are linearly dependent. Unfortunately, the constraints do not intersect, and selecting only two of these constraints leads to an infeasible solution. In general, we can make the following observation:

Observation 6.2.2 *The optimization problem of minimizing $J(\overline{w})$ subject to $A\overline{w} \leq \overline{b}$ is much simpler to solve when the rows of A are linearly independent. One can simply use projected gradient descent by first making an unconstrained gradient descent step, identifying the violated constraints, and projecting to the closest point that satisfies the violated constraints with strict equality.*

In other words, we start from a feasible point $\overline{w}_0$, and then make the following gradient-descent steps starting with $t = 0$:

1. Make the step $\overline{w}'_{t+1} \Leftarrow \overline{w}_t - \alpha_t \nabla J(\overline{w}_t)$. Here, α_t is the step-size.

2. Extract the violated constraints $A_v \overline{w} \leq \overline{b}_v$. We assume that the rows of A_v are linearly independent because the rows of A are linearly independent.

3. Update $\overline{w}_{t+1} \Leftarrow \overline{w}'_{t+1} + A_v^T (A_v A_v^T)^{-1} [\overline{b}_v - A_v \overline{w}'_{t+1}]$. Note that $A_v \overline{w}_{t+1}$ can be shown to be exactly equal to $\overline{b}_v$ by multiplying both sides of the above equation by A_v. This

update can also be derived by applying an origin translation to $\overline{w}'_{t+1}$ in order to use the right-inverse results of Section 2.8 in Chapter 2; then one can add back $\overline{w}'_{t+1}$. We need to translate the origin to $\overline{w}'_{t+1}$ because we want to find the closest point to $\overline{w}'_{t+1}$ on $A_v \overline{w} = \overline{b}_v$, whereas the right-inverse in Section 2.8 finds the most concise solution to $A_v \overline{w} = \overline{b}_v$ (i.e., closest point to the origin). However, translating the origin in this way transforms the vector $\overline{b}_v$ to $[\overline{b}_v - A_v \overline{w}'_{t+1}]$, and therefore the weight vector in translated space is $A_v^T (A_v A_v^T)^{-1} [\overline{b}_v - A_v \overline{w}'_{t+1}]$. Adding back $\overline{w}'_{t+1}$ yields the update.

4. Increment t by 1 and go back to step 1.

These steps are iterated to convergence. Here, a key point is that the projection step does not result in violation of the other (already satisfied) constraints. This is because the nearest point in a convex set is guaranteed to lie on the intersection of all the violated constraints, when the constraints are linearly independent.

A key question arises as to how one can use the approach when the rows of the matrix A are not linearly independent. Here, an important observation is that we only need each violated set A_v to contain linearly independent rows rather than the much stronger criterion of requiring this from the full set A. Therefore, the approach will often work even in cases where there is a modest level of linear dependence between rows of A, and one never encounters any matrix A_v containing linearly dependent rows. One way of discouraging the rows of A_v to be linearly independent is to use line search on α_t, and restrict the step-size so that the violated constraints are never linearly dependent. With this modification, the aforementioned approach can be used directly. However, convergence to an optimal solution is not guaranteed by such an approach, although the approach tends to work well in practice.

6.2.2.3 Sequential Linear Programming

So far this section has only considered the case where we have *linear* inequality constraints of the form $A\overline{w} \leq \overline{b}$. However, what happens in cases where the constraints are not linear, but they might be arbitrary, convex constraints of the form $f_i(\overline{w}) \leq 0$ for $i \in \{1 \ldots m\}$. The objective function $F(\overline{w})$ is assumed to be convex. In such a case, one can linearize not only the objective function, but also the constraints. In other words, we use the first-order Taylor expansion of *both* the objective function and the constraints. Therefore, if the current feasible solution to the problem is $\overline{w}_t$, then one can pose the following linearized model for solving the problem:

$$\overline{w}_{t+1} = \text{argmin}_{\overline{w}} \; \underbrace{F(\overline{w}_t) + [\nabla F(\overline{w}_t)] \cdot [\overline{w} - \overline{w}_t]}_{\text{First-order Taylor expansion}}$$

subject to:

$$\underbrace{f_i(\overline{w}_t) + \nabla f_i(\overline{w}_t)[\overline{w} - \overline{w}_t]}_{\text{Taylor expansion}} \leq 0, \quad \forall i \in \{1 \ldots m\}$$

One problem with this approach is that the linear constraints need not be a bounded convex region. For example, if the constraint is of the form $\overline{w}^2 \leq 1$ (which is a bounded circle of radius one), then its linearized approximation is $\overline{w}_t^2 + 2\overline{w}_t(\overline{w} - \overline{w}_t) \leq 1$. In other words, the linearized constraint is simply the tangent to the concentric circle passing through $\overline{w}_t$ and the side containing the center of the circle (which is the origin in this case) is included as the feasible space. Depending on the nature of the objective function, the solution to the subproblem might be unbounded because of feasible region on one side of the tangent is unbounded. One can handle this issue in several ways, such as adding additional box constraints in order to limit the step-size. However, even adding box constraints might sometimes result in a value of $\overline{w}_{t+1}$ that does not satisfy the original constraints. In such cases, one possible solution is to perform a linear search on the region between $\overline{w}_t$ and $\overline{w}_{t+1}$ and reduce the step size, so that the solution stays feasible. There are, however, many other ways in which these issues are handled, and we refer the reader to [99] for a detailed discussion.

6.2.3 Sequential Quadratic Programming

Sequential quadratic programming is the natural generalization of sequential linear programming, in which the second-order Taylor expansion is used at each point, instead of the first-order Taylor expansion in the objective function. Furthermore, the constraints are *linearized* in order to keep the problem reasonably simple. The solution to quadratic programs with linear constraints is relatively simple, if one were to use techniques for Lagrangian relaxation. Such methods are discussed later in this chapter, and Exercise 7 provides a path to applying these techniques in quadratic programs.

Consider an optimization problem in which we are trying to minimize the convex function $F(\overline{w})$, subject to the convex constraints $f_i(\overline{w}) \leq 0$ for $i \in \{1 \ldots m\}$. Also assume that we have equality constraints of the form $h_i(\overline{w}) = 0$ for $i \in \{1 \ldots k\}$. Then, the second-order approximation to the problem is as follows:

$$\overline{w}_{t+1} = \operatorname{argmin}_{\overline{w}} \underbrace{F(\overline{w}_t) + [\nabla F(\overline{w}_t)] \cdot [\overline{w} - \overline{w}_t] + [\overline{w} - \overline{w}_t]^T H_F^t [\overline{w} - \overline{w}_t]}_{\text{Second-order Taylor expansion}}$$

subject to:

$$\underbrace{f_i(\overline{w}_t) + \nabla f_i(\overline{w}_t)[\overline{w} - \overline{w}_t]}_{\text{First-order Taylor expansion}} \leq 0, \quad \forall i \in \{1 \ldots m\}$$

$$\underbrace{h_i(\overline{w}_t) + \nabla h_i(\overline{w}_t)[\overline{w} - \overline{w}_t]}_{\text{First-order Taylor expansion}} = 0, \quad \forall i \in \{1 \ldots k\}$$

Here, H_F^t represents the Hessian of $F(\cdot)$ at the point $\overline{w}_t$. This Hessian is positive semidefinite, since we are only dealing with convex functions. If the Hessian H_F^t is positive definite, the problem will have a bounded global minimum even without constraints. Although quadratic programs are harder to solve as subproblems than linear programs, they are much easier to solve than many other linear programs (see Exercise 7). Many of the methods discussed in later sections (such as Lagrangian relaxation) can be used for solving convex quadratic programs effectively. The main issue is that the solution to the linearized problem may not be feasible for the original constraints to the problem. We refer the reader to [21, 99] for a detailed discussion of solution methods. In particular, a practical line-search method discussed by [99] is very useful in this context.

6.3 Primal Coordinate Descent

The coordinate descent method is discussed in Section 4.10 of Chapter 4. The basic idea in coordinate descent is to perform the optimization one variable at a time. Consider an objective function $F(\overline{w})$, which is a function of a d-dimensional vector of variables. In

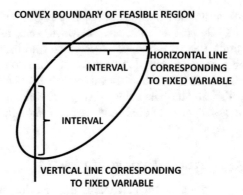

Figure 6.5: Fixing variables results in an interval constraint over remaining variable when the feasible region is convex

coordinate descent, we optimize a single variable w_i from the vector $\overline{w}$, while holding all the other parameters fixed to their values $\overline{w}^t$ in the tth iteration. This leads to the following update in the tth iteration:

$$\overline{w}^{t+1} = \text{argmin}_{[i\text{th component of } \overline{w} \,]} \; F(\overline{w}) \quad [\text{All parameters except } w_i \text{ are fixed to } \overline{w}^t]$$

Here, i is the index of the ith variable, and other variables are fixed to the corresponding values in $\overline{w}^t$. One cycles through the variables one at a time, until convergence is achieved. For example, if no improvement occurs during a cycle of optimizing each variable, then it means that the solution is a global optimum. In block coordinate descent, a block of variables is optimized at a given time, and one cycles through the different blocks one at a time.

Coordinate descent is particularly suitable for constrained optimization. This is because the variable-at-a-time optimization significantly simplifies the structure of the resulting sub-problem; in fact, the problem reduces to the univariate case. Although block coordinate descent does not yield univariate optimization problems, it still results in significant simplification. Very often, the constraints that tie together different variables can be dropped in an iteration, since some of the variable values are fixed in an iteration. A specific example of this situation is the k-means algorithm discussed in Section 4.10.3 of Chapter 4.

6.3.1   Coordinate Descent for Convex Optimization Over Convex Set

Coordinate descent reduces a multivariate optimization problem into a sequence of univariate optimization problems. When using coordinate descent over a convex set, a very useful observation is that *any univariate convex set is a continuous interval*, and the corresponding variable w can be expressed in the form of the *box constraint* $l_i \leq w \leq u_i$. This fact follows from the fact that a convex set is defined as any set such that any line passing through it must have exactly one continuous region belonging to the set. Therefore, if a horizontal or vertical line is passed through a convex set, as shown in Figure 6.5, the feasible region already corresponds to a continuous interval.

For example, consider the case where we are trying to optimize some function $F(w_1, w_2, w_3)$, over a feasible region in 3-dimensions. This 3-dimensional region is defined by the following constraints:

$$w_1^2 - w_1 \cdot w_2 + w_2^2/4 + 3w_2 \cdot w_3 + 4w_3^2 \leq 4$$
$$2w_1 + w_2 - 3w_3 \leq 4$$

Note that the constraints are both quadratic and linear, and therefore the problem is more complex than the linear constraints considered in the previous section. Now consider the case in which one is performing coordinate descent, and we are trying to compute the optimum value w_1 so that $F(w_1, w_2, w_3)$ is minimized (while holding w_2 and w_3 fixed). The values of w_2 and w_3 are set to 2 and 0, respectively. Plugging in these values of w_2 and w_3, we obtain the following pair of constraints:

$$w_1^2 - 2w_1 - 3 = (w_1 - 3)(w_1 + 1) \leq 0$$
$$w_1 \leq 1$$

Note that the first constraint implies that $w_1 \in [-1, 3]$ and the second constraint implies that $w_1 \in (-\infty, 1]$. Therefore, by combining the constraints, we obtain the fact that the variable w_1 must lie in $[-1, +1]$. Furthermore, the objective function can be simplified to $G(w_1) = F(w_1, 2, 0)$. Therefore, the subproblem reduces to optimizing a univariate convex function $G(w_1)$ over an interval.

How does one optimize a univariate convex function over an interval? One possibility is to simply set the derivative of the convex function (with respect to the only variable w being optimized) to 0, and obtain a value of the variable w by solving the resulting equation. At this point, one must check the two ends of the interval in order to check whether the optimum lies at one of the two ends. The reason that one is able to use this simple approach is because of the convexity of the optimization function. Alternatively, one can use the line search methods discussed in Section 4.4.3 of Chapter 4. One cycles through the variables using this iterative approach, until convergence is reached.

Depending on the structure of the objective function and optimization variables, the univariate subproblem in coordinate descent often has a very simple structure. Therefore, even when one is faced with an arbitrarily complex problem, it is worthwhile trying ideas from coordinate descent for the purposes of optimization. In some cases, coordinate descent can even provide good heuristic solutions to difficult optimization problems like *mixed integer programs*. This is because the subproblems are often much easier to solve than the original formulation. A specific example is the case of the k-means algorithm, which has integer constraints on the variables (cf. Section 4.10.3 of Chapter 4). However, there are also cases in which coordinate descent fails (see Exercise 19).

6.3.2 Machine Learning Application: Box Regression

The box regression problem is an enhancement of the linear regression problem in which constraints are added to the regression variables. As evident from Problem 6.2.3, the box regression problem can be addressed using projected gradient descent methods. In this section, we address this problem with the use of coordinate descent.

The linear regression problem with box constraints can be posed as follows:

$$\text{Minimize } J = \frac{1}{2}\|D\overline{w} - \overline{y}\|^2 + \frac{\lambda}{2}\|\overline{w}\|^2$$
$$\text{subject to:}$$
$$l_i \leq w_i \leq u_i, \quad \forall i \in \{1 \ldots d\}$$

Here, D is an $n \times d$ matrix of feature values, $\overline{w} = [w_1, \ldots, w_d]^T$ is a d-dimensional vector of coefficients, and $\overline{y} = [y_1 \ldots y_n]^T$ is an n-dimensional vector of response values.

In the case of unconstrained linear regression, the value of w_i is updated using the following formula (cf. Problem 4.10.1 of Chapter 4):

$$w_i \Leftarrow \frac{w_i \|\overline{d}_i\|^2 + \overline{d}_i^T \overline{r}}{\|\overline{d}_i\|^2 + \lambda}$$

Here, $\overline{r} = \overline{y} - D\overline{w}$ is the n-dimensional vector of residuals. In this case, the only difference is that we use the additional *truncation operator* $T_i(\cdot)$ after each coordinate descent step in order to bring the variable back into the relevant bounds.

$$w_i \Leftarrow T_i \left[\frac{w_i \|\overline{d}_i\|^2 + \overline{d}_i^T \overline{r}}{\|\overline{d}_i\|^2 + \lambda} \right]$$

Here, the truncation operator $T_i(\cdot)$ is defined as follows:

$$T_i(x) = \begin{cases} l_i & x < l_i \\ x & l_i \leq x \leq u_i \\ u_i & u_i < x \end{cases}$$

In other words, each coordinate is immediately truncated to its lower and upper bounds after the coordinate update. We also make the following observation:

Observation 6.3.1 *Nonnegative least-squares regression is a special case of box regression in which all coefficients have a lower bound of zero, but no upper bound. Nonnegative regression can be directly implemented as a special case of the above algorithm.*

6.4 Lagrangian Relaxation and Duality

Lagrangian relaxation is an approach whereby the constraints of an optimization problem are relaxed, while penalizing their violation within the objective function. The magnitudes of the penalties depend on factors referred to as the *Lagrange multipliers*. For a minimization problem, the Lagrangian relaxation always provides a lower bound on the optimal solution, no matter what the value of the Lagrange multipliers might be. A key point is that for certain types of optimization problems, such as convex objective functions with convex constraints, the exact optimal solution to the original solution can be obtained with the appropriate choice of multipliers on the relaxed problem.

Consider a minimization problem of the following form:

$$P = \text{Minimize } F(\overline{w})$$

$$\text{subject to:}$$

$$f_i(\overline{w}) \leq 0, \quad \forall i \in \{1 \ldots m\}$$

This problem is referred to as the *primal problem* in optimization parlance, and we introduce the notation P to denote its optimal solution. The Lagrangian relaxation methodology is particularly useful when the functions $F(\overline{w})$ and each $f_i(\overline{w})$ are convex. The Lagrangian relaxation is defined with the use of *nonnegative* Lagrangian multipliers $\overline{\alpha} = [\alpha_1 \ldots \alpha_m]^T$:

$$L(\overline{\alpha}) = \text{Minimize}_{\overline{w}} \; F(\overline{w}) + \sum_{i=1}^{m} \alpha_i f_i(\overline{w})$$

subject to:

No constraints on $\overline{w}$

We have introduced the notation $L(\overline{\alpha})$ to indicate the solution to the relaxed problem at any particular value of the parameter vector $\overline{\alpha}$. Note that the minimization is *only* with respect to the parameters in $\overline{w}$ and not the parameters in $\overline{\alpha}$, which is fixed (and therefore a part of the argument of $L(\overline{\alpha})$). It is important to note that each α_i is nonnegative to ensure that violations of the constraints are penalized. When a constraint is violated, we will have $f_i(\overline{w}) > 0$, and the penalty $\alpha_i f_i(\overline{w})$ will also be nonnegative. Although $L(\overline{\alpha})$ is defined over any value of $\overline{\alpha}$, it makes sense to consider only nonnegative values of $\overline{\alpha}$. For example, if the value of α_i is negative, then violation of the ith constraint will be rewarded.

In the case of equality constraints, the Lagrange multipliers do not have any nonnegativity constraints. Consider the following equality-constrained optimization problem:

Minimize $F(\overline{w})$

subject to:

$$f_i(\overline{w}) = 0, \quad \forall i \in \{1 \ldots m\}$$

Each equality constraint can be converted to a pair of inequality constraints $f_i(\overline{w}) \leq 0$ and $-f_i(\overline{w}) \leq 0$ with nonnegative Lagrangian multipliers $\alpha_{i,1}$ and $\alpha_{i,2}$, respectively. Then, the Lagrangian relaxation contains terms of the form $f_i(\overline{w})(\alpha_{i,1} - \alpha_{i,2})$. One can instead treat $\alpha_i = \alpha_{i,1} - \alpha_{i,2}$ as the *sign unconstrained* Lagrange multiplier. Most of the discussion in this chapter will, however, be centered around inequality constraints.

Let us examine why the Lagrangian relaxation problem provides a lower bound on the solution to the original optimization problem. Let $\overline{w}^*$ be the optimal solution to the original optimization problem, and $\overline{\alpha}$ be any nonnegative vector of Lagrangian parameters. Since $\overline{w}^*$ is also a feasible solution to the original problem, it follows that each $f_i(\overline{w}^*)$ is no larger than zero. Therefore, the "penalty" $\alpha_i f_i(\overline{w}^*) \leq 0$. In other words, the penalties can become rewards for primal-feasible solutions like w^*, if the penalties are non-zero. Therefore, we have:

$$L(\overline{\alpha}) = \text{Minimize}_{\overline{w}} \; F(\overline{w}) + \sum_{i=1}^{m} \alpha_i f_i(\overline{w})$$

$$\leq F(\overline{w}^*) + \underbrace{\sum_{i=1}^{m} \alpha_i f_i(\overline{w}^*)}_{\leq 0} \quad [w^* \text{ might not be optimal for relaxation}]$$

$$\leq F(\overline{w}^*) = P$$

In other words, the value of $L(\overline{\alpha})$ for *any* nonnegative vector $\overline{\alpha}$ is always no larger than the *optimal* solution to the primal. One can tighten this bound by maximizing $L(\overline{\alpha})$ over all nonnegative $\overline{\alpha}$ and formulating the *dual problem* with objective function D:

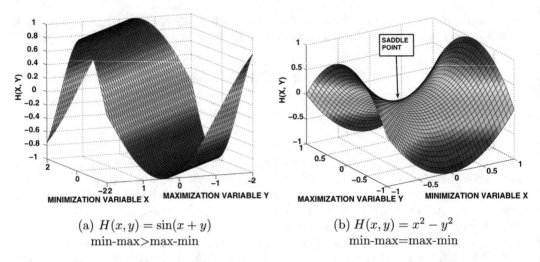

(a) $H(x, y) = \sin(x + y)$
min-max>max-min

(b) $H(x, y) = x^2 - y^2$
min-max=max-min

Figure 6.6: Examples of two minimax functions with a single minimization variable and a single maximization variable. The first is neither concave nor convex in either variable. The second is convex in the minimization variable and concave in the maximization variable, and has a well-defined saddle point

$$D = \text{Maximize } _{\overline{\alpha} \geq 0} \ L(\overline{\alpha})$$

$$= \text{Maximize} _{\overline{\alpha} \geq 0} \text{Minimize} _{\overline{w}} \left[F(\overline{w}) + \sum_{i=1}^{m} \alpha_i f_i(\overline{w}) \right]$$

We summarize the relationship between the primal and the dual as follows:

$$D = L(\overline{\alpha}^*) \leq P$$

This result is referred to as that of *weak duality*. It is noteworthy that the Lagrangian optimization problem is a *minimax* problem containing disjoint minimization and maximization variables. The minimization and maximization is done in a specific order. *The ordering of the minimization and maximization for any minimax optimization problem does matter.*

Problem 6.4.1 *Consider the 2-dimensional function* $G(x, y) = sin(x + y)$. *Show that* $min_x max_y G(x, y) = 1$ *and* $max_y min_x G(x, y) = -1$.

The ordering effects of minimization and maximization in minimax problems can be formalized in terms of John von Neumann's *minimax theorem* [37] in mathematics. It states that "min-max" is an upper bound on "max-min" of a function containing both minimization and maximization variables. Furthermore, strict equality occurs when the function is convex in its minimization variables and also concave in the maximization variables. For example, the function $H(x, y) = \sin(x + y)$ is neither concave nor convex in either x or y. The corresponding plot is shown in Figure 6.6(a). As shown in Problem 6.4.1, the order of minimization and maximization matters in this case. On the other hand, the function $H(x, y) = x^2 - y^2$ is convex in the minimization variable x and concave in the maximization variable y. This function is shown in Figure 6.6(b). Therefore, this function has a single *saddle point*, which is the optimal solution to both minimax problems.

Armed with this understanding of the importance of ordering of minimization and maximization in minimax problems, we revisit the effect of this ordering on the Lagrangian relaxation. We denote the minimax optimization function of Lagrangian relaxation as $H(\overline{w}, \overline{\alpha})$:

$$H(\overline{w}, \overline{\alpha}) = F(\overline{w}) + \sum_{i=1}^{m} \alpha_i f_i(\overline{w}) \tag{6.6}$$

Here, $\overline{w}$ contains the minimization variables and $\overline{\alpha}$ contains the maximization variables. While the dual computes $\max_{\overline{\alpha} \geq 0} \min_{\overline{w}} H(\overline{w}, \overline{\alpha})$ (which is a lower bound on the primal), *reversing the order to $\min_{\overline{w}} \max_{\overline{\alpha} \geq 0} H(\overline{w}, \overline{\alpha})$ always yields the original (primal) optimization problem irrespective of whether the original problem has a convex objective function or convex constraints.* We summarize this result below:

Lemma 6.4.1 (Minimax Primal Formulation) *Let $H(\overline{w}, \overline{\alpha})$ of Equation 6.6 represent the Lagrangian relaxation of the unrelaxed primal formulation with constraints. Then, the unconstrained minimax problem $\min_{\overline{w}} \max_{\overline{\alpha} \geq 0} H(\overline{w}, \overline{\alpha})$ is equivalent to the original, unrelaxed primal formulation irrespective of the convexity structure of the original problem.*

Proof: Consider the Lagrangian objective function $H(\overline{w}, \overline{\alpha})$ of Equation 6.6. Then, the value of $\max_{\overline{\alpha} \geq 0} H(\overline{w}, \overline{\alpha})$ is ∞ at any fixed value of $\overline{w}$ that violates one or more of the original primal constraints. This is achieved by setting the corresponding α_i of the violated constraint to ∞. Therefore, the primal problem of $\min_{\overline{w}} \max_{\overline{\alpha} \geq 0} H(\overline{w}, \overline{\alpha})$ will never yield a solution for $\overline{w}$ at (minimax) optimality that violates constraints of the form $f_i(\overline{w}) \leq 0$. In other words, minimax optimality of $\min_{\overline{w}} \max_{\overline{\alpha} \geq 0} H(\overline{w}, \overline{\alpha})$ always yields solutions for $\overline{w}$ satisfying each $f_i(\overline{w}) \leq 0$.

For any value of $\overline{w}$ satisfying each $f_i(\overline{w}) \leq 0$, the contribution of the penalty term to $H(\overline{w}, \overline{\alpha})$ is non-positive because $\alpha_i f_i(\overline{w}) \leq 0$ for each i. Therefore, for any such fixed value of $\overline{w}$ satisfying primal constraints, the function $H(\overline{w}, \overline{\alpha})$ will be maximized with respect to $\overline{\alpha}$ only when the value of α_i is set to zero for each i satisfying $f_i(\overline{w}) < 0$. This ensures that the corresponding value of $\alpha_i f_i(\overline{w})$ is zero, and therefore the contribution of the penalty term $\sum_{i=1}^{m} \alpha_i f_i(\overline{w})$ to $H(\overline{w}, \overline{\alpha})$ is 0 at minimax optimality.

The above two facts imply that the optimization of $F(\overline{w})$ with respect to the primal constraints is the same problem as $\min_{\overline{w}} \max_{\overline{\alpha}} H(\overline{w}, \overline{\alpha})$. At optimality of the second problem, the primal constraints are satisfied, and the objective function is the same as well (since the penalty contribution drops to 0). ∎

We make some key observations about the Lagrangian relaxation $H(\overline{w}, \overline{\alpha})$ of Equation 6.6:

1. **Dual is a minimax problem:** The dual problem of Lagrangian optimization is based on the relaxation of Equation 6.6 in which the minimax optimization is done in a specific order:

$$D = \max_{\overline{\alpha} \geq 0} \min_{\overline{w}} H(\overline{w}, \overline{\alpha}) \tag{6.7}$$

2. **Primal is a minimax problem of the same objective function as dual (but in different order):** The unrelaxed primal formulation with constraints can also be expressed in terms of minimax optimization of the function $H(\overline{w}, \overline{\alpha})$ of Equation 6.6, but in a different order than the dual:

$$P = \min_{\overline{w}} \max_{\overline{\alpha} \geq 0} H(\overline{w}, \overline{\alpha}) \tag{6.8}$$

3. **Duality results of Lagrangian relaxation can be derived from the more general minimax theorem in mathematics:** The weak duality result that $D \leq P$ can

also be derived from John von Neumann's minimax theorem of optimization [37]. The minimax theorem of optimization is designed for general minimax functions containing a disjoint set of minimization and maximization variables (of which the Lagrangian relaxation is a special case). The theorem states that max-min is always bounded above by min-max, which implies that $D \leq P$. Furthermore, the minimax theorem also states that strict equality $D = P$ occurs when the optimization function is convex in the minimization (primal) variables and concave in the maximization (dual) variables.

What types of optimization problems are such that their Lagrangian relaxations show strict equality between primal and dual solutions? First, the function $H(\overline{w}, \overline{\alpha})$ is linear in the maximization variables, and therefore concavity with respect to maximization variables is always satisfied. Second, the function $H(\overline{w}, \overline{\alpha})$ is a sum of $F(\overline{w})$ and nonnegative multiples of the various $f_i(\overline{w})$ for $i \in \{1 \ldots m\}$. Therefore, if $F(\overline{w})$ and each of $f_i(\overline{w})$ are convex in $\overline{w}$, then $H(\overline{w}, \overline{\alpha})$ will be convex in the minimization variables. This is the primary pre-condition for **strong duality**:

Lemma 6.4.2 (Strong Duality) *Consider the following optimization problem:*

$$P = Minimize\ F(\overline{w})$$
$$subject\ to:$$
$$f_i(\overline{w}) \leq 0, \qquad \forall i \in \{1 \ldots m\}$$

Let $F(\overline{w})$ and each $f_i(\overline{w})$ be convex functions. Then, the optimal objective function value of the dual problem created using Lagrangian relaxation is **almost always** *the same as that of the primal.*

We use the qualification "almost always," because we also need a relatively weak condition referred to as *Slater's condition*, which states that at least one strictly feasible point exists satisfying $f_i(\overline{w}) < 0$ for each i. For most machine learning problems, these conditions hold by default. For simplicity in presentation, we will drop this condition in the subsequent exposition. Many optimization problems in machine learning such as support vector machines and logistic regression satisfy strong duality.

6.4.1   Kuhn-Tucker Optimality Conditions

We start by repeating the primal and dual minimax optimization problems:

$$P = \min_{\overline{w}} \max_{\overline{\alpha} \geq 0} H(\overline{w}, \overline{\alpha}) \quad (OP1)$$
$$D = \max_{\overline{\alpha} \geq 0} \min_{\overline{w}} H(\overline{w}, \overline{\alpha}) \quad (OP2)$$

We refer to these primal and dual optimization problems as OP1 and OP2, respectively. We make the following observation, which is true irrespective of the convexity structure of the primal optimization problem:

> For a solution $(\overline{w}, \overline{\alpha})$ to be optimal to the primal minimax problem (OP1), $\overline{w}$ must be a feasible solution satisfying $f_i(\overline{w}) \leq 0$ for each i (see Lemma 6.4.1). Furthermore, if any constraint $f_i(\overline{w}) \leq 0$ is satisfied with strict inequality, then setting $\alpha_i = 0$ ensures maximization of (OP1) with respect to $\overline{\alpha}$. This ensures that we have $\alpha_i f_i(\overline{w}) = 0$ for each i for any optimal solution to (OP1).

The condition $\alpha_i f_i(\overline{w}) = 0$ is referred to as the *complementary slackness condition*. The (general) minimax theorem of mathematics tells us that the optimal pairs $(\overline{w}, \overline{\alpha})$ are the same in the two cases of the primal and the dual minimax problems [i.e., (OP1) and (OP2)], when the function $H(\overline{w}, \overline{\alpha})$ is convex in $\overline{w}$ and concave in $\overline{\alpha}$. Although we have shown the complementary slackness condition only for (OP1), any solution $(\overline{w}, \overline{\alpha})$ that is optimal for (OP1) must also be optimal for (OP2) and vice versa for problems with convex structure. Therefore, the complementary slackness condition must hold for both (OP1) and (OP2) in such problems. The primal constraints are of the form $f_i(\overline{w}) \leq 0$ and the corresponding dual constraints are of the form $\alpha_i \geq 0$. The complementary slackness condition implies that at most one of these complementary conditions can be "slack" (i.e., satisfied at strict inequality).

Another important condition that needs to be satisfied is that the gradient of $H(\overline{w}, \overline{\alpha})$ with respect to the primal variables $\overline{w}$ need to be set to 0 in the dual because we are minimizing this objective function at each fixed value of $\overline{\alpha}$. This leads to the *stationarity conditions*:

$$\nabla_{\overline{w}} H(\overline{w}, \overline{\alpha}) = \nabla F(\overline{w}) + \sum_{i=1}^{m} \alpha_i \nabla f_i(\overline{w}) = \overline{0}$$

The Kuhn-Tucker conditions are obtained by combining the primal feasibility conditions, dual feasibility conditions, complementary slackness conditions, and stationarity conditions. For convex objective functions, these represent the first-order conditions that are both necessary and sufficient for optimality:

Theorem 6.4.1 (Kuhn-Tucker Optimality Conditions) *Consider an optimization problem in which we wish to minimize the convex objective function $F(\overline{w})$, subject to convex constraints of the form $f_i(\overline{w}) \leq 0$ for $i \in \{1 \ldots m\}$. Then, a solution $\overline{w}$ is optimal for the primal and a solution $\overline{\alpha}$ is optimal for the dual, if and only if:*

- **Feasibility:** *$\overline{w}$ is feasible for the primal by satisfying each $f_i(\overline{w}) \leq 0$ and $\overline{\alpha}$ is feasible for the dual by being nonnegative.*

- **Complementary slackness:** *We have $\alpha_i f_i(\overline{w}) = 0$ for each $i \in \{1 \ldots m\}$.*

- **Stationarity:** *The primal and dual variables are related as follows:*

$$\nabla F(\overline{w}) + \sum_{i=1}^{m} \alpha_i \nabla f_i(\overline{w}) = 0$$

Note that one does not have to worry about second-order optimality conditions in the case of convex optimization problems. The Kuhn-Tucker optimality conditions are useful because they provide an alternative approach to solving the optimization problem by simply finding a feasible solution to a set of constraints as follows:

Observation 6.4.1 *For a convex optimization problem, any pair $(\overline{w}, \overline{\alpha})$ that satisfies primal feasibility $f_i(\overline{w}) \leq 0$, dual feasibility $\alpha_i \geq 0$, complementary slackness $\alpha_i f_i(\overline{w}) = 0$, and the stationarity conditions is an optimal solution to the original optimization problem.*

The stationarity conditions relate the primal and dual variables, and therefore they are often useful for eliminating primal variables from the Lagrangian. We will also refer to them as *primal-dual (PD) constraints*, because they relate primal and dual variables at optimality. The stationarity conditions are often used to formulate the minimax dual purely in terms of the dual variable (and therefore create a pure maximization problem). We discuss this general procedure in the next section.

6.4.2   General Procedure for Using Duality

The general procedure for using duality in constrained optimization is somewhat similar across problems. The first step is to formulate $L(\overline{\alpha})$ which is the objective function of the dual problem (OP2), after eliminating primal variables:

$$L(\overline{\alpha}) = \min_{\overline{w}} H(\overline{w}, \overline{\alpha}) \qquad (6.9)$$

The primal variables $\overline{w}$ can often be eliminated from $L(\overline{\alpha})$ by setting the gradients of $H(\overline{w}, \overline{\alpha})$ with respect to the primal variables $\overline{w}$ to zero. Setting the gradient with respect to primal variables to zero will result in exactly as many conditions as the number of primal variables. These are exactly the stationarity conditions of the previous section, which represent a subset of the Kuhn-Tucker optimality conditions. We also refer to these conditions as primal-dual (PD) constraints, because they relate the primal and dual variables. The (PD) constraints can be used to substitute for (and eliminate) the primal variables $\overline{w}$, and obtain a pure maximization objective function $L(\overline{\alpha})$, which is expressed in terms of $\overline{\alpha}$. In some cases, the feasibility and complementary slackness conditions are also used in the elimination process. At the end of the day, the process of generating the dual from the primal is almost purely a mechanical and algebraic process based on the Kuhn-Tucker conditions. While the specific mechanics might vary somewhat at the detailed level, the basic principle remains the same across different problems. In Section 6.4.3, we will provide an example of this procedure with the L_1-loss support vector machine. Furthermore, guided exercises (i.e., exercises broken up into simpler steps), are also available on the L_2-loss SVM and logistic regression, and the reader is advised to work them out in the same sequence as they occur.

6.4.2.1   Inferring the Optimal Primal Solution from Optimal Dual Solution

One needs to compute the optimal primal variables in order to have an interpretable solution. Therefore, a natural question arises as to how one can infer an optimal primal solution $\overline{w}$ from the optimal dual solution $\overline{\alpha}$. In this context, the (PD) constraints (i.e., the stationarity conditions) are very helpful, because they can be used to substitute in the values of the optimal dual variables and solve for the primal variables (although the algebraic approach might vary slightly across problems).

6.4.3   Application: Formulating the SVM Dual

In order to illustrate how duality is used in machine learning, we will revisit the support vector machine (SVM). We have already shown how the primal stochastic gradient descent approach can be used for the SVM in Section 4.8.2 of Chapter 4. We repeat the objective function of Equation 4.51:

$$J = \frac{1}{\lambda} \sum_{i=1}^{n} \max\{0, (1 - y_i[\overline{W} \cdot \overline{X}_i^T])\} + \frac{1}{2}\|\overline{W}\|^2 \quad \text{[Hinge-loss SVM]}$$

Note that this objective function is cosmetically different from Equation 4.51 by the scaling factor of $1/\lambda$. We have made this cosmetic adjustment because one often uses the notation corresponding to the slack penalty $C = 1/\lambda$ in the literature on dual SVM optimization, which is what we will use in subsequent restatements of this formulation. In order to create the dual, we would like to reformulate the problem as a constrained optimization problem, while simplifying the objective function without the maximization operator. This is achieved with the use of *slack variables* $\xi_1 \ldots \xi_n$ as follows:

$$\text{Minimize } J = \frac{1}{2}\|\overline{W}\|^2 + C\sum_{i=1}^{n}\xi_i$$

subject to:

$$\xi_i \geq 1 - y_i[\overline{W} \cdot \overline{X}_i^T] \quad \forall i \in \{1\ldots n\} \quad \text{[Margin Constraints]}$$
$$\xi_i \geq 0 \quad \forall i \in \{1\ldots n\} \quad \text{[Nonnegativity Constraints]}$$

Ideally, we would like $\xi_i = \max\{0, (1 - y_i[\overline{W} \cdot \overline{X}_i^T])\}$. Note that the constraints do allow values of ξ_i larger than $\max\{0, (1 - y_i[\overline{W} \cdot \overline{X}_i^T])\}$, but such values can never be optimal. The first set of constraints is referred to as the set of "margin" constraints, because they define the margins for the predicted values of y_i beyond which points are not penalized. For example, if $\overline{W} \cdot \overline{X}_i^T$ has the same sign as y_i and its absolute value is "sufficiently" positive by a margin of 1, ξ_i will drop to 0. Therefore, the point is not penalized. Strictly speaking, the constraints need to be converted to "$\leq$" form by multiplying with -1, but we can take care of it during the relaxation by multiplying the penalties with -1. We introduce the Lagrangian multiplier α_i for the ith of n margin constraints and the multiplier γ_i for the ith nonnegativity constraint on ξ_i. With these notations, the Lagrangian relaxation is as follows:

$$L_D(\overline{\alpha}, \overline{\gamma}) = \text{Minimize } J_r = \frac{1}{2}\|\overline{W}\|^2 + C\sum_{i=1}^{n}\xi_i - \underbrace{\sum_{i=1}^{n}\alpha_i(\xi_i - 1 + y_i(\overline{W} \cdot \overline{X}_i^T))}_{\text{Relax margin constraint}} - \underbrace{\sum_{i=1}^{n}\gamma_i\xi_i}_{\text{Relax } \xi_i \geq 0}$$

Here, J_r is the relaxed objective function. Since the relaxed constraints are inequalities, it follows that both α_i and γ_i must be nonnegative for the relaxation to make sense. Therefore, when we optimize over the dual variables such as α_i and γ_i, the optimization problem has a box constraint structure, which makes it somewhat simpler to solve. In this type of dual problem, one first minimizes over primal variables (with dual variables fixed) to obtain $L_D(\overline{\alpha}, \overline{\gamma})$ and then maximizes $L_D(\overline{\alpha}, \overline{\gamma})$ over the dual variables, while imposing box constraints on them. One can express this type of minimax optimization problem as follows:

$$L_D^* = \max_{\alpha_i, \gamma_i \geq 0} L_D(\overline{\alpha}, \overline{\gamma}) = \max_{\alpha_i, \gamma_i \geq 0} \min_{\overline{W}, \xi_i} J_r$$

As discussed in the previous section, the general approach to solving the dual is to use the (PD) constraints to eliminate the primal variables in order to create a pure maximization problem in terms of the dual variables. The (PD) constraints are obtained by setting the gradient of the minimax objective with respect to the primal variables to 0. This gives us exactly as many constraints as the number of primal variables, which is precisely what we need for eliminating all of them:

$$\frac{\partial J_r}{\partial \overline{W}} = \overline{W} - \sum_{i=1}^{n}\alpha_i y_i \overline{X}_i^T = \overline{0}, \quad \text{[Gradient with respect to } \overline{W} \text{ is 0]} \tag{6.10}$$

$$\frac{\partial J_r}{\partial \xi_i} = C - \alpha_i - \gamma_i = 0, \quad \forall i \in \{1\ldots n\} \tag{6.11}$$

The equations resulting from the partial derivatives with respect to ξ_i are independent of ξ_i, but the resulting equations are still useful in eliminating ξ_i from J_r. This is because the coefficient of ξ_i in J_r is $(C - \alpha_i - \gamma_i)$, which turns out to be 0 based on Equation 6.11. The ability to drop ξ_i is a direct result of the linearity of the J_r in ξ_i; the linear coefficient of ξ_i in J_r is also its derivative, which is set to 0 as an optimality condition. Furthermore, based on Equation 6.10, we can substitute $\overline{W} = \sum_{i=1}^{n} \alpha_i y_i \overline{X}_i^T$ everywhere it occurs in J_r. By dropping the terms involving ξ_i and substituting for $\overline{W}$, J_r is simplified as follows:

$$J_r = \frac{1}{2}\|\overline{W}\|^2 + \sum_{i=1}^{n} \alpha_i(1 - y_i(\overline{W} \cdot \overline{X}_i^T)), \quad \text{[Dropping terms with } \xi_i]$$

$$= \frac{1}{2}\|\sum_{j=1}^{n} \alpha_j y_j \overline{X}_j^T\|^2 + \sum_{i=1}^{n} \alpha_i(1 - y_i \sum_{j=1}^{n} \alpha_j y_j \overline{X}_i \cdot \overline{X}_j), \quad \text{[Substituting } \overline{W} = \sum_{j=1}^{n} \alpha_j y_j \overline{X}_j^T]$$

$$= \sum_{i=1}^{n} \alpha_i - \frac{1}{2}\sum_{i=1}^{n}\sum_{j=1}^{n} \alpha_i \alpha_j y_i y_j \overline{X}_i \cdot \overline{X}_j, \quad \text{[Algebraic simplification]}$$

This objective function is expressed purely in terms of the dual variables. Furthermore, the variable γ_i has dropped out of the optimization formulation. Nevertheless, the constraint $\gamma_i \geq 0$ also needs to be modified by substituting γ_i as $C - \alpha_i$ (cf. Equation 6.11):

$$\gamma_i = C - \alpha_i \geq 0$$

Therefore, the variables α_i satisfy the box constraints $0 \leq \alpha_i \leq C$. We can multiply the objective function by -1 in order to turn the maximization problem into a minimization problem:

$$\text{Minimize}_{0 \leq \overline{\alpha} \leq C} \frac{1}{2}\sum_{i=1}^{n}\sum_{j=1}^{n} \alpha_i \alpha_j y_i y_j \overline{X}_i \cdot \overline{X}_j - \sum_{i=1}^{n} \alpha_i$$

Beyond the fact that the dual problem (in minimization form) is *always* convex (see Exercise 12), one can show that the leading term in the quadratic is of the form $\overline{\alpha}^T H \overline{\alpha}$, where H is a positive semidefinite matrix of similarities between points. This makes the dual problem convex. To this effect, we assert the following result:

Observation 6.4.2 *The quadratic term $\sum_{i=1}^{n}\sum_{j=1}^{n} \alpha_i \alpha_j y_i y_j \overline{X}_i \cdot \overline{X}_j$ in the dual SVM can be expressed in the form $\overline{\alpha}^T B B^T \overline{\alpha}$, where B is an $n \times d$ matrix in which the ith row of B contains $y_i \overline{X}_i$. In other words, the ith row of B simply contains the ith data instance, after multiplying it with the class label $y_i \in \{-1, +1\}$.*

This result can be shown by simply expanding the (i, j)th term of $\overline{\alpha}^T B B^T \overline{\alpha}$. As shown in Lemma 3.3.14 of Chapter 3, matrices of the form $B B^T$ are always positive semidefinite. Therefore, this is a convex optimization problem.

6.4.3.1   Inferring the Optimal Primal Solution from Optimal Dual Solution

As discussed in Section 6.4.2.1, the (PD) constraints can be used to infer the primal variables from the dual variables. In the particular case of the SVM, the constraints correspond to Equations 6.10–6.11. Among these constraints, Equation 6.10 is in a particularly useful form, because it directly yields all the primal variables in terms of the dual variables:

$$\overline{W} = \sum_{i=1}^{n} \alpha_i y_i \overline{X}_i^T$$

One can obtain the slack variables ξ_i by using the constraints among the primal variables and substituting the inferred value of $\overline{W}$.

6.4.4 Optimization Algorithms for the SVM Dual

The dual is a constrained optimization problem, albeit a simple one because of the use of box constraints. The dual can be solved using almost all the primal optimization techniques discussed earlier in this chapter. Therefore, we still need the primal algorithms for constrained optimization, even though we are working with the dual! In the following, we provide some examples of computational algorithms.

6.4.4.1 Gradient Descent

We state the dual problem in minimization form with box constraints:

$$\text{Minimize } L_D = \frac{1}{2}\sum_{i=1}^{n}\sum_{j=1}^{n}\alpha_i\alpha_j y_i y_j \overline{X}_i \cdot \overline{X}_j - \sum_{i=1}^{n}\alpha_i$$

subject to:

$$0 \le \alpha_i \le C \quad \forall i \in \{1\ldots n\}$$

The partial derivative of L_D with respect to α_k is as follows:

$$\frac{\partial L_D}{\partial \alpha_k} = y_k \sum_{s=1}^{n} y_s \alpha_s \overline{X}_k \cdot \overline{X}_s - 1 \quad \forall k \in \{1\ldots n\} \tag{6.12}$$

One can use the standard gradient-descent procedure:

$$\overline{\alpha} \Leftarrow \overline{\alpha} - \eta\left[\frac{\partial L_D}{\partial \overline{\alpha}}\right]$$

One problem is that an update might lead to some of the values of α_k violating the feasibility constraints. In such a case, we project such infeasible components of $\overline{\alpha}$ to the feasible box, as shown in Figure 6.3. In other words, the value of each α_k is reset to 0 if it becomes negative, and it is reset to C if it exceeds C. Therefore, one starts by setting the vector of Lagrangian parameters $\overline{\alpha} = [\alpha_1\ldots\alpha_n]$ to an n-dimensional vector of 0s and uses the following update steps with learning rate η:

repeat
    Update $\alpha_k \Leftarrow \alpha_k + \eta\left[1 - y_k\sum_{s=1}^{n}y_s\alpha_s\overline{X}_k \cdot \overline{X}_s\right]$ for each $k \in \{1\ldots n\}$;
    $\left\{$ Update is equivalent to $\overline{\alpha} \Leftarrow \overline{\alpha} - \eta\left[\frac{\partial L_D}{\partial \overline{\alpha}}\right]\right\}$
    for each $k \in \{1\ldots n\}$ **do begin**
        $\alpha_k \Leftarrow \min\{\alpha_k, C\}$;
        $\alpha_k \Leftarrow \max\{\alpha_k, 0\}$;
    endfor;
until convergence

It is noteworthy that the gradient-descent procedure updates *all* the components $\alpha_1\ldots\alpha_n$ at a time. This is the main difference from coordinate descent, which updates a *single component* at a time, and it chooses a specific learning rate for that component, so that that particular value of α_k is optimized. This is the point of discussion in the next section.

6.4.4.2   Coordinate Descent

In coordinate descent, the update for α_k should be such that the updated value is optimized. In other words, the partial derivative of the dual objective function L_D with respect to α_k should be set to 0. By using Equation 6.12 to set the partial derivative with respect to α_k to 0, we obtain the following condition:

$$y_k \sum_{s=1}^{n} y_s \alpha_s \overline{X}_k \cdot \overline{X}_s - 1 = 0$$

On bringing all the terms involving α_k to one side, we obtain:

$$\alpha_k \|\overline{X}_k\|^2 y_k^2 = 1 - y_k \sum_{s \neq k} y_s \alpha_s \overline{X}_k \cdot \overline{X}_s$$

We can set $y_k^2 = 1$ because each $y_k \in \{-1, +1\}$:

$$\alpha_k = \frac{1 - y_k \sum_{s \neq k} y_s \alpha_s \overline{X}_k \cdot \overline{X}_s}{\|\overline{X}_k\|^2} = \alpha_k + \frac{1 - y_k \sum_{s=1}^{n} y_s \alpha_s \overline{X}_k \cdot \overline{X}_s}{\|\overline{X}_k\|^2}$$

In the very final simplification, we added and subtracted α_k on the right-hand side. One can simply treat the above as an iterative update (like gradient-descent) in which α_k is updated at learning rate $\eta_k = 1/\|\overline{X}_k\|^2$.

$$\alpha_k \Leftarrow \alpha_k + \eta_k \left[1 - y_k \sum_{s=1}^{n} y_s \alpha_s \overline{X}_k \cdot \overline{X}_s \right]$$

$$\alpha_k \Leftarrow \alpha_k - \eta_k \left[\frac{\partial L_D}{\partial \alpha_k} \right] \quad \text{[Equivalent update]}$$

In other words, the update for coordinate descent looks just like gradient-descent, except that it is done in component-wise fashion with a component-specific learning rate:

> **repeat**
>   **for** each $k \in \{1 \ldots n\}$ **do begin**
>   Update $\alpha_k \Leftarrow \alpha_k + \eta_k \left[1 - y_k \sum_{s=1}^{n} y_s \alpha_s \overline{X}_k \cdot \overline{X}_s) \right]$;
> $\left\{ \text{Update is equivalent to } \alpha_k \Leftarrow \alpha_k - \eta_k \left[\frac{\partial L_D}{\partial \alpha_k} \right] \right\}$
> $\alpha_k \Leftarrow \min\{\alpha_k, C\}$;
> $\alpha_k \Leftarrow \max\{\alpha_k, 0\}$;
>   **endfor**;
> **until** convergence

It is instructive to compare the pseudocode for coordinate descent with the pseudocode for gradient descent in the previous section. It is evident that the main difference is that all components of $\overline{\alpha}$ are updated in gradient descent (with the learning rate heuristically chosen), whereas updates are performed one component at a time in coordinate descent (with the learning rate specifically chosen to ensure optimality). The coordinate descent procedure always yields faster convergence than gradient descent. Furthermore, *block* coordinate descent, in which more than one variable is chosen at a time, is even more efficient. In fact, Platt's popular sequential minimal optimization (SMO) [102] is an example of block coordinate descent. We also provide a series of practice problems for the L_2-SVM, which provide the systematic steps for formulating its dual and solving it. We strongly advise the reader to work out the practice problems below using the provided solution for the hinge-loss SVM as a guideline. Working out these practice problems will provide the reader a better feel for the way in which dual optimization problems are solved.

Problem 6.4.2 (Relaxation of L_2-SVM) *Consider the following formulation for the L_2-SVM:*

$$\text{Minimize } J = \frac{1}{2}\|\overline{W}\|^2 + C\sum_{i=1}^{n}\xi_i^2$$

subject to:

$$\xi_i \geq 1 - y_i[\overline{W}\cdot\overline{X}_i^T], \quad \forall i \in \{1\ldots n\}$$

In comparison with the hinge-loss SVM, the parameter ξ is squared in the objective function, and the nonnegativity constraints on ξ_i have been dropped. Discuss why dropping of nonnegativity constraints on ξ_i does not affect the optimal solution in this case. Write the minimax Lagrangian relaxation containing both primal and dual variables. Use the Lagrange parameter α_i for the ith slack constraint to enable comparison with the hinge-loss SVM.

Problem 6.4.3 (Primal-Dual Constraints of L_2-SVM) *Let α_i be the Lagrange parameter associated with the ith slack constraint. Show that setting the gradients of the Lagrangian relaxation to 0 (with respect to primal variables) yields the following primal-dual constraints:*

$$\overline{W} = \sum_{i=1}^{n}\alpha_i y_i \overline{X}_i^T$$

$$\xi_i = \alpha_i/2C$$

Problem 6.4.4 (Dual Formulation of L_2-SVM) *Use the Lagrangian relaxation and the primal-dual constraints in the previous two exercises to eliminate the primal variables from the minimax formulation. Show that the dual problem of the L_2-SVM is as follows:*

$$\text{Maximize}_{\overline{\alpha}\geq 0}\sum_{i=1}^{n}\alpha_i - \frac{1}{2}\sum_{i=1}^{n}\sum_{j=1}^{n}\alpha_i\alpha_j y_i y_j(\overline{X}_i\cdot\overline{X}_j + \delta_{ij}/2C)$$

Here, δ_{ij} is 1 if $i = j$, and 0, otherwise. Note that the main difference from the dual formulation of the hinge-loss SVM is the addition of $\delta_{ij}/2C$ to the dot product $\overline{X}_i\cdot\overline{X}_j$, in order to constrain the magnitudes of α_i^2 in a soft way rather than the explicit constraint $\alpha_i \leq C$.

Problem 6.4.5 (Optimization Algorithm for L_2-SVM Dual) *Carefully examine the gradient-descent and coordinate-descent pseudo-codes for the hinge-loss SVM in Sections 6.4.4.1 and 6.4.4.2. The actual updates of each α_k always contain terms with $\overline{X}_k\cdot\overline{X}_s$ as a multiplicative factor for each s. Show that the gradient descent and coordinate descent algorithms for the dual L_2-SVM are exactly the same as the hinge-loss SVM, except that the dot product $\overline{X}_k\cdot\overline{X}_s$ within each update equation is substituted with $[\overline{X}_k\cdot\overline{X}_s + (\delta_{ks}/2C)]$. The value of δ_{ks} is 1 if $k = s$, and 0, otherwise. Furthermore, the values of α_i are not reset to C when they are larger than C.*

6.4.5 Getting the Lagrangian Relaxation of Unconstrained Problems

The Lagrangian relaxation is naturally designed for constrained problems, and the Lagrange multipliers automatically yield the dual variables. A natural question arises as to how one can create the dual in cases where the optimization problem is unconstrained to begin with. There are several approaches for achieving this goal, one of which uses Lagrangian relaxation. For example, a dual approach for logistic regression uses a parametrization approach to construct the dual [68]. We refer the reader to the bibliographic notes for discussions of other forms of duality.

Here, it is important to understand that an optimization problem need not be formulated in a unique way. An unconstrained optimization problem can always be recast as a constrained problem by simply introducing additional variables for various terms in the objective function, and defining those variables within the constraints. The way in which the dual was generated for the hinge-loss SVM already provides a hint for the kinds of formulations that are more friendly to creating dual problems. For example, the SVM formulation in Section 4.8.2 of Chapter 4 does not use slack variables, whereas the dual SVM of the previous section introduces slack variables for specific portions of the objective function, and then defines those slack variables within the constraints. This approach of generating additional variables for specific terms within the objective function provides a natural way to create a Lagrangian relaxation. Therefore, we summarize the basic approach for creating a Lagrangian relaxation of an unconstrained problem:

> Introduce new variables in lieu of specific parts of the objective function, and define those variables within the constraints.

Here, it is important to understand that there is more than one way in which one might choose ways of defining the new variables. Correspondingly, one would obtain a different dual, and the structure of some might be more friendly than others to optimization. Learning to define the correct variables and constraints is often a matter of skill and experience.

Consider the following simple 2-variable optimization problem without constraints:

$$\text{Minimize } J = (x - 1)^2/2 + (y - 2)^2/2$$

One can easily solve this problem in any number of ways, including the use of gradient descent, or by simply setting each partial derivative to 0. In either case, one obtains an optimal solution $x = 1$, and $y = 2$ with a corresponding objective function value of 0. However, it is instructive to formulate the dual of this optimization problem. In this case, we choose to introduce two new variables $\xi = x - 1$ and $\beta = y - 2$. The resulting optimization problem is as follows:

$$\text{Minimize } J = \xi^2/2 + \beta^2/2$$
$$\text{subject to:}$$
$$\xi = x - 1$$
$$\beta = y - 2$$

It is noteworthy that the constraints are equality constraints, and therefore the Lagrange multipliers would not have nonnegativity constraints either. We introduce the Lagrange multiplier α_1 with the first constraint and the multiplier α_2 with the second constraint. The corresponding Lagrangian relaxation then becomes the following:

$$L(\alpha_1, \alpha_2) = \text{Minimize } {}_{\xi,\beta,x,y,} \xi^2/2 + \beta^2/2 + \alpha_1(\xi - x + 1) + \alpha_2(\beta - y + 2)$$

Note that the minimization is performed only over the primal variables, and $L(\alpha_1, \alpha_2)$ needs to be maximized over the dual variables. In order to eliminate the four primal variables, we need to set the partial derivative with respect to each to zero, and obtain four stationarity constraints, which we also refer to as (PD) constraints. However, in this particular case, the (PD) constraints have a simple form:

$$\frac{\partial J}{\partial \xi} = \xi + \alpha_1, \quad \frac{\partial J}{\partial \beta} = \beta + \alpha_2$$

$$\frac{\partial J}{\partial x} = -\alpha_1, \quad \frac{\partial J}{\partial y} = -\alpha_2$$

Setting the first two derivatives with respect to ξ and β to 0 allows us to replace ξ and β with $-\alpha_1$ and $-\alpha_2$, respectively. However, setting the second two derivatives with respect to x and y to 0 yields $\alpha_1 = \alpha_2 = 0$, which allows us to drop the penalty portions of the objective function. However, we need to include[2] the constraints that are independent of the primal variables (i.e., $\alpha_1 = \alpha_2 = 0$) within the dual formulation. This yields the following trivial dual problem:

$$\text{Maximize } \alpha_1^2 + \alpha_2^2$$
$$\text{subject to:}$$
$$\alpha_1 = 0, \quad \alpha_2 = 0$$

In this case, the feasible space contains only one point with an objective function value of 0. Therefore, the optimal dual objective function value is 0 at $\alpha_1 = \alpha_2 = 0$. Furthermore, since ξ and β are equal to $-\alpha_1$ and $-\alpha_2$ (according to the stationarity constraints), it follows that we have $\xi = x - 1 = 0$ and $\beta = y - 2 = 0$. Note that this solution of $x = 1$ and $y = 2$ can be obtained by simply setting the derivative of the primal objective function to 0.

6.4.5.1 Machine Learning Application: Dual of Linear Regression

Another example of an unconstrained optimization problem is linear regression. The training data contains n feature-value pairs $(\overline{X}_i, y_i)$, and the target $\hat{y}_i$ is predicted using $\hat{y}_i \approx \overline{W} \cdot \overline{X}_i^T$. Each $\overline{X}_i$ is a row of the $n \times d$ data matrix D. The column vector of response variables is denoted by $\overline{y} = [y_1 \ldots y_n]^T$. The objective function minimizes the sum-of-squared errors over all training instances:

$$J = \frac{1}{2} \sum_{i=1}^{n} (y_i - \overline{W} \cdot \overline{X}_i^T)^2 + \frac{\lambda}{2} \|\overline{W}\|^2 \tag{6.13}$$

This is again an unconstrained problem, but we somehow want to create the Lagrangian relaxation for it in order to generate the dual. In order to do so, we create new variables and new constraints by introducing a new variable $\xi_i = y_i - \overline{W} \cdot \overline{X}_i^T$ for the error of each data point. The corresponding optimization problem is as follows:

$$\text{Minimize } J = \frac{1}{2} \sum_{i=1}^{n} \xi_i^2 + \frac{\lambda}{2} \|\overline{W}\|^2$$
$$\text{subject to:}$$
$$\xi_i = y_i - \overline{W} \cdot \overline{X}_i^T, \quad \forall i \in \{1 \ldots n\}$$

---

[2]As discussed in the previous section, this situation also arose with the hinge-loss SVM when the constraint $C - \alpha_i - \gamma_i = 0$ contains only dual variables. In that case, the constraint $C - \alpha_i - \gamma_i = 0$ was implicitly included in the formulation by using it to eliminate γ_i from the dual.

We introduce the dual variable α_i for the ith constraint, which results in the following dual objective function:

$$L(\overline{\alpha}) = \text{Minimize}_{\overline{W}, \xi_i}\, J = \frac{1}{2}\sum_{i=1}^{n}\xi_i^2 + \frac{\lambda}{2}\|\overline{W}\|^2 + \sum_{i=1}^{n}\alpha_i(-\xi_i + y_i - (\overline{W}\cdot\overline{X}_i^T))$$

Next, we will generate the primal-dual (PD) constraints by differentiating the objective function with respect to all the primal variables and setting it to zero.

$$\frac{\partial J}{\partial \overline{W}} = \lambda\overline{W} - \sum_{i=1}^{n}\alpha_i\overline{X}_i^T = \overline{0}$$

$$\frac{\partial J}{\partial \xi_i} = \xi_i - \alpha_i = 0, \quad \forall i \in \{1\ldots n\}$$

Substituting $\xi_i = \alpha_i$ and $\overline{W} = \sum_{j=1}^{n}\alpha_j\overline{X}_j^T/\lambda$, we obtain the following for $L(\overline{\alpha})$ purely in terms of only the dual variables:

$$L(\overline{\alpha}) = \frac{1}{2}\sum_{i=1}^{n}\alpha_i^2 + \frac{1}{2\lambda}\sum_{i=1}^{n}\sum_{j=1}^{n}\alpha_i\alpha_j\overline{X}_i\cdot\overline{X}_j + \sum_{i=1}^{n}\alpha_i\left(-\alpha_i + y_i - \overline{X}_i^T\cdot[\sum_{j=1}^{n}\alpha_j\overline{X}_j^T]/\lambda\right)$$

$$= \sum_{i=1}^{n}\alpha_i y_i - \sum_{i=1}^{n}\alpha_i^2/2 - \frac{1}{2\lambda}\sum_{i=1}^{n}\sum_{j=1}^{n}\alpha_i\alpha_j\overline{X}_i\cdot\overline{X}_j$$

One can rewrite the above objective function in matrix form by replacing the d-dimensional row vectors $\overline{X}_1\ldots\overline{X}_n$ with a single $n \times d$ matrix D whose rows contain these vectors in the same order. Furthermore, the scalar variables are converted to vector forms such as $\overline{\alpha} = [\alpha_1\ldots\overline{\alpha}_n]^T$ and $\overline{y} = [y_1\ldots y_n]^T$:

$$L(\overline{\alpha}) = \overline{\alpha}^T\overline{y} - \frac{1}{2}\|\overline{\alpha}\|^2 - \frac{1}{2\lambda}\overline{\alpha}^T DD^T\overline{\alpha}$$

$$= \overline{\alpha}^T\overline{y} - \frac{1}{2\lambda}\overline{\alpha}^T(DD^T + \lambda I)\overline{\alpha}$$

One can simply set the gradient of the objective function to 0 in order to solve for $\overline{\alpha}$ in closed form. By using matrix calculus to compute the gradient of the objective function, we obtain the following:

$$(DD^T + \lambda I)\overline{\alpha} = \lambda\overline{y}$$

$$\overline{\alpha} = \lambda(DD^T + \lambda I)^{-1}\overline{y}$$

It now remains to relate the optimal dual variables to the optimal primal variables by using the primal-dual constraints. From the (PD) constraints, we already know that $\overline{W} = \sum_{j=1}^{n}\alpha_j\overline{X}_j^T/\lambda = D^T\overline{\alpha}/\lambda$. This yields the following optimal solution for primal variable $\overline{W}$:

$$\overline{W} = D^T(DD^T + \lambda I_n)^{-1}\overline{y} \tag{6.14}$$

Here, I_n is the $n \times n$ identity matrix. It is helpful to compare this solution with that obtained by setting the gradient of the primal loss function to zero. The resulting solution is described in Section 4.7 of Chapter 4. We repeat Equation 4.39 from that section here:

$$\overline{W} = (D^T D + \lambda I_d)^{-1}D^T\overline{y} \tag{6.15}$$

At first glance, this solution seems to be different. However, the two solutions are really equivalent, and one can derive this result from the push-through identity (cf. Problem 1.2.13 of Chapter 1). Specifically, the following can be shown:

$$D^T(DD^T + \lambda I_n)^{-1} = (D^T D + \lambda I_d)^{-1} D^T \tag{6.16}$$

Another example of an unconstrained problem is logistic regression, which is discussed in Section 4.8.3 of Chapter 4. The following sequence of problems provides a step-by-step guide to how one can formulate the dual of logistic regression [140]. Since logistic regression is a fundamental problem in machine learning, it is to advised to work out this sequence of problems for better insights.

Problem 6.4.6 (Relaxation of Logistic Regression) *Logistic regression is an unconstrained optimization problem, as evident from its objective function in Equation 4.56 of Chapter 4. Consider the following formulation for logistic regression:*

$$Minimize\ J = \frac{1}{2}\|\overline{W}\|^2 + C\sum_{i=1}^{n} log(1 + exp[\xi_i])$$

subject to:

$$\xi_i = -y_i(\overline{W} \cdot \overline{X}_i^T)$$

Discuss why this objective function is the same as Equation 4.56 with an appropriate choice of C. Assume that the other notations are the same as Equation 4.56. Formulate a Lagrangian relaxation of this problem, where α_i is the dual variable used for the ith constraint associated with $\overline{X}_i$.

Since the Lagrange multiplier is sign-unconstrained in this case, and the constraints are equality constraints, one could obtain either of two possible answers to the previous problem with different signs of α_i. This issue is also applicable to the next problem, where you might get the results in the statement of the exercise with the sign of α_i flipped.

Problem 6.4.7 (Primal-Dual Constraints of Logistic Regression) *Let α_i be the Lagrange parameter associated with the ith slack constraint. Show that setting the gradients of the Lagrangian relaxation to 0 (with respect to primal variables) yields the following primal-dual constraints:*

$$\overline{W} = \sum_{i=1}^{n} y_i \alpha_i \overline{X}_i^T$$

$$\alpha_i = \frac{C}{1 + exp(-\xi_i)}$$

Now discuss why α_i must lie in the range $(0, C)$ based on the primal dual constraints (just like the hinge-loss SVM).

The similarity of the logistic dual with the hinge-loss SVM dual is not particularly surprising, given the fact that we have shown the similarity of the primal logistic regression objective function with that of the hinge-loss SVM, especially for the critical, difficult-to-classify points (see Section 4.8.4 of Chapter 4).

Problem 6.4.8 *Show that the dual of logistic regression can be expressed in* **minimization form** *as follows:*

$$Minimize_{\overline{\alpha}} \frac{1}{2} \sum_{i=1}^{n} \sum_{j=1}^{n} \alpha_i \alpha_j y_i y_j (\overline{X}_i \cdot \overline{X}_j) + \sum_{i=1}^{n} \alpha_i log(\alpha_i) + \sum_{i=1}^{n} (C - \alpha_i) log(C - \alpha_i)$$

Note that the objective function of logistic regression only makes sense for $\alpha_i \in (0, C)$ because the logarithm function can only have positive arguments. In practice, one explicitly adds the constraints $\alpha_i \in (0, C)$ to avoid an undefined objective function. This makes the entire formulation very similar to the hinge-loss SVM dual, and the pseudo-code in Section 6.4.4.1 can be used directly, but with stronger box-constraint updates to *strictly* within $(0, C)$. Another difference is that α_k is updated as follows:

$$\alpha_k \Leftarrow \alpha_k + \eta \left[log \frac{C - \alpha_k}{\alpha_k} - y_k \sum_{s=1}^{n} y_s \alpha_s \overline{X}_k \cdot \overline{X}_s \right]$$

The term $log([C - \alpha_k]/\alpha_k)$ replaces 1 in the pseudo-code, and it tries to keep α_k in the middle of the range $(0, C)$.

6.5 Penalty-Based and Primal-Dual Methods

The Lagrangian relaxation methods formulate the dual of the optimization problem by relaxing primal constraints in terms of the penalty variables. The idea is that the relaxed version of the problem always satisfies weak duality. Therefore, if we can come up with a relaxed version (with appropriate values of the penalty multipliers) that satisfies the primal constraints, then the resulting solution is also an optimal solution to the original problem (i.e., unrelaxed primal problem with constraints). This type of approach requires us to successively modify the penalty variables and perform gradient descent on the relaxed problem until the primal constraints are satisfied. In all cases, the solution to the relaxed problem provides us hints as to whether the penalty variables should be increased or decreased. For example, if the primal constraints are violated on solving the relaxed problem then the penalty variables for the violated primal constraints need to be increased. Otherwise, if the constraints are not satisfied tightly, the penalty variables for the primal constraints can be decreased. The form of the penalty is sometimes different from a traditional Lagrangian relaxation, and in other cases can be shown to be exactly or almost equivalent. In the latter case, these methods are sometimes referred to as *primal-dual methods* because they simultaneously learn primal and dual variables. Even in cases where the form of the penalty constraint is not the same as a traditional Lagrangian relaxation, the broader principle is quite similar. In order to understand this point, we will first work with an example of an optimization problem with a single constraint as a motivating idea.

6.5.1 Penalty Method with Single Constraint

Consider the following optimization problem in which we wish to minimize a convex function subject to a distance constraint. In other words, we wish to find the optimum point of the convex objective function $F(\overline{w})$ subject to the constraint that the distance between vector $\overline{w}$ and constant vector $\overline{a}$ is at most δ. Note that this problem arises frequently in machine learning, when using trust-region optimization in conjunction with the Newton method (cf. Section 5.6.3.1 of Chapter 5). This problem is stated as follows:

$$\text{Minimize } F(\overline{w})$$

$$\text{subject to:}$$

$$\|\overline{w} - \overline{a}\|^2 \le \delta^2$$

The first step is to solve the optimization problem while ignoring the constraint. If the optimal solution already satisfies the constraint (in spite of the fact that it was not used), then we need to do nothing else. We can simply terminate. On the other hand, if the constraint is violated, then we formulate the following relaxed version of the problem with penalty parameter $\alpha > 0$:

$$\text{Minimize } F(\overline{w}) + \alpha \left(\max\{\|\overline{w} - \overline{a}\|^2 - \delta^2, 0\} \right)^2$$

Note that there is no penalty or gain when the constraint is satisfied. This ensures that the objective function value of the relaxed problem is the same as that of the original problem as long as one operates in the feasible space. Choosing very small values of α might result in violation of the constraints. On the other hand, choosing large enough values of α will always result in feasible solutions, in which the penalty does not contribute anything to the objective function. An important observation about penalty functions is as follows:

Observation 6.5.1 *Consider a penalty-based variation of a constrained optimization problem in which violation of constraints is penalized and added to the objective function. Furthermore, feasible points have zero penalties (or gains). If the optimal solution to the penalty-based relaxation is feasible for the constraints in the original problem, then that solution is also optimal for the original problem.*

The above observation is the key to the success of penalty-based methods. We simply need to start with small enough values of α and gradually test successively large values of α until the relaxation yields a feasible solution. One can solve this problem by starting at $\alpha = 1$ and solving the optimization problem. If the constraints are satisfied, we terminate and report the corresponding value of the parameter vector $\overline{w}$ as optimal. If the solution is not feasible, one can double the value of α and perform gradient descent again to find the best value of the parameter vector $\overline{w}$ with gradient descent. One can use the parameter vector $\overline{w}$ at the end of an iteration as the starting point for gradient descent in the next iteration (with increased α). This reduces the work in the next iteration. This approach of increasing α is continued until no constraints are violated. It is also noteworthy that the relaxed objective function is convex when the objective function and the constraints are convex.

6.5.2 Penalty Method: General Formulation

The general formulation of convex optimization problems is as follows:

$$\text{Minimize } F(\overline{w})$$

$$\text{subject to:}$$

$$f_i(\overline{w}) \le 0, \quad \forall i \in \{1 \ldots m\}$$

$$h_i(\overline{w}) = 0, \quad \forall i \in \{1 \ldots k\}$$

To ensure convexity of this problem, the function $F(\overline{w})$ is convex, all the functions $f_i(\cdot)$ must be convex, and all the functions $h_i(\cdot)$ are linear. Note that the penalty method can be used even in the cases where these conditions are not met; however, in those cases, one might not be able to obtain the global optimum. Then, the relaxed objective function of this problem is as follows:

$$\text{Minimize } R(\overline{w}, \alpha) = F(\overline{w}) + \frac{\alpha}{2} \left(\sum_{i=1}^{m} \max\{0, f_i(\overline{w})\}^2 + \sum_{i=1}^{k} h_i(\overline{w})^2 \right) \qquad (6.17)$$

Note the difference between how equality and inequality constraints are treated. The penalty parameter α is always greater than zero. We make the following observation:

Observation 6.5.2 (Convexity of Relaxation) *If $F(\overline{w})$ is convex, each $f_i(\overline{w})$ is convex, and each $h_i(\overline{w})$ is linear, then the relaxed objective function of Equation 6.17 is convex for $\alpha > 0$.*

The gradient of this objective function with respect to $\overline{w}$ can be computed as follows:

$$\nabla_{\overline{w}} R(\overline{w}, \alpha) = \nabla F(\overline{w}) + \alpha \sum_{i=1}^{m} \max\{f_i(\overline{w}), 0\} \nabla f_i(\overline{w}) + \alpha \sum_{i=1}^{k} h_i(\overline{w}) \nabla h_i(\overline{w})$$

As in the case of single-variable penalty methods, we perform gradient descent at a fixed value of α. In the event that the resulting solution $\overline{w}$ at termination is feasible, we use this parameter vector as the optimal solution. Otherwise, we increase α and repeat the process. In the next iteration, we can start with the vector $\overline{w}$ obtained from the previous iteration as a starting point.

A natural question arises as to why one should not start with the largest possible value of α to begin with. After all, choosing large values of α ensures feasibility of the solution with respect to the constraints of the original problem. The main problem with using very large values of α is that it often leads to ill-conditioning during intermediate stages of the gradient descent, where the gradient is much more sensitive to some directions than others. As we have seen in Chapter 5, this type of situation can cause problems (such as "bouncing" behavior) in gradient descent, and convergence will not occur. This is the reason that one should generally start with smaller values of α, and increase it over time. This ensures that the algorithm will show good convergence behavior.

6.5.3 Barrier and Interior Point Methods

Penalty-based methods do not yield intermediate values of $\overline{w}$ that are feasible. In contrast, barrier methods always maintain values of $\overline{w}$ that are not only feasible but *strictly* feasible. The notion of strict feasibility makes sense only for inequality constraints. Therefore, barrier methods are designed only for inequality constraints of the form $f_i(\overline{w}) \geq 0$. Note that we have flipped the direction of the inequality for notational ease. A point $\overline{w}$ is strictly feasible if and only if we have $f_i(\overline{w}) > 0$ for each constraint. Obviously, such a point exists only when the feasible region has non-zero volume in the space. This is the reason that barrier methods are not designed for the case of equality constraints. Consider the following optimization problem:

$$\text{Minimize } F(\overline{w})$$
$$\text{subject to:}$$
$$f_i(\overline{w}) \geq 0, \quad \forall i \in \{1 \dots m\}$$

Then, the *barrier function* $B(\overline{w}, \alpha)$ is well-defined only for feasible values of the parameter vector $\overline{w}$, and it is defined as follows:

$$B(\overline{w}, \alpha) = F(\overline{w}) - \alpha \sum_{i=1}^{m} \log(f_i(\overline{w}))$$

This is an example of the use of the *logarithmic barrier function*, although other choices (such as the inverse barrier function) exist. One observation is that the barrier function is convex as long as $F(\overline{w})$ is convex, and each $f_i(\overline{w})$ is concave. This is because the logarithm[3] of a concave function is concave, and the negative logarithm is therefore convex. The sum of convex functions is convex, and therefore the barrier function is convex. Note that we require each $f_i(\overline{w})$ to be concave (rather than convex) because our inequality constraints are of the form $f_i(\overline{w}) \geq 0$ rather than $f_i(\overline{w}) \leq 0$.

A key point is that each $f_i(\overline{w})$ must be *strictly* greater than zero even for the objective function to be meaningfully evaluated at a given step; one cannot compute the logarithm of zero or negative values. Therefore, barrier methods start with feasible solutions $\overline{w}$ in the interior of the data. Furthermore, unlike penalty methods, one starts with large values of α in early iterations, and this value is reduced over time. At any fixed value of α, gradient-descent is performed on $\overline{w}$ to optimize the weight vector. Smaller values of α allow $\overline{w}$ to approach closer to the boundary of the feasible region defined by the constraints. This is because the barrier function always approaches ∞ near the boundary irrespective of the value of α, but small values of α allow a closer approach. However, small values of α also result in sharp ill-conditioning, and using small values of α early is bad for convergence. For example, using high values of α in the initial phases is helpful in maintaining strict feasibility of the weight vector $\overline{w}$.

In cases where the true optimal solution is not near the boundary of the feasible region, one will often approach the optimal solution quickly, and convergence is smooth. In these cases, the constraints might even be redundant, and the unconstrained version of the problem will yield the same solution. In more difficult cases, the optimal weight vector might lie near the boundary of the feasible region. As the feasible weight vector $\overline{w}$ approaches close enough to the boundary $f_i(\overline{w}) \geq 0$, the penalty contribution increases rapidly like a "barrier" and increases to ∞ when one reaches the boundary $f_i(\overline{w}) = 0$. Therefore, we only need relatively small values of α in order to ensure feasibility. However, at small values of α, the function becomes ill-conditioned near the boundary. Therefore, the barrier method starts with large values of α and gradually reduces it, while performing gradient descent with respect to $\overline{w}$ and fixed α. The optimal vector $\overline{w}$ at the end of a particular iteration is used as a starting point for the next iteration (with a smaller value of α).

For gradient descent, the gradient of the objective function is as follows:

$$\nabla_{\overline{w}} B(\overline{w}, \alpha) = \nabla F(\overline{w}) - \alpha \sum_{i=1}^{m} \frac{\nabla f_i(\overline{w})}{f_i(\overline{w})}$$

---

[3]Since the logarithm is concave, we know that:

$$\log[\lambda f_i(\overline{w}_1) + (1 - \lambda) f_i(\overline{w}_2)] \geq \lambda \log[f_i(\overline{w}_1)] + (1 - \lambda) \log[f_i(\overline{w}_2)] \qquad (6.18)$$

At the same time, we know that $f_i(\lambda \overline{w}_1 + (1 - \lambda)\overline{w}_2) \geq \lambda f_i(\overline{w}_1) + (1 - \lambda) f_i(\overline{w}_2)$ because $f_i(\cdot)$ is concave. Since, the logarithm is an increasing function, we can take the logarithm of both sides to show the result that $\log[f_i(\lambda \overline{w}_1 + (1 - \lambda)\overline{w}_2)] \geq \log[\lambda f_i(\overline{w}_1) + (1 - \lambda) f_i(\overline{w}_2)]$. Combining this inequality with Equation 6.18 using transitivity, we can show that $\log[f_i(\lambda \overline{w}_1 + (1 - \lambda)\overline{w}_2)] \geq \lambda \log[f_i(\overline{w}_1)] + (1 - \lambda) \log[f_i(\overline{w}_2)]$. In other words, $\log(f_i(\cdot))$ is concave. More generally, we just went through all the steps required to show that the composition $g(f(\cdot))$ of two concave functions is concave as long as $g(\cdot)$ is non-decreasing. Closely related results are available in Lemma 4.3.2.

Setting this gradient to zero yields the optimality condition. It is instructive to compare this optimality condition with the primal-dual (PD) constraint of the Lagrangian $L(\overline{w}, \overline{\alpha}) = F(\overline{w}) - \sum_i \alpha_i f_i(\overline{w})$:

$$\nabla_{\overline{w}} L(\overline{w}, \overline{\alpha}) = \nabla F(\overline{w}) - \sum_{i=1}^{m} \alpha_i \nabla f_i(\overline{w}) = \overline{0}$$

Here, we are using $\alpha_1 \ldots \alpha_k$ as the Lagrangian parameters, which can be distinguished from the penalty parameter α by virtue of having a subscript. Furthermore, since the Lagrangian relaxation is computed using the "$\leq$" form of the constraint (which is $-f_i(\overline{w}) \leq 0$), we have a negative sign in front of each penalty term. Note that the value of $\alpha / f_i(\overline{w})$ is an estimate of the Lagrangian multiplier α_i, if one were to use the traditional Lagrangian relaxation $L(\overline{w}, \overline{\alpha}) = F(\overline{w}) - \sum_i \alpha_i f_i(\overline{w})$. Interestingly, this means that we have $\alpha_i f_i(\overline{w}) = \alpha$. Note that this is almost the complementary-slackness condition of Lagrangian relaxation, except that we have substituted 0 with a small value α. Therefore, at small values of α, the optimality conditions of the (traditional) dual relaxation are nearly satisfied when one views the barrier function as a Lagrangian relaxation. The barrier method belongs to the class of interior point methods that approach the optimal solution from the interior of the feasible space. Therefore, one benefit of such methods is that they yield estimates of the Lagrangian dual variables in addition to yielding the primal values.

6.6   Norm-Constrained Optimization

The use of eigenvectors in norm-constrained optimization is discussed in Section 3.4.5 of Chapter 3. This problem appears repeatedly in different types of machine learning problems, such as principal component analysis, singular value decomposition, and spectral clustering. We revisit an optimization problem introduced in Section 3.4.5:

$$\text{Minimize} \sum_{i=1}^{k} \overline{x}_i^T A \overline{x}_i$$

$$\text{subject to:}$$

$$\|\overline{x}_i\|^2 = 1, \quad \forall i \in \{1 \ldots k\}$$

$$\overline{x}_1 \ldots \overline{x}_k \text{ are mutually orthogonal}$$

Here, A is a *symmetric* $d \times d$ matrix, and $\overline{x}_1 \ldots \overline{x}_k$ correspond to the d-dimensional vectors containing the optimization variables. The symmetric nature of A is important in this case, because it simplifies the handling of the orthogonality constraints. This problem essentially tries to find the top-k orthogonal vectors in d dimensions, such that the sum of $\overline{x}_i^T A \overline{x}_i$ over all i is as small as possible. It is assumed that the value of k is less than or equal to d, or else the problem will not have feasible solutions. One difference from the problem discussed in Section 3.4.5 is that we are explicitly trying to minimize the objective function in this case, whereas the problem of Section 3.4.5 is stated more generally in terms of either minimization or maximization. Although one can deal with the maximization in an exactly analogous way, we deal only with minimization in order to create a crisp and unambiguous Lagrangian. It is also noteworthy that the orthogonality constraints can be restated as $\binom{k}{2}$ constraints of the form $\overline{x}_i \cdot \overline{x}_j = 0$ for all $i < j \leq k$. We introduce the Lagrangian multiplier $-\alpha_i$ with each constraint of the form $\|\overline{x}_i\|^2 = 1$. However, we do not choose to relax the orthogonality constraints. This is an example of the fact that Lagrangian relaxations can choose not to relax all the constraints, although one can obtain an equivalent solution by relaxing all constraints. Note that the Lagrangian multipliers are not constrained to be nonnegative because we are relaxing equality constraints rather than inequality constraints. We also add a negative sign in front of the multipliers for algebraic interpretability of the Lagrangian multipliers as eigenvalues (as we will show later). Correspondingly, one can write the Lagrangian relaxation as follows:

$$L(\overline{\alpha}) = \text{Minimize}_{\overline{x}_1 \ldots \overline{x}_k \text{ are orthogonal}} \sum_{i=1}^{k} \overline{x}_i^T A \overline{x}_i - \sum_{i=1}^{k} \alpha_i (\|\overline{x}_i\|^2 - 1)$$

Setting the gradient of the Lagrangian with respect to each $\overline{x}_i$ to 0, one obtains the following:

$$A\overline{x}_i = \alpha_i \overline{x}_i, \quad \forall i \in \{1 \ldots k\}$$

As discussed earlier, we need to use the primal-dual (PD) constraints to eliminate the primal variables, and obtain an optimization problem in terms of the dual variables. Note that the constrains $A\overline{x}_i = \alpha_i \overline{x}_i$ implies that the feasible space for α_i is restricted to the d eigenvalues of A. Note that the orthogonality constraints on the vectors $\overline{x}_1 \ldots \overline{x}_k$ are automatically satisfied because the eigenvectors of the symmetric matrix A are orthonormal. Using the (PD) constraints to substitute $A\overline{x}_i = \alpha_i \overline{x}_i$ within the Lagrangian relaxation, we obtain the following:

$$L(\overline{\alpha}) = \text{Minimize}_{[\overline{x}_1 \ldots \overline{x}_k \text{ are orthogonal}]} \sum_{i=1}^{k} \alpha_i \overline{x}_i^T \overline{x}_i - \sum_{i=1}^{k} \alpha_i (\|\overline{x}_i\|^2 - 1)$$

$$= \text{Minimize}_{[\text{Eigenvalues of } A]} \sum_{i=1}^{k} \alpha_i$$

Clearly, the above objective function is minimized over the smallest eigenvalues of A. Therefore, one obtains the following trivial dual problem:

$$\text{Maximize } L(\overline{\alpha}) = \sum_{i=1}^{k} \alpha_i$$

$$\text{subject to:}$$

$$\alpha_1 \ldots \alpha_k \text{ are smallest eigenvalues of } A$$

Note that the dual problem has a single point in its feasible solution. The primal solutions $\overline{x}_1 \ldots \overline{x}_k$, correspond to the smallest eigenvectors of A because of the (PD) constraints $A\overline{x}_i = \alpha_i \overline{x}_i$. A key point is that even though we assumed that the matrix A is symmetric, we did not assume that it is positive semi-definite. Therefore, the objective function might not be convex. In other words, strong duality is not guaranteed, and there might be a gap between the primal and dual solutions. One way of checking optimality of the derived primal solution is to explicitly check if a gap exists. In other words, we substitute the derived primal solution into the primal objective function and compare it with the dual objective function value at optimality. On making this substitution, we find that the primal objective function is also the sum of the smallest k eigenvalues. Therefore, there is no gap between the derived primal and dual solutions. The result of this section, therefore, provides an example of how it is sometimes possible to use Lagrangian relaxation even in the case of objective functions that are not convex. This section also provides a detailed proof of the norm-constrained optimization problem introduced in Section 3.4.5.

The maximization variant of this problem is very similar:

$$\text{Maximize } \sum_{i=1}^{k} \overline{x}_i^T A \overline{x}_i$$

$$\text{subject to:}$$

$$\|\overline{x}_i\|^2 = 1, \quad \forall i \in \{1 \ldots k\}$$

$$\overline{x}_1 \ldots \overline{x}_k \text{ are mutually orthogonal}$$

As in the case of the minimization version of the problem, it is important for the matrix A to be symmetric (because of orthogonality constraints). The approach to the maximization variant of the problem is very similar, and one can show that the best solution is obtained by choosing the largest eigenvectors of A. We leave the proof of this result as an exercise for the reader.

Problem 6.6.1 *Show that the optimal solution to the maximization variant of norm-constrained optimization with objective function $\sum_{i=1}^{k} \overline{x}_i^T A \overline{x}_i$ corresponds to the largest k eigenvectors of the symmetric matrix A.*

6.7   Primal Versus Dual Methods

A natural question arises as to whether primal methods or dual methods are desirable in terms of performance. For example, in the case of the support vector machine, dual methods are used so universally that it has sometimes led to the impression that it is the only reasonable way to solve the optimization problem. Interestingly, many machine learning problems like the SVM can be posed as purely unconstrained problems in the primal (cf. Section 4.8.2 of Chapter 4), as long as we allow functions like maximization (e.g., $\max\{x, 0\}$) within the objective function; therefore, complicated techniques for gradient-descent are often not required in the primal. Even in cases where the primal contains constraints, one can use techniques like (primal) projected gradient descent. Interestingly, to create the dual problem, we actually *add constraints and variables* to the primal, so that a Lagrangian relaxation can be created (cf. Section 6.4.5). This fascination of the machine learning community with the dual has been pointed out in a seminal paper [28]:

> "The vast majority of text books and articles introducing support vector machines (SVMs) first state the primal optimization problem, and then go directly to the dual formulation. A reader could easily obtain the impression that this is the only possible way to train an SVM."

An incorrect perception among some data scientists is that the dual is useful for solving the *kernel* SVM using similarities between points (rather than feature values), whereas the primal can be solved using only the feature values. Here, one observation is that the primal optimization problem for an $n \times d$ data matrix D is often posed in terms of the *scatter matrices* $D^T D$, whereas the dual optimization problem is often posed in terms of *similarity matrices* $D D^T$. Note that all the dual optimization problems posed in this chapter contain the dot-product similarity $\overline{X}_i \cdot \overline{X}_j$ within the objective function; therefore, one can write the objective function in terms of only the similarities between the ith and jth points. This observation is useful in cases where one wants to use arbitrary similarities between points in lieu of their feature representations. In some cases, one might want to use a domain-specific similarity, another kernel-based similarity (cf. Chapter 9), or a heuristic similarity function between objects that are not inherently multidimensional. Such techniques are referred to as *kernel methods*. However, the idea that dual objective functions are *essential* for the use of kernel methods is a widespread misconception. As we will see in Chapter 9, there is a systematic way in which every primal objective function discussed in this chapter and the previous chapters can be recast in terms of similarities. This approach uses a fundamental idea in linear algebra, known as the *representer theorem*. Note that the dual problems are often constrained optimization problems like the primal (albeit with simple box constraints). Therefore, all that the dual formulation achieves is to provide another perspective to the problem, which *might* have (relatively minor) benefits.

For example, consider the issue of computational efficiency for a problem with n data points and d dimensions. The scatter matrix (used in the primal) has $O(d^2)$ entries, whereas the similarity matrix (used in the dual) has $O(n^2)$ entries. Therefore, the primal is often cheaper to solve when the dimensionality is smaller than the number of points. This situation is quite common. On the other hand, if the number of points is smaller than the dimensionality, the dual methods can be cheaper. However, some principles like the *representer theorem* (cf. Chapter 9) enable techniques for the primal, which are of similar complexity as the dual.

Another point to be kept in mind is that most gradient descent methods arrive at an approximately optimal objective function value. After all, there are many practical challenges associated with computational optimization, and one often arrives at a numerically approximate solution. However, the primal has the advantage that the level of final approximation is guaranteed, because we are directly optimizing the objective function we wanted in the first place. On the other hand, the final dual solution needs to be mapped to a primal solution via the primal-dual constraints. For example, on computing the dual variables $\alpha_1 \ldots \alpha_n$ in the hinge-loss SVM, the primal solution $\overline{W}$ is computed as $\overline{W} = \sum_i \alpha_i y_i \overline{X}_i^T$. Optimizing the dual objective function approximately *might* provide an arbitrarily poor solution for the primal. Although the primal and dual objective function values are exactly the same at optimality (for convex objective functions like the SVM), this is not the case for *approximately* optimal solutions; the approximate dual objective function value (which is a function of $\alpha_1 \ldots \alpha_n$) might be quite different from the final objective function value when translated to the primal solution. Finally, intermediate primal solutions are more interpretable than dual solutions. This interpretability has an advantage from a practical point of view, and early termination is easier in the event of computational constraints. The dual approach has been historically favored in models like support vector machines. However, there is no inherent reason to so so, given the vast number of simple methods available for primal optimization. Our recommendation is to *always use a primal method where possible*.

6.8 Summary

Many optimization problems have constraints in them, which makes the solution methodology more challenging. Several methods for handing constrained optimization were discussed in this chapter, such as projected gradient descent, coordinate descent, and Lagrangian relaxation. Penalty-based and barrier methods combine ideas from primal and dual formulations. Among these methods, primal methods have some advantages because of their better interpretability. Nevertheless, dual problems can also work well in some settings, where the number of points is fewer than the number of variables.

6.9   Further Reading

The dual algorithm for SVMs was introduced in the original paper by Cortes and Vapnik [30]. The formulation of the dual for logistic regression is discussed in [68, 140], and various numerical algorithms are compared in [93]. Techniques for kernel logistic regression based on the representer theorem are presented in [142]. Detailed discussions of dual methods for SVMs are provided in [31]. Dual coordinate descent methods for the SVM and logistic regression are proposed in [64, 136]. Although the Lagrangian relaxation is the most common approach for formulating the dual of a problem, it is not the only way to do so. As long as we can parameterize a problem with additional variables, so that its minimax solution provides the true optimum, it can be used to formulate a dual problem. An example of such an approach is that for logistic regression [68].

6.10   Exercises

1. Suppose you want to find the largest area of rectangle that can be inscribed in a circle of radius 1. Formulate a 2-variable optimization problem with constraints to solve this problem. Discuss how you can convert this problem into a single-variable optimization problem without constraints.

2. Consider the following optimization problem:

$$\text{Minimize } x^2 + 2x + y^2 + 3y$$
$$\text{subject to:}$$
$$x + y = 1$$

Suppose that (x_0, y_0) is a point satisfying the constraint $x + y = 1$. Compute the projected gradient at (x_0, y_0).

3. Use the method of Gaussian elimination to eliminate both the constraint and variable y in Exercise 2. Compute the optimal solution of the resulting unconstrained problem. What is the optimal objective function value?

4. Compute the dual of the objective function in Exercise 2. Compute the optimal solution as well as the resulting objective function value.

5. Implement a gradient-descent algorithm for linear regression with box constraints. Use Python or any other programming language of your choice.

6. **Linear programming dual:** Consider the following linear programming optimization problem with respect to primal variables $\overline{w} = [w_1, w_2, \ldots w_d]^T$:

$$\text{Minimize } \sum_{i=1}^{d} c_i w_i$$
$$\text{subject to:}$$
$$A\overline{w} \leq \overline{b}$$

Here, A is an $n \times d$ matrix, and $\overline{b}$ is an n-dimensional column vector. Formulate the dual of this optimization problem by using the Lagrangian relaxation only in terms of dual variables. Are there any conditions under which strong duality holds?

7. **Quadratic programming dual:** Consider the following quadratic programming optimization problem with respect to primal variables $\overline{w} = [w_1, w_2, \ldots w_d]^T$:

$$\text{Minimize } \frac{1}{2}\overline{w}^T Q\overline{w} + \sum_{i=1}^{d} \overline{c}^T \overline{w}$$

subject to:

$$A\overline{w} \leq \overline{b}$$

Here, Q is a $d \times d$ matrix, A is an $n \times d$ matrix, $\overline{c}$ is a d-dimensional column vector, and $\overline{b}$ is an n-dimensional column vector. Formulate the dual of this optimization problem by using the Lagrangian relaxation only in terms of dual variables. Assume that Q is invertible. Are there any conditions under which strong duality holds?

8. Consider the SVM optimization problem where we explicitly allow a bias variable b. In other words, the primal SVM optimization problem is stated as follows:

$$J = \sum_{i=1}^{n} \max\{0, (1 - y_i[\overline{W} \cdot \overline{X}_i^T] + b)\} + \frac{\lambda}{2}\|\overline{W}\|^2$$

Compute the dual of this optimization formulation by using analogous steps to those discussed in the chapter. How would you handle the additional constraint in the dual formulation during gradient descent?

9. As you will learn in Chapter 9, the primal formulation for least-squares regression can be recast in terms of similarities s_{ij} between pairs of data points as follows:

$$J = \frac{1}{2}\sum_{i=1}^{n}(y_i - \sum_{p=1}^{n}\beta_p s_{pi})^2 + \frac{\lambda}{2}\sum_{i=1}^{n}\sum_{j=1}^{n}\beta_i\beta_j s_{ij}$$

Here, s_{ij} is the similarity between points i and j. Convert this unconstrained optimization problem into a constrained problem, and formulate the dual of the problem in terms of s_{ij}.

10. Let $\overline{z} \in \mathcal{R}^d$ lie outside the ellipsoid $\overline{x}^T A\overline{x} + \overline{b}^T \overline{x} + c \leq 0$, where A is a $d \times d$ positive semi-definite matrix and $\overline{x} \in \mathcal{R}^d$. We want to find the closest projection of $\overline{z}$ on this convex ellipsoid to enable projected gradient descent. Use Lagrangian relaxation to show that the projection point $\overline{z}_0$ must satisfy the following:

$$\overline{z} - \overline{z}_0 \propto 2A\overline{z}_0 + \overline{b}$$

Interpret this condition geometrically in terms of the tangent to the ellipsoid.

11. Consider the following optimization problem:

$$\text{Minimize } x^2 - y^2 - 2xy + z^2$$

subject to:

$$x^2 + y^2 + z^2 \leq 2$$

Imagine that we are using coordinate descent in which we are currently optimizing the variable x, when y and z are set to 1 and 0, respectively. Solve for x. Then, solve for y by setting x and z to their current values. Finally, solve for z in the same way. Perform another full cycle of coordinate descent to confirm that coordinate descent cannot improve further. Provide an example of a solution with a better objective function value. Discuss why coordinate descent was unable to find an optimal solution.

12. Consider the dual objective function in Lagrangian relaxation, as a function of only the dual variables:

$$L(\overline{\alpha}) = \text{Minimize}_{\overline{w}} \left[F(\overline{w}) + \sum_{i=1}^{m} \alpha_i f_i(\overline{w}) \right]$$

The notations here for $F(\cdot)$ and $f_i(\cdot)$ are the same as those used in Section 6.4. Show that $L(\overline{\alpha})$ is always concave in $\overline{\alpha}$, *irrespective of the convexity structure of the original optimization problem.*

13. **Nonnegative box regression:** Formulate the Lagrangian dual (purely in terms of dual variables) for L_2-regularized linear regression $D\overline{w} \approx \overline{y}$ with $n \times d$ data matrix D, regressand vector $\overline{y}$, and with nonnegativity constraints $\overline{w} \geq 0$ on the parameter vector.

14. **Hard Regularization:** Consider the case where instead of Tikhonov regularization, you solve the linear regression problem of minimizing $\|A\overline{x}-\overline{b}\|^2$ subject to the spherical constraint $\|\overline{x}\| \leq r$. Formulate the Lagrangian dual of the problem with variable $\alpha \geq 0$. Show that the primal and dual variables are related at optimality in a similar way to Tikhonov regularization:

$$\overline{x} = (A^T A + \alpha I)^{-1} A^T \overline{b}$$

Under what conditions is α equal to 0? If α is non-zero, show that it is equal to the solution to the following *secular equation*:

$$\overline{b}^T A (A^T A + \alpha I)^{-2} A^T \overline{b} = r^2$$

15. Propose a (primal) gradient-descent algorithm for the hard regularization model of the previous exercise. Use the projected gradient-descent method. The key point is in knowing how to perform the projection step.

16. **Best subset selection:** Consider an $n \times d$ data matrix D in which you want to find the best subset of k features that are related to the n-dimensional regressand vector $\overline{y}$. Therefore, the following mixed integer program is formulated with d-dimensional real vector $\overline{w}$, d-dimensional binary vector $\overline{z}$, and an a priori (constant) upper bound M on each coefficient in $\overline{w}$. The optimization problem is to minimize $\|D\overline{w} - \overline{y}\|^2$ subject to the following constraints:

$$\overline{z} \in \{0,1\}^d, \quad \overline{w} \leq M\overline{z}, \quad \overline{1}^T \overline{z} = k$$

The notation $\overline{1}$ denotes a d-dimensional vector of 1s. Propose an algorithm using block coordinate descent for this problem, where each optimized block contains just two integer variables and all real variables.

17. **Duality Gap:** Suppose that you are running the dual gradient descent algorithm for the SVM, and you have the (possibly suboptimal) dual variables $\alpha_1 \ldots \alpha_n$ in the current iteration. Propose a quick computational procedure to estimate an upper bound on how far this dual solution is from optimality. [Hint: The current dual solution can be used to construct a primal solution.]

18. State whether the following minimax functions $f(x, y)$ satisfy John von Neumann's strong duality condition, where x is the minimization variable and y is the maximization variable: (i) $f(x, y) = x^2 + 3xy - y^4$, (ii) $f(x, y) = x^2 + xy + y^2$, (iii) $f(x, y) = \sin(y - x)$, and (iv) $f(x, y) = \sin(y - x)$ for $0 \leq x \leq y \leq \pi/2$.

19. Failure of coordinate descent: Consider the problem of minimizing $x^2 + y^2$, subject to $x + y \geq 1$. Show using Lagrangian relaxation that the optimal solution is $x = y = 0.5$. Suppose that you start coordinate descent for this problem at $x = 1$ and $y = 0$. Discuss why coordinate descent will fail.

20. Propose a linear variable transformation for Exercise 19, so that coordinate descent will work on the reformulated problem.

21. Formulate a variation of an SVM with hinge loss, in which the binary target (drawn from -1 or $+1$) is known to be non-negatively correlated with each feature based on prior knowledge. Propose a variation of the gradient descent method by using only feasible directions.

Chapter 7

Singular Value Decomposition

"The SVD is absolutely a high point of linear algebra."– Gilbert Strang and Kae
Borre

7.1 Introduction

In Chapter 3, we learned that certain types of matrices, which are referred to as positive
semidefinite matrices, can be expressed in the following form:

$$A = V\Delta V^T$$

Here, V is a $d \times d$ matrix with orthonormal columns, and Δ is a $d \times d$ diagonal matrix
with *nonnegative* eigenvalues of A. The orthogonal matrix V can also be viewed as a rota-
tion/reflection matrix, the diagonal matrix Δ as a nonnegative scaling matrix along axes
directions, and the matrix V^T is the inverse of V. By *factorizing* the matrix A into simpler
matrices, we are expressing a linear transform as a sequence of simpler linear transformations (such as rotation and scaling). This chapter will study the generalization of this type
of factorization to arbitrary matrices. This generalized form of factorization is referred to
as *singular value decomposition*.

Singular value decomposition generalizes the factorization approach to arbitrary matrices that might not even be square. Given an $n \times d$ matrix B, singular value decomposition
decomposes it as follows:

$$B = Q\Sigma P^T$$

Here, B is an $n \times d$ matrix, Q is an $n \times n$ matrix with orthonormal columns, Σ is an
$n \times d$ *rectangular* diagonal matrix with *nonnegative* entries, and P is a $d \times d$ matrix with
orthonormal columns. The notion of a rectangular diagonal matrix is discussed in Figure 1.3
of Chapter 1 in which only entries with indices of the form (i, i) (i.e., with the same row
and column indices) are non-zero. The columns of Q and the columns of P are referred to
as *left singular vectors* and *right singular vectors*, respectively. The entries of Σ are referred
to as *singular values*, and they are arranged in non-increasing order (by convention). We
emphasize that the diagonal matrix Σ is nonnegative.

© Springer Nature Switzerland AG 2020
C. C. Aggarwal, *Linear Algebra and Optimization for Machine Learning*,
https://doi.org/10.1007/978-3-030-40344-7_7

Singular value decomposition has some insightful linear algebra properties in terms of enabling the discovery of all four fundamental subspaces of the matrix B. Furthermore, if exact decomposition is not essential, singular value decomposition provides the ability to approximate B very well with small portions of the factor matrices Q, P, and Σ. This is an *optimization-centric* view of singular value decomposition. The optimization-centric view naturally generalizes to the broader concept of *low-rank matrix factorization*, which lies at the heart of many machine learning applications (cf. Chapter 8).

We will first approach singular value decomposition simply from a linear algebra point of view, as a way of exploring the row and column spaces of a matrix. This view is, however, incomplete because it does not provide an understanding of the compression-centric properties of singular value decomposition. Therefore, we will also present singular value decomposition in terms of the optimization-centric view together with its natural applications to compression and dimensionality reduction.

This chapter is organized as follows. In the next section, we will introduce singular value decomposition from the point of view of linear algebra. An optimization-centric view of singular value decomposition is presented in Section 7.3. Both these views expose somewhat different properties of singular value decomposition. Singular value decomposition (SVD) has numerous applications in machine learning, and an overview is provided in Section 7.4. Numerical algorithms for singular value decomposition are introduced in Section 7.5. A summary is given in Section 7.6.

7.2 SVD: A Linear Algebra Perspective

Singular value decomposition (SVD) is a generalization of the concept of diagonalization, which is discussed in Chapter 3. While diagonalization with nonnegative eigenvalues and orthogonal eigenvectors is only assured for square, symmetric, and positive semidefinite matrices, singular value decomposition is assured for any matrix, irrespective of its size or other properties. Since we have already explored the diagonalization of square matrices, we will first study the singular value decomposition of square matrices in order to show how singular value decomposition is a natural generalization of diagonalization. Then, we will generalize these ideas to rectangular matrices.

7.2.1 Singular Value Decomposition of a Square Matrix

In this section, we will discuss the existence of a singular value decomposition of a square $m \times m$ matrix B. First, we note that the matrices $B^T B$ and BB^T are positive semidefinite and symmetric (cf. Lemma 3.3.14 of Chapter 3). Therefore, these matrices are diagonalizable with orthonormal eigenvectors and nonnegative eigenvalues. In the following, we show that these matrices share eigenvalues, and their eigenvectors are also related.

Lemma 7.2.1 *Let B be a square, $m \times m$ matrix. Then, the following results are true:*

1. *If $\overline{p}$ is a unit eigenvector of $B^T B$ with non-zero eigenvalue λ, then $B\overline{p}$ is an eigenvector of BB^T with the same eigenvalue λ. Furthermore, the norm of $B\overline{p}$ is $\sqrt{\lambda}$.*

2. *If $\overline{q}$ is a unit eigenvector of BB^T with non-zero eigenvalue λ, then $B^T\overline{q}$ is an eigenvector of $B^T B$ with the same eigenvalue λ. Furthermore, the norm of $B^T\overline{q}$ is $\sqrt{\lambda}$.*

Proof: We only show the first part of the above result, because the proof of the second part is exactly identical by working with B^T instead of B throughout the proof. If $\overline{p}$ is an eigenvector of $B^T B$ with eigenvalue λ, we have the following:

$$B^T B \overline{p} = \lambda \overline{p}$$

$$BB^T [B\overline{p}] = \lambda [B\overline{p}] \quad \{ \text{ Pre-multiplying with } B \}$$

In other words, $B\overline{p}$ is an eigenvector of BB^T with eigenvalue λ.

The *squared* norm of $B\overline{p}$ may be computed as follows:

$$\|B\overline{p}\|^2 = [p^T B^T][B\overline{p}] = p^T \underbrace{[B^T B\overline{p}]}_{\lambda \overline{p}}$$

$$= \overline{p}^T [\lambda \overline{p}] = \lambda \|\overline{p}\|^2 = \lambda$$

The last of these equalities follows from the fact that $\overline{p}$ is a unit eigenvector. Since the squared norm of $B\overline{p}$ is λ, it follows that the norm of $B\overline{p}$ is $\sqrt{\lambda}$. ∎

The pairing of the eigenvectors/eigenvalues of $B^T B$ and BB^T can also be expressed as the following corollary:

Corollary 7.2.1 (Eigenvector Pairing) *Let B be a square, $m \times m$ matrix. Then, the matrices $B^T B$ and BB^T have the same set of m eigenvalues $\lambda_1 \ldots \lambda_m$. Let the m orthonormal eigenvectors of the symmetric matrix $B^T B$ be denoted by $\overline{p}_1 \ldots \overline{p}_m$ with eigenvalues $\lambda_1 \ldots \lambda_m$. Then, it is possible to find m orthonormal eigenvectors $\overline{q}_1 \ldots \overline{q}_m$ of BB^T, such that the following holds:*

$$\overline{q}_i \sqrt{\lambda_i} = B\overline{p}_i$$

Proof: This proof works by defining each $\overline{q}_i$ as a function of $\overline{p}_i$. Let there be $r \leq m$ non-zero eigenvalues. In the case when $\overline{p}_i$ is associated with a non-zero eigenvalue, we define $\overline{q}_i = B\overline{p}_i/\sqrt{\lambda_i}$, and Lemma 7.2.1 ensures that each $\overline{q}_i$ is a unit eigenvector of BB^T. The extracted eigenvectors $\overline{q}_1 \ldots \overline{q}_r$ for non-zero eigenvalues are orthogonal to one another:

$$\overline{q}_i^T \overline{q}_j = (B\overline{p}_i)^T (B\overline{p}_j)/\lambda = \overline{p}_i^T ([B^T B]\overline{p}_j)/\lambda = \overline{p}_i^T \overline{p}_j = 0$$

Next, we focus on the remaining $(m - r)$ zero eigenvectors of both $B^T B$ and BB^T. Any zero eigenvector $\overline{q}_i$ of BB^T and any zero eigenvector of $B^T B$ trivially satisfies $\overline{q}_i \sqrt{\lambda_i} = B\overline{p}_i$ because both sides evaluate to zero. The key point is that $B^T B\overline{p}_i = \overline{0}$ implies that $B\overline{p}_i = \overline{0}$ (see Exercise 2 of Chapter 2). Therefore, we can pair the zero eigenvectors of $B^T B$ and BB^T arbitrarily. ∎

Corollary 7.2.1 provides a way of pairing the eigenvectors of $B^T B$ and BB^T in such a way that the condition $\overline{q}_i \sqrt{\lambda_i} = B\overline{p}_i$ is always satisfied for any pair of eigenvectors $(\overline{p}_i, \overline{q}_i)$. This observation can be used to write these paired relationships in matrix form. This way of expressing the pairing is referred to as singular value decomposition.

Theorem 7.2.1 (Existence of SVD) *Let the columns of the $m \times m$ matrix P contain the m orthonormal eigenvectors of the $m \times m$ matrix $B^T B$, and let Σ be an $m \times m$ diagonal matrix with diagonal entries containing the square-root of the corresponding eigenvalues. By convention, the columns of P and Σ are ordered, so that the singular values are in non-increasing order. Then, it is possible to find an $m \times m$ orthogonal matrix Q containing the orthonormal eigenvectors of BB^T, such that the following holds:*

$$B = Q\Sigma P^T$$

Proof: Corollary 7.2.1 ensures that for any ordered set $\bar{p}_1 \ldots \bar{p}_m$ of eigenvectors of $B^T B$, an ordered set $\bar{q}_1 \ldots \bar{q}_m$ of eigenvectors of BB^T exists, so that the following is satisfied for each $i \in \{1 \ldots m\}$:

$$\bar{q}_i \sqrt{\lambda} = B\bar{p}_i$$

One can write the m vector-centric relationships as a single matrix-centric relationship:

$$[\bar{q}_1, \ldots, \bar{q}_m]\Sigma = B[\bar{p}_1 \ldots \bar{p}_m]$$

Here, Σ is an $m \times m$ diagonal matrix whose (i, i)th entry is $\sqrt{\lambda_i}$. One can write the above relationship in the following form:

$$Q\Sigma = BP$$

Here, P is an $m \times m$ orthogonal matrix with columns containing $\bar{p}_1 \ldots \bar{p}_m$, and Q is an $m \times m$ orthogonal matrix with columns containing $\bar{q}_1 \ldots \bar{q}_m$. Post-multiplication of both sides with P^T and setting $PP^T = I$ yields $Q\Sigma P^T = B$. Therefore, a singular value decomposition of a square matrix B always exists. ∎

Consider the following matrix B and its derived scatter matrix $B^T B$:

$$B = \begin{bmatrix} 14 & 8 & -6 \\ 21 & 11 & 14 \\ 16 & -6 & 2 \end{bmatrix}, \quad B^T B = \begin{bmatrix} 893 & 247 & 242 \\ 247 & 221 & 94 \\ 242 & 94 & 236 \end{bmatrix}$$

On performing the eigendecomposition of $B^T B$ we obtain eigenvectors proportional to $[3, 1, 1]^T$, $[1, -1, -2]^T$, and $[1, -7, 4]^T$ (although the vectors need to be unit normalized to create P). The corresponding eigenvalues are 1052, 162, and 232, and the square-roots of these eigenvalues are the singular values, which can be used to create the diagonal matrix Σ. Since we have $B = Q\Sigma P^T$, the matrix Q can then be obtained as $BP\Sigma^{-1}$, which is as follows:

$$Q = \underbrace{\begin{bmatrix} 14 & 8 & -6 \\ 21 & 11 & 14 \\ 16 & -6 & 2 \end{bmatrix}}_{B} \underbrace{\begin{bmatrix} 3/\sqrt{11} & 1/\sqrt{6} & 1/\sqrt{66} \\ 1/\sqrt{11} & -1/\sqrt{6} & -7/\sqrt{66} \\ 1/\sqrt{11} & -2/\sqrt{6} & 4/\sqrt{66} \end{bmatrix}}_{P} \underbrace{\begin{bmatrix} 4\sqrt{66} & 0 & 0 \\ 0 & 9\sqrt{2} & 0 \\ 0 & 0 & 2\sqrt{33} \end{bmatrix}^{-1}}_{\Sigma^{-1}}$$

Upon performing this multiplication, we obtain a matrix Q whose columns are proportional to $[1, 2, 1]^T$, $[1, -1, 1]^T$, and $[-1, 0, 1]^T$, although the matrix Q is obtained in terms of unit normalized columns. Therefore, the SVD of matrix B can be expressed as $Q\Sigma P^T$ as follows:

$$\underbrace{\begin{bmatrix} 1/\sqrt{6} & 1/\sqrt{3} & -1/\sqrt{2} \\ 2/\sqrt{6} & -1/\sqrt{3} & 0 \\ 1/\sqrt{6} & 1/\sqrt{3} & 1/\sqrt{2} \end{bmatrix}}_{Q} \underbrace{\begin{bmatrix} 4\sqrt{66} & 0 & 0 \\ 0 & 9\sqrt{2} & 0 \\ 0 & 0 & 2\sqrt{33} \end{bmatrix}}_{\Sigma} \underbrace{\begin{bmatrix} 3/\sqrt{11} & 1/\sqrt{6} & 1/\sqrt{66} \\ 1/\sqrt{11} & -1/\sqrt{6} & -7/\sqrt{66} \\ 1/\sqrt{11} & -2/\sqrt{6} & 4/\sqrt{66} \end{bmatrix}^{T}}_{P^T}$$

One important point is that we *derived* Q from P, rather than independently diagonalizing BB^T and $B^T B$, and doing the latter *might* lead to incorrect results because of sign dependence between Q and P. For example, one could use $-Q$ and $-P$ as the decomposition matrices without changing the product of the matrices. However, we cannot use $-Q$ and P to create an SVD. The signs of matching pairs of singular vectors are also interdependent.

SVD also decomposes non-diagonalizable matrices, such as the following:

$$\underbrace{\begin{bmatrix} 0 & -7 \\ 0 & 0 \end{bmatrix}}_{} = \underbrace{\begin{bmatrix} -1 & 0 \\ 0 & 1 \end{bmatrix}}_{Q} \underbrace{\begin{bmatrix} 7 & 0 \\ 0 & 0 \end{bmatrix}}_{\Sigma} \underbrace{\begin{bmatrix} 0 & 1 \\ 1 & 0 \end{bmatrix}}_{P^T}$$

Note that the above matrix has no diagonalization, since it is nilpotent (see Exercise 26 of Chapter 3). However, it has a valid singular value decomposition. Furthermore, even though this matrix only has zero eigenvalues, it has a non-zero singular value of 7, containing one of the key *scaling factors* of the transformation. In fact, SVD has the neat property of relating arbitrary (square) matrices to positive semidefinite ones with the use of *polar decomposition*, which explicitly separates out the rotreflection matrix from the scaling (positive semidefinite) matrix:

Lemma 7.2.2 (Polar Decomposition) *Any square matrix can be expressed in the form* US, *where* U *is an orthogonal matrix, and* S *is a symmetric positive semidefinite matrix.*

Proof: One can write the SVD of a square matrix as $Q\Sigma P^T = (QP^T)(P\Sigma P^T)$. The matrix QP^T can be set to U, and it is orthogonal because of the closure of orthogonal matrices under multiplication (cf. Chapter 2). Furthermore, S can be set to $P\Sigma P^T$, which is positive semidefinite because of the nonnegativity of Σ. ∎

The polar decomposition is geometrically insightful, because it tells us that *every matrix multiplication causes an anisotropic scaling along orthogonal directions with nonnegative scale factors, followed by rotreflection*. When the rotreflection component is missing, the resulting matrix is positive semidefinite. The matrix U is also the *nearest orthogonal matrix* to B, just as $[\cos(\theta), \sin(\theta)]^T$ is the nearest unit vector to the polar coordinates $r[\cos(\theta), \sin(\theta)]^T$.

Problem 7.2.1 *Let* B *be a symmetric and square matrix, which is negative semidefinite. Show that the singular value decomposition of* B *is of the form* $B = Q\Sigma P^T$, *where* $Q = -P$.

The important point of the previous exercise is to emphasize the fact that the singular values need to be nonnegative. We provide another exercise to emphasize this fact:

Problem 7.2.2 *Suppose that somebody gave you an* $m \times m$ *matrix* B *and a decomposition of the form* $B = Q\Sigma P^T$, *where* Q *and* P *are both orthogonal matrices of size* $m \times m$, *and* Σ *is an* $m \times m$ *diagonal matrix. However, you are told that some of the entries of* Σ *are negative. Discuss how you would adjust the decomposition in order to convert it into a standard form of singular value decomposition.*

Problem 7.2.3 *Suppose that the eigendecomposition of a* 3×3 *symmetric matrix* A *can be written as follows:*

$$A = V\Delta V^T = \begin{bmatrix} v_{11} & v_{12} & v_{13} \\ v_{21} & v_{22} & v_{23} \\ v_{31} & v_{32} & v_{33} \end{bmatrix} \begin{bmatrix} 5 & 0 & 0 \\ 0 & -2 & 0 \\ 0 & 0 & -3 \end{bmatrix} \begin{bmatrix} v_{11} & v_{21} & v_{31} \\ v_{12} & v_{22} & v_{32} \\ v_{13} & v_{23} & v_{33} \end{bmatrix}$$

What is the singular value decomposition of this matrix?

The number of non-zero singular values yields the rank of the original matrix.

Lemma 7.2.3 *Let* B *be an* $m \times m$ *matrix with rank* $k \leq m$. *Let the singular value decomposition of* B *be* $B = Q\Sigma P^T$, *where* Q, Σ, *and* P^T *are* $m \times m$ *matrices. Then, exactly* $m - k$ *singular values must be zeros.*

Proof: As discussed in Corollary 2.6.3, multiplication with a non-singular (or orthogonal) matrix does not change the rank of a matrix. Therefore, the rank of $B = Q\Sigma P^T$ is the same as that of Σ. Since the rank of Σ is equal to the number of non-zero singular values, the result follows.   ∎

7.2.2 Square SVD to Rectangular SVD via Padding

Consider the special case in which the matrix B is obtained by padding an $n \times d$ matrix D with additional rows or columns of zero values, so that we have a square matrix B with $m = max\{n, d\}$ rows and columns. This type of padding leads to natural way of performing SVD of rectangular matrices because portions of the (unnecessarily large) factored matrices of the padded matrix can be extracted to create a decomposition of the original matrix (without the padding). For example, while working with an $n \times d$ matrix denoted by D, one can factorize it into a sequence of an $n \times n$ orthogonal matrix, an $n \times d$ rectangular diagonal matrix, and a $d \times d$ orthogonal matrix. These three (smaller) matrices can be extracted directly as portions of the three (larger) factors of the $m \times m$ matrix B. Consider a situation where an $n \times d$ matrix D is padded with zeros (in either rows or columns) in order to obtain the square matrix B. In such cases, it can be shown that singular value decomposition has one of the following two types of block diagonal structures of factor matrices:

Lemma 7.2.4 (Block Diagonal Structure of Padded SVD) *Let B be an $m \times m$ matrix obtained by padding the $n \times d$ matrix D with either zero rows or zero columns, where $m = max\{n, d\}$. Then, depending on whether n or d is greater, a singular value decomposition $B = Q\Sigma P^T$ exists with one of the following two types of forms:*

$$B = [D\ 0] = Q \underbrace{\begin{bmatrix} \Sigma_1 & 0 \\ 0 & 0 \end{bmatrix}}_{\Sigma} \underbrace{\begin{bmatrix} P_1 & 0 \\ 0 & P_2 \end{bmatrix}^T}_{P^T}, \quad \textit{[When } d < n]$$

$$B = \begin{bmatrix} D \\ 0 \end{bmatrix} = \underbrace{\begin{bmatrix} Q_1 & 0 \\ 0 & Q_2 \end{bmatrix}}_{Q} \underbrace{\begin{bmatrix} \Sigma_1 & 0 \\ 0 & 0 \end{bmatrix}}_{\Sigma} P^T, \quad \textit{[When } n < d]$$

The matrices, Q, P, and Σ are all of sizes $m \times m$, as is normally the case for square SVD. The matrix P_1 is of size $d \times d$, and Q_1 is of size $n \times n$. The matrices P_2 and Q_2 are of sizes $(m - d) \times (m - d)$ and $(m - n) \times (m - n)$, respectively. The matrix Σ_1 is of size $min\{n, d\} \times min\{n, d\}$.

Proof Sketch: Consider the first case above where $B = [D\ 0]$ and $d < n$. In such a case, $B^T B$ will only have a single non-zero block of size $d \times d$ in the upper-left corner. As a result, it will have at most d non-zero eigenvalues, the square-roots of which can be used to create the $d \times d$ diagonal matrix Σ_1. The eigenvectors of its upper-left block will be contained in the $d \times d$ matrix P_1. Let the $(n - d) \times (n - d)$ matrix P_2 be created by stacking up any set of $(n - d)$ orthonormal column vectors in $\mathcal{R}^{(n-d)}$. It remains to show that if matrix P and Σ are constructed using P_1, P_2, and Σ_1 using the block structure shown on the right-hand side of the first relationship above, then (i) P will contains both the non-zero and zero eigenvectors of $B^T B$, and (ii) the matrix Σ^2 contains the eigenvalues of $B^T B$. This can be achieved by showing that the ith column of P is a right-eigenvector of $B^T B$ with the corresponding eigenvalue contained in the ith diagonal entry of Σ^2. The result holds because for $i \leq d$, the eigenvectors and eigenvalues are inherited from eigenvectors of the upper-left block of $B^T B$ with size $d \times d$. These eigenvectors are contained in P_1 and the padding simply adds $(n - d)$ zero values both to the ith column of $B^T B$ and to the ith column of P. For $i > d$, any n-dimensional vector with zero values in the first d components can be shown to be an eigenvector of $B^T B$ (with 0 eigenvalue) because of the block structure of $B^T B$. Furthermore, the matrix P can be shown to be orthogonal because both of its blocks are orthogonal matrices. The matrix Q can be extracted from B, Σ, and P using the methods discussed in the proof of Theorem 7.2.1. Therefore, one can create an SVD respecting the block diagonal structure in the first case of the statement of the lemma (when $n > d$). The second case for $n < d$ can be proven using a similar argument. ∎

Instead of using singular value decomposition on the padded matrix B, one can directly decompose the matrix D by pulling out portions of the block structure of padded SVD:

$$D = Q \begin{bmatrix} \Sigma_1 \\ 0 \end{bmatrix} P_1^T, \quad \text{[When } d < n]$$

$$D = Q_1 [\Sigma_1 \ 0] P^T, \quad \text{[When } n < d]$$

Both Q and P are square, and only the $n \times d$ diagonal matrix Σ is rectangular in both relationships. The square submatrix Σ_1 is of size $\min\{n, d\} \times \min\{n, d\}$, and the $n \times d$ matrix Σ is obtained by padding it with $|n - d|$ zero rows or columns. Unlike the SVD of B, the right singular vectors and left singular vectors of D are no longer of the same dimensionality. The left singular vector matrix is always of size $n \times n$, whereas the right singular vector matrix is always of size $d \times d$. This is the standard form of rectangular singular value decomposition. However, other variations of singular value decomposition are even more economical, and will be discussed in the next section.

7.2.3 Several Definitions of Rectangular Singular Value Decomposition

We start with a formal summary of the rectangular SVD derived in the previous section:

Definition 7.2.1 (Singular Value Decomposition) *Consider an $n \times d$ matrix D with real-valued entries. Such a matrix can always be factorized into three matrices as follows:*

$$D = Q \Sigma P^T$$

Here, Q is an $n \times n$ matrix with orthonormal columns containing the left singular vectors, Σ is an $n \times d$ rectangular "diagonal" matrix with diagonal entries containing the nonnegative singular values in non-increasing order, and P is a $d \times d$ matrix with orthonormal columns containing the right singular vectors.

We present a number of important properties of the right singular vectors and left singular vectors below. These properties follow directly from the discussion in the previous section:

1. The n columns of Q, which are referred to as the left singular vectors, correspond to the n eigenvectors of the $n \times n$ matrix DD^T. Note that these eigenvectors are orthonormal because DD^T is a symmetric matrix.

2. The d columns of P, which correspond to the right singular vectors, correspond to the d eigenvectors of the $d \times d$ matrix $D^T D$. These eigenvectors are orthonormal because $D^T D$ is a symmetric matrix.

3. The diagonal entries of the $n \times d$ rectangular diagonal matrix Σ contain the singular values, which are the square-roots of the $\min\{n, d\}$ largest eigenvalues of $D^T D$ or DD^T.

4. By convention, the columns of Q, P, and Σ are ordered by non-increasing singular value.

The above form of singular value decomposition is also referred to as *full* singular value decomposition. Note that either Q or P will be larger than the original matrix D when $n \neq d$, and the $n \times d$ matrix Σ is of the same size as the original matrix. In fact, the larger of Q and P will contain $|n - d|$ *unmatched* eigenvectors that are not represented in the $\min\{n, d\}$ diagonal entries of Σ. This would seem wasteful.

A more economical form of the decomposition is *economy* singular value decomposition, which can be derived from the *spectral decomposition* of the matrix. Let σ_{rr} be the (r, r)th entry of Σ, $\overline{q}_r$ be the rth column of Q, and $\overline{p}_r$ be the rth column of P. Then, the matrix product $Q\Sigma P^T$ can be decomposed into the sum of rank-1 matrices:

$$D = Q\Sigma P^T = \sum_{r=1}^{\min\{n,d\}} \sigma_{rr}\overline{q}_r\overline{p}_r^T \tag{7.1}$$

The right-hand side of the above result is obtained by simply applying one of the fundamental ways of characterizing matrix multiplication (cf. Lemma 1.2.1 of Chapter 1) to the product of the matrices $(Q\Sigma)$ and P^T. The above form of the decomposition is also referred to as the spectral decomposition of the matrix D. Each of the $\min\{n, d\}$ terms (i.e., the $n \times d$ matrix $\sigma_{rr}\overline{q}_r\overline{p}_r^T$) in the above summation is referred to as a *latent component* of the original $n \times d$ matrix D. This term is referred to as a *latent component*, because it represents the independent, hidden (or *latent*) pieces of the matrix D. Note that each $\overline{q}_r^T\overline{p}_r$ is a rank-1 matrix of size $n \times d$, because it is obtained from the product of an n-dimensional column vector with a d-dimensional row vector. The above form of the spectral decomposition provides the insight necessary to propose a form of SVD, referred to as *economy* singular value decomposition. The idea is that each term of Equation 7.1 can be used to create one of the $p = \min\{n, d\}$ columns of each of the decomposed matrices:

Definition 7.2.2 (Economy Singular Value Decomposition) *Consider an $n \times d$ matrix D with real-valued entries, where $p = min\{n, d\}$. Such a matrix can always be factorized into three matrices as follows:*

$$D = Q\Sigma P^T$$

Here, Q is an $n \times p$ matrix with orthonormal columns containing the left-singular vectors, Σ is an $p \times p$ diagonal matrix with diagonal entries containing nonnegative singular values in non-increasing order, and P is a $d \times p$ matrix with orthonormal columns containing the right-singular vectors.

One of the two matrices Q and P may no longer be square, as we are shedding unmatched singular vectors from the larger of the two matrices in full singular value decomposition.

One can further reduce the size of the decomposition by observing that some of the $\min\{n, d\}$ values of σ_{rr} might be zero. Such a situation will occur in the case of a matrix D with rank k that is strictly smaller than $\min\{n, d\}$. In such cases, one can keep only the $k < \min\{n, d\}$ *strictly positive* singular values without affecting the sum. Assume that the singular values are ordered by non-increasing value, so that $\sigma_{11} \geq \sigma_{22} \geq \ldots \geq \sigma_{kk}$. In such a case, we can write the above decomposition as follows:

$$D = \sum_{r=1}^{k} \sigma_{rr} \bar{q}_r \bar{p}_r^T \qquad (7.2)$$

Note that the above summation uses *all* the k *strictly positive* singular values. This leads to a slightly different form of singular value decomposition, which is referred to as *compact* singular value decomposition or *reduced* singular value decomposition. Compact singular value decomposition is defined as follows:

Definition 7.2.3 (Compact Singular Value Decomposition) *Consider an $n \times d$ matrix D with real-valued entries, which has rank $k \leq min\{n, d\}$. Such a matrix can always be factorized into three matrices as follows:*

$$D = Q\Sigma P^T$$

Here, Q is an $n \times k$ matrix with orthonormal columns containing the left-singular vectors, Σ is an $k \times k$ diagonal matrix with diagonal entries containing all the **positive** *singular values in non-increasing order, and P is a $d \times k$ matrix with orthonormal columns containing the right-singular vectors.*

The compact version of singular value decomposition can factorize a matrix into much smaller matrices, especially if $k \ll min\{n, d\}$. The number of entries in D is $n \cdot d$, whereas the total number of entries in the three factorized matrices is $(n + d + k) \cdot k$. The latter value can often be much smaller. If one is willing to take this argument further and lose some representation accuracy, further reduction in the sizes of the factorized matrices can be achieved with *truncated* singular value decomposition. Truncated singular value decomposition is, in fact, the primary way in which SVD is used in real applications.

7.2.4 Truncated Singular Value Decomposition

In many real applications, it suffices to be able to reconstruct the matrices *approximately*. Consider the spectral decomposition of the matrix D based on the discussion in the previous section:

$$D = Q\Sigma P^T = \sum_{r=1}^{min\{n,d\}} \sigma_{rr} \bar{q}_r \bar{p}_r^T \qquad (7.3)$$

Instead of only dropping the additive components for which $\sigma_{rr} = 0$, we might also drop those components for which σ_{rr} is very small. In other words, we keep the top-k values of σ_{rr} in the decomposition (like compact SVD), except that k might be smaller than the number of non-zero singular values. In such a case, we obtain an *approximation D_k* of the original matrix D, which is also referred to as the *rank-k* approximation of the $n \times d$ matrix D:

$$D \approx D_k = \sum_{r=1}^{k} \sigma_{rr} \bar{q}_r \bar{p}_r^T \qquad (7.4)$$

Note that Equation 7.4 for truncated singular value decomposition is the same as that for compact singular value decomposition (cf. Equation 7.2); the only difference is that the value of k is no longer chosen to ensure zero information loss. Consequently, we can express truncated singular value decomposition as a matrix factorization as follows:

$$D \approx D_k = Q_k \Sigma_k P_k^T \qquad (7.5)$$

Here, Q_k is an $n \times k$ matrix with columns containing the top-k left singular vectors, Σ_k is a $k \times k$ diagonal matrix containing the top-k singular values, and P_k is a $d \times k$ matrix with columns containing the top-k right singular vectors. It is not difficult to see that the matrix D_k is of rank-k, and therefore it is viewed as a *low-rank approximation* of D.

Almost all forms of matrix factorization, including singular value decomposition, are low-rank approximations of the original matrix. Truncated singular value decomposition can retain a surprisingly large level of accuracy using values of k that are much smaller than $\min\{n, d\}$. This is because only a very small proportion of the singular values are large in real-world matrices. In such cases, D_k becomes an excellent approximation of D by retaining the few singular vectors that are large.

A useful property of truncated singular value decomposition is that it is also possible to create a lower dimensional representation of the data by changing the basis to P_k, so that each d-dimensional data point is now represented in only k dimensions. In other words, we change the axes so that the basis vectors correspond to the columns of P_k. This transformation is achieved by post-multiplying the data matrix D with P_k to obtain the $n \times k$ matrix U_k. By post-multiplying Equation 7.5 with P_k and using $P_k^T P_k = I_k$, we obtain the following:

$$U_k = DP_k = Q_k\Sigma_k \tag{7.6}$$

Each row of U_k contains a reduced k-dimensional representation of the corresponding row in D. Therefore, we can obtain a reduced representation of the data either by post-multiplying the data matrix with the matrix containing the dominant right singular vectors (i.e., using DP_k), or we can simply scale the dominant left singular vectors with the singular values (i.e., using $Q_k\Sigma_k$). Both these types of methods are used in real applications, depending on whether n or d is larger.

The reduction in dimensionality can be very significant in some domains such as images and text. Image data are often represented by matrices of numbers corresponding to pixels. For example, an image corresponding to an 807×611 matrix of numbers is illustrated in Figure 7.1(a). Only the first 75 singular values are represented in Figure 7.1(b). The remaining $611 - 75 = 536$ singular values are not shown because they are very small. The

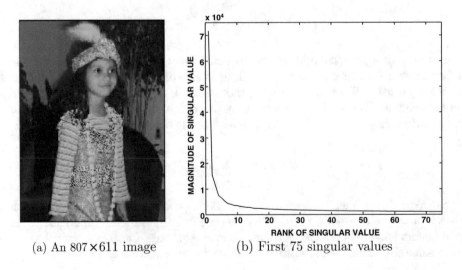

(a) An 807×611 image   (b) First 75 singular values

Figure 7.1: The rapid decay in singular values for an 807×611 image

rapid decay in singular values is quite evident in the figure. It is this rapid decay that enables effective truncation without loss of accuracy. In the text domain, each document is represented as a row in a matrix with as many dimensions as the number of words. The value of each entry is the frequency of the word in the corresponding document. Note that this matrix is sparse, which is a standard use-case for SVD. The word-frequency matrix D might have $n = 10^6$ and $d = 10^5$. In such cases, truncated SVD might often yield excellent approximations of the matrix by using $k \approx 400$. This represents a drastic level of reduction in the dimensionality of representation. The use of SVD in text is also referred to as *latent semantic analysis* because of its ability to discover latent (hidden) topics represented by the rank-1 matrices of the spectral decomposition.

7.2.4.1 Relating Truncation Loss to Singular Values

A natural question arises as to the amount of accuracy loss caused by truncation. Here, it is important to understand that the spectral decomposition of SVD expresses a matrix as a sum of matrices that are *Frobenius* orthogonal in terms of their *Frobenius inner product*:

Definition 7.2.4 (Frobenius Inner Product and Orthogonality) *The Frobenius inner product* $\langle A, B \rangle_F$ *between* $n \times d$ *matrices* $A = [a_{ij}]$ *and* $B = [b_{ij}]$ *is equal to* $\sum_i \sum_j a_{ij} b_{ij}$, *which is equal to the trace of* $A^T B$:

$$\langle A, B \rangle_F = \langle B, A \rangle_F = tr(A^T B) = tr(AB^T)$$

Two matrices are Frobenius orthogonal if their Frobenius inner product is 0.

The squared Frobenius norm is a special case of the Frobenius inner product. The Frobenius orthogonality of matrices can be viewed in a similar way to the pairwise orthogonality of vectors by simply converting each matrix into a vector representation. One simply flattens all the entries of each matrix into a vector and computes the dot product between them. Many of the norm properties of sums of pairwise orthogonal vectors are also inherited by matrices. This is not particularly surprising because one can view the set of all $n \times d$ matrices as a vector space in $\mathcal{R}^{n \times d}$ and an inner product that behaves similarly to the dot product. For example, the Frobenius inner product also satisfies the Pythagorean theorem:

Lemma 7.2.5 *Let A and B be two $n \times d$ matrices that are Frobenius orthogonal. Then, the squared Frobenius norm of $(A + B)$ can be expressed in terms of the Frobenius norms of A and B as follows:*

$$\|A + B\|_F^2 = \|A\|_F^2 + \|B\|_F^2$$

Proof: The above result is relatively easy to show by expressing the Frobenius norm in terms of the trace of the matrix:

$$\|A + B\|_F^2 = tr[(A + B)^T (A + B)] = tr(A^T A) + \underbrace{tr(A^T B) + tr(B^T A)}_{=0} + tr(B^T B)$$

$$= \|A\|_F^2 + \|B\|_F^2$$

Note that we used Frobenius orthogonality to set some of the terms to 0. ∎
One can generalize the above result to the sum of any number of matrices by recursively applying the above lemma.

Corollary 7.2.2 *Let $A_1 \ldots A_k$ be any set of k matrices of the same size that are all Frobenius orthogonal to one another. Then, the squared Frobenius norm of the sum of these matrices can be expressed in terms of the Frobenius norms of the individual matrices as follows:*

$$\| \sum_{i=1}^{k} A_i \|_F^2 = \sum_{i=1}^{k} \| A_i \|_F^2$$

One can generalize the above result to the case where a weighted sum of the matrices is used. We leave the proof of the generalized result as an exercise:

Corollary 7.2.3 *Let $A_1 \ldots A_k$ be any set of k matrices of the same size that are all Frobenius orthogonal to one another. Then, the Frobenius norm of a linear combination of these matrices can be expressed in terms of the Frobenius norms of the individual matrices as follows:*

$$\| \sum_{i=1}^{k} \sigma_i A_i \|_F^2 = \sum_{i=1}^{k} \sigma_i^2 \| A_i \|_F^2$$

Here, each σ_i is a scalar weight.

Next, we will show that the rank-1 matrices of spectral decomposition are all Frobenius orthogonal to one another:

Lemma 7.2.6 *Let $\overline{q}_i$ and $\overline{q}_j$ be orthogonal to one another, and also let $\overline{p}_i$ and $\overline{p}_j$ be orthogonal. Then, the rank-1 matrices $D_i = \overline{q}_i \overline{p}_i^T$ and $D_j = \overline{q}_j \overline{p}_j^T$ are Frobenius orthogonal.*

Proof: One can show that the matrices D_i and D_j are Frobenius orthogonal by showing that the trace of $D_i^T D_j$ is 0. Therefore, we have:

$$\text{tr}(D_i^T D_j) = \text{tr}([\overline{p}_i \overline{q}_i^T][\overline{q}_j \overline{p}_j^T]) = \text{tr}(\overline{p}_i \underbrace{[\overline{q}_i^T \overline{q}_j]}_{0} \overline{p}_j^T) = 0$$

Note that we used the orthogonality of $\overline{q}_i$ and $\overline{q}_j$ in the above proof, but we did not use the orthogonality of $\overline{p}_i$ and $\overline{p}_j$. This lemma can be shown to be true under the weaker conditions that *either* of the vector pairs $(\overline{q}_i, \overline{q}_j)$ and $(\overline{p}_i, \overline{p}_j)$ are orthogonal. ∎

The matrix $\overline{q}_i \overline{p}_i^T$ in the spectral decomposition is the outer-product of two vectors with unit norm. The Frobenius norm of such a matrix can be shown to be 1.

Lemma 7.2.7 *Let $\overline{p}_i$ and $\overline{q}_i$ be a pair of vectors with unit norm. The Frobenius norm of the rank-1 matrix of the form $D_i = \overline{q}_i \overline{p}_i^T$ is 1.*

Proof: The Frobenius norm of D_i can be expressed in terms of the trace as follows:

$$\| D_i \|_F^2 = \text{tr}(D_i^T D_i) = \text{tr}(\overline{p}_i \underbrace{[\overline{q}_i^T \overline{q}_i]}_{=1} \overline{p}_i^T) = \text{tr}(\overline{p}_i \overline{p}_i^T) = \text{tr}(\underbrace{\overline{p}_i^T \overline{p}_i}_{=1}) = 1$$

∎

Let us now take a moment to examine the spectral decomposition of the matrix created by truncated SVD. We replicate the spectral decomposition of rank-k truncated SVD from Equation 7.4 here:

$$D \approx D_k = Q_k \Sigma_k P_k^T = \sum_{r=1}^{k} \sigma_{rr} \overline{q}_r \overline{p}_r^T \tag{7.7}$$

Here, it is evident that the spectral decomposition on the right-hand side contains a bunch of Frobenius orthogonal matrices. Each of these matrices has a Frobenius norm of 1, but they are weighted by σ_{rr}. Therefore, taking the Frobenius norm of all expressions in Equation 7.7, we obtain the following (based on Corollary 7.2.3):

$$\|D\|_F^2 \approx \|D_k\|_F^2 = \|\sum_{r=1}^{k} \sigma_{rr} \bar{q}_r \bar{p}_r^T\|_F^2 = \sum_{r=1}^{k} \sigma_{rr}^2 \underbrace{\|\bar{q}_r \bar{p}_r^T\|_F^2}_{=1} = \sum_{r=1}^{k} \sigma_{rr}^2$$

Therefore, we obtain the result that the squared Frobenius norm of the rank-k approximation is equal to the sum of the squares of the top-k singular values. The squared Frobenius norm of a matrix is referred to as its *energy* (cf. Section 1.2.6 of Chapter 1). Therefore, the lost energy is equal to the sum of the squares of the *smallest* singular values (excluding the top-k singular values), which is also a measure of the squared error of the approximation. In fact, Section 7.3 shows that SVD provides a rank-k approximation of the matrix D, which has the smallest squared error among the universe of *all possible rank-k approximations*.

7.2.4.2 Geometry of Rank-k Truncation

The rank-k truncation reduces the dimensionality of the data, because the rank-k approximation $D_k = Q_k \Sigma_k P_k^T$ no longer requires d dimensions for representation. Rather, it can be expressed in $k \ll d$ dimensions, which translates to significant space savings. One can simply rotate the truncated representation to a k-dimensional basis without further loss of accuracy. As discussed in Equation 7.6, one can obtain an $n \times k$ reduced representation matrix U_k as follows:

$$U_k = DP_k = Q_k \Sigma_k \tag{7.8}$$

Each row of $U_k = DP_k$ contains a k-dimensional representation of the corresponding row in D. The k columns of P_k contain the top eigenvectors of the scatter matrix $D^T D$, and they preserve the directions with the largest possible scatter among all possible directions. We will explicitly show this result in Section 7.3, which provides an optimization-centric view of SVD. This situation is illustrated for a 3-dimensional data set in Figure 7.2, in which most of the energy is retained in one or two eigenvectors with the largest scatter. Therefore, by projecting the data onto this new axis system, most of the scatter of the data about the origin (i.e., energy) can be preserved in one or two dimensions.

In order to understand the geometric effect of the dimensionality reduction achieved by singular value decomposition, consider a large data set in which all points are normally distributed in a perfect ellipsoid centered at the origin, and the standard deviation along the ith axis of the ellipsoid is β_i. Singular value decomposition will find all the axes of this ellipsoid as the right singular vectors, and the ith singular value will be $\sigma_i = \beta_i$. An example of an origin-centered ellipsoid and its three axes directions is shown in Figure 7.3. These three axes directions are the right singular vectors. The left singular vectors may be obtained by applying the same approach to the transpose of the data set.

7.2.4.3 Example of Truncated SVD

We provide an example of truncated SVD with the use of a toy text collection, which has 6 documents and 6 words. The (i, j)th in the data matrix D is the frequency of word j in document i. The 6×6 data matrix D is defined over the following vocabulary:

lion, tiger, cheetah, jaguar, porsche, ferrari

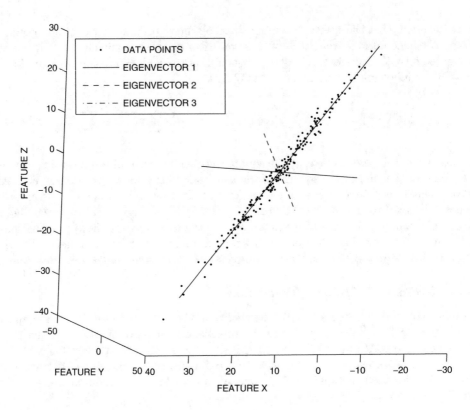

Figure 7.2: Most of energy of the data is retained in the projection along the one or two largest eigenvectors of the 3×3 matrix $D^T D$

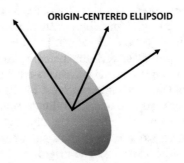

Figure 7.3: SVD models the data to be distributed in an ellipsoid centered at the origin

The frequencies of the words in each document of the data matrix D are illustrated below:

$$D = \begin{pmatrix} & \text{lion} & \text{tiger} & \text{cheetah} & \text{jaguar} & \text{porsche} & \text{ferrari} \\ \text{Document-1} & 2 & 2 & 1 & 2 & 0 & 0 \\ \text{Document-2} & 2 & 3 & 3 & 3 & 0 & 0 \\ \text{Document-3} & 1 & 1 & 1 & 1 & 0 & 0 \\ \text{Document-4} & 2 & 2 & 2 & 3 & 1 & 1 \\ \text{Document-5} & 0 & 0 & 0 & 1 & 1 & 1 \\ \text{Document-6} & 0 & 0 & 0 & 2 & 1 & 2 \end{pmatrix}$$

Note that this matrix represents topics related to both cars and cats. The first three documents are primarily related to cats, the fourth is related to both, and the last two are primarily related to cars. The word "jaguar" is ambiguous because it could correspond to either a car or a cat. We perform an SVD of rank-2 to capture the two latent components in the collection, which is as follows:

$$D \approx Q_2 \Sigma_2 P_2^T$$

$$\approx \begin{pmatrix} -0.41 & 0.17 \\ -0.65 & 0.31 \\ -0.23 & 0.13 \\ -0.56 & -0.20 \\ -0.10 & -0.46 \\ -0.19 & -0.78 \end{pmatrix} \begin{pmatrix} 8.4 & 0 \\ 0 & 3.3 \end{pmatrix} \begin{pmatrix} -0.41 & -0.49 & -0.44 & -0.61 & -0.10 & -0.12 \\ 0.21 & 0.31 & 0.26 & -0.37 & -0.44 & -0.68 \end{pmatrix}$$

$$= \begin{pmatrix} 1.55 & 1.87 & 1.67 & 1.91 & 0.10 & 0.04 \\ 2.46 & 2.98 & 2.66 & 2.95 & 0.10 & -0.03 \\ 0.89 & 1.08 & 0.96 & 1.04 & 0.01 & -0.04 \\ 1.81 & 2.11 & 1.91 & 3.14 & 0.77 & 1.03 \\ 0.02 & -0.05 & -0.02 & 1.06 & 0.74 & 1.11 \\ 0.10 & -0.02 & 0.04 & 1.89 & 1.28 & 1.92 \end{pmatrix}$$

The reconstructed matrix is a very good approximation of the original data matrix D. One can also obtain a 2-dimensional embedding of each row of D as $DP_2 = Q_2\Sigma_2$:

$$DP_2 = Q_2\Sigma_2 \approx \begin{pmatrix} -3.46 & 0.57 \\ -5.44 & 1.03 \\ -1.95 & 0.41 \\ -4.74 & -0.66 \\ -0.83 & -1.49 \\ -1.57 & -2.54 \end{pmatrix}$$

It is clear that the reduced representations of the first three rows are quite similar, which is not surprising. After all the corresponding documents belong to similar topics. At the same time, the reduced representations of the last two rows are also similar. The fourth row seems to be somewhat different because it contains a combination of two topics. Therefore, the latent components seem to capture the hidden "concepts" in the data matrix. In this case, these hidden concepts correspond to cats and cars.

7.2.5 Two Interpretations of SVD

In this section, we will discuss two interpretations of SVD, which correspond to the *data-centric* and the *transformation-centric* interpretations of SVD.

In the data-centric interpretation, SVD is viewed as a way of providing an orthogonal basis for both the row space and the column space of the data matrix D. Note that QR decomposition (cf. Section 2.7.2 of Chapter 2) can provide an orthonormal basis for either the row space or the column space (depending on whether it is performed on the matrix or its transpose), but not both simultaneously. Consider the compact SVD of the $n \times d$ data matrix D of rank k:

$$D = Q\Sigma P^T$$

The $d \times k$ matrix P contains the d-dimensional basis vectors of the (transposed) rows of D in its columns; there are k of them because the original data matrix has rank k. The $n \times k$

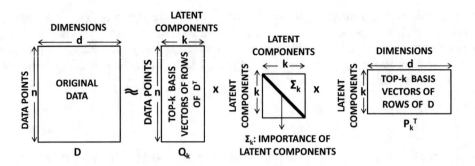

Figure 7.4: Interpretation of SVD in terms of the basis vectors of rows and columns of D

matrix Q contains the n-dimensional basis vectors of the columns of D in its columns. In other words, *SVD simultaneously finds the basis sets of both the (transposed) rows and the columns of the data matrix.* The square of the ith diagonal entry of the matrix Σ provides a quantification of the energy of the 1-dimensional data set $D\overline{p}_i$ obtained by projecting it along the ith right singular vector. Directions with larger scatter obviously retain larger information about the data set. For example, when the singular value σ_{ii} is small, each value in $D\overline{p}_i$ tends to be close to zero. When truncated SVD is used instead of compact SVD, we are restricting ourselves to finding *approximate* basis sets rather than exact basis sets. In other words, we can use these basis sets to represent all the rows in the data matrix approximately, but not exactly. This ability of truncated SVD to simultaneously find approximate bases for the row space and column space is shown in Figure 7.4. Note that each of the k pieces $\sigma_{ii}\overline{q}_i\overline{p}_i^T$ represents a portion of D corresponding to a latent (or hidden) component of the matrix. Truncated SVD, therefore, represents a matrix in terms of its dominant hidden components.

SVD can also be interpreted from a transformation-centric point of view, especially when it is performed on square matrices. Consider a square $d \times d$ matrix A, which is used to transform the d-dimensional rows of the $n \times d$ data matrix D into the d-dimensional rows of the $n \times d$ matrix DA. One can replace A with its SVD $Q\Sigma P^T$, which corresponds to a sequence of rotation/reflection, anisotropic scaling, and another rotation/reflection. This seems very similar to what happens in diagonalization of positive semidefinite matrices. The only difference is that the two rotations/reflections cancel each other out in positive semidefinite matrices, whereas they do not cancel each other out in SVD. SVD implies that *any linear transformation can be expressed as a combination of rotation/reflection and scaling.* Another way of viewing this point is that if we have an $n \times d$ data matrix D, whose scatter plot is an origin-centered ellipsoid in d-dimensions, and we multiply it with an *arbitrary* $d \times d$ matrix A to create the matrix DA, the resulting scatter plot will still be a re-scaled and re-oriented ellipsoid! Both the left and right singular vectors will affect the final orientation, and the singular values will affect the scaling. An example of a transformation of a 2-dimensional scatter plot is illustrated in Figure 7.5.

Both the aforementioned interpretations are rooted in linear algebra. SVD can also be interpreted from an optimization-centric point of view, wherein it tries to find an approximate factorization that preserves the maximum energy from the data set. In Section 7.3, we will explore this optimization-centric interpretation, which is a gateway to more general forms of matrix factorization (cf. Chapter 8).

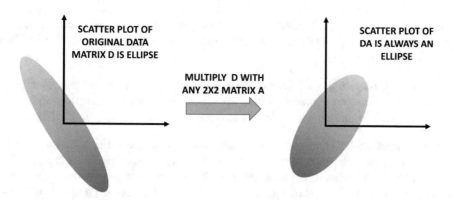

Figure 7.5: The transformation-centric interpretation of SVD as a pair of rotations/reflections and a distortion

7.2.6 Is Singular Value Decomposition Unique?

Given a data matrix, the SVD is a relatively restricted form of decomposition compared to other types of decompositions like the QR method. For example, the QR decomposition varies drastically according to the order in which one processes the different vectors that are orthogonalized. However, the SVD is much more specialized, and can sometimes be close to unique (irrespective of the numerical algorithm used). In Section 3.3.3, you learned that the diagonalization of a square matrix is unique (after imposing sign and normalization conventions) if there are no repeated eigenvalues. Singular value decomposition can be viewed as a generalization of the diagonalization of positive semidefinite matrices (with orthonormal *eigenvectors*) to matrices that are not symmetric or even square (with orthonormal *singular vectors*). Interestingly, the condition for uniqueness of singular value decomposition is also similar to that of diagonalization — the non-zero singular values need to be distinct.

We first consider the singular value decomposition of a square matrix B. The singular value decomposition is almost unique, if and only if all the eigenvalues of $B^T B$ and BB^T are distinct. In such cases, singular value decomposition is unique up to multiplication of any column of Q with -1, and multiplication of any column of P by -1. Note that if we multiply the ith column of Q by -1 and we also multiply the ith column of P by -1, the product $Q\Sigma P^T$ remains unchanged. Throughout this chapter, the definition of the word "uniqueness" is slightly relaxed to allow for this type of reflection.

Lemma 7.2.8 (Condition for Uniqueness) *Consider a square matrix B of size $m \times m$, which is such that all the eigenvalues of $B^T B$ (and BB^T) are distinct. Then, the singular value decomposition of B is unique up to the multiplication of the singular vectors by -1.*

Note that if the singular values are not distinct, then one can choose any orthonormal basis of the eigenspace of the tied eigenvalues of $B^T B$ as the corresponding right singular vectors in P^T. The corresponding left singular vectors are obtained by pre-multiplying each of the these right-singular vectors with B and scaling the result to unit norm (cf. Lemma 7.2.1). In fact, there are an infinite number of possible (orthonormal) basis systems to choose from in the subspace corresponding to the tied eigenvectors (by simply selecting any basis of the tied eigenvectors). Therefore, ties in the singular values always ensure that singular value decomposition is not unique in a very fundamental way.

The above discussion pertains only to the singular value decomposition of square matrices. What about the singular value decomposition of rectangular matrices? One can generalize the uniqueness result of Lemma 7.2.8 to rectangular singular value decomposition, as long as we use the *compact* variant of singular value decomposition in which only non-zero singular values are included.

Lemma 7.2.9 (Uniqueness of Compact SVD) *Consider an $n \times d$ matrix D, which is such that all the non-zero singular values are distinct. Then, the compact singular value decomposition of D is unique up to the multiplication of the singular vectors by -1.*

In addition, truncated SVD will also be unique, as long as the retained singular values in the decomposition are distinct. Truncated singular value decomposition is very likely to be unique in real applications, because most of the (exact or approximate) ties in singular values often occur at the lower-order singular values at or near zero. The truncation process often removes most of these singular values.

7.2.7   Two-Way Versus Three-Way Decompositions

Singular value decomposition is inherently defined as a three-way factorization $Q\Sigma P^T$, in which the leftmost factor Q provides a basis for the column space, the rightmost factor P^T provides a basis for the row space, and the diagonal matrix Σ provides a quantification of the relative importance of the different basis vectors. Although this division of labor is elegant, two-way decompositions are often more popular in the literature on matrix factorization. In the two-way decomposition, an $n \times d$ matrix D is factorized into an $n \times k$ matrix U and a $d \times k$ matrix V, where k is the rank of the decomposition:

$$D \approx UV^T \tag{7.9}$$

If the original matrix D has rank larger than k, the above decomposition is only approximate (like truncated SVD). One can convert any three-way factorization like SVD into a two-way factorization as follows:

$$D \approx \underbrace{(Q\Sigma)}_{U} \underbrace{P^T}_{V^T}$$

In the case of SVD, it is natural to absorb the diagonal matrix within Q, because $U = Q\Sigma$ provides the coordinates of the data point in the k-dimensional basis space corresponding to the columns of $V = P$. When converting a three-way decomposition into a two-way decomposition, the general preference is to keep the normalization of the right factor and absorb the diagonal matrix in the left factor. However, the reality is that the 2-way decomposition has a much lower level of uniqueness as compared to 3-way decomposition. For example, one could absorb Σ in V^T instead of U. Furthermore, one could scale U and V in all sorts of ways without affecting the product UV^T. For example, if we multiply each entry of U by 2, we can divide each entry of V by 2 to get the same product UV^T. Furthermore, we can apply this trick to just a particular (say, rth) column of each of U and V to get the same result. In this sense, two-way factorizations are often ambiguously defined, unless one takes care to have clear normalization rules for one of the factors. Nevertheless, two-way factorizations are extremely useful in other forms of dimensionality reduction (like nonnegative matrix factorization) because of the simplicity in working with only two matrices in optimization formulations. Many forms of factorization use optimization models over two factors, which are relatively simple from the perspective of optimization algorithms like gradient descent. The good news that two-way factorizations can always be converted to a standardized three-way factorization like SVD by using the procedure discussed below.

In singular value decomposition, the (r, r)th diagonal entry is chosen in such a way that the rth columns of the left-most factor matrix Q and the right-most factor matrix P become normalized to unit norm. In other words, the diagonal matrix contains the scaling factors which create the ambiguity in 2-way factorization in terms of their distribution between U and V. Consider a two-way matrix factorization $D \approx UV^T$ into $n \times k$ and $d \times k$ matrices U and V, respectively. We can convert it into a near-unique (ignoring column reflection) three-way matrix factorization of the following form:

$$D \approx Q\Sigma P^T \tag{7.10}$$

Here, Q is a *normalized* $n \times k$ matrix (derived from U), P is a *normalized* $d \times k$ matrix (derived from V), and Σ is a $k \times k$ diagonal matrix in which the diagonal entries contain the nonnegative normalization factors for the k concepts. Each of the columns of Q and P satisfy the constraint that its L_2-norm (or L_1-norm) is one unit. It is common to use L_2-normalization in methods like singular value decomposition and L_1-normalization in some variations of *nonnegative matrix factorization* (discussed in Chapter 8). For the purpose of discussion, let us assume that we use L_2-normalization. Then, the conversion from two-way factorization to three-way factorization can be achieved as follows:

1. For each $r \in \{1 \ldots k\}$, divide the rth column $\overline{U}_r$ of U with its L_2-norm $\|\overline{U}_r\|$. The resulting matrix is denoted by Q.

2. For each $r \in \{1 \ldots k\}$, divide the rth column $\overline{V}_r$ of V with its L_2-norm $\|\overline{V}_r\|$. The resulting matrix is denoted by P.

3. Create a $k \times k$ diagonal matrix Σ, in which the (r, r)th diagonal entry is the nonnegative value $\|\overline{U}_r\| \cdot \|\overline{V}_r\|$.

It is easy to show that the matrices Q, Σ, and P satisfy the following relationship:

$$Q\Sigma P^T = UV^T \tag{7.11}$$

It is noteworthy that all diagonal entries of Σ are always nonnegative because of how the normalization is done. The optimization-centric view of SVD, which is discussed in the next section, uses two-way factorization in order to create compact optimization formulations. In general, two-way decompositions are more common in optimization-centric matrix factorization, because it is simpler to work with fewer matrices (and optimization variables).

7.3 SVD: An Optimization Perspective

The previous section provides a linear algebra perspective of singular value decomposition. While it provides insights about the existence/uniqueness of *full* SVD or even *compact* SVD, it makes no claim on the comparative accuracy of *truncated* SVD to the best possible low-rank approximation of the matrix. Another important point is that linear algebra can be used to derive full SVD, but it does not work for other forms of matrix factorization. In many cases, one may want to have constraints on the factors that move them away from properties of vector spaces. For example, if we want to put arbitrary constraints on the factors (such as nonnegative factors), it suddenly becomes very difficult to use techniques from linear algebra. The problem is that the space of nonnegative vectors is not even a vector space, and therefore the principles of linear algebra no longer apply. As we will see in the next chapter, many forms of matrix factorization use different choices of objective functions and constraints in order to control the properties of the factorization. Controlling the properties of the factorization is the key to being able to use them in different types of machine learning models, and these properties will be explored in Chapter 8. The optimization perspective is useful in all these cases. The most important result that arises from optimization-centric analysis is the following:

> Truncated SVD provides the best possible rank-k approximation of a matrix in terms of squared error.

An important point is that SVD also happens to provide a factorization $D \approx UV^T = Q\Sigma P^T$, which is such that the columns of each of U and V are orthogonal. However, even if we allow factorizations $D \approx UV^T$ in which the columns of each of U and V are not necessarily orthogonal, one would not gain anything from this relaxation in terms of accuracy. In other words, even for the optimization problem of minimizing the squared error of *unconstrained* low-rank factorization of D into U and V^T, one of the alternative optima is a pair of matrices U and V, such that the columns of each of the matrices are orthogonal. This section will show this beautiful property of SVD by approaching it from an optimization perspective.

In the following exposition, we will consistently work with the two-way factorization $D \approx UV^T$ rather than the three-way factorization $D \approx Q\Sigma P^T$. Here, D is an $n \times d$ matrix, U is an $n \times k$ matrix, and V is a $d \times k$ matrix. The hyperparameter k is the rank of the factorization. In such a case, the columns of each of U and V are mutually orthogonal, although there is some ambiguity in how these columns are scaled. Therefore, we will make the assumption that the columns of V are scaled to unit norm.

7.3.1   A Maximization Formulation with Basis Orthogonality

First, we present an optimization model that assumes orthonormality of the columns of the matrix V. In such a case, we have $V^TV = I$, and therefore the reduced representation of D can be obtained as $U = DV$. Therefore, one way of formulating singular value decomposition is to maximize the energy of the matrix $U = DV$ as follows:

$$\text{Maximize}_V \|DV\|_F^2 \quad (OP)$$
$$\text{subject to:}$$
$$V^TV = I_k$$

We refer to this optimization problem as (OP). Here, V is a $d \times k$ matrix, and the $n \times k$ matrix $U = DV$ is not included in the optimization formulation. *The objective function of this problem (in minimization form) is not convex even for simple versions of this problem such as $k = 1$.* Nevertheless, it can still be solved optimally because of the specialized structure of the problem. It is important to note that one can decompose $\|DV\|_F^2$ in terms of the sums of L_2-norms of the k columns of DV. Therefore, if $\overline{V}_r$ is the rth column of V, one can simplify the objective function as follows:

$$\|DV\|_F^2 = \sum_{r=1}^{k} \|D\overline{V}_r\|^2 = \sum_{r=1}^{k} \overline{V}_r^T [D^TD] \overline{V}_r$$

Note that this optimization problem is the same as the norm-constrained optimization problem introduced in Section 6.6 of Chapter 6. The solution to this problem corresponds to the top-k eigenvectors of D^TD. Recall from the previous section that the eigenvalues of D^TD are $\sigma_{11}^2 \ldots \sigma_{rr}^2$, which are the same as the squares of the singular values of D.

Furthermore, the energy retained in DV is equal to $\sum_{r=1}^{k} \sigma_{rr}^2$ based on the discussion in Section 6.6 of Chapter 6. This is consistent with the energy retained by truncated singular value decomposition (cf. Section 7.2.4). We have, therefore, just shown that the energy retained by truncated SVD (cf. Section 7.2.4) is as large as possible among all possible orthonormal basis systems V. We summarize this result as follows:

Lemma 7.3.1 *The optimal solution V for the optimization problem (OP) is obtained by setting the columns of V to the largest eigenvectors in $D^T D$.*

We can also show that the transformed representation $U = DV$ contains the (scaled) eigenvectors of DD^T.

Lemma 7.3.2 *Let $U = DV$ be the transformed representation of the data, when V is obtained using (OP). Then U contains the scaled eigenvectors of DD^T.*

Proof: Let the n-dimensional column vector $\overline{U}_r$ contain the rth column of DV. This is equal to $D\overline{V}_r$, where $\overline{V}_r$ contains the rth column of V. In other words, we have:

$$\overline{U}_r = D\overline{V}_r$$

Multiplying both sides with DD^T, we obtain the following:

$$DD^T\overline{U}_r = (DD^T)D\overline{V}_r = D\underbrace{[(D^TD)\overline{V}_r]}_{\propto \overline{V}_r} \propto D\overline{V}_r = \overline{U}_r$$

In other words, $\overline{U}_1 \ldots \overline{U}_k$ are the eigenvectors of DD^T. The only difference is that the columns of V are scaled to unit norm, whereas those of U are not. ∎

Since DD^T is a symmetric matrix, its eigenvectors $\overline{U}_1 \ldots \overline{U}_k$ will be mutually orthogonal as well. Note that this optimization model only uses the assumption that the columns of V are orthogonal, and we were able to automatically derive the fact that the columns of $U = DV$ are mutually orthogonal.

7.3.2 A Minimization Formulation with Residuals

The aforementioned optimization model tries to *maximize* the *retained* energy $\|U\|_F^2 = \|DV\|_F^2$ in the projected matrix DV. An alternative approach is to *minimize* the *lost* energy, which is $\|D - UV^T\|_F^2$. The matrix $R = (D - UV^T)$ is commonly referred to as the *residual matrix* from an approximate factorization $D \approx UV^T$.

Consider the following *unconstrained* optimization problem, which is obtained by dropping the orthonormality constraints on the columns of V:

$$\text{Minimize}_{U,V} \; J = \|D - UV^T\|_F^2$$

The optimization problem is also referred to as *unconstrained matrix factorization*. Here, U is an $n \times k$ matrix, and V is a $d \times k$ matrix. *This objective function is not convex, but can nevertheless be optimized easily.* This optimization problem is an example of the fact that not all non-convex problems are impossible to solve.

First, note that even though this problem is unconstrained, we can find at least one optimal V with orthonormal columns. This is because we can replace any optimal solution pair $(U, V) = (U^0, V^0)$ with the pair $(U^0 R^T, Q)$, where $V^0 = QR$ is the QR-decomposition of V^0, and Q, R are $d \times k$ and $k \times k$ matrices, respectively. Both solutions have the same objective function value, since the product of both pairs of matrices is $U^0 R^T Q^T$. Furthermore, as shown in Figure 8.1 of Chapter 8, a necessary condition for optimality of this matrix factorization problem is as follows:

$$DV - UV^T V = 0$$

The solution with orthonormal columns of V (obtained via QR decomposition of any optimal V^0), satisfies $V^T V = I$, and, therefore, the condition simplifies to $U = DV$. Substituting for U in the optimization formulation, the unconstrained matrix factorization problem has the same objective function value as that of minimizing $\|D - UV^T\|^2 = \|D - DVV^T\|_F^2$ subject to $V^T V = I_k$. The sum of the squared Frobenius norms of DV and $D - DVV^T$ can be shown[1] to be the constant $\|D\|_F^2$, and therefore this minimization problem reduces to the maximization of the Frobenius norm of DV. This is exactly the problem (OP) of the previous section. Therefore, the *unconstrained minimization formulation with residuals also yields the top eigenvectors of DD^T and $D^T D$ for U and V, respectively, as one of the alternate optima.* In other words, we have the following important result:

Theorem 7.3.1 *Truncated singular value decomposition provides one of the alternate optima to unconstrained matrix factorization.*

7.3.3   Generalization to Matrix Factorization Methods

The formulation contained in the previous section is the most basic form of optimization-centric matrix factorization. By changing the objective function and the constraints, other forms of matrix factorization can be supported. All matrix factorization methods have the following general form:

Maximize similarity between entries of D and UV^T

subject to:

Constraints on U and V

For example, *probabilistic* matrix factorization methods use a log-likelihood function rather than the Frobenius norm as the optimization function. Similarly, various types of nonnegative matrix factorization impose nonnegativity constraints on U and V. Logistic matrix factorization methods apply a logistic function on the entries of UV^T in order to materialize the probability that a particular entry is 1. Such an approach works well for matrices with binary entries. Therefore, the optimization framework of unconstrained matrix factorization provides a starting point for factorizations with different properties. These methods will be discussed in detail in Chapter 8. *Most matrix factorization formulations are not convex.* Nevertheless, gradient descent works quite well in these cases.

7.3.4   Principal Component Analysis

Principal component analysis (PCA) is very closely related to SVD. SVD tries to find a k-dimensional subspace, so that projecting the data points in that subspace maximizes their aggregate squared distances about the *origin*; in contrast, principal component analysis tries to preserve the aggregate squared distances about the *data mean*. The aggregate squared distances about the data mean are captured by the variance (albeit in *averaged* form). As a result, given a data set D, the relationship between SVD and PCA is as follows:

---

[1] DV and DVV^T have the same energy (see Exercise 18), and the latter is Frobenius orthogonal to $(D - DVV^T)$. Therefore, the sum of the squared Frobenius norms of DV and $D - DVV^T$ is simply $\|D\|_F^2$.

PCA performs exactly the same dimensionality reduction as SVD on a *mean-centered* data set D.

When the data is not mean-centered up front, PCA and SVD will yield different results. In PCA, we first mean-center the data set by subtracting the d-dimensional mean-vector of the full data set D from each row as follows:

$$M = D - \underbrace{\overline{1}\,\overline{\mu}}_{n \times d}$$

Here, $\overline{1}$ is a column vector of n ones, and $\overline{\mu}$ is a d-dimensional row vector containing the mean values of each of the d dimensions. Therefore, $\overline{1}\,\overline{\mu}$ is an $n \times d$ matrix in which each row is the mean vector $\overline{\mu}$. We compute the covariance matrix C as follows:

$$C = \frac{M^T M}{n}$$

The covariance matrix C is a $d \times d$ matrix, in which the (i, j)th entry is simply the covariance between the dimensions i and j. The diagonal entries are the dimension-specific variances. Like the scatter matrix $D^T D$ in SVD, the covariance matrix in SVD is also positive semidefinite. The covariance matrix may be approximately diagonalized at rank-k as follows:

$$C \approx V \Delta V^T$$

Here, V is a $d \times k$ matrix with columns containing the top-k eigenvectors, and Δ is a $k \times k$ diagonal matrix with the diagonal entries containing the top-k eigenvalues (which are always nonnegative for the positive semidefinite matrix $C \propto M^T M$). The (r, r)th diagonal entry is therefore denoted by the nonnegative value λ_r^2, and it represents the rth eigenvalue. As we will see later, the value of λ_r^2 is equal to the variance of the rth column of the k-dimensional projection DV of the matrix D. Instead of referring to the eigenvectors as singular vectors (as in SVD), they are referred to as *principal components* in PCA. Note that if one were to perform singular value decomposition on the mean-centered matrix M, the right singular vectors are the PCA eigenvectors, and the rth singular value σ_{rr} of SVD is related to the eigenvalue λ_r^2 of PCA as follows:

$$\lambda_r^2 = \frac{\sigma_{rr}^2}{n}$$

The additional factor of n in the denominator comes from dividing $M^T M$ by n to obtain the covariance matrix. The $n \times k$ matrix U containing the k-dimensional representation of the n rows of D is defined by projecting the rows of M on the columns of V:

$$U = MV$$

We make the following observations about PCA:

1. The matrix U is mean-centered just like the mean-centered data set M. In other words, the reduced representation of the data is also mean-centered. Note that the sum of the rows of U is given by $\overline{1}U = \overline{1}[MV] = \underbrace{[\overline{1}M]}_{\overline{0}}V$.

2. The covariance of the matrix U is the diagonal matrix Δ. Consider the case in which the matrix V contains the k columns $\overline{v}_1 \ldots \overline{v}_k$. Since the matrix U is mean-centered,

its covariance matrix is given by $U^T U/n$, which can be simplified as follows:

$$\frac{U^T U}{n} = V^T \frac{[M^T M]}{n} V = [\overline{v}_1 \ldots \overline{v}_k]^T (C[\overline{v}_1 \ldots \overline{v}_k])$$

$$= [\overline{v}_1 \ldots \overline{v}_k]^T [\lambda_1^2 \overline{v}_1 \ldots \lambda_k^2 \overline{v}_k] = \Delta$$

In the above simplification, we used the fact that each $\overline{v}_i$ is an eigenvector of the covariance matrix C, and that these k vectors are orthonormal. Therefore, $\overline{v}_i \cdot \overline{v}_j$ is 1 when $i = j$, and 0, otherwise. As a result, the diagonal entries of Δ will contain $\lambda_1^2 \ldots \lambda_k^2$.

3. The retained variance in the data is given by $\sum_{i=1}^{k} \lambda_i^2$. This is easy to show because the covariance matrix of U is Δ. Therefore, the sum of its diagonal entries, which is $\sum_{i=1}^{k} \lambda_i^2$, yields the retained variance.

All of the above results show that PCA has very similar properties to SVD. In order to completely reconstruct the data from U and V^T, one also needs to store the mean vector $\overline{\mu}$, which was used to mean-center the data. In other words, the original (uncentered) data set can be reconstructed by using the following approach:

$$D \approx D_{pca} = UV^T + \overline{1}\,\overline{\mu} \tag{7.12}$$

The amount of overhead for storing $\overline{\mu}$ is small, and it asymptotically vanishes for large data sets.

The mean-centering of PCA helps in improving the accuracy of the approximation. In order to understand this point, we have shown an example of a 3-dimensional data set that is not originally mean-centered in Figure 7.6. Most of the data is distributed near a plane far

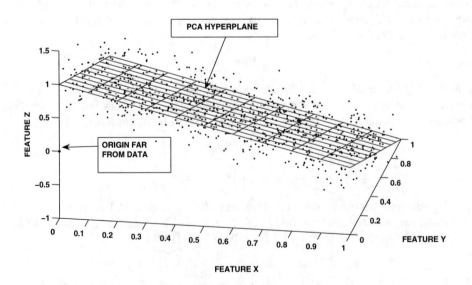

Figure 7.6: PCA for data that is not originally mean-centered

away from the origin (before preprocessing or mean-centering). In this case, a 2-dimensional hyperplane can approximate the data quite well, where the mean-centering process ensures that the PCA hyperplane passed through the mean of the original data set. This is not the case for SVD, which will struggle to approximate the data without using all the three dimensions. It can be explicitly shown that the accuracy of PCA is at least as good as that of SVD for the same number of eigenvectors.

Problem 7.3.1 *Consider an $n \times d$ data set D, whose rank-k approximations using truncated SVD and PCA are D_{svd} and D_{pca}, respectively (see Equation 7.12). Then, the information loss in PCA can never be larger that that in SVD:*

$$\|D - D_{pca}\|_F^2 \le \|D - D_{svd}\|_F^2$$

For mean-centered data, the accuracy of the two methods is identical because $D_{pca} = D_{svd}$.

The geometric intuition for the above exercise is that PCA finds a k-dimensional hyperplane that must pass through the *mean of the data*, whereas SVD finds the k-dimensional hyperplane passing through the *origin*. The former provides better reconstruction. However, as the next exercise shows, the difference is usually not too large.

Problem 7.3.2 *Show that the squared error of SVD at a truncation rank of $(k + 1)$ is no larger than the squared error of PCA at a truncation rank of k for any $k \ge 1$.*

A hint for solving the above problem is to show using Lemma 2.6.2 of Chapter 2 that the mean-corrected reconstruction D_{pca} (cf. Equation 7.12) has rank at most $(k + 1)$. The SVD of D at rank-$(k + 1)$ will provide a better rank-$(k + 1)$ reconstruction because of its optimality properties.

7.4 Applications of Singular Value Decomposition

Singular value decomposition has numerous applications in machine learning. The following will provide an overview of some of the key applications of singular value decomposition.

7.4.1 Dimensionality Reduction

The most widely used application of singular value decomposition and principal component analysis is dimensionality reduction. Given a $d \times k$ basis matrix V, both PCA and SVD transform the $n \times d$ data matrix D to the $n \times k$ data matrix DV. In other words, each d-dimensional row in D is transformed to a k-dimensional row in DV.

The above dimensionality reduction can be performed with either PCA or with SVD. What types of data sets are more suitable for PCA, and which ones are suitable for SVD? PCA is often used for numerical data of modest dimensionality that is not sparse, whereas SVD is often used for sparse and high-dimensional data. A classical example of a data domain that is more suitable to the use of SVD as opposed to PCA is text data. Note that if one attempted to use PCA on text data, the mean-centering process would destroy the sparsity of the data. Destroying the sparsity of the data results in dense matrices that are computationally difficult to process from a practical point of view. When SVD is used for text data, it is referred to as *latent semantic analysis* or *LSA*. A detailed discussion of latent semantic analysis for text is provided in [2].

Singular value decomposition is often used for compression of images. An image can be represented as a matrix of pixels, which can be compressed using SVD. In cases where there

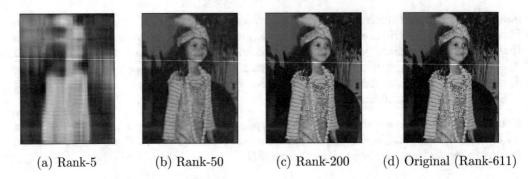

(a) Rank-5   (b) Rank-50   (c) Rank-200   (d) Original (Rank-611)

Figure 7.7: SVD reconstruction at different ranks. The reconstruction at rank-200 is nearly identical to that of the full-rank image

are multiple colors in the image, each color channel is processed as a separate matrix. An image matrix is often of full rank, although the lower ranks have very small singular values. Figure 7.7 illustrates the case of an image of size 807×611 in which the 611th singular value is non-zero. The rank of the image matrix is therefore 611, and the full-rank reconstruction of Figure 7.7(d) is identical to the original image. Obviously, there are no space advantages of full-rank reconstruction, and one must use truncation. Using a rank that is too low, such as 5, loses a lot of information, and the resulting image does not show too many useful details (cf. Figure 7.7(a)). An SVD of rank-50 loses only a small amount of detail, as shown in Figure 7.7(b). Furthermore, an SVD of rank-200 is virtually indistinguishable from the original image (cf. Figure 7.7(c)).

With certain types of images, noisy artifacts of the image can even be removed by the SVD truncation at intermediate values of the rank. This is because the dropping of the lower-order components leads to the discarding of the grainy noise components rather than the informative portions of the image. Therefore, the "lossiness" of the low-rank reconstruction is sometimes useful. This is an issue, which will be discussed in the next section.

7.4.2   Noise Removal

One interesting side effect of dimensionality reduction is that it often reduces the amount of noise in the data. For example, if image data is corrupted with some amount of noise, it is often helpful to reconstruct it with truncated SVD. The basic intuition is that a minor amount of noise is often independent of the aggregate patterns in the data. Therefore, this noise often shows up in the lower-order components of SVD, which are largely independent of the dominant patterns in the higher-order components. This type of behavior is also exhibited in text data, where singular value decomposition tends to improve the retrieval accuracy. In the particular case of text, singular value decomposition reduces the noise and ambiguity effects inherent in languages; two examples of such ambiguity effects are *synonymy* and *polysemy*. For example, the fact that a word might have multiple meanings might be viewed as a kind of noise in the lower order components of SVD. The higher-order components of SVD tend to focus on the correlations, and therefore they do a much better job at disambiguating a word based on its context. A detailed discussion of the noise removal effects of SVD may be found in [7, 33]. This type of behavior is also observed in the case of image data reconstruction. In many cases, the quality of reconstruction of blurry images is higher with the use of intermediate values of the ranks.

7.4.3 Finding the Four Fundamental Subspaces in Linear Algebra

The four fundamental subspaces in linear algebra are the row space, the column space, the right null space, and the left null space (cf. Section 2.4 of Chapter 2). Consider an $n \times d$ matrix D with rank $r \leq \min\{n, d\}$. Let the full SVD of D be given by $D = Q\Sigma P^T$. Then, the four fundamental subspaces of linear algebra are given by the following:

1. The r non-zero right singular vectors of D define an orthogonal basis for the row space of D. This is because the vector $D^T\overline{x} = P\Sigma^T[Q^T\overline{x}] = [P\Sigma^T]\overline{y}$ can always be shown to be a linear combination of the *non-zero* right singular vectors [non-zero columns of $P\Sigma^T$] for any $\overline{x} \in \mathcal{R}^n$.

2. The r non-zero left singular vectors of D define an orthogonal basis for the column space of D. This is because the vector $D\overline{x} = Q\Sigma[P^T\overline{x}] = [Q\Sigma]\overline{z}$ can always be shown to be a linear combination of the *non-zero* left singular vectors [non-zero columns of $Q\Sigma$] for any $\overline{x} \in \mathcal{R}^d$.

3. The $(d-r)$ zero right singular vectors contained in the columns of P define an orthogonal basis for the right null space of D, because the right null space is the orthogonal complementary space to the row space of D.

4. The $(n-r)$ zero left singular vectors contained in the columns of Q define an orthogonal basis for the left null space of D. This is because the left null space is the orthogonal complementary space to the column space of D.

Problem 7.4.1 *In Chapter 2, we showed that the row rank of a matrix is the same as its column rank. This value is referred to as the matrix rank, which is used throughout this chapter. Discuss why the existence of SVD provides an alternative proof that the row rank of a matrix is the same as its column rank.*

7.4.4 Moore-Penrose Pseudoinverse

The Moore-Penrose pseudoinverse can be used for solving systems of linear equations, and for providing a solution to the problem of linear regression (cf. Section 4.7.1.1 of Chapter 4). SVD can be used to efficiently compute the Moore-Penrose pseudoinverse. Consider the compact SVD of an $n \times d$ matrix D of rank $k \leq \min\{n, d\}$:

$$D = Q\Sigma P^T$$

Here, Q is an $n \times k$ matrix, Σ is a $k \times k$ diagonal matrix with only positive diagonal entries, and P is a $d \times k$ matrix. Note that all diagonal entries of Σ are positive and Σ is a square matrix, because we are using the compact SVD rather than the full SVD. Then, the pseudoinverse D^+ of D is given by the following:

$$D^+ = \lim_{\lambda \to 0^+}(D^T D + \lambda I_d)^{-1}D^T = \lim_{\lambda \to 0^+}P(\Sigma^2 + \lambda I_k)^{-1}\Sigma Q^T$$
$$= P\underbrace{[\lim_{\lambda \to 0^+}(\Sigma^2 + \lambda I_k)^{-1}\Sigma]}_{\Sigma^{-1}}Q^T = P\Sigma^{-1}Q^T$$

The matrix Σ^{-1} is obtained by replacing each diagonal entry in Σ by its reciprocal.

7.4.4.1 Ill-Conditioned Square Matrices

As discussed in Section 2.9 of Chapter 2, a singular (square) matrix might occasionally appear to be non-singular because of computational errors in fixed-precision computers while arriving at that matrix. Inverting such matrices directly can cause numerical overflows. In such cases, one wants to detect this situation, and compute the pseudoinverse of the nearest singular approximation, rather than the direct inverse. This situation is detected using the *condition number* of the $d \times d$ matrix D, which is defined as the ratio of the largest value of $\|D\overline{x}\|/\|\overline{x}\|$ to the smallest value of $\|D\overline{x}\|/\|\overline{x}\|$ over all vectors $\overline{x} \in \mathcal{R}^d$ (cf. Definition 2.9.1 of Chapter 2). An interesting result is that the condition number can be expressed in terms of the ratio of the singular values of the matrix:

Lemma 7.4.1 *The condition number of a matrix is the ratio of its largest singular value to its smallest singular value.*

The above lemma can be shown by showing that the scaling ratio $\|D\overline{x}\|/\|\overline{x}\|$ of vector $\overline{x}$ is maximized by choosing $\overline{x}$ to be the largest right singular vector of the matrix, and the scaling ratio is minimized by choosing the smallest right singular vector. As a result, the condition number is the ratio of the two quantities.

Lemma 7.4.2 *The maximum value of $\|D\overline{x}\|$ for $m \times m$ matrix D is σ_1 for unit vector $\overline{x}$, where σ_1 is the largest singular value. A similar result holds for the minimum value of $\|D\overline{x}\|$.*

Proof: One can express $Q\Sigma P^T$, where Q and P are rotation matrices. Therefore, $P^T\overline{x}$ can be written as another unit vector $\overline{x}'$. Let $x_1 \ldots x_m$ be the components of $\overline{x}'$. The value $\Sigma\overline{x}'$ has the norm of σ_1, when the first component of the unit vector $\overline{x}'$ is 1 and all other components are 0. This vector $\overline{x}'$ is obtained by choosing $\overline{x}$ to be the largest right singular vector of D. Furthermore, the squared norm of $\Sigma\overline{x}'$ for any other vector $\overline{x}' = [f_1 \ldots f_m]$ is $\sum_{i=1}^{m} f_i^2\sigma_i^2$, where $\sum_{i=1}^{m} f_i^2 = 1$. This squared norm is the weighted average of $\sigma_1^2 \ldots \sigma_m^2$, which can be no larger than σ_1^2. The result follows. ∎

Problem 7.4.2 *Show that a real eigenvalue of a square matrix can never be larger in magnitude than the largest singular value of the matrix. Also show that a real eigenvalue can never be smaller in magnitude than the smallest singular value of the matrix.*

Poorly conditioned matrices will become obvious in SVD when all singular values are positive but are different by orders of magnitude. How can one invert such a matrix without causing overflows? Note that one need not even use *all* the positive singular values in SVD. Rather, one can use truncated SVD, and remove the smaller singular values. In other words, if the truncated SVD of rank-k is $Q_k\Sigma_k P_k^T$, then the resulting inverse is $P_k\Sigma_k^{-1}Q_k^T$. Surprisingly, truncating very small singular values often has a beneficial effect in terms of prediction accuracy, when using the Moore-Penrose pseudoinverse in machine learning applications like least-squares regression. This is because it helps in reducing the effect of computational errors caused by the numerical inexactness of floating point representation in computers. Such computational errors might cause singular values that ought to be zero to take on small values like 10^{-6}. This approach can also help with solving ill-conditioned systems of equations. Furthermore, truncating small singular values has a regularization effect, when the Moore-Penrose pseudoinverse is used to compute the solution to machine learning problems like linear regression. This improves the performance on out-of-sample predictions. This issue is discussed in the next section, and is an example of the fact that the goals of optimization in machine learning are often different from those of traditional optimization (see Section 4.5.3 of Chapter 4).

7.4.5   Solving Linear Equations and Linear Regression

The problem of solving homogeneous linear equations is that of finding the solution to the homogeneous system of equations $A\overline{x} = 0$, where A is an $n \times d$ matrix and $\overline{x}$ is a d-dimensional column vector. In other words, we want to discover the right null space of A. Singular value decomposition is a natural way to achieve this goal because the right singular vectors contain the basis of the rows of A, and the zero singular values correspond to the null space. If $\overline{x}$ is a right-singular vector of A with zero singular value, then the eigenvector properties of A also yield $A\overline{x} = 0$. What happens when there is no zero singular value? In such a case, there is no solution to this system of linear equations, although one can find a (unit normalized) solution $\overline{x}$ in which $\|A\overline{x}\|^2$ is minimized. As shown in Lemma 7.4.2, *the minimum value is the smallest singular value.* Furthermore, this optimum value of $\overline{x}$ is the right singular vector. We leave this problem as an exercise:

Problem 7.4.3 *The unit-normalized solution to the problem of minimizing $\|A\overline{x}\|^2$ is the smallest right-singular vector of A.*

Singular value decomposition can also be used for solving the system of equations $A\overline{x} = \overline{b}$. Here, $\overline{b}$ is an n-dimensional column vector. Solving the system equations $A\overline{x} = \overline{b}$ is a special case of the (more general) problem of minimizing $\|A\overline{x} - \overline{b}\|^2$. This is identical to the problem of linear regression. One can use the Moore-Penrose pseudoinverse is used to compute the solution to least-squares regression (cf. Section 4.7.1.1 of Chapter 4):

$$\overline{x} = A^+\overline{b} \tag{7.13}$$

Using truncated SVD instead of SVD is beneficial from the perspective of regularization. In the event that the original data matrix A is mean-centered, and truncated SVD is used to compute the pseudoinverse, the resulting solution is referred to as *principal components regression.*

7.4.6   Feature Preprocessing and Whitening in Machine Learning

Principal component analysis is used for feature preprocessing in machine learning, by first reducing the dimensionality of the data and then normalizing the newly transformed features, so that the variance along each transformed direction is the same. Let V_k be the $d \times k$ matrix containing the top-k eigenvectors found by principal component analysis. Then, the first step is to transform the mean-centered data matrix D to the k-dimensional representation U_k as follows:

$$U_k = DV_k$$

The next step is to divide each column of U_k by its standard deviation. As a result, the original data distribution becomes roughly spherical in shape. This type of approach is referred to as *whitening.*

This type of data distribution works much more effectively with gradient-descent algorithms. This is because widely varying variances along different directions also lead to loss functions in which different directions have different levels of curvature. Examples of two loss functions with different levels of curvature are illustrated in Figure 5.2 of Chapter 5. A loss function like Figure 5.2(a) tends to be more easily optimized with gradient descent algorithms. Normalizing the data to have unit variance in all directions tends to reduce obvious forms of ill-conditioning in the loss function. As a result, gradient-descent tends to become much faster. Furthermore, the normalization of the data in this way sometimes prevents some subsets of features from having undue influence on the final results.

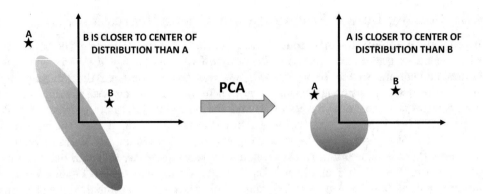

Figure 7.8: An example of the whitening of an ellipsoidal data distribution by principal component analysis and its use in outlier detection

This type of preprocessing is also used in unsupervised applications like outlier detection. In fact, whitening is arguably more important in unsupervised applications because one does not have labels to provide guidance about the relative importance of different directions in the data. An example of the whitening of an ellipsoidal data distribution is illustrated in Figure 7.8. The resulting data distribution has a spherical shape.

7.4.7 Outlier Detection

The whitening approach described in the previous section is also used for outlier detection. The resulting technique is also referred to as *soft PCA* or the Mahalanobis method. The overall approach uses two steps:

1. First, the $n \times d$ data matrix D is mean-centered, and PCA is used to transform it to an $n \times k$ data matrix $U_k = DV_k$. Here, V_k contains the top-k eigenvectors of the covariance matrix. Each column of U_k is normalized to unit variance. This type of approach will tend to increase the absolute distance of outliers from the data mean when they deviate along low-variance directions. In fact, for low-dimensional data, the value of the rank, k, might be the full dimensionality but the distortion of the ellipsoidal data distribution to a spherical distribution will change the relative propensity of different points to be considered outliers.

2. The squared distance of each point from the data mean is reported as its outlier score.

Although some of the low-variance principal components may be dropped (in order to avoid directions in which variances are caused by computational errors), the primary goal of whitening is to change the relative importance of the independent directions, so as to emphasize *relative* variations along the principal directions. It is the distortion of the shape of the data distribution that is the key to the discovery of non-obvious outliers. For example, the point A is further from the center of the original data distribution, as compared to point B. However, the point A is aligned along the elongated axis of the data distribution, and therefore it is much more consistent with overall shape of the distribution. This pattern becomes more obvious when we apply principal component analysis to the data distribution. This tends to separate B from the data distribution and the distance from the center of the data distribution provides an outlier score that is larger for point B as compared to point A. The resulting method is referred to as soft PCA because its uses soft distortions of the data distribution rather than truncation of low-variance directions. In fact, low-variance directions are more important in this case for discovering outliers. This approach is also referred to as the Mahalanobis method because the distance of each point from the center of the data distribution after PCA-based normalization is equivalent to the Mahalanobis distance. Intuitively, the Mahalanobis distance is the exponent of the Gaussian distribution, which assumes that the original data has an ellipsoidal shape. The whitening along principal component directions simply discovers which points are unlikely to belong to this Gaussian distribution.

Definition 7.4.1 (Mahalanobis Distance) *Let $\overline{X}$ be a d-dimensional row vector from the data set and $\overline{\mu}$ be the mean (row vector) of a data set. Let C be the $d \times d$ covariance matrix of a d-dimensional data set in which the (i, j)th entry is the covariance between the dimensions i and j. Then, the squared Mahalanobis distance of the point $\overline{X}$ is given by the following:*

$$Maha(\overline{X}, \overline{\mu})^2 = (\overline{X} - \overline{\mu})C^{-1}(\overline{X} - \overline{\mu})^T \qquad (7.14)$$

At first glance, the Mahalanobis distance seems to have little to do with PCA or to normalization of points by the standard deviations along principal components. However, the key point is that the covariance matrix can be expressed as $V\Delta V^T$, where the columns of V contain the eigenvectors. Then, the Mahalanobis distance can be expressed in terms of the eigenvectors as follows:

$$Maha(\overline{X}, \overline{\mu})^2 = (\overline{X} - \overline{\mu})C^{-1}(\overline{X} - \overline{\mu})^T \qquad (7.15)$$

$$= (\overline{X} - \overline{\mu})[V\Delta V^T]^{-1}(\overline{X} - \overline{\mu})^T \qquad (7.16)$$

$$= \underbrace{[(\overline{X} - \overline{\mu})V]}_{\text{Basis Change}} \Delta^{-1}[(\overline{X} - \overline{\mu})V]^T \qquad (7.17)$$

Note that postmultiplying $(\overline{X} - \overline{\mu})$ with V results in mean-centering with $\overline{\mu}$ and then transformation to the orthonormal basis system in the columns of V. Multiplying with $\Delta^{-1/2}$ simply scales each dimension with the inverse of the standard deviation. Consider the following definition of the row vector $\overline{Z}$ containing the scaled coordinates of the data points:

$$\overline{Z} = (\overline{X} - \overline{\mu})V\Delta^{-1/2}$$

Note that the matrix $\Delta^{-1/2}$ is a diagonal matrix containing the inverse of the standard deviation along each principal component. This approach transforms the mean-centered row vector $(\overline{X} - \overline{\mu})$ to a new basis and normalizes it with the standard deviation along each principal component. It is not difficult to see that the row vector $\overline{Z}$ is the whitened coordinate representation of the row vector $\overline{X}$, and the squared Mahalanobis distance is given by $\|\overline{Z}\|^2$. Furthermore, one can easily verify that $\|\overline{Z}\|^2 = \overline{Z}\,\overline{Z}^T$ simplifies to the covariance-based definition of Mahalanobis distance in Equation 7.17. A detailed discussion of the Mahalanobis method is provided in [4]. One can even combine the Mahalanobis method with feature engineering in order to discover non-obvious outliers. One example of such a feature engineering approach is discussed in the next section.

7.4.8 Feature Engineering

Singular value decomposition uses the eigenvectors of the dot product similarity matrix DD^T as the left singular vectors, and the eigenvectors of the scatter matrix D^TD are the right singular vectors. The decomposition $D = Q\Sigma P^T$ provides the left singular vectors in the columns of Q and the right singular vectors in the columns of P. The (scaled) left singular vectors $Q\Sigma$ provide the embedding, whereas the right singular vectors P provide the basis. Either of the two matrices can be used to compute the transformed data in the standard version of SVD. While the direct extraction of the right singular vectors is more common because of the intuitive appeal of a basis, the left singular vectors can *directly* provide the embeddings (without worrying about a basis). In some application-centric settings, no multidimensional representation of the data is available, but only a similarity matrix S is available. For example, S might represent the pairwise similarities between a set of small graph objects. In such cases, one can assume that the provided similarity matrix corresponds to DD^T for some unknown $n \times n$ matrix D, whose rows contain the multidimensional representations of the graph objects. Note that the matrix D might have as many as n dimensions because any set of n objects (together with the origin) always defines an n-dimensional plane. In such cases, one can simply diagonalize S as follows:

$$S = DD^T = Q\Sigma^2 Q^T = (Q\Sigma)(Q\Sigma)^T$$

The $n \times n$ matrix $Q\Sigma$ is provides the multidimensional embeddings of the points in its rows. If the similarity matrix S was derived by using dot products on multidimensional data, then the resulting representation will provide the vanilla SVD embedding of D. Note that any rotated representation DV of D will provide the same embedding, because $(DV)(DV)^T = D(V^T V)D = DD^T$. We cannot control the basis in which the unknown matrix D is represented in the final embedding $Q\Sigma$; singular value decomposition happens to choose the basis in which the columns of the embedded representation are orthogonal. This type of approach works only when the similarity matrix S is positive semidefinite, because the eigenvalues in Σ^2 need to be nonnegative. Such similarity matrices are referred to as *kernel matrices*. You will learn more about kernel matrices in Chapter 9.

This type of approach is referred to as *feature engineering*, because we can convert any arbitrary object (e.g., graph) to a multidimensional representation by using the pairwise similarities between them. For example, one can combine the Mahalanobis method for outlier detection (see previous section) with the feature engineering approach discussed in this section. Consider a set of n graphs with an $n \times n$ similarity matrix S. We wish to identify the graphs that should be labeled as outliers. One can extract the embedding $Q\Sigma$ from the diagonalization $S = Q\Sigma^2 Q^T$. By whitening the representation, one obtains the embedding Q in which each column has unit variance. The distance of each row in Q from the mean of the rows of Q provides the *kernel* Mahalanobis outlier score. Even for multidimensional data, one can extract more insightful features by replacing the dot products in DD^T with other similarity functions between points. In Chapter 9, we will provide specific examples of such similarity functions.

7.5   Numerical Algorithms for SVD

In this section, we will discuss some simple algorithms for singular value decomposition. These algorithms are not optimized for efficiency, but they provide the basic ideas for some of the advanced algorithms. Given an $n \times d$ matrix D, the simplest (and most naive) approach is to find the eigenvectors of $D^T D$ or DD^T using the approach discussed in Section 3.5 of Chapter 3. The choice of using either $D^T D$ or DD^T depends on which matrix is smaller. Using the former helps in finding the right singular vectors, whereas using the latter helps in finding the left singular vectors. In most cases, we only need to find the top-k singular vectors, where $k \ll \text{mbox}\{n, d\}$. If we use $D^T D$ to compute the $d \times k$ matrix P containing the right singular vectors, the left singular vectors will be contained in the $n \times k$ matrix $Q = DP\Sigma^{-1}$. The $k \times k$ matrix Σ is computed by placing the square root of the eigenvalues on the diagonal of the matrix. We can assume that we are only interested in non-zero singular vectors, and therefore Σ is invertible. On the other hand, if the left singular vectors are found by diagonalizing DD^T, then the matrix P can be computed as $P = D^T Q\Sigma^{-1}$.

This approach is inefficient for sparse matrices. For example, consider a text data set in which each row contains about 100 non-zero values, but the dimensionality d of the row is 10^5. Similarly, the collection contains 10^6 documents, which is not large by modern standards. The number of non-zero entries in D is 10^8, which is much smaller than the total number of entries in D. Therefore, this is a *sparse* matrix for which special data structures can be used. On the other hand, $D^T D$ is a *dense* matrix that contains 10^{10} entries. Therefore, it is inefficient to work with $D^T D$ as compared to D. In the following, we present the generalization of the power method discussed in Section 3.5 that works with D rather than $D^T D$. It is noteworthy that this method is not optimized for efficiency, but it provides the starting points for understanding some efficient methods such as the Lanczos algorithm [52]. In recent years, methods based on QR decomposition have become more popular. A specific example is the Golub and Kahan algorithm [52].

The Power Method

The power method can find the dominant eigenvector of any matrix (like $D^T D$) by first initializing it to a random d-dimensional column vector $\bar{p}_1$ and then repeatedly pre-multiplying with $D^T D$ and scaling to unit norm. However, $D^T D$ is dense, and one has to be careful about how these operations are performed. To reduce the number of operations, it makes sense to compute the operations in the order dictated by the brackets in $[D^T (D\bar{p})]$. Therefore, we repeat the following step to convergence:

$$\bar{p}_1 \Leftarrow \frac{[D^T (D\bar{p}_1)]}{\|[D^T (D\bar{p}_1)]\|}$$

The projection of the data matrix D on the vector $\bar{p}_1$ has an energy that is equal to the square of the first singular value. Therefore, the first singular value σ_{11} is obtained by using the L_2-norm of the vector $D\bar{p}_1$. The first column $\bar{q}_1$ of Q is obtained by a single execution of the following step:

$$\bar{q}_1 \Leftarrow \frac{D\bar{p}_1}{\sigma_{11}} \tag{7.18}$$

The above result is a 1-dimensional simplification of $Q = DP\Sigma^{-1}$. This completes the determination of the first set of singular vectors and singular values. The next eigenvector and eigenvalue pair is obtained by making use of the spectral decomposition of Equation 7.1. One possibility is to remove the rank-1 component contributed by the first set of singular vectors by adjusting the data matrix as follows:

$$D \Leftarrow D - \sigma_{11}\bar{q}_1\bar{p}_1^T \tag{7.19}$$

Once the impact of the first component has been removed, we can repeat the process to obtain the second set of singular vectors from the modified matrix. The main problem with this approach is that the removal of spectral components hurts the sparsity of D.

Therefore, in order to avoid hurting the sparsity of D, one need not explicitly remove the rank-1 matrix $\overline{q}_1 \overline{p}_1^T$ from D. Rather, the original matrix D is used, and the second set of singular vectors can be computed by using the following iterative step (that removes the effect of the first component within the iterations):

$$\overline{p}_2 \Leftarrow (D^T - \sigma_{11} \overline{p}_1 \overline{q}_1^T)([D - \sigma_{11} \overline{q}_1 \overline{p}_1^T] \overline{p}_2)$$

$$\overline{p}_2 \Leftarrow \frac{\overline{p}_2}{\|\overline{p}_2\|}$$

When computing a quantity like $[D - \sigma_{11} \overline{q}_1 \overline{p}_1^T] \overline{p}_2$, one computes $D\overline{p}_2$ and $\overline{q}_1 [\overline{p}_1^T \overline{p}_2]$ separately. Note that the order of operations in $\overline{q}_1 [\overline{p}_1^T \overline{p}_2]$ is preferred to the order $[\overline{q}_1 \overline{p}_1^T] \overline{p}_2$ to ensure that one never has to store large and dense matrices. Therefore, the associativity property of matrix multiplication comes in handy to ensure that one is always working with multiplications between vectors, or between vectors and sparse matrices. This basic idea can be generalized to finding the kth singular vector:

$$\overline{p}_k \Leftarrow (D^T - \sum_{r=1}^{k-1} \sigma_{rr} \overline{p}_r \overline{q}_r^T)([D - \sum_{r=1}^{k-1} \sigma_{rr} \overline{q}_r \overline{p}_r^T] \overline{p}_k)$$

$$\overline{p}_k \Leftarrow \frac{\overline{p}_k}{\|\overline{p}_k\|}$$

In each case, it is possible to control the order of multiplication of matrices, so that one never has to work with dense matrices. The singular value σ_{kk} is the norm of $D\overline{p}_k$. The kth left singular vector can be derived from the kth right singular vector as follows:

$$\overline{q}_k \Leftarrow \frac{D\overline{p}_k}{\sigma_{kk}} \tag{7.20}$$

The entire process is repeated m times to obtain the rank-m singular value decomposition.

7.6   Summary

Singular value decomposition is one of the most fundamental techniques among a class of methods called matrix factorization. We present a linear algebra and an optimization perspective on singular value decomposition. These perspectives provide different insights:

- The linear algebra perspective is helpful is showing that a singular value decomposition exists, and the singular vectors are the eigenvectors of DD^T and $D^T D$.

- The optimization perspective shows that singular value decomposition provides a matrix factorization with the least error. The optimization perspective can be generalized to other forms of matrix factorization, which is the subject of the next chapter.

Singular value decomposition has numerous applications in machine learning applications like least-squares regression. The basic ideas in singular value decomposition also provide the foundations for kernel methods, which are discussed in Chapter 9.

7.7   Further Reading

A discussion of SVD is provided in various linear algebra books [77, 122, 123, 130]. The fact that SVD provides an optimal solution to unconstrained matrix factorization was first pointed out in the Eckart-Young theorem [41]. Different types of numerical algorithms for singular value decomposition are discussed in [52, 130]. The noise reduction properties of singular value decomposition are discussed in [7]. The use of singular value decomposition methods in outlier detection is discussed in [4].

7.8 Exercises

1. Use SVD to show the push-through identity of Problem 1.2.13 for any $n \times d$ matrix D and scalar $\lambda > 0$:

$$(\lambda I_d + D^T D)^{-1} D^T = D^T (\lambda I_n + DD^T)^{-1}$$

This exercise is almost the same as Problem 1.2.13 in Chapter 1.

2. Let D be an $n \times d$ data matrix, and $\overline{y}$ be an n-dimensional column vector containing the dependent variables of linear regression. The Tikhonov regularization solution to linear regression (cf. Section 4.7.1 of Chapter 4) predicts the dependent variables of a test instance $\overline{Z}$ using the following equation:

$$\text{Prediction}(\overline{Z}) = \overline{Z}\,\overline{W} = \overline{Z}(D^T D + \lambda I)^{-1} D^T \overline{y}$$

Here, the vectors $\overline{Z}$ and $\overline{W}$ are treated as $1 \times d$ and $d \times 1$ matrices, respectively. Show using the result of Exercise 1, how you can write the above prediction purely in terms of similarities between training points or between $\overline{Z}$ and training points.

3. Suppose that you are given a truncated SVD $D \approx Q\Sigma P^T$ of rank-k. Show how you can use this solution to derive an alternative rank-k decomposition $Q'\Sigma' P'^T$ in which the unit columns of Q (or/and P) might not be mutually orthogonal and the truncation error is the same.

4. Suppose that you are given a truncated SVD $D \approx Q\Sigma P^T$ of rank-k. Two of the non-zero singular values are identical. The corresponding right singular vectors are $[1, 0, 0]^T$ and $[0, 1, 0]^T$. Show how you can use this solution to derive an alternative rank-k SVD $Q'\Sigma' P'^T$ for which the truncation error is the same. At least some columns of matrices Q' and P' need to be non-trivially different from the corresponding columns in Q and P (i.e., the ith column of Q' should not be derivable from the ith column of Q by simply multiplying with either -1 or $+1$). Give a specific example of how you might manipulate the right singular vectors to obtain a non-trivially different solution.

5. Suppose that you are given a particular solution $\overline{x} = \overline{x}_0$ that satisfies the system of equations $A\overline{x} = \overline{b}$. Here, A is an $n \times d$ matrix, $\overline{x}$ is a d-dimensional vector of variables, and $\overline{b}$ is an n-dimensional vector of constants. Show that *all* possible solutions to this system of equations are of the form $\overline{x}_0 + \overline{v}$, where $\overline{v}$ is any vector drawn from a vector space $\mathcal{V}$. Show that $\mathcal{V}$ can be found easily using SVD. [Hint: Think about the system of equations $A\overline{x} = 0$.]

6. Consider the $n \times d$ matrix D. Construct the $(n + d) \times (n + d)$ matrix B as follows:

$$B = \begin{bmatrix} 0 & D^T \\ D & 0 \end{bmatrix}$$

Note that the matrix B is square and symmetric. Show that diagonalizing B yields all the information needed for constructing the SVD of D. [Hint: Relate the eigenvectors of B to the singular vectors of SVD.]

7. Consider the following matrix A whose SVD is given by the following:

$$A = \begin{bmatrix} -1/\sqrt{2} & 1/\sqrt{2} \\ 1/\sqrt{2} & 1/\sqrt{2} \end{bmatrix} \begin{bmatrix} 4 & 0 \\ 0 & 2 \end{bmatrix} \begin{bmatrix} 1 & 0 \\ 0 & 1 \end{bmatrix}^T$$

Compute the inverse of A without explicitly materializing A.

8. Consider the following 2-way factorization of the matrix A:

$$A = UV^T = \begin{bmatrix} 4 & 1 \\ 3 & 2 \end{bmatrix} \begin{bmatrix} 1 & 2 \\ 1 & 1 \end{bmatrix}^T$$

Convert this factorization into a 3-way factorization $Q\Sigma P^T$ in each of the following ways:

   (a) The L_2-norm of each column of Q and P is 1.
   (b) The L_1-norm of each column of Q and P is 1.

The second form of decomposition is used for nonnegative factorizations with probabilistic interpretability.

9. Suppose that you add a small amount of noise to each entry of an $n \times d$ matrix D with rank $r \ll d$ and $n \gg d$. The noise is drawn from a Gaussian distribution, whose variance $\lambda > 0$ is much smaller than the smallest non-zero singular value of D. The non-zero singular values of D are $\sigma_{11} \ldots \sigma_{rr}$. What do you expect the rank of the modified matrix D' to become?

10. Consider the unconstrained optimization problem of minimizing the Frobenius norm $\|D - UV^T\|_F^2$, which is equivalent to SVD. Here, D is an $n \times d$ data matrix, U is an $n \times k$ matrix, and V is a $d \times k$ matrix.

   (a) Use differential calculus to show that the optimal solution satisfies the following conditions:

$$DV = UV^TV$$
$$D^TU = VU^TU$$

   (b) Let $E = D - UV^T$ be a matrix of errors from the current solutions U and V. Show that an alternative way to solve this optimization problem is by using the following gradient-descent updates:

$$U \Leftarrow U + \alpha EV$$
$$V \Leftarrow V + \alpha E^TU$$

   Here, $\alpha > 0$ is the step-size.

   (c) Will the resulting solution necessarily contain mutually orthogonal columns in U and V?

11. Suppose that you change the objective function of SVD in Exercise 10 to add penalties on large values of the parameters. This is often done to reduce overfitting and improve generalization power of the solution. The new objective function to be minimized is as follows:

$$J = \|D - UV^T\|_F^2 + \lambda(\|U\|_F^2 + \|V\|_F^2)$$

Here, $\lambda > 0$ defines the penalty. How would your answers to Exercise 10 change?

12. Recall from Chapter 3 that the determinant of a square matrix is equal to the product of its eigenvalues. Show that the determinant of a square matrix is also equal to the product of its singular values *but only in absolute magnitude*. Show that the Frobenius norm of the inverse of a $d \times d$ square matrix A is equal to the sum of squared inverses of the singular values of A.

13. Show using SVD that a square matrix A is symmetric (i.e., $A = A^T$) if and only if $AA^T = A^T A$.

14. Suppose that you are given the following valid SVD of a matrix:

$$D = \begin{bmatrix} 1 & 0 & 0 \\ 0 & 1/\sqrt{2} & 1/\sqrt{2} \\ 0 & 1/\sqrt{2} & -1/\sqrt{2} \end{bmatrix} \begin{bmatrix} 2 & 0 & 0 \\ 0 & 1 & 0 \\ 0 & 0 & 1 \end{bmatrix} \begin{bmatrix} 1 & 0 & 0 \\ 0 & 1/\sqrt{2} & -1/\sqrt{2} \\ 0 & 1/\sqrt{2} & 1/\sqrt{2} \end{bmatrix}$$

Is the SVD of this matrix unique? You may ignore multiplication of singular vectors by -1 as violating uniqueness. If the SVD is unique, discuss why this is the case. If the SVD is not unique, provide an alternative SVD of this matrix.

15. State a simple way to find the SVD of (a) a diagonal matrix with both positive and negative entries that are all different; and (b) an orthogonal matrix. Is the SVD unique in these cases?

16. Show that the largest singular value of $(A + B)$ is at most the sum of the largest singular values of each of A and B. Also show that the largest singular value of AB is at most the product of the largest singular values of A and B. Finally, show that the largest singular value of a matrix is a convex function of the matrix entries.

17. If A is a square matrix, use SVD to show that AA^T and $A^T A$ are similar. What happens when A is rectangular?

18. The Frobenius norm of a matrix A is defined as the trace of either AA^T or $A^T A$. Let P be a $d \times k$ matrix with orthonormal columns. Let D be an $n \times d$ data matrix. Show that the squared Frobenius norm of DP is the same as that of DPP^T. Interpret the matrices DP and DPP^T in terms of their relationship with D, when P contains the top-k right singular vectors of the SVD of D.

19. Consider two data matrices D_1 and D_2 that share the same scatter matrix $D_1^T D_1 = D_2^T D_2$ but are otherwise different. We aim to show that the columns of one are rotre-flections of the other and vice versa. Show that a *partially shared* (full) singular value decomposition can be found for D_1 and D_2, so that $D_1 = Q_1 \Sigma P^T$ and $D_2 = Q_2 \Sigma P^T$. Use this fact to show that $D_2 = Q_{12} D_1$ for some orthogonal matrix Q_{12}.

20. Let $A = \bar{a} \bar{b}^T$ be a rank-1 matrix for vectors $\bar{a}, \bar{b} \in \mathcal{R}^n$. Find the non-zero eigenvectors, eigenvalues, singular vectors, and singular values of A.

21. What are the singular values of (i) a $d \times d$ Givens rotation matrix, (ii) a $d \times d$ Householder reflection matrix, (iii) a $d \times d$ projection matrix of rank r, (iv) a 2×2 shear matrix $A = [a_{ij}]$ with 1s along the diagonal, and a value of $a_{12} = 2$ in the upper right corner.

22. Consider an $n \times d$ matrix A with linearly independent columns and non-zero singular values $\sigma_1 \ldots \sigma_d$. Find the non-zero singular values of $A^T(AA^T)^5$, $A^T(AA^T)^5 A$, $A(A^T A)^{-2} A^T$, and $A(A^T A)^{-1} A^T$. Do you recognize the last of these matrices? Which of these matrices have economy SVDs with zero singular values in addition to the non-zero singular values?

23. Suppose that you have the $n \times 3$ scatterplot matrix D of an ellipsoid in 3-dimensions, whose three axes have lengths 3, 2, and 1, respectively. The axes directions of this ellipsoid are $[1, 1, 0]$, $[1, -1, 0]$, and $[0, 0, 1]$. You multiply the scatter plot matrix D with a 3×3 transformation matrix A to obtain the scatter plot $D' = DA$ of a new ellipsoid, in which the axes $[1, 1, 1]$, $[1, -2, 1]$, and $[1, 0, -1]$ have lengths 12, 6, and 5, respectively. Write the singular value decompositions of two possible matrices that can perform the transformation. You should be able to write down the SVDs with very little numerical calculation. [The answer to this question is not unique, as the specific mapping of points between the two ellipsoids is not known. For example, an axis direction in the original ellipsoid may or may not match with an axis direction in the transformed ellipsoid.]

24. **Regularization impact:** Consider the regularized least-squares regression problem of minimizing $\|A\overline{x} - \overline{b}\|^2 + \lambda \|\overline{x}\|^2$ for d-dimensional optimization vector $\overline{x}$, n-dimensional vector $\overline{b}$, nonnegative scalar λ, and $n \times d$ matrix A. There are several ways of showing that the norm of the optimum solution $\overline{x} = \overline{x}^*$ is non-increasing with increasing λ (and this is also intuitively clear from the nature of the optimization formulation). Use SVD to show that the optimum solution $\overline{x}^* = (A^T A + \lambda I_d)^{-1} A^T \overline{b}$ has non-increasing norm with increasing λ.

25. The function $f(\lambda)$ arises commonly in spherically constrained least-squares regression:

$$f(\lambda) = \overline{b}^T A (A^T A + \lambda I)^{-2} A^T \overline{b}$$

Here, A is an $n \times d$ matrix of rank-r, $\overline{b}$ is an n-dimensional column vector, and $\lambda > 0$ is an optimization parameter. Furthermore, $A = Q\Sigma P^T$ is the reduced SVD of A with $n \times r$ matrix Q, $d \times r$ matrix P, and $r \times r$ diagonal matrix Σ. The diagonal elements of Σ are $\sigma_{11} \ldots \sigma_{rr}$. Show that $f(\lambda)$ can be written in scalar form as follows:

$$f(\lambda) = \sum_{i=1}^{r} \left(\frac{\sigma_{ii} c_i}{\sigma_{ii}^2 + \lambda} \right)^2$$

Here, c_i is the ith component of $Q^T \overline{b}$.

26. **Pseudoinverse properties:** Show using SVD that $AA^+ A = A$ and $A^+ AA^+ = A^+$. Also show using SVD that AA^+ is a symmetric and idempotent matrix (which is an alternative definition of a projection matrix).

27. Compute the compact SVD of the matrix A along with the Moore-Penrose pseudoinverse:

$$A = \begin{bmatrix} 2 & 1 & 3 \\ 1 & 2 & 0 \end{bmatrix}$$

28. **Generalized singular value decomposition:** The generalized singular value decomposition of an $n \times d$ matrix D is given by $D = Q\Sigma P^T$, where $Q^T S_1 Q = I$ and

$P^T S_2 P = I$. Here, S_1 and S_2 are (given) $n \times n$ and $d \times d$ positive definite matrices, and therefore the singular vectors in Q and P are orthogonal from the perspective of the generalized definition of inner products. Show how to reduce generalized singular value decomposition to a singular value decomposition on a modified version of D. [Hint: Pre-multiply and post-multiply D with appropriate square-root matrices.]

Chapter 8

Matrix Factorization

"He who knows only his own side of the case knows little of that. His reasons may be good, and no one may have been able to refute them. But if he is equally unable to refute the reasons on the opposite side, if he does not so much as know what they are, he has no ground for preferring either opinion."–John Stuart Mill

8.1 Introduction

Just as multiplication can be generalized from scalars to matrices, the notion of factorization can also be generalized from scalars to matrices. Exact matrix factorizations need to satisfy the size and rank constraints that are imposed on matrix multiplication. For example, when an $n \times d$ matrix A is factorized into two matrices B and C (i.e., $A = BC$), the matrices B and C must be of sizes $n \times k$ and $k \times d$ for some constant k. For exact factorization to occur, the value of k must be equal to at least the rank of A. This is because the rank of A is at most equal to the minimum of the ranks of B and C. In practice, it is common to perform *approximate* factorization with much smaller values of k than the rank of A.

As in scalars, the factorization of a matrix is not unique. For example, the scalar 12 can be factorized into 2 and 6, or it can be factorized into 3 and 4. If we allow real factors, there are an infinite number of possible factorizations of a given scalar. The same is true of matrices, where even the sizes of the factors might vary. For example, consider the following factorizations of the same matrix:

$$\begin{bmatrix} 3 & 6 \\ 3 & 6 \end{bmatrix} = \begin{bmatrix} 1 \\ 1 \end{bmatrix} \begin{bmatrix} 3 & 6 \end{bmatrix} = \begin{bmatrix} 1 & 1 \\ 1 & 1 \end{bmatrix} \begin{bmatrix} 2 & 4 \\ 1 & 2 \end{bmatrix}$$

It is clear that a given matrix can be factorized in an unlimited number of ways. However, factorizations with certain types of properties are more useful than others. There are two types of properties that are commonly desired in decompositions:

1. *Linear algebra properties with exact decomposition:* In these cases, one tries to create decompositions in which the individual components of the factorization have specific linear algebra/geometric properties such as orthogonality, triangular nature of the

© Springer Nature Switzerland AG 2020

C. C. Aggarwal, *Linear Algebra and Optimization for Machine Learning*,

https://doi.org/10.1007/978-3-030-40344-7_8

matrix, and so on. These types of properties are useful for various linear algebra applications like basis construction. All the decompositions that we have seen so far, such as LU decomposition, QR decomposition, and SVD, have linear algebra properties.

2. *Optimization and compression properties with approximate decomposition:* In these cases, one is attempting to factorize a much larger matrix into two or more smaller matrices. Truncated SVD is an example of this type of factorization. Consider the $n \times d$ matrix D, which is truncated to rank-k to create the following factorization:

$$D \approx Q_k \Sigma_k P_k^T \tag{8.1}$$

Here, Q_k is an $n \times k$ orthogonal matrix, Σ_k is a $k \times k$ diagonal matrix with nonnegative entries, and P_k is a $d \times k$ orthogonal matrix. The total number of entries in all three matrices is $(n+d+k)k$, which is often much smaller than the nd entries in the original matrix for large values of n and d. For example, if $n = d = 10^6$ and $k = 1000$, the number of entries in D is 10^{12}, whereas the total number of entries in the factorized matrices is approximately 2×10^9, which is only 0.2% of the original number of entries.

Singular value decomposition is one of the few factorizations that is useful *both* in terms of its linear algebra properties (when used in exact form), and in terms of its compression properties (when used in truncated form). The value k is referred to as the *rank* of the factorization. The optimization view of matrix factorization $D \approx UV^T$ is particularly useful in machine learning by instantiating D, U, and V as follows:

1. When D is a *document-term matrix* containing frequencies of words (columns of D) in documents (rows of D), the rows of U provide latent representations of documents, whereas the rows of V provide latent representations of words.

2. A rating is a numerical score that a user gives to an item (e.g., movie). Recommender systems collect ratings of users for items in order to make predictions of ratings for items they have not yet evaluated. When D is a user-item matrix of ratings, the rows correspond to users and the columns correspond to items. The entries of D contain ratings. Matrix factorization decomposes the *incomplete* matrix $D \approx UV^T$ using only observed ratings. The rows of U provide latent representations of users, whereas the rows of V are the latent representations of items. The matrix UV^T reconstructs the entire ratings matrix (including predictions for missing ratings).

3. Let $D \approx UV^T$ be a graph adjacency matrix, so that the (i,j)th entry of D contains the weight of edge between nodes i and j. In such a case, the rows of both U and V are the latent representations of nodes. The latent representations of U and V can be used for applications like clustering and *link prediction* (cf. Chapters 9 and 10).

In the optimization-centric view, one can impose specific properties on the decomposed matrices as constraints of the optimization problem (such as nonnegativity of matrix entries). These specific properties are often useful in various types of applications.

This chapter is organized as follows. The next section provides an overview of the optimization-centric view to matrix factorization. Unconstrained matrix factorization methods are discussed in Section 8.3. Nonnegative matrix factorization methods are introduced in Section 8.4. Weighted matrix factorization methods are introduced in Section 8.5. Logistic and maximum-margin matrix factorizations are discussed in Section 8.6. Generalized low-rank models are introduced in Section 8.7. Methods for shared matrix factorization are discussed in Section 8.8. Factorization machines are discussed in Section 8.9. A summary is given in Section 8.10.

8.2 Optimization-Based Matrix Factorization

The optimization-centric view of matrix factorization is at the core of its usefulness in machine learning applications. The optimization-centric view creates a *compressed* representation of the matrix, which is always helpful in getting rid of the random artifacts and generalizing predictions of missing values from seen data to unseen data. After all, repeated patterns in the data, which are useful for predictions of missing values in new data instances, are retained in a compressed representation.

In the following, we discuss a two-way factorization of an $n \times d$ matrix D into an $n \times k$ matrix U and a $d \times k$ matrix V, although any two-way factorization can be converted into a three-way factorization $Q\Sigma P^T$ (like SVD) using the approach in Section 7.2.7 of Chapter 7. The main goal of the factorization is to create an objective function so that UV^T can be used to reconstruct the original matrix D. Most forms of optimization-centric matrix factorization are special cases of the following optimization model over matrices U and V:

$$\text{Maximize similarity between entries of } D \text{ and } UV^T$$

$$\text{subject to:}$$

$$\text{Constraints on } U \text{ and } V$$

Constraints are used to ensure specific properties of the factor matrices. A commonly used constraint is that of nonnegativity of the matrices U and V. The simplest possible objective function, which is used in SVD, is $\|D - UV^T\|_F^2$. Other objective functions such as *log-likelihood* and *I-divergence* are also used to create probabilistic models. Most matrix factorization formulations are not convex; nevertheless, gradient descent works quite well in these cases.

In some cases, it is possible to weight specific matrix entries in the objective function. In fact, for certain types of matrices, it makes more sense to interpret the entry of the matrix as a weight. This is common in the case of *implicit feedback data* in recommender systems, where all entries are assumed to be binary, and the values of non-zero entries are treated as weights. For example, a matrix containing the quantities of sales of products (column identifiers) to various users (row identifiers) is always assumed to be binary depending on whether or not users have bought the products. This approach is also sometimes used with frequency matrices in the text domain [101].

Logistic matrix factorization methods apply a logistic function on the entries of UV^T in order to materialize the probability that a particular entry is 1. Such an approach works well for matrices in which the nonnegative values should be treated as frequencies of binary values. The basic idea here is to assume that the entries of D are frequencies obtained by repeatedly sampling each entry in the matrix with probabilities present in the matrix $P = \text{Sigmoid}(UV^T)$. The sigmoid function is defined as follows:

$$\text{Sigmoid}(x) = \frac{1}{1 + \exp(-x)}$$

Note that D and P are two matrices of the same size, and the frequencies in D will be roughly proportional to the entries in P:

$$D \sim \text{Instantiation of frequencies obtained by sampling from } P$$

The optimization model maximizes a log-likelihood function based on this probabilistic model. A surprisingly large number of applications in machine learning can be shown to be special cases of matrix factorization, especially if one is willing to incorporate complex objective functions and constraints in the factorization. Matrix factorization is used for feature engineering, clustering, kernel methods, link prediction, and recommendations. In each case, the secret is to choose an appropriate objective function and corresponding constraints for the problem at hand. As a specific example, we will show that the k-means algorithm is a special case of matrix factorization, albeit with some special constraints.

8.2.1   Example: K-Means as Constrained Matrix Factorization

The k-means algorithm determines a set of k centroids, so that the sum-of-squared errors of each row in an $n \times d$ data matrix D from the closest centroid is minimized. This algorithm can be shown to be a special case of matrix factorization if one is willing to incorporate additional constraints in the factorization. Consider the following optimization problem with $n \times k$ and $d \times k$ factor matrices U and V, respectively:

$$\text{Minimize}_{U,V} \, \|D - UV^T\|_F^2$$
$$\text{subject to:}$$
$$\text{Columns of } U \text{ are mutually orthogonal}$$
$$u_{ij} \in \{0, 1\}$$

This is a *mixed integer matrix factorization problem*, because the entries of U are constrained to be binary values. An equivalent optimization formulation is given in Section 4.10.3 of Chapter 4. In this case, each row of U can be shown to contain exactly a single 1, corresponding to the cluster membership of that row. Each column of V contains the d-dimensional centroid of one of the k clusters. As discussed in Section 4.10.3 of Chapter 4, this optimization problem can be solved using block coordinate descent, *which is identical to the k-means algorithm*. The fact that k-means is a special case of matrix factorization is an example of the fact that the family of matrix factorization methods is extremely expressive in its relationship to a wide variety of machine learning methods. This chapter will, therefore, explore multiple ways of performing matrix factorization together with their applications.

8.3   Unconstrained Matrix Factorization

The problem of unconstrained matrix factorization is defined as follows:

$$\text{Minimize}_{U,V} \, J = \frac{1}{2}\|D - UV^T\|_F^2$$

Here, D, U, and V are matrices of sizes n, d, and k, respectively. The value of k is typically much smaller than the rank of the matrix D. This problem, which is discussed in Section 7.3.2 of Chapter 7, can be shown to provide the same solution as SVD. As discussed in Chapter 7, the top eigenvectors of $D^T D$ provide the columns of V, and the top eigenvectors of DD^T provide the columns of U. The columns of V are unit normalized, whereas those of U are normalized so that the norm of the ith column is equal to the ith singular value of D.

However, many alternative optima are possible. For example, even the normalization of the columns is not unique. Instead of normalizing columns of V to unit norm, one could easily normalize the columns of U to unit norm, and adjust the normalization of each column of V appropriately. The main point is that the product of the norms of the ith columns of U and V needs to be the ith singular value of D. Furthermore, the columns of U and V need not be orthonormal sets for an optimum solution to exist. Given an optimum pair (U_0, V_0), one could change the basis of the column space of V_0 to a non-orthogonal one and adjust U_0 to the corresponding coordinates in the non-orthogonal basis system, so that the product $U_0 V_0^T$ does not change. To understand this point, we recommend the reader to solve the following problem:

Problem 8.3.1 *Let $D \approx Q_k \Sigma_k P_k^T$ be the rank-k SVD of D. The results in Chapter 7 show that $(U, V) = (Q_k \Sigma_k, P_k)$ represents an optimal pair of rank-k solution matrices to the unconstrained matrix factorization problem that is posed in this section. Show that $(U, V) = (Q_k \Sigma_k R_k^T, P_k R_k^{-1})$ is an alternative optimal solution to the unconstrained matrix factorization problem for any $k \times k$ invertible matrix R_k.*

8.3.1 Gradient Descent with Fully Specified Matrices

In this section, we will investigate a method that finds a solution to the unconstrained optimization problem with the use of gradient descent. This approach does not guarantee the orthogonal solutions provided by singular value decomposition; however, the formulation is equivalent and should (ideally) lead to a solution with the same objective function value. The approach also has the advantage that it can easily adapted to more difficult settings such as the presence of missing values in the matrix. A natural application of this type of approach is that of matrix factorization in recommender systems. Recommender systems use the same optimization formulation as SVD; however, the resulting basis vectors of the factorization are not guaranteed to be orthogonal.

In order to perform gradient descent, we need to compute the derivative of the unconstrained optimization problem with respect to the parameters in the matrices $U = [u_{iq}]$ and $V = [v_{jq}]$. The simplest approach is to compute the derivative of the objective function J with respect to each parameter in the matrices U and V. First, the objective function is expressed in terms of the individual entries in the various matrices. Let the (i, j)th entry of the $n \times d$ matrix D be denoted by x_{ij}. Then, the objective function can be restated in terms of the entries of the matrices D, U, and V as follows:

$$\text{Minimize } J = \frac{1}{2} \sum_{i=1}^{n} \sum_{j=1}^{d} \left(x_{ij} - \sum_{s=1}^{k} u_{is} \cdot v_{js} \right)^2$$

The quantity $e_{ij} = x_{ij} - \sum_{s=1}^{k} u_{is} \cdot v_{js}$ is the error of the factorization for the (i, j)th entry. Note that the objective function J minimizes the sum of squares of e_{ij}. One can compute the partial derivative of the objective function with respect to the parameters in the matrices U and V as follows:

$$\frac{\partial J}{\partial u_{iq}} = \sum_{j=1}^{d} \left(x_{ij} - \sum_{s=1}^{k} u_{is} \cdot v_{js} \right) (-v_{jq}) \quad \forall i \in \{1 \ldots n\}, q \in \{1 \ldots k\}$$

$$= \sum_{j=1}^{d} (e_{ij})(-v_{jq}) \quad \forall i \in \{1 \ldots n\}, q \in \{1 \ldots k\}$$

$$\frac{\partial J}{\partial v_{jq}} = \sum_{i=1}^{n} \left(x_{ij} - \sum_{s=1}^{k} u_{is} \cdot v_{js} \right) (-u_{iq}) \quad \forall j \in \{1 \ldots d\}, q \in \{1 \ldots k\}$$

$$= \sum_{i=1}^{n} (e_{ij})(-u_{iq}) \quad \forall j \in \{1 \ldots d\}, q \in \{1 \ldots k\}$$

One can also express these derivatives in terms of matrices. Let $E = [e_{ij}]$ be the $n \times d$ matrix of errors. In the denominator layout of matrix calculus, the derivatives can be expressed as follows:

$$\frac{\partial J}{\partial U} = -(D - UV^T)V = -EV$$

$$\frac{\partial J}{\partial V} = -(D - UV^T)^T U = -E^T U$$

The above matrix calculus identity can be verified by using the relatively tedious process of expanding the (i, q)th and (j, q)th entries of each of the above matrices on the right-hand side, and showing that they are equivalent to the (corresponding) scalar derivatives $\frac{\partial J}{\partial u_{iq}}$ and $\frac{\partial J}{\partial v_{jq}}$. An alternative approach that directly uses the matrix calculus identities of Chapter 4 is given in Figure 8.1. The reader may choose to skip over this derivation without loss of continuity.

The optimality conditions for this optimization problem are therefore obtained by setting these derivatives to 0. Therefore, we obtain the optimality conditions $DV = UV^TV$ and $D^TU = VU^TU$. These optimality conditions can be shown to hold for the solution obtained from SVD $U = Q_k \Sigma_k$ and $V = P_k$.

Problem 8.3.2 *Let $Q_k \Sigma_k P_k^T$ be the rank-k truncated SVD of matrix D. Show that the solution $U = Q_k \Sigma_k$ and $V = P_k$ satisfies the optimality conditions $DV = UV^TV$ and $D^TU = VU^TU$.*

A useful hint for solving the above problem is to use the spectral decomposition of SVD as a sum of rank-1 matrices.

Although the optimality condition leads to the standard SVD solution $D \approx [Q_k \Sigma_k] P_k^T$, one can also find an optimal solution by using gradient descent. The updates for gradient descent are as follows:

$$U \Leftarrow U - \alpha \frac{\partial J}{\partial U} = U + \alpha EV$$

$$V \Leftarrow V - \alpha \frac{\partial J}{\partial V} = V + \alpha E^T U$$

Here, $\alpha > 0$ is the learning rate.

The optimization model is identical to that of SVD. If the aforementioned gradient descent method is used (instead of the power iteration method of the previous chapter), one will typically obtain solutions that are equally good in terms of objective function value, but for which the columns of U (or V) are not mutually orthogonal. The power iteration methods yields solutions with orthogonal columns. Although the standardized SVD solution with orthonormal columns is typically not obtained by gradient descent, the k columns of U will span[1] the same subspace as the columns of Q_k, and the columns of V will span the same subspace as the columns of P_k.

The gradient-descent approach can be implemented efficiently when the matrix D is sparse by sampling entries from the matrix for making updates. This is essentially a *stochastic* gradient descent method. In other words, we sample an entry (i, j) and compute its error

---

[1]This will occur under the assumption that the top-k eigenvalues of $D^T D$ are distinct. Tied eigenvalues result in a non-unique solution for SVD, which might sometimes result in some differences in the subspace corresponding to the smallest eigenvalue within the rank-k solution.

Consider the following objective function the $n \times d$ matrix D with rank-k matrices U and V:

$$J = \frac{1}{2}\|D - UV^T\|_F^2$$

Matrix calculus can be used for computing derivatives with respect to U and V after decomposing the Frobenius norm into row-wise vector norms or column-wise vector norms, depending on whether we wish to compute the derivative of J with respect to U or V. Let $\overline{X}_i$ be the ith row of D (row vector), $\overline{d}_j$ be the jth column of D (column vector), $\overline{u}_i$ be the ith row of U (row vector), and $\overline{v}_j$ be the jth row of V (row vector). Then, the Frobenius norm can be decomposed in row-wise fashion as follows:

$$J = \frac{1}{2}\sum_{i=1}^{n}\|\overline{X}_i - \overline{u}_iV^T\|^2 = \underbrace{\frac{1}{2}\sum_{i=1}^{n}\overline{X}_i\overline{X}_i^T}_{\text{Constant}} - \sum_{i=1}^{n}\overline{X}_iV\overline{u}_i^T + \frac{1}{2}\sum_{i=1}^{n}\overline{u}_iV^TV\overline{u}_i^T$$

To compute the derivative of the two non-constant terms with respect to $\overline{u}_i$, we can use identities (i) and (ii) of Table 4.2(a) in Chapter 4. This yields the following:

$$\frac{\partial J}{\partial \overline{u}_i^T} = -V^T\overline{X}_i^T + V^TV\overline{u}_i^T$$

$$\frac{\partial J}{\partial [\overline{u}_1^T \ldots \overline{u}_n^T]} = -V^T[\overline{X}_1^T \ldots \overline{X}_n^T] + V^TV[\overline{u}_1^T \ldots \overline{u}_n^T]$$

$$\frac{\partial J}{\partial U^T} = -V^TD^T + V^TVU^T$$

$$\frac{\partial J}{\partial U} = -DV + UV^TV = -(D - UV^T)V$$

In order to compute the derivative with respect to V, one will need to decompose the squared Frobenius norm in J in column-wise fashion as follows:

$$J = \frac{1}{2}\sum_{j=1}^{d}\|\overline{d}_j - U\overline{v}_j^T\|^2 = \underbrace{\frac{1}{2}\sum_{j=1}^{d}\overline{d}_j^T\overline{d}_j}_{\text{Constant}} - \sum_{j=1}^{d}\overline{d}_j^TU\overline{v}_j^T + \frac{1}{2}\sum_{j=1}^{d}\overline{v}_jU^TU\overline{v}_j^T$$

One can again use identities (i) and (ii) of Table 4.2(a) to show the following:

$$\frac{\partial J}{\partial \overline{v}_j^T} = -U^T\overline{d}_j + U^TU\overline{v}_j^T$$

As in the previous case, one can put together the derivatives for different rows of V to obtain the following:

$$\frac{\partial J}{\partial V} = -D^TU + VU^TU = -(D - UV^T)^TU$$

Figure 8.1: Alternative derivation of factorization gradients using matrix calculus

e_{ij}. Subsequently, we make the following updates to the ith row $\overline{u}_i$ of U and the jth row $\overline{v}_j$ of V, which are also referred to as *latent factors*:

$$\overline{u}_i \Leftarrow \overline{u}_i + \alpha e_{ij} \overline{v}_j$$
$$\overline{v}_j \Leftarrow \overline{v}_j + \alpha e_{ij} \overline{u}_i$$

One cycles through the sampled entries of the matrix (making the above updates) until convergence. The fact that we can sample entries of the matrix for updates means that we do not need fully specified matrices in order to learn the latent factors. This basic idea forms the foundations of recommender systems.

Problem 8.3.3 (Regularized Matrix Factorization) *Let D be an $n \times d$ matrix that we want to factorize using a rank-k decomposition into U and V. Suppose that we add the regularization terms $\frac{\lambda}{2}(\|U\|^2 + \|V\|^2)$ to the objective function $\frac{1}{2}\|D - UV^T\|_F^2$. Show that the gradient descent updates need to be modified as follows:*

$$U \Leftarrow U(1 - \alpha\lambda) + \alpha EV$$
$$V \Leftarrow V(1 - \alpha\lambda) + \alpha E^T U$$

The entries of the matrices U and V can be initialized as follows. First, all $n \times k$ entries in U are independently sampled from a standard normal distribution, and then each column is normalized to the unit vector. Note that the matrix U contains roughly orthogonal columns, if n is large (see Exercise 18 of Chapter 1). The matrix V is selected to be $D^T U$. This approach ensures that UV^T yields $UU^T D$, where UU^T is (roughly) a projection matrix because of the approximate orthogonality of U. Thus, the initialized product is already closely related to the target matrix.

8.3.2   Application to Recommender Systems

Let D be an $n \times d$ ratings matrix representing the ratings of n users for d items. The (i,j)th entry in the matrix D is denoted by x_{ij}, and it represents the rating of user i for item j. The key distinguishing point of a recommendation application is that the vast majority of ratings are missing. This is because users do not specify the ratings of the vast majority of the items in collaborative filtering applications. An example of a ratings matrix with missing entries is shown in Figure 8.2. In other words, the value of x_{ij} is observed (known) for only a small subset of the entries. The goal of the recommendation problem is to predict the missing ratings from the known ones.

Let S represent the set of indices of all the observed ratings. Therefore, we have:

$$S = \{(i,j) : x_{ij} \text{ is observed}\} \tag{8.2}$$

As in the case of traditional matrix factorization, we would like to factorize the *incomplete* ratings matrix D with the use of only the entries in S. In the terminology of recommender systems, the $n \times k$ matrix U is referred to as the *user factor* matrix, and the $d \times k$ matrix V is referred to as the *item factor matrix*. Regularization is particularly important in the case of the collaborative filtering application because of the paucity of observed data. Therefore, an additional term $\frac{\lambda}{2}(\|U\|_F^2 + \|V\|_F^2)$ is added to the objective function.

Once the user and item factor matrices have been learned, the entire ratings matrix can be reconstructed as UV^T. In practice, we only need to reconstruct the (i,j)th entry of matrix D as follows:

$$\hat{x}_{ij} = \sum_{s=1}^{k} u_{is} \cdot v_{js} \tag{8.3}$$

	GLADIATOR	GODFATHER	BEN-HUR	GOODFELLAS	SCARFACE	SPARTACUS
TOM	1			5		2
JIM		5			4	
JOE	5	3		1		
ANN			3			4
JILL					3	5
SUE	5		4			

Figure 8.2: A ratings matrix with missing ratings

Note the "hat" symbol (i.e., circumflex) on the rating on the left-hand side to indicate that it is a predicted value rather than an observed value. The error e_{ij} of the prediction is $e_{ij} = x_{ij} - \hat{x}_{ij}$ for ratings that are observed.

One can then formulate the objective function in terms of the *observed* entries in D as follows:

$$\text{Minimize } J = \frac{1}{2} \sum_{(i,j) \in S} \left(x_{ij} - \sum_{s=1}^{k} u_{is} \cdot v_{js} \right)^2 + \frac{\lambda}{2} \sum_{i=1}^{n} \sum_{s=1}^{k} u_{is}^2 + \frac{\lambda}{2} \sum_{j=1}^{d} \sum_{s=1}^{k} v_{js}^2$$

The main difference from the objective function in the previous section is the use of only observed entries in S for squared error computation, and the use of regularization. As in the previous section, we can compute the partial derivative of the objective function with respect to the various parameters as follows:

$$\frac{\partial J}{\partial u_{iq}} = \sum_{j:(i,j) \in S} (e_{ij})(-v_{jq}) + \lambda u_{iq} \quad \forall i \in \{1 \ldots n\}, q \in \{1 \ldots k\}$$

$$\frac{\partial J}{\partial v_{jq}} = \sum_{i:(i,j) \in S} (e_{ij})(-u_{iq}) + \lambda v_{jq} \quad \forall j \in \{1 \ldots d\}, q \in \{1 \ldots k\}$$

One can also define these errors in matrix calculus notation. Let E be an $n \times d$ error matrix, which is defined to be e_{ij} for each observed entry $(i,j) \in S$ and 0 for each missing entry in the ratings matrix. Note that (unlike vanilla SVD), the error matrix E is already sparse because the vast majority of entries are not specified.

$$\frac{\partial J}{\partial U} = -EV + \lambda U$$

$$\frac{\partial J}{\partial V} = -E^T U + \lambda V$$

Note that the form of the derivative is exactly identical to traditional SVD except for the regularization term and the difference in how the error matrix is defined (to account for missing ratings). Then, the gradient-descent updates for the matrices U and V are as follows:

$$U \Leftarrow U - \alpha \frac{\partial J}{\partial U} = U(1 - \alpha\lambda) + \alpha E V$$

$$V \Leftarrow V - \alpha \frac{\partial J}{\partial V} = V(1 - \alpha\lambda) + \alpha E^T U$$

Here, $\alpha > 0$ is the learning rate. The matrix E can be explicitly materialized as a sparse error matrix, and the above updates can be achieved using only sparse matrix multiplications. Although this approach is referred to as singular value decomposition in the literature on recommender systems (because of the relationship of unconstrained matrix factorization with the SVD optimization model), one will typically not obtain orthogonal columns of U and V with this approach.

8.3.2.1 Stochastic Gradient Descent

Stochastic gradient descent is introduced in Section 4.5.2 of Chapter 4. From the perspective of matrix factorization, the idea is to sample observed entries in S in order to perform the updates one at a time. Note that in the above updates, the (i, j)th error entry e_{ij} influences only the ith rows of U and V respectively. Stochastic gradient descent simply pulls out these updates on an entry-wise basis, rather than summing up the updates over all entries. Let $\overline{u}_i$ be a k-dimensional row vector containing the ith row of U and $\overline{v}_j$ be a k-dimensional row vector containing the jth row of V. Then, stochastic gradient descent cycles through each entry $(i, j) \in S$ in random order, and performs the following updates:

$$\overline{u}_i \Leftarrow \overline{u}_i(1 - \alpha\lambda) + e_{ij}\overline{v}_j$$
$$\overline{v}_j \Leftarrow \overline{v}_j(1 - \alpha\lambda) + e_{ij}\overline{u}_i$$

Here, $\alpha > 0$ is the learning rate. Note that exactly $2k$ entries in the matrices U and V are updated for each observed entry in S. Therefore, a single cycle of stochastic gradient descent through all the observed ratings will make exactly $2k|S|$ updates. One starts by initializing the matrices U and V to uniform random values in $(0, M/\sqrt{k})$, where M is the maximum value of a rating. This type of initialization enures that the initial product UV^T yields values in a similar order of magnitude as the original ratings matrix. One then performs the aforementioned updates to convergence. Stochastic gradient descent tends to converge faster than gradient descent, and is often the method of choice in recommender systems.

8.3.2.2 Coordinate Descent

The method of coordinate descent is introduced in Section 4.10 of Chapter 4. The basic idea in coordinate descent methods is to optimize one parameter at a time. In the case of matrix factorization, this amounts to optimizing a single parameter in U and V.

In the following, we will use the objective function J used in recommender systems (with incomplete matrices), because it is the more general form of the objective function. All notations such as the observed entry set S and the error e_{ij} are the same as those used in the previous section. By setting the partial derivative of the objective function J with respect to u_{iq} to 0, we obtain the following for each $i \in \{1 \ldots n\}$ and $q \in \{1 \ldots k\}$:

$$\frac{\partial J}{\partial u_{iq}} = \sum_{j:(i,j)\in S} (e_{ij})(-v_{jq}) + \lambda u_{iq} = 0$$

$$u_{iq}(\lambda + \sum_{j:(i,j)\in S} v_{jq}^2) = \sum_{j:(i,j)\in S} (e_{ij} + u_{iq}v_{jq})v_{jq}$$

$$u_{iq} = \frac{\sum_{j:(i,j)\in S}(e_{ij} + u_{iq}v_{jq})v_{jq}}{\lambda + \sum_{j:(i,j)\in S} v_{jq}^2}$$

In the second step of the above algebraic manipulation, we added the quantity $\sum_{j:(i,j)\in S} u_{iq}v_{jq}^2$ to both sides in order to create a stable form of the update. The final form of the algebraic equation contains u_{iq} on both sides, and therefore it provides an iterative update. One can also derive a similar iterative update for each v_{jq}. The updates for the various values of u_{iq} and v_{jq} need to be performed sequentially as follows:

$$u_{iq} \Leftarrow \frac{\sum_{j:(i,j)\in S}(e_{ij} + u_{iq}v_{jq})v_{jq}}{\lambda + \sum_{j:(i,j)\in S} v_{jq}^2} \quad \forall i, q$$

$$v_{jq} \Leftarrow \frac{\sum_{i:(i,j)\in S}(e_{ij} + u_{iq}v_{jq})u_{iq}}{\lambda + \sum_{i:(i,j)\in S} u_{iq}^2} \quad \forall j, q$$

One simply starts with random values of the parameters in the matrices U and V, and performs the above updates. One cycles through the $(m + n) \cdot k$ parameters in U and V with these updates until convergence is reached.

8.3.2.3 Block Coordinate Descent: Alternating Least Squares

Alternating least squares is a form of block coordinate descent, which is introduced in Section 4.10.2 of Chapter 4. This approach with an initial pair of matrices U and V, and then *alternates* between updating U and V, while keeping the other matrix fixed. Therefore, the process works as follows:

1. Keeping U fixed, we solve for each of the d rows of V by treating the problem as a least-squares regression problem. Only the observed ratings in S can be used for building the least-squares model in each case. Let $\overline{v}_j$ be the jth row of V. In order to determine the optimal vector $\overline{v}_j$, we wish to minimize $\sum_{i:(i,j)\in S}(x_{ij} - \sum_{s=1}^k u_{is}v_{js})^2$, which is a least-squares regression problem in $v_{j1} \ldots v_{jk}$. The terms $u_{i1} \ldots u_{ik}$ are treated as constant values, whereas $v_{j1} \ldots v_{jk}$ are treated as optimization variables. Therefore, the k-dimensional vector $\overline{v}_j$ for the jth item are determined with least-squares regression. A total of d such least-squares problems need to be executed, and each least-squares problem has k variables. Because the least-squares problem for each item is independent, this step can be parallelized easily.

2. Keeping V fixed, solve for each of the n rows of U by treating the problem as a least-squares regression problem. Only the specified ratings in S can be used for building the least-squares model in each case. Let $\overline{u}_i$ be the ith row of U. In order to determine the optimal vector $\overline{u}_i$, we wish to minimize $\sum_{j:(i,j)\in S}(x_{ij} - \sum_{s=1}^k u_{is}v_{js})^2$, which is a least-squares regression problem in $u_{i1} \ldots u_{ik}$. The terms $v_{j1} \ldots v_{jk}$ are treated as constant values, whereas $u_{i1} \ldots u_{ik}$ are treated as optimization variables. A total of n such least-squares problems need to be executed, and each least-squares problem has k variables. Because the least-squares problem for each user is independent, this step can be parallelized easily.

Least-squares regression is discussed in Section 4.7 of Chapter 4.

8.4   Nonnegative Matrix Factorization

Nonnegative matrix factorization expresses a matrix as the product of nonnegative factor matrices. Since the product of two nonnegative matrices is nonnegative, it makes sense for the original matrix to be nonnegative as well. Nonnegativity is often satisfied by many real-world matrices containing frequency counts, such as the following:

1. The quantities of items bought by a user are a nonnegative values. The rows of the matrix correspond to users, and the columns correspond to items. The (i, j)th entry corresponds to the number of units bought by user i for item j.

2. The frequencies of various words in a document is a nonnegative value. In this case, the rows correspond to the documents, and the columns correspond to the entire set of words in the lexicon. The (i, j)th entry corresponds to frequency of the jth lexicon word in the ith document.

3. In graph applications, a square adjacency matrix might contain nonnegative weights associated with edges. For example, in the adjacency matrix of a publication network between authors, the entry (i, j) represents the number of times that authors i and j have collaborated with one another.

Why is the nonnegativity of factors useful? As we will see later, the nonnegativity of the factor matrices results in a very high level of interpretability of the factorization. Secondly, the nonnegativity of the factor matrices plays a role in regularizing the factorization. While the error of the factorization always increases by adding constraints such as nonnegativity, the predictions obtained from the factorization often improve for out-of-sample data (such as making predictions with missing data). This is an example of how the goals of optimization in machine learning are often different from those of traditional optimization (cf. Section 4.5.3 of Chapter 4).

8.4.1   Optimization Problem with Frobenius Norm

The most common formulation for nonnegative matrix factorization uses the Frobenius norm as the objective function and imposes a nonnegativity constraints on the factor matrices. Let $D = [x_{ij}]$ be an $n \times d$ data matrix in which the entries are nonnegative. Let U and V be the corresponding $n \times k$ and $d \times k$ factor matrices, so that $D \approx UV^T$. Then, the nonnegative matrix factorization problem is formulated as follows:

$$\text{Minimize } J = \frac{1}{2}\|D - UV^T\|_F^2 + \frac{\lambda}{2}\|U\|_F^2 + \frac{\lambda}{2}\|V\|_F^2$$
$$\text{subject to:}$$
$$U \geq 0, \quad V \geq 0$$

It is evident that this problem differs from unconstrained matrix factorization only in terms of the addition of the nonnegativity constraints. These are *box constraints*, which are particularly easy to address in the context of constrained optimization (cf. Section 6.3.2 of Chapter 6).

8.4.1.1 Projected Gradient Descent with Box Constraints

The optimization problem only has box constraints, which makes it particularly easy to solve. The basic idea is to use the same gradient-descent method as used in unconstrained optimization. Subsequently, the optimization variables are reset to their nonnegative values. In other words, any entry of the matrices U and V, which is negative, is set to 0.

As discussed in Sections 8.3.1 and 8.3.2 on unconstrained matrix factorization, the gradients of the objective function J with respect to the factor matrices U and V are as follows:

$$\frac{\partial J}{\partial U} = -(D - UV^T)V + \lambda U$$

$$\frac{\partial J}{\partial V} = -(D - UV^T)^T U + \lambda V$$

Therefore, the gradient-descent updates (without worrying about the nonnegativity constraints) are as follows:

$$U \Leftarrow U - \alpha \frac{\partial J}{\partial U} = U(1 - \alpha\lambda) + \alpha(D - UV^T)V$$

$$V \Leftarrow V - \alpha \frac{\partial J}{\partial V} = V(1 - \alpha\lambda) + \alpha(D - UV^T)^T U$$

The main difference is that we add two steps to the updates to ensure nonnegativity of each matrix entry:

$$U \Leftarrow \max\{U, 0\}, \qquad V \Leftarrow \max\{V, 0\}$$

This procedure is based on the ideas discussed in Section 6.3.2 on box constraints. In practice, projected gradient descent is used rarely for nonnegative matrix factorization.

8.4.2 Solution Using Duality

In this section, we will provide a solution to the nonnegativity problem using Lagrangian relaxation (cf. Section 6.4 of Chapter 6). We convert the inequality constraints to the form $-U \leq 0$ and $-V \leq 0$ in order to be consistent with the conventions used in Section 6.4. Note that there is a constraint for each entry in the matrices U and V, and therefore we need as many Lagrange multipliers as the number of matrix entries. For the (i, s)th entry u_{is} in U, we introduce the Lagrange multiplier $\alpha_{is} \geq 0$, whereas for the (j, s)th entry v_{js} in V, we introduce the Lagrange multiplier $\beta_{js} \geq 0$. One can create a vector $(\overline{\alpha}, \overline{\beta})$ of dimensionality $(n + d) \cdot k$ by putting together all the Lagrangian parameters into a vector. Then, the Lagrangian relaxation is stated as follows:

$$L = \frac{1}{2}\|D - UV^T\|_F^2 + \frac{\lambda}{2}\|U\|_F^2 + \frac{\lambda}{2}\|V\|_F^2 - \sum_{i=1}^{n}\sum_{r=1}^{k} u_{ir}\alpha_{ir} - \sum_{j=1}^{d}\sum_{r=1}^{k} v_{jr}\beta_{jr} \qquad (8.4)$$

The minimax problem of Lagrangian optimization is stated as follows:

$$\text{Max}_{\overline{\alpha} \geq 0, \overline{\beta} \geq 0} \text{Min}_{U, V} \, L \qquad (8.5)$$

As discussed in Section 6.4, the first step is to compute the gradient of the Lagrangian relaxation with respect to the (minimization) optimization variables u_{is} and v_{js}. Therefore, we have:

$$\frac{\partial L}{\partial u_{is}} = -(DV)_{is} + (UV^TV)_{is} + \lambda u_{is} - \alpha_{is} \qquad \forall i \in \{1, \ldots, n\}, s \in \{1, \ldots, k\} \quad (8.6)$$

$$\frac{\partial L}{\partial v_{js}} = -(D^TU)_{js} + (VU^TU)_{js} + \lambda v_{js} - \beta_{js} \qquad \forall j \in \{1, \ldots, d\}, s \in \{1, \ldots, k\} \quad (8.7)$$

These partial derivatives are set to zero in order to obtain the following conditions:

$$-(DV)_{is} + (UV^TV)_{is} + \lambda u_{is} - \alpha_{is} = 0 \qquad \forall i \in \{1, \ldots, n\}, s \in \{1, \ldots, k\} \quad (8.8)$$

$$-(D^TU)_{js} + (VU^TU)_{js} + \lambda v_{js} - \beta_{js} = 0 \qquad \forall j \in \{1, \ldots, d\}, s \in \{1, \ldots, k\} \quad (8.9)$$

We would like to eliminate the Lagrangian parameters and set up the optimization conditions purely in terms of U and V. In this context, the complementary slackness components of the Kuhn-Tucker optimality conditions turn out to be very helpful. These conditions are $u_{is}\alpha_{is} = 0$ and $v_{js}\beta_{js} = 0$ over all parameters. By multiplying Equation 8.8 with u_{is} and multiplying Equation 8.9 with v_{js}, one obtains a condition purely in terms of the primal variables:

$$-(DV)_{is}u_{is} + (UV^TV)_{is}u_{is} + \lambda u_{is}^2 - \underbrace{\alpha_{is}u_{is}}_{0} = 0 \qquad \forall i \in \{1, \ldots, n\}, s \in \{1, \ldots, k\}$$

$$(8.10)$$

$$-(D^TU)_{js}v_{js} + (VU^TU)_{js}v_{js} + \lambda v_{js}^2 - \underbrace{\beta_{js}v_{js}}_{0} = 0 \qquad \forall j \in \{1, \ldots, d\}, s \in \{1, \ldots, k\}$$

$$(8.11)$$

One can rewrite these optimality conditions, so that a single parameter occurs on one side of the condition:

$$u_{is} = \frac{[(DV)_{is} - \lambda u_{is}]u_{is}}{(UV^TV)_{is}} \qquad \forall i \in \{1, \ldots, n\}, s \in \{1, \ldots, k\} \quad (8.12)$$

$$v_{js} = \frac{[(D^TU)_{js} - \lambda v_{js}]v_{js}}{(VU^TU)_{js}} \qquad \forall j \in \{1, \ldots, d\}, s \in \{1, \ldots, k\} \quad (8.13)$$

The aforementioned conditions can be used in order to perform iterative updates. A small value of ϵ is typically added to the denominator to avoid ill conditioning. Therefore, the iterative approach starts by initializing the parameters in U and V to nonnegative random values in $(0, 1)$ and then uses the following updates:

$$u_{is} \Leftarrow \frac{[(DV)_{is} - \lambda u_{is}]u_{is}}{(UV^TV)_{is} + \epsilon} \qquad \forall i \in \{1, \ldots, n\}, s \in \{1, \ldots, k\} \quad (8.14)$$

$$v_{js} \Leftarrow \frac{[(D^TU)_{js} - \lambda v_{js}]v_{js}}{(VU^TU)_{js} + \epsilon} \qquad \forall j \in \{1, \ldots, d\}, s \in \{1, \ldots, k\} \quad (8.15)$$

These iterations are then repeated to convergence. Improved initialization provides significant advantages [76].

As in all other forms of matrix factorization, it is possible to convert the factorization UV^T into the three-way factorization $Q\Sigma P^T$ by using the approach discussed in Section 7.2.7 of Chapter 7. For a nonnegative factorization, it makes sense to use L_1-normalization on each column of U and V, so that the columns of the resulting matrices Q and P each sum to 1. This type of normalization makes nonnegative factorization similar to a closely related factorization known as *Probabilistic Semantic Analysis (PLSA)*. The main difference between PLSA and nonnegative matrix factorization is that the former uses a maximum likelihood optimization function (or *I-divergence* objective) whereas nonnegative matrix factorization (typically) uses the Frobenius norm. Refer to Section 8.4.5.

8.4.3 Interpretability of Nonnegative Matrix Factorization

Nonnegative matrix factorization is a highly interpretable form of decomposition. In order to understand this point, we will consider the case in which the matrix D is an $n \times d$ document-term matrix; correspondingly, the corpus contains n documents and d words. The (i, j)th entry is the frequency of term j in document i. The $n \times k$ matrix U can be viewed as the factors of documents, whereas the $d \times k$ matrix V can be viewed as the factors of words. Each of the k factors can be viewed as a *topic* or a cluster of related documents. The rth columns U_r and V_r of each of U and V respectively contain document- and word-membership information about the rth topic (or cluster) in the data. The n entries in U_r correspond to the nonnegative components (coordinates) of the n documents along the rth topic. If a document strongly belongs to topic r, then it will have a very positive coordinate in U_r. Otherwise, its coordinate will be zero or mildly positive (representing noise). Similarly, the rth column V_r of V provides the frequent vocabulary of the rth cluster. Terms that are highly related to a particular topic will have large components in V_r. The k-dimensional representation of each document is provided by the corresponding row of U. This approach allows a document to belong to multiple clusters, because a given row in U might have multiple positive coordinates. For example, if a document discusses both science and history, it will have components along latent components with science-related and history-related vocabularies. This provides a more realistic "sum-of-parts" decomposition of the corpus along various topics, which is primarily enabled by the nonnegativity of U and V. In fact, one can create a decomposition of the document-term matrix into k different rank-1 document-term matrices corresponding to the k topics captured by the decomposition. Let us treat U_r as an $n \times 1$ matrix and V_r as a $d \times 1$ matrix. If the rth component is related to science, then $U_r V_r^T$ is an $n \times d$ document-term matrix containing the science-related portion of the original corpus. Then the decomposition of the document-term matrix is defined as the sum of the following components:

$$D \approx \sum_{r=1}^{k} U_r V_r^T \tag{8.16}$$

This decomposition represents matrix multiplication as a sum of outer-products; in this particular case, the nonnegativity of each additive component $U_r V_r^T$ makes it interpretable as a "document-term matrix."

8.4.4 Example of Nonnegative Matrix Factorization

In order to illustrate the semantic interpretability of nonnegative matrix factorization, let us revisit the same example used in Section 7.2.4.3 of Chapter 7, and create a decomposition in terms of nonnegative matrix factorization:

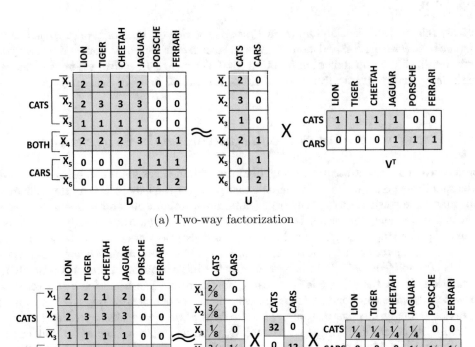

(a) Two-way factorization

(b) Three-way factorization by applying L_1-normalization to (a) above

Figure 8.3: The highly interpretable decomposition of nonnegative matrix factorization

$$
D = \begin{pmatrix}
 & \text{lion} & \text{tiger} & \text{cheetah} & \text{jaguar} & \text{porsche} & \text{ferrari} \\
\text{Document-1} & 2 & 2 & 1 & 2 & 0 & 0 \\
\text{Document-2} & 2 & 3 & 3 & 3 & 0 & 0 \\
\text{Document-3} & 1 & 1 & 1 & 1 & 0 & 0 \\
\text{Document-4} & 2 & 2 & 2 & 3 & 1 & 1 \\
\text{Document-5} & 0 & 0 & 0 & 1 & 1 & 1 \\
\text{Document-6} & 0 & 0 & 0 & 2 & 1 & 2
\end{pmatrix}
$$

This matrix represents topics related to both cars and cats. The first three documents are related to cats, the fourth is related to both, and the last two are related to cars. The *polysemous* word "jaguar" is present in documents of both topics.

A highly interpretable nonnegative factorization of rank-2 is shown in Figure 8.3(a). We have shown an approximate decomposition containing only integers for simplicity, although the optimal solution would (almost always) be dominated by floating point numbers in practice. It is clear that the first latent concept is related to cats and the second latent concept is related to cars. Furthermore, documents are represented by two non-negative coordinates indicating their affinity to the two topics. Correspondingly, the first three documents have strong positive coordinates for cats, the fourth has strong positive coordinates in both, and

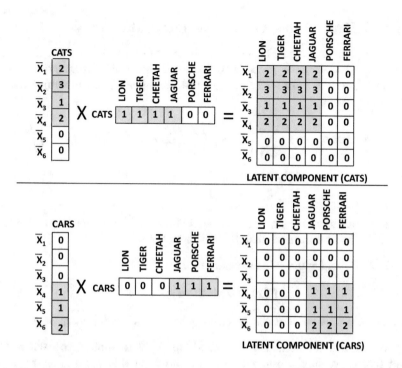

Figure 8.4: The highly interpretable "sum-of-parts" decomposition of the document-term matrix into rank-1 matrices representing different topics

the last two belong only to cars. The matrix V tells us that the vocabularies of the various topics are as follows:

Cats: lion, tiger, cheetah, jaguar
Cars: jaguar, porsche, ferrari

It is noteworthy that the polysemous word "jaguar" is included in the vocabulary of both topics, and its usage is automatically inferred from its context (i.e., other words in document) during the factorization process. This fact becomes especially evident when we decompose the original matrix into two rank-1 matrices according to Equation 8.16. This decomposition is shown in Figure 8.4 in which the rank-1 matrices for cats and cars are shown. It is particularly interesting that the occurrences of the polysemous word "jaguar" are nicely divided up into the two topics, which roughly correspond with their usage in these topics.

As discussed in Section 7.2.7 of Chapter 7, any two-way matrix factorization can be converted into a standardized three-way factorization. In the case of nonnegative matrix factorization, it is common to use L_1-normalization, rather than L_2-normalization (which is used in SVD). The three-way normalized representation is shown in Figure 8.3(b), and it tells us a little bit more about the relative frequencies of the two topics. Since the diagonal entry in Σ is 32 for cats in comparison with 12 for cars, it indicates that the topic of cats is more dominant than cars. This is consistent with the observation that more documents and terms in the collection are associated with cats as compared to cars.

8.4.5   The I-Divergence Objective Function

The previous section used the Frobenius norm in order to ensure that D is as close to UV^T as possible. However, it is possible to use a different objective function, referred to as the I-divergence function, in order to achieve the same goal [79]. This objective function is formulated as follows:

$$\text{Minimize } _{U,V} \sum_{i=1}^{n} \sum_{j=1}^{d} \left(D_{ij} \log \left\{ \frac{D_{ij}}{(UV^T)_{ij}} \right\} - D_{ij} + (UV^T)_{ij} \right)$$

$$\text{subject to:}$$
$$U \geq 0, \quad V \geq 0$$

This formulation takes on its minimum value $D = UV^T$. The reader is advised to solve the following problem to obtain more insight on this point:

Problem 8.4.1 *Consider the following function $F(x)$:*

$$F(x) = a \cdot log(a/x) - a + x$$

Here, a is a constant. Show that the function achieves its minimum value $x = a$.

In nonnegative matrix factorization, the function $F(x)$ is applied to each reconstructed entry x and (corresponding) observed entry a, and then this value is aggregated over all matrix entries. In the case of the Frobenius norm, the function $\|x - a\|^2$ is used instead of $F(x)$. In both cases, the objective function tries to make x as close to a as possible. The model requires the following iterative solution for $U = [u_{is}]$ and $V = [v_{js}]$:

$$u_{is} \Leftarrow u_{is} \frac{\sum_{j=1}^{d} [D_{ij} v_{js}/(UV^T)_{ij}]}{\sum_{j=1}^{d} v_{js}} \quad \forall i, s$$

$$v_{js} \Leftarrow v_{js} \frac{\sum_{i=1}^{n} [D_{ij} u_{is}/(UV^T)_{ij}]}{\sum_{i=1}^{n} u_{is}} \quad \forall j, s$$

The two-way factorization can be converted into a normalized three-way factorization using the approach discussed in Section 7.2.7 of Chapter 7. The three-way factorization can be interpreted from the perspective of a *probabilistic generative model*, which is *identical to probabilistic latent semantic analysis*.

8.5   Weighted Matrix Factorization

In weighted matrix factorization, weights are associated with the individual entries in the matrix, because errors on some entries are considered more important than others. The optimization model is similar to the unconstrained matrix factorization model discussed in Section 8.3. Furthermore, the application of matrix factorization to incomplete data (cf. Section 8.3.2) is a relatively trivial special case of weighted matrix factorization; in this case, the weights of observed entries are set to 1 and the weights of missing entries are set to 0. However, the most important (and somewhat nontrivial) use case, which will be discussed in this section, turns out to be an alternative to nonnegative matrix factorization for sparse matrices containing frequency counts. A surprising number of real-world applications, such as implicit feedback data and graph adjacency matrices, belong to this category.

In weighted matrix factorization, we have a weight w_{ij} associated with the (i,j)th entry of the $n \times d$ matrix $D = [x_{ij}]$ to be factorized. As in the case of unconstrained matrix factorization, we assume that the two factor matrices are the $n \times k$ matrix U and the $d \times k$ matrix V. Then, the objective function of weighted matrix factorization is as follows:

$$\text{Minimize } J = \frac{1}{2} \sum_{i=1}^{n} \sum_{j=1}^{d} w_{ij} \left(x_{ij} - \sum_{s=1}^{k} u_{is} \cdot v_{js} \right)^2 + \frac{\lambda}{2} \sum_{i=1}^{n} \sum_{s=1}^{k} u_{is}^2 + \frac{\lambda}{2} \sum_{j=1}^{d} \sum_{s=1}^{k} v_{js}^2$$

Note that this objective function is different from that of unconstrained matrix factorization only in terms of the weights w_{ij} of the entries. The partial derivative of the objective function with respect to the various parameters can be expressed in terms of the error $e_{ij} = x_{ij} - \hat{x}_{ij}$ of the factorization as follows:

$$\frac{\partial J}{\partial u_{iq}} = \sum_{i=1}^{n} \sum_{j=1}^{d} (w_{ij} e_{ij})(-v_{jq}) + \lambda u_{iq} \quad \forall i \in \{1 \ldots n\}, q \in \{1 \ldots k\}$$

$$\frac{\partial J}{\partial v_{jq}} = \sum_{i=1}^{n} \sum_{j=1}^{d} (w_{ij} e_{ij})(-u_{iq}) + \lambda v_{jq} \quad \forall j \in \{1 \ldots d\}, q \in \{1 \ldots k\}$$

The main difference from unconstrained matrix factorization is in terms of weighting the errors with w_{ij}. In order to express the aforementioned derivatives in matrix form, we define E as an $n \times d$ error matrix in which the (i,j)th entry is e_{ij}. Furthermore $W = [w_{ij}]$ is an $n \times d$ matrix containing the weights of various entries.

$$\frac{\partial J}{\partial U} = -(W \odot E)V + \lambda U$$

$$\frac{\partial J}{\partial V} = -(W \odot E)^T U + \lambda V$$

Here, the notation $\odot$ indicates elementwise multiplication between two matrices of exactly the same size. The weight matrix W essentially controls the importance of the errors of the individual entries in gradient descent. One can, therefore, express the gradient descent updates for the matrices U and V as follows:

$$U \Leftarrow U - \alpha \frac{\partial J}{\partial U} = U(1 - \alpha\lambda) + \alpha(W \odot E)V$$

$$V \Leftarrow V - \alpha \frac{\partial J}{\partial V} = V(1 - \alpha\lambda) + \alpha(W \odot E)^T U$$

Here, $\alpha > 0$ is the learning rate.

8.5.1 Practical Use Cases of Nonnegative and Sparse Matrices

Interestingly, weighted matrix factorization is often used in nonnegative and sparse matrices like implicit feedback data, graph adjacency matrices, and various text-centric matrices. In these cases, the weight matrix W and factorized matrix D are both defined as functions of the original raw *quantity* matrix Q. The matrix Q might correspond to the number of occurrences of customer buying behavior, link-link interactions, or document-word interactions. Although it is possible to use $D = Q$, the common approach is to modify the entries in Q in some way (such as setting them to 0-1 values, depending on whether they are non-zero or

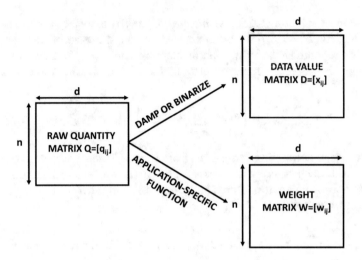

Figure 8.5: Deriving the data value and weight matrices from raw quantity matrix

not). Setting the entries to binary values makes sense in cases where the final prediction of the factorization is intended to be binary (e.g., recommend an item or link). In such a case, a new *binary* data matrix D is used in lieu of Q, where the (i, j)th entry x_{ij} of D is set to 1 when the corresponding entry of $Q = [q_{ij}]$ is non-zero. In other applications, the values of the *raw* data matrix Q are "damped" before factorization. In other words, each raw entry q_{ij} is replaced with the damped value $x_{ij} = f(q_{ij})$, where $f(\cdot)$ is a damping function like the square-root or logarithm. An example of such an approach is the GloVe embedding [101] for factorization of matrices derived from text (cf. Section 8.5.5). The weight matrix W is also derived as a function of the quantity matrix. This overall process is illustrated in Figure 8.5.

The choice of the weight matrix $W = [w_{ij}]$ is, however, more application-specific. In some cases, the weight matrix is set to the entries in D when they are non-zero. However, the zero entries also need to be set to specific weights. Typically, the weight of a zero entry is either set to a constant value or it is set to a column-specific value. Allowing non-zero weights on zero entries amounts to using *negative sampling* in the context of stochastic gradient descent. As we will see later, this type of negative sampling is important in most applications. While the weight matrix is technically dense because of the non-zero weights on zero entries, it can still be represented in compressed form. This is because all zero entries in a column have the same weight, and therefore one only needs to store the column-specific weight.

Why is this type of weighted matrix factorization more desirable than vanilla nonnegative matrix factorization? The reason is that the data matrix Q is sparse, and the vast majority of entries are 0s. In such cases, the fact that an entry is non-zero is more important than the specific magnitude of that value. Factorizing the values of the matrix Q might sometimes cause problems when the different entries of the matrix vary by orders of magnitude. This type of situation can occur in word matrices with large variations in word frequencies, or in graphs with power-law frequency distributions. If one simply performs value-based factorization, the preponderance of the (relatively unimportant) zero entries and the large magnitudes of a very small number of entries might play too large a role in the factorization. As a result, the modeling of most of the important entries in the matrix will be poor. As discussed in [65] in the case of ratings matrices, the general principle for treating raw numerical values is as follows:

> "The numerical value of explicit feedback [value in a dense matrix] indicates preference, whereas the numerical value of implicit feedback [value in a sparse matrix] indicates confidence."

Of course, a zero value in a sparse matrix does not necessarily indicate zero confidence, which is why one must resort to setting some non-zero weights on default values. In this section, we will provide several application-specific examples of scenarios in which sparse values should be treated as weights. Another useful property of weighted matrix factorization is that it allows a very efficient trick for parameter learning when most of the entries are zeros.

8.5.2 Stochastic Gradient Descent

The stochastic gradient descent procedure of Section 8.3.2.1 samples each entry of the matrix with equal probability in order to perform the updates. In weighted matrix factorization, the entries are sampled with probabilities that are proportional to their weights. In the (deterministic) gradient descent updates discussed earlier in this section, the weight of each entry is explicitly multiplied with the error of each entry in the update. Stochastic gradient descent replaces explicit weighting with weighted sampling, while keeping the form of the update unchanged from the unweighted case. Aside from this difference in how error entries are sampled, the algebraic form of the updates remains unchanged from the unweighted case:

$$\overline{u}_i \Leftarrow \overline{u}_i(1 - \alpha\lambda) + e_{ij}\overline{v}_j$$
$$\overline{v}_j \Leftarrow \overline{v}_j(1 - \alpha\lambda) + e_{ij}\overline{u}_i$$

Here, $\alpha > 0$ is the learning rate. Here, $\overline{u}_i$ represents the ith row of the $n \times k$ matrix U and $\overline{v}_j$ represents the jth row of the $d \times k$ matrix V. Note that exactly $2k$ entries in the matrices U and V are updated for each entry in the matrix.

This type of weighted matrix factorization is particularly efficient when the vast majority of the (raw) entries in the matrix are zeros. In many applications, the number of entries in the $n \times d$ matrix D might be very large, but the number of non-zero entries is several orders of magnitude lower. This is common in applications such as graphs in which a $10^6 \times 10^6$ adjacency matrix might have only 10 non-zero entries in each row. Therefore, the weight matrix is also sparse, and one only needs to keep track of positive sampling probabilities (i.e., non-zero entries in the original matrix D). All the zero weights are aggregated into a single negative sampling probability. The stochastic gradient descent procedure works as follows:

1. A coin is tossed with probability equal to the negative sampling rate. If the coin toss is a success, a random entry is treated as a negative entry, and an update is performed with the random entry (assuming that the observed value of the random entry is zero).

2. If the coin toss in the previous step is a failure, then a positive entry is sampled with probability proportional to its weight. Subsequently, stochastic gradient descent is performed with the positive entry.

This process of randomly sampling entries in proportion to their weights is iterated to convergence.

8.5.2.1   Why Negative Sampling Is Important

It is noteworthy that using negative sampling is particularly important, especially in cases where there is not much variation among the positive entries. For example, if one applied this procedure to a sparse binary matrix while setting the negative sampling rate to 0, one possible "optimal" solution is to obtain U and V as matrices in which each entry is $1/\sqrt{k}$. It is not difficult to see that UV^T will be a matrix of 1s, in which there are no errors on the positive entries and drastic errors on the negative entries. A negative sampling rate of 0 implicitly defines the objective function only over the positive entries, and therefore an optimal solution of zero error can be reached even when the quality of predictions is poor. One can view this situation as a type of overfitting. *In general, the weights w_{ij} of zero entries in sparse settings should always be set to non-zero values.*

8.5.3   Application: Recommendations with Implicit Feedback Data

Consider a case in which the original data set contains the amount that the user i bought for item j. Let $Q = [q_{ij}]$ be the original matrix of quantities bought by the user. In such a case, we create an $n \times d$ data matrix $D = [x_{ij}]$ as the binary indicator matrix of the quantities that were bought by the user. Therefore, the entries of the matrix D are defined as follows:

$$x_{ij} = \begin{cases} 1 & q_{ij} > 0 \\ 0 & q_{ij} = 0 \end{cases}$$

Therefore, the $n \times d$ matrix D is now a binary indicator matrix. Furthermore, the work in [65] suggests the following heuristic to select the weight w_{ij} of entry (i, j):

$$w_{ij} = 1 + \theta \cdot q_{ij} \tag{8.17}$$

The above weighting scheme ensures that *zero entries in the raw matrix Q have a non-zero weight*. These non-zero weights of zero entries define the negative sampling probability, when added over the various zero entries. It is suggested in [65] to use a value of $\theta = 40$. Then, the weighted matrix factorization of D is used to find the factor matrices U and V. The entries (i, j) with large values of $(UV^T)_{ij}$ are suggested recommendations of item j for user i.

8.5.4   Application: Link Prediction in Adjacency Matrices

Link prediction is a recommendation problem in every sense of the word. Instead of user-item matrices, we have the $n \times n$ node-node adjacency matrix $Q = [q_{ij}]$. The value q_{ij} of the entry (i, j) is the weight of edge (i, j). Such graphs are sparse, and therefore most values of q_{ij} are 0s. For example, in a bibliographic network, this weight might correspond to the number of publications between authors i and j. As in the case of the user-item recommendation application, the *binarized* version of the raw $n \times n$ node-node matrix is defined as follows:

$$x_{ij} = \begin{cases} 1 & q_{ij} > 0 \\ 0 & q_{ij} = 0 \end{cases}$$

The setting of the weight matrix requires additional thought, compared to the user-item recommendation application. One issue is that the weights of the edges in these graphs have a very high level of variation. Therefore, if care is not taken in selecting the weights, a few weights can dominate the factorization. This is obviously undesirable. Therefore, one possibility is to define w_{ij} with logarithmic damping:

$$w_{ij} = 1 + \theta \cdot \log(1 + q_{ij})$$

The value of θ can be tuned by testing its accuracy over a set of entries that are excluded from the sampling process in stochastic gradient descent.

8.5.5 Application: Word-Word Context Embedding with GloVe

The acronym GloVe stands for Global Vectors for Word Representation. The goal is to create multidimensional embeddings of words based on the other words in their context windows. Therefore, words that are similar in terms of the distribution of words in the window-based locality of these words will tend to have similar embeddings. The matrix Q in GloVe is a $d \times d$ word-context matrix. The (i, j)th entry in the matrix is the number of times that the word j occurs within a pre-defined distance δ of the word i in sentences of the document. The value of δ is typically a small quantity such as 4 or 5. The rows of the factor matrices U and V can be concatenated (or even added) to create the embeddings of the individual words. Note that there are exactly as many rows of these matrices as the number of words. These types of embeddings tend to have greater linguistic and semantic significance those obtained from methods like latent semantic analysis.

The matrix Q is then damped to create the data matrix $D = [x_{ij}]$, which contains the values to be factorized:

$$x_{ij} = \log(1 + q_{ij})$$

Note that a binary data matrix D is no longer used in this application. The weight w_{ij} is defined as follows:

$$w_{ij} = \begin{cases} \min\left\{1, \frac{c_{ij}}{M}\right\}^\alpha & q_{ij} > 0 \\ 0 & q_{ij} = 0 \end{cases}$$

The values of M and α are recommended to be 100 and 3/4, respectively, based on empirical considerations. It is possible to enhance this basic model in a number of ways, such as by the use of bias variables.

Note that GloVe sets the negative sampling probability to zero, and therefore depends almost entirely on the variation among the different *non-zero* values of x_{ij}. This is a unusual and controversial design choice, and it is very different from almost all other known techniques for weighted matrix factorization. It is significant that GloVe does not try to extract binary values of x_{ij} from the original quantity matrix $Q = q_{ij}$. Trying to use binary values of x_{ij} would be disastrous in the case of GloVe (cf. Exercise 9). A directly competing method, referred to as *word2vec*, plays great emphasis on negative sampling in order to achieve high-quality results [91, 92]. If the values of q_{ij} do not vary significantly in a given collection, it is possible for GloVe to provide overfitted results (cf. Section 8.5.2.1). This does not seem to be the case in practice, as GloVe seems to provide reasonably good results (based on independent evaluations by researchers and practitioners). This might possibly be a result of the fact that there is sufficient variation among the non-zero frequency counts (even after damping) in word-word context matrices. This property might not be the case in other domains, and therefore one should generally be cautious of factorizations that do not use some type of negative sampling.

8.6   Nonlinear Matrix Factorizations

All the models discussed so far in this chapter factorize a matrix as $D \approx UV^T$, which is linear in both U and V. Logistic and maximum margin matrix factorizations deviate from this linearity by using an elementwise function of UV^T to derive D. Although these methods are naturally designed for binary and sparse matrices, the general principle of nonlinearity can be extended to real-valued matrices with the appropriate choice of prediction function.

For the binary matrices in this section, the (i,j)th binary value is associated with a weight w_{ij}. Such factorizations are also designed for sparse matrices, where the values of the entries are treated as weights. These types of factorizations can be used for any of the applications that are used in conjunction with weighted matrix factorization. For example, all the applications for recommendations, link prediction, and text processing (discussed in Sections 8.5.3–8.5.5) can be directly supported by logistic and maximum-margin matrix factorization.

8.6.1   Logistic Matrix Factorization

Unlike weighted matrix factorization, logistic matrix factorization uses a nonlinear prediction function that can be interpreted as a probabilistic model. Let U and V be the $n \times k$ and $d \times k$ factor matrices for the original $n \times d$ matrix $Q = [q_{ij}]$ of sparse frequency counts. Logistic matrix factorization applies the logistic sigmoid function $F(x)$ to each entry of UV^T in order to generate a probability matrix P:

$$P = F(UV^T)$$

Here, the function $F(\cdot)$ is applied in an entry-wise fashion, and is defined as follows:

$$F(x) = \frac{1}{1 + \exp(-x)}$$

Each entry of the $n \times d$ matrix P is a probability value drawn from $(0, 1)$, and the goal is to maximize the log-likelihood of the observed data matrix based on these probabilities. As in the case of weighted matrix factorization, we create the binarized matrix $D = [x_{ij}]$ as follows:

$$x_{ij} = \begin{cases} 1 & q_{ij} > 0 \\ 0 & q_{ij} = 0 \end{cases}$$

We utilize a user-driven parameter m, which is the ratio of the *aggregate* weight of the negative entries to the aggregate weight of the positive entries. The weight w_{ij} of each entry of the matrix is defined as follows:

$$w_{ij} = \begin{cases} q_{ij} & q_{ij} > 0 \\ m(\sum_{s=1}^{d} q_{is})/d & q_{ij} = 0 \end{cases}$$

The value of m is set in a domain-specific way, and it is often a small integer such as 5. Under the assumption that the matrix is sparse, the sum of the negative entries in each row is roughly $m(\sum_{s=1}^{d} q_{is})$. Therefore, the weights of the negative entries are m times the weights of the positive entries, where m is a user-driven parameter. Implicitly, the negative entries are underweighted by this approach, because a sparse matrix will often contain negative entries that are hundreds of times the number of positive entries, whereas m is a small value such as 5.

A key point in logistic matrix factorization is what we would like the (learned) probability matrix $P = p_{ij}$ to have a large value of p_{ij} when x_{ij} is 1, and a small value of p_{ij} when x_{ij} is 0. This can be achieved with a log-likelihood objective function, which is defined as follows:

$$J = -\sum_{i=1}^{n}\sum_{j=1}^{d} w_{ij}\left[x_{ij}\log(p_{ij}) + (1 - x_{ij})\log(1 - p_{ij})\right]$$

Here, it is evident that this loss function is always nonnegative, and it takes[2] on its minimum value of 0 when $p_{ij} = x_{ij}$. Recall that each p_{ij} is the (i, j)th entry of $F(UV^T)$, and it is defined as follows:

$$p_{ij} = \frac{1}{1 + \exp(-\overline{u}_i \cdot \overline{v}_j)}$$

Here, $\overline{u}_i$ is the ith row of the $n \times k$ matrix U, and $\overline{v}_j$ is the jth row of the $d \times k$ matrix V. Therefore, one can substitute this value of p_{ij} in the objective function in order to obtain the following loss function for logistic matrix factorization:

$$J = -\sum_{i=1}^{n}\sum_{j=1}^{d} w_{ij}\left[x_{ij}\log\left(\frac{1}{1 + \exp(-\overline{u}_i \cdot \overline{v}_j)}\right) + (1 - x_{ij})\log\left(\frac{1}{1 + \exp(\overline{u}_i \cdot \overline{v}_j)}\right)\right]$$

Now that we have set up the objective function of logistic matrix factorization, it remains to derive the gradient descent steps.

8.6.1.1 Gradient Descent Steps for Logistic Matrix Factorization

In order to perform the gradient-descent updates, we need to compute the gradient of the objective function with respect to the k-dimensional vectors $\overline{u}_i$ and $\overline{v}_j$. This is best achieved using the chain rule in matrix calculus:

$$\frac{\partial J}{\partial \overline{u}_i} = \sum_{j=1}^{d} \frac{\partial J}{\partial (\overline{u}_i \cdot \overline{v}_j)}\frac{\partial (\overline{u}_i \cdot \overline{v}_j)}{\partial \overline{u}_i} = \sum_{j=1}^{d} \frac{\partial J}{\partial (\overline{u}_i \cdot \overline{v}_j)}\overline{v}_j$$

$$\frac{\partial J}{\partial \overline{v}_j} = \sum_{i=1}^{n} \frac{\partial J}{\partial (\overline{u}_i \cdot \overline{v}_j)}\frac{\partial (\overline{u}_i \cdot \overline{v}_j)}{\partial \overline{v}_j} = \sum_{i=1}^{n} \frac{\partial J}{\partial (\overline{u}_i \cdot \overline{v}_j)}\overline{u}_i$$

Note that the partial derivative of $\overline{u}_i \cdot \overline{v}_j$ with respect to either $\overline{u}_i$ or $\overline{v}_j$ is obtained using identity (v) of Table 4.2(a). It is relatively easy to compute the partial derivative of J with respect to $\overline{u}_i \cdot \overline{v}_j$ because the objective function is defined as a function of this quantity. By computing this derivative and substituting in the above equations, we obtain the following:

$$\frac{\partial J}{\partial \overline{u}_i} = -\sum_{j=1}^{d} \frac{w_{ij}x_{ij}\overline{v}_j}{1 + \exp(\overline{u}_i \cdot \overline{v}_j)} + \sum_{j=1}^{d} \frac{w_{ij}(1 - x_{ij})\overline{v}_j}{1 + \exp(-\overline{u}_i \cdot \overline{v}_j)}$$

$$\frac{\partial J}{\partial \overline{v}_j} = -\sum_{i=1}^{n} \frac{w_{ij}x_{ij}\overline{u}_i}{1 + \exp(\overline{u}_i \cdot \overline{v}_j)} + \sum_{i=1}^{n} \frac{w_{ij}(1 - x_{ij})\overline{u}_i}{1 + \exp(-\overline{u}_i \cdot \overline{v}_j)}$$

---

[2]Strictly speaking, the objective function is not defined when p_{ij} is 0 or 1. However, the loss is zero when $p_{ij} \to x_{ij}$ in the limit. A logistic function will never yield values of exactly 0 or 1 for p_{ij}.

With these derivatives, a straightforward gradient-descent procedure can be applied at learning rate $\alpha > 0$:

$$\overline{u}_i \Leftarrow \overline{u}_i - \alpha \frac{\partial J}{\partial \overline{u}_i} \quad \forall i$$

$$\overline{v}_j \Leftarrow \overline{v}_j - \alpha \frac{\partial J}{\partial \overline{v}_j} \quad \forall j$$

The gradient-descent approach computes the exact derivative of the objective function. What about stochastic gradient descent? It turns out that mini-batch stochastic gradient descent is particularly popular in the case of logistic matrix factorization. Since the total weight of negative entries is m times the number of positive entries, it is particularly common to sample one positive entry together with m negative entries for stochastic gradient descent. Interestingly, when this type of stochastic gradient descent is used for updates and it is applied to word-word context matrices with weight set to the frequencies of contexts, the resulting approach is *identical* to the backpropagation-based updates of the *word2vec* algorithm. The weights of the negative entries in the ith row is $K_i(\sum_{s=1}^{d} q_{is})^{\alpha}$, where $\alpha = 3/4$. The value of K_i is chosen to ensure that the weights of the negative entries in the ith row sum to roughly $m(\sum_{s=1}^{d} q_{is})$. Therefore, *word2vec* is simply an instantiation of logistic matrix factorization, and it provides an alternative to the GloVe algorithm discussed earlier in this chapter.

Problem 8.6.1 (Regularized Logistic Matrix Factorization) *Write an objective function for logistic matrix factorization that uses L_2-regularization. Derive the gradient-descent steps.*

8.6.2   Maximum Margin Matrix Factorization

Just as logistic regression is closely related to support vector machines (cf. Figure 4.9 of Chapter 4), logistic matrix factorization is closely related to maximum margin matrix factorization. The following exposition is roughly based on [120], although we simplify the algorithm in many respects while allowing the use of weights. As in the case of logistic matrix factorization, a *binary* $n \times d$ data matrix $D = [x_{ij}]$ and an $n \times d$ weight matrix $W = [w_{ij}]$ is derived from the raw quantity matrix $Q = [q_{ij}]$. The original work in [120] does not discuss the use of weights. In practice, methods like logistic matrix factorization and maximum-margin matrix factorization are most useful when weights are allowed. This is because many sparse matrices are associated with small counts on the non-zero entries (in real settings of implicit feedback). Furthermore, while the work in [120] provides a dual learning algorithm, we provide a much simpler primal algorithm.

Note that the 0-1 entry x_{ij} from the data matrix D can be converted to a value y_{ij} drawn from $\{-1, +1\}$ by using the following transformation:

$$y_{ij} = 2x_{ij} - 1$$

One can also introduce the matrix $Y = [y_{ij}]$, which is defined as $Y = 2D - \overline{1}_n \overline{1}_d^T$. Here, $\overline{1}_k$ is a column vector of k ones, and therefore $\overline{1}_n \overline{1}_d^T$ is an $n \times d$ matrix of 1s. We choose to use y_{ij} instead of x_{ij} in the objective function below because it easier to show similarity with the SVM objective function by doing so. Let U and V be the $n \times k$ and $d \times k$ factor matrices, respectively. The ith rows of U and V are denoted by $\overline{u}_i$ and $\overline{v}_i$ respectively. Then, the predicted value of the (i, j)th entry, denoted by $\hat{y}_{ij}$, is defined as follows:

$$\hat{y}_{ij} = \overline{u}_i \cdot \overline{v}_j$$

Unlike logistic matrix factorization, the predicted value $\hat{y}_{ij}$ is intended to match a quantity y_{ij} from $\{-1, +1\}$, rather than a value from $\{0, 1\}$. An important point here is that entries with large absolute values of $\hat{y}_{ij}$ are not penalized, as long as their sign is correct. This is because this factorization predicts the original entries by using the sign function on UV^T instead of using the absolute deviation from observed values:

$$Y \approx \text{sign}(UV^T)$$

This type of approach is exactly analogous to the prediction approach in an SVM. As in the case of SVMs, the hinge loss is used on the individual entries:

$$\text{Hinge}(i, j) = \max\{0, 1 - y_{ij}\hat{y}_{ij}\} = \max\{0, 1 - y_{ij}[\overline{u}_i \cdot \overline{v}_j]\}$$

This is a margin-based objective function, because an entry is not penalized only when its predicted value matches the sign of the original binary value with a sufficient margin of 1. Then, the overall objective function of maximum margin factorization (without regularization) can be expressed as follows:

$$J = \sum_{i=1}^{n} \sum_{j=1}^{d} w_{ij} \max\{0, 1 - y_{ij}[\overline{u}_i \cdot \overline{v}_j]\}$$

As in the case of logistic matrix factorization, one can use the chain rule to compute the derivative.

$$\frac{\partial J}{\partial \overline{u}_i} = \sum_{j=1}^{d} \frac{\partial J}{\partial(\overline{u}_i \cdot \overline{v}_j)} \overline{v}_j = - \sum_{j:y_{ij}(\overline{u}_i \cdot \overline{v}_j)<1} w_{ij} y_{ij} \overline{v}_j$$

$$\frac{\partial J}{\partial \overline{v}_j} = \sum_{i=1}^{n} \frac{\partial J}{\partial(\overline{u}_i \cdot \overline{v}_j)} \overline{u}_i = - \sum_{i:y_{ij}(\overline{u}_i \cdot \overline{v}_j)<1} w_{ij} y_{ij} \overline{u}_i$$

It is common to use L_2-regularization, in which case the gradients above are adjusted by $\lambda \overline{u}_i$ and $\lambda \overline{v}_j$, respectively. Here, $\lambda > 0$ is the regularization parameter. Therefore, at learning rate $\alpha > 0$, the gradient-descent updates of maximum-margin matrix factorization are as follows:

$$\overline{u}_i \Leftarrow \overline{u}_i(1 - \alpha\lambda) + \alpha \sum_{j:y_{ij}(\overline{u}_i \cdot \overline{v}_j)<1} w_{ij} y_{ij} \overline{v}_j \quad \forall i$$

$$\overline{v}_j \Leftarrow \overline{v}_j(1 - \alpha\lambda) + \alpha \sum_{i:y_{ij}(\overline{u}_i \cdot \overline{v}_j)<1} w_{ij} y_{ij} \overline{u}_i \quad \forall j$$

Just as SVMs and logistic regression provide very similar results for classification of binary labels, logistic matrix factorization and maximum margin matrix factorization provide similar results for factorization of binary matrices.

8.7 Generalized Low-Rank Models

The application of specialized forms of matrix factorization (like logistic and maximum margin factorization) for binary data leads to some interesting questions. What happens if the original data matrix contains entries of different types? In machine learning, it is common

Table 8.1: A demographic data set containing heterogeneous data types in different columns

Age (Numerical)	Gender (Binary)	Zip Code (Categorical)	Race (Categorical)	Education Level (Ordinal)
32	F	10598	Caucasian	Bachelors
41	M	10532	African American	Bachelors
36	M	10562	Filipino	High School
32	F	10532	Hispanic	Masters
29	F	10532	Native American	Doctorate

to encounter data matrices in which different features of the matrix might be numerical, binary, categorical, ordinal, and so on. Table 8.1 illustrates a table of demographic data containing *heterogeneous* data types in which the different columns correspond to different data types. A natural question that arises is how one might possibly create a factorization of a table with such bewilderingly different data types.

In this case, we have an $n \times d$ matrix $D = [x_{ij}]$ of heterogeneous data types and $W = [w_{ij}]$ is an $n \times d$ matrix of weights. An important distinction from the scenarios we have seen so far is that the data type of x_{ij} depends on the column index j. In order to perform the factorization, we use an $n \times k$ matrix U and an $r \times k$ matrix V. In most forms of factorization, the number of rows in V is equal to the number of columns d in D, whereas we have $r > d$ in this case. Why is $r > d$? The reason is that some data types like categorical data require multiple rows for a single column of D, which is not the case in any of the models we have seen so far. Therefore, the "reconstructed matrix" UV^T does not have the same size as the original matrix D; the reconstruction has the same number of rows as D, but it might have a much larger number of columns. Therefore, a one-to-one correspondence of columns between D and UV^T is no longer possible, and it is sensitive to the data type at hand. The jth column in D is associated with multiple columns in UV^T, and it is assumed that these columns are located consecutively in UV^T with column indices in the range $[l_j, h_j]$. When the column index j of D corresponds to a numeric, ordinal, or binary variable, we will have $l_j = h_j$ and therefore a single column of UV^T corresponds to a single column of D. However, for some data types like categorical data, we will have $h_j > l_j$. For each column j in D, we define a loss function that is specific to the column at hand, and it uses $h_j - l_j + 2$ arguments. The loss function $\mathcal{L}_j(\cdot)$ of the jth column of D is defined as follows:

1. The first argument of the loss function for any entry (i, j) of the jth column of D is the observed value of the entry x_{ij}.

2. The remaining $h_j - l_j + 1$ arguments of the loss function use the $r = h_j - l_j + 1$ values $z_{i,l_j} \ldots z_{i,l_j+r}$, where we have $z_{iq} = \overline{u}_i \cdot \overline{v}_q$ for each $q \in \{l_j \ldots l_j + r - 1\}$.

3. The loss value L_{ij} for the (i, j)th entry of D is defined using the loss function $\mathcal{L}_j$ specific to column j:

$$L_{ij} = \mathcal{L}_j(x_{ij}, z_{i,l_j} \ldots z_{i,l_j+r})$$

The nature of the loss function depends heavily on the data type at hand. We have already seen some examples of loss functions for binary and numerical variables. In the following, we will also introduce some loss functions for categorical and ordinal variables.

The overall objective function of the factorization can be expressed as a function of entry-specific weights and additional regularization:

$$\text{Minimize } J = \sum_{i=1}^{n} \sum_{j=1}^{d} w_{ij} L_{ij} + \frac{\lambda}{2} \left(\|U\|_F^2 + \|V\|_F^2 \right)$$

We have already seen how the loss functions for numerical and binary factorization were derived directly from their counterparts in linear regression and binary classification. Correspondingly, we can also derive the loss functions of categorical and ordinal values from their counterparts in multinomial logistic regression and ordinal regression.

8.7.1 Handling Categorical Entries

The key point is that the modeling of categorical entries requires exactly as many entries in UV^T as the number of distinct values of the categorical attribute. Consider the jth column of D, which can take on $s_j = h_j - l_j + 1$ possible values $a_1 \ldots a_{s_j}$. Then, the model of multinomial logistic regression computes the probability of the (i, j)th entry taking on the value a_r as follows:

$$P_{ij}(a_r) = \frac{\exp(z_{i,l_j+r-1})}{\sum_{s=1}^{r} \exp(z_{i,l_j+s-1})}$$

Then, the loss from this entry can be defined as follows:

$$L_{ij} = -\log \left[P_{ij}(x_{ij}) \right]$$

This is the straightforward log-likelihood model of multinomial logistic regression discussed in Section 4.9.2 of Chapter 4.

8.7.2 Handling Ordinal Entries

Ordinal entries are those in which there are fixed number of ordered values that need to be predicted. However, the distances among these different entries are unknown. For example, Table 8.1 contains four possible education levels that are ordered. However, it is not easily possible to know the distances among the different education levels. Let T_j be the number of possible distinct values for the jth column of D (which happens to be an ordinal variable). We define $m = T_j - 1$ different ordered thresholds denoted by $y_1 \ldots y_m$ (in increasing order). These thresholds, which are referred to as *intercepts*, also need to be learned in a data-driven manner, although they are not part of the matrices U and V. The matrix V contains a single row $\overline{v}_{o_j}$ with index o_j for this ordinal column. In traditional matrix factorization, the number of rows of V is equal to the number of columns in D because there is a one-to-one correspondence between the rows of V and columns of D. This is not the case for heterogeneous data tables containing categorical data types. Although ordinal data types require only one column in V, the value of o_j may be larger than j because some other types of data (like categorical data) in the same matrix require more than one row in V; this can cause a persistent mismatch in the indices of the original matrix D and latent vector matrix V. The prediction of the (i, j)th entry of D is obtained by computing $\overline{u}_i \cdot \overline{v}_{o_j}$ and then predicting it as follows:

$$\hat{x}_{ij} = \begin{cases} \text{1st ordinal value} & z_{i,o_j} \leq y_1 \\ q\text{th ordinal value} & q \in [2, m], y_{q-1} \leq z_{i,o_j} \leq y_q \\ (m+1)\text{th ordinal value} & z_{i,o_j} > y_m \end{cases}$$

In other words, we use the ordered thresholds $y_1 \ldots y_m$ to define $(m+1)$ buckets on the real line. The (i, j)th entry is mapped to an ordinal value depending on which bucket it falls in on the real line. In the following discussion, we also assume (for notational convenience) that $y_0 = -\infty$ and $y_{m+1} = +\infty$. Although these (trivial) end-point intercepts do not need to be learned, they help in reducing unnecessary case-wise analysis. For example, the prediction $\hat{x}_{ij}$ can now be collapsed into a single case as follows:

$$\hat{x}_{ij} = \begin{cases} q\text{th ordinal value} & q \in [1, m+1], y_{q-1} \leq z_{i,o_j} \leq y_q \end{cases}$$

There are many possible ways in which one can set up the loss function for ordinal entries. One possible way is to use the *proportional odds model* in which we view the ordinal prediction model as that of summing the losses of m different binary predictions for the (i, j)th entry– the qth prediction checks whether x_{ij} and z_{i,o_j} end up on the same side of y_q. Note that this is the same approach used in binary logistic matrix factorization, except that we have to learn multiple intercepts $y_1 \ldots y_m$ in this case. Then, we compute the probability that the (i, j)th entry lies on either side of y_b as follows:

$$P_{ij}(x_{ij} \leq y_b) = \frac{1}{1 + \exp(z_{i,o_j} - y_b)}$$

$$P_{ij}(x_{ij} > y_b) = \frac{1}{1 + \exp(-z_{i,o_j} + y_b)}$$

It is easy to verify that the sum of the above two probabilities is 1. Note that larger values of y_b will increase the probability $P_{ij}(x_{ij} \leq y_b)$, which makes sense in this case. At $b = 0$ and $b = m+1$, the values of y_b are fixed to $-\infty$ and $+\infty$, respectively. In such cases, it can be easily verified that each of the aforementioned probabilities is either a 0 or 1.

Suppose that the observed value of the ordinal variable x_{ij} lies between the current values of y_s and y_{s+1} for some $s \in \{0, \ldots, m\}$. Then, we would like $P(x_{ij} > y_b)$ to be as large as possible for $b \leq s$ and we would like $P(x_{ij} \leq y_b)$ to be as large as possible for $b > s$. This is achieved with the use of the following loss function:

$$L_{ij} = -\sum_{b=1}^{s} \log[P_{ij}(x_{ij} > y_b)] - \sum_{b=s+1}^{m} \log[P_{ij}(x_{ij} \leq y_b)]$$

Note that if s is 0, the first set of terms vanish. Similarly, if s is m, the second set of terms vanish. This loss function is very similar to binary logistic prediction; the main difference from binary logistic modeling is that we have m different binary predictions corresponding to each threshold y_s, and we want to reward predictions on the correct side of each of m thresholds. The loss function contains the sum of m different (negative) rewards. Here, it is important to note that each y_s is a variable. Therefore, the gradient-descent procedure not only has to update the factor matrices, but it also has to update the thresholds $y_1 \ldots y_m$.

In all problems we have seen so far, one can always substitute hinge loss wherever logistic loss is used. This is because of the similarity of these loss functions (cf. Figure 4.9 of Chapter 4). Suppose that the ordinal variable x_{ij} lies between the current values of y_s and y_{s+1} for some $s \in \{0, \ldots, m\}$. As in the case of logistic model, we can view the loss function as the sum of m different losses for the m different binary predictions (one for each non-trivial threshold y_b). The loss function penalizes cases in which z_{i,o_j} either lies on the wrong side of each y_b, or it lies on the correct side (but without sufficient margin). This is achieved by defining the loss function L_{ij} as follows:

$$L_{ij} = \sum_{b=1}^{s} \max(1 - z_{i,o_j} + y_b, 0) + \sum_{b=s+1}^{m} \max(1 + z_{i,o_j} - y_b, 0)$$

The hinge loss has the advantage of having a simpler derivative. This model is also available in the Julia package discussed in [128].

8.8 Shared Matrix Factorization

Shared matrix factorization is used in order to factorize multiple matrices simultaneously. Furthermore, the matrices have at least one modality in common. For example, if we have a graph of objects in which each node (object) of the graph also contains a document, then we have two matrices corresponding to the graph connectivity structure and the documents associated with all the nodes. Furthermore, since there is one-to-one correspondence between each node and a document, it is possible to create two matrices in which one of the dimensions is common between the two matrices. This sharing of a data modality is critical in being able to perform shared matrix factorization.

The basic idea in shared matrix factorization is to perform the factorization of two matrices with a *shared dimension* by using a *shared factor matrix*. Consider two matrices D and M that are of sizes $n \times d$ and $n \times m$, respectively. Because of the shared modality, the numbers of rows in the two matrices are the same; each row of D corresponds to a row in M. For example, D might correspond to a document-term matrix (which contains frequencies of words in documents) and M might correspond to a document-user matrix (which contains binary information on users placing a "like" rating on one or more documents). Therefore, the values are either 0 or 1 depending on whether or not a user has placed a like rating on an item. This is a classical example of implicit feedback data. An important property of the two matrices is that the ith row of D and the ith row of M correspond to the same object (which is a document in this case). Note that this one-to-one correspondence is essential for shared matrix factorization to work. Ideally, one would like to create a latent representation of documents based on *both* the content-based and the feedback data. This type of feature-engineered representation is helpful in a variety of tasks such as recommendations. For example, a similarity search using such an engineered representation will provide an output that takes both the topical and the interest behavior into account.

We introduce the shared $n \times k$ factor matrix U for documents, the $d \times k$ factor matrix V for words, and the $m \times k$ factor matrix W for users. Then, we would like to perform the following set of shared factorizations:

$$D \approx UV^T \quad \text{Document-Word Factorization}$$
$$M \approx UW^T \quad \text{Document-User Factorization}$$

One can then set up the objective function of the factorization as follows:

$$\text{Minimize } J = \frac{1}{2}\|D - UV^T\|_F^2 + \frac{\beta}{2}\|M - UW^T\|_F^2 + \underbrace{\frac{\lambda}{2}(\|U\|_F^2 + \|V\|_F^2 + \|W\|_F^2)}_{\text{Regularization}}$$

Here, $\beta > 0$ provides the relative weights of the two factorizations, and $\lambda > 0$ is the regularization parameter. The value of β is often chosen based on application-specific considerations.

8.8.1   Gradient Descent Steps for Shared Factorization

In the following, we will derive the gradient-descent steps for the matrix factorization model discussed in the previous section. We compute the gradient of J with respect to the entries in U, V, and W. For any current values of U, V, and W, let e_{ij}^D represent the (i,j)th entry of the error matrix $(D - UV^T)$, and e_{ij}^M represent the (i,j)th entry of the error matrix $(M - UW^T)$. The partial derivatives of J are as follows:

$$\frac{\partial J}{\partial u_{iq}} = -\sum_{j=1}^{d} e_{ij}^D v_{jq} - \beta \sum_{p=1}^{m} e_{ip}^M w_{pq} + \lambda u_{iq} \quad \forall i \in \{1\ldots n\}, \ \forall q \in \{1\ldots k\}$$

$$\frac{\partial J}{\partial v_{jq}} = -\sum_{i=1}^{n} e_{ij}^D u_{iq} + \lambda v_{jq} \quad \forall j \in \{1\ldots d\}, \ \forall q \in \{1\ldots k\}$$

$$\frac{\partial J}{\partial w_{pq}} = -\beta \sum_{i=1}^{n} e_{ip}^M u_{iq} + \lambda w_{pq} \quad \forall p \in \{1\ldots m\}, \ \forall q \in \{1\ldots k\}$$

These gradients can be used to update the entire set of $(n + m + d)k$ parameters with a step-size of α. This approach corresponds to vanilla gradient descent. It is also possible to use *stochastic* gradient descent, which effectively computes the gradients with respect to residual errors in randomly sampled *entries* of the matrices. One can sample any entry in *either* the document-term matrix or the adjacency matrix, and then perform the gradient-descent step with respect to the error in this single entry:

    Randomly sample any entry from either D or M;
    Perform a gradient-descent step with respect to entry-specific loss;

The probability of sampling each entry is fixed irrespective of which matrix it is drawn from. Consider a case in which the (i,j)th entry in the document-term matrix is sampled with error e_{ij}^D. Then, the following updates are executed for each $q \in \{1\ldots k\}$ and step-size α:

$$u_{iq} \Leftarrow u_{iq}(1 - \alpha \cdot \lambda/2) + \alpha e_{ij}^D v_{jq} \quad \forall q \in \{1\ldots k\}$$

$$v_{jq} \Leftarrow v_{jq}(1 - \alpha \cdot \lambda) + \alpha e_{ij}^D u_{iq} \quad \forall q \in \{1\ldots k\}$$

On the other hand, if the (i,p)th entry in the adjacency matrix is sampled, then the following updates are performed for each $q \in \{1\ldots k\}$ and step-size α:

$$u_{iq} \Leftarrow u_{iq}(1 - \alpha \cdot \lambda/2) + \alpha \beta e_{ip}^M w_{pq} \quad \forall q \in \{1\ldots k\}$$

$$w_{pq} \Leftarrow w_{pq}(1 - \alpha \cdot \lambda) + \alpha \beta e_{ip}^M u_{iq} \quad \forall q \in \{1\ldots k\}$$

These steps are repeated to convergence.

8.8.2   How to Set Up Shared Models in Arbitrary Scenarios

Shared matrix factorization is useful in any scenario where the matrices are drawn from heterogeneous domains, but they share at least some connections with one another. The connections can themselves be expressed as matrices in some cases. For example, one might have completely independent matrices for feature representations of documents and images, and a third matrix might contain hyperlinks from documents to images. In such a case, one would have an $n_1 \times d_1$ matrix D_1 for document-term representation, an $n_2 \times d_2$ matrix D_2 for image-(visual word) representation, and an $n_1 \times n_2$ matrix A containing connections between documents and visual words. Corresponding, we need two factor matrices U_1 and U_2 for documents and images, and two factor matrices V_1 and V_2 for text-words and visual-words. Therefore, one would like to perform the factorization as follows:

$$D_1 \approx U_1 V_1^T \quad \text{Document-Word Factorization}$$
$$D_2 \approx U_2 V_2^T \quad \text{Image-(Visual Word) Factorization}$$
$$A \approx U_1 U_2^T \quad \text{Connection-Matrix Factorization}$$

One can set up an objective function that minimizes the sum of the squares of the errors over all three matrices. It is even possible to weight the different types of errors differently, depending on the application at hand. We leave the derivation of the gradient descent steps as an exercise for the reader.

Problem 8.8.1 *Write down the sum-of-squared objective function for the factorization of the matrices D_1, D_2, and A, as discussed above. Derive the gradient descent steps for the entries in these matrices. You may introduce any notation as needed for this problem.*

All the settings used for shared matrix factorization are very similar; *we have a set of matrices in which some of the modalities are shared*, and we wish to extract latent representations of the shared relationships implicit in these matrices. The key in this entire process is to use *shared latent factors* between different modalities so that they are able to incorporate the impact of these relationships in an indirect (i.e., latent) way within the extracted embedding. A single set of factors is introduced for each shared modality, and each matrix is factorized. The sum-of-squared objective function is used to determine the gradient-descent updates.

8.9 Factorization Machines

Factorization machines are closely related to shared matrix factorization methods, and are particularly suitable when each data instance contains features from multiple domains. For example, consider an item that is tagged with particular keywords by a user and also rated by that user. In such a case, the feature set corresponds to all the item identifiers, all the possible keywords, and the user identifiers. The feature values of the user identifier, item identifier, and the relevant keywords are set to 1, whereas all other feature values are set to 0. The dependent variable is equal to the value of the rating.

Factorization machines are polynomial regression techniques, in which strong regularization conditions are imposed on the regression coefficients in order to handle the challenges of sparsity. Sparsity is common in short-text domains, such as the social content on bulletin boards, social network datasets, and chat messengers. It is also common in recommender systems.

An example of a data set drawn from the recommendation domain is illustrated in Figure 8.6. It is evident that there are three types of attributes corresponding to user attributes, item attributes, and tagging keywords. Furthermore, the rating corresponds to the dependent variable, which is also the regressand. At first sight, this data set seems to be no different from a traditional multidimensional data set to which one might apply least-squares regression in order to model the rating as a linear function of the regressors.

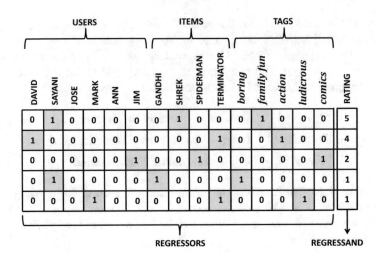

Figure 8.6: An example of a sparse regression modeling problem with heterogeneous attributes

Unfortunately, the sparsity of the data in Figure 8.6 ensures that a least-squares regression method does rather poorly. For example, each row might contain only three or four non-zero entries. In such cases, linear regression may not be able to model the dependent variable very well, because the presence of a small number of non-zero entries provides little information. Therefore, a second possibility is to use higher-order interactions between the attributes in which we use the simultaneous presence of multiple entries for modeling. As a practical matter, one typically chooses to use second-order interactions between attributes, which corresponds to second-order polynomial regression. However, as we will discuss below, an attempt to do so leads to overfitting, which is exacerbated by the sparse data representation.

Let $d_1 \ldots d_r$ be the number of attributes in each of the r data modalities such as text, images, network data and so on. Therefore, the total number of attributes is given by $p = \sum_{k=1}^{r} d_k$. We represent the variables of the row by $x_1 \ldots x_p$, most of which are 0s, and a few might be non-zero. In many natural applications in the recommendation domain, the values of x_i might be binary. Furthermore, it is assumed that a target variable is available for each row. In the example of Figure 8.6, the target variable is the rating associated with each row, although it could be any type of dependent variable in principle.

Consider the use of a regression methodology in this setting. For example, the simplest possible prediction would be use linear regression with the variables $x_1 \ldots x_p$.

$$\hat{y}(\overline{x}) = b + \sum_{i=1}^{p} w_i x_i \qquad (8.18)$$

Here, b is the bias variable and w_i is the regression coefficient of the ith attribute. This is in an almost identical form to the linear regression discussed in Chapter 4, except that we have explicitly used a global bias variable b. Although this form can provide reasonable results in some cases, it is often not sufficient for sparse data in which a lot of information is captured by the correlations between various attributes. For example, in a recommender system, the co-occurrence of a user-item pair is far more informative than the separate coefficients of users and items. Therefore, the key is to use a *second-order* regression coefficient s_{ij}, which captures the coefficient of the interaction between the ith and jth attribute.

$$\hat{y}(\overline{x}) = b + \sum_{i=1}^{p} w_i x_i + \sum_{i=1}^{p} \sum_{j=i+1}^{p} s_{ij} x_i x_j \qquad (8.19)$$

Note that one could also include the second-order term $\sum_{i=1}^{p} s_{ii} x_i^2$, although x_i is often drawn from sparse domains with little variation in non-zero values of x_i, and the addition of such a term is not always helpful. For example, if the value of x_i is binary (as is common), the coefficient of x_i^2 would be redundant with respect to that of x_i.

One observation is that the above model is very similar to what one would obtain with the use of kernel regression with a second-order polynomial kernel. In sparse domains like text, such kernels often overfit the data, especially when the dimensionality is large and the data is sparse. Even for an application in a single domain (e.g., short-text tweets), the value of d is greater than 10^5, and therefore the number of second-order coefficients is more than 10^{10}. With any training data set containing less than 10^{10} points, one would perform quite poorly. This problem is exacerbated by sparsity, in which pairs of attributes co-occur rarely in the training data, and may not generalize to the test data. For example, in a recommender application, a particular user-item pair may occur only once in the entire training data, and it will not occur in the test data if it occurs in the training data. In fact, all the user-item pairs that occur in the test data will not have occurred in the training data. How, then, does one learn the interaction coefficients s_{ij} for such user-item pairs? Similarly, in a short-text mining application, the words "movie" and "film" may occur together, and the words "comedy" and "film" may also occur together, but the words "comedy" and "movie" might never have occurred together in the training data. What does one do, if the last pair occurs in the test data?

A key observation is that one can use the learned values of s_{ij} for the other two pairs (i.e., "comedy"/"film" and "movie"/"film") in order to make some inferences about the interaction coefficient for the pair "comedy" and "movie." How does one achieve this goal? The key idea is to assume that the $d \times d$ matrix $S = [s_{ij}]$ of second-order coefficients has a *low-rank* structure for some $d \times k$ matrix $V = [v_{is}]$:

$$S = VV^T \qquad (8.20)$$

Here, k is the rank of the factorization. Intuitively, one can view Equation 8.20 as a kind of regularization constraint on the (massive number of) second-order coefficients in order to prevent overfitting. Therefore, if $\overline{v_i} = [v_{i1} \ldots v_{ik}]$ is the k-dimensional row vector representing the ith row of V, we have:

$$s_{ij} = \overline{v_i} \cdot \overline{v_j} \qquad (8.21)$$

By substituting Equation 8.21 in the prediction function of Equation 8.19, one obtains the following:

$$\hat{y}(\overline{x}) = b + \sum_{i=1}^{p} w_i x_i + \sum_{i=1}^{p} \sum_{j=i+1}^{p} (\overline{v_i} \cdot \overline{v_j}) x_i x_j \qquad (8.22)$$

The variables to be learned are b, the different values of w_i, and each of the vectors $\overline{v_i}$. Although the number of interaction terms might seem large, most of them will evaluate to zero in sparse settings in Equation 8.22. This is one of the reasons that factorization machines are designed to be used only in sparse settings where most of the terms of Equation 8.22 evaluate to 0. A crucial point is that we only need to learn the $O(d \cdot k)$ parameters represented by $\overline{v_1} \ldots \overline{v_k}$ in lieu of the $O(d^2)$ parameters in $[s_{ij}]_{d \times d}$.

A natural approach to solve this problem is to use the stochastic gradient-descent method, in which one cycles through the observed values of the dependent variable to compute the gradients with respect to the error in the observed entry. The update step with respect to any particular model parameter $\theta \in \{b, w_i, v_{is}\}$ depends on the error $e(\overline{x}) = y(\overline{x}) - \hat{y}(\overline{x})$ between the predicted and observed values:

$$\theta \Leftarrow \theta(1 - \alpha \cdot \lambda) + \alpha \cdot e(\overline{x}) \frac{\partial \hat{y}(\overline{x})}{\partial \theta} \tag{8.23}$$

Here, $\alpha > 0$ is the learning rate, and $\lambda > 0$ is the regularization parameter. The partial derivative in the update equation is defined as follows:

$$\frac{\partial \hat{y}(\overline{x})}{\partial \theta} = \begin{cases} 1 & \text{if } \theta \text{ is } b \\ x_i & \text{if } \theta \text{ is } w_i \\ x_i \sum_{j=1}^{p} v_{js} \cdot x_j - v_{is} \cdot x_i^2 & \text{if } \theta \text{ is } v_{is} \end{cases} \tag{8.24}$$

The term $L_s = \sum_{j=1}^{p} v_{js} \cdot x_j$ in the third case is noteworthy. To avoid redundant effort, this term can be pre-stored while evaluating $\hat{y}(\overline{x})$ for computation of the error term $e(\overline{x}) = y(\overline{x}) - \hat{y}(\overline{x})$. This is because Equation 8.22 can be algebraically rearranged as follows:

$$\hat{y}(\overline{x}) = b + \sum_{i=1}^{p} w_i x_i + \frac{1}{2} \sum_{s=1}^{k} \left([\sum_{j=1}^{p} v_{js} \cdot x_j]^2 - \sum_{j=1}^{p} v_{js}^2 \cdot x_j^2 \right)$$

$$= b + \sum_{i=1}^{p} w_i x_i + \frac{1}{2} \sum_{s=1}^{k} \left(L_s^2 - \sum_{j=1}^{p} v_{js}^2 \cdot x_j^2 \right)$$

Furthermore, the parameters $\overline{v_i}$ and w_i do not need to be updated when $x_i = 0$. This allows for an efficient update process in sparse settings, which is linear in both the number of non-zero entries and the value of k.

Factorization machines can be used for any (massively sparse) classification or regression task; ratings prediction in recommender systems is only one example of a natural application. Although the model is inherently designed for regression, binary classification can be handled by applying the logistic function on the numerical predictions to derive the probability whether $\hat{y}(\overline{x})$ is +1 or −1. The prediction function of Equation 8.22 is modified to a form used in logistic regression:

$$P[y(\overline{x}) = 1] = \frac{1}{1 + \exp(-[b + \sum_{i=1}^{p} w_i x_i + \sum_{i=1}^{p} \sum_{j=i+1}^{p} (\overline{v_i} \cdot \overline{v_j}) x_i x_j])} \tag{8.25}$$

This form is the same as the logistic regression approach discussed in Chapter 4. The main difference is that we are also using second-order interactions within the prediction function. A log-likelihood criterion can be optimized to learn the underlying model parameters with a gradient-descent approach [47, 107, 108].

The description in this section is based on second-order factorization machines that are popularly used in practice. In third-order polynomial regression, we would have $O(p^3)$ additional regression coefficients of the form w_{ijk}, which correspond to interaction terms of the form $x_i x_j x_k$. These coefficients would define a massive third-order *tensor*, which can be compressed with tensor factorization. Although higher-order factorization machines have also been developed, they are often impractical because of greater computational complexity and overfitting. A software library, referred to as *libFM* [108], provides an excellent set of factorization machine implementations. The main task in using *libFM* is an initial feature engineering effort, and the effectiveness of the model mainly depends on the skill of the analyst in extracting the correct set of features. Other useful libraries include *fastFM* [11] and[3] *libMF* [144], which have some fast learning methods for factorization machines.

8.10 Summary

Matrix factorization is one of the most fundamental tools in machine learning, which is exploited both for the useful linear algebra and also for the compression properties of the underlying factors. One of the most fundamental forms of factorization is singular value decomposition in which the columns of the different factor matrices are mutually orthogonal. More general forms of the matrix factorization modify the optimization model to allow different types of objective functions, constraints, and data types. Certain types of constraints like nonnegativity have a regularization effect, and they help in creating more interpretable matrix factorizations. Methods like logistic matrix factorization, maximum margin factorization, and generalized low-rank models are designed to deal with different data types. Shared matrix factorization and factorization machines are designed to factorize multiple matrices. In general, the broader theme of matrix factorization provides a very wide variety of tools that can be harnessed for various machine learning scenarios.

8.11 Further Reading

Discussions of SVD and unconstrained matrix factorization may be found in many books on linear algebra [77, 122, 123, 130]. The use of unconstrained matrix factorization for recommender systems is discussed in detail in [3, 75]. The use of coordinate descent for matrix factorization in recommender systems is discussed in [137]. Alternating least-squares methods are discussed in [69, 141].

Nonnegative matrix factorization is introduced in [79], and its probabilistic counterpart, PLSA, is discussed in [63]. The relationship between the I-divergence objective function and PLSA is discussed in [35, 50]. The importance of deriving separate weight and value matrices for factorization from implicit feedback data is discussed in [65]. The application of various types of sparse factorization models to text and graph feature engineering is discussed in [2, 55, 91, 92, 101, 103]. The use of logistic matrix factorization for implicit feedback data is discussed in [70], and that of maximum margin matrix factorization is discussed in [120]. However, the presentation of maximum margin matrix factorization is much more simplified in this chapter. Generalized low-rank models are introduced in [128]. Regression models for ordinal data are introduced in [90]. An overview of different types of shared matrix factorization models is provided in [2, 3, 117]. Factorization machines are discussed in [107, 108].

8.12 Exercises

1. **Biased matrix factorization:** Consider the factorization of an incomplete $n \times d$ matrix D into an $n \times k$ matrix U and a $d \times k$ matrix V:

$$D \approx UV^T$$

---

[3]The libraries *libFM* and *libMF* are different.

Suppose you add the constraint that all entries of the penultimate column of U and the final column of V are fixed to 1. Discuss the similarity of this model to that of the addition of bias to classification models. How is gradient descent modified?

2. In the scenario of Exercise 1, will the Frobenius norm on observed ratings be better optimized with or without constraints on the final columns of U and V? Why might it be desirable to add such a constraint during the estimation of missing entries?

3. Suppose that you have a symmetric $n \times n$ matrix D of similarities, which has missing entries. You decide to recover the missing entries by using the *symmetric factorization* $D \approx UU^T$. Here, U is an $n \times k$ matrix, and k is the rank of the factorization.

   (a) Write the objective function for the optimization model using the Frobenius norm and L_2-regularization.

   (b) Derive the gradient-descent steps in terms of matrix-centric updates.

   (c) Discuss the conditions under which an exact factorization will not exist, irrespective of how large a value of k is used for the factorization.

4. Derive the gradient-descent updates for L_1-loss matrix factorization in which the objective function is $J = \|D - UV^T\|_1$.

5. Derive the gradient-descent updates for L_2-loss matrix factorization in which L_1-regularization is used on the factors.

6. In SVD, it is easy to compute the representation of out-of-sample matrices because of the orthonormality of the basis $d \times k$ matrix V. If the SVD factorization of the $n \times d$ matrix D is $D \approx UV^T$, then one can compute the representation of an out-of-sample $m \times d$ matrix D_o as $D_o V$. Show how you can efficiently compute a similar out-of-sample representation of D_o, when you are given a non-orthonormal factorization $D = UV^T$. Assume that m and k are much smaller than n and d.

7. Show that the k-means formulation in Section 4.10.3 of Chapter 4 is identical to the formulation of Section 8.2.1. [Hint: Propose a one-to-one mapping of optimization variables in the two problems. Show that the constraints and the objective functions are equivalent in the two cases.]

8. **Orthogonal Nonnegative Matrix Factorization:** Consider a nonnegative $n \times d$ data matrix D in which we try to approximately factorize D as UV^T with the Frobenius norm as the objective function. Suppose you add nonnegativity constraints on U and V along with the constraint $U^TU = I$. How many entries in each row of U will be non-zero? Discuss how you can extract a clustering from this factorization. Show that this approach is closely related to the k-means optimization formulation.

9. Suppose that you use GloVe on a quantity matrix $Q = [q_{ij}]$ in which each count q_{ij} is either 0 or 10000. A sizeable number of counts are 0s. Show that GloVe can discover a trivial factorization with zero error in which each word has the same embedded representation.

10. Derive the gradient update equations for using factorization machines in binary classification with logistic loss and hinge loss.

11. Suppose you want to perform the rank-k factorization $D \approx UV^T$ of the $n \times d$ matrix D using gradient descent. Propose an initialization method for U and V using QR decomposition of k randomly chosen columns of D.

12. Suppose that you have a sparse non-negative matrix D of size $n \times d$. What can you say about the dot product of any pair of columns as a consequence of sparsity? Use this fact along with the intuition derived from the previous exercise to initialize U using k randomly sampled columns of D for *non-negative matrix factorization*. In this case, the initialized matrices U and V need to be non-negative.

13. **Nonlinear matrix factorization of positive matrices:** Consider a nonlinear model for matrix factorization of positive matrices $D = [x_{ij}]$, where $D = F(UV^T)$, and $F(x) = x^2$ is applied in element-wise fashion. The vectors $\overline{u}_i$ and $\overline{v}_j$ represent the ith and jth rows of U and V, respectively. The loss function is $\|D - F(UV^T)\|_F^2$. Show that the gradient descent steps are as follows:

$$\overline{u}_i \Leftarrow \overline{u}_i + \alpha \sum_j (\overline{u}_i \cdot \overline{v}_j)(x_{ij} - F(\overline{u}_i \cdot \overline{v}_j))\overline{v}_j$$

$$\overline{v}_j \Leftarrow \overline{v}_j + \alpha \sum_i (\overline{u}_i \cdot \overline{v}_j)(x_{ij} - F(\overline{u}_i \cdot \overline{v}_j))\overline{u}_i$$

14. **Out-of-sample factor learning:** Suppose that you learn the optimal matrix factorization $D \approx UV^T$ of $n \times d$ matrix D, where U, V are $n \times k$ and $d \times k$ matrices, respectively. Now you are given a new out-of-sample $t \times d$ data matrix D_o with rows collected using the same methodology as the rows of D (and with the same d attributes). You are asked to quickly factorize this out-of-sample data matrix into $D_o \approx U_o V^T$ with the objective of minimizing $\|D_o - U_o V^T\|_F^2$, where V is *fixed* to the matrix learned from the earlier in-sample factorization. Show that the problem can be decomposed into t linear regression problems, and the optimal solution U_o is given by:

$$U_o^T = V^+ D_o^T$$

Here, V^+ is the pseudoinverse of V. Show that the rank-k approximation of $D_o \approx U_o V^T$ is given by $D_o P_v$, where $P_v = V(V^TV)^{-1}V^T$ is the $d \times d$ projection matrix induced by V. Propose a fast solution approach using QR decomposition of V and back-substitution with a triangular equation system. How does this problem relate to the alternating minimization approach?

15. **Out-of-sample factor learning:** Consider the same scenario as Exercise 14, where you are trying to learn the out-of-sample factor matrix U_o for in-sample data matrix $D \approx UV^T$ and out-of-sample data matrix D_o. The factor matrix V is fixed from in-sample learning. Closed-form solutions, such as the one in Exercise 14, are rare in most matrix factorization settings. Discuss how the gradient-descent updates discussed in this chapter can be modified so that U_o can be learned directly. Specifically discuss the case of (i) unconstrained matrix factorization, (ii) nonnegative matrix factorization, and (iii) logistic matrix factorization.

16. Suppose that you have a user-item ratings matrix with numerical/missing values. Furthermore, users have rated each other's trustworthiness with binary/missing values.

    (a) Show how you can use shared matrix factorization for estimating the rating of a user on an item that they have not already rated.

    (b) Show how you can use factorization machines to achieve similar goals as (a).

17. Propose an algorithm for finding outlier entries in a matrix with the use of matrix factorization.

18. Suppose that you are given the linkage of a large Website with n pages, in which each page contains a bag of words drawn from a lexicon of size d. Furthermore, you are given information on how m users have rated each page on a scale of 1 to 5. The ratings data is incomplete. Propose a model to create an embedding for each Webpage by combining all three pieces of information. [Hint: This is a shared matrix factorization problem.]

19. **True or false:** A zero error non-negative matrix factorization (NMF) UV^T of an $n \times d$ non-negative matrix D always exists, where U is an $n \times k$ matrix and V is a $d \times k$ matrix, as long as k is chosen large enough. At what value of k can you get an exact NMF of the following matrix?

$$D = \begin{bmatrix} 1 & 1 \\ 1 & 0 \end{bmatrix}$$

20. **True or false:** Suppose you have the exact non-negative factorization (NMF) UV^T of a matrix D, so that each column of V is constrained to sum to 1. Subject to this normalization rule, the NMF of D is unique.

21. Discuss why the following algorithm will work in computing the matrix factorization $D_{n \times d} \approx UV^T$ after initializing $U_{n \times k}$ and $V_{d \times k}$ randomly:

   repeat; $U \Leftarrow DV^+$; $V \Leftarrow D^T U^+$; **until convergence;**

22. Derive the gradient-descent updates of unconstrained matrix factorization with L_1-regularization. You may assume that the regularization parameter is $\lambda > 0$.

23. **Alternating nonnegative least-squares:** Propose an algorithm for nonnegative matrix factorization using the alternating least-squares method [Hint: See nonnegative regression in Chapter 6.]

24. **Bounded matrix factorization:** In bounded matrix factorization, the entries of U and V in the factorization $D \approx UV^T$ are bounded above and below by specific values. Propose a computational algorithm for bounded matrix factorization using (i) gradient descent, and (ii) alternating least-squares.

25. Suppose that you have a very large and dense matrix D of low rank that you cannot hold in memory, and you want to factorize it as $D \approx UV^T$. Propose a method for factorization that uses only sparse matrix multiplication. [Hint: Read the section on recommender systems.]

26. **Temporal matrix factorization:** Consider a sequence of $n \times d$ matrices $D_1 \ldots D_t$ that are slowly evolving over t time stamps. Show how one can create an optimization model to infer a single $n \times k$ *static* factor matrix that does not change over time, and multiple $d \times k$ *dynamic* factor matrices, each of which is time-specific. Derive the gradient descent steps to find the factor matrices.

Chapter 9

The Linear Algebra of Similarity

"The worst form of inequality is to try to make unequal things equal." – Aristotle

9.1 Introduction

A dot-product similarity matrix is an alternative way to represent a multidimensional data set. In other words, one can convert an $n \times d$ data matrix D into an $n \times n$ similarity matrix $S = DD^T$ (which contains n^2 pairwise dot products between points). One can use S instead of D for machine learning algorithms. The reason is that the similarity matrix contains almost the same information about the data as the original matrix. This equivalence is the genesis of a large class of methods in machine learning, referred to as *kernel methods*. This chapter builds the linear algebra framework required for understanding this important class of methods in machine learning. The real utility of such methods arises when the similarity matrix is chosen differently from the use of dot products (and the data matrix is sometimes not even available).

This chapter is organized as follows. The next section discusses how similarity matrices are alternative representations of data matrices. The efficient recovery of data matrices from similarity matrices is discussed in Section 9.3. The different types of linear algebra operations on similarity matrices are discussed in Section 9.4. The implementation of machine learning algorithms with similarity matrices is discussed in Section 9.5. The representer theorem is discussed in Section 9.6. The choice of similarity matrices that promote linear separation is discussed in Section 9.7. A summary is given in Section 9.8.

9.2 Equivalence of Data and Similarity Matrices

This section will establish the rough equivalence between data matrices and similarity matrices. Therefore, the next subsection will concretely show how one can convert a data matrix into a similarity matrix and vice versa.

© Springer Nature Switzerland AG 2020
C. C. Aggarwal, *Linear Algebra and Optimization for Machine Learning*,
https://doi.org/10.1007/978-3-030-40344-7_9

9.2.1   From Data Matrix to Similarity Matrix and Back

Consider an $n \times d$ data matrix D, in which the ith row is denoted by $\overline{X}_i = [x_{i1}, x_{i2}, \ldots, x_{id}]$, and it corresponds to the ith object in the data set. Then, a *symmetric* $n \times n$ similarity matrix $S = [s_{ij}]$ can be defined among the n objects as follows:

$$s_{ij} = \overline{X}_i \cdot \overline{X}_j = \sum_{k=1}^{d} x_{ik} x_{jk}$$

One can write the above similarity relationship in matrix form as well:

$$S = DD^T$$

How does one recover the original data set D from the similarity matrix? First, note that the recovery can never be unique. This is because dot products are invariant to rotation and reflection. Therefore, rotating a data set about the origin or reflecting it along any axis will result in the same similarity matrix. For example, consider a $d \times d$ matrix P with orthonormal columns, which is essentially a rotation/reflection matrix. Then, the rotated/reflected version of D is as follows:

$$D' = DP$$

Then, the similarity matrix S' using D' can be shown to be equal to S as follows:

$$S' = D'D'^T = (DP)(DP)^T = D \underbrace{(PP^T)}_{I} D^T = S$$

In other words, the similarity matrices using D and D' are the same.

It is noteworthy that both $(DP)(DP)^T$ and DD^T represent *symmetric factorizations* of the similarity matrix S. A symmetric factorization of an $n \times n$ matrix is a factorization of S into two $n \times k$ matrices of the form $S = UU^T$. For exact factorization, the value of k will be equal to the rank of the similarity matrix S. The ith row of U in any symmetric factorization UU^T of S yields a valid set of features of the ith data point.

The simplest way to perform symmetric factorization of a similarity matrix is with the use of *eigendecomposition*. First, note that if S was indeed created using dot products on the data matrix D, it is of the form DD^T and is therefore positive semidefinite (cf. Lemma 3.3.14 of Chapter 3). Therefore, one can diagonalize it with nonnegative eigenvalues of which at most $\min\{n, d\}$ are non-zero. To emphasize the nonnegativity of the eigenvalues, we will represent the diagonal matrix as Σ^2:

$$S = Q\Sigma^2 Q^T = \underbrace{(Q\Sigma)}_{U}(Q\Sigma)^T$$

Therefore, $Q\Sigma$ is the extracted representation from the similarity matrix S, and it will contain at most $\min\{n, d\}$ non-zero columns. Specifically, the ith row of $Q\Sigma$ contains the embedded representation of the ith data point (based on the ordering of rows/columns of the similarity matrix S). Note that the eigenvectors and eigenvalues of $DD^T = Q\Sigma^2 Q^T$ are (respectively) the left singular vectors and squared singular values of D.

The eigendecomposition of the similarity matrix provides one of an infinite number of possible embeddings obtained from factorization of the similarity matrix, and it is one of the most compact ones in terms of the number of non-zero columns. The compactness can be improved further by dropping smaller eigenvectors. As another example, one can extract a symmetric Cholesky factorization $S = LL^T$, and use the rows of matrix L as the engineered representations of the points (see Section 3.3.9), although a small positive value might be needed to be added to each diagonal entry of S to make it positive definite. Another example is the symmetric *square-root* matrix, which can also be extracted from the eigendecomposition as $S = Q\Sigma^2 Q^T = (Q\Sigma Q^T)(Q\Sigma Q^T)^T = (\sqrt{S})^2$. Choosing any particular embedding among these will not affect the predictions of any machine learning algorithm that relies on dot products (or Euclidean distances), because they remain the same whether we use eigendecomposition, Cholesky factorization, or the square-root matrix.

Problem 9.2.1 (Alternative Embeddings Are Orthogonally Related) *Show that if the rank-k similarity matrix S of size $n \times n$ can be expressed as either $U_1 U_1^T$ or $U_2 U_2^T$ with $n \times k$ matrices, then (i) a full SVD of each of U_1 and U_2 can be constructed so that the* **left** *singular vectors and the singular values are the same in the two cases, and (ii) an orthogonal matrix P_{12} can be found so that $U_2 = U_1 P_{12}$.*

9.2.2   When Is Data Recovery from a Similarity Matrix Useful?

The above discussion simply creates a dot-product similarity matrix from a data matrix and then recovers a rotated/reflected version of the data set from it. At face value, this does not seem like a useful exercise. However, the real usefulness of this type of data recovery approach arises when the similarity matrix is constructed from the data using a method *different* from dot products. In fact, the original data set might not even be a multidimensional data type (and dot products are not possible). Rather, it might represent a set of structural data objects, such as small graph objects (e.g., chemical compounds), time-series, or discrete sequences. The similarity matrix S might have been created using a domain-specific similarity function on these objects. In such cases, the matrix $Q\Sigma$ of the scaled eigenvectors of similarity matrix $S = Q\Sigma^2 Q^T$ contains *engineered features* of the objects in its rows. *In fact, the most common approach for feature engineering of arbitrary object types is the extraction of eigenvectors from the similarity matrix of the objects.*

Why is it useful to create multidimensional embeddings? One reason is that it is hard to apply machine learning algorithms on many data types such as discrete sequences or graph-centric chemical compounds. However, extracting a multidimensional embedding of each object opens the door to the use of many machine learning algorithms like SVMs or logistic regression that work with multidimensional data.

At most d eigenvectors of the $n \times n$ similarity matrix are non-zero, when the similarity matrix was truthfully created using dot products on a multidimensional data of dimensionality d. However, similarity matrices are rarely created using dot products on multidimensional data (and there is no practical use of doing so). For arbitrary similarity matrices created using domain-specific similarity functions on different types of objects, it is possible for all n eigenvalues of the $n \times n$ similarity matrix to be non-zero. One can interpret this result from the point of view that any n points in an embedding (and the origin) lie on an n-dimensional plane passing through the origin, although they might define an even lower dimensional plane if they are linearly dependent. The rows of matrix U in the factorization $S = UU^T$ contain the coordinates with respect to an n-dimensional orthogonal basis of this plane. When the $n \times n$ similarity matrix is extracted from the use of dot products on a multidimensional data set of dimensionality $d \ll n$, the linear dependence among the n embedded vectors ensures that the dimensionality of the hyperplane defined by these n vectors is no larger than d. However, this is not the case if one uses similarity functions other than the dot product on the original data set, which results in feature engineering from the data set.

9.2.3   What Types of Similarity Matrices Are "Valid"?

Similarity matrices are alternative representations of data sets (ignoring rotation and reflection). Even when the original data is not multidimensional (e.g., graph objects), one can simply assume that the similarities represent dot products between (fictitious) multidimensional objects. However, this assumption of the existence of a fictitious embedding needs to meet an important mathematical test of validity. As a small example, diagonal entries of similarity matrices are *non-negative* squared norms of embedded objects. Therefore, a similarity matrix with a negative diagonal entry could not have possibly been created using dot products even on a fictitious embedding.

A multidimensional embedding can be extracted from a similarity matrix S, if and only if it can be expressed in the form UU^T. Any matrix S expressible in this form must be positive semidefinite (cf. Lemma 3.3.14 of Chapter 3). *In other words, a similarity matrix needs to be positive semidefinite in order for a valid embedding to exist.*

Unfortunately, if an $n \times n$ similarity matrix S is extracted using a domain-specific similarity function, there is no guarantee that it will be positive semidefinite. What can be done in such cases? It turns out that it is always possible to repair any similarity matrix (without significantly changing the interpretation of the similarities) so that a valid embedding exists. The idea is to add $\delta > 0$ to each diagonal entry of the similarity matrix S, where δ is the magnitude of the most negative eigenvalue of the matrix S. In such a case, it can be shown that the resulting matrix $S' = S + \delta I$ is positive semidefinite.

$$S' = S + \delta I = Q\Delta Q^T + \delta I = Q\underbrace{(\Delta + \delta I)}_{\geq 0} Q^T$$

In this case, the embedding can be extracted as $Q\sqrt{\Delta + \delta I}$. The modification of the similarity matrix is often not a significant one from the perspective of application-centric interpretability. By doing so, we are only translating the (less important) self-similarity values to make them sufficiently large, while keeping the pairwise similarity values unchanged. Intuitively, the self-similarities among points are always larger than those across different points (on the average), when working with dot products.

Problem 9.2.2 *Let $\overline{X}$ and $\overline{Y}$ be two d-dimensional points. Show that the average of the two dot-product self-similarities $\overline{X} \cdot \overline{X}$ and $\overline{Y} \cdot \overline{Y}$ is at least as large as the pairwise dot-product similarity $\overline{X} \cdot \overline{Y}$.*

Stated differently, the above problem implies that if we have a 2×2 symmetric similarity matrix in which the sum of diagonal entries is smaller than the sum of off-diagonal entries, the matrix would not be positive semidefinite.

Problem 9.2.3 *Show that the sum of the entries in a similarity matrix S can be expressed as $\overline{y}^T S \overline{y}$ for appropriately chosen column vector $\overline{y}$. What can you infer about the sign of the sum of the values in a similarity matrix that is positive semidefinite?*

Problem 9.2.4 *Let $\overline{y}$ be an n-dimensional column vector. Show that the expression $\overline{y}^T S \overline{y}$ represents the squared norm of some vector in the multidimensional space containing the embedding induced by the $n \times n$ similarity matrix S.*

It is noteworthy that even though increasing the diagonal entries of S by δ affects the embedding matrix $Q\sqrt{\Delta + \delta I}$, it does not affect the *normalized* embedding matrix Q. In fact, several forms of feature engineering (such as *spectral embeddings*) work with normalized embeddings. Although the importance of positive semidefiniteness is often emphasized in the machine learning literature on kernel methods, the reality is that this requirement is a lot less important (from a practical perspective) than appears at first glance – the similarity matrix can always be repaired by increasing the self-similarity entries along the diagonal (which are semantically less significant anyway).

9.2.4 Symmetric Matrix Factorization as an Optimization Model

The extraction of embeddings from similarity matrices is a special case of symmetric matrix factorization. However, it is not necessary to perform exact factorization. When one is looking to generate a k-dimensional embedding for $k \ll n$, the $n \times k$ embedding U defines a representation in which $\|S - UU^T\|_F^2$ is minimized. Therefore, one can pose the problem of finding a k-dimensional embedding as an unconstrained matrix factorization problem for an $n \times k$ matrix of variables U:

$$\text{Minimize}_U \; J = \frac{1}{2}\|S - UU^T\|_F^2$$

The top-k (scaled) eigenvectors $Q_k \Sigma_k$ of this similarity matrix $S = Q_k \Sigma_k^2 Q_k^T$ represent one of the solutions to this optimization problem. A particular property of this solution is that the columns of $U = Q_k \Sigma_k$ are mutually orthogonal. However, other alternative solutions to the optimization problem are possible. To understand this point, we recommend the reader to work out the following exercise:

Problem 9.2.5 *It is known that $U = Q_k \Sigma_k$ is one of the optimal solutions to the optimization problem presented above. Here, Q_k is an $n \times k$ matrix containing the top-k eigenvectors of S in its columns, and Σ_k is a diagonal matrix containing the square-root of the corresponding eigenvalues in its diagonal entries. Show that any solution of the form $U' = Q_k \Sigma_k R_k$ is also a solution to this optimization problem. Here, R_k is any $k \times k$ orthogonal matrix. Discuss why the columns of U' need not necessarily be mutually orthogonal.*

A hint for solving the last part of the above problem is to compute the expression for $U'^T U'$ and show that it will typically not be diagonal other than in some very special cases. As discussed in Section 9.3, it is possible to discover these alternative embeddings efficiently by using methods such as column sampling and stochastic gradient descent.

9.2.5 Kernel Methods: The Machine Learning Terminology

The linear algebra operations on similarity matrices are commonly used in kernel methods for machine learning. Therefore, we connect these linear algebra concepts with the terminology used in machine learning. The multidimensional representation extracted from a similarity matrix (using eigendecomposition) is referred to as a *data-specific Mercer kernel map*. More generally, these representations are referred to as *kernel feature spaces* or simply *feature spaces*.

Definition 9.2.1 (Kernel Feature Spaces) *The multidimensional data space obtained by diagonalizing the positive semidefinite similarity matrix between objects is referred to as a data-specific kernel feature space, or simply a data-specific feature space.*

We would like to emphasize that the definition of a kernel feature space here is a "data-specific" version, where a similarity matrix of finite size is already given, which contains *samples* of similarity values. The specification of a similarity matrix of finite size ensures that the feature space is bounded above by the size of the similarity matrix. However, it is also possible to (implicitly) specify a similarity matrix of infinite size, by defining similarities between each pair of objects from an infinite domain *as a function in closed form*. For example, consider a situation where one wants to engineer new features $\Phi(\overline{x})$ from multidimensional vectors $\overline{x} \in \mathcal{R}^d$. In such a case, one can define a similarity function between multidimensional objects $\overline{x}$ and $\overline{y}$ (different from the dot product) by the following simplified *Gaussian kernel* with unit variance:

$$K(\overline{x}, \overline{y}) = \Phi(\overline{x}) \cdot \Phi(\overline{y}) = \exp(-\|\overline{x} - \overline{y}\|^2/2)$$

By providing a closed-form expression, one has effectively defined a similarity matrix between all pairs of objects $\overline{x}$ and $\overline{y}$ in the infinite set $\mathcal{R}^d$. Since the dimensionality of the eigenvectors increases with similarity matrix size, it is possible[1] for the resulting eigenvectors to also be infinite dimensional. Such spaces of infinite-dimensional vectors are natural generalizations of finite-dimensional Euclidean spaces, and are referred to as *Hilbert spaces*. However, even in these abstract cases of infinite-dimensional representations, it is possible to represent a specific data set of finite size containing n objects in n-dimensional space— the key point is that an n-dimensional projection of this infinite-dimensional space always exists that contains all n objects (and the origin). After all, any set of n vectors defines an (at most) n-dimensional *subspace*. The eigendecomposition of the *sample matrix* of size $n \times n$ discovers precisely this *subspace*, which is the *data-specific* feature space. For most machine learning problems, only the data-specific feature space is needed. We emphasize this point below:

> For a finite similarity matrix of size $n \times n$, one can always extract an (at most) n-dimensional engineered representation using eigendecomposition of the similarity matrix. This is true even when the dimensionality of the true feature space induced by a (closed-form) similarity function over an infinite domain of points is much larger.

The key point is that as long as we do not need to know the representations of points outside our finite data set of n points, one can restrict the dimensionality of representation to an n-dimensional subspace (and much lower in many cases). In this chapter, whenever we refer to kernel feature space, we refer to the data-specific feature space whose dimensionality is bounded above by the number of points. The eigendecomposition of the similarity matrix to extract features is also referred to as *kernel SVD*.

Definition 9.2.2 (Kernel SVD) *The embedding $Q\Sigma$ extracted by the eigendecomposition $S = Q\Sigma^2 Q^T$ of an $n \times n$ positive semidefinite similarity matrix S is referred to as* kernel SVD. *The $n \times n$ matrix $Q\Sigma$ contains the embedding of the ith data point in its ith row for each $i \in \{1 \ldots n\}$. When S already contains dot products between points from $\mathcal{R}^d$, the approach specializes to standard SVD.*

All kernel methods in machine learning *implicitly* transform the data using kernel SVD via a method referred to as the "kernel trick." However, we will revisit some traditional applications of kernels like SVMs and show how to implement them using *explicit* eigendecomposition of the similarity matrix. Although this approach is unusual, it is instructive and has some advantages over the alternative that avoids this eigendecomposition.

---

[1]For some closed-form functions like the dot product, only d components of the eigenvectors will be non-zero, whereas for others like the Gaussian kernel, the entire set of infinite components will be needed.

9.3 Efficient Data Recovery from Similarity Matrices

The most basic approach for embedding extraction is the materialization of an $n \times n$ similarity matrix S from the n objects by computing all pairwise similarities, and then extracting large eigenvectors of S. However, this can be difficult to achieve in practice. Imagine a situation in which one has a billion data objects (e.g., chemical compounds), and therefore $n = 10^9$. By modern standards, a data set containing a billion objects is not considered extraordinarily large. In such a case, the number of entries in the similarity matrix is 10^{18}, which can be difficult to even materialize explicitly. In many cases, a closed-form function might be available to compute the similarities between each pair of objects, although one does not want to be forced to compute this function 10^{18} times. For example, if 1 cycle of a computer were required to compute a single pairwise similarity, then a 10 GHz computer will require 10^8 seconds to compute all similarities (which is more than three years). The space required for explicitly storing the similarity matrix is of order of 10^6 TB. The vanilla method for embedding extraction can, therefore, be impractical.

It is possible to extract embeddings approximately by materializing only a subset of the entries in the similarity matrix $S = [s_{ij}]$. The key point is that there are large correlations between the entries in S, as a result of which the similarity matrix has a lot of built-in redundancy. For example, if both s_{ij} and s_{ik} are very large, then it is often the case that s_{jk} is very large as well. In mathematical terms, this observation also amounts to the fact that the matrix S often has much lower rank than its physical dimensionality n. With such redundancies, it is often possible to extract a lower-dimensional embedding of S approximately with sampling methods — this is a form of compression. We will discuss two solutions for achieving this goal. The first solution is a row-wise sampling approach that modifies the eigendecomposition technique already discussed. The second solution is a stochastic gradient-descent method that builds on the matrix factorization models of Chapter 8.

9.3.1 Nyström Sampling

The Nyström approach speeds up the embedding process by subsampling a subset of data objects and constructing the similarity matrix only on this (small) subset. The prototype embedding model is created using only this subset, and then it is generalized to out-of-sample points using some tricks from linear algebra [133].

The first step is to sample a set of p objects, and an in-sample similarity matrix S_{in} of size $p \times p$ is constructed in which the (i, j)th entry is the similarity between the ith and jth in-sample objects. Similarly, an $n \times p$ similarity matrix S_a is constructed in which the (i, j)th entry is the similarity between the ith object with the jth *in-sample* object. Note that the matrix S_{in} is contained within S_a, since each row of S_{in} is also a row of S_a. Then, the following pair of steps is used to first generate the embeddings of the in-sample points and then generalize the in-sample embeddings to all points (including out-of-sample points):

- **(In-sample embedding):** Diagonalize $S_{in} = Q\Sigma^2 Q^T$. If there are fewer than p non-zero eigenvectors, then extract all $k < p$ non-zero eigenvectors in the $n \times k$ matrix Q_k and $k \times k$ diagonal matrix Σ_k. This step requires $O(p^2 \cdot k)$ time and $O(p^2)$ space. Since p is typically a small constant of the order of a few thousand, this step is extremely fast and space-efficient irrespective of the base number of objects.

- **(Universal embedding):** Let U_k denote the unknown $n \times k$ matrix containing the k-dimensional representation of the all n points in its rows. Although we already know the embeddings of the in-sample points, we will use the properties of the similarity matrix in transformed space to derive all rows in a uniform way. The dot products of the n points in U_k and in-sample points in $Q_k \Sigma_k$ can be computed as the matrix product of U_k and $(Q_k \Sigma_k)^T$. This set of $n \times p$ dot products is contained in the matrix S_a, because it is assumed that S_a contains the dot products of embedded representations of all points and in-sample points. Therefore, we have the following:

$$S_a \approx \underbrace{U_k (Q_k \Sigma_k)^T}_{\text{Dot Products}} \tag{9.1}$$

The approximation is caused by the fact that the embedding of *all* points might require n-dimensional space, whereas we have restricted ourselves to at most p dimensions defined by the in-sample points. By postmultiplying each side with $Q_k \Sigma_k^{-1}$ and using $Q_k^T Q_k = I_k$, we obtain the following:

$$U_k \approx S_a Q_k \Sigma_k^{-1} \tag{9.2}$$

Therefore, we have an embedding of all n points in k-dimensional space. This step requires a simple matrix multiplication in time $O(n \cdot p \cdot k)$, which is linear in the number of objects in the data set.

It is noteworthy that the p in-sample rows in U_k are the same as the p rows in $Q_k \Sigma_k$.

It is interesting to note that we are able to represent n points in at most p dimensions, whereas the data-specific feature space for the full data might have dimensionality as large as n. What we have effectively done is to use the fact that a hyperplane defined by p points (and the origin) in feature space is an at most p-dimensional projection of the n-dimensional data-specific feature space, and it can be represented in at most $k \leq p$ coordinates. Therefore, we first find the exact k-dimensional representation of a subset of p points (where $k \leq p$); then, we project the remaining $(n - p)$ points from n-dimensional feature space to the k-dimensional subspace in which the p points lie. Therefore, the remaining $(n - p)$ points lose some accuracy of representation, which is expected in a sampling method. In fact, it is even possible to drop some of the smaller non-zero eigenvectors from the in-sample embedding for better efficiency.

9.3.2   Matrix Factorization with Stochastic Gradient Descent

A second approach is to use stochastic gradient descent in conjunction with the optimization model introduced in Section 9.2.4. Stochastic gradient descent performs the updates by sampling entries from the similarity matrix. Since the entries in the similarity matrix are highly correlated, it means that the matrix is often of approximately low rank. In such cases, is possible to learn the top-k components of the embedding, by minimizing the squared sum of the residual (noise) entries. This is a similar approach to matrix factorization in recommender systems (cf. Section 8.3.2 of Chapter 8), where a small subset of matrix entries is sufficient to learn the factor matrices.

For ease in discussion, we will assume that a a tiny subset of entries of the similarity matrix is materialized up front in the same way as a tiny subset of entries is available in a ratings matrix in recommender systems (cf. Section 8.3.2 of Chapter 8). In practice, one can always compute the similarity values on-the-fly for stochastic gradient descent, although fixing the "observed" entries up front also allows the use of vanilla gradient descent with sparse matrix multiplications. Let $S = [s_{ij}]$ be an $n \times n$ similarity matrix, in which only a subset O of entries are observed:

$$O = \{(i,j) : s_{ij} \text{ is "observed" }\} \quad (9.3)$$

One can assume that the matrix S is symmetric, and therefore the observed set of similarities O can be grouped into symmetric pairs of entries satisfying $s_{ij} = s_{ji}$. It is desired to learn an $n \times k$ embedding U for user-specified rank k, so that for any observed entry (i,j) the dot product of the ith row of U and the jth row of U is as close as possible to the (i,j)th entry, s_{ij}, of S. In other words, the value of $\|S - UU^T\|_F^2$ should be as small as possible for the observed entries in S. This problem can be formulated only over the "observed" entries in O as follows:

$$\text{Minimize } J = \frac{1}{2} \sum_{(i,j) \in O} \left(s_{ij} - \sum_{p=1}^{k} u_{ip} u_{jp} \right)^2 + \frac{\lambda}{2} \sum_{i=1}^{n} \sum_{p=1}^{k} u_{ip}^2$$

Therefore, we have changed the optimization model of Section 9.2.4, so that it is formulated only over a tiny subset of entries in S. Furthermore, regularization becomes particularly important in these cases as the subset of entries to be used is small. This problem is similar to the determination of factors in recommendation problems, and is a natural candidate for gradient-descent methods. The main difference is that the factorization is symmetric.

Let $e_{ij} = s_{ij} - \sum_{p=1}^{k} u_{ip} u_{jp}$ be the error of any entry (i,j) from set O at a particular value of the parameter matrix U. On computing the partial derivative of J with respect to u_{im}, one obtains the following:

$$\frac{\partial J}{\partial u_{im}} = \sum_{j:(i,j) \in O} \left(s_{ij} + s_{ji} - 2 \cdot \sum_{p=1}^{k} u_{ip} u_{jp} \right)(-u_{jm}) + \lambda u_{im} \quad \forall i \in \{1 \ldots n\}, m \in \{1 \ldots k\}$$

$$= \sum_{j:(i,j) \in O} (e_{ij} + e_{ji})(-u_{jm}) + \lambda u_{im} \quad \forall i \in \{1 \ldots n\}, m \in \{1 \ldots k\}$$

$$= -2 \sum_{j:(i,j) \in O} e_{ij} u_{jm} + \lambda u_{im} \quad \forall i \in \{1 \ldots n\}, m \in \{1 \ldots k\}$$

Note that s_{ij} and s_{ji} are either both present or both absent from the observed entries because of the symmetric assumption. It is possible to express these partial derivatives in matrix form. Let $E = [e_{ij}]$ be an error matrix, in which (i,j)th entry is set to the error for any observed entry (i,j) in O, and 0, otherwise. When a small number of entries are observed, this matrix is a sparse matrix. It is not difficult to see that the entire $n \times k$ matrix of partial derivatives $\left[\frac{\partial J}{\partial u_{im}} \right]_{n \times k}$ is given by $-2EU$. This suggests that one should randomly initialize the matrix U of parameters, and use the following gradient-descent steps:

$$U \Leftarrow U(1 - \alpha\lambda) + 2\alpha EU \quad (9.4)$$

Here, $\alpha > 0$ is the step size, which one can follow through to convergence. Note that the error matrix E is sparse, and therefore it makes sense to compute only those entries that are present in O before converting to a sparse data structure.

To determine the optimal rank k of the factorization, one can hold out a small subset $O_1 \subset O$ of the observed entries, which are not used for learning U. These entries are used to test the squared error $\sum_{(i,j) \in O_1} e_{ij}^2$ of the matrix U learned using various values of k. The value of k at which the error of the held out entries is minimized is used. Furthermore, one can also use the held-out entries to determine the stopping criterion for the gradient-descent approach. The gradient-descent procedure is terminated when the error on the held-out entries begins to rise. The recovered matrix U provides a k-dimensional embedding of the data, which can be used in conjunction with machine learning algorithms.

The use of a fixed set of pre-computed entries in O allows the leveraging of gradient-descent methods. On the other hand, if we use *stochastic* gradient descent, we can simply sample any position in S and compute the similarity value on the fly. This type of approach does have the advantage that one does not have to cycle through the same set of entries in O. Presumably, the number of entries in the similarity matrix is so large that even when one samples as many entries as possible (with replacement) for stochastic gradient descent, most entries would not be visited more than once (or at all). Therefore, the stochastic gradient descent step boils down to the following step, which is executed repeatedly:

> Randomly sample index pair $[i, j]$ and compute similarity value s_{ij};
> Compute the error $e_{ij} = s_{ij} - \sum_{p=1}^{k} u_{ip} u_{jp}$;
> Update $u_{im}^{+} \Leftarrow u_{im}(1 - \alpha\lambda) + 2e_{ij}u_{jm}$ for all $m \in \{1 \ldots k\}$;
> Update $u_{jm}^{+} \Leftarrow u_{jm}(1 - \alpha\lambda) + 2e_{ij}u_{im}$ for all $m \in \{1 \ldots k\}$;
> Update $u_{im} \Leftarrow u_{im}^{+}$ and $u_{jm} \Leftarrow u_{jm}^{+}$;

The similarity values are computed on the fly, as entries are sampled. The algorithm can be used even for similarity matrices that are not positive semidefinite. The diagonal entries will be learned automatically to create the closest positive semidefinite approximation.

Problem 9.3.1 *Let S be an $n \times n$ symmetric matrix that is not positive semidefinite. It has $r \ll n$ negative eigenvalues of sizes $\lambda_1 \ldots \lambda_r$. Show that the objective function $J = \|S - UU^T\|_F^2$ is always at least $\sum_{p=1}^{r} \lambda_p^2$, irrespective of the value of k in the $n \times k$ matrix U. What is the minimum value of k at which this error is guaranteed?*

9.3.3   Asymmetric Similarity Decompositions

The decomposition $S = UU^T$ of the similarity matrix S is a symmetric one. However, it is also possible to use the *asymmetric* decomposition $S \approx UV^T$. Here, S is an $n \times n$ matrix, whereas U and V are both $n \times k$ matrices. In such a case, one can use some combination of the ith row of U and the ith row of V to the create the embedding of the ith data point. For example, one can concatenate the ith row of U and the ith row of V to create a $2k$-dimensional embedding of the ith data point. In such a case, the updates become similar to recommender systems (cf. Chapter 8), with the error matrix defined as $E = S - UV^T$, and the updates defined as follows:

$$U \Leftarrow U + \underbrace{\alpha EV}_{\Delta U}$$

$$V \Leftarrow V + \underbrace{\alpha E^T U}_{\Delta V}$$

How does one use the decomposition components to create the embedding? There are several choices; for example, one can use only V to create the embedding. However, one can also concatenate the ith row of U and the ith row of V to create a $2k$-dimensional embedding of the ith object. Using both the matrices U and V recognizes the fact that rows and columns capture different aspects of the similarity between objects. For example, in an asymmetric follower-followee link matrix, similarities in terms of followers is different from similarity in terms of followees. Alice and Bob might both be movie stars and be similar in terms of their followers, whereas Alice and John might belong to the same family and be similar in terms of who they follow (i.e., followees). Using both U and V is helpful in accounting for both types of similarities.

Another way of performing asymmetric decompositions is truncated SVD of rank-k:

$$S \approx Q_k \Sigma_k P_k^T \tag{9.5}$$

Here, Q_k is analogous to U and P_k is analogous to V, if the scaling factors in Σ_k are ignored.

Finally, if S is diagonalizable with real eigenvectors/eigenvalues, one can use straightforward eigendecomposition to extract the embedding:

$$S = U \Delta U^{-1} \tag{9.6}$$

The columns of U contain the eigenvectors, and they are not necessarily orthonormal. In such a case, one can extract the top-k columns of U (corresponding to the largest eigenvalues) to create a k-dimensional embedding. Asymmetric decompositions are particularly useful for asymmetric similarity matrices that arise in a number of real-world applications:

1. In a social network, one user might follow another (or like another user), but the "similarity" relationship might not be reciprocated. Similarly, hyperlinks between Webpages can be viewed as directed indicators of similarity.

2. Even an undirected graph might have an asymmetric similarity network, if the edge weights are normalized in an asymmetric way. As we will see in Chapter 10, an adjacency matrix of an undirected graph can be converted into a *stochastic transition matrix* by normalizing each row to sum to 1, and the right eigenvectors of this transition matrix provide an embedding, referred to as the Shi-Malik embedding [115]. On the other hand, the symmetric decomposition of a symmetric normalization of the same adjacency matrix leads to a related embedding known as the Ng-Jordan-Weiss embedding [98]. Both embeddings are different forms of spectral decomposition, which are used for applications like *spectral clustering* (see Section 10.5.1 of Chapter 10).

Asymmetric decompositions can be computed using any of the methods discussed in Chapter 8. Most of this chapter will focus on symmetric embeddings.

9.4 Linear Algebra Operations on Similarity Matrices

Many machine learning applications use basic statistical and geometric operations on a data matrix such as computing the mean/variance of a data matrix, centering it, normalizing the data points, computing pairwise Euclidean distances (instead of dot product similarities), and so on. These operations are relatively easy to perform, if one already had access to the multidimensional data set. However, what if one was only provided the similarities? Would it be possible to perform these operations *indirectly* by modifying or using the similarity matrix (rather than the points)? These types of basic operations often turn out to be very useful in various machine learning applications.

Consider an $n \times n$ similarity matrix S based on n objects $o_1 \ldots o_n$. These objects could be arbitrary types of objects such as time-series, sequences, and so on. Each object o_i has a multidimensional embedding $\Phi(o_i)$, so that the (i, j)th entry s_{ij} in matrix S is defined by the following dot product:

$$s_{ij} = \Phi(o_i) \cdot \Phi(o_j)$$

With these notations, we will define the basic operations between two points:

9.4.1   Energy of Similarity Matrix and Unit Ball Normalization

The squared norm of the multidimensional representation $\Phi(o_i)$ of the ith object o_i, is computed in terms of similarities as follows:

$$\|\Phi(o_i)\|^2 = \Phi(o_i) \cdot \Phi(o_i) = s_{ii}$$

The above computation is defined for the *squared* norm. The norm is simply $\sqrt{s_{ii}}$. The total energy $E(S)$ of the data set is the sum of the squared norms of the points, which is simply the trace of the similarity matrix:

$$E(S) = \sum_{i=1}^{n} s_{ii} = \text{tr}(S)$$

In other words, the total energy in the data set is equal to the sum of the diagonal entries of the similarity matrix!

The norm can be used to normalize a similarity matrix, so that all engineered points $\Phi_n(o_i)$ lie on a unit ball. Note that dot products become *cosine* similarities for unit normalized points. Unlike dot products, cosine similarities are invariant to normalization. Consider the case where we have an unnormalized similarity matrix $S = [s_{ij}]$ corresponding to engineered representation $\Phi(\cdot)$, and we want to normalize these points to $\Phi_n(\cdot)$ on the unit ball.

$$\Phi_n(o_i) \cdot \Phi_n(o_j) = \text{cosine}[\Phi_n(o_i), \Phi_n(o_j)] = \frac{\Phi(o_i) \cdot \Phi(o_j)}{\|\Phi(o_i)\| \cdot \|\Phi(o_j)\|} = \frac{s_{ij}}{\sqrt{s_{ii}}\sqrt{s_{jj}}}$$

Each entry s_{ij} is replaced with the above normalized value. Note that a normalized similarity matrix will contain only 1s along the diagonal. This is because one has effectively normalized the data-specific kernel features to lie on a unit ball in $\mathcal{R}^n$.

9.4.2   Norm of the Mean and Variance

It is possible to compute the norm of the *mean* of a data set as follows:

$$\|\overline{\mu}\|^2 = \|\sum_{i=1}^{n} \Phi(o_i)/n\|^2 = \sum_{i=1}^{n}\sum_{j=1}^{n} \Phi(o_i) \cdot \Phi(o_j)/n^2 = \sum_{i=1}^{n}\sum_{j=1}^{n} s_{ij}/n^2$$

In other words, the squared norm of the mean is equal to average value of the entries in the similarity matrix. This value is always nonnegative according to Problem 9.2.3.

The total variance $\sigma^2(S)$ of the data set (over all dimensions) in the embedded space (induced by similarity matrix S) is obtained by subtracting the squared norm of the mean from the normalized energy (i.e., energy averaged over number of dimensions):

$$\sigma^2(S) = \text{Energy}(S)/n - \overline{\mu}^2 = \sum_{i=1}^{n} s_{ii}/n - \sum_{i=1}^{n}\sum_{j=1}^{n} s_{ij}/n^2$$

Note that the variance is the difference between the average diagonal entry and average matrix entry.

Problem 9.4.1 *The variance of a data set containing n points can be shown to be proportional to the sum of squared pairwise distances between points (over all $\binom{n}{2}$ pairs). Use this result to show that the variance $\sigma^2(S)$ in the data induced by similarity matrix S can be expressed in the following form for appropriately chosen n-dimensional vectors $\overline{y}_r$ for $r \in \{1, 2, \ldots, n(n-1)/2\}$:*

$$\sigma^2(S) \propto \sum_{r=1}^{n(n-1)/2} \overline{y}_r^T S \overline{y}_r$$

The above problem also makes it evident why the variance will be nonnegative, given that the similarity matrix S is positive semidefinite.

9.4.3 Centering a Similarity Matrix

In some applications like PCA, it is assumed that the data is mean-centered. Unfortunately, there is no guarantee that the embedding induced by an arbitrary similarity matrix is mean-centered. Consider the case, where the data set $D = Q\Sigma$ can be extracted from the similarity matrix $S = Q\Sigma^2 Q^T$. Therefore, the similarity matrix S can also be expressed as DD^T. Let M be an $n \times n$ matrix of 1s. Then, the centered version D_c of D can be expressed as follows:

$$D_c = (I - M/n)D \tag{9.7}$$

Then, the centered version S_c of similarity matrix S is given by $S_c = D_c D_c^T$. This similarity matrix can be expressed in terms of S as follows:

$$\begin{aligned} S_c = D_c D_c^T &= [(I - M/n)D][(I - M/n)D]^T \\ &= (I - M/n)\underbrace{(DD^T)}_{S}(I - M/n)^T = (I - M/n)S(I - M/n) \end{aligned}$$

Data matrices are often centered in machine learning as a preprocessing step to various tasks. A specific example is kernel PCA. It is also possible to recognize when a similarity matrix is mean-centered as follows:

Observation 9.4.1 (Recognizing Mean-Centered Similarities) *The sum of the ith row (or column) of a similarity matrix is the dot product between the embedding of the ith point and the sum of all vectors in the embedding. Therefore, all rows and columns of a **mean-centered** similarity matrix sum to 0.*

9.4.3.1 Application: Kernel PCA

The above approach provides a route to kernel PCA [112]. In kernel PCA, the $n \times n$ similarity matrix S is first centered to $S_c = (I - M/n)S(I - M/n)$, and subsequently it is diagonalized as $S_c = Q\Sigma^2 Q^T$, which can be further expressed as the symmetric factorization $(Q\Sigma)(Q\Sigma)^T$. The embedding matrix is, therefore, given by $Q\Sigma$:

Definition 9.4.1 (Kernel PCA) *Let S be a positive semidefinite similarity matrix of size $n \times n$, which can be centered to $S_c = (I - M/n)S(I - M/n)$. Here, M is an $n \times n$ matrix of 1s. The embedding $Q\Sigma$ extracted by the eigendecomposition $S_c = Q\Sigma^2 Q^T$ is referred to as the kernel PCA embedding of the points. The matrix $Q\Sigma$ contains the kernel PCA embedding of the ith point in its ith row.*

Therefore, kernel PCA differs from kernel SVD in terms of the preprocessing of the similarity matrix, just as PCA differs from SVD in the preprocessing of the data matrix. The reader is advised to carefully compare the above definition with Definition 9.2.2.

9.4.4   From Similarity Matrix to Distance Matrix and Back

In many applications of machine learning, distance functions are available instead of similarity functions. Therefore, a natural question arises as to how (dot product) similarity matrices can be converted into distance matrices and back.

It is easier to convert similarity matrices into distance matrices rather than the reverse. In order to understand this point, note that the Euclidean distance is closely related to the dot product similarity as follows:

$$\|\overline{X} - \overline{Y}\|^2 = \overline{X} \cdot \overline{X} + \overline{Y} \cdot \overline{Y} - 2\overline{X} \cdot \overline{Y}$$

Therefore, given a dot product matrix, it is relatively easy to create a squared Euclidean distance matrix using the above relationship for each entry. If δ_{ij} is the Euclidean distance between points i and j, then one can express δ_{ij} in terms of the entries of $S = [s_{ij}]$ as follows:

$$\delta_{ij}^2 = s_{ii} + s_{jj} - 2s_{ij}$$

The above relationship can also be expressed in matrix form. Let $\overline{1}_n$ be an n-dimensional column vector of 1s, and $\overline{z} = [s_{11}, s_{22}, \ldots s_{nn}]^T$. Let the matrix of squared distances be denoted by $\Delta = [\delta_{ij}^2]$. Then, Δ can be expressed in terms of similarities as follows:

$$\Delta = \overline{1}_n \overline{z}^T + \overline{z} \overline{1}_n^T - 2S \tag{9.8}$$

Note that the first two matrices on the right-hand side of the above summation are defined by outer products of vectors.

One can also compute the squared distance matrix from the similarity matrix. Let $\Delta = [\delta_{ij}^2]$ be the squared distance matrix. Then, the similarity between any pair of points $\overline{X}$ and $\overline{Y}$ can be expressed in terms of distances as follows:

$$\overline{X} \cdot \overline{Y} = \frac{1}{2} \left(\|\overline{X}\|^2 + \|\overline{Y}\|^2 - \|\overline{X} - \overline{Y}\|^2 \right) \tag{9.9}$$

The squared norms of the individual points $\overline{X}$ and $\overline{Y}$ create a challenge for expressing the problem in terms of distances. The squared norms represent squared distances from the origin and the distance matrix will typically not contain any information about distances of points from the origin. Here, it is important to understand that the pairwise distances are invariant to origin translation, whereas dot product similarity is not. In other words, the similarity matrix will depend on which point we choose as the origin! A natural choice is to assume that the similarity matrix to be extracted is mean centered. In such a case, the squared norm of a data point becomes the squared distance of the point from the mean. Therefore, we make the following claim:

Lemma 9.4.1 *Let $\Delta = [\delta_{ij}^2]$ be an $n \times n$ squared distance matrix between embedded data points. Then, the mean-centered similarity matrix of dot products is given by the following:*

$$S = -\frac{1}{2} \left(I - \frac{M}{n} \right) \Delta \left(I - \frac{M}{n} \right)$$

Here, M is an $n \times n$ matrix of 1s.

Proof: When the similarity matrix S is mean-centered, Observation 9.4.1 implies that $MS = SM = 0$, when M is a matrix of 1s. Therefore, one can show the following:

$$S = \left(I - \frac{M}{n} \right) S \left(I - \frac{M}{n} \right) \tag{9.10}$$

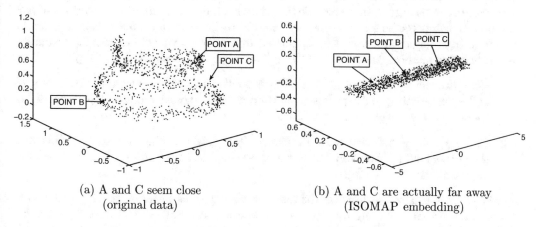

(a) A and C seem close
(original data)

(b) A and C are actually far away
(ISOMAP embedding)

Figure 9.1: Impact of *ISOMAP* embedding on distances

One can left-multiply and right-multiply both sides of Equation 9.8 with $(I - M/n)$ to obtain the following:

$$\left(I - \frac{M}{n}\right) \Delta \left(I - \frac{M}{n}\right) = \left(I - \frac{M}{n}\right) [\bar{1}_n \bar{z}^T + \bar{z} \bar{1}_n^T - 2S] \left(I - \frac{M}{n}\right)$$

Note that $M\bar{1} = n\bar{1}$. As a result, it is easy to show that $\left(I - \frac{M}{n}\right)\bar{1}_n = \bar{0}$ and $\bar{1}_n^T \left(I - \frac{M}{n}\right) = \bar{0}^T$. We can use these results to simplify the above equation as follows:

$$\left(I - \frac{M}{n}\right) \Delta \left(I - \frac{M}{n}\right) = -2\left(I - \frac{M}{n}\right) [S] \left(I - \frac{M}{n}\right)$$

$$= -2S, \qquad \text{[Using Equation 9.10]}$$

One can divide both sides by -2 to obtain the desired result. ∎

The conversion of distance matrices to similarity matrices is more useful because it enables the use of kernel methods when distances are available. For example, in time-series data, some domain-specific methods provide distances instead of similarities to begin with. In such cases, the technique of *multidimensional scaling (MDS)* is used to create an embedding. Starting with the squared distance matrix Δ, it is converted to a centered similarity matrix using the approach discussed above. The large eigenvectors of this similarity matrix are used to create the embedding.

Problem 9.4.2 (Almost Negative Semi-definite Distance Matrix) *We know that a valid similarity matrix using dot products must be positive semidefinite. A matrix is* **almost** *negative semidefinite, if it satisfies $\bar{y}^T S \bar{y} \leq 0$ for any* **mean-centered** *vector $\bar{y}$. Show using this fact and Lemma 9.4.1 that any valid squared distance matrix in Euclidean space is almost negative semidefinite.*

9.4.4.1 Application: ISOMAP

The *ISOMAP* approach is a great technique for straightening out curved manifolds in multidimensional space [126]. It can also compute *geodesic* distances and similarities, which correspond to distances (and similarities) along a curved manifold rather than straight line distances. It can be argued that geodesic distances are more accurate representations of true distances as compared to straight-line distances in real applications. Such distances can be computed by using an approach that is derived from a non-linear dimensionality reduction and embedding method, known as *ISOMAP*. The approach consists of two steps:

1. Compute the k-nearest neighbors of each point. Construct a weighted graph G with nodes representing data points, and edge weights (costs) representing distances of these k-nearest neighbors.

2. For any pair of points $\overline{X}$ and $\overline{Y}$, report $Dist(\overline{X}, \overline{Y})$ as the shortest path between the corresponding nodes in G. Any graph-theoretic algorithm, such as the Dijkstra algorithm can be used [8].

Subsequently, a squared distance matrix is constructed. This distance matrix is converted into a similarity matrix using the approach of this section. Subsequently, the eigenvectors of this matrix are used to create the *ISOMAP* embedding. A 3-dimensional example is illustrated in Figure 9.1(a), in which the data is arranged along a spiral. In this figure, data points A and C seem much closer to each other than data point B. However, in the *ISOMAP* embedding of Figure 9.1(b), the data point B is much closer to each of A and C. This example shows how *ISOMAP* has a drastically different view of similarity and distances, as compared to the pure use of Euclidean distances.

9.5   Machine Learning with Similarity Matrices

There are two ways in which machine learning algorithms may be used, when a similarity matrix is provided instead of the data matrix. These two ways are as follows:

1. The similarity matrix S can be decomposed as $S = Q\Sigma^2 Q^T$, and the embedding $Q\Sigma$ can be extracted. In some cases, only the top eigenvectors are retained, and the representation may be otherwise processed (e.g., whitening) to improve its quality. Subsequently, off-the-shelf machine learning algorithms are applied to the extracted representation.

2. Some algorithms in machine learning can be directly expressed in terms of similarities between points. An example is the SVM, in which the dual can be expressed in terms of dot products between points (cf. Section 6.4.4.1 of Chapter 6). In these cases, one can simply substitute the appropriate entry from the similarity matrix within the optimization function. This approach is referred to as the *kernel trick*.

The two choices provide equivalent solutions. Which one is preferable in practice? The general tendency in the machine learning community is to prefer the kernel trick. The reason is that the kernel trick is more space-efficient. However, explicit feature engineering also has a number of advantages. One can post-process the extracted features and discard the irrelevant ones. The lower-order features (i.e., smaller eigenvectors) can sometimes contain irrelevant noise, and explicit feature engineering methods tend to extract only the higher-order features. When using the kernel trick, one is effectively using *all* the features without any improvement/change including the irrelevant ones. When using feature engineering, tricks like whitening can also be used on the extracted features. In some problems like outlier detection, whitening is absolutely essential to create a high-quality implementation. The additional flexibility of feature engineering over the kernel trick is illustrated in Figure 9.2.

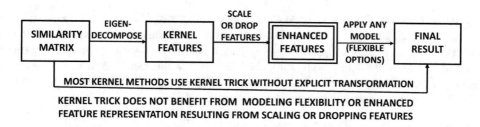

Figure 9.2: Two choices in algorithm design with kernel methods

As regards the issue of space- or time-efficiency, Nyström sampling or stochastic gradient descent can be used to efficiently extract the dominant features. In many cases, only the dominant features need to be extracted because lower-order features are not informative in most[2] applications. It is noteworthy that dominant feature extraction is far more efficient than the extraction of all features. The following sections discuss both feature engineering and the kernel trick.

9.5.1 Feature Engineering from Similarity Matrix

In this section, we will discuss algorithms for clustering, classification, and outlier detection with the use of feature engineering. One advantage of this approach is that it is very general, and one is not restricted to the use of a specific algorithm for clustering and classification.

9.5.1.1 Kernel Clustering

Imagine a setting, where you have an $n \times n$ similarity matrix S over n objects (e.g., chemical compounds). You would like to cluster these objects into similar groups. The broader approach of explicit feature engineering works by diagonalizing an $n \times n$ similarity matrix $S = Q\Sigma^2 Q^T$ as follows:

Diagonalize $S = Q\Sigma^2 Q^T$;
Extract the n-dimensional embeddings in rows of $Q\Sigma$;
Drop any zero columns from $Q\Sigma$ to create $Q_0\Sigma_0$;
Apply any existing clustering algorithm on rows of $Q_0\Sigma_0$;

The columns of Q_0 contain the non-zero eigenvectors, and the n rows of $Q_0\Sigma_0$ contain the embeddings of the n points. Note that Q_0 is an $n \times r$ matrix and Σ_0 is an $r \times r$ matrix, since the zero rows *and* columns of Σ are removed to create Σ_0. It is noteworthy that all n eigenvectors are extracted and only the zero eigenvectors are dropped. Such zero eigenvectors show up as zero columns in $Q\Sigma$. The embedding dimensionality can be as large as the number of points n, if no dimensions are dropped. The *space requirements* of such an approach can therefore be $O(n^2)$. Furthermore, the running time requirement for extracting all n eigenvectors is $O(n^3)$, which can be prohibitive. In many cases, one can use a dimensionality of the embedding that is far less than n. Furthermore, some implementations use the matrix Q to generate the embedding rather than $Q\Sigma$. Such an approach can be viewed as an indirect form of whitening (cf. Section 7.4.7 of Chapter 7). A specific example of a kernel clustering method that uses this form of whitening is *spectral clustering*, which is discussed in Section 10.5 of Chapter 10.

---

[2]An exception is outlier detection.

Since lower-order eigenvectors are often dropped anyway, it is possible to use any sampling method that preserves information only about dominant eigenvectors. A specific example is Nyström sampling, which subsamples a set of s objects in order to create an s-dimensional representation. Typically, the value of s is independent of the data set size, although it depends on the complexity of the underlying data distribution (e.g., number of clusters). Then, the approach proceeds as follows:

> Draw a subsample of s objects from the data set;
> Use the Nyström method (cf. Section 9.3.1) to create an s-dimensional
>   representation of all objects denoted by the $n \times s$ matrix U_s;
> Apply any existing clustering algorithm on U_s;

It is also possible to use stochastic gradient descent (cf. Section 9.3.2) to extract the embedding matrix U_s. Furthermore, sampling-based methods can often be repeated to create multiple models. The averaged model from these multiple models is referred to as an *ensemble*, and it provides superior results.

9.5.1.2   Kernel Outlier Detection

The Mahalanobis method discussed in Section 7.4.7 can be generalized to the *kernel* Mahalanobis method. In the kernel Mahalanobis method, the Mahalanobis method is applied to an engineered representation of the data [5]. Note that the feature engineering approach already extracts a normalized SVD from the similarity matrix. Technically, the Mahalanobis method requires centering of the similarity matrix up front, although it makes no practical difference even if the uncentered matrix is used. Given an $n \times n$ similarity matrix S, the kernel Mahalanobis method works as follows:

> Diagonalize $S = Q\Sigma^2 Q^T$;
> Extract the n-dimensional embeddings in rows of $Q\Sigma$;
> Drop any zero columns from $Q\Sigma$ to create $Q_0\Sigma_0$;
> Report the outlier score of each row of Q_0 as the Euclidean distance of
>   that row from the mean computed over all rows of Q_0;

Here, it is noteworthy that we are using Q_0 rather than $Q_0\Sigma_0$ in order to compute the outlier score of each point. This is particularly important in outlier detection, because outliers are often hidden in the deviations along lower-order singular vectors. Multiplying with Σ_0 would de-emphasize such outliers. These forms of whitening and feature postprocessing are not possible when using methods like the kernel trick (which is equivalent to always using $Q\Sigma$). Interestingly, some of the methods that do not use this type of whitening, such as the one-class SVM [113], are known to have weak performance [42].

Problem 9.5.1 *Write the pseudocode for kernel outlier detection with Nyström sampling.*

9.5.1.3   Kernel Classification

Consider the case where we want to implement a kernel SVM. Assume that the $n \times n$ similarity matrix on the training objects is denoted by S. In addition, we have t test objects, and therefore we have the $t \times n$ matrix of test-training similarities denoted by S_t. Therefore, each row of S_t contains the similarity of a test object to all training objects. The columns of S_t are sorted in the same order of training objects as S.

The similarity matrix of training data is diagonalized as $S = Q\Sigma^2 Q^T$. One can drop the zero columns of $Q\Sigma$ to yield the $n \times r$ matrix $U_0 = Q_0\Sigma_0$ with $r \leq n$ dimensions. Note that Q_0 is an $n \times r$ matrix, whereas Σ_0 is an $r \times r$ diagonal matrix with only non-zero diagonal entries (singular values) of S. Therefore, Σ_0 is invertible. The rows of U_0 contain the *explicit* transformations of the training objects. The t *out-of-sample* objects in the test data can also be projected into this r-dimensional representation U_{test} by using the same trick as used in Nyström sampling:

$$\underbrace{S_t}_{t \times n} = U_{test} U_0^T = \underbrace{U_{test}}_{t \times r} \underbrace{(Q_0 \Sigma_0)^T}_{r \times n} \tag{9.11}$$

The above relationship is a result of the fact that the dot product of the rows in U_{test} and U_0 correspond to test-training similarities. Multiplying both sides with $Q_0 \Sigma_0^{-1}$ and using $Q_0^T Q_0 = I$ on the left-hand side, we obtain:

$$U_{test} = S_t Q_0 \Sigma_0^{-1} \tag{9.12}$$

The matrix U_{test} contains the engineered representations of the test objects. Therefore, we present the algorithm for kernel SVMs as follows:

Diagonalize $S = Q \Sigma^2 Q^T$;
Extract the n-dimensional embedding in rows of $Q\Sigma$;
Drop any zero eigenvectors from $Q\Sigma$ to create $Q_0 \Sigma_0$;
{ The n rows of $Q_0 \Sigma_0$ and their class labels constitute training data }
Apply linear SVM on $Q_0 \Sigma_0$ and class labels to learn model $\mathcal{M}$;
Convert test-train similarity matrix S_t to representation matrix U_{test} using Equation 9.12;
Apply $\mathcal{M}$ on each row of U_{test} to yield predictions;

The above implementation is *identical* to the kernel SVM that is implemented using the kernel trick (cf. Section 9.5.2.1). *One can substitute the SVM with any learning algorithm like logistic regression or least-squares classification*, which is one of the advantages of explicit feature engineering. One can also use this approach in conjunction with Nyström sampling in order to improve efficiency.

Problem 9.5.2 *Show how the kernel SVM approach discussed in this section can be efficiently implemented with Nyström sampling.*

9.5.2 Direct Use of Similarity Matrix

The direct use of similarity matrices to implement machine learning algorithms (without extracting the embedding as an intermediate step) is referred to as the *kernel trick*. Although the kernel trick is often touted as the only practical way of implementing these algorithms, this is not precisely an accurate view. Explicit feature engineering has several benefits, the most important of which is the fact that one can modify or normalize features in intermediate steps. In some applications like outlier detection, this is so important that the variants using the kernel trick are not quite as effective [5, 42].

9.5.2.1 Kernel K-Means

Let $S = [s_{ij}]$ be an $n \times n$ similarity matrix, which contains the pairwise similarity information between the objects $o_1 \ldots o_n$. These objects may be of any arbitrary type (e.g., sequences or chemical compound graphs). Assume that the embedding implied by the kernel similarity matrix is denoted by $\Phi(\cdot)$ so that $s_{ij} = \Phi(o_i) \cdot \Phi(o_j)$.

The kernel k-means algorithm proceeds as follows. We start with a random assignment of points to the k clusters, denoted by $\mathcal{C}_1 \ldots \mathcal{C}_k$. The usual implementation of the k-means algorithm determines the centroids of the clusters as the representatives of the next iteration. The kernel k-means algorithm computes the dot product of each point to the various cluster centroids *in transformed space* and re-assigns each point to its closest centroid in the next iteration. How can one compute the dot product between a embedded object $\Phi(o_i)$ and the centroid $\overline{\mu}_j$ of $\mathcal{C}_j$ (in transformed space)? This can be achieved as follows:

$$\Phi(o_i) \cdot \overline{\mu}_j = \Phi(o_i) \cdot \frac{(\sum_{q \in \mathcal{C}_j} \Phi(o_q))}{|\mathcal{C}_j|} = \frac{\sum_{q \in \mathcal{C}_j} \Phi(o_i) \cdot \Phi(o_q)}{|\mathcal{C}_j|} = \sum_{q \in \mathcal{C}_j} \frac{s_{iq}}{|\mathcal{C}_j|}$$

Therefore, for any given object o_i, we only need to compute its average kernel similarity to all points in that cluster. Instead of the centroids, the approach does require the explicit maintenance of assignments of each point to various clusters in order to recompute the assignments for the next iteration. As in all k-means algorithms, the approach is iterated to convergence. For a data set containing n points, the approach requires $O(n^2)$ time in each iteration of the k-means algorithm, which can be quite costly for large data sets. The approach also requires the computation of the entire kernel matrix, which might require $O(n^2)$ storage. However, if the similarity function can be computed efficiently, then one does not need to store the kernel matrix a priori, but simply recompute individual entries on the fly when they are needed.

This algorithm is *identical* to the approach discussed in Section 9.5.1.1, when the embedding $Q_0 \Sigma_0$ is used in Section 9.5.1.1 and the k-means algorithm is used in the final step. However, a disadvantage of the kernel trick is that it can be paired with only a restricted subset of clustering algorithms (e.g., k-means) that use similarity functions between points. Not all clustering algorithms are equally friendly to the use of the kernel trick. Furthermore, one can perform no further engineering or normalization of the extracted features, if they are being used only indirectly via the kernel trick.

9.5.2.2 Kernel SVM

As in other problems of this section, we assume that each object o_i has an engineered representation $\Phi(o_i)$. The similarities between the n objects are contained in the $n \times n$ similarity matrix S. As consistently used throughput this book, the class labels are denoted by $y_1 \ldots y_n \in \{-1, +1\}$ for the n training instances. We simply copy the dual problem of the SVM from Section 6.4.4.1 as follows (in minimization form):

$$\text{Minimize } L_D = \frac{1}{2} \sum_{i=1}^{n} \sum_{j=1}^{n} \alpha_i - \left\{ \sum_{i=1}^{n} \alpha_i \alpha_j y_i y_j \underbrace{\Phi(o_i) \cdot \Phi(o_j)}_{s_{ij}} \right\}$$

subject to:
$$0 \leq \alpha_i \leq C \quad \forall i \in \{1 \ldots n\}$$

The notations of this problem are the same as those in Section 6.4.4.1. Each α_i is the ith dual variable. The quantity C is the slack penalty. The *only difference* from the objective function of Section 6.4.4.1 is that we have replaced the training point $\overline{X}_i$ with an engineered point $\Phi(o_i)$. However, the dot product between $\Phi(o_i)$ and $\Phi(o_j)$ is simply s_{ij}, and therefore the engineered points can be made to disappear from the formulation and be replaced by similarities. The partial derivative of L_D with respect to α_k is as follows:

$$\frac{\partial L_D}{\partial \alpha_k} = y_k \sum_{q=1}^{n} y_q \alpha_q s_{kq} - 1 \quad \forall k \in \{1 \ldots n\} \tag{9.13}$$

This is a convex optimization problem with box constraints. Therefore, one starts by setting the vector of Lagrangian parameters $\overline{\alpha} = [\alpha_1 \ldots \alpha_n]$ to an n-dimensional vector of 0s and uses the following update steps with learning rate η:

> **repeat**
>   Update $\alpha_k \Leftarrow \alpha_k + \eta \left[1 - y_k \sum_{q=1}^n y_q \alpha_q s_{kq} \right]$ for each $k \in \{1 \ldots n\}$;
> $\left\{ \text{ Update is equivalent to } \overline{\alpha} \Leftarrow \overline{\alpha} - \eta \left[\frac{\partial L_D}{\partial \overline{\alpha}} \right] \right\}$
>   **for** each $k \in \{1 \ldots n\}$ **do begin**
> $\alpha_k \Leftarrow \min\{\alpha_k, C\}$;
> $\alpha_k \Leftarrow \max\{\alpha_k, 0\}$;
>   **endfor**;
> **until** convergence

After the variables $\alpha_1 \ldots \alpha_n$ have been learned, the test objects are predicted using their similarities with training objects. This is because the classification of an unseen test instance $\overline{Z}$ is given by the sign of $\overline{W} \cdot \Phi(\overline{Z})$, where our analysis in Chapter 6 shows that $\overline{W} = \sum_{j=1}^n \alpha_j \overline{y}_j \Phi(\overline{X}_j)$. Here, $\Phi(\overline{X}_j)$ is the engineered representation of the jth training point. Therefore, the prediction of $\overline{Z}$ is given by $\sum_{j=1}^n \alpha_j \overline{y}_j \Phi(\overline{X}_j) \cdot \Phi(\overline{Z})$. This is the simply the weighted sum of the similarities of the training instances with the test instance, where the weight of the jth similarity is $y_j \alpha_j$.

One can also express this result in terms of training-test similarity matrices. Let S_t be the $t \times n$ similarity matrix of training-test similarities between test objects and training objects. Let $\overline{\gamma}$ be an n-dimensional column vector in which the jth component is $y_j \alpha_j$. Then, the analysis of the previous paragraph shows that the prediction of the t test instances is given by the sign of each element in the t-dimensional vector $S_t \overline{\gamma}$.

The optimization model discussed in this section is *identical* to the one discussed in Section 9.5.1.3 (although the computational procedures are very different in the two cases). However, the approach in Section 9.5.1.3 is more flexible, because it cleanly decouples feature engineering from model building. Therefore, it can be used more easily with any off-the-shelf classification model or computational procedure.

9.6 The Linear Algebra of the Representer Theorem

In this section, we will drop the use of $\Phi(\cdot)$ for brevity; the training points $\overline{X}_1 \ldots \overline{X}_n$ and the test point $\overline{Z}$ already represent the engineered representations. Note that we could make the same arguments by adding the function $\Phi(\cdot)$ to each data point, but it will unnecessarily make our mathematical formulas cumbersome.

When using the kernel trick, it becomes essential to identify a computational procedure that uses similarities rather than individual data points. The *representer theorem* is a useful principle from linear algebra that provides a boiler plate method to convert many optimization formulations into one that uses similarities. The approach is applicable to optimization models satisfying the following two properties:

1. The parameters of the optimization problem can be expressed as one or more vectors in the same multidimensional space as the individual data points. For example, the weight vector $\overline{W}$ in problems like the SVM lies in the same multidimensional space as the data points.

2. The objective function of the optimization problem can be expressed as a function of one or more of (i) dot products between points, (ii) dot products between the

parameter vectors and points, and (iii) dot products between the parameter vectors themselves (e.g., L_2-regularizer).

Under these circumstances, a representer theorem can be used to transform any machine learning problem on multidimensional vectors into a formulation that uses only similarities between points. Interestingly, all the linear classification models we have seen so far satisfy this property. In other words, the optimization formulation can be converted into one that uses only similarities between objects.

Consider the L_2-regularized form of all linear models discussed in Chapter 4 over the training pairs $(\overline{X}_1, y_1) \ldots (\overline{X}_n, y_n)$, where each $\overline{X}_i$ is a row vector. Furthermore, the prediction of y_i is done as $\hat{y}_i = f(\overline{W} \cdot \overline{X}_i^T)$ for some function $f(\cdot)$ that depends on the nature of the target variable (e.g., numeric, binary, or categorical). The loss function can be written as $L(y_i, \overline{W} \cdot \overline{X}_i^T)$ in each case, because the objective function compares $\overline{W} \cdot \overline{X}_i^T$ to y_i in each case to decide the loss. The overall objective function, including the regularizer, may be written as follows:

$$\text{Minimize } J = \sum_{i=1}^{n} L(y_i, \overline{W} \cdot \overline{X}_i^T) + \frac{\lambda}{2} \|\overline{W}\|^2 \tag{9.14}$$

Consider a situation in which the training data points have dimensionality d, but all of them lie on a 2-dimensional plane. Note that the optimal linear separation of points on this plane can always be achieved with the use of a 1-dimensional line on this 2-dimensional plane. Furthermore, this separator is more concise than any higher dimensional separator and will therefore be preferred by the L_2-regularizer. A 1-dimensional separator of training points lying on a 2-dimensional plane is shown in Figure 9.3(a). Although it is also possible to get the same separation of training points using any 2-dimensional plane (e.g., Figure 9.3(b)) passing through the 1-dimensional separator of Figure 9.3(a), such a separator would not be preferred by an L_2-regularizer because of its lack of conciseness. In other words, given a set of training data points (row vectors from a data matrix) denoted by $\overline{X}_1 \ldots \overline{X}_n$, the separator $\overline{W}$, defined as a column vector, always lies in the space spanned by these vectors (after converting them to column vectors). We state this result below, which is a very simplified version of the representer theorem, and is specific to linear models with L_2-regularizers.

Theorem 9.6.1 (Simplified Representer Theorem) *Let J be any optimization problem of the following form:*

$$\text{Minimize } J = \sum_{i=1}^{n} L(y_i, \overline{W} \cdot \overline{X}_i^T) + \frac{\lambda}{2} \|\overline{W}\|^2$$

Then, any optimum solution $\overline{W}^*$ to the aforementioned problem lies in the subspace spanned by the training points $\overline{X}_1^T \ldots \overline{X}_n^T$. In other words, there must exist real values $\beta_1 \ldots \beta_n$ such that the following is true:

$$\overline{W}^* = \sum_{i=1}^{n} \beta_i \overline{X}_i^T$$

Proof: Suppose that $\overline{W}^*$ cannot be expressed in the subspace spanned by the training points. Then, let us decompose $\overline{W}^*$ into the portion $\overline{W}_{\parallel} = \sum_{i=1}^{n} \beta_i \overline{X}_i^T$ spanned by the training points and an additional orthogonal residual $\overline{W}_{\perp}$. In other words, we have:

$$\overline{W}^* = \overline{W}_{\parallel} + \overline{W}_{\perp} \tag{9.15}$$

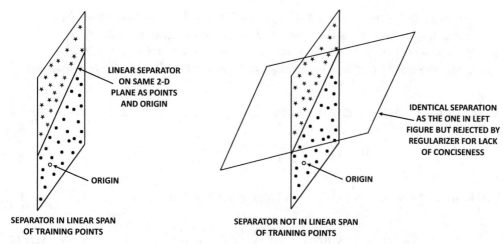

(a) Separator in subspace spanned spanned by training points

(b) Separator not in subspace spanned by training points

Figure 9.3: Both the linear separators in (a) and (b) provide exactly the same separation of training points, except that the one in (a) can be expressed as a linear combination of the training points. The separator in (b) will always be rejected by the regularizer. The key point of the representer theorem is that a separator $\overline{W}$ can always be found in the plane (subspace) of the training points with an identical separation to one that does not

Then, it suffices to show that $\overline{W}^*$ can be optimal only when $\overline{W}_\perp$ is the zero vector.

Each $(\overline{W}_\perp \cdot \overline{X}_i)$ has to be 0, because $\overline{W}_\perp$ is orthogonal to the subspace spanned by the various training points. The optimal objective J^* can be written as follows:

$$J^* = \sum_{i=1}^{n} L(y_i, \overline{W}^* \cdot \overline{X}_i^T) + \frac{\lambda}{2}\|\overline{W}^*\|^2 = \sum_{i=1}^{n} L(y_i, (\overline{W}_\| + \overline{W}_\perp) \cdot \overline{X}_i^T) + \frac{\lambda}{2}\|\overline{W}_\| + \overline{W}_\perp\|^2$$

$$= \sum_{i=1}^{n} L(y_i, \overline{W}_\| \cdot \overline{X}_i^T + \underbrace{\overline{W}_\perp \cdot \overline{X}_i^T}_{0}) + \frac{\lambda}{2}\|\overline{W}_\|\|^2 + \frac{\lambda}{2}\|\overline{W}_\perp\|^2$$

$$= \sum_{i=1}^{n} L(y_i, \overline{W}_\| \cdot \overline{X}_i^T) + \frac{\lambda}{2}\|\overline{W}_\|\|^2 + \frac{\lambda}{2}\|\overline{W}_\perp\|^2$$

It is noteworthy that $\|\overline{W}_\perp\|^2$ must be 0, or else $\overline{W}_\|$ will be a better solution than $\overline{W}^*$. Therefore, $\overline{W}^* = \overline{W}_\|$ lies in the subspace spanned by the training points. ∎

Intuitively, the representer theorem states that for a particular family of loss functions, one can always find an optimal linear separator within the subspace spanned by the training points (see Figure 9.3), and the regularizer ensures that this is the concise way to do it. After all, even though the embedding of an object might be infinite dimensional, each data object only lies in an n-dimensional projection of this space for a data set of size n. This n-dimensional projection is defined by the span of the n vectors $\overline{X}_1^T \ldots \overline{X}_n^T$. The parameter vector $\overline{W}$ also lies in the span of this n-dimensional subspace; this is the essence of the representer theorem.

The representer theorem provides a boilerplate method to create an optimization model that is expressed as a function of dot products:

For any given optimization model of the form of Equation 9.14 plug in $\overline{W} = \sum_{i=1}^{n} \beta_i \overline{X}_i^T$ to obtain a new optimization problem parameterized by $\beta_1 \ldots \beta_n$, and expressed only in terms of dot products between training points. Furthermore, the same approach is also used while evaluating $\overline{W} \cdot \overline{Z}^T$ for test instance $\overline{Z}$.

Consider what happens when one evaluates $\overline{W} \cdot \overline{X}_i^T$ in order to plug it into the loss function:

$$\overline{W} \cdot \overline{X}_i^T = \sum_{p=1}^{n} \beta_p \overline{X}_p^T \cdot \overline{X}_i^T = \sum_{p=1}^{n} \beta_p \overline{X}_p \cdot \overline{X}_i \qquad (9.16)$$

Furthermore, the regularizer $\|\overline{W}\|^2$ can be expressed as follows:

$$\|\overline{W}\|^2 = \sum_{i=1}^{n} \sum_{j=1}^{n} \beta_i \beta_j \overline{X}_i \cdot \overline{X}_j \qquad (9.17)$$

In order to kernelize the problem, all we have to do is to substitute the dot product with the similarity value $s_{ij} = \overline{X}_i \cdot \overline{X}_j$ from the $n \times n$ similarity matrix S. Note that each $\overline{X}_i$ is really the embedded representation $\Phi(o_i)$ of an object. Therefore, one obtains the following optimization objective function:

$$J = \sum_{i=1}^{n} L(y_i, \sum_{p=1}^{n} \beta_p s_{pi}) + \frac{\lambda}{2} \sum_{i=1}^{n} \sum_{j=1}^{n} \beta_i \beta_j s_{ij} \quad \text{[General form]}$$

In other words, all we need to do is to substitute each $\overline{W} \cdot \overline{X}_i^T$ in the loss function with $\sum_p \beta_p s_{pi}$. Therefore, one obtains the following form for least-squares regression:

$$J = \frac{1}{2} \sum_{i=1}^{n} (y_i - \sum_{p=1}^{n} \beta_p s_{pi})^2 + \frac{\lambda}{2} \sum_{i=1}^{n} \sum_{j=1}^{n} \beta_i \beta_j s_{ij} \quad \text{[Least-squares regression]}$$

By substituting $\overline{W} \cdot \overline{X}_i^T = \sum_p \beta_p s_{pi}$ into the loss functions of various classifiers for binary data, one can obtain corresponding optimization formulations:

$$J = \sum_{i=1}^{n} \max\{0, 1 - y_i \sum_{p=1}^{n} \beta_p s_{pi}\} + \frac{\lambda}{2} \sum_{i=1}^{n} \sum_{j=1}^{n} \beta_i \beta_j s_{ij} \quad \text{[SVM]}$$

$$J = \sum_{i=1}^{n} \log(1 + \exp(-y_i \sum_{p=1}^{n} \beta_p s_{pi})) + \frac{\lambda}{2} \sum_{i=1}^{n} \sum_{j=1}^{n} \beta_i \beta_j s_{ij} \quad \text{[Logistic Regression]}$$

These *unconstrained* optimization problems are conveniently expressed in terms of pairwise similarities, and parameterized by $\overline{\beta} = [\beta_1 \ldots \beta_n]^T$. Any of the optimization procedures discussed in Chapter 4 can be used to learn these parameters.

How does one perform prediction with the use of similarities? Predictions in linear classifiers are done by using the dot products between test instances and $\overline{W}$. Consider the case in which we have a $t \times n$ matrix S_t of similarities between test-training pairs. The dot product of each test instance with $\overline{W}$ amounts to the dot product between the corresponding row of S_t and $\overline{\beta}^T$, as in the case of training instances (cf. Equation 9.16). This is because

$\overline{W}$ can be expressed as the summation $\sum_i \beta_i \overline{X}_i^T$ over all engineered training instances, and the dot product with the engineered test instance $\overline{Z}^T$ simply extracts the corresponding row from S_t; the entries in that row are $\overline{Z}^T \cdot \overline{X}_i^T$. Therefore, the entire set of t test instances can be predicted as the t-dimensional vector $S_t \overline{\beta}$. In the case of the classification problem, one needs to use the sign of each element of this vector as the predicted class label.

9.7 Similarity Matrices and Linear Separability

So far, we have justified the use of similarity matrices as an avenue for feature engineering of data types that are not multidimensional (e.g., chemical compounds). What happens when we have a multidimensional data set? Surely, it would not make sense to compute a dot product matrix and then extract features from the matrix via eigendecomposition — by doing so, one would only obtain a rotreflection of the original data set. Here, the key point is that changing the dot product to a more carefully designed similarity function has the effect of changing the data representation so that a simple model with inherent limitations (e.g., the limitation of being a linear model) works much more effectively. This is because by changing the similarity function, we are also changing its underlying multidimensional (engineered) representation to a nonlinear function of the original representation.

In classification and regression applications, one is often looking for feature representations such that the dependent variable is linearly related to the features. For example, in the case of an SVM, we are looking for the linear separator $\overline{W} \cdot \Phi(\overline{X}) = 0$, where $\Phi(\overline{X})$ is the engineered representation obtained by symmetric decomposition of the similarity matrix. Since $\Phi(\overline{X})$ depends on the similarity matrix, a natural question arises as to whether some similarity matrices are better than others in achieving linear separation between the classes. In many applications, an analyst might have little control over the similarity or distance matrix provided by a given application. For example, in a time-series application, the distance matrix may be defined by a dynamic time-warping application, which might be converted to a similarity matrix (cf. Section 9.4.4). However, the analyst still has the ability to *post-process* the matrix, so that its decomposition yields better features. In general, *higher embedding dimensionalities tend to make linear separation more likely*. The effective embedding dimensionality of an $n \times n$ similarity matrix is often much less than n because many eigenvalues are extremely small. The embedding dimensionality of a positive semidefinite similarity matrix can often be increased by applying a *superlinear function* on its entries. In other words, we apply an *element-wise superlinear function* $F(s_{ij})$ to each element s_{ij} of the similarity matrix S. The simplest elementwise function is the polynomial function:

$$F(s_{ij}) = (c + s_{ij})^h$$

In this case, c and h are nonnegative hyperparameters. For example, choosing $c = 1$ and $h = 2$ corresponds to adding 1 to each value in the similarity matrix and squaring it. *For certain types of functions, such as the above, the positive semidefiniteness of the similarity matrix is not lost by this change.*

If the original embedded data (implied by the similarity matrix) is such that one of the two classes lies inside the ellipse $x^2 + 4y^2 \leq 10$ and the other lies outside it, then performing this type of quadratic operation on the similarity matrix will lead to at least one two-dimensional projection in the embedding, where the elliptical boundary becomes a linear separator. This situation is shown in Figure 9.4, where two of the embedded dimensions are shown. In order to understand this point, we recommend the reader to work out the practice Problem 9.7.1 below.

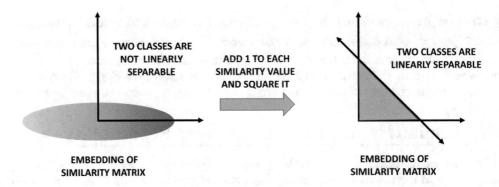

Figure 9.4: Applying an element-wise superlinear function on a similarity matrix often creates higher-dimensional embeddings in which points are linearly separable in carefully chosen projections: the relevant 2-dimensional projection of the embedding on the right is shown

Problem 9.7.1 *Consider two points (x_1, y_1) and (x_2, y_2) in 2-dimensional space and the dot product similarity $s = x_1 \cdot x_2 + y_1 \cdot y_2$. Now imagine that you modify the similarity to the superlinear function $s' = (1 + s)^2$. Show that s' can be expressed as the dot product between $(x_1^2, y_1^2, x_1 y_1 \sqrt{2}, x_1 \sqrt{2}, y_1 \sqrt{2}, 1)$ and $(x_2^2, y_2^2, x_2 y_2 \sqrt{2}, x_2 \sqrt{2}, y_2 \sqrt{2}, 1)$.*

Now consider a situation where an 100000×100000 similarity matrix S has two non-zero eigenvalues, and the 2-dimensional embedding $[x_1, x_2]$ extracted by eigendecomposition of S exhibits the property that all members of one class are lie inside the ellipse $x_1^2 + 4y_1^2 \leq 10$, whereas all members of the second class lie outside this ellipse. Discuss why the two classes will become linearly separable if the embedding is extracted by eigendecomposition of a modified similarity matrix S' in which we add 1 to each similarity entry and then square it. What is the dimensionality of the modified embedding (i.e., number of non-zero eigenvectors of S')?

There are two other functions that are commonly used for increasing the embedding dimensionality and capturing nonlinearity:

$$F(s_{ij}) = \tanh(\kappa s_{ij} - \delta) \quad \text{[Sigmoid function]}$$
$$F(s_{ij}) = \exp(s_{ij}/\sigma^2) \quad \text{[Gaussian function]}$$

Of course, applying a superlinear function does not always help, because it could lead to overfitting. The level of sensitivity of the superlinear function depends on the parameters (such as the bandwidth, σ, of the Gaussian function above), which are often chosen in a data-driven manner. For example, one can test the classification accuracy on out-of-sample data in order to select σ^2. A critical fact about many of these functions is that they do not destroy the positive semidefinite nature of the underlying similarity matrix. In Section 9.7.1, we will discuss some of these transformations.

The above ideas are used frequently for multidimensional data, where the similarity value s_{ij} is often set to a superlinear function of the dot product. Let $\overline{X}_1 \ldots \overline{X}_n$ be the n points, and the similarity s_{ij} be defined by the *kernel function* $K(\overline{X}_i, \overline{X}_j)$. Then, the common kernel functions used for multidimensional data are defined in Table 9.1.

Table 9.1: Table of common kernel functions

Function	Form
Linear Kernel	$K(\overline{X}_i, \overline{X}_j) = \overline{X}_i \cdot \overline{X}_j$
Gaussian Radial Basis Kernel	$K(\overline{X}_i, \overline{X}_j) = \exp(-\|\overline{X}_i - \overline{X}_j\|^2/(2 \cdot \sigma^2))$
Polynomial Kernel	$K(\overline{X}_i, \overline{X}_j) = (\overline{X}_i \cdot \overline{X}_j + c)^h, \ c \geq 0$
Sigmoid Kernel	$K(\overline{X}_i, \overline{X}_j) = \tanh(\kappa \overline{X}_i \cdot \overline{X}_j - \delta)$

Each of the kernels in Table 9.1 has parameters associated with it, which need to be learned in a data-driven manner. Note that the above kernels are similar to the techniques discussed for modifying the similarity matrix. These modifications improve the level of separation among different classes. The dimensionality of the embedding depends on the nature of the kernel function. For example, the Gaussian kernel leads to an infinite-dimensional embedding to represent all possible data pairs in $\mathcal{R}^n \times \mathcal{R}^n$, although the data-specific embedding is always n-dimensional and can be materialized for a data set containing n points (using the eigendecomposition methods discussed earlier in this chapter).

9.7.1 Transformations That Preserve Positive Semi-definiteness

It is useful to understand the nature of transformations of a positive semidefinite matrix that preserve the positive semidefiniteness property. All the results below apply to the $n \times n$ positive semidefinite matrix $S = [s_{ij}]$, unless otherwise mentioned:

1. The matrix aS is positive semidefinite for $a > 0$. [Note that $\overline{x}^T(aS)\overline{x} = a[\overline{x}^T S\overline{x}] \geq 0$.]

2. If S_1 and S_2 are positive semidefinite, then $S_1 + S_2$ is positive semidefinite. [This is easy to prove by showing $\overline{x}^T(S_1 + S_2)\overline{x} \geq 0$ for all $\overline{x}$.]

3. An $n \times n$ matrix C containing the constant non-negative value c in each entry is positive semidefinite because $\overline{x}^T C\overline{x} = c(\sum_i x_i)^2 \geq 0$.

4. If S_1 and S_2 are positive semidefinite matrices of the same size, then $S_1 \odot S_2$ is positive semidefinite. Here, $\odot$ indicates entry-wise product. This result is referred to as *Schur's product theorem*, and the proof is nonobvious (see Problem 9.7.2).

5. The matrix $\overbrace{S \odot S \odot \ldots \odot S}^{k \text{ times}}$ is positive semidefinite. [This is easy to show by applying the previous result recursively.]

6. Let $f(x)$ be a polynomial function with nonnegative coefficients that is applied to each entry of S. Then, the resulting matrix is positive semidefinite. [This is easy to show by combining four of the above results.]

7. The matrix $\exp(aS)$ is positive semidefinite for $a > 0$. Here $\exp(\cdot)$ refers to entry-wise exponentiation of the matrix. [This is easy to show by observing that an exponentiation can be expressed as an infinite polynomial with nonnegative coefficients using the Taylor expansion (cf. Equation 1.31 of Chapter 1). Therefore, each entry in the matrix is an infinite polynomial, and one reverts to the polynomial case above.]

8. Let $\delta_1 \ldots \delta_n$ be n real values. Then, the scaled similarity matrix in which the (i,j)th entry is $\delta_i s_{ij} \delta_j$ is positive semidefinite. [Since $\overline{x}^T S\overline{x}$ is nonnegative for any $\overline{x}$, so is the value $[\overline{x} \odot \overline{\delta}]^T S[\overline{x} \odot \overline{\delta}]$. Here, we have $\overline{\delta} = [\delta_1 \ldots \delta_n]^T$.]

Next, we list the Schur's product theorem as a practice exercise in a step-by-step manner, because it was used in one of the above results to show that $S_1 \odot S_2$ is positive semidefinite, when S_1 and S_2 are positive semidefinite.

Problem 9.7.2 (Schur's Product Theorem) *Let $S_1 = AA^T$ and $S_2 = BB^T$ be two positive semidefinite matrices. Let $\overline{a}_i$ the ith row of A and $\overline{b}_i$ be the ith row of B.*

- *Show that for any vector $\overline{x}$, one can express $\overline{x}^T(S_1 \odot S_2)\overline{x}$ in the following form:*

$$\overline{x}^T(S_1 \odot S_2)\overline{x} = \sum_i \sum_j x_i x_j [\overline{a}_i \overline{a}_j^T][\overline{b}_i \overline{b}_j^T]$$

- *Suppose that qth components of $\overline{a}_i$ and $\overline{b}_i$ are a_{iq} and b_{iq} respectively. Show that one can simplify the above expression to the following:*

$$\overline{x}^T(S_1 \odot S_2)\overline{x} = \sum_i \sum_j x_i x_j [\sum_k a_{ik} a_{jk}][\sum_l b_{il} b_{jl}]$$

- *Show that one can simplify the above expression as follows:*

$$\overline{x}^T(S_1 \odot S_2)\overline{x} = \sum_k \sum_l [\sum_i x_i a_{ik} b_{il}][\sum_j x_j a_{jk} b_{jl}]$$

Discuss why this expression is always nonnegative, and therefore the matrix $S_1 \odot S_2$ is positive semidefinite.

Problem 9.7.3 *Show that adding a non-negative value c to each entry of a positive semidefinite matrix does not affect its positive semidefinite property.*

The aforementioned list of properties of positive semidefinite matrices map to properties of positive semidefinite kernels (in closed form as in Table 9.1). First, we define the notion of a (closed-form) positive semidefinite kernel function:

Definition 9.7.1 *A kernel function is positive semidefinite if and only if all possible matrices created by samples of the arguments of that function are positive semidefinite.*

For example, in order to show that the polynomial kernel of Table 9.1 is positive semidefinite, we will have to show that *any* $n \times n$ similarity matrix $P = [p_{ij}]$ created from *arbitrary* $\overline{X}_1 \ldots \overline{X}_n \in \mathcal{R}^d$ using the function $p_{ij} = (c + \overline{X}_i \cdot \overline{X}_j)^h$ and $c \geq 0$ is positive semidefinite. The value of n can also be arbitrary, whereas h is a positive integer.

Lemma 9.7.1 (Polynomial Kernel Is Positive Semidefinite) *The $n \times n$ similarity matrix $P = [p_{ij}]$ defined by the polynomial kernel $p_{ij} = (\overline{X}_i \cdot \overline{X}_j + c)^h$ for any $\overline{X}_1 \ldots \overline{X}_n \in \mathcal{R}^d$ and $c \geq 0$ is positive semidefinite.*

Proof: Let $S = [s_{ij}]$ be an $n \times n$ matrix in which $s_{ij} = \overline{X}_i \cdot \overline{X}_j$. We already know that the matrix S is positive semidefinite because it is a dot product (Gram) matrix. Let C be an $n \times n$ matrix containing c in each entry. Since $c \geq 0$, it follows that C is positive semidefinite, and the matrix $C + S$ is positive semidefinite as well. The polynomial kernel P can be expressed in the following form:

$$P = (C + S) \odot (C + S) \odot \ldots \odot (C + S)$$

From Schur's product theorem, the matrix P is positive semidefinite as well. ∎
From Definition 9.7.1, this means that the polynomial kernel is positive semidefinite. One can also show that the Gaussian kernel is positive semidefinite.

Lemma 9.7.2 (Gaussian Kernel Is Positive Semidefinite) *The $n \times n$ similarity matrix $G = [g_{ij}]$ defined by the Gaussian kernel $g_{ij} = exp(-\|\overline{X}_i - \overline{X}_j\|^2/(2 \cdot \sigma^2))$ for any $\overline{X}_1 \ldots \overline{X}_n \in \mathcal{R}^d$ is positive semidefinite.*

Proof: Let $h_{ij} = \frac{\overline{X}_i}{\sigma} \cdot \frac{\overline{X}_j}{\sigma}$. Then, the matrix $H = [h_{ij}]$ is positive semidefinite, since it is a Gram matrix of dot products. Therefore, the matrix defined by $s_{ij} = \exp(g_{ij})$ is also positive semidefinite, since the (element-wise) exponentiation operation does not affect positive semidefiniteness (see list of properties at the beginning of the section). Define $\delta_i = \exp(-\|\overline{X}_i\|^2/(2\sigma^2))$. Then, it is easy to show the following:

$$\delta_i s_{ij} \delta_j = \exp(-\|\overline{X}_i - \overline{X}_j\|^2/(2 \cdot \sigma^2)) = g_{ij}$$

Since the matrix $S = [s_{ij}]$ is positive semidefinite and scaling does not affect positive semidefiniteness, it follows that the matrix $G = [g_{ij}]$ is positive semidefinite as well. ∎
Interestingly, the sigmoid kernel is not always positive semi-definite, but works well in practice. One can also create more complicated kernel functions by combining multiple kernels.

Problem 9.7.4 *Use the matrix transformation properties preserving positive semidefiniteness (listed at the beginning of this section) together with Property 9.7.1 to show that (i) The kernel function defined by the sum of two kernel functions is positive semidefinite, (ii) the kernel function defined by the product of two kernel functions is positive semidefinite, and (iii) the kernel function defined by any polynomial function of a kernel function with nonnegative coefficients is positive semidefinite.*

9.8   Summary

Many forms of data are not multidimensional, and examples include discrete sequences and graphs. In such cases, one might have similarities available between objects, but one might not have any multidimensional representation of the data. Eigendecomposition methods from linear algebra help in converting such similarity matrices to multidimensional embeddings. This chapter discusses the use of the similarity matrix in lieu of the multidimensional representation in order to implement machine learning algorithms with similarities rather than multidimensional representations. Similarity-based representations also allow the flattening of nonlinear relationships in the data, so that linear learners become more effective.

9.9   Further Reading

This chapter discusses the linear algebra of similarity matrices and kernel methods. The basics of kernel methods follow from Mercer's theorem [118]. The book by Schölkopf and Smola [118] provides a lot of detail about mathematical properties of kernels and their use in machine learning. The kernel PCA technique is discussed in [112]. The ISOMAP method is introduced in [126]. The kernel outlier detection technique is discussed in [5]. Kernel one-class SVMs are discussed in [113]. The representer theorem is due to Wahba [129].

9.10   Exercises

1. Suppose that you are given a 10×10 *binary* matrix of similarities between objects. The similarities between all pairs of objects for the first four objects is 1, and also

between all pairs of objects for the next six objects is 1. All other similarities are 0. Derive an embedding of each object.

2. Suppose that you have two non-disjoint sets of objects A and B. The set $A \cap B$ is a modestly large sample of objects. You are given all similarities between pairs of objects, one drawn from each of the two sets. Discuss how you can efficiently approximate the entire similarity matrix over the entire set $A \cup B$. It is known that the similarity matrix is symmetric. [Hint: Think of the connections of this setting with matrix factorization. After all, all embeddings are extracted as symmetric matrix factorizations. Here, we are given only a block of the similarity matrix.]

3. Suppose that S_1 and S_2 are $n \times n$ positive semidefinite matrices of ranks k_1 and k_2, respectively, where $k_2 > k_1$. Show that $S_1 - S_2$ can never be positive semidefinite.

4. Suppose you are given a binary matrix of similarities between objects, in which most entries are 0s. Discuss how you can adapt the logistic matrix factorization approach of Chapter 8 to make it more suitable to symmetric matrix factorization.

5. Suppose that you were given an incomplete matrix of similarities between objects belonging to two sets A and B that are completely disjoint (unlike Exercise 2). Discuss how you can find an embedding for each of the objects in the two sets. Are the embeddings of the objects in set A comparable to those in the set B?

6. A centered vector is one whose elements sum to 0. Show that for any valid (squared) distance matrix $\Delta = [\delta_{ij}^2]$ defined on a Euclidean space, the following must be true for any *centered* d-dimensional vector $\overline{y}$:

$$\overline{y}^T \Delta \overline{y} \leq 0$$

   (a) Suppose that you are given a symmetric matrix Δ in which all entries along the diagonal are 0s, and it always satisfies $\overline{y}^T \Delta \overline{y} \leq 0$ for all centered $\overline{y}$. Show that all entries of Δ must be nonnegative by using an appropriate choice of vector $\overline{y}$.

   (b) Discuss why a distance matrix Δ of (squared) Euclidean distances is always indefinite, unless it is a trivial matrix of 0s.

7. You have an $n \times n$ (dot-product) similarity matrix between training points and a $t \times n$ similarity matrix S_t between test and training points. The n-dimensional column vector of class variables is $\overline{y}$. Furthermore, the true $n \times d$ data matrix is D (i.e., $S = DD^T$), but you are not shown this matrix. As discussed in Chapter 4, the d-dimensional coefficient vector $\overline{W}$ of linear regression is given by the following:

$$\overline{W} = (D^T D + \lambda I)^{-1} D^T \overline{y}$$

Here, λ is the regularization parameter. Then, show the following results:

   (a) Let $\overline{p}$ be the t-dimensional vector of predictions for test instances. Show the following using the push-through identity of Problem 1.2.13:

$$\overline{p} = S_t(S + \lambda I)^{-1} \overline{y}$$

   (b) The previous exercise performs differentiation with respect to the weight vector. Show the result of (a) using the representer-based loss function discussed in this chapter, and differentiating with respect to $\overline{\beta}$.

(c) Take a moment to examine the coefficient vector obtained using the dual approach in Equation 6.14 of Chapter 6 and compare it to one in this exercise. What do you observe?

8. Derive the gradient descent steps for the primal formulation of logistic regression using the similarity matrix S and the representer theorem.

9. A student is given a square and symmetric similarity matrix S that is not positive semidefinite. The student computes the following new matrix:

$$S' = I - S + S^2$$

Is the new similarity matrix S' always positive semidefinite? If it is positive semidefinite, provide a proof. Otherwise, provide a counterexample.

10. A student used three different experimental ways to estimate $n \times n$ similarity matrices S_1, S_2, and S_3 among a set of n objects. These similarity matrices were all positive semidefinite. The student then computed the composite similarity matrix S as follows:

$$S = S_1 \odot S_2 + S_2 \odot S_3 + S_3 \odot S_1$$

Is the composite similarity matrix positive semidefinite?

11. Suppose $S(\overline{X}_1, \overline{X}_2) = S(\overline{X}_2, \overline{X}_1)$ is a symmetric similarity function between vectors $\overline{X}_1$ and $\overline{X}_2$, which is not necessarily a valid kernel. Then, is the similarity function $K(\overline{X}_1, \overline{X}_2) = S(\overline{X}_1, \overline{X}_2)^2$ a valid kernel? Either provide a proof or a counter-example.

12. Suppose that S is a positive semidefinite kernel, and a sub-linear element-wise function $f(\cdot)$ is applied to each element of S to create the new matrix $f(S)$. In each case, either show that $f(S)$ is positive semidefinite or provide a counter-example: (i) $f(x)$ is the natural logarithmic function, and S originally contains positive entries, and (ii) $f(x)$ is the non-negative square-root function, and S is originally nonnegative.

13. **Symmetric nonnegative factorization:** Consider a symmetric and nonnegative $n \times n$ matrix S that is factorized as $S \approx UU^T$, where U is a **nonnegative** $n \times k$ matrix for some $k < n$. The errors on the diagonal entries are ignored with the use of the objective function $\|W \odot (S - UU^T)\|^2$. Here, W is an $n \times n$ binary weight matrix that is set to 1 for all entries other than the diagonal ones. Derive a projected gradient-descent update for this box-constrained optimization problem. Discuss why the factor matrix U is more interpretable in the nonnegative case.

14. Show that at least one symmetric factorization $S = UU^T$ exists of a positive semidefinite matrix S, so that U is symmetric as well.

15. Express the loss function of the regularized L_2-loss SVM (cf. Chapter 5) using the representer theorem in terms of a similarity matrix. Here, we will convert the regularized Newton update of Chapter 5 to a representer update. The Newton update of Chapter 5 (using the same notations as the chapter) is as follows:

$$\overline{W} \Leftarrow (D_w^T D_w + \lambda I_d)^{-1} D_w^T \overline{y}$$

Here, the $n \times d$ matrix $D_w = \Delta_w D$ is a partial copy of data matrix D in **feature space**, except that it has zero rows for margin-satisfying rows of D. Δ_w is a binary diagonal matrix in which the ith diagonal entry is 1 only if the ith training instance of D is margin-violating. How would you compute Δ_w with representer coefficients? Use the push-through identity to show that this update is equivalent to the following with representer coefficients $\overline{\beta}$:

$$D^T\overline{\beta} \Leftarrow D^T \Delta_w (S_w + \lambda I_n)^{-1} \overline{y}$$

Note that one can implicitly implement this update using the following:

$$\overline{\beta} \Leftarrow \Delta_w (S_w + \lambda I_n)^{-1} \overline{y}$$

Here, $S_w = D_w D_w^T$ is a similarity matrix.

16. Consider loss functions of the following form (same notations as text):

$$\text{Minimize } J = \sum_{i=1}^{n} L(y_i, \overline{W} \cdot \overline{X_i}) + \frac{\lambda}{2}||\overline{W}||^2$$

Show that the gradient descent update is as follows:

$$\overline{W} \Leftarrow \overline{W}(1 - \alpha\lambda) - \sum_{i=1}^{n} \frac{\partial L(y_i, \overline{W} \cdot \overline{X_i})}{\partial(\overline{W} \cdot \overline{X_i})} \overline{X}_i^T$$

Now imagine that you only had similarities $S = [s_{ij}]$ available to you. Show that $\overline{W}$ can be updated indirectly by updating its representer coefficients $\overline{\beta}$ as follows:

$$\beta_i \Leftarrow \beta_i(1 - \alpha\lambda) - \alpha\frac{\partial L(y_i, t_i)}{\partial t_i} \quad \forall i \in \{1 \ldots n\}$$

Here, we define $t_i = \sum_{p=1}^{n} s_{ip}$.

Chapter 10

The Linear Algebra of Graphs

"If people do not believe that mathematics is simple, it is only because they do not realize how complicated life is." – John von Neumann

10.1 Introduction

Graphs are encountered in many real-world settings, such as the Web, social networks, and communication networks. Furthermore, many machine learning applications are conceptually represented as optimization problems on graphs. Graph matrices have a number of useful algebraic properties, which can be leveraged in machine learning. There are close connections between kernels and the linear algebra of graphs; a classical application that naturally belongs to both fields is *spectral clustering* (cf. Section 10.5).

This chapter is organized as follows. The next section introduces the basics of graphs and representations with adjacency matrices. The structural properties of the powers of adjacency matrices are discussed in Section 10.3. The eigenvectors and eigenvalues of graph matrices are discussed in Section 10.4. The linear algebra of graph clustering is explored in Section 10.5, whereas the linear algebra of graph ranking algorithms is explored in Section 10.6. The linear algebra of graphs with poor connectivity properties is discussed in Section 10.7. Machine learning applications of graphs are discussed in Section 10.8. A summary is given in Section 10.9.

10.2 Graph Basics and Adjacency Matrices

A graph, which is sometimes also referred to as a *network*, is a structure that is used to represent "relationships" among objects. The objects may be of any type, such as Web pages, social network actors, or chemical elements. Similarly, the relationships may be of a (corresponding) application-relevant type, such as Web links, social network friendships, or chemical bonds. For example, the chemical compound *Acetaminophen* and its associated graph structure are illustrated in Figure 10.1(a) and (b), respectively. A graph representing a social network is illustrated in Figure 10.1(c).

© Springer Nature Switzerland AG 2020
C. C. Aggarwal, *Linear Algebra and Optimization for Machine Learning*,
https://doi.org/10.1007/978-3-030-40344-7_10

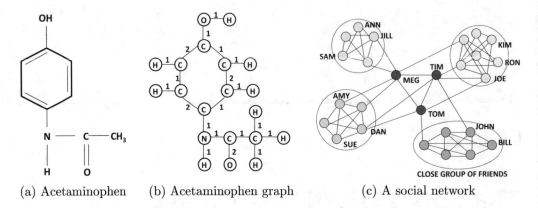

(a) Acetaminophen   (b) Acetaminophen graph   (c) A social network

Figure 10.1: Examples of undirected graphs

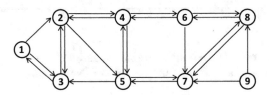

Figure 10.2: A directed graph

The objects in a graph are referred to as *vertices*, and the relationships among them are referred to as *edges*. A vertex is also sometimes referred to as a *node*. Throughout this book, we use the term "vertex" and "node" interchangeably. A graph G is denoted by the pair (V, E), where V is a set of vertices (nodes), and E is a set of edges. If the graph contains n vertices, it is assumed that the vertices are $V = \{1 \dots n\}$. Similarly, each edge $(i, j) \in E$ represents a connection between the vertices i and j.

Graphs may be *directed* or *undirected*. In directed graphs, each edge has a direction. For example, Web links have a direction from the source page to the destination page. The source of an edge is referred to as its *tail* and the destination is referred to as its *head*. Therefore, edges are shown using arrows, where the head corresponds to the end containing an arrowhead. An example of a directed graph is illustrated in Figure 10.2. On the other hand, edges do not have direction in undirected graphs. For example, a Facebook friendship link or a chemical bond does not have direction. All the graphs illustrated in Figure 10.1 are undirected graphs. An undirected graph may be converted into a directed graph by replacing each undirected edge with a pair of directed edges in opposite directions.

Finally, graphs may be *unweighted* or *weighted*. In an unweighted graph, an edge may be present or absent between two vertices, and there is no "strength" associated with a specific edge. In algebraic terms, the representation is binary and the relationship between a pair of vertices has a value of either 1 or 0, depending on whether or not an edge is present between the pair. On the other hand, in many applications, the relationship might have a weight associated with it. For example, a chemical bond has a strength corresponding to the number of shared electrons. Correspondingly, weights are shown in Figure 10.1(b). In an email network, the weight of an edge from one participant to another might correspond to the number of messages sent along that edge. Since weighted graphs are more general, the graphs discussed in this chapter are always associated with nonnegative weights.

In an undirected graph, the *degree* of a vertex is defined as the number of incident edges at that vertex. For example, in Figure 10.1(c), the degree of the vertex corresponding to Sam is 4. Since every edge is incident on two vertices, the sum of the degrees of the vertices in an undirected graph with m edges is always equal to $2m$. In the case of a directed graph, it makes sense to talk about the *indegree* and the *outdegree* of a vertex. The indegree of a vertex is the number of incoming edges at a vertex, whereas the outdegree is the number of outgoing edges. For example, the indegree of vertex 1 in Figure 10.2 is 1, whereas its outdegree is 2. The sum of the indegrees over all vertices and the sum of the outdegrees over all vertices are both equal to the number of edges m. This is because each edge is incident on exactly one vertex as an incoming edge, and it is incident on one vertex as an outgoing edge. All definitions of vertex degrees can be generalized to the weighted case by adding the weights of the edges instead of using a default weight of 1 for each edge.

Basic Structures in Graphs

A *walk* is any sequence $i_1, i_2, \ldots i_k$ of vertices, so that an edge exists from each i_r to i_{r+1}. In the case of directed graphs, the tail of the edge must be at i_r and the head must be at i_{r+1}. In undirected graphs, an edge can be traversed in both directions. There is no restriction on repetition of vertices within a walk. In Figure 10.2, the sequence $2, 3, 1, 2, 4$ of vertices is a walk. A *path* is any sequence of vertices $i_1, i_2, \ldots i_k$ of vertices, so that an edge exists from each i_r to i_{r+1} and there is no repetition of vertices. In directed graphs, the direction of the edges must be from i_r to i_{r+1}. Therefore, every path is a walk, but not vice versa. In Figure 10.2, the sequence $3, 1, 2, 4$ is a path. A cycle is any sequence of vertices $i_1 i_2 \ldots i_k$, so that an edge exists between each successive pair of vertices, $i_1 = i_k$, and there is no other repeating vertex. In other words, a cycle is a closed and directed "loop" of vertices in a directed graph. A cycle is also a special case of a walk. For undirected graphs, a cycle is simply a closed loop of undirected edges. In a directed graph, the direction of all edges in a cycle must be the same. Directed graphs that do not contain cycles are referred to as *directed acyclic graphs*. For example, consider a directed graph containing three vertices $\{1, 2, 3\}$, and the directed edges $(1, 2)$, $(1, 3)$, and $(2, 3)$. This graph does not contain any directed cycle and is therefore a directed acyclic graph.

A *subgraph* of a graph is any subset of vertices and edges in the graph. Note that if an edge is included in the subgraph, its end points must be included as well. The subgraph of a graph $G = (V, E)$ *induced* by a set of vertices $V' \subseteq V$ is the graph $G' = (V', E')$, in which $E' \subseteq E$ contains all edges between vertices in V'.

Connectivity and Diameter

An undirected graph is referred to as connected, if a path exists between each pair of vertices. An undirected graph that is not connected can be divided into a number of *connected components*. A connected component is a subset of vertices from the original graph, so that the subgraph induced by that vertex set is connected. Examples of both connected and unconnected undirected graphs are shown in Figure 10.3(a) and (b), respectively. The graph in Figure 10.3(b) has two connected components.

A directed graph is referred to as *strongly* connected, if a directed path exists between each pair of vertices in either direction. In other words, for any given pair of vertices $[i, j]$, a path must exist from i to j, and a path must also exist from j to i. For example, a graph corresponding to a single cycle of vertices is strongly connected. On the other hand, a single directed path of vertices or a directed acyclic graph is not strongly connected

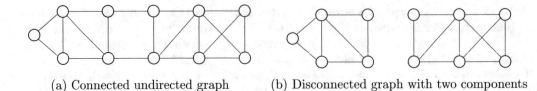

(a) Connected undirected graph   (b) Disconnected graph with two components

Figure 10.3: Examples of connected and disconnected undirected graphs

(because *directed* paths do not exist between specific *ordered* pairs of vertices). The graph in Figure 10.2 is not strongly connected, because a directed path does not exist from vertex 7 to vertex 9. As we will see later, strongly connected graphs have useful algebraic properties.

The *distance* or *shortest path* between a pair of vertices in a directed graph is defined as the least number of edges on a directed path between them. The *diameter* of a directed graph is defined as the largest distance between two vertices in the graph. Note that the distance from vertex i to vertex j might be different from that from vertex j to vertex i. Therefore, one needs to compute the distances between all $n(n-1)$ *ordered* pairs of vertices in the graph and compute the largest among them in order to compute the graph diameter. If no directed path exists between a particular pair of vertices, then the diameter of the graph is ∞. Therefore, a directed graph needs to be strongly connected in order for its diameter to be finite. For example, the diameter of the directed graph in Figure 10.2 is ∞ because no directed path exists from vertex 7 to vertex 9.

In *undirected* graphs, the shortest path distance from vertex i to j is the same as that from vertex j to i. If no path exists between a pair of vertices, it means that the graph is *disconnected*, and the distance between this vertex pair is ∞. The diameter of an undirected graph is the maximum of the shortest path distances between each pair of vertices. The diameter of a disconnected graph [like Figure 10.3(b)] is ∞.

Graph Adjacency Matrix

The adjacency matrix of an undirected graph is a special case of that of a directed graph, because each undirected edge can be replaced with two directed edges in opposite directions of equal weight. Therefore, we will first discuss the more general case of directed graphs.

For a directed graph containing n vertices and m edges, a square $n \times n$ matrix $A = [a_{ij}]$ is defined in which the value of a_{ij} is the weight of the edge from vertex i to vertex j. If no edge exists from vertex i to vertex j, the value of that entry is 0. Therefore, the adjacency matrix of a directed graph with m edges will contain m non-zero entries. In the case of unweighted graphs, all entries in the matrix are 0s or 1s. It is common for the diagonal entries of an adjacency matrix to be 0s, because self-loops are extremely uncommon in graphs. The adjacency matrix of a directed graph is usually asymmetric because a_{ij} is typically not the same as a_{ji}. On the other hand, an undirected graph with m edges will have $2m$ non-zero entries, and it will be symmetric because a_{ij} has the same value as a_{ji}. The symmetric nature of undirected graph adjacency matrices simplifies their linear algebra, because they have real-valued and orthonormal eigenvectors.

Normalized Adjacency Matrices

There are several ways in which graph adjacency matrices are normalized. The goal of normalization is to prevent a few vertices with many incident edges from dominating the

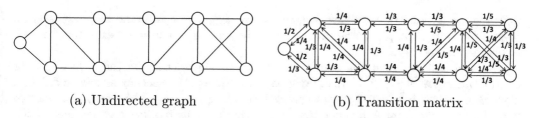

(a) Undirected graph   (b) Transition matrix

Figure 10.4: An undirected graph and its random walk graph. Note that asymmetric normalization makes a symmetric adjacency matrix asymmetric

algebraic properties of the graph. Most real-world graphs have *power-law degree distributions* [43], as a result of which the sum of the degrees of a tiny fraction of the vertices often form the vast majority of the sum of the degrees of all vertices in the full graph. As a result, the structure of the edges incident on these vertices dominate any type of analysis or the results of a machine learning algorithm applied to the entire network. This is undesirable because the structural behavior of high-degree nodes is often caused by spam and other irrelevant/noisy edges.

Some forms of normalization have a probabilistic interpretation, which are useful in real applications. The first type of normalization is *asymmetric* normalization, in which every row is normalized to sum to 1 unit. Therefore, we sum the elements of each row, and divide each element in that row by this sum. The result of this type of normalization is to create a *stochastic transition matrix*, that converts the adjacency matrix into the transition matrix of a *Markov chain*. This transition matrix defines a random walk at the graph, where the outgoing probabilities at each vertex define the probability of traversing along that edge. The resulting graph is referred to as a *random walk graph*. Note that this type of normalization results in asymmetric weights even for an undirected graph (for which the unnormalized adjacency matrix is symmetric). An example of asymmetric normalization is shown in Figure 10.4, where the original graph with binary edge weights is shown on the left, whereas the normalized graph (i.e., random walk graph) is shown on the right. It is noteworthy that this type of asymmetric normalization can also be applied to a directed graph. In such a case, the weight of each edge is divided by the sum of the weights of the *outgoing* edges at a vertex. The goal is again to interpret each edge weight as a random walk probability *out* of a given vertex.

Symmetric normalization is generally defined for undirected graphs. Therefore, one starts with a symmetric adjacency matrix and the goal is to preserve its symmetry in the normalization process. In symmetric normalization, we sum up the nonnegative entries of the ith row to create the sum δ_i. Since the matrix is symmetric, the sum of the elements of the ith column is also δ_i. In other words, we have the following:

$$\delta_i = \sum_{j=1}^{n} a_{ij} = \sum_{j=1}^{n} a_{ji}$$

In symmetric normalization, we divide each entry with the *geometric mean* of its row and column sums. The resulting similarity value s_{ij} is defined as follows:

$$s_{ij} \Leftarrow \frac{a_{ij}}{\sqrt{\delta_i \delta_j}}$$

Note that in asymmetric normalization, we always use $p_{ij} \Leftarrow a_{ij}/\delta_i$, and therefore the sum of each row is 1. Here, p_{ij} represents the probability of transition to vertex j from vertex i in the random walk graph.

One can also represent the above normalizations algebraically in the form of matrix multiplication. Let $A = [a_{ij}]$ be the original $n \times n$ (undirected) adjacency matrix, and Δ be a diagonal $n \times n$ matrix in which the ith diagonal entry is $\delta_i = \sum_j a_{ij}$. The matrix Δ is referred to as the *degree matrix* of A. It is noteworthy that the degree matrix incorporates information about the weights a_{ij} of edges; the values on the diagonal of Δ will be the aggregate *weights* of incident edges rather than the *number* of incident edges. We occasionally refer to the matrix Δ as the *weighted* degree matrix, although referring to it simply as "degree matrix" is more common. Let $P = [p_{ij}]$ be the asymmetrically normalized stochastic transition matrix, and $S = [s_{ij}]$ be the symmetrically normalized adjacency matrix. Then, the asymmetrically and symmetrically normalized matrices are defined as follows:

$$P = \Delta^{-1}A \qquad\qquad \text{[Asymmetric Normalization]}$$
$$S = \Delta^{-1/2}A\Delta^{-1/2} \qquad\qquad \text{[Symmetric Normalization]}$$

As we will see, these two related matrices play an important role in many network applications including clustering, classification, and *PageRank* computation.

10.3   Powers of Adjacency Matrices

Consider a binary adjacency matrix A of a directed graph in which each entry is either 0 or 1. The powers of A are related to the number of walks of specific lengths:

Property 10.3.1 *Let A be a binary adjacency matrix of a graph. The value of the (i,j)th entry in A^k is equal to the number of walks from i to j of length exactly k.*

This result can be shown using induction. It is easy to see that $A^1 = A$ contains all walks of length exactly 1. Now, if the (i,j)th entry of A^k contains the number of walks of length k from i to j, then the number of walks of length $(k+1)$ can be obtained by summing up the number of walks of length k from i to all vertices connected to j (i.e., all vertices with a value of 1 in the jth column of A). This value is precisely given by the (i,j)th entry of $A^k A = A^{k+1}$. What happens when the edges are weighted? In such a case, the value of A^k is equal to the number of weighted walks, where the contribution of each walk is equal to the product of the values of a_{ij} on that walk.

The number of walks of length *at most* t from i to j is contained in the matrix $W(t) = \sum_{k=0}^{t} A^t$. Note that A^0 is the identity matrix, and we are including it in the summation because every vertex is reachable from itself with a walk of length 0. In many cases, the zeroth order term is omitted because walks are assumed to have non-zero length. In general, this summation does not converge as t increases. For example, if A corresponds to a cycle of two vertices, it is easy to show the following:

$$A^r = \begin{bmatrix} 0 & 1 \\ 1 & 0 \end{bmatrix} \quad \text{[Odd } r]$$
$$A^r = \begin{bmatrix} 1 & 0 \\ 0 & 1 \end{bmatrix} \quad \text{[Even } r]$$

This is because there is only one way of walking from a vertex to itself in a specified number of even steps, and only one way of walking across the two vertices in a specified number of odd steps. An infinite summation over these matrices will not yield a converging summation. Furthermore, the entries of A^r could themselves blow up over an infinite summation (in some types of matrices).

However, by allowing for a decay factor $\gamma < 1$, it is possible for this summation to converge. From a semantic perspective, this means that a walk of length r is weighted by γ^r. Even though there might be more walks of greater length, the decay factor ensures that the infinite summation eventually converges. Clearly, the choice of γ required for convergence depends on the structural properties of the graph; interestingly, these structural properties can be captured by the eigendecomposition of the underlying adjacency matrix. Choosing γ less than the reciprocal of the largest absolute eigenvalue of A ensures that the powers of $(\gamma A)^k$ do converge over an infinite summation. In other words, we are causing a decay at the multiplicative factor of $\gamma < 1$, and then summing up the weights of all walks between a pair of vertices.

This result follows from the fact that if λ is an eigenvalue of A, then $\gamma\lambda$ is an eigenvalue of the matrix $A_\gamma = \gamma A$. This is because $\det(A_\gamma - \lambda\gamma I) = \gamma^n \det(A - \lambda I)$. As a result, by choosing γ to be less than the reciprocal of the largest absolute magnitude of the eigenvalues in A, the largest absolute magnitude of the eigenvalues of the matrix $A_\gamma = \gamma A$ is strictly less than 1. The following result can be easily shown:

Lemma 10.3.1 *Let A be a matrix for which all eigenvalues have absolute magnitude less than $1/\gamma$ for $\gamma > 0$. Then, the following can be shown:*

$$lim_{r \to \infty}(\gamma A)^r = 0$$

Proof Sketch: We denote γA with the matrix A_γ, and all its eigenvalues have absolute magnitude less than 1. Any matrix can be converted into Jordan normal form (cf. Section 3.3.3) with possibly complex eigenvalues as follows:

$$A_\gamma = VJV^{-1}$$

Here, J is an upper-triangular matrix in Jordan normal form in which the diagonal entries contain the (possibly complex) eigenvalues, each with magnitude less than 1. In such a case, we can show the following:

$$A_\gamma^r = VJ^rV^{-1}$$

As r goes to ∞, the matrix J^r can be shown[1] to go to 0. Therefore, the matrix A_γ^r converges to the zero matrix as well. ∎

This result provides an approach for computing the decay-weighted sum of walks between any pair of vertices.

Lemma 10.3.2 *Given a directed graph with adjacency matrix A in which the largest eigenvalue is less than $1/\gamma$, the weighted sum of all decayed walks between each pair of vertices is contained in the matrix $(I - \gamma A)^{-1} - I$.*

$$\sum_{r=1}^{\infty}(\gamma A) = (I - \gamma A)^{-1} - I$$

---

[1] The diagonal entries in the upper-triangular matrix J^r are powers of those in J, which will go to 0. Such a strictly triangular matrix is known to be nilpotent, and, therefore, J^r will eventually go to 0.

Proof Sketch: All eigenvalues of γA have magnitude less than 1. Each eigenvalue of γA has a corresponding eigenvalue in $(I - \gamma A)$, and the two eigenvalues sum to 1 with the same eigenvector. Therefore all eigenvalues of $(I - \gamma A)$ are non-zero, and the matrix is non-singular. In other words, we can multiply both sides of the above equation with $(I - \gamma A)$ without affecting the correctness of the above result. Multiplying each side with $(I - \gamma A)$ yields the matrix γA for both sides. This proves the result.   ∎

The matrix $(I - \gamma A)^{-1}$ is very useful because, the (i,j)th entry tells us about the level of indirect connectivity from vertex i to j even when this pair of vertices is not directly connected. This matrix contains all the *Katz measures* between pairs of vertices. The Katz measure is generally used for undirected graphs, although it can also be applied to directed graphs.

Definition 10.3.1 (Katz Measure) *Given the adjacency matrix A of an undirected graph, the Katz measure between vertices i and j is the (i,j)th entry of $(I - \gamma A)^{-1} - I$. Here, γ is a decay parameter used for computing the Katz measure.*

The Katz measure can be used for *link prediction*. In this problem, the goal is to discover pairs of vertices between which links are likely to form in the future in a graph of interest (e.g., social network). Clearly, if many (short) walks exist between a pair of vertices, links are more likely to form between them in the future. For example, one is more likely to form links with the friends of one's friends in a social networks.

This problem of link prediction can be viewed in a similar manner to that of the problem of recommendations with implicit feedback. To solve this problem, we first set a small subset of the non-zero entries in A to 0 to obtain A', and save those edges as a *validation set A_v*. We also add some of the zero edges in A to A_v. One can compute the inverse of $(I - \gamma A')$ at different values of γ and select the one at which the link predictions on the validation set A_v are the most accurate. Once the value of γ has been determined, we compute the matrix $(I - \gamma A)^{-1}$ (with all edges included) in order to rank the edges in the order of the likelihood of forming links.

The powers of the adjacency matrix can also be used to characterize the diameter and connectivity of both undirected and directed graphs. Note that if the diameter of a graph is d, then a path of length at most d exists between each pair of vertices. In other words, if $r_{ij} \leq d$ be the length of the shortest path between an arbitrary pair of vertices i and j, then the (i,j)th entry of $A^{r_{ij}}$ will be non-zero. This implies the following way of defining the diameter of a graph in terms of the powers of the adjacency matrix.

Property 10.3.2 *The diameter of a (directed or undirected) graph with adjacency matrix A is the smallest value of d for which all entries of the matrix $\sum_{k=0}^{d} A^k$ are non-zero. For any connected, undirected graph or strongly connected, directed graph, the value of d is at most $n - 1$.*

When an undirected graph is not connected or a directed graph is not strongly connected, the diameter of the graph is ∞. Since all graphs with finite diameter have a diameter value at most $n - 1$, this provides a simple approach for testing the connectivity of a (directed or undirected) graph with the use of powers of the adjacency matrix.

Lemma 10.3.3 (Connectivity of Undirected Graph) *An undirected graph with $n \times n$ adjacency matrix A is connected if and only if every entry of the matrix $\sum_{k=0}^{n-1} A^k$ is non-zero.*

Lemma 10.3.4 (Strong Connectivity of Directed Graph) *A directed graph with $n \times n$ adjacency matrix A is strongly connected if and only if every entry of the matrix $\sum_{k=0}^{n-1} A^k$ is non-zero.*

A simple approach for testing connectivity is to compute the matrix containing all Katz measures and checking if all nondiagonal entries are non-zero.

Undirected and directed graphs that are not connected/strongly connected will have missing edges between specific sets of vertices in the matrix $\sum_{k=0}^{n-1} A^k$. The missing edges in $\sum_{k=0}^{n-1} A^k$ will result in a specific type of block structure of the non-zero entries. For example, consider a directed graph in which the vertices can be divided into sets 1 and 2. Assume that the vertices in each of the sets 1 and 2 are strongly connected. Furthermore, edges exist from vertices in set 1 to vertices in set 2, but no edges exist in the other direction. This type of graph is not strongly connected and its adjacency matrix can always be converted into a *block upper-triangular form* by appropriately reordering the vertices:

$$A = \begin{bmatrix} A_{11} & A_{12} \\ 0 & A_{22} \end{bmatrix}$$

Note that the blocks along the diagonal are always square, but the blocks above the diagonal might not be square. It is not very difficult to verify that no matter how many times we exponentiate the matrix A, the lower block of zeros will stay as zeros. Similarly, a disconnected, undirected adjacency matrix can be converted into *block diagonal form* like the one below:

$$A = \begin{bmatrix} A_{11} & 0 \\ 0 & A_{22} \end{bmatrix}$$

It is relatively easy to verify that exponentiating this matrix any number of times will only result in the individual blocks being exponentiated:

$$A^k = \begin{bmatrix} A_{11}^k & 0 \\ 0 & A_{22}^k \end{bmatrix}$$

The two blocks A_{11} and A_{22} do not interact with one another in matrix multiplication.

10.4 The Perron-Frobenius Theorem

In the following, we will show some key properties of the eigenvectors of adjacency matrices in both directed and undirected graphs. The Perron-Frobenius theorem applies to the general case of *directed* graphs, with some additional simplifications for undirected graphs. All graph adjacency matrices (directed or undirected) have nonnegative entries, as a result of which the underlying eigenvectors have special properties. Undirected graphs have more interesting properties than directed graphs, because the symmetry in their adjacency matrices assures orthonormal eigenvectors and real eigenvalues. The adjacency matrices of directed graphs may contain both real and complex eigenvalues. This family of results is generally referred to as the *Perron-Frobenius theorem*.

Property 10.4.1 *The adjacency matrices of directed graphs might have one or more complex eigenvalues. Furthermore, a directed graph is not even guaranteed to have a diagonalizable adjacency matrix even after allowing for complex eigenvalues.*

As an example of non-diagonalizability, consider a directed graph of four nodes, which has bidirectional edges between each *consecutive* node pair $[i, i + 1]$ and a single unidirectional edge from node 1 to node 4:

$$N = \begin{bmatrix} 0 & 1 & 0 & 1 \\ 1 & 0 & 1 & 0 \\ 0 & 1 & 0 & 1 \\ 0 & 0 & 1 & 0 \end{bmatrix}$$

Without the unidirectional edge, the graph is symmetric and diagonalizable with real and orthonormal eigenvectors. However, the addition of a single edge from node 1 to node 4 makes the adjacency matrix non-diagonalizable. On computing $\det(N - \lambda I)$, it can be shown that this matrix has the characteristic polynomial $\lambda^4 - 3\lambda^2$, which corresponds to the eigenvalues $\{-\sqrt{3}, \sqrt{3}, 0, 0\}$. Therefore, the eigenvalue 0 is repeated. However, there is only one eigenvector $[0, 1, 0, -1]^T$ with eigenvalue 0, since the matrix has rank 3. Therefore, this adjacency matrix is not diagonalizable.

As an example of a directed graph with a diagonalizable adjacency matrix and complex eigenvalues, consider the following adjacency matrix, which contains a directed cycle of three vertices:

$$A = \begin{bmatrix} 0 & 0 & 1 \\ 1 & 0 & 0 \\ 0 & 1 & 0 \end{bmatrix} \tag{10.1}$$

Note that this matrix does have one real-valued eigenvector $[1, 1, 1]^T$ with an eigenvalue of 1. The other two eigenvalues can be shown to be $(-1 + i\sqrt{3})/2$ and $(-1 - i\sqrt{3})/2$, which are obviously complex. The corresponding eigenvectors are also complex. All eigenvalues can be shown to be the real and complex cube roots of unity because the characteristic polynomial is $\lambda^3 - 1$. It is noteworthy that we did get at least one real eigenvector-eigenvalue pair. Furthermore, this eigenvector is the largest eigenvector in absolute[2] magnitude, although the other two eigenvectors also have a magnitude of 1. This is not a coincidence. It can be shown that the adjacency matrix of any strongly connected directed graph will have at least one real eigenvector-eigenvalue pair, which is also the dominant pair. The adjacency matrix of a strongly connected graph is said to be *irreducible*.

Definition 10.4.1 (Irreducible Matrix) *The adjacency matrix of a directed graph is said to be irreducible, if and only if the underlying graph is strongly connected.*

An adjacency matrix that is not irreducible is said to be *reducible*. Note that if the graph is not strongly connected (i.e., its adjacency matrix is reducible), then either the graph is completely disconnected, or its vertices can be partitioned into two sets (i.e., a *cut* can be created) such that edges between the two sets point in only one direction. In other words, the matrix can be expressed in the following *block upper-triangular form*:

$$A = \begin{bmatrix} A_{11} & A_{12} \\ 0 & A_{22} \end{bmatrix} \tag{10.2}$$

In terms of walks on the directed graph, it means that once one moves from block 1 to block 2 via edges in A_{12}, it is impossible to come back to any vertex in block 1. In the above matrix it is assumed that the vertices are ordered so that all vertices in one component occur before all vertices in another component.

---

[2]The magnitude of the complex number $a + ib$ is $\sqrt{a^2 + b^2}$.

The Perron-Frobenius theorem applies only to strongly connected graphs, which are represented by irreducible adjacency matrices. The primary focus of this result is on the eigenvector of largest *magnitude*, which is also referred to as the *principal eigenvector*. Note that the left eigenvectors and right eigenvectors of an asymmetric matrix (like a directed graph) are different. The general version of the Perron-Frobenius theorem is stated as follows:

Theorem 10.4.1 (Perron-Frobenius Theorem for Directed Graphs) *Let A be a square, irreducible adjacency matrix with nonnegative entries for a directed graph containing n vertices. Then, one of the largest eigenvalues of A (in absolute magnitude) is always real-valued and positive (denoted by λ_{max}), and the multiplicity of this (positive) eigenvalue is 1. However, other complex or negative eigenvalues could exist with absolute magnitude λ_{max}. The following results also hold true:*

- *The unique left eigenvector and unique right eigenvector corresponding to the real and positive eigenvalue λ_{max} contains only real and strictly positive entries.*

- *The real and positive eigenvalue λ_{max} satisfies the following:*

$$average_i \sum_j a_{ij} \le \lambda_{max} \le max_i \sum_j a_{ij}$$

$$average_j \sum_i a_{ij} \le \lambda_{max} \le max_j \sum_i a_{ij}$$

 This means that for unweighted matrices, the largest eigenvalue lies between the average and maximum indegree (and outdegree). Therefore, the minimum of the maximum indegree and maximum outdegree can be used to provide an upper bound on λ_{max}.

- **Special case of stochastic transition matrices:** *Since stochastic transition matrices have a weighted outdegree of 1 for each vertex, the largest eigenvalue is 1 according to the above results. The corresponding right eigenvector of a stochastic transition matrix P is the n-dimensional column vector of 1s because each row of P sums to 1. The corresponding left eigenvector, referred to as the **PageRank** vector, is the solution to $\overline{\pi}^T P = \overline{\pi}^T$.*

It is noteworthy that even though the largest eigenvalue in magnitude is positive, complex or negative eigenvalues might exist with the same absolute magnitude λ_{max}. As a specific example, consider the directed cycle of three vertices in the adjacency matrix of Equation 10.1. The eigenvalues of this matrix are the three real or complex cube-roots of 1, which are 1, $(-1 + i\sqrt{3})/2$ and $(-1 - i\sqrt{3})/2$. All three roots have an absolute magnitude of 1. In general, it can be shown that a directed cycle of n vertices has n eigenvalues corresponding to the n real and complex nth roots of 1. These values can be shown to be $\exp(2i\pi t/n) = \cos(2\pi t/n) + i\sin(2\pi t/n)$ for $t \in \{0 \ldots n-1\}$. All real and complex eigenvalues have an absolute magnitude of 1. If the graph is reducible, there might be multiple eigenvectors corresponding to the largest (positive) eigenvalue. Furthermore, the principal eigenvector is no longer guaranteed to contain only positive entries in such a case.

A simplified version of the Perron-Frobenius theorem also applies to undirected graphs. In undirected graphs, the adjacency matrix is symmetric and therefore all eigenvectors are real. Furthermore, the graph is always strongly connected, when viewed as a directed network (with two directed edges replacing each undirected edge). Correspondingly, one can state the Perron-Frobenius theorem for undirected graphs as follows:

Corollary 10.4.1 (Perron-Frobenius Theorem for Undirected Graphs) *Let A be a nonnegative adjacency matrix for an undirected, connected graph containing n vertices. Then, the largest eigenvalue λ_{max} of A (in absolute magnitude) is always real-valued and positive. There is only one eigenvalue with value λ_{max}, although it is possible for a negative eigenvalue of the same magnitude to exist. The following results are true:*

- *All eigenvectors and eigenvalues are real. If the graph does not have self-loops, the sum of the eigenvalues is equal to the trace of A, which is 0. Therefore, some eigenvalues will always be negative. The eigenvalue λ_{max} is always at least as large as the most negative eigenvalue (in absolute magnitude).*

- *There is a single eigenvector corresponding to the eigenvalue λ_{max}. The eigenvector only has strictly positive entries (after multiplying with -1 if necessary). The uniqueness means that the solution to $A\overline{\pi} = \lambda_{max}\overline{\pi}$ is unique to within scaling.*

- *The largest eigenvalue satisfies the following:*

$$average_i \sum_j a_{ij} \leq \lambda_{max} \leq max_i \sum_j a_{ij}$$

This means that for unweighted matrices, the largest eigenvalue lies between the average and maximum degree.

Finally, the stochastic transition matrices of undirected graphs are not symmetric; however, they inherit some of the properties of undirected (symmetric) adjacency matrices from which they are derived. For example, the stochastic transition matrices of undirected graphs continue to have real eigenvectors and real eigenvalues like their symmetric counterparts.

Corollary 10.4.2 (Stochastic Transition Matrices of Undirected Graphs) *Let $P = \Delta^{-1}A$ be the normalized transition matrix of an undirected, connected graph with $n \times n$ adjacency matrix A and degree matrix Δ. The following results are true:*

- *All eigenvectors and eigenvalues are real. The largest eigenvalue in absolute magnitude is always unique and has a value of 1, although it is also possible to have an eigenvalue with value -1.*

- *The single right eigenvector with eigenvalue of 1 corresponds to an n-dimensional column of 1s. This vector is a valid right eigenvector because each row sums to 1 in the stochastic transition matrix.*

- *The left eigenvector solution to $\overline{\pi}^T P = \overline{\pi}^T$ is unique to within scaling. This vector is referred to as the **PageRank** vector and all components are strictly positive.*

Note that even though the eigenvector with largest eigenvalue of 1 is unique, other eigenvectors could exist with an eigenvalue of -1. Therefore, the eigenvector in largest *magnitude* is not unique. As a specific example, consider the stochastic transition matrix of an undirected graph with two vertices and a single edge between them. Both the adjacency matrix A and the stochastic transition matrix P have the following form:

$$A = P = \begin{bmatrix} 0 & 1 \\ 1 & 0 \end{bmatrix}$$

This graph has the same adjacency matrix as a directed cycle of two vertices. It has eigenvalues of -1 and $+1$.

The eigenvectors of normalized adjacency matrices are used extensively in machine learning for applications like spectral clustering and ranking. The left eigenvectors are used for applications like ranking nodes, whereas the right eigenvectors are used for spectral clustering. This will be the focus of subsequent sections.

10.5 The Right Eigenvectors of Graph Matrices

In this section, we will discuss the linear algebra of right eigenvectors of *undirected* graphs, which are used for a specific type of graph clustering, referred to as *spectral clustering*. Throughout this section, our assumption is that we are working with an undirected graph with a symmetric adjacency matrix, which ensures real eigenvalues and eigenvectors.

Consider an undirected graph with adjacency matrix A. Before applying any form of graph clustering, a preprocessing step is applied to the adjacency matrix A. The goal of this preprocessing step is to break the low-weight links in the graph and set their weights to 0. Doing so tends to differentially remove the inter-cluster links, and makes the clustering much cleaner. There are several ways in which this can be done, both of which require the use of a threshold parameter to decide which links to remove.

1. All non-zero entries in the adjacency matrix A that are less than a threshold value of ϵ are set to 0. In other words, if a_{ij} is less than ϵ, it is set to 0.

2. For each vertex i, we set all those entries a_{ij} to 0 if j is not among the κ-nearest neighbors of i for some user-driven parameter κ. However, this change will make the matrix asymmetric. Therefore, we prefer the use of *mutual κ-nearest* neighbors over the use of κ-nearest neighbors. In other words, we set a_{ij} and a_{ji} to 0 if and only if i and j are both κ-nearest neighbors of each other.

These types of methods are simply heuristics. One can use almost any reasonable heuristic to break weak links, as long as the final adjacency matrix A is symmetric. Keeping a symmetric adjacency matrix is important for ensuring real-valued eigenvectors and eigenvalues in spectral clustering. *Throughout this section, we will assume that the matrix A refers to an adjacency matrix in which the weak links have been removed.* An adjacency matrix is simply a form of a similarity matrix (or *kernel*) although it is not positive semidefinite because the diagonal elements are zeros. It is noteworthy that one often converts multidimensional data sets to graphs by treating data points as objects and the weights of edges as similarity values based on the Gaussian kernel (cf. Chapter 9). In fact, if we use such a graph without breaking the weak links, the resulting clustering is very similar to that of kernel k-means (with only small detail-oriented differences associated with feature selection and normalization).

The removal of weak links makes each entry in the adjacency matrix dependent on the values of other entries. This type of similarity function is referred to as a *data-dependent* kernel. Unlike a pure Gaussian kernel (cf. Chapter 9), a data-dependent kernel cannot be computed purely as a function of its two arguments and depends on the remainder of the similarity matrix. In this section, we provide three equivalent views of spectral clustering, which correspond to the kernel view, the Laplacian view, and the matrix factorization view.

10.5.1 The Kernel View of Spectral Clustering

Before discussing spectral clustering, we first discuss the most obvious kernel method obtained by treating the symmetric adjacency matrix of an undirected graph as a similarity matrix. This discussion sets the stage for understanding spectral methods as variants of this kernel method. We assume that the adjacency matrix (after removing the weak links) is denoted by $A = [a_{ij}]$. The $n \times n$ adjacency matrix A of an undirected graph is a symmetric similarity matrix, although it is not positive semidefinite because the sum of its eigenvalues (i.e., matrix trace) is 0. Nevertheless, one can always diagonalize the symmetric adjacency matrix as $A = Q\Lambda Q^T$ and simply use the top-k columns of Q as the engineered representation. Any clustering algorithm like k-means can be applied on the embedding (cf. Section 9.5.1.1). The lack of positive semidefiniteness of the similarity matrix (i.e., the adjacency matrix) might lead one to assume that this is not a kernel method. However, this is not quite correct; we could also condition the matrix A by adding the absolute value $\gamma > 0$ of the most negative eigenvalue to each diagonal entry. The matrix $A + \gamma I = Q(\Lambda + \gamma I)Q^T$ is a positive semidefinite matrix *with exactly the same eigenvectors*. The eigenvalues are not used. This is slightly different from kernel k-means (cf. Section 9.5.2.1), which implicitly scales the eigenvectors with the square-root of the eigenvalues via the kernel trick. Spectral clustering ignores the scaling effect of eigenvalues, and works with *whitened* representations (cf. Section 7.4.6 of Chapter 7). Furthermore, it only uses the top-k eigenvectors, thereby replacing soft eigenvalue weighting with discrete selection. These types of normalization and feature selection differences always occur in cases where one uses feature engineering (as in spectral methods) rather than the kernel trick.

The approach discussed above is a kernel method (related to spectral clustering), but it is not precisely spectral clustering. One problem with using adjacency matrices (as similarity matrices) directly is that the entries of the matrix are dominated by a few vertices. Most real-world graphs satisfy *power-law degree distributions* [43] in which a tiny fraction (typically less than 1%) of the vertices account for most of the edges in the graph. As a result, the embedding is dominated by the topological structure of a small fraction of vertices, which is undesirable. Spectral clustering solves this problem with *vertex degree normalization*.

In symmetric normalization, we compute the degree matrix Δ, which is an $n \times n$ diagonal matrix containing the degree $\delta_i = \sum_j a_{ij}$ on the ith diagonal entry. Each entry a_{ij} is divided by the geometric mean of δ_i and δ_j in order to reduce the influence of high-degree vertices. As discussed in Section 10.2, the symmetrically normalized similarity matrix S is defined as follows:

$$S = \Delta^{-1/2} A \Delta^{-1/2} \tag{10.3}$$

We can diagonalize this similarity matrix $S = Q_{(s)}\Lambda Q_{(s)}^T$, and then use the top-$k$ columns of $Q_{(s)}$ (i.e., largest eigenvectors) as the $n \times k$ matrix $Q_{(s),k}$ containing the embedding. We subscript the matrix $Q_{(s),k}$ containing the embedding with '(s)' to emphasize symmetric normalization. The ith row of $Q_{(s),k}$ contains the k-dimensional embedding of the ith vertex. This embedding is referred to as the *Ng-Jordan-Weiss* embedding [98]. Any clustering algorithm like k-means can be applied to this embedding. It is also common to normalize each *row* of the $n \times k$ matrix $Q_{(s),k}$ to unit norm just before applying the k-means algorithm. Note that normalizing each row of $Q_{(s),k}$ will result in an embedding matrix in which the columns are no longer normalized.

A related variation of this approach is the *Shi-Malik algorithm* [115], which uses the stochastic transition matrix instead of symmetrically normalized matrix. In other words, the normalization is asymmetric:

$$P = \Delta^{-1} A$$

One can also view P as a similarity matrix; however, the main problem is that it is asymmetric; therefore, it does not even make sense to talk about positive semidefiniteness.

As discussed in Section 9.3.3 of Chapter 9, it is still possible to extract embeddings from asymmetric similarity matrices if one is willing to relax the definition of similarity to allow asymmetry. In this case, asymmetric decomposition is used:

$$P = Q_{(a)} \Lambda Q_{(a)}^{-1}$$

Note that this decomposition always contain real-valued eigenvectors and eigenvalues according to the Perron-Frobenius theorem for stochastic transition matrices (cf. Corollary 10.4.2). The columns of $Q_{(a)}$ contain the right eigenvectors of P. We subscript the matrix with '(a)' to emphasize the fact that it is extracted from an asymmetric similarity matrix. The top-k columns of $Q_{(a)}$ are extracted in order to create an $n \times k$ embedding matrix $Q_{(a),k}$. The ith row of this matrix contains the k-dimensional embedding of the ith vertex, and it is referred to as the Shi-Malik embedding. Any off-the-shelf clustering algorithm can be applied to this embedding. We make an observation about the top eigenvector in this embedding, which is also stated in the Perron-Frobenius result for stochastic matrices (cf. Corollary 10.4.2):

Property 10.5.1 *The largest eigenvector in the Shi-Malik embedding is a column of 1s.*

This particular eigenvector is not very informative from a clustering point of view and is sometimes discarded (although including it does not seem to make much of a difference).

10.5.1.1 Relating Shi-Malik and Ng-Jordan-Weiss Embeddings

The Shi-Malik and Ng-Jordan-Weiss embeddings are almost identical– in fact, one can obtain either embedding from the other with a simple post-processing step. This is because the symmetrically normalized matrix S and transition matrix P are related as $S = \Delta^{1/2} P \Delta^{-1/2}$, which makes the matrices similar. Therefore, their eigenvalues are the same and the eigenvectors are related.

Lemma 10.5.1 *Let Δ be the degree matrix of adjacency matrix A. Let S and P be the symmetrically normalized and the stochastic transition matrices of A. Then, $\overline{x}$ is an eigenvector of P with eigenvalue λ, if and only if $\sqrt{\Delta}\overline{x}$ is an eigenvector of S with eigenvalue λ.*

Proof: We show that both of the above statements are true if and only if $\overline{x}$ is a generalized eigenvector of A satisfying $A\overline{x} = \lambda\Delta\overline{x}$.

First, we note that $P\overline{x} = \lambda\overline{x}$ is true if and only if $\Delta^{-1}A\overline{x} = \lambda\overline{x}$, which is the same as saying that $A\overline{x} = \lambda\Delta\overline{x}$.

Second, we note that $S[\sqrt{\Delta}\overline{x}] = \lambda[\sqrt{\Delta}\overline{x}]$ is true if and only if $\Delta^{-1/2}A\overline{x} = \lambda[\sqrt{\Delta}\overline{x}]$, which is the same as saying that $A\overline{x} = \lambda\Delta\overline{x}$. This completes the proof. ∎

Since the first eigenvector of the Shi-Malik embedding is a column of 1s, it follows that the first eigenvector of the Ng-Jordan-Weiss embedding is proportional to $\sqrt{\Delta}[1, 1, \ldots 1]^T$, which is the same as $[\sqrt{\delta_1}, \sqrt{\delta_2}, \ldots, \sqrt{\delta_n}]^T$.

Corollary 10.5.1 *The first eigenvector of the symmetrically normalized adjacency matrix $S = \Delta^{-1/2}A\Delta^{-1/2}$ is proportional to an n-dimensional vector containing the square-roots of the weighted vertex degrees.*

An important point is that the Ng-Jordan-Weiss embedding normalizes the rows before applying k-means, whereas the Shi-Malik approach does not normalize the rows before applying k-means. In other words, the former is *row-normalized* before applying k-means, whereas the latter is *column-normalized* before applying k-means. This implies the following unified extraction of both embeddings using the symmetric kernel matrix $S = \Delta^{-1/2}A\Delta^{-1/2}$:

Lemma 10.5.2 *Let R be the $n \times k$ matrix containing the top-k (unit normalized) eigenvectors of the symmetric matrix $S = \Delta^{-1/2} A \Delta^{-1/2}$ in its columns. Then, both the Shi-Malik and Ng-Jordan-Weiss embeddings can be obtained from R using the following postprocessing steps:*

- *The Ng-Jordan-Weiss embedding is obtained by normalizing each row of the $n \times k$ matrix R to unit norm.*

- *The Shi-Malik embedding is obtained by normalizing each column of the $n \times k$ matrix $\Delta^{-1/2} R$ to unit norm.*

This observation shows that both methods are very similar, and differ only in terms of the minor post-processing steps of scaling/normalizing the rows/columns of the same matrix.

10.5.2   The Laplacian View of Spectral Clustering

The Laplacian view of spectral clustering is the most popular way of presenting it in most textbooks and survey papers [84]. We have not chosen this presentation of spectral clustering as the primary one because the relationship of spectral clustering to kernel methods is more fundamental. Nevertheless, we will briefly present the Laplacian view and discuss its relationship to the kernel view. An additional contribution of the Laplacian view (beyond the kernel view) is an unnormalized variant of the algorithm (although it is rarely used).

10.5.2.1   Graph Laplacian

Let A be the adjacency matrix of an undirected graph, and Δ be its degree matrix. The notations P and S correspond to the stochastic transition matrix and the symmetrically normalized adjacency matrix, respectively. The Laplacian is almost always used in undirected graph applications, although one can easily extend the definitions below to directed graphs. There are three types of Laplacians that are commonly used in graph applications. The first is the *unnormalized* Laplacian L:

$$L = \Delta - A \tag{10.4}$$

The Laplacian of an undirected graph is symmetric because both Δ and A are symmetric. The Laplacian is always a singular matrix because the sum of each row is 0. Therefore, one of the eigenvalues of the Laplacian is 0.

The asymmetrically normalized Laplacian L_a applies a one-sided normalization to the matrix L introduced above:

$$L_a = \Delta^{-1} L = \Delta^{-1}(\Delta - A) = I - P \tag{10.5}$$

Here, P is the stochastic transition matrix $P = \Delta^{-1} A$, which was introduced slightly earlier. Therefore, the asymmetrically normalized Laplacian is closely connected to the stochastic transition matrix P. This Laplacian is not symmetric, because P is not symmetric. The symmetrically normalized Laplacian L_s is defined in a similar way, except that two-sided normalization is applied to L:

$$L_s = \Delta^{-1/2} L \Delta^{-1/2} = \Delta^{-1/2}(\Delta - A)\Delta^{-1/2} = I - S \tag{10.6}$$

Therefore, the symmetrically normalized Laplacian is closely related to the symmetric similarity matrix S.

The graph Laplacian has some interesting interpretations in terms of embedding vertices in multidimensional space. In the simplest case, consider a setting in which each vertex i is embedded to the real number x_i. Therefore, one can create a normalized vector $\overline{x} = [x_1 \ldots x_n]^T$, so that $\overline{x}^T \overline{x} = 1$. First, let us examine the unnormalized Laplacian L of Equation 10.4. One can show the following result:

Lemma 10.5.3 *If L is the unnormalized Laplacian of an undirected adjacency matrix, then $\overline{x}^T L \overline{x}$ is proportional to the weighted sum of square distances of each pair (x_i, x_j) with weight a_{ij}. In other words, we have the following:*

$$\overline{x}^T L \overline{x} = \frac{1}{2} \sum_{i=1}^{n} \sum_{j=1}^{n} a_{ij} (x_i - x_j)^2$$

Proof: In order to show this result, we can expand the expressions on both sides of the equation in the statement of the lemma and examine the coefficient of each term on both sides. For $i \neq j$, it can be shown that the coefficient of any term of the form $x_i x_j$ is $-(a_{ij} + a_{ji})$ on both sides. On the other hand, for any i, it can be shown that the coefficient of x_i^2 is $\frac{1}{2} \sum_{j=1}^{n} (a_{ij} + a_{ji})$. ∎

There are several observations that one can make from the above result. These observations are used in various ways in machine learning applications on graphs:

1. An immediately obvious observation is that $\overline{x}^T L \overline{x}$ is nonnegative for any $\overline{x}$ according to Lemma 10.5.3. Therefore, it is positive semidefinite.

2. The fact that the unnormalized graph Laplacian is positive semidefinite can be easily extended to the symmetrically normalized graph Laplacian (see Problem 10.5.1). Even though the notion of positive semidefiniteness is defined only for symmetric graphs, it can also be shown that the asymmetrically normalized graph Laplacian has nonnegative eigenvalues and satisfies $\overline{x}^T L_a \overline{x} \geq 0$ for all $\overline{x}$. In other words, it satisfies an extended notion of positive semidefiniteness for asymmetric graphs. Each type of Laplacian always has one of its eigenvalues as 0, because Laplacians are singular matrices with a null space of rank at least 1. The null space has rank exactly 1 if the graph is connected. Since Laplacians always have nonnegative eigenvalues, it follows that 0 is the smallest eigenvalue, which is unique for connected graphs.

3. Finding a unit vector $\overline{x}$ that minimizes $\overline{x}^T L \overline{x}$ will find a 1-dimensional embedding of the vertices, so that vertex pairs $[i, j]$ connected by edges (i, j) of heavy weight a_{ij} are close together in terms of the value of $(x_i - x_j)^2$. This can be easily inferred from the weighted sum-of-squares interpretation of $\overline{x}^T L \overline{x}$ in Lemma 10.5.3. In other words, this type of objective function finds an embedding that is friendly to clustering, which is why it is useful for spectral clustering.

The following result can be shown relatively easily by using Lemma 10.5.3 and some properties of determinants.

Problem 10.5.1 *Let L, $L_a = \Delta^{-1} L$, and $L_s = \Delta^{-1/2} L \Delta^{-1/2}$ be the unnormalized, asymmetrically normalized, and symmetrically normalized Laplacians of an undirected graph with nonnegative weights on edges. Show the following: (i) The value of $\overline{x}^T L_s \overline{x}$ is always nonnegative for any $\overline{x}$; (ii) the eigenvalues of L_s and L_a must be the same; and (iii) The value of $\overline{x}^T L_a \overline{x}$ is always nonnegative.*

The overall approach for spectral clustering works as follows:

1. Select one of the three Laplacians L, L_a, and L_s. Find the **smallest** k eigenvectors of the Laplacian, and create an $n \times k$ matrix Q_k, whose columns contain these eigenvectors after scaling them to unit norm. Each row of Q_k corresponds to a k-dimensional embedding of a vertex.

2. In the event that the symmetric Laplacian L_s was chosen in the first step, perform the additional step of normalizing each row of Q_k to unit norm.

3. Apply a k-means algorithm to the embedding representations of the different vertices.

An important point is that we use the large eigenvectors of similarity matrices, whereas we use the *small* eigenvectors of the Laplacian. This is not surprising because of the relationship between the two. For examples, the symmetrically normalized adjacency matrix S is related to the symmetric Laplacian L_s as $L_s = I - S$. As a result, *the eigenvectors of the two matrices are identical and the corresponding eigenvalues sum to 1.* We summarize these results in terms of the equivalence of the kernel view and the Laplacian view of spectral clustering.

Lemma 10.5.4 (Equivalence of Kernel and Laplacian View) *Let A be an $n \times n$ adjacency matrix, let P be its stochastic transition matrix, and S be its symmetrically normalized adjacency matrix. Let $L_a = I - P$ and $L_s = I - S$ be the asymmetric and symmetric Laplacians, respectively. Then, the following are true:*

- *The **largest** k eigenvectors of the stochastic transition matrix P are the same as the **smallest** k eigenvectors of asymmetric Laplacian L_a.*

- *The **largest** k eigenvectors of the symmetrically normalized adjacency matrix S are the same as the **smallest** k eigenvectors of symmetric Laplacian L_s.*

Proof: We show the result in the case of the symmetric Laplacian S. The proof for the asymmetric Laplacian is similar. Thd eigenvector of any matrix S with eigenvalue λ is also an eigenvector of the matrix $(I - S)$ with eigenvalue $1 - \lambda$. This is because $S\overline{x} = \lambda \overline{x}$ is true if and only if $(I - S)\overline{x} = (1 - \lambda)\overline{x}$ is true. Therefore, large eigenvectors of S correspond to small eigenvectors of $L_s = (I - S)$.   ∎

It is noteworthy that the small eigenvectors of the *unnormalized* Laplacian L are not exactly the same as the large eigenvectors of the adjacency matrix A. In any case, the use of unnormalized variants is relatively uncommon in most practical applications.

10.5.2.2   Optimization Model with Laplacian

The fact that the small eigenvectors of the Laplacian are used for embedding can also be derived with the use of an optimization model involving the Laplacian. In fact, this approach is the popular presentation of most textbooks and surveys [84]. We recap the result from Lemma 10.5.3, according to which $\overline{x}^T L \overline{x}$ provides the weighted sum-of-square distances between pairs of data points:

$$\overline{x}^T L \overline{x} = \frac{1}{2} \sum_{i=1}^{n} \sum_{j=1}^{n} a_{ij} (x_i - x_j)^2$$

Here, $\overline{x}$ is an n-dimensional vector, which contains one coordinate for each vertex. Note that one can easily extend this result to a k-dimensional embedding of each vertex by using

k vectors $\overline{x}_1 \ldots \overline{x}_k$. In this case, the n-dimensional vector $\overline{x}_i$ contains the ith coordinate of the n different vertices. In such a case, the sum over the k different values of $\overline{x}_i^T L \overline{x}_i$ provides the weighted sum-of-square Euclidean distances for the embedding. Therefore, a clustering-friendly embedding may be defined by the following optimization model:

$$\text{Minimize} \ \sum_{i=1}^{k} \overline{x}_i^T L \overline{x}_i$$

subject to:

$$\|\overline{x}_i\|^2 = 1 \ \ \forall i \in \{1 \ldots k\}$$
$$\overline{x}_1 \ldots \overline{x}_k \ \text{are mutually orthogonal}$$

This optimization model tries to find a k-dimensional embedding of each vertex, so that the weighted sum-of-square Euclidean distances is minimized. The above model is a special case of the norm-constrained optimization problem discussed in Section 3.4.5 of Chapter 3. As discussed in Section 3.4.5, the smallest k eigenvectors of L provide a solution to this optimization problem. Note that one can use exactly the same model for the symmetrically normalized Laplacian by replacing L with L_s in the above model. The case of the asymmetric Laplacian is slightly different, because the problem boils down to the following:

$$\text{Minimize} \ \sum_{i=1}^{k} \overline{x}_i^T L \overline{x}_i$$

subject to:

$$\overline{x}_1 \ldots \overline{x}_k \ \text{are} \ \Delta\text{-orthonormal}$$

The only difference from the original optimization problem is that the vectors are Δ-orthonormal. The notion of Δ-orthonormality is defined as follows:

$$\overline{x}_i^T \Delta \overline{x}_j = \begin{cases} 1 & i = j \\ 0 & i \neq j \end{cases}$$

Imposing Δ-orthonormality de-emphasizes the impact of high-degree vertices. The optimal solution can be shown to be the smallest k eigenvectors of $\Delta^{-1} L$. We leave the proof of this result as a practice problem.

Problem 10.5.2 *Show that the optimum solution to the optimization model of asymmetric spectral clustering corresponds to the smallest eigenvector of $L_a = \Delta^{-1} L$.*

The key hint in the above problem is to use a variable transformation, wherein each $\overline{x}_i = \sum_{j=1}^{n} \beta_{ij} \overline{p}_j$ is expressed as a linear combination of the basis system of unit eigenvectors of $\Delta^{-1} L$. Subsequently, one has to solve for the coefficients β_{ij}, while transforming the objective function and the constraints in terms of these coefficient variables. Show that the eigenvectors $\overline{p}_1 \ldots \overline{p}_n$ are Δ-orthogonal, and use it to simplify the objective function in terms of the different values of β_{ij}. It can be shown that all values of β_{ij} will be either $1/\sqrt{\Delta}$ or 0.

10.5.3   The Matrix Factorization View of Spectral Clustering

Spectral clustering is a form of matrix factorization. Both asymmetric and symmetric spectral clusterings are obtained using the following factorizations:

$$S = \underbrace{Q_{(s)}}_{U} \underbrace{\Lambda Q_{(s)}^T}_{V^T} \quad \text{[Symmetric]}$$

$$P = \underbrace{Q_{(a)}}_{U} \underbrace{\Lambda Q_{(a)}^{-1}}_{V^T} \quad \text{[Asymmetric]}$$

In other words, both factorizations can be expressed in the form UV^T. We can also approximately factorize either S or P into UV^T using the gradient-descent methods of Chapter 8. Note that both U and V are $n \times k$ matrices, where k is the rank of the factorization. Subsequently, the k-dimensional rows of U and V can be concatenated in order to create the $2k$-dimensional embedded representations of each vertex. The result will not be exactly the same as that of spectral clustering, but this generalized approach will often provide similar results to spectral clustering. This generalized view is useful in cases where the direct application of spectral clustering is not possible. For example, in the case of directed graphs, the adjacency matrix may not be diagonalizable with real-valued eigenvectors and eigenvalues. In such cases, we can use generalized forms of the factorization on the directed graph.

Let A be an adjacency matrix of a directed graph. As in the case of undirected adjacency matrices, weak links are removed by keeping only those edges, which are *both* among the top-k incoming edges and the top-k outgoing edges of the two vertices at their end points. Note that the matrix A will not be symmetric either before or after removal of the weak links (since the graph is directed to begin with). Let δ_i^{in} be the weighted indegree of the ith vertex, which is obtained by adding the (possibly non-binary) elements of the ith column of A. Similarly, let δ_i^{out} be the weighted outdegree of the ith vertex, which is obtained by summing the values in the ith row of A. One can create corresponding $n \times n$ diagonal matrices denoted by Δ_{in} and Δ_{out}, whose diagonal elements are the δ_i^{in} and δ_i^{out}, respectively. Then, each edge (i, j) is normalized using the geometric mean of outdegree at i and the indegree at j:

$$a_{ij} \Leftarrow \frac{a_{ij}}{\sqrt{\delta_i^{out}} \sqrt{\delta_j^{in}}}$$

One can also write this relationship in matrix form with the use of the normalized matrix N:

$$N = \Delta_{out}^{-1/2} A \Delta_{in}^{-1/2}$$

Once the matrix N has been computed, we can factorize it as $N \approx UV^T$ using any of the methods discussed in Chapter 8. Here, U and V are $n \times k$ matrices, where k is the rank of the factorization. The ith row of U provides the *outgoing factor* of the ith vertex (or *sender* factor), and is related to some of the centrality measures discussed in the next section. The ith row of V provides the *incoming factor* of the ith vertex (or *receiver* factor). One can concatenate these representations to create a $2k$-dimensional representation of each vertex. Subsequently, this representation can be used for clustering.

10.5.3.1   Machine Learning Application: Directed Link Prediction

One can use the aforementioned approach for directed link prediction. This type of approach can be useful for link prediction in directed follower-followee networks like Twitter. Once the normalized matrix N has been factorized as $N \approx UV^T$, the reconstruction UV^T predicts the directed links of the adjacency matrix. Normally, link prediction is required in undirected graphs, although some applications may also require the prediction of links in directed graphs. It is important to note that the (i, j)th entry in UV^T may not be the same as the (j, i)th entry in UV^T. In other words, the prediction of links is direction-sensitive. After all, the probability of a teenager following a famous rock star on Twitter is not the same as that of the rock star following the teenager.

10.5.4 Which View of Spectral Clustering Is Most Informative?

This section provides multiple presentations of spectral clustering. In the following, we provide a broader perspective on these different views of spectral clustering:

1. **The kernel view with adjacency matrices:** The *normalized* variations of spectral clustering can be viewed as special cases of the similarity-based clustering methods with explicit feature engineering (cf. Section 9.5.1.1 of Chapter 9). The main difference between spectral clustering and this family of methods is only in terms of how the matrix is preprocessed to remove weak links. The main advantage of the kernel view is that it provides a unified view with all the other kernel methods we have seen so far in Chapter 9. A spectral method is simply an equal citizen of the vast family of kernel methods such as kernel k-means – nothing more and nothing less. The main distinguishing characteristic is the heuristic sparsening/normalization of the similarity matrix and the normalization (whitening) of engineered features.

2. **The Laplacian view:** The use of the Laplacian is the dominant treatment in popular expositions of spectral clustering [84]. Because of the different nature of this treatment as compared to kernel or factorization methods, spectral clustering is often viewed in a very different light than other related embedding methods. The Laplacian view is often interpreted as an elegant *discrete* optimization problem of finding minimum cuts in a normalized graph [115]. However, the actual optimization problem of spectral clustering is only a *continuous approximation* of this problem, which could be an arbitrarily poor approximation of the discrete version.

3. **The matrix factorization view and its extensions:** Asymmetric forms of spectral clustering can be viewed in the context of matrix factorization. Most importantly, this point of view provides a generalization of the approach to clustering *directed* graphs, which is not possible with the vanilla version of spectral clustering.

An important principle in understanding a method well is to understand its relationship with other similar methods. This type of understanding promotes useful extensions and applications. Spectral clustering is simply *a special case of a kernel clustering method with explicit feature engineering and a data-dependent kernel.* Because of its need to postprocess the features extracted from the kernel before applying k-means on the features, it cannot be used in conjunction with the kernel trick (like kernel k-means).

10.6 The Left Eigenvectors of Graph Matrices

As discussed in the previous section, the right eigenvectors of the stochastic transition matrix of the adjacency matrix A are used to create the Shi-Malik embedding for clustering. In this section, we will explore the left eigenvectors of the stochastic transition matrix.

However, there are several differences from the exposition in the previous section. First, we will examine the characteristics of only the *principal* (i.e., largest) eigenvector, which has an eigenvalue of 1 according to the Perron-Frobenius theorem. Second, since we will be examining applications associated with the graph structure of the Web (rather than clustering), the focus will be on directed graphs rather than undirected graphs. After all, Web page linkage structure is not symmetric. Finally, the clustering application discussed in the previous section always modifies the adjacency matrix in order to remove weak links. This modification is not made for the applications discussed in this section. Rather a different type of modification is used, which focuses on making the directed graph strongly connected.

In many applications such as the Web, one is looking for vertices with a high level of *prestige*. Intuitively, a Web page has a high level of prestige if many Web pages point to it. However, simply using the number of Web pages pointing to a page might be a deceptive indicator of its prestige, because the pages pointing to it might be of low quality themselves. Therefore, one typically wants to discover pages that are pointed to by other high prestige pages. One can model this type of recursive relationship by using the notion of *random walks* in the graph.

Imagine a directed graph with adjacency matrix A, weighted degree matrix Δ, and stochastic transition matrix $P = \Delta^{-1}A$. One can interpret each entry p_{ij} in P as the probability of a transition of random surfer from Web page i to Web page j, under the assumption that the surfer selects the vertices outgoing from i using the vector of transition probabilities $[p_{i1}, p_{i2}, \ldots p_{in}]$. Note that these transition probabilities sum to 1. The *PageRank*-based prestige of a vertex (Web page) is defined as the steady-state probability of a random surfer visiting that vertex (Web page). Interestingly, the process that we just described is precisely the transition process of a *Markov chain*. A Markov chain consists of a set of states (vertices) along with a set of transitions (edges with probabilities) among them. Therefore, a Markov chain is perfectly described by the graph structures introduced in this section. Furthermore, a Markov chain has a well-defined procedure for finding steady state probabilities in terms of eigenvectors of its transition matrix.

One question that arises is as to the conditions under which a Markov chain has steady-state probabilities that are independent of the starting vertex of the random surfer. Such Markov chains are referred to as *ergodic* and they must satisfy the following condition:

Definition 10.6.1 (Ergodic Markov Chain) *A Markov chain is defined as ergodic, if its transition matrix is strongly connected.*

Graphs that are not strongly connected might have steady-state probabilities for a walk that depend on the starting point of the walk. For example, consider the directed graph in Figure 10.5(b), which is not strongly connected. In this case, starting a random walk at vertex 1 will result in steady-state probabilities that are distributed only among the vertices in the component A1. However, starting the walk at vertex 6 could lead to either of the components A1 or A2.

An immediate problem that arises because of this requirement is that many real applications do not result in directed graphs that are strongly connected. For example, the Web is certainly not a directed graph that is strongly connected. When you set up a new Website, the chances are that you might point to many other Websites, but no one might know of your Website or point to it. This can be a problem in terms of computing steady-state probabilities.

This problem is solved with the use of *restart probabilities*. In each step, the random surfer is allowed to reset to a completely random vertex in the network with probability

$\alpha < 1$ and continue with the random walk with probability $(1 - \alpha)$. The value of α is a hyper-parameter that is chosen in an application-specific manner. The transition matrix P is obtained from the old transition matrix P_o as follows:

$$P \Leftarrow (1 - \alpha)P_o + \alpha M/n$$

Here, M is an $n \times n$ matrix containing 1s. The matrix M/n is a transition matrix in which one can move from any vertex to another with probability $1/n$. Therefore, this is a strongly-connected restart matrix. The final transition matrix P is a weighted combination of the original transition matrix and the restart matrix. Henceforth, we will always assume that the transition matrix is strongly connected, and the restart matrix has been incorporated as a preprocessing step.

10.6.1 PageRank as Left Eigenvector of Transition Matrix

Let π_i be the steady-state probability of vertex i under the random walk model. Then, the probability π_i of visiting a particular vertex i is given by the sum of the probabilities of transitioning into that vertex from each of its incoming vertices. The probability of transitioning into vertex i from vertex j is given by $\pi_j p_{ji}$. We can write this relationship mathematically for each vertex $i \in \{1, \ldots n\}$ as follows:

$$\pi_i = \sum_{j=1}^{n} \pi_j p_{ji} \quad \forall i \in \{1 \ldots n\}$$

Note that we have n equations in n variables $\pi_1 \ldots \pi_n$. We can denote the vector of n variables by $\overline{\pi} = [\pi_1, \pi_2, \ldots \pi_n]^T$. The above system of equations can then be written in vector form as follows:

$$\overline{\pi}^T = \overline{\pi}^T P$$

This is exactly a left-eigenvector equation at an eigenvalue of 1. From the Perron-Frobenius theorem, we already know that a unique eigenvector with eigenvalue 1 exists for the stochastic transition matrices associated with strongly connected graphs.

One can use the power method (cf. Section 3.5.2 of Chapter 3) to solve for $\overline{\pi}$ since it is the principal eigenvector of the stochastic transition matrix. The approach works by initializing $\overline{\pi}$ to a vector of random positive values between 0 and 1, and then scaling all values to sum to 1. Subsequently, the following iterative process is repeated:

1. $\overline{\pi}^T \Leftarrow \overline{\pi}^T P$

2. Normalize $\overline{\pi}$ so that its elements sum to 1.

This approach is applied to convergence.

Problem 10.6.1 (Efficient Computation) *One problem with the use of restart is that it makes the transition matrix P dense, even when the underlying graph is sparse. Discuss how you can treat the restart component of P more carefully, so as to be able to compute $\overline{\pi}$ using only sparse matrix operations.*

A hint for solving the above problem is that the final transition matrix can be represented in terms of the original transition matrix P_o as $(1 - \alpha)P_o + \alpha \overline{1}_n \overline{1}_n^T/n$. Here, $\overline{1}_n$ is an n-dimensional column vector of 1s.

10.6.2   Related Measures of Prestige and Centrality

The *PageRank* algorithm can be easily applied to undirected networks by replacing each undirected edge with two directed edges in opposite directions. When applied to directed networks, *PageRank* is a *prestige* measure, and when applied to undirected networks, *PageRank* is a *centrality measure*. Prestige computes an asymmetric measure of importance, and is high when a vertex is reachable from other vertices. However, if a vertex points to many vertices but is reachable from very few vertices, its *PageRank* might be quite low. Because of the symmetric nature of undirected networks, incoming and outgoing reachability cannot be distinguished. Therefore, it is considered a centrality measure. The *PageRank* is only one of many mechanisms used for centrality and prestige computation in networks. The first among them is *eigenvector* centrality, which uses the principal left eigenvector of the (unnormalized) adjacency matrix, as opposed to the stochastic transition matrix:

Definition 10.6.2 (Eigenvector Centrality) *Let A be the $n \times n$ adjacency matrix of an undirected graph. Then, the n components of the principal left eigenvector define the eigenvector centrality of the vertices. In other words, the eigenvector centrality of each of the n vertices is contained in the n-dimensional vector $\overline{\pi}^T A = \lambda \overline{\pi}^T$. Here, λ is the largest eigenvalue of A.*

According to the Perron-Frobenius theorem, the largest eigenvalue lies between the maximum and average degree of the matrix. Like *PageRank*, eigenvector centrality can be computed using the power method. For directed graphs, one can also compute the notion of *eigenvector prestige*.

Definition 10.6.3 (Eigenvector Prestige) *Let A be the $n \times n$ adjacency matrix of a directed graph. Then, the n components of the principal left eigenvector define the eigenvector prestige of the vertices. In other words, the eigenvector prestige of each of the n vertices is contained in the n-dimensional vector $\overline{\pi}^T A = \lambda \overline{\pi}^T$. Here, λ is the largest eigenvalue of A.*

As in the case of *PageRank*, the largest eigenvector may not be unique if the graph is not strongly connected. Therefore, the matrix A may need to be averaged with a restart matrix M/n in order to make it strongly connected. Here, M is a matrix of 1s. As in *PageRank*, the averaging is done in a weighted way with smoothing parameter α.

Finally, it is interesting to explore the left eigenvectors of symmetrically normalized matrices. Let A be the adjacency matrix of an undirected network, and Δ be its weighted degree matrix in which the ith diagonal entry is the degree δ_i. Then, the symmetrically normalized matrix is given by the following:

$$S = \Delta^{-1/2} A \Delta^{-1/2}$$

Since the matrix S is symmetric, its left and right eigenvectors are the same. The principal eigenvector of this matrix contains the square-roots of the vertex degrees. This result is given in Corollary 10.5.1. The degree of a vertex is its *degree centrality*.

Since degree centrality can be computed trivially from the adjacency matrix, it does not make sense to use eigenvector computation. Nevertheless, it is interesting that all eigenvectors relate in one way or another to centrality measures. The corresponding notion of *degree prestige* is defined by the indegree of each vertex in a directed graph. It is possible to relate degree prestige to the eigenvectors of the adjacency matrix after careful normalization, although it is not practically useful. Importantly, the normalization can be done only with weighted indegrees (i.e., column sums) of A rather than outdegrees (i.e., row sums). Note

Table 10.1: Relationship of different eigenvectors to prestige and centrality measures

	Centrality (undirected)	Prestige (directed)
Unnormalized matrix A	Eigenvector Centrality $\lambda\overline{\pi}^T = \overline{\pi}^T A$	Eigenvector Prestige $\lambda\overline{\pi}^T = \overline{\pi}^T A$
Left-normalized matrix $P = \Delta^{-1}A$	PageRank Centrality $\overline{\pi}^T = \overline{\pi}^T P$	PageRank Prestige $\overline{\pi}^T = \overline{\pi}^T P$
Bi-normalized matrix Undirected: $S = \Delta^{-1/2}A\Delta^{-1/2}$ Directed: $S_d = \Delta_{in}^{-1/2}A\Delta_{in}^{-1/2}$	Degree Centrality $\overline{\pi}^T = \overline{\pi}^T S$	Degree Prestige $\overline{\pi}^T = \overline{\pi}^T S_d$

that S will not be symmetric, because A is not symmetric to begin with. Just as directed graphs associate *prestige* measures with vertex *indegrees*, it is also possible to associate *gregariousness* measures with vertex *outdegrees*. Intuitively, the gregariousness of a vertex is defined by the propensity of that vertex to easily reach other vertices via directed paths. This is a complementary idea to prestige measures, in which vertices with high prestige are likely to be easily reached by other vertices via directed paths. In other words, the key difference is in terms of the direction of the edges in the two cases. We leave the modeling of this problem as a practice exercise.

Problem 10.6.2 (Gregariousness) *Each of the three prestige measures (i.e., eigenvector, PageRank, and degree) has a corresponding gregariousness measure. The difference is that prestige measures value in-linking vertices, whereas gregariousness measures value out-linking vertices. For example, the degree gregariousness is the outdegree of a vertex. How would you define the different gregariousness measures of a directed graph in terms of the eigenvectors of an appropriately chosen matrix? This matrix should be defined as a function of the (directed) adjacency matrix A, outdegree matrix Δ_{out}, and indegree matrix Δ_{in}.*

10.6.3 Application of Left Eigenvectors to Link Prediction

The left eigenvector of a stochastic transition matrix can also be used for link prediction. In link prediction, we attempt to find pairs of vertices that do not have a link between them, but are likely to become connected in the future (based on the fact that these vertices are connected to similar vertices). In the following, we assume the case of undirected link prediction although the approach can also be used for directed link prediction (cf. Problem 10.6.3). Given a vertex i, we would like to find all vertices to which it is likely to become connected in the future. The main difference from *PageRank* is in terms of how the restart probabilities are set. Instead of restarting at any vertex, we allow a restart at only a vertex of interest i. Therefore, we first compute an $n \times n$ restart matrix $R = [r_{ij}]$. The matrix R is also a stochastic transition matrix in which all entries in the ith column are set to 1s and the remaining entries are 0s. Then, the current transition matrix P of the undirected graph is modified as follows with the use of restart probability $\alpha \in (0,1)$:

$$P \Leftarrow (1-\alpha)P + \alpha R$$

Subsequently, the components of the left eigenvectors of P provide *personalized PageRank* values for vertex i. The value of α provides the trade-off between the level of personalization and the core social popularity of a vertex.

Problem 10.6.3 (Directed Link Prediction) *Propose a method for using the left eigenvectors of appropriately chosen matrices to predict (i) incoming links, and (ii) outgoing links, of a given vertex.*

10.7   Eigenvectors of Reducible Matrices

The Perron-Frobenius theorem incorporates a strong connectivity assumption. This section examines cases where this assumption does not hold.

10.7.1   Undirected Graphs

Consider an undirected graph with an $n \times n$ adjacency matrix A, degree matrix Δ, and stochastic transition matrix $P = \Delta^{-1}A$. If the undirected graph is not connected, its adjacency matrix and stochastic transition matrix can both be represented in block diagonal form. Specifically, the stochastic transition matrix of a graph containing two connected components can be represented as follows:

$$P = \begin{bmatrix} P_{11} & 0 \\ 0 & P_{22} \end{bmatrix}$$

In general, a graph having r connected components will have r square blocks along the diagonal. Each block P_{ii} along the diagonal is a stochastic transition matrix within that subset of vertices.

   The Perron-Frobenius theorem states that the largest eigenvector of the stochastic transition matrix of a connected graph is unique, and has an eigenvalue of 1. The uniqueness of the largest eigenvector of the adjacency matrix no longer holds when the graph is not connected. For example, in the case of the two-component adjacency matrix, an eigenvector containing 1s for the first component and 0s for the second component has an eigenvalue of 1. This is because the block P_{11} is a transition matrix in its own right, and it is the only part of P that will interact with the non-zero part of the eigenvector. Similarly, one can create an eigenvector containing 0s for the first component and 1s for the second component. This eigenvector also has an eigenvalue of 1. In other words, the dimensionality of the *eigenspace* corresponding to an eigenvalue of 1 is equal to the number of connected components. The eigenvectors corresponding to different connected components form a basis of this eigenspace.

Property 10.7.1 *The dimensionality of the vector space corresponding to the largest eigenvectors (i.e., eigenvectors with eigenvalue 1) of the stochastic transition matrix of an undirected graph is equal to the number of connected components in it.*

We further note that one can compute the number of connected components in a graph using linear algebra.

Property 10.7.2 *Let P be the stochastic transition matrix of an undirected graph. Then, the dimensionality of the null space of $(P - I)$ yields the number of connected components in the graph.*

It is evident from Property 10.7.1 that the eigenspace of the matrix P (for the eigenvalue of 1) is the null space of $(P - I)$. The null space of a matrix can be easily computed using SVD.

10.7.2   Directed Graphs

The notion of connectivity is more complex in directed graphs, because the direction of an edge matters in the definition of concepts like strong connectivity. As in the case of undirected graphs, we will assume that we are working with the stochastic transition matrix.

First, the creation of a stochastic transition matrix from a directed adjacency matrix is not always possible because some vertices might not have any *outgoing* edges. In other words, the corresponding row in the adjacency matrix contains 0s, although its column does contain non-zero entries. It is impossible to normalize such rows to sum to 1. Therefore, the following analysis will assume that such dead-end nodes do not exist. When such dead-end nodes do exist, one can add a self-loop with probability of 1 to facilitate the following analysis. This type of addition of self-loops is often performed in numerous machine learning algorithms like *PageRank* and collective classification (in the presence of dead-end nodes). Another assumption that we make to facilitate analysis is that the graph is fully connected if the directions of edges are ignored.

When a graph is not strongly connected, the matrix will be *block upper-triangular*, in which the diagonal contains square blocks of strongly connected components with some additional non-zero entries above these diagonal blocks. These additional entries only allow unidirectional walks between blocks (sets) of vertices. It is easiest to understand this type of block structure in terms of random walks. Strongly connected graphs result in *ergodic Markov chains* in which the steady-state probabilities do not depend on where the random walk is started. Furthermore, all vertices have non-zero probability of being reached in steady state.

Graphs that are not strongly connected do not satisfy these properties. Such graphs containing vertices of two types:

1. The first type is the *transient vertex set* in the graph. The transient vertex set is a maximal set of vertices, such that the direction of the edges between vertices belonging to the transient set and all other vertices are always outgoing from vertices belonging to the transient set. Note that the vertices in the transient set may or may not be connected/strongly connected to one another, and therefore the edge structure on the subgraph of transient vertices may be arbitrary. These vertices are referred to as transient because a random walk will never visit these vertices in steady-state. Once a walk exits this component, it is no longer possible for the walk to return to this component in steady state. For example, in the case of Figure 10.5(a), vertex 9 is the only transient vertex. On the other hand, in Figure 10.5(b), the vertices 1, 4, 5, 6, and 7 are transient vertices. The transient vertices are labeled by 'T' in these figures. The reader should take a moment to verify that a random walk starting at any of these vertices will eventually reach a vertex outside this set so that it becomes impossible to ever visit any of these vertices again. The transient component is an essential property of reducible graphs. If a connected graph does not have a transient component, then it is also strongly connected.

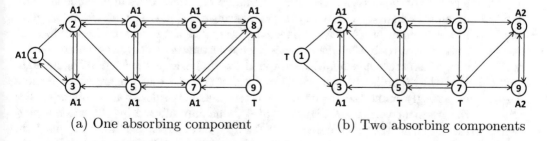

(a) One absorbing component   (b) Two absorbing components

Figure 10.5: Examples of directed graphs that are not strongly connected. Transient vertices are labeled as 'T' and vertices belonging to absorbing components are labeled as 'A1' and 'A2'

2. In addition to the transient vertex set, the graph contains l vertex-disjoint components, which are referred to as *absorbing components*. Each absorbing component is strongly connected in terms of the subgraph induced by its vertex set. Furthermore, an absorbing component only has incoming edges from transient vertices, is not connected to other absorbing components, and has no outgoing edges. Each absorbing component has a non-zero probability to be visited in steady-state from a random walk that starts at a random vertex in the network. However, the steady-state probability depends on the vertex at which the random walk starts. In Figure 10.5(a), there is a single absorbing component containing all vertices except vertex 9. On the other hand, Figure 10.5(b) contains two absorbing components, which are labeled by 'A1' and 'A2.' Note that starting the random walk at vertex 1 will always reach absorbing component A1, but will never reach absorbing component A2. Starting the walk at vertex 4 will allow both A1 and A2 to be reached.

Without loss of generality, we assume that the vertices of such a graph are ordered as follows. The first block of this graph contains all the transient vertices. For this set a_{ij} can be non-zero for each transient vertex i and j can be any value from 1 to n. All other l absorbing components are arranged in block diagonal form, as in the case of undirected matrices. For example, a graph with a transient set and three absorbing components will have the following block structure of the stochastic transition matrix on appropriate reordering the vertex indices (to put the transient vertices first and the vertices of the absorbing components contiguously in succession):

$$P = \begin{bmatrix} P_{11} & P_{12} & P_{13} & P_{14} \\ 0 & P_{22} & 0 & 0 \\ 0 & 0 & P_{33} & 0 \\ 0 & 0 & 0 & P_{44} \end{bmatrix}$$

It is important to note that the square blocks P_{22}, P_{33}, and P_{44} correspond to the edges within absorbing components, and are complete stochastic transition matrices. This matrix has three eigenvectors in the eigenspace belonging to eigenvalue 1. Each of these eigenvectors can be defined by an absorbing component. It is easier to discuss the left eigenvectors. First one can compute the principal left eigenvectors of each of the blocks P_{22}, P_{33}, and P_{44} (corresponding to the absorbing components) separately. Each such eigenvector of the absorbing submatrix can be used to define a left eigenvector of P by setting the remaining components of the larger eigenvector to zeros. Therefore, the Markov chains defined by such reducible graphs have more than one solution to the steady-state equation $\overline{\pi}^T P = \overline{\pi}^T$.

Although it is possible for directed graphs to be completely disconnected (i.e., disconnected irrespective of the direction of edges), this situation rarely arises in machine learning applications. In this chapter, we ignore this case, because it is not very interesting from an application-centric point of view. All connected graphs that are not reducible will have at least some transient vertices that have a one-way connection to strongly connected components. The basic structure of a reducible graph always appears in the form of Figure 10.6, assuming that the graph is not completely disconnected. Note that the transient vertices need not be strongly connected (or even connected) when considered as a subgraph. For example, the transient vertices in Figure 10.5(b) (which are labeled by 'T') are not connected to one another, when considered as a subgraph. Each of the satellite components in Figure 10.6 is an absorbing component. If a random walk is performed on the full graph, and the random walk happens to enter an absorbing component, the walk will never exit the component. It is possible for a reducible network to contain only one absorbing component,

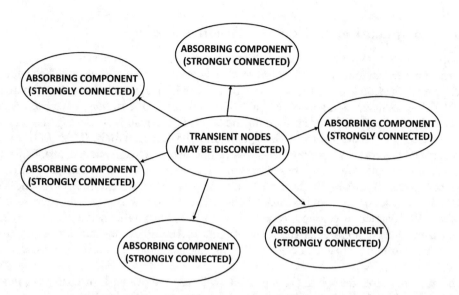

Figure 10.6: If a graph is not strongly connected, it will have at least some transient vertices with one-way connections to strongly connected components

as long as some of the vertices are transient. In fact, an adjacency matrix for a directed, connected graph is reducible if and only if transient vertices exist in the graph.

We provide a number of extended properties of these types of reducible matrices. Let P be the $n \times n$ stochastic transition matrix of a connected (but not strongly connected), directed graph with at least one transient vertex and l absorbing components. We provide an intuitive interpretation of the principal eigenspace in P:

1. The total rank of the eigenspace corresponding to the principal eigenvalue of 1 is always l, where l is the number of absorbing components.

2. It is possible to construct an eigenspace basis of the principal left eigenvectors, so that each eigenvector "belongs" to an absorbing component. The ith component of the left eigenvector belonging to the jth absorbing component contains the steady-state probability of vertex i in a random walk starting in the jth absorbing component.

3. It is possible to construct an eigenspace basis of principal right eigenvectors, so that each eigenvector "belongs" to an absorbing component. The ith component of the right eigenvector belonging to the jth absorbing component contains the probability that a walk starting at vertex i will terminate in absorbing component j. This probability can be fractional for transient nodes and is 1 for nodes belonging to component j. Note that the sum of these l eigenvectors is itself an eigenvector of 1s (which is what we obtain in the irreducible case). This basis of l eigenvectors is very useful in applications like *collective classification* (see next section and Problem 10.8.1).

10.8 Machine Learning Applications

In this section, we will discuss a number of machine learning applications of graphs.

10.8.1　Application to Vertex Classification

Reducible matrices can be used for applications to vertex classification in graphs. The problem of vertex classification is also referred to as *collective classification*. The problem of collective classification is suited to undirected graphs with a symmetric adjacency matrix A. In collective classification, a subset of the vertices are labeled with one of k labels, denoted by $\{1, \ldots, k\}$. Some of the vertices may not be labeled. The goal is to classify the unlabeled vertices based on the structure of the graph and the known labels. This problem often arises in social networks, where one attempts to find actors with particular properties (e.g., interest in a particular product), based on other (known) actors with these properties. The known actors can, therefore, the labeled with their properties. An example of an undirected graph with a subset of vertices that are labeled either 'A' or 'B' is shown in Figure 10.7(a). Therefore, this particular example corresponds to the binary label setting with $k = 2$. For ease in discussion, assume that the edge weights of the adjacency matrix A are binary. Many vertices are not labeled in Figure 10.7(a), and the goal is to classify precisely these vertices.

The basic principle for solving this problem is to use the principle of *homophily* in social networks. The idea is that vertices tend to be connected with vertices that have similar properties. Therefore, a random walk starting at an unlabeled vertex is more likely to first reach a labeled vertex, whose label value matches its own. Therefore, a probabilistic approach for solving this problem is as follows:

> Given an unlabeled vertex, perform a random walk by using the stochastic transition matrix of the adjacency matrix until a labeled vertex is reached. Output the observed label of the destination (labeled) vertex as the predicted class label of the source (unlabeled) vertex from which the random walk begins.

For better robustness, one can compute the *probability* that a vertex of each class is reached. The intuition for this approach is that the walk is more likely to terminate at labeled vertices in the proximity of the starting vertex i. Therefore, when many vertices of a particular class are located in its proximity, then the vertex i is more likely to be labeled with that class. In the particular case of Figure 10.7(a), any random walk starting from test vertex X will always reach label 'A' first rather than label 'B' because of the topology of the graph. However, it is also possible to select test vertices, where there is a non-zero probability of reaching either the label 'A' or the label 'B.' For example, if one starts the random walk at test vertex Y, then a vertex corresponding to either label 'A' or label 'B' could be reached first.

An important assumption is that the graph needs to be *label connected*. In other words, every unlabeled vertex needs to be able to reach a labeled vertex in the random walk. For undirected graphs, this means that every connected component of the graph needs to contain at least one labeled vertex. In the following discussion, it will be assumed that the entire undirected graph is connected; any undirected graph with only one connected component will always lead to a modified transition matrix that is label connected.

Since the approach is based on random walks, the first step is to create the (directed) stochastic transition matrix P from the undirected $n \times n$ adjacency matrix A. Let Δ be the (diagonal) degree matrix that contains the weighted degree $\delta_i = \sum_j a_{ij} = \sum_j a_{ji}$ on the ith entry of its main diagonal. As in the case of all other applications of this chapter, the adjacency matrix is converted into the stochastic transition matrix by using left-normalization:

$$P = \Delta^{-1}A \qquad (10.7)$$

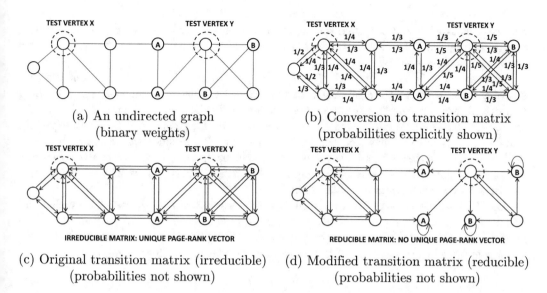

(a) An undirected graph
(binary weights)

(b) Conversion to transition matrix
(probabilities explicitly shown)

(c) Original transition matrix (irreducible)
(probabilities not shown)

(d) Modified transition matrix (reducible)
(probabilities not shown)

Figure 10.7: Creating directed transition graphs from undirected graph

Since this transition matrix is derived from an undirected graph, it will always be an irreducible matrix. The corresponding strongly connected graph is illustrated in Figure 10.7(b), and the probabilities on the various edges are explicitly shown. The same illustration is shown in Figure 10.7(c), but without the probabilities on the edges (to avoid clutter).

Although we can use this transition matrix to model the random walks in the graph, such an approach will not provide the *first stopping point* of the walk. In other words, we need to model the random walks in such a way that they always terminate at their *first arrival* at labeled vertices. This can be achieved by removing outgoing edges from labeled vertices and replacing them with self-loops. This results in a singleton absorbing component containing only one vertex, which we refer to as an *absorbing vertex*. Such vertices are referred to as absorbing vertices because they trap the random walk after an incoming transition. The goal of creating such absorbing vertices is to ensure that a random walk is trapped by the first labeled node it reaches.

The stochastic transition matrix P needs to be modified to account for the effect of absorbing vertices. For each absorbing vertex i, the ith row of P is replaced with the ith row of the identity matrix. Henceforth, we will assume that the matrix denoted by notation P incorporates this modification (and is therefore not exactly equal to $\Delta^{-1}A$ according to Equation 10.7). An example of the final transition graph is illustrated in Figure 10.7(d). The resulting matrix is no longer irreducible because the resulting graph is no longer strongly connected; no other vertex can be reached from an absorbing vertex. Note that this graph has exactly the structure of Figure 10.6 because each of the absorbing components is a singleton (labeled) vertex, and all unlabeled vertices are transient. Since the transition matrix P is reducible, it does not have a unique eigenvector with eigenvalue 1. Rather, it has as many principal eigenvectors with eigenvalue 1 as the number of absorbing vertices.

For any given starting vertex i, the steady-state probability distribution has positive values only at labeled vertices. This is because a random walk will eventually reach an absorbing vertex in a label-connected graph, and it will never emerge from that vertex. Therefore, if one can estimate the steady-state probability distribution of labeled nodes for a starting unlabeled vertex i, then the probability values of the labeled vertices in each class can be aggregated. The class with the highest probability is reported as the relevant label of the unlabeled vertex i.

Note that the (i, j)th entry of P^r yields the probability of a random walk of length r starting at any vertex i to terminate at any vertex j. Because of the self-loops at absorbing vertices, all walks of length less than r are automatically included in the probability (as the remaining steps can be completed inside the self-loop). Therefore, P^∞ is the steady-state matrix of probabilities, in which the (i, j)th entry of P^∞ provides the probability that a walk starting a vertex i terminates at vertex j. For each row, we would like to aggregate the probabilities of the classes belonging to the different labels. As we will see below, this can be achieved with a simple matrix multiplication.

Let Y be an $n \times k$ matrix in which the (i, c)th entry is 1, if the ith vertex is labeled and it belongs to the class $c \in \{1, \ldots, k\}$. Then, the aggregation of the probabilities of labeled vertices in each row of P^∞ is given by the matrix $P^\infty Y$. In other words, we can obtain an $n \times k$ matrix Z of probabilities of the classes for the various vertices as follows:

$$Z = P^\infty Y \tag{10.8}$$

The class with the maximum probability in Z for unlabeled vertex (row) i may be reported as its class label. This approach is also referred to as the *rendezvous approach* to label propagation [9]. How is P^∞ computed? One possibility is keep multiplying P with itself in order to compute P^∞. One can use eigendecomposition tricks to speed up the process. However, we do not want to work with large matrices (like P^∞) of size $n \times n$. Therefore, a more efficient (but equivalent) approach is *iterative label propagation* [143].

In iterative label propagation, we initialize $Z^{(0)} = Y$ and then repeatedly use the following update for increasing value of iteration index t:

$$Z^{(t+1)} = P Z^{(t)} \tag{10.9}$$

It is easy to see that $Z^{(\infty)}$ is the same as the value of Z in Equation 10.8. Furthermore, each column of Z is a principal right eigenvector of P at convergence. Each principal eigenvector corresponds to a label with multiple absorbing components rather than a single absorbing component. Furthermore, the absorbing components are singleton vertices in this case. The following exercise generalizes this idea to finding principal eigenvectors of absorbing components with no special structure.

Problem 10.8.1 (Right Principal Eigenvectors) *Consider a directed graph with l absorbing components, which are already demarcated. Discuss how you can use the ideas in iterative label propagation to create a basis of l principal right eigenvectors satisfying the following property. The ith component of the jth principal eigenvector should be equal to the probability that a walk starting at node i ends in absorbing component j.*

10.8.2   Applications to Multidimensional Data

Graphs can also be applied to various machine learning applications on multidimensional data. This is because any type of multidimensional data set can be modeled as a *similarity graph*. Consider a multidimensional data set with n data instances denoted by $\overline{X}_1 \ldots \overline{X}_n$. The first step is to compute the k-nearest neighbors of each data point. The similarity between any pair of points $\overline{X}_i$ and $\overline{X}_j$ can be computed with the use of a Gaussian kernel:

$$K(\overline{X}_i, \overline{X}_j) = e^{-\|\overline{X}_i - \overline{X}_j\|^2 / (2 \cdot \sigma^2)} \tag{10.10}$$

Here, σ is the bandwidth of the Gaussian kernel (cf. Table 9.1 of Chapter 9). For supervised applications, the value of σ is often tuned using out-of-sample data. For unsupervised applications, the value of σ is chosen to be of the order of the median of all pairwise Euclidean distances between points.

A graph is constructed, where the ith vertex corresponds to the data point $\overline{X}_i$. The similarity value can be used to compute the mutual k-nearest neighbor graph, as discussed in Section 10.5. The weight of an edge is set to the kernel similarity value introduced in Equation 10.10. Subsequently any of the applications such as clustering or classification (discussed in this chapter) can be used on this graph. In the case of classification, the graph is constructed on both the labeled and the unlabeled vertices. Therefore, unlike traditional classification, the classification benefits from unlabeled samples in the data. This type of approach can be helpful when the number of labeled instances is small. The use of unlabeled instances for better classification is referred to as *semi-supervised classification*. Furthermore, one can use this approach not only for multidimensional data, but for any type of data where a similarity function can be computed between the objects. After all the various forms of graph embeddings (e.g., spectral embeddings) can be viewed as special cases of kernel methods. Like any kernel method, a similarity graph and its embeddings can be used for any data type.

10.9 Summary

This chapter discusses the linear algebra of graphs. Many important structural properties of graphs such as the walks between vertices and connectivity can be inferred from the linear algebra of the adjacency matrix. The fundamental result underpinning the primary results on spectral analysis of graphs is the Perron-Frobenius theorem. A particular type of matrix associated with graphs is the stochastic transition matrix, whose principal eigenvector has an eigenvalue of 1. The dominant left eigenvector and the dominant right eigenvector(s) of the stochastic transition matrix have different types of applications. The dominant left eigenvector is used for ranking, whereas the right eigenvectors are used for clustering. One can also use a symmetrically normalized adjacency matrix for clustering, which is roughly equivalent to the use of kernel methods. Different forms of normalization of the adjacency matrix help in extracting different centrality and prestige measures. One can also compute the eigenvectors of modified versions of stochastic transition matrices in order to perform collective classification of graphs. All graph-based machine learning applications can be generalized to arbitrary data types with the construction of suitable similarity graphs over the underlying objects.

10.10 Further Reading

The basics of spectral graph theory are available in [25, 29]. A more application-centric approach may be found in [40]. The basics of graph Laplacians and spectral clustering may be found in the well-known survey by Luxburg [84]. The symmetric and asymmetric variations of spectral clustering were proposed in [98, 115]. Both the variants are roughly equivalent, although the symmetric variant is easier to relate to a kernel method. The original *PageRank* algorithm was proposed in [24]. A discussion of different types of centrality measures may be found in [138]. Different presentations of the label propagation algorithm are provided in [9, 143]. Surveys on vertex classification in graphs may be found in [17, 82].

10.11 Exercises

1. Consider the $n \times n$ adjacency matrices A_1 and A_2 of two graphs. Suppose that the graphs are known to be *isomorphic*. Two graphs are said to be isomorphic, if one graph can be obtained from the other by reordering its vertices. Show that isomorphic graphs have the same eigenvalues. [Hint: What is the nature of the relationship between their adjacency matrices in algebraic form? You may introduce any new matrices as needed.]

2. Suppose that you were given the eigenvectors and eigenvalues of the stochastic transition matrix P of an undirected graph. Discuss how you can quickly compute P^∞ using these eigenvectors and eigenvalues.

3. Let Δ be the weighted degree of matrix of an $n \times n$ (undirected) adjacency matrix A, and $\bar{e}_1 \ldots \bar{e}_n$ be the n eigenvectors of the stochastic transition matrix $P = \Delta^{-1} A$. Show that any pair of eigenvectors $\bar{e}_i$ and $\bar{e}_j$ are Δ-orthogonal. In other words, any pair of eigenvectors $\bar{e}_i$ and $\bar{e}_j$ must satisfy the following:
$$\bar{e}_i^T \Delta \bar{e}_j = 0$$

4. Show that all eigenvectors (other than the first eigenvector) of the stochastic transition matrix of a connected, undirected graph will have both positive and negative components.

5. Consider the adjacency matrix A of an $n \times n$ undirected graph, which is also bipartite. In a bipartite graph, the n vertices can be divided into two vertex sets V_1 and V_2 of respectively n_1 and n_2 vertices, so that all edges occur between vertices of V_1 and vertices of V_2. The adjacency matrix of such a graph always has the following form for an $n_1 \times n_2$ matrix B:
$$A = \begin{bmatrix} 0 & B \\ B^T & 0 \end{bmatrix}$$
Even though A is symmetric, B might not be symmetric. Given the eigenvectors and eigenvalues of A, show how you can perform the SVD of B quickly (and vice versa).

6. A complete directed graph is defined on n vertices and it contains all $n(n-1)$ possible edges in both directions between each pair of vertices (other than self-loops). Each edge weight is 1.

   (a) Give a short reason why all eigenvalues must be real.

   (b) Give a short reason why the eigenvalues must sum to 0.

   (c) Show that this graph has one eigenvalue of $(n-1)$ and $(n-1)$ eigenvalues are -1. [Express adjacency matrix as $\bar{1}\bar{1}^T - I$.]

7. A complete bipartite graph (see Exercise 5) is defined on 4 vertices, where 2 vertices are contained in each partition. A edge of weight 1 exists in both directions between each pair of vertices drawn from the two partitions. Find the eigenvalues of this graph. Can you generalize this result to the case of a complete bipartite graph containing $2n$ vertices, where n vertices are contained in each partition?

8. Suppose you create a symmetrically normalized adjacency matrix $S = \Delta^{-1/2} A \Delta^{-1/2}$ for an undirected adjacency matrix A. You decide that some vertices are "important" and they should get relative weight $\gamma > 1$ in an embedding that is similar to that in spectral clustering, whereas other vertices only get a weight to 1.

(a) Propose a weighted matrix factorization model that creates an embedding in which the "important" vertices have a relative weight of γ in the objective function. The matrix factorization model should yield the same embedding at $\gamma = 1$ as symmetric spectral clustering.

(b) Show how you can create a class-sensitive embedding with this approach, if some vertices in a graph are labeled.

(c) You are given a black-box classifier that works with multidimensional data. Show how you can select γ appropriately and use it for collective classification of the unlabeled vertices of a partially labeled graph.

9. Propose an embedding-based algorithm for outlier detection in multidimensional data that uses the concept of the similarity graph proposed in Section 10.8.2. Discuss the choice of an appropriate dimensionality of the embedding, and how this choice is different from the case of the clustering problem.

10. Suppose you are given a very large graph for which the symmetric similarity matrix S (for spectral clustering) cannot be materialized in the disk space available to you. Discuss why the data-dependent nature of this kernel matrix makes Nyström sampling difficult. Propose a rough approximation of Nyström sampling for spectral clustering. [The answer to the last part of the question is not unique.]

11. Provide an example of a 2×2 adjacency matrix of a directed graph that is not diagonalizable.

12. A bipartite graph is defined as a graph $G = (V_1 \cup V_2)$ with a partitioned vertex set $V_1 \cup V_2$, so that no edges in E exist within vertices of a single partition. In other words, for all $(i, j) \in E$, both i, j cannot be drawn from V_1, and both i, j cannot be drawn from V_2. Show that if λ is the eigenvalue of the adjacency matrix of an undirected bipartite graph, then $-\lambda$ is an eigenvalue as well.

13. **Degree conditioning:** Suppose that you have the adjacency matrix A of an undirected graph. Discuss using the Perron-Frobenius theorem why adding the (weighted) degree of each node i to the ith diagonal entry results in a positive semidefinite matrix. The text of the chapter already shows that negating each entry of A and adding the degree matrix results in a positive semidefinite Laplacian. Also provide an alternate proof of the positive semidefiniteness of the unnormalized Laplacian using the Perron-Frobenius theorem.

14. Suppose that you are given the symmetric factorization $A \approx UU^T$ of an undirected (and possibly weighted) graph adjacency matrix in which all entries of A are well approximated by UU^T (with the possible exception of the zero diagonal entries). Furthermore, all entries of the $n \times k$ matrix U are nonnegative. Discuss how you can use this factorization to express A as the sum of k adjacency matrices, each of which is a rank-1 matrix. Discuss how you can use this decomposition to create a possibly overlapping clustering of the graph by inspection of U. Discuss the interpretability advantages of nonnegative matrix factorization.

15. Let P be the stochastic transition matrix of an undirected and connected graph. Show that all left eigenvectors of P other than the principal left eigenvector (i.e., *PageRank* vector) have vector components that sum to 0. [Hint: What are the angles between left eigenvectors and right eigenvectors of a matrix?]

16. Let S be the symmetrically normalized adjacency matrix of spectral clustering of an undirected graph. In some cases, the clusters do not clearly separate out by applying the k-means algorithm on the features obtained from eigenvector extraction on S. Use the kernel intuition from Chapter 9 to discuss the advantages of using $(C+S) \odot (C+S)$ instead of S for eigenvector extraction in such cases. Here, $C \geq 0$ is a matrix of constant values. [Hint: The k-means algorithm works best with linearly separable clusters. Check Figure 9.4.]

17. Consider two $n \times n$ symmetric matrices A and B, such that B is also positive definite. Show that BA need not be symmetric, but it is diagonalizable with real eigenvalues. [Hint: This is a generalization of the proof that stochastic transition matrices have real eigenvalues by setting B to the inverse degree matrix and A to the adjacency matrix.]

18. Suppose that A is the 20×20 binary adjacency matrix of a directed graph of 20 nodes. Interpret the matrix $(I - A^{20})(I - A)^{-1}$ in terms of walks in the graph. Will this matrix have any special properties for a strongly connected graph? Argue algebraically why the following is true:

$$(I - A^{20})(I - A)^{-1} = (I - A)^{-1}(I - A^{20})$$

19. Exercise 13 of the previous chapter introduces symmetric non-negative matrix factorization, which can also be used to factorize the symmetrically normalized adjacency matrix $S \approx UU^T$, which is used in spectral clustering. Here, U is an $n \times k$ non-negative factor matrix. Discuss why the top-r components of each column of U (in magnitude) directly provide clustered bags of nodes of size r in the graph.

20. Find the *PageRank* of each node in (i) an undirected cycle of n nodes, and (ii) a single central node connected with an undirected edge to each of $(n-1)$ nodes. In each case, compute the *PageRank* at a restart probability of 0.

21. **Signed network embedding:** Suppose that you have a graph with both positive and negative weights on edges. Propose modifications of the algorithms used to remove "weak edges" and to symmetrically normalize the graph for spectral clustering. Will the resulting graph be diagonalizable with orthogonal eigenvectors and real eigenvalues? Is there anything special about the first eigenvector? [This is an open-ended question with multiple solutions.]

22. **Heterogeneous network embedding:** Consider a social network graph with directed/undirected edges of multiple types (e.g., undirected friendship links, directed messaging links, and directed "like" links). Propose a shared matrix factorization algorithm (cf. Chapter 8) to extract an embedding of each node. How would you tune the parameters? [This is an open-ended question with multiple solutions.]

Chapter 11

Optimization in Computational Graphs

"Science is the differential calculus of the mind. Art the integral calculus; they may be beautiful when apart, but are greatest only when combined."– Ronald Ross

11.1 Introduction

A computational graph is a network of connected nodes, in which each node is a unit of computation and stores a variable. Each edge joining two nodes indicates a relationship between the corresponding variables. The graph may be either directed or undirected. In a directed graph, a node computes its associated variable as a function of the variables in the nodes that have edges incoming to it. In an undirected graph, the functional relationship works in both directions. Most practical computational graphs (e.g., conventional neural networks) are *directed acyclic graphs*, although many undirected probabilistic models in machine learning can be implicitly considered computational graphs with cycles. Similarly, the variables at the nodes might be continuous, discrete, or probabilistic, although most real-world computational graphs work with continuous variables.

In many machine learning problems, parameters may be associated with the edges, which are used as additional arguments to the functions computed at nodes connected to these edges. These parameters are learned in a *data-driven* manner so that variables in the nodes mirror relationships among attribute values in data instances. Each data instance contains both *input* and *target* attributes. The variables in a subset of the *input* nodes are fixed to input attribute values in data instances, whereas the variables in all other nodes are *computed* using the node-specific functions. The variables in some of the computed nodes are compared to observed *target* values in data instances, and edge-specific parameters are modified to match the observed and computed values as closely as possible. By learning the parameters along the edges in a data-driven manner, one can learn a function relating the input and target attributes in the data.

In this chapter, we will primarily focus on directed acyclic graphs with continuous, deterministic variables. A *feed-forward neural network* is an important special case of this type

447

C. C. Aggarwal, *Linear Algebra and Optimization for Machine Learning*,
https://doi.org/10.1007/978-3-030-40344-7_11

of computational graph. The inputs often correspond to the features in each data point, whereas the output nodes might correspond to the target variables (e.g., class variable or regressand). The optimization problem is defined over the edge parameters so that the predicted variables match the observed values in the corresponding nodes as closely as possible. In other words, the *loss function* of a computational graph might penalize differences between predicted and observed values. In computational graphs with continuous variables, one can use gradient descent for optimization. *Almost all machine learning problems that we have seen so far in this book, such as linear regression, logistic regression, SVMs, SVD, PCA, and recommender systems, can be modeled as directed acyclic computational graphs with continuous variables.*

This chapter is organized as follows. The next section will introduce the basics of computational graphs. Section 11.3 discusses optimization in directed acyclic graphs. Applications to neural networks are discussed in Section 11.4. A general view of computational graphs is provided in Section 11.5. A summary is given in Section 11.6.

11.2   The Basics of Computational Graphs

We will define the notion of a directed acyclic computational graph (without cycles).

Definition 11.2.1 (Directed Acyclic Computational Graph) *A directed acyclic computational graph contains nodes, so that each node is associated with a variable. A set of directed edges connect nodes, which indicate functional relationships among nodes. Edges* **might** *be associated with learnable parameters. A variable in a node is either fixed externally (for input nodes with no incoming edges), or it is computed as a function of the variables in the tail ends of edges incoming into the node and the learnable parameters on the incoming edges.*

It is technically possible to define computational graphs with cycles, although we do not consider this rare possibility. The computational graph contains three types of nodes, which are the input, output, and hidden nodes. The input nodes contain the external inputs to the computational graph, and the output node(s) contain(s) the final output(s). The hidden nodes contain intermediate values. Each hidden and output node computes a relatively simple **local** function of its incoming node variables. The cascading effect of the computations over the whole graph implicitly defines a **global** vector-to-vector function from input to output nodes. The variable in each input node is fixed to an externally specified input value. Therefore, no function is computed at an input node. The node-specific functions also use parameters associated with their incoming edges, and the inputs along those edges are scaled with the weights. By choosing weights appropriately, one can control the (global) function defined by the computational graph. This global function is often *learned* by feeding the computational graph input-output pairs (training data) and adjusting the weights so that predicted outputs matched observed outputs.

An example of a computational graph with two weighted edges is provided in Figure 11.1. This graph has three inputs, denoted by x_1, x_2, and x_3. Two of the edges have weights w_2 and w_3. Other than the input nodes, all nodes perform a computation such as addition, multiplication, or evaluating a function like the logarithm. In the case of weighted edges, the values at the tail of the edge are scaled with the weights before computing the node-specific function. The graph has a single output node, and computations are cascaded in

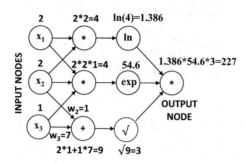

Figure 11.1: Examples of computational graph with two weighted edges

the forward direction from the input to the output. For example, if the weights w_2 and w_3 are chosen to be 1 and 7, respectively, the global function $f(x_1, x_2, x_3)$ is as follows:

$$f(x_1, x_2, x_2) = \ln(x_1 x_2) \cdot \exp(x_1 x_2 x_3) \cdot \sqrt{x_2 + 7 x_3}$$

For $[x_1, x_2, x_3] = [2, 2, 1]$, the cascading sequence of computations is shown in the figure with a final output value of approximately 227.1. However, if the *observed* value of the output is only 100, it means that the weights need to be readjusted to change the computed function. In this case, one can observe from inspection of the computational graph that reducing either w_2 or w_3 will help reduce the output value. For example, if we change the weight w_3 to -1, while keeping $w_2 = 1$, the computed function becomes the following:

$$f(x_1, x_2, x_2) = \ln(x_1 x_2) \cdot \exp(x_1 x_2 x_3) \cdot \sqrt{x_2 - x_3}$$

In this case, for the same set of inputs $[x_1, x_2, x_3] = [2, 2, 1]$, the computed output becomes 75.7, which is much closer to the true output value of 100. Therefore, it is clear that one must use the mismatch of predicted values with observed outputs to adjust the computational function, so that there is a better matching between predicted and observed outputs across the data set. Although we adjusted w_3 here by inspection, such an approach will not work in very large computational graphs containing millions of weights.

The goal in machine learning is to learn parameters (like weights) using examples of input-output pairs, while adjusting weights with the help of the observed data. The key point is to convert the problem of adjusting weights into an optimization problem. The computational graph may be associated with a loss function, which typically penalizes the differences in the *predicted* outputs from *observed* outputs, and adjusts weights accordingly. Since the outputs are functions of inputs and edge-specific parameters, the loss function can also be viewed as a complex function of the inputs and edge-specific parameters. The goal of learning the parameters is to minimize the loss, so that the input-output pairs in the computational graph mimic the input-output pairs in the observed data. It should be immediately evident that the problem of learning the weights is likely to challenging, if the underlying computational graph is large with a complex topology.

The choice of loss function depends on the application at hand. For example, one can model least-squares regression by using as many input nodes as the number of input variables (regressors), and a single output node containing the predicted regressand. Directed edges exist from each input node to this output node, and the parameter on each such edge

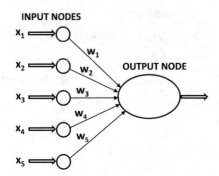

Figure 11.2: A single-layer computational graph that can perform linear regression

corresponds to the weight associated with that input variable (cf. Figure 11.2). The output
node computes the following function of the variables $x_1 \ldots x_d$ in the d input nodes:

$$\hat{o} = f(x_1, x_2, \ldots, x_d) = \sum_{i=1}^{d} w_i x_i$$

If the observed regressand is o, then the loss function simply computes $(o - \hat{o})^2$, and adjusts
the weights $w_1 \ldots w_d$ so as to reduce this value. Typically, the derivative of the loss is
computed with respect to each weight in the computational graph, and the weights are
updated by using this derivative. One processes each training point one-by-one and updates
the weights. *The resulting algorithm is identical to using stochastic gradient descent in the
linear regression problem* (cf. Section 4.7 of Chapter 4). In fact, by changing the nature of
the loss function at the output node, it is possible to model both logistic regression and the
support vector machine.

Problem 11.2.1 (Logistic Regression with Computational Graph) *Let o be an observed* **binary** *class label drawn from $\{-1, +1\}$ and $\hat{o}$ be the predicted* **real** *value by the
neural architecture of Figure 11.2. Show that the loss function $log(1 + exp(-o\hat{o}))$ yields the
same loss function for each data instance as the logistic regression model in Equation 4.56
of Chapter 4. Ignore the regularization term in Chapter 4.*

Problem 11.2.2 (SVM with Computational Graph) *Let o be an observed* **binary**
class label drawn from $\{-1, +1\}$ and $\hat{o}$ be the predicted **real** *value by the neural architecture
of Figure 11.2. Show that the loss function $max\{0, 1 - o\hat{o}\}$ yields the same loss function for
each data instance as the L_1-loss SVM in Equation 4.51 of Chapter 4.*

In the particular case of Figure 11.2, the choice of a computational graph for model representation does not seem to be useful because a single computational node is rather rudimentary
for model representation; indeed, one can directly compute gradients of the loss function
with respect to the weights without worrying about computational graphs at all! The main
usefulness of computational graphs is realized when the topology of computation is more
complex.

The nodes in the directed acyclic graph of Figure 11.2 are arranged in *layers*, because
all paths from an input node to any node in the network have the same length. This type
of architecture is common in computational graphs. Nodes that are reachable by a path of
a particular length i from input nodes are assumed to belong to layer i. At first glance,

Figure 11.2 looks like a two-layer network. However, such networks are considered single-layer networks, because the non-computational input layer is not counted among the number of layers.

11.2.1 Neural Networks as Directed Computational Graphs

The real power of computational graphs is realized when one uses multiple layers of nodes. Neural networks represent the most common use case of a multi-layer computational graph. The nodes are (typically) arranged in layerwise fashion, so that all nodes in layer-i are connected to nodes in layer-$(i + 1)$ (and no other layer). The vector of variables in each layer can be written as a vector-to-vector function of the variables in the previous layer. A pictorial illustration of a multilayer neural network is shown in Figure 11.3(a). In this case, the network contains three computational layers in addition to the input layer. For example, consider the first hidden layer with output values $h_{11} \ldots h_{1r} \ldots h_{1,p_1}$, which can be computed as a function of the input nodes with variables $x_1 \ldots x_d$ in the input layer as follows:

$$h_{1r} = \Phi(\sum_{i=1}^{d} w_{ir} x_i) \quad \forall r \in \{1, \ldots, p_1\}$$

The value p_1 represents the number of nodes in the first hidden layer. Here, the function $\Phi(\cdot)$ is referred to as an *activation* function. The final numerical value of the variable in a particular node (i.e., h_{1r} in this case) for a particular input is also sometimes referred to as its *activation* for that input. In the case of linear regression, the activation function is missing, which is also referred to as using the *identity* activation function or *linear* activation function. However, computational graphs primarily gain better expressive power by using nonlinear activation functions such as the following:

$$\Phi(v) = \frac{1}{1 + e^{-v}} \qquad \qquad \text{[Sigmoid function]}$$

$$\Phi(v) = \frac{e^{2v} - 1}{e^{2v} + 1} \qquad \qquad \text{[Tanh function]}$$

$$\Phi(v) = \max\{v, 0\} \qquad \qquad \text{[ReLU: Rectified Linear Unit]}$$

$$\Phi(v) = \max\{\min[v, 1], -1\} \qquad \qquad \text{[Hard tanh]}$$

It is noteworthy that these functions are nonlinear, and nonlinearity is essential for greater expressive power of networks with increased depth. Networks containing only linear activation functions are not any more powerful than single-layer networks.

In order to understand this point, consider a two-layer computational graph (not counting the input layer) with 4-dimensional input vector $\overline{x}$, 3-dimensional hidden-layer vector $\overline{h}$, and 2-dimensional output-layer vector $\overline{o}$. Note that we are creating a column vector from the node variables in each layer. Let W_1 and W_2 be two matrices of sizes 3×4 and 2×3 so that $\overline{h} = W_1 \overline{x}$ and $\overline{o} = W_2 \overline{h}$. The matrices W_1 and W_2 contain the weight parameters of each layer. Note that one can express $\overline{o}$ directly in terms of $\overline{x}$ without using $\overline{h}$ as $\overline{o} = W_2 W_1 \overline{x} = (W_2 W_1) \overline{x}$. One can replace the matrix $W_2 W_1$ with a single 2×4 matrix W without any loss of expressive power. In other words, this is a single-layer network! It is not possible to use this type of approach to (easily) eliminate the hidden layer in the case of nonlinear activation functions without creating extremely complex functions at individual nodes (thereby increasing node-specific complexity). This means that increased depth results in increased complexity only when using nonlinear activation functions.

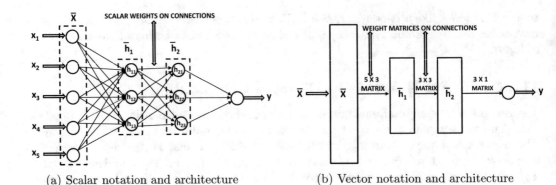

(a) Scalar notation and architecture   (b) Vector notation and architecture

Figure 11.3: A feed-forward network with two hidden layers and a single output layer

In the case of Figure 11.3(a), the neural network contains three layers. Note that the input layer is often not counted, because it simply transmits the data and no computation is performed in that layer. If a neural network contains $p_1 \ldots p_k$ units in each of its k layers, then the (column) vector representations of these outputs, denoted by $\overline{h}_1 \ldots \overline{h}_k$ have dimensionalities $p_1 \ldots p_k$. Therefore, the number of units in each layer is referred to as the *dimensionality* of that layer. It is also possible to create a computational graph in which the variables in nodes are vectors, and the connections represent vector-to-vector functions. Figure 11.3(b) creates a computational graph in which the nodes are represented by *rectangles* rather than *circles*. Rectangular representations of nodes correspond to nodes containing vectors. The connections now contain matrices. The sizes of the corresponding *connection* matrices are shown in Figure 11.3(b). For example, if the input layer contains 5 nodes and the first hidden layer contains 3 nodes, the *connection matrix* is of size 5×3. However, as we will see later, the *weight* matrix has size that is the transpose of the connection matrix (i.e., 3×5) in order to facilitate matrix operations. Note that the computational graph in the vector notation has a simpler structure, where the entire network contains only a single path. The weights of the connections between the input layer and the first hidden layer are contained in a *matrix* W_1 with size $p_1 \times d$, whereas the weights between the rth hidden layer and the $(r+1)$th hidden layer are denoted by the $p_{r+1} \times p_r$ matrix denoted by W_r. If the output layer contains s nodes, then the final matrix W_{k+1} is of size $s \times p_k$. Note that the weight matrix has transposed dimensions with respect to the connection matrix. The d-dimensional input vector $\overline{x}$ is transformed into the outputs using the following recursive equations:

$$\overline{h}_1 = \Phi(W_1 \overline{x}) \qquad \text{[Input to Hidden Layer]}$$
$$\overline{h}_{p+1} = \Phi(W_{p+1} \overline{h}_p) \ \ \forall p \in \{1 \ldots k-1\} \qquad \text{[Hidden to Hidden Layer]}$$
$$\overline{o} = \Phi(W_{k+1} \overline{h}_k) \qquad \text{[Hidden to Output Layer]}$$

Here, the activation functions are applied in *element-wise* fashion to their vector arguments. Here, it is noteworthy that the final output is a **recursively nested composition function of the inputs**, which is as follows:

$$\overline{o} = \Phi(W_{k+1}(\Phi(W_k \Phi(W_{k-1} \ldots))))$$

This type of neural network is harder to train than single-layer networks because one must compute the derivative of a **nested** composition function with respect to each weight. In particular, the weights of earlier layers lie inside the recursive nesting, and are harder to learn with gradient descent, because the methodology for computation of the gradient of weights in the inner portions of the nesting (i.e., earlier layers) is not obvious, especially when the computational graph has a complex topology. It is also noticeable that the **global** input-to-output function computed by the neural network is harder to express in closed form neatly. The recursive nesting makes the closed-form representation look extremely cumbersome. A cumbersome closed-form representation causes challenges in derivative computation for parameter learning.

11.3 Optimization in Directed Acyclic Graphs

The optimization of loss functions in computational graphs requires the computation of gradients of the loss functions with respect to the network weights. This computation is done using *dynamic programming* (cf. Section 5.8.4 of Chapter 5). Dynamic programming is a technique from optimization that can be used to compute all types of path-centric functions in *directed acyclic graphs*.

In order to train computational graphs, it is assumed that we have training data corresponding to input-output pairs. The number of input nodes is equal to the number of input attributes and the number of output nodes is equal to the number of output attributes. The computational graph can predict the outputs using the inputs, and compare them to the observed outputs in order to check whether the function computed by the graph is consistent with the training data. If this is not the case, the weights of the computational graph need to be modified.

11.3.1 The Challenge of Computational Graphs

A computational graph naturally evaluates compositions of functions. Consider a variable x at a node in a computational graph with only three nodes containing a path of length 2. The first node applies the function $g(x)$, whereas the second node applies the function $f(\cdot)$ to the result. Such a graph computes the function $f(g(x))$, and it is shown in Figure 11.4. The example shown in Figure 11.4 uses the case when $f(x) = \cos(x)$ and $g(x) = x^2$. Therefore, the overall function is $\cos(x^2)$. Now, consider another setting in which both $f(x)$ and $g(x)$ are set to the same function, which is the sigmoid function:

$$f(x) = g(x) = \frac{1}{1 + \exp(-x)}$$

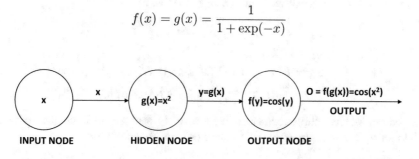

Figure 11.4: A simple computational graph with an input node and two computational nodes

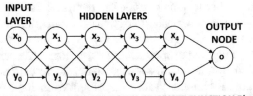

Figure 11.5: The awkwardness of recursive nesting caused by a computational graph

Then, the global function evaluated by the computational graph is as follows:

$$f(g(x)) = \frac{1}{1 + \exp\left[-\frac{1}{1+\exp(-x)}\right]} \tag{11.1}$$

This simple graph already computes a rather awkward composition function. Trying the find the derivative of this composition function becomes increasingly tedious with increasing complexity of the graph.

Consider a case in which the functions $g_1(\cdot), g_2(\cdot) \ldots g_k(\cdot)$ are the functions computed in layer m, and they feed into a particular layer-$(m + 1)$ node that computes the multivariate function $f(\cdot)$ that uses the values computed in the previous layer as arguments. Therefore, the layer-$(m+1)$ function computes $f(g_1(\cdot), \ldots g_k(\cdot))$. This type of multivariate composition function already appears rather awkward. As we increase the number of layers, a function that is computed several edges downstream will have as many layers of nesting as the length of the path from the source to the final output. For example, if we have a computational graph which has 10 layers, and 2 nodes per layer, the overall composition function would have 2^{10} nested "terms". This makes the handling of closed-form functions of deep networks unwieldy and impractical.

In order to understand this point, consider the function in Figure 11.5. In this case, we have two nodes in each layer other than the output layer. The output layer simply sums its inputs. Each hidden layer contains two nodes. The variables in the ith layer are denoted by x_i and y_i, respectively. The input nodes (variables) use subscript 0, and therefore they are denoted by x_0 and y_0 in Figure 11.5. The two computed functions in the ith layer are $F(x_{i-1}, y_{i-1})$ and $G(x_{i-1}, y_{i-1})$, respectively.

In the following, we will write the expression for the variable in each node in order to show the increasing complexity with increasing number of layers:

$$x_1 = F(x_0, y_0)$$
$$y_1 = G(x_0, y_0)$$
$$x_2 = F(x_1, y_1) = F(F(x_0, y_0), G(x_0, y_0))$$
$$y_2 = G(x_1, y_1) = G(F(x_0, y_0), G(x_0, y_0))$$

We can already see that the expressions have already started looking unwieldy. On computing the values in the next layer, this becomes even more obvious:

$$x_3 = F(x_2, y_2) = F(F(F(x_0, y_0), G(x_0, y_0)), G(F(x_0, y_0), G(x_0, y_0)))$$
$$y_3 = G(x_2, y_2) = G(F(F(x_0, y_0), G(x_0, y_0)), G(F(x_0, y_0), G(x_0, y_0)))$$

An immediate observation is that the complexity and length of the closed-form function increases *exponentially* with the path lengths in the computational graphs. This type of complexity further increases in the case when optimization parameters are associated with the edges, and one tries to express the outputs/losses in terms of the inputs and the parameters on the edges. This is obviously a problem, if we try to use the boilerplate approach of first expressing the loss function in closed form in terms of the optimization parameters on the edges (in order to compute the derivative of the closed-form loss function).

11.3.2 The Broad Framework for Gradient Computation

The previous section makes it evident that differentiating closed-form expressions is not practical in the case of computational graphs. Therefore, one must somehow *algorithmically* compute gradients with respect to edges by using the topology of the computational graph. The purpose of this section is to introduce this broad algorithmic framework, and later sections will expand on the specific details of individual steps.

To learn the weights of a computational graph, an input-output pair is selected from the training data and the error of trying to predict the observed output with the observed input with the current values of the weights in the computational graph is quantified. When the errors are large, the weights need to be modified because the current computational graph does not reflect the observed data. Therefore, a loss function is computed as a function of this error, and the weights are updated so as to reduce the loss. This is achieved by computing the gradient of the loss with respect to the weights and performing a gradient-descent update. The overall approach for training a computational graph is as follows:

1. Use the attribute values from the input portion of a training data point to fix the values in the input nodes. Repeatedly select a node for which the values in all incoming nodes have already been computed and apply the node-specific function to also compute its variable. Such a node can be found in a directed acyclic graph by processing the nodes in order of increasing distance from input nodes. Repeat the process until the values in all nodes (including the output nodes) have been computed. If the values on the output nodes do not match the observed values of the output in the training point, compute the loss value. This phase is referred to as the *forward phase*.

2. Compute the gradient of the loss with respect to the weights on the edges. This phase is referred to as the *backwards phase*. The rationale for calling it a "backwards phase" will become clear later, when we introduce an algorithm that works backwards along the topology of the (directed acyclic) computational graph from the outputs to the inputs.

3. Update the weights in the negative direction of the gradient.

As in any stochastic gradient descent procedure, one cycles through the training points repeatedly until convergence is reached. A single cycle through all the training points is referred to as an *epoch*.

The main challenge is in computing the gradient of the loss function with respect to the weights in a computational graph. It turns out that *the derivatives of the node variables with respect to one another can be easily used to compute the derivative of the loss function with respect to the weights on the edges*. Therefore, in this discussion, we will focus on the computation of the derivatives of the variables with respect to one another. Later, we will show how these derivatives can be converted into gradients of loss functions with respect to weights.

11.3.3   Computing Node-to-Node Derivatives Using Brute Force

As discussed in an earlier section, one can express the function in a computational graph in terms of the nodes in early layers using an awkward closed-form expression that uses nested compositions of functions. If one were to indeed compute the derivative of this closed-form expression, it would require the use of the *chain rule of differential calculus* in order to deal with the repeated composition of functions. However, a blind application of the chain rule is rather wasteful in this case because many of the expressions in different portions of the inner nesting are identical, and one would be repeatedly computing the same derivative. The key idea in *automatic differentiation over computational graphs* is to recognize the fact that structure of the computational graph already provides all the information about which terms are repeated. We can avoid repeating the differentiation of these terms by using the structure of the computational graph itself to store intermediate results (by working backwards starting from output nodes to compute derivatives)! This is a well-known idea from dynamic programming, which has been used frequently in control theory [26, 71]. In the neural network community, this same algorithm is referred to as *backpropagation* (cf. Section 11.4). It is noteworthy that the applications of this idea in control theory were well-known to the traditional optimization community in 1960 [26, 71], although they remained unknown to researchers in the field of artificial intelligence for a while (who coined the term "backpropagation" in the 1980s to independently propose and describe this idea in the context of neural networks).

The simplest version of the chain rule is defined for a univariate composition of functions:

$$\frac{\partial f(g(x))}{\partial x} = \frac{\partial f(g(x))}{\partial g(x)} \cdot \frac{\partial g(x)}{\partial x} \tag{11.2}$$

This variant is referred to as the *univariate chain rule*. Note that each term on the right-hand side is a *local gradient* because it computes the derivative of a *local* function with respect to its immediate argument rather than a recursively derived argument. The basic idea is that a composition of functions is applied on the input x to yield the final output, and the gradient of the final output is given by the product of the local gradients along that path. Each local gradient only needs to worry about its specific input and output, which simplifies the computation. An example is shown in Figure 11.4 in which the function $f(y)$ is $\cos(y)$ and $g(x) = x^2$. Therefore, the composition function is $\cos(x^2)$. On using the univariate chain rule, we obtain the following:

$$\frac{\partial f(g(x))}{\partial x} = \underbrace{\frac{\partial f(g(x))}{\partial g(x)}}_{-\sin(g(x))} \cdot \underbrace{\frac{\partial g(x)}{\partial x}}_{2x} = -2x \cdot \sin(x^2)$$

Note that we can annotate each of the above two multiplicative components on the two *connections* in the graph, and simply compute the product of these values. Therefore, *for a computational graph containing a single path, the derivative of one node with respect to another is simply the product of these annotated values on the connections between the two nodes.* The example of Figure 11.4 is a rather simple case in which the computational graph is a single path. In general, a computational graph with good expressive power will not be a single path. Rather, a single node may feed its output to multiple nodes. For example, consider the case in which we have a single input x, and we have k independent computational nodes that compute the functions $g_1(x), g_2(x), \ldots g_k(x)$. If these nodes are connected to a single output node computing the function $f()$ with k arguments, then the

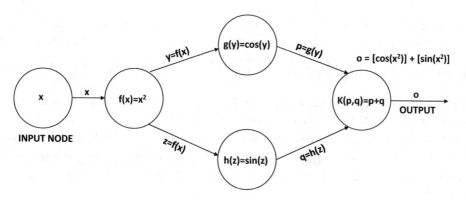

Figure 11.6: A simple computational function that illustrates the chain rule

resulting function that is computed is $f(g_1(x), \ldots g_k(x))$. In such cases, the *multivariate chain rule* needs to be used. The multivariate chain rule is defined as follows:

$$\frac{\partial f(g_1(x), \ldots g_k(x))}{\partial x} = \sum_{i=1}^{k} \frac{\partial f(g_1(x), \ldots g_k(x))}{\partial g_i(x)} \cdot \frac{\partial g_i(x)}{\partial x} \tag{11.3}$$

It is easy to see that the multivariate chain rule of Equation 11.3 is a simple generalization of that in Equation 11.2.

One can also view the multivariate chain rule in a path-centric fashion rather than a node-centric fashion. *For any pair of source-sink nodes, the derivative of the variable in the sink node with respect to the variable in the source node is simply the sum of the expressions arising from the univariate chain rule being applied to all paths existing between that pair of nodes.* This view leads to a direct expression for the derivative between any pair of nodes (rather than the recursive multivariate rule). However, it leads to an excessive computation, because the number of paths between a pair of nodes is exponentially related to the path length. In order to show the repetitive nature of the operations, we work with a very simple closed-form function with a single input x:

$$o = \sin(x^2) + \cos(x^2) \tag{11.4}$$

The resulting computational graph is shown in Figure 11.6. In this case, the multivariate chain rule is applied to compute the derivative of the output o with respect to x. This is achieved by summing the results of the univariate chain rule for each of the two paths from x to o in Figure 11.6:

$$\frac{\partial o}{\partial x} = \underbrace{\frac{\partial K(p,q)}{\partial p}}_{1} \cdot \underbrace{g'(y)}_{-\sin(y)} \cdot \underbrace{f'(x)}_{2x} + \underbrace{\frac{\partial K(p,q)}{\partial q}}_{1} \cdot \underbrace{h'(z)}_{\cos(z)} \cdot \underbrace{f'(x)}_{2x}$$

$$= -2x \cdot \sin(y) + 2x \cdot \cos(z)$$

$$= -2x \cdot \sin(x^2) + 2x \cdot \cos(x^2)$$

In this simple example, there are two paths, both of which compute the function $f(x) = x^2$. As a result, the function $f(x)$ is differentiated *twice*, once for each path. This type of repetition can have severe effects for large multilayer networks containing many shared nodes, where the same function might be differentiated hundreds of thousands of times as a portion of the nested recursion. It is this *repeated and wasteful* approach to the computation of the derivative, that it is impractical to express the global function of a computational graph in closed form and explicitly differentiating it.

One can summarize the path-centric view of the multivariate chain rule as follows:

Lemma 11.3.1 (Pathwise Aggregation Lemma) *Consider a directed acyclic compu-tational graph in which the ith node contains variable $y(i)$. The local derivative $z(i, j)$ of the directed edge (i, j) in the graph is defined as $z(i, j) = \frac{\partial y(j)}{\partial y(i)}$. Let a non-null set of paths $\mathcal{P}$ exist from a node s in the graph to node t. Then, the value of $\frac{\partial y(t)}{\partial y(s)}$ is given by computing the product of the local gradients along each path in $\mathcal{P}$, and summing these products over all paths in $\mathcal{P}$.*

$$\frac{\partial y(t)}{\partial y(s)} = \sum_{P \in \mathcal{P}} \prod_{(i,j) \in P} z(i, j) \tag{11.5}$$

This lemma can be easily shown by applying the multivariate chain rule (Equation 11.3) recursively over the computational graph. Although the use of the pathwise aggregation lemma is a wasteful approach for computing the derivative of $y(t)$ with respect to $y(s)$, it enables a simple and intuitive exponential-time algorithm for derivative computation.

An Exponential-Time Algorithm

The pathwise aggregation lemma provides a natural exponential-time algorithm, which is roughly similar to the steps one would go through by expressing the computational function in closed form with respect to a particular variable and then differentiating it. Specifically, the pathwise aggregation lemma leads to the following exponential-time algorithm to com-pute the derivative of the output o with respect to a variable x in the graph:

1. Use computational graph to compute the value $y(i)$ of each node i in a forward phase.

2. Compute the local partial derivatives $z(i, j) = \frac{\partial y(j)}{\partial y(i)}$ on each edge in the computational graph.

3. Let $\mathcal{P}$ be the set of all paths from an input node with value x to the output o. For each path $P \in \mathcal{P}$ compute the product of each local derivative $z(i, j)$ on that path.

4. Add up these values over all paths in $\mathcal{P}$.

In general, a computational graph will have an exponentially increasing number of paths with depth and one must add the product of the local derivatives over all paths. An example is shown in Figure 11.7, in which we have five layers, each of which has only two units. Therefore, the number of paths between the input and output is $2^5 = 32$. The jth hidden unit of the ith layer is denoted by $h(i, j)$. Each hidden unit is defined as the product of its inputs:

$$h(i, j) = h(i - 1, 1) \cdot h(i - 1, 2) \quad \forall j \in \{1, 2\} \tag{11.6}$$

In this case, the output is x^{32}, which is expressible in closed form, and can be differentiated easily with respect to x. In other words, we do not really need computational graphs in order to perform the differentiation. However, we will use the exponential-time algorithm to elucidate its workings. The derivatives of each $h(i, j)$ with respect to its two inputs are

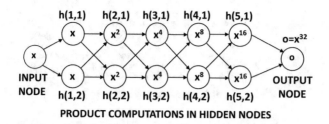

PRODUCT COMPUTATIONS IN HIDDEN NODES

Figure 11.7: The chain rule aggregates the product of local derivatives along $2^5 = 32$ paths

the values of the complementary inputs, because the partial derivative of the multiplication of two variables is the complementary variable:

$$\frac{\partial h(i,j)}{\partial h(i-1,1)} = h(i-1,2), \quad \frac{\partial h(i,j)}{\partial h(i-1,2)} = h(i-1,1)$$

The pathwise aggregation lemma implies that the value of $\frac{\partial o}{\partial x}$ is the product of the local derivatives (which are the complementary input values in this particular case) along all 32 paths from the input to the output:

$$\frac{\partial o}{\partial x} = \sum_{j_1,j_2,j_3,j_4,j_5 \in \{1,2\}^5} \prod \underbrace{h(1,j_1)}_{x} \underbrace{h(2,j_2)}_{x^2} \underbrace{h(3,j_3)}_{x^4} \underbrace{h(4,j_4)}_{x^8} \underbrace{h(5,j_5)}_{x^{16}}$$

$$= \sum_{\text{All 32 paths}} x^{31} = 32x^{31}$$

This result is, of course, consistent with what one would obtain on differentiating x^{32} directly with respect to x. However, an important observation is that it requires 2^5 aggregations to compute the derivative in this way for a relatively simple graph. More importantly, *we repeatedly differentiate the same function computed in a node* for aggregation. For example, the differentiation of the variable $h(3,1)$ is performed 16 times because it appears in 16 paths from x to o.

Obviously, this is an inefficient approach to compute gradients. For a network with 100 nodes in each layer and three layers, we will have a million paths. *Nevertheless, this is exactly what we do in traditional machine learning when our prediction function is a complex composition function.* Manually working out the details of a complex composition function is tedious and impractical beyond a certain level of complexity. It is here that one can apply dynamic programming (which is guided by the structure of the computational graph) in order to store important intermediate results. By using such an approach, one can minimize repeated computations, and achieve polynomial complexity.

11.3.4 Dynamic Programming for Computing Node-to-Node Derivatives

In graph theory, computing all types of path-aggregative values over directed acyclic graphs is done using dynamic programming. Consider a directed acyclic graph in which the value $z(i,j)$ (interpreted as local partial derivative of variable in node j with respect to variable

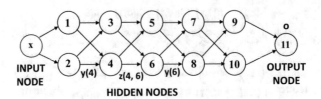

Figure 11.8: Edges are labeled with local partial derivatives such as $z(4,6) = \frac{\partial y(6)}{\partial y(4)}$

in node i) is associated with edge (i, j). In other words, if $y(p)$ is the variable in the node p, we have the following:

$$z(i, j) = \frac{\partial y(j)}{\partial y(i)} \tag{11.7}$$

An example of such a computational graph is shown in Figure 11.8. In this case, we have associated the edge $(2, 4)$ with the corresponding partial derivative. We would like to compute the product of $z(i, j)$ over each path $P \in \mathcal{P}$ from source node s to output node t and then add them in order to obtain the partial derivative $S(s, t) = \frac{\partial y(t)}{\partial y(s)}$:

$$S(s, t) = \sum_{P \in \mathcal{P}} \prod_{(i,j) \in P} z(i, j) \tag{11.8}$$

Let $A(i)$ be the set of nodes at the end points of outgoing edges from node i. We can compute the aggregated value $S(i, t)$ for each intermediate node i (between source node s and output node t) using the following well-known dynamic programming update:

$$S(i, t) \Leftarrow \sum_{j \in A(i)} S(j, t) z(i, j) \tag{11.9}$$

This computation can be performed backwards starting from the nodes directly incident on o, since $S(t, t) = \frac{\partial y(t)}{\partial y(t)}$ is already known to be 1. This is because the partial derivative of a variable with respect to itself is always 1. Therefore one can describe the pseudocode of this algorithm as follows:

Initialize $S(t, t) = 1$;
repeat
  Select an unprocessed node i such that the values of $S(j, t)$ all of its outgoing
    nodes $j \in A(i)$ are available;
  Update $S(i, t) \Leftarrow \sum_{j \in A(i)} S(j, t) z(i, j)$;
until all nodes have been selected;

Note that the above algorithm always selects a node i for which the value of $S(j, t)$ is available for all nodes $j \in A(i)$. Such a node is always available in directed acyclic graphs, and the node selection order will always be in the backwards direction starting from node t. Therefore, the above algorithm will work only when the computational graph does not have cycles, and it is referred to as the *backpropagation algorithm.*

The algorithm discussed above is used by the network optimization community for computing all types of path-centric functions between *source-sink* node pairs (s, t) on directed acyclic graphs, which would otherwise require exponential time. For example, one can even use a variation of the above algorithm to find the longest path in a directed acyclic graph [8].

Interestingly, *the aforementioned dynamic programming update is exactly the multivariate chain rule of Equation 11.3, which is repeated in the backwards direction starting at the output node where the local gradient is known.* This is because we derived the path-aggregative form of the loss gradient (Lemma 11.3.1) using this chain rule in the first place. The main difference is that we apply the rule in a particular order in order to minimize computations. We emphasize this important point below:

> Using dynamic programming to efficiently aggregate the product of local gradients along the exponentially many paths in a computational graph results in a dynamic programming update that is identical to the multivariate chain rule of differential calculus. The main point of dynamic programming is to apply this rule in a particular order, so that the derivative computations at different nodes are not repeated.

This approach is the backbone of the backpropagation algorithm used in neural networks. We will discuss more details of neural network-specific enhancements in Section 11.4. In the case where we have multiple output nodes $t_1, \ldots t_p$, one can initialize each $S(t_r, t_r)$ to 1, and then apply the same approach for each t_r.

11.3.4.1 Example of Computing Node-to-Node Derivatives

In order to show how the backpropagation approach works, we will provide an example of computation of node-to-node derivatives in a graph containing 10 nodes (see Figure 11.9). A variety of functions are computed in various nodes, such as the sum function (denoted by '+'), the product function (denoted by '*'), and the trigonometric sine/cosine functions. The variables in the 10 nodes are denoted by $y(1) \ldots y(10)$, where the variable $y(i)$ belongs to the ith node in the figure. Two of the edges incoming into node 6 also have the weights w_2 and w_3 associated with them. Other edges do not have weights associated with them. The functions computed in the various layers are as follows:

Layer 1: $y(4) = y(1) \cdot y(2), \quad y(5) = y(1) \cdot y(2) \cdot y(3), \quad y(6) = w_2 \cdot y(2) + w_3 \cdot y(3)$
Layer 2: $y(7) = \sin(y(4)), \quad y(8) = \cos(y(5)), \quad y(9) = \sin(y(6))$
Layer 3: $y(10) = y(7) \cdot y(8) \cdot y(9)$

We would like to compute the derivative of $y(10)$ with respect to each of the inputs $y(1)$, $y(2)$, and $y(3)$. One possibility is to simply express the $y(10)$ in closed form in terms of the inputs $y(1)$, $y(2)$ and $y(3)$, and then compute the derivative. By recursively using the above relationships, it is easy to show that $y(10)$ can be expressed in terms of $y(1)$, $y(2)$, and $y(3)$ as follows:

$$y(10) = \sin(y(1) \cdot y(2)) \cdot \cos(y(1) \cdot y(2) \cdot y(3)) \cdot \sin(w_2 \cdot y(2) + w_3 \cdot y(3))$$

As discussed earlier computing the closed-form derivative is not practical for larger networks. Furthermore, since one needs to compute the derivative of the output with respect to each and every node in the network, such an approach would also required closed-form expressions in terms of upstream nodes like $y(4)$, $y(5)$ and $y(6)$. All this tends to increase the amount of repeated computation. Luckily, backpropagation frees us from this repeated computation, since the derivative in $y(10)$ with respect to each and every node is computed by the backwards phase. The algorithm starts, by initializing the derivative of the output $y(10)$ with respect to itself, which is 1:

$$S(10, 10) = \frac{\partial y(10)}{\partial y(10)} = 1$$

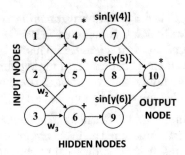

Figure 11.9: Example of node-to-node derivative computation

Subsequently, the derivatives of $y(10)$ with respect to all the variables on its incoming nodes are computed. Since $y(10)$ is expressed in terms of the variables $y(7)$, $y(8)$, and $y(9)$ incoming into it, this is easy to do, and the results are denoted by $z(7, 10)$, $z(8, 10)$, and $z(9, 10)$ (which is consistent with the notations used earlier in this chapter). Therefore, we have the following:

$$z(7, 10) = \frac{\partial y(10)}{\partial y(7)} = y(8) \cdot y(9)$$

$$z(8, 10) = \frac{\partial y(10)}{\partial y(8)} = y(7) \cdot y(9)$$

$$z(9, 10) = \frac{\partial y(10)}{\partial y(9)} = y(7) \cdot y(8)$$

Subsequently, we can use these values in order to compute $S(7, 10)$, $S(8, 10)$, and $S(9, 10)$ using the recursive backpropagation update:

$$S(7, 10) = \frac{\partial y(10)}{\partial y(7)} = S(10, 10) \cdot z(7, 10) = y(8) \cdot y(9)$$

$$S(8, 10) = \frac{\partial y(10)}{\partial y(8)} = S(10, 10) \cdot z(8, 10) = y(7) \cdot y(9)$$

$$S(9, 10) = \frac{\partial y(10)}{\partial y(9)} = S(10, 10) \cdot z(9, 10) = y(7) \cdot y(8)$$

Next, we compute the derivatives $z(4, 7)$, $z(5, 8)$, and $z(6, 9)$ associated with all the edges incoming into nodes 7, 8, and 9:

$$z(4, 7) = \frac{\partial y(7)}{\partial y(4)} = \cos[y(4)]$$

$$z(5, 8) = \frac{\partial y(8)}{\partial y(5)} = -\sin[y(5)]$$

$$z(6, 9) = \frac{\partial y(9)}{\partial y(6)} = \cos[y(6)]$$

These values can be used to compute $S(4, 10)$, $S(5, 10)$, and $S(6, 10)$:

$$S(4, 10) = \frac{\partial y(10)}{\partial y(4)} = S(7, 10) \cdot z(4, 7) = y(8) \cdot y(9) \cdot \cos[y(4)]$$

$$S(5, 10) = \frac{\partial y(10)}{\partial y(5)} = S(8, 10) \cdot z(5, 8) = -y(7) \cdot y(9) \cdot \sin[y(5)]$$

$$S(6, 10) = \frac{\partial y(10)}{\partial y(6)} = S(9, 10) \cdot z(6, 9) = y(7) \cdot y(8) \cdot \cos[y(6)]$$

In order to compute the derivatives with respect to the input values, one now needs to compute the values of $z(1, 3)$, $z(1, 4)$, $z(2, 4)$, $z(2, 5)$, $z(2, 6)$, $z(3, 5)$, and $z(3, 6)$:

$$z(1, 4) = \frac{\partial y(4)}{\partial y(1)} = y(2)$$

$$z(2, 4) = \frac{\partial y(4)}{\partial y(2)} = y(1)$$

$$z(1, 5) = \frac{\partial y(5)}{\partial y(1)} = y(2) \cdot y(3)$$

$$z(2, 5) = \frac{\partial y(5)}{\partial y(2)} = y(1) \cdot y(3)$$

$$z(3, 5) = \frac{\partial y(5)}{\partial y(3)} = y(1) \cdot y(2)$$

$$z(2, 6) = \frac{\partial y(6)}{\partial y(2)} = w_2$$

$$z(3, 6) = \frac{\partial y(6)}{\partial y(3)} = w_3$$

These partial derivatives can be backpropagated to compute $S(1, 10)$, $S(2, 10)$, and $S(3, 10)$:

$$S(1, 10) = \frac{\partial y(10)}{\partial y(1)} = S(4, 10) \cdot z(1, 4) + S(5, 10) \cdot z(1, 5)$$

$$= y(8) \cdot y(9) \cdot \cos[y(4)] \cdot y(2) - y(7) \cdot y(9) \cdot \sin[y(5)] \cdot y(2) \cdot y(3)$$

$$S(2, 10) = \frac{\partial y(10)}{\partial y(2)} = S(4, 10) \cdot z(2, 4) + S(5, 10) \cdot z(2, 5) + S(6, 10) \cdot z(2, 6)$$

$$= y(8) \cdot y(9) \cdot \cos[y(4)] \cdot y(1) - y(7) \cdot y(9) \cdot \sin[y(5)] \cdot y(1) \cdot y(3) +$$
$$+ y(7) \cdot y(8) \cdot \cos[y(6)] \cdot w_2$$

$$S(3, 10) = \frac{\partial y(10)}{\partial y(3)} = S(5, 10) \cdot z(3, 5) + S(6, 10) \cdot z(3, 6)$$

$$= -y(7) \cdot y(9) \cdot \sin[y(5)] \cdot y(1) \cdot y(2) + y(7) \cdot y(8) \cdot \cos[y(6)] \cdot w_3$$

Note that the use of a backward phase has the advantage of computing the derivative of $y(10)$ (output node variable) with respect to all the hidden and input node variables. These different derivatives have many sub-expressions in common, although the derivative computation of these sub-expressions is not repeated. This is the advantage of using the backwards phase for derivative computation as opposed to the use of closed-form expressions.

Because of the tedious nature of the closed-form expressions for outputs, the algebraic expressions for derivatives are also very long and awkward (no matter how we compute them). One can see that this is true even for the simple, ten-node computational graph of this section. For example, if one examines the derivative of $y(10)$ with respect to each of nodes $y(1)$, $y(2)$ and $y(3)$, the algebraic expression wraps into multiple lines. Furthermore, one cannot avoid the presence of repeated subexpressions within the algebraic derivative. This is counter-productive because our original goal in the backwards algorithm was to avoid the repeated computation endemic to traditional derivative evaluation with closed-form expressions. Therefore, one does not *algebraically* compute these types of expressions in real-world networks. One would first *numerically* compute all the node variables for a *specific* set of numerical inputs from the training data. Subsequently, one would *numerically* carry the derivatives backward, so that one does not have to carry the large algebraic expressions (with many repeated sub-expressions) in the backwards direction. The advantage of carrying numerical expressions is that multiple terms get consolidated into a single numerical value, which is specific to a particular input. By making the numerical choice, one must repeat the backwards computation algorithm *for each training point*, but it is still a better choice than computing the (massive) symbolic derivative in one shot and substituting the values in different training points. This is the reason that such an approach is referred to as *numerical differentiation* rather than *symbolic differentiation*. In much of machine learning, one first computes the algebraic derivative (which is symbolic differentiation) before substituting numerical values of the variables in the expression (for the derivative) to perform gradient-descent updates. This is different from the case of computational graphs, where the backwards algorithm is *numerically* applied to each training point.

11.3.5 Converting Node-to-Node Derivatives into Loss-to-Weight Derivatives

Most computational graphs define loss functions with respect to output node variables. One needs to compute the derivatives with respect to weights on *edges* rather than the node variables (in order to update the weights). In general, the node-to-node derivatives can be converted into loss-to-weight derivatives with a few additional applications of the univariate and multivariate chain rule.

Consider the case in which we have computed the node-to-node derivative of output variables in nodes indexed by $t_1, t_2, \ldots t_p$ with respect to the variable in node i using the dynamic programming approach in the previous section. Therefore, the computational graph has p output nodes in which the corresponding variable values are $y(t_1) \ldots y(t_p)$ (since the indices of the output nodes are $t_1 \ldots t_p$). The loss function is denoted by $L(y(t_1), \ldots y(t_p))$. We would like to compute the derivative of this loss function with respect to *the weights in the incoming edges of i*. For the purpose of this discussion, let w_{ji} be the weight of an edge from node index j to node index i. Therefore, we want to compute the derivative of the loss function with respect to w_{ji}. In the following, we will abbreviate $L(y_{t_1}, \ldots y_{t_p})$ with L for compactness of notation:

$$\frac{\partial L}{\partial w_{ji}} = \left[\frac{\partial L}{\partial y(i)}\right] \frac{\partial y(i)}{\partial w_{ji}} \qquad \text{[Univariate chain rule]}$$

$$= \left[\sum_{k=1}^{p} \frac{\partial L}{\partial y(t_k)} \frac{\partial y(t_k)}{\partial y(i)}\right] \frac{\partial y(i)}{\partial w_{ji}} \qquad \text{[Multivariate chain rule]}$$

Here, it is noteworthy that the loss function is typically a closed-form function of the variables in the node indices $t_1 \ldots t_p$, which is often either is least-squares function or a logarithmic loss function (like the examples in Chapter 4). Therefore, each derivative of the loss L with respect to $y(t_i)$ is easy to compute. Furthermore, the value of each $\frac{\partial y(t_k)}{\partial y(i)}$ for $k \in \{1 \ldots p\}$ can be computed using the dynamic programming algorithm of the previous section. The value of $\frac{\partial y_i}{\partial w_{ji}}$ is a derivative of the **local** function at each node, which usually has a simple form. Therefore, the loss-to-weight derivatives can be computed relatively easily, once the node-to-node derivatives have been computed using dynamic programming.

Although one can apply the pseudocode of page 460 to compute $\frac{\partial y(t_k)}{\partial y(i)}$ for each $k \in \{1 \ldots p\}$, it is more efficient to collapse all these computations into a single backwards algorithm. In practice, one initializes the derivatives at the output nodes to the loss derivatives $\frac{\partial L}{\partial y(t_k)}$ for each $k \in \{1 \ldots p\}$ rather than the value of 1 (as shown in the pseudocode of page 460). Subsequently, the entire loss derivative $\Delta(i) = \frac{\partial L}{\partial y(i)}$ is propagated backwards. Therefore, the modified algorithm for computing the loss derivative with respect to the *node variables* as well as the *edge variables* is as follows:

> Initialize $\Delta(t_r) = \frac{\partial L}{\partial y(t_k)}$ for each $k \in \{1 \ldots p\}$;
> **repeat**
>   Select an unprocessed node i such that the values of $\Delta(j)$ all of its outgoing
>   nodes $j \in A(i)$ are available;
>   Update $\Delta(i) \Leftarrow \sum_{j \in A(i)} \Delta(j) z(i,j)$;
> **until** all nodes have been selected;
> **for each edge** (j,i) with weight w_{ji} **do** compute $\frac{\partial L}{\partial w_{ji}} = \Delta(i) \frac{\partial y(i)}{\partial w_{ji}}$;

In the above algorithm, $y(i)$ denotes the variable at node i. The key difference of this algorithm from the algorithm on page 460 is in the nature of the initialization and the addition of a final step computing the edge-wise derivatives. However, the core algorithm for computing the node-to-node derivatives remains an integral part of this algorithm. In fact, one can convert all the weights on the edges into additional "input" nodes containing weight parameters, and also add computational nodes that multiply the weights with the corresponding variables at the tail nodes of the edges. Furthermore, a computational node can be added that computes the loss from the output node(s). For example, the architecture of Figure 11.9 can be converted to that in Figure 11.10. Therefore, a computational graph with *learnable weights* can be converted into an unweighted graph with *learnable node variables* (on a subset of nodes). Performing only node-to-node derivative computation in Figure 11.10 from the loss node to the weight nodes is equivalent to loss-to-weight derivative computation. In other words, *loss-to-weight derivative computation in a weighted graph is equivalent to node-to-node derivative computation in a modified computational graph*. The derivative of the loss with respect to each weight can be denoted by the vector $\frac{\partial L}{\partial \overline{W}}$ (in matrix calculus notation), where $\overline{W}$ denotes the weight vector. Subsequently, the standard gradient descent update can be performed:

$$\overline{W} \Leftarrow \overline{W} - \alpha \frac{\partial L}{\partial \overline{W}} \tag{11.10}$$

Here, α is the learning rate. This type of update is performed to convergence by repeating the process with different inputs in order to learn the weights of the computational graph.

11.3.5.1 Example of Computing Loss-to-Weight Derivatives

Consider the case of Figure 11.9 in which the loss function is defined by $L = \log[y(10)^2]$, and we wish to compute the derivative of the loss with respect to the weights w_2 and w_3. In such a case, the derivative of the loss with respect to the weights is given by the following:

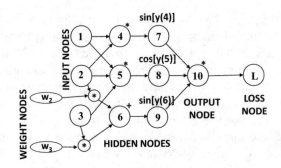

Figure 11.10: Converting loss-to-weight derivatives into node-to-node derivative computation based on Figure 11.9. Note the extra weight nodes and an extra loss node

$$\frac{\partial L}{\partial w_2} = \frac{\partial L}{\partial y(10)} \frac{\partial y(10)}{\partial y(6)} \frac{\partial y(6)}{\partial w_2} = \left[\frac{2}{y(10)}\right] [y(7) \cdot y(8) \cdot \cos[y(6)]]\, y(2)$$

$$\frac{\partial L}{\partial w_3} = \frac{\partial L}{\partial y(10)} \frac{\partial y(10)}{\partial y(6)} \frac{\partial y(6)}{\partial w_3} = \left[\frac{2}{y(10)}\right] [y(7) \cdot y(8) \cdot \cos[y(6)]]\, y(3)$$

Note that the quantity $\frac{\partial y(10)}{\partial y(6)}$ has been obtained using the example in the previous section on node-to-node derivatives. In practice, these quantities are not computed algebraically. This is because the aforementioned algebraic expressions can be extremely awkward for large networks. Rather, for each numerical input set $\{y(1), y(2), y(3)\}$, one computes the different values of $y(i)$ in a forward phase. Subsequently, the derivatives of the loss with respect to each node variable (and incoming weights) are computed in a backwards phase. Again, these values are computed numerically for a specific input set $\{y(1), y(2), y(3)\}$. The numerical gradients can be used in order to update the weights for learning purposes.

11.3.6 Computational Graphs with Vector Variables

The previous section discuss the simple case in which each node of a computational graph contains a single scalar variable, whereas this section allows vector variables. In other words, the ith node contains the vector variable $\overline{y}_i$. Therefore, the **local** functions applied at the computational nodes are also vector-to-vector functions. For any node i, its local function uses an argument which corresponds to all the vector components of all its incoming nodes. From the input perspective of this local function, this situation is not too different from the previous case, where the argument is a vector corresponding to all the scalar inputs. However, the main difference is that the *output* of this function is a vector rather than a scalar. One example of such a vector-to-vector function is the *softmax* function (cf. Equation 4.64 of Chapter 4), which takes k real values as inputs and outputs k probabilities. In general, the number of inputs of the function need not be the same as the number of outputs in vector-to-vector functions. An important observation is the following:

> One can compute vector-to-vector derivatives in a computational graph using the vector-centric chain rule (cf. Section 4.6.3 of Chapter 4) rather than the scalar chain rule.

As discussed in Equation 4.19 of Chapter 4, a vector-to-vector derivative is a *matrix*. Consider two vectors, $\overline{v} = [v_1 \ldots v_d]^T$ and $\overline{h} = [h_1 \ldots h_m]^T$, which occur somewhere in the computational graph shown in Figure 11.11(a). There might be nodes incoming into $\overline{v}$ as

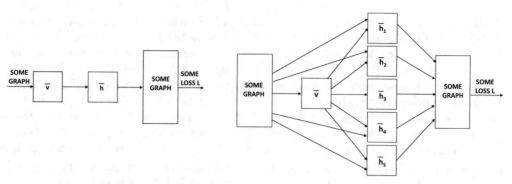

(a) Vector-centric graph with single path   (b) Vector-centric graph with multiple paths

Figure 11.11: Examples of vector-centric computational graphs

well as a loss L computed in a later layer. Then, using the denominator layout of matrix calculus, the vector-to-vector derivative is the transpose of the *Jacobian* matrix, which was introduced in Chapter 4:

$$\frac{\partial \overline{h}}{\partial \overline{v}} = \text{Jacobian}(\overline{h}, \overline{v})^T$$

The (i, j)th entry of the above vector-to-vector derivative is simply $\frac{\partial h_j}{\partial v_i}$. Since $\overline{h}$ is an m-dimensional vector and $\overline{v}$ is a d-dimensional vector, the vector derivative is a $d \times m$ matrix. As discussed in Section 4.6.3 of Chapter 4, the chain rule over a single vector-centric path looks almost identical to the univariate chain rule over scalars, when one substitutes local partial derivatives with Jacobians. In other words, we can derive the following vector-valued chain rule for the single path of Figure 11.11(a):

$$\frac{\partial L}{\partial \overline{v}} = \underbrace{\frac{\partial \overline{h}}{\partial \overline{v}}}_{d \times m} \underbrace{\frac{\partial L}{\partial \overline{h}}}_{m \times 1} = \text{Jacobian}(\overline{h}, \overline{v})^T \frac{\partial L}{\partial \overline{h}}$$

Therefore, once the gradient of the loss is available with respect to a layer, it can be backpropagated by multiplying it with the transpose of a Jacobian! Here the *ordering of the matrices is important*, since matrix multiplication is not commutative.

The above provides the chain rule only for the case where the computational graph is a single path. What happens when the computational graph has an arbitrary structure? In such a case, we might have a situation where we have multiple nodes $\overline{h}_1 \ldots \overline{h}_s$ between node $\overline{v}$ and a network in later layers, as shown in Figure 11.11(b). Furthermore, there are connections between alternate layers, which are referred to as *skip connections*. Assume that the vector $\overline{h}_i$ has dimensionality m_i. In such a case, the partial derivative turns out to be a simple generalization of the previous case:

$$\frac{\partial L}{\partial \overline{v}} = \sum_{i=1}^{s} \underbrace{\frac{\partial \overline{h}_i}{\partial \overline{v}}}_{d \times m_i} \underbrace{\frac{\partial L}{\partial \overline{h}_i}}_{m_i \times 1} = \sum_{i=1}^{s} \text{Jacobian}(\overline{h}_i, \overline{v})^T \frac{\partial L}{\partial \overline{h}_i}$$

In most layered neural networks, we only have a single path and we rarely have to deal with the case of branches. Such branches might, however, arise in the case of neural networks with skip connections [see Figures 11.11(b) and 11.13(b)]. However, even in complicated network architectures like Figures 11.11(b) and 11.13(b), *each node only has to worry about its local outgoing edges during backpropagation.* Therefore, we provide a very general vector-based algorithm below that can work even in the presence of skip connections.

Consider the case where we have p output nodes containing vector-valued variables, which have indices denoted by $t_1 \ldots t_p$, and the variables in it are $\overline{y}(t_1) \ldots \overline{y}(t_p)$. In such a case, the loss function L might be function of all the components in these vectors. Assume that the ith node contains a column vector of variables denoted by $\overline{y}(i)$. Furthermore, in the denominator layout of matrix calculus, each $\overline{\Delta}(i) = \frac{\partial L}{\partial \overline{y}(i)}$ is a column vector with dimensionality equal to that of $\overline{y}(i)$. It is this *vector* of loss derivatives that will be propagated backwards. The vector-centric algorithm for computing derivatives is as follows:

> Initialize $\overline{\Delta}(t_k) = \frac{\partial L}{\partial \overline{y}(t_k)}$ for each output node t_k for $k \in \{1 \ldots p\}$;
> **repeat**
>   Select an unprocessed node i such that the values of $\overline{\Delta}(j)$ all of its outgoing
>   nodes $j \in A(i)$ are available;
>   Update $\overline{\Delta}(i) \Leftarrow \sum_{j \in A(i)} \text{Jacobian}(\overline{y}(j), \overline{y}(i))^T \overline{\Delta}(j)$;
> **until** all nodes have been selected;
> **for** the vector $\overline{w}_i$ of edges incoming to each node i **do** compute $\frac{\partial L}{\partial \overline{w}_i} = \frac{\partial \overline{y}(i)}{\partial \overline{w}_i} \overline{\Delta}(i)$;

In the final step of the above pseudocode, the derivative of vector $\overline{y}(i)$ with respect to the vector $\overline{w}_i$ is computed, which is itself the transpose of a Jacobian matrix. This final step converts a vector of partial derivatives with respect to node variables into a vector of partial derivatives with respect to weights incoming at a node.

11.4 Application: Backpropagation in Neural Networks

In this section, we will describe how the generic algorithm based on computational graphs can be used in order to perform the backpropagation algorithm in neural networks. The key idea is that specific variables in the neural networks need to be defined as nodes of the computational-graph abstraction. The same neural network can be represented by different types of computational graphs, depending on which variables in the neural network are used to create computational graph nodes. The precise methodology for performing the backpropagation updates depends heavily on this design choice.

Consider the case of a neural network that first applies a linear function with weights w_{ij} on its inputs to create the pre-activation value $a(i)$, and then applies the activation function $\Phi(\cdot)$ in order to create the output $h(i)$:

$$h(i) = \Phi(a(i))$$

The variables $h(i)$ and $a(i)$ are shown in Figure 11.12. In this case, it is noteworthy that there are several ways in which the computational graph can be created. For example, one might create a computational graph in which each node contains the post-activation value $h(i)$, and therefore we are implicitly setting $y(i) = h(i)$. A second choice is to create a computational graph in which each node contains the pre-activation variable $a(i)$ and therefore we are setting $y(i) = a(i)$. It is even possible to create a decoupled computational graph containing both $a(i)$ and $h(i)$; in the last case, the computational graph will have twice as many nodes as the neural network. In all these cases, a relatively straightforward special-case/simplification of the pseudocodes in the previous section can be used for learning the gradient:

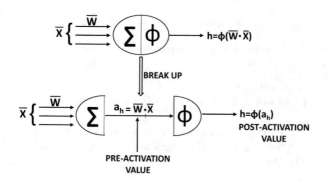

Figure 11.12: Pre- and post-activation values within a neuron

1. The post-activation value $y(i) = h(i)$ could represent the variable in the ith computational node in the graph. Therefore, each computational node in such a graph *first* applies the linear function, *and then* applies the activation function. The post-activation value is shown in Figure 11.12. In such a case, the value of $z(i,j) = \frac{\partial y(j)}{\partial y(i)} = \frac{\partial h(j)}{\partial h(i)}$ in the pseudocode of page 465 is $w_{ij}\Phi'_j$. Here, w_{ij} is the weight of the edge from i to j and $\Phi'_j = \frac{\partial \Phi(a(j))}{\partial a(j)}$ is the local derivative of the activation function at node j with respect to its argument. The value of each $\Delta(t_r)$ at output node t_r is simply the derivative of the loss function with respect to $h(t_r)$. The final derivative with respect to the weight w_{ji} (in the final line of the pseudocode on page 465) is equal to $\Delta(i)\frac{\partial h(i)}{\partial w_{ji}} = \Delta(i)h(j)\Phi'_i$.

2. The pre-activation value (after applying the linear function), which is denoted by $a(i)$, could represent the variable in each computational node i in the graph. Note the subtle distinction between the work performed in computational nodes and neural network nodes. Each *computational* node *first* applies the activation function to each of its inputs before applying a linear function, whereas these operations are performed in the reverse order in a neural network. The structure of the computational graph is roughly similar to the neural network, except that the first layer of computational nodes do not contain an activation. In such a case, the value of $z(i,j) = \frac{\partial y(j)}{\partial y(i)} = \frac{\partial a(j)}{\partial a(i)}$ in the pseudocode of page 465 is $\Phi'_i w_{ij}$. Note that $\Phi(a(i))$ is being differentiated with respect to its argument in this case, rather than $\Phi(a(j))$ as in the case of the post-activation variables. The value of the loss derivative with respect to the pre-activation variable $a(t_r)$ in the rth output node t_r needs to account for the fact that it is a pre-activation value, and therefore, we cannot directly use the loss derivative with respect to post-activation values. Rather the post-activation loss derivative needs to be *multiplied with the derivative* Φ'_{t_r} *of the activation function at that node*. The final derivative with respect to the weight w_{ji} (final line of pseudocode on page 465) is equal to $\Delta(i)\frac{\partial a(i)}{\partial w_{ji}} = \Delta(i)h(j)$.

The use of pre-activation variables for backpropagation is more common than the use of post-activation variables. Therefore, we present the backpropagation algorithm in a crisp pseudocode with the use of pre-activation variables. Let t_r be the index of the rth output node. Then, the backpropagation algorithm with pre-activation variables may be presented as follows:

Initialize $\Delta(t_r) = \frac{\partial L}{\partial y(t_r)} = \Phi'(a(t_r))\frac{\partial L}{\partial h(t_r)}$ for each output node t_r with $r \in \{1 \dots k\}$;

repeat

    Select an unprocessed node i such that the values of $\Delta(j)$ all of its outgoing

          nodes $j \in A(i)$ are available;

    Update $\Delta(i) \Leftarrow \Phi'_i \sum_{j \in A(i)} w_{ij}\Delta(j)$;

until all nodes have been selected;

for each edge (j, i) with weight w_{ji} **do** compute $\frac{\partial L}{\partial w_{ji}} = \Delta(i)h(j)$;

It is also possible to use both pre-activation and post-activation variables as separate nodes of the computational graph. In the next section, we will combine this approach with a vector-centric representation.

11.4.1 Derivatives of Common Activation Functions

It is evident from the discussion in the previous section that backpropagation requires the computation of derivatives of activation functions. Therefore, we discuss the computation of the derivatives of common activation functions in this section:

1. *Sigmoid activation:* The derivative of sigmoid activation is particularly simple, when it is expressed in terms of the *output* of the sigmoid, rather than the input. Let o be the output of the sigmoid function with argument v:

$$o = \frac{1}{1 + \exp(-v)} \tag{11.11}$$

Then, one can write the derivative of the activation as follows:

$$\frac{\partial o}{\partial v} = \frac{\exp(-v)}{(1 + \exp(-v))^2} \tag{11.12}$$

The key point is that this sigmoid can be written more conveniently in terms of the outputs:

$$\frac{\partial o}{\partial v} = o(1 - o) \tag{11.13}$$

The derivative of the sigmoid is often used as a function of the output rather than the input.

2. *Tanh activation:* As in the case of the sigmoid activation, the tanh activation is often used as a function of the output o rather than the input v:

$$o = \frac{\exp(2v) - 1}{\exp(2v) + 1} \tag{11.14}$$

One can then compute the derivative as follows:

$$\frac{\partial o}{\partial v} = \frac{4 \cdot \exp(2v)}{(\exp(2v) + 1)^2} \tag{11.15}$$

One can also write this derivative in terms of the output o:

$$\frac{\partial o}{\partial v} = 1 - o^2 \tag{11.16}$$

3. *ReLU and hard tanh activations:* The ReLU takes on a partial derivative value of 1 for non-negative values of its argument, and 0, otherwise. The hard tanh function takes on a partial derivative value of 1 for values of the argument in $[-1, +1]$ and 0, otherwise.

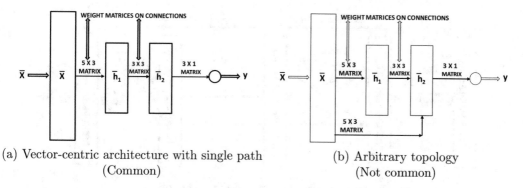

(a) Vector-centric architecture with single path
(Common)

(b) Arbitrary topology
(Not common)

Figure 11.13: Most neural networks have a layer-wise architecture, and therefore the vector-centric architecture has a single path. However, if there are shortcuts across layers, it is possible for the topology of the vector-centric architecture to be arbitrary

11.4.2 Vector-Centric Backpropagation

As illustrated in Figure 11.3, any *layer-wise* neural architecture can be represented as a computational graph of vector variables *with a single path*. We repeat the vector-centric illustration of Figure 11.3(b) in Figure 11.13(a). Note that the architecture corresponds to a single path of vector variables, which can be further decoupled into linear layers and activation layers. Although it is possible for the neural network to have an arbitrary architecture (with paths of varying length), this situation is not so common. Some variations of this idea have been explored recently in the context of a specialized[1] neural network for image data, referred to as *ResNet* [6, 58]. We illustrate this situation in Figure 11.13(b), where there is a shortcut between alternate layers.

Since the layerwise situation of Figure 11.13(a) is more common, we discuss the approach used for performing backpropagation in this case. As discussed earlier, a node in a neural network performs a combination of a linear operation and a nonlinear activation function. In order to simplify the gradient evaluations, the linear computations and the activation computations are decoupled as separate "layers," and one separately backpropagates through the two layers. Therefore, one can create a neural network in which activation layers are alternately arranged with linear layers, as shown in Figure 11.14. Activation layers (usually) perform one-to-one, elementwise computations on the vector components with the activation function $\Phi(\cdot)$, whereas linear layers perform all-to-all computations by multiplying with the coefficient matrix W. Then, if $\overline{g}_i$ and $\overline{g}_{i+1}$ be the loss gradients in the ith and $(i+1)$th layers, and J_i be the Jacobian matrix between the ith and $(i+1)$th layers, the update is as follows: Let J be the matrix whose elements are J_{kr}. Then, it is easy to see that the backpropagation update from layer to layer can be written as follows:

$$\overline{g}_i = J_i^T \overline{g}_{i+1} \tag{11.17}$$

Writing backpropagation equations as matrix multiplications is often beneficial from an implementation-centric point of view, such as acceleration with Graphics Processor Units, which work particularly well with vector and matrix operations.

---

[1] *ResNet* is a convolutional neural network in which the structure of the layer is spatial, and the operations correspond to convolutions.

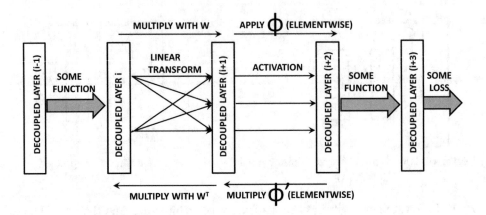

Figure 11.14: A decoupled view of backpropagation

First, the forward phase is performed on the inputs in order to compute the activations in each layer. Subsequently, the gradients are computed in the backwards phase. For each pair of matrix multiplication and activation function layers, the following forward and backward steps need to be performed:

1. Let $\overline{z}_i$ and $\overline{z}_{i+1}$ be the column vectors of activations in the forward direction when the matrix of linear transformations from the ith to the $(i+1)$th layer is denoted by W. Each element of the gradient $\overline{g}_i$ is the partial derivative of the loss function with respect to a hidden variable in the ith layer. Then, we have the following:

$$\overline{z}_{i+1} = W\overline{z}_i \qquad \text{[Forward Propagation]}$$
$$\overline{g}_i = W^T \overline{g}_{i+1} \qquad \text{[Backward Propagation]}$$

2. Now consider a situation where the activation function $\Phi(\cdot)$ is applied to each node in layer $(i+1)$ to obtain the activations in layer $(i+2)$. Then, we have the following:

$$\overline{z}_{i+2} = \Phi(\overline{z}_{i+1}) \qquad\qquad \text{[Forward Propagation]}$$
$$\overline{g}_{i+1} = \overline{g}_{i+2} \odot \Phi'(\overline{z}_{i+1}) \quad \text{[Backward Propagation]}$$

Here, $\Phi(\cdot)$ and its derivative $\Phi'(\cdot)$ are applied in element-wise fashion to vector arguments. The symbol $\odot$ indicates elementwise multiplication.

Note the extraordinary simplicity once the activation is decoupled from the matrix multiplication in a layer. The forward and backward computations are shown in Figure 11.14. Examples of different types of backpropagation updates for various forward functions are shown in Table 11.1. Therefore, the backward propagation operation is just like forward propagation. Given the vector of gradients in a layer, one only has to apply the operations shown in the final column of Table 11.1 to obtain the gradients of the loss with respect to the previous layer. In the table, the vector indicator function $I(\overline{x} > 0)$ is an *element-wise* indicator function that returns a binary vector of the sam size as $\overline{x}$; the ith output component is set to 1 when the ith component of $\overline{x}$ is larger than 0. The notation $\overline{1}$ denotes a column vector of 1s.

Table 11.1: Examples of different functions and their backpropagation updates between layers i and $(i + 1)$. The hidden values and gradients in layer i are denoted by $\overline{z}_i$ and $\overline{g}_i$. Some of these computations use $I(\cdot)$ as the binary indicator function

Function	Type	Forward	Backward
Linear	Many-Many	$\overline{z}_{i+1} = W\overline{z}_i$	$\overline{g}_i = W^T\overline{g}_{i+1}$
Sigmoid	One-One	$\overline{z}_{i+1} =$sigmoid$(\overline{z}_i)$	$\overline{g}_i = \overline{g}_{i+1} \odot \overline{z}_{i+1} \odot (1 - \overline{z}_{i+1})$
Tanh	One-One	$\overline{z}_{i+1} =$tanh$(\overline{z}_i)$	$\overline{g}_i = \overline{g}_{i+1} \odot (1 - \overline{z}_{i+1} \odot \overline{z}_{i+1})$
ReLU	One-One	$\overline{z}_{i+1} = \overline{z}_i \odot I(\overline{z}_i > 0)$	$\overline{g}_i = \overline{g}_{i+1} \odot I(\overline{z}_i > 0)$
Hard Tanh	One-One	Set to ± 1 ($\notin [-1, +1]$) Copy ($\in [-1, +1]$)	Set to 0 ($\notin [-1, +1]$) Copy ($\in [-1, +1]$)
Max	Many-One	Maximum of inputs	Set to 0 (non-maximal inputs) Copy (maximal input)
Arbitrary function $f_k(\cdot)$	Anything	$\overline{z}_{i+1}^{(k)} = f_k(\overline{z}_i)$	$\overline{g}_i = J_i^T\overline{g}_{i+1}$ J_i is Jacobian$(\overline{z}_{i+1}, \overline{z}_i)$

Initialization and Final Steps

The gradient of the final (output) layer is initialized to the vector of derivatives of the loss with respect to the various outputs in the output layer. This is generally a simple matter, since the loss in a neural network is generally a closed-form function of the outputs. Upon performing the backpropagation, one only obtains the loss-to-node derivatives but not the loss-to-weight derivatives. Note that the elements in $\overline{g}_i$ represent gradients of the loss with respect to the *activations* in the ith layer, and therefore an additional step is needed to compute gradients with respect to the *weights*. The gradient of the loss with respect to a weight between the pth unit of the $(i - 1)$th layer and the qth unit of ith layer is obtained by multiplying the pth element of $\overline{z}_{i-1}$ with the qth element of $\overline{g}_i$. One can also achieve this goal using a vector-centric approach, by simply computing the *outer product* of $\overline{g}_i$ and $\overline{z}_{i-1}$. In other words, the entire matrix M of derivatives of the loss with respect to the weights in the $(i - 1)$th layer and the ith layer is given by the following:

$$M = \overline{g}_i \overline{z}_{i-1}^T$$

Since M is given by the product of a column vector and a row vector of sizes equal to two successive layers, it is a matrix of exactly the same size as the weight matrix between the two layers. The (q, p)th element of M yields the derivative of the loss with respect to the weight between the pth element of $\overline{z}_{i-1}$ and qth element of $\overline{z}_i$.

11.4.3 Example of Vector-Centric Backpropagation

In order to explain vector-specific backpropagation, we will use an example in which the linear layers and activation layers have been decoupled. Figure 11.15 shows an example of a neural network with two computational layers, but they appear as four layers, since the activation layers have been decoupled as separated layers from the linear layers. The vector for the input layer is denoted by the 3-dimensional column vector $\overline{x}$, and the vectors for the computational layers are $\overline{h}_1$ (3-dimensional), $\overline{h}_2$ (3-dimensional), h_3 (1-dimensional), and output layer o (1-dimensional). The loss function is $L = -\log(o)$. These notations are annotated in Figure 11.15. The input vector $\overline{x}$ is $[2, 1, 2]^T$, and the weights of the edges in the two linear layers are annotated in Figure 11.15. Missing edges between $\overline{x}$ and $\overline{h}_1$

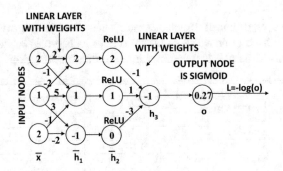

Figure 11.15: Example of decoupled neural network with vector layers $\overline{x}$, $\overline{h}_1$, $\overline{h}_2$, h_3, and o: variable values are shown within the nodes

are assumed to have zero weight. In the following, we will provide the details of both the forward and the backwards phase.

Forward phase: The first hidden layer $\overline{h}_1$ is related to the input vector $\overline{x}$ with the weight matrix W as $\overline{h}_1 = W\overline{x}$. We can reconstruct the weights matrix W and then compute $\overline{h}_1$ for forward propagation as follows:

$$W = \begin{bmatrix} 2 & -2 & 0 \\ -1 & 5 & -1 \\ 0 & 3 & -2 \end{bmatrix}; \quad \overline{h}_1 = W\overline{x} = \begin{bmatrix} 2 & -2 & 0 \\ -1 & 5 & -1 \\ 0 & 3 & -2 \end{bmatrix} \begin{bmatrix} 2 \\ 1 \\ 2 \end{bmatrix} = \begin{bmatrix} 2 \\ 1 \\ -1 \end{bmatrix}$$

The hidden layer $\overline{h}_2$ is obtained by applying the ReLU function in element-wise fashion to $\overline{h}_1$ during the forward phase. Therefore, we obtain the following:

$$\overline{h}_2 = \text{ReLU}(\overline{h}_1) = \text{ReLU} \begin{bmatrix} 2 \\ 1 \\ -1 \end{bmatrix} = \begin{bmatrix} 2 \\ 1 \\ 0 \end{bmatrix}$$

Subsequently, the 1×3 weight matrix $W_2 = [-1, 1, -3]$ is used to transform the 3-dimensional vector $\overline{h}_2$ to the 1-dimensional "vector" h_3 as follows:

$$h_3 = W_2\overline{h}_2 = [-1, 1, -3] \begin{bmatrix} 2 \\ 1 \\ 0 \end{bmatrix} = -1$$

The output o is obtained by applying the sigmoid function to h_3. In other words, we have the following:

$$o = \frac{1}{1 + \exp(-h_3)} = \frac{1}{1 + e} \approx 0.27$$

The point-specific loss is $L = -\log_e(0.27) \approx 1.3$.

Backwards phase: In the backward phase, we first start by initializing $\frac{\partial L}{\partial o}$ to $-1/o$, which is $-1/0.27$. Then, the 1-dimensional "gradient" g_3 of the hidden layer h_3 is obtained by using the backpropagation formula for the sigmoid function in Table 11.1:

$$g_3 = o(1 - o) \underbrace{\frac{\partial L}{\partial o}}_{-1/o} = o - 1 = 0.27 - 1 = -0.73 \qquad (11.18)$$

The gradient $\overline{g}_2$ of the hidden layer $\overline{h}_2$ is obtained by multiplying g_3 with the transpose of the weight matrix $W_2 = [-1, 1, -3]$:

$$\overline{g}_2 = W_2^T g_3 = \begin{bmatrix} -1 \\ 1 \\ -3 \end{bmatrix} (-0.73) = \begin{bmatrix} 0.73 \\ -0.73 \\ 2.19 \end{bmatrix}$$

Based on the entry in Table 11.1 for the ReLU layer, the gradient $\overline{g}_2$ can be propagated backwards to $\overline{g}_1 = \frac{\partial L}{\partial \overline{h}_1}$ by copying the components of $\overline{g}_2$ to $\overline{g}_1$, when the corresponding components in $\overline{h}_1$ are positive; otherwise, the components of $\overline{g}_1$ are set to zero. Therefore, the gradient $\overline{g}_1 = \frac{\partial L}{\partial \overline{h}_1}$ can be obtained by simply copying the first and second components of $\overline{g}_2$ to the first and second components of $\overline{g}_1$, and setting the third component of $\overline{g}_1$ to 0. In other words, we have the following:

$$\overline{g}_1 = \begin{bmatrix} 0.73 \\ -0.73 \\ 0 \end{bmatrix}$$

Note that we can also compute the gradient $\overline{g}_0 = \frac{\partial L}{\partial \overline{x}}$ of the loss with respect to the input layer $\overline{x}$ by simply computing $\overline{g}_0 = W^T \overline{g}_1$. However, this is not really needed for computing loss-to-weight derivatives.

Computing loss-to-weight derivatives: So far, we have only shown how to compute loss-to-node derivatives in this particular example. These need to be converted to loss-to-weight derivatives with the additional step of multiplying with a hidden layer. Let M be the loss-to-weight derivatives for the weight matrix W between the two layers. Note that there is a one-to-one correspondence between the positions of the elements of M and W. Then, the matrix M is defined as follows:

$$M = \overline{g}_1 \overline{x}^T = \begin{bmatrix} 0.73 \\ -0.73 \\ 0 \end{bmatrix} [2, 1, 2] = \begin{bmatrix} 1.46 & 0.73 & 1.46 \\ -1.46 & -0.73 & -1.46 \\ 0 & 0 & 0 \end{bmatrix}$$

Similarly, one can compute the loss-to-weight derivative matrix M_2 for the 1×3 matrix W_2 between $\overline{h}_2$ and h_3:

$$M_2 = g_3 \overline{h}_2^T = (-0.73)[2, 1, 0] = [-1.46, -0.73, 0]$$

Note that the size of the matrix M_2 is identical to that of W_2, although the weights of the missing edges should not be updated.

11.5 A General View of Computational Graphs

Although the use of directed acyclic graphs on continuous-valued data is extremely common in machine learning (with neural networks being a prominent use case), other variations of such graphs exist. For example, it is possible for a computational graph to define probabilistic functions on edges, to have discrete-valued variables, and also to have cycles in the graph. In fact, the entire field of probabilistic graphical models is devoted to these types of computational graphs. Although the use of cycles in computational graphs is not common in feed-forward neural networks, they are extremely common in many advanced variations

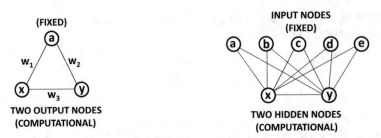

(a) An undirected computational graph   (b) More input states than hidden states

Figure 11.16: Examples of undirected computational graphs

of neural networks like Kohonen self-organizing maps, Hopfield networks, and Boltzmann machines. Furthermore, these neural networks use discrete and probabilistic data types as variables within their nodes (implicitly or explicitly).

Another important variation is the use of undirected computational graphs. In undirected computational graphs, each node computes a function of the variables in nodes incident on it, and there is no direction to the links. This is the only difference between an undirected computational graph and a directed computational graph. As in the case of directed computational graphs, one can define a loss function on the observed variables in the nodes. Examples of undirected computational graphs are shown in Figure 11.16. Some nodes are fixed (for observed data) whereas others are computational nodes. The computation can occur in both directions of an edge as long as the value in the node is not fixed externally.

It is harder to learn the parameters in undirected computational graphs, because the presence of cycles creates additional constraints on the values of variables in the nodes. In fact, it is not even necessary for there to be a set of variable values in nodes that satisfy all the functional constraints implied by the computational graph. For example, consider a computational graph with two nodes in which the variable of each node is obtained by adding 1 to the variable on the other node. It is impossible to find a pair of values in the two nodes that can satisfy both constraints (because both variable values cannot be larger than the other by 1). Therefore, one would have to be satisfied with a *best-fit* solution in many cases. This situation is different from a directed acyclic graph, where appropriate variables values can always be defined over all values of the inputs and parameters (as long as the function in each node is computable over its inputs).

Undirected computational graphs are often used in all types of unsupervised algorithms, because the cycles in these graphs help in relating other hidden nodes to the input nodes. For example, if the variables x and y are assumed to be hidden variables in Figure 11.16(b), this approach learns weights so that the two hidden variables correspond to the compressed representations of 5-dimensional data. The weights are often learned to minimize a loss function (or energy function) that rewards large weights when connected nodes are highly correlated in a positive way. For example, if variable x is heavily correlated with input a in a positive way, then the weight between these two nodes should be large. By learning these weights, one can compute the hidden representation of any 5-dimensional point by providing it as an input to the network.

The level of difficulty in learning the parameters of a computational graph is regulated by three characteristics of the graph. The first characteristic is the structure of the graph itself. It is generally much easier to learn the parameters of computational graphs without cycles (which are always directed). The second characteristic is whether the variable in a node is continuous or discrete. It is much easier to optimize the parameters of a computational graph with continuous variables with the use of differential calculus. Finally, the function computed at a node of can be either probabilistic or deterministic. The parameters of deterministic computational graphs are almost always much easier to optimize with observed data. All these variations are important, and they arise in different types of machine learning applications. Some examples of different types of computational graphs in machine learning are as follows:

1. **Hopfield networks:** Hopfield networks are *undirected* computational graphs, in which the nodes always contain discrete, binary values. Since the graph is undirected, it contains cycles. The discrete nature of the variables makes the problem harder to optimize, because it precludes the use of simple techniques from calculus. In many cases, the optimal solutions to undirected graphs with discrete-valued variables are known[2] to be NP-hard [49]. For example, a special case of the Hopfield network can be used to solve the *traveling salesman problem*, which is known to be NP-hard. Most of the algorithms for such types of optimization problems are iterative heuristics.

2. **Probabilistic graphical models:** Probabilistic graphical models [74] are graphs representing the structural dependencies among random variables. Such dependencies may be either undirected or directed; directed dependencies may or may not contain cycles. The main distinguishing characteristic of a probabilistic graphical model from other types of graphical computational models is that the variables are probabilistic in nature. In other words, a variable in the computational graph corresponds to the *outcome* resulting from sampling from a *probability distributions that is conditioned on the variables in the incoming nodes.* Among all classes of models, probabilistic graphical models are the hardest to solve, and often require computationally intensive procedures like *Markov chain Monte Carlo sampling.* Interestingly, a generalization of Hopfield networks, referred to as *Boltzmann machines*, represents an important class of probabilistic graphical models.

3. **Kohonen self-organizing map:** A Kohonen self-organizing map uses a 2-dimensional *lattice-structured graph* on the hidden nodes. The activations on hidden nodes are analogous to the centroids in a k-means algorithm. This type of approach is a *competitive learning algorithm.* The lattice structure ensures that hidden nodes that are close to one another in the graph have similar values. As a result, by associating data points with their closest hidden nodes, one is able to obtain a 2-dimensional visualization of the data.

Table 11.2 shows several variations of the computational graph paradigm in machine learning, and their specific properties. It is evident that the methodology used for a particular problem is highly dependent on the structure of the computational graph, its variables, and the nature of the node-specific function. We refer the reader to [6] for the neural architectures of the basic machine learning models discussed in this book (like linear regression, logistic regression, matrix factorization, and SVMs).

---

[2]When a problem is NP-hard, it means that a polynomial-time algorithm for such a problem is not known (although it is unknown whether one exists). More specifically, finding a polynomial-time algorithm for such a problem would automatically provide a polynomial-time algorithm for thousands of related problems for which no one has been able to find a polynomial-time algorithm. The inability to find a polynomial-time problem for a large class of related problems is generally assumed to be evidence of the fact that the entire set of problems is hard to solve, and a polynomial time algorithm *probably* does not exist for any of them.

Table 11.2: Types of computational graphs for different machine learning problems. The properties of the computational graph vary according to the application at hand

Model	Cycles?	Variable	Function	Methodology
SVM Logistic Regression Linear Regression SVD Matrix Factorization	No	Continuous	Deterministic	Gradient Descent
Feedforward Neural Networks	No	Continuous	Deterministic	Gradient Descent
Kohonen Map	Yes	Continuous	Deterministic	Gradient Descent
Hopfield Networks	Yes (Undirected)	Discrete (Binary)	Deterministic	Iterative (Hebbian Rule)
Boltzmann Machines	Yes (Undirected)	Discrete (Binary)	Probabilistic	Monte Carlo Sampling + Iterative (Hebbian)
Probabilistic Graphical Models	Varies	Varies	Probabilistic (largely)	Varies

11.6   Summary

This chapter introduces the basics of computational graphs for machine learning applications. Computational graphs often have parameters associated with their edges, which need to be learned. Learning the parameters of a computational graph from observed data provides a route to learning a function from observed data (whether it can be expressed in closed form or not). The most commonly used type of computational graph is a directed acyclic graph. Traditional neural networks represent a class of models that is a special case of this type of graph. However, other types of undirected and cyclic graphs are used to represent other models like Hopfield networks and restricted Boltzmann machines.

11.7   Further Reading

Computational graphs represent a fundamental way of defining the computations associated with many machine learning models such as neural networks or probabilistic models. Detailed discussions of neural networks may be found in [6, 53], whereas detailed discussions of probabilistic graphical models may be found in [74]. Automatic differentiation in computational graphs has historically been used extensively in control theory [26, 71]. The backpropagation algorithm was first proposed in the context of neural networks by Werbos [131], although it was forgotten. Eventually, the algorithm was popularized in the paper by Rumelhart *et al.* [110]. The Hopfield network and the Boltzmann machine are both discussed in [6]. A discussion of Kohonen self-organizing maps may also be found in [6].

11.8   Exercises

1. Problem 11.2.2 proposes a loss function for the L_1-SVM in the context of a computational graph. How would you change this loss function, so that the same computational graph results in an L_2-SVM?

2. Repeat Exercise 1 with the changed setting that you want to simulate Widrow-Hoff learning (least-squares classification) with the same computational graph. What will be the loss function associated with the single output node?

3. The book discusses a vector-centric view of backpropagation in which backpropagation in linear layers can be implemented with matrix-to-vector multiplications. Discuss how you can deal with *batches* of training instances at a time (i.e., mini-batch stochastic gradient descent) by using matrix-to-matrix multiplications.

4. Let $f(x)$ be defined as follows:

$$f(x) = \sin(x) + \cos(x)$$

Consider the the function $f(f(f(f(x))))$. Write this function in closed form to obtain an appreciation of the awkwardly long function. Evaluate the derivative of this function at $x = \pi/3$ radians by using a computational graph abstraction.

5. Suppose that you have a computational graph with the constraint that specific sets of weights are always constrained to be at the same value. Discuss how you can compute the derivative of the loss function with respect to these weights. [Note that this trick is used frequently in the neural network literature to handle shared weights.]

6. Consider a computational graph in which you are told that the variables on the edges satisfy k linear equality constraints. Discuss how you would train the weights of such a graph. How would your answer change, if the variables satisfied box constraints. [The reader is advised to refer to the chapter on constrained optimization for answering this question.]

7. Discuss why the dynamic programming algorithm for computing the gradients will not work in the case where the computational graph contains cycles.

8. Consider the neural architecture with connections between alternate layers, as shown in Figure 11.13(b). Suppose that the recurrence equations of this neural network are as follows:

$$\overline{h}_1 = \text{ReLU}(W_1 \overline{x})$$
$$\overline{h}_2 = \text{ReLU}(W_2 \overline{x} + W_3 \overline{h}_1)$$
$$y = W_4 \overline{h}_2$$

Here, W_1, W_2, W_3, and W_4 are matrices of appropriate size. Use the vector-centric backpropagation algorithm to derive the expressions for $\frac{\partial y}{\partial h_2}$, $\frac{\partial y}{\partial h_1}$, and $\frac{\partial y}{\partial \overline{x}}$ in terms of the matrices and activation values in intermediate layers.

9. Consider a neural network that has hidden layers $\overline{h}_1 \ldots \overline{h}_t$, inputs $\overline{x}_1 \ldots \overline{x}_t$ into each layer, and outputs $\overline{o}$ from the final layer $\overline{h}_t$. The recurrence equation for the pth layer is as follows:

$$\overline{o} = U\overline{h}_t$$
$$\overline{h}_p = \tanh(W\overline{h}_{p-1} + V\overline{x}_p) \quad \forall p \in \{1 \ldots t\}$$

The vector output $\overline{o}$ has dimensionality k, each $\overline{h}_p$ has dimensionality m, and each $\overline{x}_p$ has dimensionality d. The "tanh" function is applied in element-wise fashion. The notations U, V, and W are matrices of sizes $k \times m$, $m \times d$, and $m \times m$, respectively. The vector $\overline{h}_0$ is set to the zero vector. Start by drawing a (vectored) computational graph for this system. Show that node-to-node backpropagation uses the following recurrence:

$$\frac{\partial \overline{o}}{\partial \overline{h}_t} = U^T$$

$$\frac{\partial \overline{o}}{\partial \overline{h}_{p-1}} = W^T \Delta_{p-1} \frac{\partial \overline{o}}{\partial \overline{h}_p} \quad \forall p \in \{2 \ldots t\}$$

Here, Δ_p is a diagonal matrix in which the diagonal entries contain the components of the vector $\overline{1} - \overline{h}_p \odot \overline{h}_p$. What you have just derived contains the node-to-node backpropagation equations of a recurrent neural network. What is the size of each matrix $\frac{\partial \overline{o}}{\partial \overline{h}_p}$?

10. Show that if we use the loss function $L(\overline{o})$ in Exercise 9, then the loss-to-node gradient can be computed for the final layer $\overline{h}_t$ as follows:

$$\frac{\partial L(\overline{o})}{\partial \overline{h}_t} = U^T \frac{\partial L(\overline{o})}{\partial \overline{o}}$$

The updates in earlier layers remain similar to Exercise 9, except that each $\overline{o}$ is replaced by $L(\overline{o})$. What is the size of each matrix $\frac{\partial L(\overline{o})}{\partial \overline{h}_p}$?

11. Suppose that the output structure of the neural network in Exercise 9 is changed so that there are k-dimensional outputs $\overline{o}_1 \ldots \overline{o}_t$ in each layer, and the overall loss is $L = \sum_{i=1}^{t} L(\overline{o}_i)$. The output recurrence is $\overline{o}_p = U\overline{h}_p$. All other recurrences remain the same. Show that the backpropagation recurrence of the hidden layers changes as follows:

$$\frac{\partial L}{\partial \overline{h}_t} = U^T \frac{\partial L(\overline{o}_t)}{\partial \overline{o}_t}$$

$$\frac{\partial L}{\partial \overline{h}_{p-1}} = W^T \Delta_{p-1} \frac{\partial L}{\partial \overline{h}_p} + U^T \frac{\partial L(\overline{o}_{p-1})}{\partial \overline{o}_{p-1}} \quad \forall p \in \{2 \ldots t\}$$

12. For Exercise 11, show the following loss-to-weight derivatives:

$$\frac{\partial L}{\partial U} = \sum_{p=1}^{t} \frac{\partial L(\overline{o}_p)}{\partial \overline{o}_p} \overline{h}_p^T, \quad \frac{\partial L}{\partial W} = \sum_{p=2}^{t} \Delta_{p-1} \frac{\partial L}{\partial \overline{h}_p} \overline{h}_{p-1}^T, \quad \frac{\partial L}{\partial V} = \sum_{p=1}^{t} \Delta_p \frac{\partial L}{\partial \overline{h}_p} \overline{x}_p^T$$

What are the sizes and ranks of these matrices?

13. Consider a neural network in which a vectored node $\overline{v}$ feeds into two distinct vectored nodes $\overline{h}_1$ and $\overline{h}_2$ computing different functions. The functions computed at the nodes are $\overline{h}_1 = \text{ReLU}(W_1\overline{v})$ and $\overline{h}_2 = \text{sigmoid}(W_2\overline{v})$. We do not know anything about the values of the variables in other parts of the network, but we know that $\overline{h}_1 = [2, -1, 3]^T$ and $\overline{h}_2 = [0.2, 0.5, 0.3]^T$, that are connected to the node $\overline{v} = [2, 3, 5, 1]^T$. Furthermore, the loss gradients are $\frac{\partial L}{\partial \overline{h}_1} = [-2, 1, 4]^T$ and $\frac{\partial L}{\partial \overline{h}_2} = [1, 3, -2]^T$, respectively. Show that the backpropagated loss gradient $\frac{\partial L}{\partial \overline{v}}$ can be computed in terms of W_1 and W_2 as follows:

$$\frac{\partial L}{\partial \overline{v}} = W_1^T \begin{bmatrix} -2 \\ 0 \\ 4 \end{bmatrix} + W_2^T \begin{bmatrix} 0.16 \\ 0.75 \\ -0.42 \end{bmatrix}$$

What are the sizes of W_1, W_2, and $\frac{\partial L}{\partial v}$?

14. **Forward Mode Differentiation:** The backpropagation algorithm needs to compute node-to-node derivatives of *output* nodes with respect to all other nodes, and therefore computing gradients in the backwards direction makes sense. Consequently, the pseudocode on page 460 propagates gradients in the backward direction. However, consider the case where we want to compute the node-to-node derivatives of all nodes with respect to *source* (input) nodes $s_1 \ldots s_k$. In other words, we want to compute $\frac{\partial x}{\partial s_i}$ for each non-input node variable x and each input node s_i in the network. Propose a variation of the pseudocode of page 460 that computes node-to-node gradients in the forward direction.

15. **All-pairs node-to-node derivatives:** Let $y(i)$ be the variable in node i in a directed acyclic computational graph containing n nodes and m edges. Consider the case where one wants to compute $S(i,j) = \frac{\partial y(j)}{\partial y(i)}$ for all pairs of nodes in a computational graph, so that at least one directed path exists from node i to node j. Propose an algorithm for all-pairs derivative computation that requires at most $O(n^2 m)$ time. [Hint: The pathwise aggregation lemma is helpful. First compute $S(i,j,t)$, which is the portion of $S(i,j)$ in the lemma belonging to paths of length exactly t. How can $S(i,k,t+1)$ be expressed in terms of the different $S(i,j,t)$?]

16. Use the pathwise aggregation lemma to compute the derivative of $y(10)$ with respect to each of $y(1)$, $y(2)$, and $y(3)$ as an algebraic expression (cf. Figure 11.9). You should get the same derivative as obtained using the backpropagation algorithm in the text of the chapter.

17. Consider the computational graph of Figure 11.8. For a particular numerical input $x = a$, you find the unusual situation that the value $\frac{\partial y(j)}{\partial y(i)}$ is 0.3 for each and every edge (i,j) in the network. Compute the numerical value of the partial derivative of the output with respect to the input x (at $x = a$). Show the computations using both the pathwise aggregation lemma and the backpropagation algorithm.

18. Consider the computational graph of Figure 11.8. The upper node in each layer computes $\sin(x+y)$ and the lower node in each layer computes $\cos(x+y)$ with respect to its two inputs. For the first hidden layer, there is only a single input x, and therefore the values $\sin(x)$ and $\cos(x)$ are computed. The final output node computes the product of its two inputs. The single input x is 1 radian. Compute the numerical value of the partial derivative of the output with respect to the input x (at $x = 1$ radian). Show the computations using both the pathwise aggregation lemma and the backpropagation algorithm.

19. **Matrix factorization with neural networks:** Consider a neural network containing an input layer, a hidden layer, and an output layer. The number of outputs is equal to the number of inputs d. Each output value corresponds to an input value, and the loss function is the sum of squared differences between the outputs and their corresponding inputs. The number of nodes k in the hidden layer is much less than d. The d-dimensional rows of a data matrix D are fed one by one to train this neural network. Discuss why this model is identical to that of unconstrained matrix factorization of rank-k. Interpret the weights and the activations in the hidden layer in the context of matrix factorization. You may assume that the matrix D has full column rank. Define weight matrix and data matrix notations as convenient.

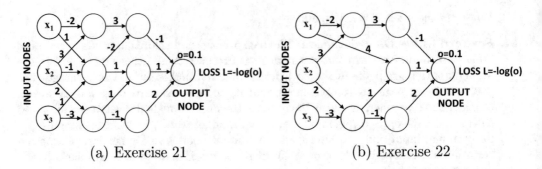

(a) Exercise 21   (b) Exercise 22

Figure 11.17: Computational graphs for Exercises 21 and 22

20. **SVD with neural networks:** In the previous exercise, unconstrained matrix factorization finds the same k-dimensional subspace as SVD. However, it does not find an orthonormal basis in general like SVD (see Chapter 8). Provide an iterative training method for the computational graph of the previous section by gradually increasing the value of k so that an orthonormal basis is found.

21. Consider the computational graph shown in Figure 11.17(a), in which the local derivative $\frac{\partial y(j)}{\partial y(i)}$ is shown for each edge (i, j), where $y(k)$ denotes the activation of node k. The output o is 0.1, and the loss L is given by $-\log(o)$. Compute the value of $\frac{\partial L}{\partial x_i}$ for each input x_i using both the path-wise aggregation lemma, and the backpropagation algorithm.

22. Consider the computational graph shown in Figure 11.17(b), in which the local derivative $\frac{\partial y(j)}{\partial y(i)}$ is shown for each edge (i, j), where $y(k)$ denotes the activation of node k. The output o is 0.1, and the loss L is given by $-\log(o)$. Compute the value of $\frac{\partial L}{\partial x_i}$ for each input x_i using both the path-wise aggregation lemma, and the backpropagation algorithm.

23. Convert the weighted computational graph of Figure 11.2 into an unweighted graph by defining additional nodes containing $w_1 \ldots w_5$ along with appropriately defined hidden nodes.

24. **Multinomial logistic regression with neural networks:** Propose a neural network architecture using the softmax activation function and an appropriate loss function that can perform multinomial logistic regression. You may refer to Chapter 4 for details of multinomial logistic regression.

25. **Weston-Watkins SVM with neural networks:** Propose a neural network architecture and an appropriate loss function that is equivalent to the Weston-Watkins SVM. You may refer to Chapter 4 for details of the Weston-Watkins SVM.

Bibliography

1. C. Aggarwal. Data mining: The textbook. *Springer*, 2015.

2. C. Aggarwal. Machine learning for text. *Springer*, 2018.

3. C. Aggarwal. Recommender systems: The textbook. *Springer*, 2016.

4. C. Aggarwal. Outlier analysis. *Springer*, 2017.

5. C. C. Aggarwal and S. Sathe. Outlier Ensembles: An Introduction. *Springer*, 2017.

6. C. Aggarwal. Neural networks and deep learning: A textbook. *Springer*, 2018.

7. C. Aggarwal. On the effects of dimensionality reduction on high dimensional similarity search. *ACM PODS Conference*, pp. 256–266, 2001.

8. R. Ahuja, T. Magnanti, and J. Orlin. Network flows: theory, algorithms, and applications. *Prentice Hall*, 1993.

9. A. Azran. The rendezvous algorithm: Multiclass semi-supervised learning with markov random walks. *ICML*, pp. 49–56, 2007.

10. M. Bazaraa, H. Sherali, and C. Shetty. Nonlinear programming: theory and algorithms. *John Wiley and Sons*, 2013.

11. I. Bayer. Fastfm: a library for factorization machines. *arXiv preprint arXiv:1505.00641*, 2015. https://arxiv.org/pdf/1505.00641v2.pdf

12. A. Beck and M. Teboulle. A fast iterative shrinkage-thresholding algorithm for linear inverse problems. SIAM journal on imaging sciences, 2(1), pp. 183–202, 2009.

13. S. Becker, and Y. LeCun. Improving the convergence of back-propagation learning with second order methods. *Proceedings of the 1988 connectionist models summer school*, pp. 29–37, 1988.

14. J. Bergstra and Y. Bengio. Random search for hyper-parameter optimization. *Journal of Machine Learning Research*, 13, pp. 281–305, 2012.

15. D. Bertsekas. Nonlinear programming. *Athena scientific*, 1999.

16. D. Bertsimas and J. Tsitsiklis. Introduction to linear optimization. *Athena Scientific*, 1997.

17. S. Bhagat, G. Cormode, and S. Muthukrishnan. Node classification in social networks. *Social Network Data Analytics*, Springer, pp. 115–148. 2011.

© Springer Nature Switzerland AG 2020
C. C. Aggarwal, *Linear Algebra and Optimization for Machine Learning*,
https://doi.org/10.1007/978-3-030-40344-7

18. C. M. Bishop. Pattern recognition and machine learning. *Springer*, 2007.

19. C. M. Bishop. Neural networks for pattern recognition. *Oxford University Press*, 1995.

20. E. Bodewig. Matrix calculus. *Elsevier*, 2014.

21. P. Boggs and J. Tolle. Sequential quadratic programming. *Acta Numerica*, 4, pp. 1–151, 1995.

22. S. Boyd and L. Vandenberghe. Convex optimization. *Cambridge University Press*, 2004.

23. S. Boyd and L. Vandenberghe. Applied linear algebra. *Cambridge University Press*, 2018.

24. S. Brin, and L. Page. The anatomy of a large-scale hypertextual web search engine. *Computer Networks*, 30(1–7), pp. 107–117, 1998.

25. A. Brouwer and W. Haemers. Spectra of graphs. *Springer Science and Business Media*, 2011.

26. A. Bryson. A gradient method for optimizing multi-stage allocation processes. *Harvard University Symposium on Digital Computers and their Applications*, 1961.

27. C. Chang and C. Lin. LIBSVM: a library for support vector machines. *ACM Transactions on Intelligent Systems and Technology*, 2(3), 27, 2011.
http://www.csie.ntu.edu.tw/~cjlin/libsvm/

28. O. Chapelle. Training a support vector machine in the primal. *Neural Computation*, 19(5), pp. 1155–1178, 2007.

29. F. Chung. Spectral graph theory. *American Mathematical Society*, 1997.

30. C. Cortes and V. Vapnik. Support-vector networks. *Machine Learning*, 20(3), pp. 273–297, 1995.

31. N. Cristianini, and J. Shawe-Taylor. An introduction to support vector machines and other kernel-based learning methods. *Cambridge University Press*, 2000.

32. Y. Dauphin, R. Pascanu, C. Gulcehre, K. Cho, S. Ganguli, and Y. Bengio. Identifying and attacking the saddle point problem in high-dimensional non-convex optimization. *NIPS Conference*, pp. 2933–2941, 2014.

33. S. Deerwester, S. Dumais, G. Furnas, T. Landauer, and R. Harshman. Indexing by latent semantic analysis. *Journal of the American Society for Information Science*, 41(6), 41(6), pp. 391–407, 1990.

34. C. Deng. A generalization of the Sherman Morrison Woodbury formula. *Applied Mathematics Letters*, 24(9), pp. 1561–1564, 2011.

35. C. Ding, T. Li, and W. Peng. On the equivalence between non-negative matrix factorization and probabilistic latent semantic indexing. *Computational Statistics and Data Analysis*, 52(8), pp. 3913–3927, 2008.

36. N. Draper and H. Smith. Applied regression analysis. *John Wiley & Sons*, 2014.

37. D. Du and P. Pardalos (Eds). Minimax and applications, *Springer*, 2013.

38. J. Duchi, E. Hazan, and Y. Singer. Adaptive subgradient methods for online learning and stochastic optimization. *Journal of Machine Learning Research*, 12, pp. 2121–2159, 2011.

39. R. Duda, P. Hart, and D. Stork. Pattern classification. *John Wiley and Sons*, 2012.

40. D. Easley, and J. Kleinberg. Networks, crowds, and markets: Reasoning about a highly connected world. *Cambridge University Press*, 2010.

41. C. Eckart and G. Young. The approximation of one matrix by another of lower rank. *Psychometrika*, 1(3), pp. 211–218, 1936.

42. A. Emmott, S. Das, T. Dietterich, A. Fern, and W. Wong. Systematic Construction of Anomaly Detection Benchmarks from Real Data. *arXiv:1503.01158*, 2015.
https://arxiv.org/abs/1503.01158

43. M. Faloutsos, P. Faloutsos, and C. Faloutsos. On power-law relationships of the internet topology. *ACM SIGCOMM Computer Communication Review*, pp. 251–262, 1999.

44. R. Fan, K. Chang, C. Hsieh, X. Wang, and C. Lin. LIBLINEAR: A library for large linear classification. *Journal of Machine Learning Research*, 9, pp. 1871–1874, 2008.
http://www.csie.ntu.edu.tw/~cjlin/liblinear/

45. R. Fisher. The use of multiple measurements in taxonomic problems. *Annals of Eugenics*, 7: pp. 179–188, 1936.

46. P. Flach. Machine learning: the art and science of algorithms that make sense of data. *Cambridge University Press*, 2012.

47. C. Freudenthaler, L. Schmidt-Thieme, and S. Rendle. Factorization machines: Factorized polynomial regression models. *GPSDAA*, 2011.

48. J. Friedman, T. Hastie, and R. Tibshirani. Sparse inverse covariance estimation with the graphical lasso. *Biostatistics*, 9(3), pp. 432–441, 2008.

49. M. Garey, and D. S. Johnson. Computers and intractability: A guide to the theory of NP-completeness. *New York, Freeman*, 1979.

50. E. Gaussier and C. Goutte. Relation between PLSA and NMF and implications. *ACM SIGIR Conference*, pp. 601–602, 2005.

51. H. Gavin. The Levenberg-Marquardt method for nonlinear least squares curve-fitting problems, 2011.
http://people.duke.edu/~hpgavin/ce281/lm.pdf

52. G. Golub and C. F. Van Loan. Matrix computations, *John Hopkins University Press*, 2012.

53. I. Goodfellow, Y. Bengio, and A. Courville. Deep learning. *MIT Press*, 2016.

54. I. Goodfellow, O. Vinyals, and A. Saxe. Qualitatively characterizing neural network optimization problems. *arXiv:1412.6544*, 2014. [Also appears in *ICLR*, 2015]
https://arxiv.org/abs/1412.6544

55. A. Grover and J. Leskovec. node2vec: Scalable feature learning for networks. *ACM KDD Conference*, pp. 855–864, 2016.

56. T. Hastie, R. Tibshirani, and J. Friedman. The elements of statistical learning. *Springer*, 2009.

57. T. Hastie, R. Tibshirani, and M. Wainwright. Statistical learning with sparsity: the lasso and generalizations. *CRC Press*, 2015.

58. K. He, X. Zhang, S. Ren, and J. Sun. Delving deep into rectifiers: Surpassing human-level performance on imagenet classification. *IEEE International Conference on Computer Vision*, pp. 1026–1034, 2015.

59. M. Hestenes and E. Stiefel. Methods of conjugate gradients for solving linear systems. *Journal of Research of the National Bureau of Standards*, 49(6), 1952.

60. G. Hinton. Connectionist learning procedures. *Artificial Intelligence*, 40(1–3), pp. 185–234, 1989.

61. G. Hinton. Neural networks for machine learning, *Coursera Video*, 2012.

62. K. Hoffman and R. Kunze. Linear algebra, Second Edition, *Pearson*, 1975.

63. T. Hofmann. Probabilistic latent semantic indexing. *ACM SIGIR Conference*, pp. 50–57, 1999.

64. C. Hsieh, K. Chang, C. Lin, S. S. Keerthi, and S. Sundararajan. A dual coordinate descent method for large-scale linear SVM. *ICML*, pp. 408–415, 2008.

65. Y. Hu, Y. Koren, and C. Volinsky. Collaborative filtering for implicit feedback datasets. *IEEE ICDM*, pp. 263–272, 2008.

66. H. Yu and B. Wilamowski. Levenberg–Marquardt training. *Industrial Electronics Handbook*, 5(12), 1, 2011.

67. R. Jacobs. Increased rates of convergence through learning rate adaptation. *Neural Networks*, 1(4), pp. 295–307, 1988.

68. T. Jaakkola, and D. Haussler. Probabilistic kernel regression models. *AISTATS*, 1999.

69. P. Jain, P. Netrapalli, and S. Sanghavi. Low-rank matrix completion using alternating minimization. *ACM Symposium on Theory of Computing*, pp. 665–674, 2013.

70. C. Johnson. Logistic matrix factorization for implicit feedback data. *NIPS Conference*, 2014.

71. H. J. Kelley. Gradient theory of optimal flight paths. *Ars Journal*, 30(10), pp. 947–954, 1960.

72. D. Kingma and J. Ba. Adam: A method for stochastic optimization. *arXiv:1412.6980*, 2014.
https://arxiv.org/abs/1412.6980

73. M. Knapp. Sines and cosines of angles in arithmetic progression. *Mathematics Magazine*, 82(5), 2009.

74. D. Koller and N. Friedman. Probabilistic graphical models: principles and techniques. *MIT Press*, 2009.

75. Y. Koren, R. Bell, and C. Volinsky. Matrix factorization techniques for recommender systems. *Computer*, 8, pp. 30–37, 2009.

76. A. Langville, C. Meyer, R. Albright, J. Cox, and D. Duling. Initializations for the nonnegative matrix factorization. *ACM KDD Conference*, pp. 23–26, 2006.

77. D. Lay, S. Lay, and J. McDonald. Linear Algebra and its applications, *Pearson*, 2012.

78. Q. Le, J. Ngiam, A. Coates, A. Lahiri, B. Prochnow, and A. Ng, On optimization methods for deep learning. *ICML Conference*, pp. 265–272, 2011.

79. D. Lee and H. Seung. Algorithms for non-negative matrix factorization. *Advances in Neural Information Processing Systems*, pp. 556–562, 2001.

80. C. J. Lin, R. C. Weng, and S. S. Keerthi. Trust region newton method for logistic regression. , 9(Apr), 627–650. *Journal of Machine Learning Research*, 9, pp. 627–650, 2008.

81. T.-Y. Liu. Learning to rank for information retrieval. *Foundations and Trends in Information Retrieval*, 3(3), pp. 225–231, 2009.

82. B. London and L. Getoor. Collective classification of network data. *Data Classification: Algorithms and Applications*, CRC Press, pp. 399–416, 2014.

83. D. Luenberger and Y. Ye. Linear and nonlinear programming, *Addison-Wesley*, 1984.

84. U. von Luxburg. A tutorial on spectral clustering. *Statistics and computing*, 17(4), pp. 395–416, 2007.

85. S. Marsland. Machine learning: An algorithmic perspective, *CRC Press*, 2015.

86. J. Martens. Deep learning via Hessian-free optimization. *ICML Conference*, pp. 735–742, 2010.

87. J. Martens and I. Sutskever. Learning recurrent neural networks with hessian-free optimization. *ICML Conference*, pp. 1033–1040, 2011.

88. J. Martens, I. Sutskever, and K. Swersky. Estimating the hessian by back-propagating curvature. *arXiv:1206.6464*, 2016.
https://arxiv.org/abs/1206.6464

89. J. Martens and R. Grosse. Optimizing Neural Networks with Kronecker-factored Approximate Curvature. *ICML Conference*, 2015.

90. P. McCullagh. Regression models for ordinal data. *Journal of the royal statistical society. Series B (Methodological)*, pp. 109–142, 1980.

91. T. Mikolov, K. Chen, G. Corrado, and J. Dean. Efficient estimation of word representations in vector space. *arXiv:1301.3781*, 2013.
https://arxiv.org/abs/1301.3781

92. T. Mikolov, I. Sutskever, K. Chen, G. Corrado, and J. Dean. Distributed representations of words and phrases and their compositionality. *NIPS Conference*, pp. 3111–3119, 2013.

93. T. Minka. A comparison of numerical optimizers for logistic regression. *Unpublished Draft*, 2003.

94. T. Mitchell. Machine learning, *McGraw Hill*, 1997.

95. K. Murphy. Machine learning: A probabilistic perspective, *MIT Press*, 2012.

96. G. Nemhauser, A. Kan, and N. Todd. Nondifferentiable optimization. *Handbooks in Operations Research and Management Sciences*, 1, pp. 529–572, 1989.

97. Y. Nesterov. A method of solving a convex programming problem with convergence rate $O(1/k^2)$. *Soviet Mathematics Doklady*, 27, pp. 372–376, 1983.

98. A. Ng, M. Jordan, and Y. Weiss. On spectral clustering: Analysis and an algorithm. *NIPS Conference*, pp. 849–856, 2002.

99. J. Nocedal and S. Wright. Numerical optimization. *Springer*, 2006.

100. N. Parikh and S. Boyd. Proximal algorithms. *Foundations and Trends in Optimization*, 1(3), pp. 127–239, 2014.

101. J. Pennington, R. Socher, and C. Manning. Glove: Global Vectors for Word Representation. *EMNLP*, pp. 1532–1543, 2014.

102. J. C. Platt. Sequential minimal optimization: A fast algorithm for training support vector machines. *Advances in Kernel Method: Support Vector Learning*, MIT Press, pp. 85–208, 1998.

103. B. Perozzi, R. Al-Rfou, and S. Skiena. Deepwalk: Online learning of social representations. *ACM KDD Conference*, pp. 701–710, 2014.

104. E. Polak. Computational methods in optimization: a unified approach. *Academic Press*, 1971.

105. B. Polyak and A. Juditsky. Acceleration of stochastic approximation by averaging. *SIAM Journal on Control and Optimization*, 30(4), pp. 838–855, 1992.

106. N. Qian. On the momentum term in gradient descent learning algorithms. *Neural networks*, 12(1), pp. 145–151, 1999.

107. S. Rendle. Factorization machines. *IEEE ICDM Conference*, pp. 995–100, 2010.

108. S. Rendle. Factorization machines with libfm. *ACM Transactions on Intelligent Systems and Technology*, 3(3), 57, 2012.

109. F. Rosenblatt. The perceptron: A probabilistic model for information storage and organization in the brain. *Psychological Review*, 65(6), 386, 1958.

110. D. Rumelhart, G. Hinton, and R. Williams. Learning internal representations by back-propagating errors. In *Parallel Distributed Processing: Explorations in the Microstructure of Cognition*, pp. 318–362, 1986.

111. T. Schaul, S. Zhang, and Y. LeCun. No more pesky learning rates. *ICML Confererence*, pp. 343–351, 2013.

112. B. Schölkopf, A. Smola, and K.-R. Müller. Nonlinear component analysis as a kernel eigenvalue problem. *Neural Computation*, 10(5), pp. 1299–1319, 1998.

113. B. Schölkopf, J. C. Platt, J. Shawe-Taylor, A. J. Smola, and R. C. Williamson. Estimating the Support of a High-Dimensional Distribution. *Neural Computation*, 13(7), pp. 1443–1472, 2001.

114. J. Shewchuk. An introduction to the conjugate gradient method without the agonizing pain. *Technical Report, CMU-CS-94-125*, Carnegie-Mellon University, 1994.

115. J. Shi and J. Malik. Normalized cuts and image segmentation. *IEEE Transactions on Pattern Analysis and Machine Intelligence*, 22(8), pp. 888–905, 2000.

116. N. Shor. Minimization methods for non-differentiable functions (Vol. 3). *Springer Science and Business Media*, 2012.

117. A. Singh and G. Gordon. A unified view of matrix factorization models. *Joint European Conference on Machine Learning and Knowledge Discovery in Databases*, pp. 358–373, 2008.

118. B. Schölkopf and A. J. Smola. Learning with kernels: support vector machines, regularization, optimization, and beyond. *Cambridge University Press*, 2001.

119. J. Solomon. Numerical Algorithms: Methods for Computer Vision, Machine Learning, and Graphics. *CRC Press*, 2015.

120. N. Srebro, J. Rennie, and T. Jaakkola. Maximum-margin matrix factorization. *Advances in neural information processing systems*, pp. 1329–1336, 2004.

121. G. Strang. The discrete cosine transform. *SIAM review*, 41(1), pp. 135–147, 1999.

122. G. Strang. An introduction to linear algebra, Fifth Edition. *Wellseley-Cambridge Press*, 2016.

123. G. Strang. Linear algebra and its applications, Fourth Edition. *Brooks Cole*, 2011.

124. G. Strang and K. Borre. Linear algebra, geodesy, and GPS. *Wellesley-Cambridge Press*, 1997.

125. G. Strang. Linear algebra and learning from data. *Wellesley-Cambridge Press*, 2019.

126. J. Tenenbaum, V. De Silva, and J. Langford. A global geometric framework for nonlinear dimensionality reduction. *Science*, 290 (5500), pp. 2319–2323, 2000.

127. A. Tikhonov and V. Arsenin. Solution of ill-posed problems. *Winston and Sons*, 1977.

128. M. Udell, C. Horn, R. Zadeh, and S. Boyd. Generalized low rank models. *Foundations and Trends in Machine Learning*, 9(1), pp. 1–118, 2016.
https://github.com/madeleineudell/LowRankModels.jl

129. G. Wahba. Support vector machines, reproducing kernel Hilbert spaces and the randomized GACV. *Advances in Kernel Methods-Support Vector Learning*, 6, pp. 69–87, 1999.

130. H. Wendland. Numerical linear algebra: An introduction. *Cambridge University Press*, 2018.

131. P. Werbos. Beyond Regression: New Tools for Prediction and Analysis in the Behavioral Sciences. *PhD thesis, Harvard University*, 1974.

132. B. Widrow and M. Hoff. Adaptive switching circuits. *IRE WESCON Convention Record*, 4(1), pp. 96–104, 1960.

133. C. Williams and M. Seeger. Using the Nyström method to speed up kernel machines. *NIPS Conference*, 2000.

134. S. Wright. Coordinate descent algorithms. *Mathematical Programming*, 151(1), pp. 3–34, 2015.

135. T. T. Wu, and K. Lange. Coordinate descent algorithms for lasso penalized regression. *The Annals of Applied Statistics*, 2(1), pp. 224–244, 2008.

136. H. Yu, F. Huang, and C. J. Lin. Dual coordinate descent methods for logistic regression and maximum entropy models. *Machine Learning*, 85(1–2), pp. 41–75, 2011.

137. H. Yu, C. Hsieh, S. Si, and I. S. Dhillon. Scalable coordinate descent approaches to parallel matrix factorization for recommender systems. *IEEE ICDM*, pp. 765–774, 2012.

138. R. Zafarani, M. A. Abbasi, and H. Liu. Social media mining: an introduction. *Cambridge University Press*, 2014.

139. M. Zeiler. ADADELTA: an adaptive learning rate method. *arXiv:1212.5701*, 2012. https://arxiv.org/abs/1212.5701

140. T. Zhang. On the dual formulation of regularized linear systems with convex risks. *Machine Learning*, 46, 1–3, pp. 81–129, 2002.

141. Y. Zhou, D. Wilkinson, R. Schreiber, and R. Pan. Large-scale parallel collaborative filtering for the Netflix prize. *Algorithmic Aspects in Information and Management*, pp. 337–348, 2008.

142. J. Zhu and T. Hastie. Kernel logistic regression and the import vector machine. *Advances in neural information processing systems*, 2002.

143. X. Zhu, Z. Ghahramani, and J. Lafferty. Semi-supervised learning using gaussian fields and harmonic functions. *ICML Conference*, pp. 912–919, 2003.

144. https://www.csie.ntu.edu.tw/~cjlin/libmf/

Index

Symbols
(PD) Constraints, 275

A
Activation, 451
AdaGrad, 214
Adam Algorithm, 215
Additively Separable Functions, 128
Adjacency Matrix, 414
Affine Transform, 42
Affine Transform Definition, 43
Algebraic Multiplicity, 110
Alternating Least-Squares Method, 197
Alternating Least Squares, 349
Anisotropic Scaling, 49
Armijo Rule, 162
Asymmetric Laplacian, 426

B
Backpropagation, 459
Barrier Function, 289
Barrier Methods, 288
Basis, 54
Basis Change Matrix, 104
BFGS, 237, 251
Binary Search, 161
Block Coordinate Descent, 197, 349
Block Diagonal Matrix, 13, 419
Block Upper-Triangular Matrix, 419
Bold-Driver Algorithm, 160
Box Regression, 269

C
Cauchy-Schwarz Inequality, 6
Cayley-Hamilton Theorem, 106
Chain Rule for Vectored Derivatives, 175
Chain Rule of Calculus, 174
Characteristic Polynomial, 105
Cholesky Factorization, 119
Closed Convex Set, 155
Clustering Graphs, 423
Collective Classification, 440
Compact Singular Value Decomposition, 307
Competitive Learning Algorithm, 477
Complementary Slackness Condition, 275
Complex Eigenvalues, 107
Computational Graphs, 34
Condition Number of a Matrix, 85, 326
Conjugate Gradient Method, 233
Conjugate Transpose, 88
Connected Component of Graph, 413
Connected Graph, 413
Constrained Optimization, 255
Convergence in Gradient Descent, 147
Convex Objective Functions, 124, 154
Convex Sets, 154
Coordinate, 2, 55
Coordinate Descent, 194, 348
Coordinate Descent in Recommenders, 348
Cosine Law, 7
Covariance Matrix, 122
Critical Points, 143
Cycle, 413

© Springer Nature Switzerland AG 2020
C. C. Aggarwal, *Linear Algebra and Optimization for Machine Learning*,
https://doi.org/10.1007/978-3-030-40344-7

D

Data-Specific Mercer Kernel Map, 383
Davidson–Fletcher–Powell, 238
Decision Boundary, 181
Decomposition of Matrices, 339
Defective Matrix, 110
Degree Centrality, 434
Degree Matrix, 416
Degree of a Vertex, 413
Degree Prestige, 434
Denominator Layout, 170
DFP, 238
Diagonal Entries of a Matrix, 13
Diameter of Graph, 414
Dimensionality Reduction, 307
Directed Acyclic Graphs, 413
Directed Acyclic Graphs, 413
Directed Graph, 412
Directed Link Prediction, 430
Discrete Cosine Transform, 77
Discrete Fourier Transform, 79, 89
Discrete Wavelet Transform, 60
Disjoint Vector Spaces, 61
Divergence in Gradient Descent, 148
Document-Term Matrix, 340
Duality, 255
Dynamic Programming, 248, 453

E

Economy Singular Value Decomposition, 306
Eigenspace, 110, 436
Eigenvalues, 104
Eigenvector Centrality, 434
Eigenvector Prestige, 434
Eigenvectors, 104
Elastic-Net regression, 244
Elementary Matrix, 22
Energy, 20, 311
Epoch, 165
Ergodic Markov Chain, 432
Euler Identity, 33, 87

F

Factorization of Matrices, 299, 339
Fat Matrix, 3
Feasible Direction method, 256
Feature Engineering, 329
Feature Preprocessing with PCA, 327
Feature Spaces, 383

Fields, 3
Finite-Difference Approximation, 159
Frobenius Inner Product, 309
Full Rank Matrix, 63
Full Row/Column Rank, 63
Full Singular Value Decomposition, 306
Fundamental Subspaces of Linear Algebra, 63, 325

G

Gaussian Elimination, 65
Gaussian Radial Basis Kernel, 405
Generalization, 165
Generalized Low-Rank Models, 365
Geometric Multiplicity, 111
Givens Rotation, 47
Global Minimum, 145
GloVe, 361
Golden-Section Search, 161
Gram-Schmidt Orthogonalization, 73
Gram Matrix, 72
Graphs, 411

H

Hard Tanh Activation, 451
Hessian, 127, 152, 217
Hessian-free Optimization, 233
Hinge Loss, 184
Homogeneous System of Equations, 327
Homophily, 440
Householder Reflection Matrix, 47
Huber Loss, 223
Hyperbolic Tangent Activation, 451

I

Idempotent Property, 83
Identity Activation, 451
Ill-Conditioned Matrices, 85
Implicit Feedback Data, 341
Indefinite Matrix, 118
Indegree of a Vertex, 413
Inflection Point, 143
Initialization, 163
Inner Product, 86, 309
Interior Point Methods, 288
Irreducible Matrix, 420
ISOMAP, 394
ISTA, 246

Iterative Label Propagation, 442
Iterative Soft Thresholding Algorithm, 246

J

Jacobian, 170, 217, 467
Jordan Normal Form, 111

K

K-Means Algorithm, 197, 342
Katz Measure, 418
Kernel Feature Spaces, 383
Kernel K-Means, 395, 397
Kernel Methods, 122
Kernel PCA, 391
Kernel SVD, 384
Kernel SVM, 396, 398
Kernel Trick, 395, 397
Kuhn-Tucker Optimality Conditions, 274

L

L-BFGS, 237, 239, 251
Lagrangian Relaxation, 270
Laplacian, 426
Latent Components, 306
Latent Semantic Analysis, 323
Learning Rate Decay, 159
Learning Rate in Gradient Descent, 33, 146
Left Eigenvector, 108
Left Gram Matrix, 73
Left Inverse, 79
Left Null Space, 63
Leibniz formula, 100
Levenberg–Marquardt Algorithm, 251
libFM, 374
LIBLINEAR, 199
Linear Activation, 451
Linear Conjugate Gradient Method, 237
Linear Independence/Dependence, 53
Linear Kernel, 405
Linearly Additive Functions, 149
Linear Programming, 257
Linear Transform as Matrix Multiplication, 9
Linear Transform Definition, 42
Line Search, 160
Link Prediction, 360
Local Minimum, 145
Logarithmic Barrier Function, 289
Loss Function, 142

Loss Functions, 28
Low-Rank Approximation, 308
Low-Rank Matrix Update, 19
Lower Triangular Matrix, 13
LSA, 323
LU Decomposition, 66, 119

M

Mahalanobis Distance, 329
Manhattan Norm, 5
Markov Chain, 432
Matrix Calculus, 170
Matrix Decomposition, 339
Matrix Factorization, 299, 339
Matrix Inversion, 67
Matrix Inversion Lemma, 18
Maximum Margin Matrix Factorization, 364
Minimax Theorem, 272
Momentum-based Learning, 212
Moore-Penrose Pseudoinverse, 81, 179, 325
Multivariate Chain Rule, 175

N

Negative Semidefinite Matrix, 118
Newton Update, 218
Nilpotent Matrix, 14
Noise Removal with SVD, 324
Non-Differentiable Optimization, 239
Nonlinear Conjugate Gradient Method, 237
Nonnegative Matrix Factorization, 350
Nonsingular Matrix, 15
Norm, 5
Normal Equation, 56, 80
Null Space, 63
Nyström Technique, 385

O

One-Sided Inverse, 79
Open Convex Set, 155
Orthogonal Complementary Subspace, 62
Orthogonal Matrix, 17
Orthogonal Vectors, 7
Orthogonal Vector Spaces, 61
Orthonormal Vectors, 7
Outdegree of a Vertex, 413
Outer Product, 10
Outlier Detection, 328
Overfitting, 166

P
Path, 413
PCA, 320
Permutation Matrix, 24
Perron-Frobenius Theorem, 421, 422
Polar Decomposition of Matrix, 303
Polynomial Kernel, 405
Polynomial of Matrix, 14
Positive Semidefinite Matrices, 117
Power Method, 133
Primal-Dual Constraints, 275
Primal-Dual Methods, 286
Principal Component Analysis, 123, 320
Principal Components, 321
Principal Components Regression, 327
Principal Eigenvector, 133, 421
Probabilistic Graphical Models, 477
Projected Gradient Descent, 256
Projection, 7
Projection Matrix, 82, 114, 259
Proportional Odds Model, 368
Proximal Gradient Method, 244
Pseudoinverse, 179, 325
Push-Through Identity, 19

Q
QR Decomposition, 74
Quadratic Programming, 124, 130, 173, 257
Quasi-Newton Methods, 237

R
Ranking Support Vector Machines, 247
Recommender Systems, 346
Rectangular Diagonal Matrix, 13
Rectangular Matrix, 3
Reduced Singular Value Decomposition, 307
Reducible Matrix, 420
Regressand, 30, 176
Regressors, 30, 176
ReLU Activation, 451
Representer Theorem, 400
Residual Matrix, 319
Response Variable, 30
Right Eigenvector, 108
Right Inverse, 79
Right Null Space, 63
Rigid Transformation, 48
RMSProp, 215
Rotreflection, 46

Row Echelon Form, 65
Row Space, 63

S
Saddle Point, 143
Saddle Points, 128, 230
Scatter Matrix, 123
Schur's Product Theorem, 405
Schur Decomposition, 112, 132
Schwarz Theorem, 152
Semi-Supervised Classification, 443
Separator, 181
Sequential Linear Programming, 266
Sequential Quadratic Programming, 267
Shear Matrix, 26
Shear Transform, 26
Sherman–Morrison–Woodbury Identity, 19
Shortest Path, 414
Sigmoid Activation, 451
Sigmoid Kernel, 405
Similarity Graph, 442
Similarity Transformation, 113
Simultaneously Diagonalizable, 115
Singular Matrix, 15
Singular Value Decomposition, 299
Span of Vector Set, 59
Sparse Matrix, 14
Spectral Decomposition, 134, 306
Spectral Theorem, 115
Square Matrix, 3
Standard Basis, 55
Stationarity Condition, 275
Steepest Descent Direction, 146
Stochastic Gradient Descent, 164
Stochastic Transition Matrix, 415
Strict Convexity, 158
Strictly Triangular Matrix, 13
Strong Duality, 274
Strongly Connected Graph, 413
Subgradient Method, 240
Subgraph, 413
Subspace of Vector Space, 52
Subtangents, 240
Super-diagonal Entries of a Matrix, 112
SVD, 299
Sylvester's Inequality, 72
Symmetric Laplacian, 426
Symmetric Matrix, 12, 115

T

Tall Matrix, 3
Taylor Expansion, 31, 217
Trace, 20, 113
Triangle Inequality, 6
Triangular Matrix, 13
Triangular Matrix Inversion, 18
Truncated SVD, 307
Trust Region Method, 232
Tuning Hyperparameters, 168

U

Undirected Graph, 412
Unitary Matrix, 89
Univariate Optimization, 142
Upper Triangular Matrix, 13

V

Vector Space, 51, 87
Vertex Classification, 440
Von Mises Iterations, 133

W

Walk, 413
Weak Duality, 272
Weighted Graph, 412
Weston-Watkins SVM, 190
Whitening with PCA, 327
Wide Matrix, 3
Widrow-Hoff Update, 183
Woodbury Identity, 19
Word2vec, 364

Printed in the United States
by Baker & Taylor Publisher Services